8th International Particle Accelerator Conference (IPAC'17)

Journal of Physics: Conference Series Volume 874

Copenhagen, Denmark
14 – 19 May 2017

ISBN: 978-1-5108-4560-2
ISSN: 1742-6588

Printed from e-media with permission by:

Curran Associates, Inc.
57 Morehouse Lane
Red Hook, NY 12571

Some format issues inherent in the e-media version may also appear in this print version.

Printed by Curran Associates, Inc. (2017)

For permission requests, please contact the Institute of Physics
at the address below.

Institute of Physics
Dirac House, Temple Back
Bristol BS1 6BE UK

Phone: 44 1 17 929 7481
Fax: 44 1 17 920 0979

techtracking@iop.org

Additional copies of this publication are available from:

Curran Associates, Inc.
57 Morehouse Lane
Red Hook, NY 12571 USA
Phone: 845-758-0400
Fax: 845-758-2633
Email: curran@proceedings.com
Web: www.proceedings.com

Table of contents

Volume 874

8th International Particle Accelerator Conference
14–19 May 2017, Copenhagen, Denmark

Accepted papers received: 3 July 2017
Published online: 20 July 2017

Preface

Circular and Linear Colliders

Photon Sources and Electron Accelerators

Novel Particle Sources and Acceleration Techniques

Hadron Accelerators

Beam Dynamics and Electromagnetic Fields

Beam Instrumentation, Controls, Feedback and Operational Aspects

Accelerator Technology

Applications of Accelerators, Technology Transfer and Industrial Relations

IOP Conf. Series: Journal of Physics: Conf. Series **874** (2017) 011001 doi:10.1088/1742-6596/874/1/011001

PREFACE

The 8th International Particle Accelerator Conference, IPAC'17, took place at the Bella Center Copenhagen, Denmark, from Sunday to Friday, 14 to 19 May, 2017.

It was attended by more than 1350 full time delegates from approximately 34 different countries. All participants considered (incl. JACOW team, grant students and exhibitors), IPAC'17 was visited by 1550 people.

Hosted by the European Spallation Source (ESS), it was supported by MAXIV and Aarhus University. It was organised under the auspices of the European Physical Society Accelerator Group (EPS-AG) and the International Union of Pure and Applied Physics (IUPAP).

The attendance of almost 100 young scientists and engineers from all over the world was made possible through the sponsorship of the following societies, institutes and laboratories (in alphabetical order):

The International Union of Pure and Applied Physics (IUPAP), The American Physical Society Division of Physics of Beams APS-DB and the United States National Science Foundation (Plasma Physics and Accelerator Science) from the Americas, the Asian Committee for Future Accelerators (ACFA) with contributions from AS, IBS, IHEP, IMP, KAERI, KEK, KIRAMS, NSRRC, PAL, RIKEN Nishina, RIKEN SPring-8, SSRF from Asia, the European Physical Society Accelerator Group (EPS-AG) with contributions from ALBA, CEA Saclay, CERN, Cockcroft Institute, DESY, Diamond, ELETTRA, ESRF, ESS, GANIL, GSI/FAIR, HZB, IN2P3, INFN, JAI, PSI, SOLEIL, STFC from Europe.
The organisers of IPAC'17 are grateful to all sponsors for their valuable support.

The scientific programme was developed by the IPAC'17 Scientific Programme Committee (SPC). It was a truly international body with members coming 50% from Europe and 50% from Asia and North America.

The conference programme spanned four and a half days, with plenary sessions on Monday to Friday mornings, and on Thursday afternoon. All other sessions were composed of two oral sessions in parallel, followed by poster sessions at the end of each afternoon during which 1455 posters were scheduled. These proceedings contain more than 1400 contributions.

A total of 124 posters were presented by young scientists and engineers in the Special Student Poster Session on Sunday, 14 May 2017.

An industrial exhibition took place during the first three days of the conference, plus the Welcome Reception on Sunday evening. Industrial exhibitors (115 companies from 16 countries) occupied the Exhibition area and presented their high technology products and services to the delegates in a superb set-up and an excellent atmosphere conducive to discussions.

During the Accelerator Awards Session, the EPS-AG best student poster prizes were awarded to: Annalisa Romano (CERN, Geneva), for her contribution entitled *Macroparticle Simulation Studies of LHC Beam Dynamics in the Presence of the E-Cloud* and to Daniel Leslie Hall (Cornell

IOP Conf. Series: Journal of Physics: Conf. Series **874** (2017) 011001 doi:10.1088/1742-6596/874/1/011001

University (CLASSE), Ithaca, New York), for his contribution *Impact of Trapped Magnetic Flux and Thermal Gradients on the Performance of Nb3Sn Cavities.*

The **Rolf Wideröe Prize** for "outstanding work in the accelerator field without age limit" was awarded to **Dr. Lyndon Evans of CERN** *for his many major professional accomplishments in the field of accelerator design, construction and operation, including his contributions to the SPS, where he was essential for converting the SPS to a proton-anti-proton collider that led to the discovery of the W and Z Bosons and the design and construction of the LHC which led to the discovery of the Higgs Boson in 2012. The LHC at CERN is the most powerful and complex collider ever build and one of the most ambitious science projects. Dr. Lyndon Evans' leadership abilities and experience were essential ingredients for the successful completion of the LHC construction.*

The **Gersh Budker Prize**, for "a recent significant, original contribution to the accelerator field, with no age limit," was awarded to **Dr. Pantaleo Raimondi of ESRF** *for the invention of the Hybrid Multi Bend Achromat HMBA-lattice, reducing the emittance of the ESRF Synchrotron Light Source by a factor of 30, while still keeping a large enough dynamic aperture. The design of the cell shows Dr. Raimondi's ability to foster new ideas, his deep understanding of accelerator physics and mastering of technological aspects. The HMBA-lattice has been adopted as the basis for the design for most future 4th generation storage ring light sources (APS-U, SPring 8-2, ALS-U, SSRF, IHEP).*

The **Frank Sacherer Prize**, for "an individual in the early part of his or her career, having made a recent significant, original contribution to the accelerator field," was awarded to **Dr. Anna Grassellino of Fermilab** *for her major impact on the field of superconducting RF technology, in particular, the improvement of the cavity quality factor Q and more recently the accelerating field gradient and quality factor combined. The technique of "nitrogen-doping" has already been applied in ongoing accelerator projects including the technology transfer to industry around the world and the recent development of "Nitrogen-Infusion" has recently resulted in a major improvement of the maximum field gradient and quality factor combined. Both methods have the potential for revolutionizing the field of SC RF and underline the importance of her contributions to the field.*

The **Bruno Touschek prize** winner, awarded to a student registered for a PhD or diploma in accelerator physics or engineering or to a trainee accelerator physicist or engineer in the educational phase of their professional career, for the quality of work and promise for the future, was awarded to **Fabrizio Guiseppe Bisesto of INFN/LNF** *for his contributions to the plasma related activities underway at SPARC_LAB exploiting the high-power laser FLAME. In particular, for his experimental work on the single shot diagnostics systems, including Electro Optical Sampling (EOS) for temporal measurement and Optical Transition Radiation (OTR) measurements for an innovative, one-shot emittance measurement.*

The EPS-AG/IPAC'17 prizes were presented by the Chair of the Prizes Selection Committee, Oliver Brüning (CERN). Mike Seidel (PSI), Chair of the SPC, presented the student poster prizes.

IOP Conf. Series: Journal of Physics: Conf. Series **874** (2017) 011001 doi:10.1088/1742-6596/874/1/011001

An additional Industrial Committee Student Poster Award was to be given to 2 students whose work, presented in the special session for students, most incorporates criteria relative or applicable to industry. The jury eventually decided to award 3 prizes; winners were announced during the Industry Session on Tuesday, 16 May. They are:

- Hiroaki Takeuchi (The University of Tokyo) for his work on *Structural Analysis and Evaluation of Actual PC Bridge Using 950 keV/3.95 MeV X-Band Linacs.*
- Yumi Lee (Korea University Sejong Campus) for her *Design Study of Drift Tube Linac for BNCT Accelerator.*
- Jean-Michel Antoine Bereder (The University of Tokyo), for his *Development of a 3.95 Mev X-Band Linac-Driven X-Ray Combined Neutron Source.*

This publication contains a fraction of the papers presented at the conference and submitted by the authors within the deadline set for the start of the peer-reviewing process. These papers were accepted following a peer-review conducted by a committee of experts nominated by the Editorial Board. The Editorial Board was composed by:

Ralph Assmann (DESY)*, Giovanni Bisoffi, (INFN)*, Marie-Emmanuelle Couprie, (SOLEIL)*, Mike Seidel (PSI)*, Gianluigi Arduini (CERN), Matt Arena (FNAL), Oliver Boine-Frankenheim (GSI), Mark Boland (AS), Paul Collier (CERN), Angeles Faus-Golfe (IFIC), Wolfram Fischer (BNL), Andreas Jansson (ESS), In Soo Ko (POSTECH), Shane Koscielniak (TRIUMF), Juliana Pranke (ESS), Qing Quin (IHEP), Tor Raubenheimer (SLAC), Linda Spentzouris (IIT), Hitoshi Tanaka (SPring-8), Thomas Vinatier (DESY), Maurizio Vretenar (CERN)
*members of the IPAC'17 pilot refeering team

The success of IPAC'17 was due in great part to the truly excellent collaboration between the international teams of the OC and the SPC, and the LOC. Membership of the LOC, under the leadership of Mats Lindroos (ESS), included the following staff:

- Håkan Danared (ESS), *Deputy LOC Chair*
- Volker Schaa (GSI), *Chief Editor*
- Garry Trahern (ESS), *Proceedings Office Coordinator*
- Juliana Pranke (ESS), *Project Leader and Scientific Secretary*
- Caroline Prabert (ESS), *Student Programme Manager*
- Inga Tejedor (ESS), *Technical Tours to ESS and Poster Manager*
- Mikael Johansson (ESS), *Presentation Manager*
- Rob Yarbray (ESS), *Industrial Exhibition Manager*
- Roger Eriksson (ESS), *Sponsoring and Industry Support*
- Henno Gous and Daniel Friis (ESS), *Website Support*
- Johan Olander (ESS), *IT Manager, App and AV coordination onsite*
- Søren Pape Moeller (Aarhus University), *Public Lecture, Tour to ASTRID2*
- Sverker Werin, *General Support*
- Ann Frisenborg Marker (DTI), *Public Lecture*
- Tutti Johansson Falk (MAXI IV), *Public Affairs and Technical Tours to MAXIV*
- Dušan Stric (ESS), *Satellite Meeting Manager*
- Alexandra Schmidli (ESS), *onsite support with Students, Posters, Sessions*

IOP Conf. Series: Journal of Physics: Conf. Series **874** (2017) 011001 doi:10.1088/1742-6596/874/1/011001

- Caroline Holgersson (ESS), *onsite support with Students, Posters, Sessions*
- Julia Öberg (ESS), *Press*
- Helle Klestrup (DIS), *Overall PCO Support*
- Peder Andersen (DIS), *Industry Exhibitor Coordination*
- Rikke Blæsberg (DIS), *Visa, Registration*
- Charlotte Nielsen (DIS), *Accommodation*
- Jørgen Nielsen (Aarhus University), *Poster Police*
- Heine Dølrath (Aarhus University), *Poster Police*

Twenty years ago the first meeting among the representatives of the three major Accelerator Conferences (Particle Accelerator Conference - PAC, European Particle Accelerator Conference - EPAC and Asian Particle Accelerator Conference - APAC) took place in Vancouver. It was the first act of a collaboration effort among the three regions that culminated in the decision to introduce a three-year cycle for the particle accelerator conferences in 2007, ten years ago, leading to the First International Particle Accelerator Conference (IPAC) held in Kyoto in 2010.

Today, the continuously increasing number of delegates, confirms that IPAC, at its third cycle, provides a unique occasion for an exciting and fruitful exchange of ideas across countries and domains, between younger and more experienced colleagues, between industry and research institutes.

This success builds on the participants, on their work and enthusiasm as well as on the continuous support of societies, institutes and laboratories worldwide!

The ninth IPAC will return to the Americas and take place in Vancouver, Canada.

Gianluigi Arduini (CERN), Chair of the IPAC'17 Organising Committee

IOP Conf. Series: Journal of Physics: Conf. Series **874** (2017) 011002 doi:10.1088/1742-6596/874/1/011002

CONFERENCE ORGANISATION

Hosted by
European Spallation Source (ESS), MAXIV and Aarhus University

Under the Auspices of
Asian Committee for Future Accelerators (ACFA)
American Physical Society Division of Physics of Beams (APS-DPB)
International Union of Pure and Applied Physics (IUPAP)
European Physical Society Accelerator Group (EPS-AG)

Supported by
The European Physical Society Accelerator Group (EPS-AG) and the International Union of Pure and Applied Physics (IUPAP)

Student Grant Program Supported by
International Union of Pure and Applied Physics (IUPAP), the American Physical Society Division of Physics of Beams APS-DB and the United States National Science Foundation (Plasma Physics and Accelerator Science) from the Americas, the Asian Committee for Future Accelerators (ACFA) with contributions from AS, IBS, IHEP, IMP, KAERI, KEK, KIRAMS, NSRRC, PAL, RIKEN Nishina, RIKEN SPring-8, SSRF from Asia, the European Physical Society Accelerator Group (EPS-AG) with contributions from ALBA, CEA Saclay, CERN, Cockroft Institute, DESY, Diamond, ELETTRA, ESRF, ESS, GANIL, GSI/FAIR, HZB, IN2P3, INFN, JAI, PSI, SOLEIL, STFC from Europe.

Chair Organising Committee	Gianluigi Arduini (CERN)
Chair Scientific Program Committee	Mike Seidel (PSI)
Chair Local Organising Committee	Mats Lindroos (ESS)

ORGANISING COMMITTEE

Gianluigi Arduini, CERN (OC Chair), Deepa Angal-Kalinin, ASTEC/STFC, Ralph Assmann, DESY, Philip Bambade, LAL Orsay, Ivan Bazarov, Cornell, Giovanni Bisoffi, INFN-LNL, Oliver Boine-Frankenheim, GSI, Mark Boland, AS, Hans Braun, PSI, Paul Collier, CERN, Marie-Emmanuelle Couprie, SOLEIL, Angeles Faus-Golfe, IFIC, Wolfram Fischer, BNL, Susanna Guiducci, INFN-LNF Mark Hogan, SLAC, Andreas Jansson, ESS, In Soo Ko, POSTECH, Shane Koscielniak, TRIUMF, Tadashi Koseki, KEK, Gwo-Huei Luo, NSRRC, Lia Merminga, SLAC, Anke-Susanne Müller, KIT, Sergei Nagaitsev, FNAL, Won Namkung, PAL, Montse Pont, ALBA/CELLS, Qing Qin, IHEP, Tor Raubenheimer, SLAC, Mike Seidel, PSI, Linda Spentzouris, IIT, Hitoshi Tanaka, SPring-8, Maurizio Vretenar, CERN, Hongwei Zhao, IMP, Frank Zimmermann, CERN

SCIENTIFIC PROGRAM COMMITTEE

Mike Seidel, PSI (SPC Chair), Ralph Assmann, DESY, Giovanni Bisoffi, INFN-LNL, Oliver Boine-Frankenheim, GSI, Mark Boland, AS, Paul Collier, CERN, Marie-Emmanuelle Couprie, SOLEIL, Angeles Faus-Golfe, IFIC, Wolfram Fischer, BNL, Andreas Jansson, ESS, In Soo Ko, POSTECH,

IOP Conf. Series: Journal of Physics: Conf. Series **874** (2017) 011002 doi:10.1088/1742-6596/874/1/011002

Shane Koscielniak, TRIUMF, Qing Qin, IHEP, Tor Raubenheimer, SLAC, Linda Spentzouris, IIT, Hitoshi Tanaka, SPring-8, Maurizio Vretenar, CERN.

LOCAL ORGANISING COMMITTEE

Mats Lindroos, ESS (LOC Chair), Håkan Danared (ESS), Garry Trahern (ESS), Juliana Pranke (ESS), Caroline Prabert (ESS), Inga Tejedor (ESS), Mikael Johansson (ESS), Rob Yarbray (ESS), Roger Eriksson (ESS), Henno Gous and Daniel Friis (ESS), Johan Olander (ESS), Søren Pape Moeller (Aarhus University), Sverker Werin (MAXIV), Ann Frisenborg Marker (DTI), Tutti Johansson Falk (MAXI IV), Dušan Stric (ESS)

SCIENTIFIC ADVISORY BOARD

AMERICAS

Biedron, Sandra, CSU
Blaskiewicz, Mike, BNL
Boulware, Chas, Niowave
Cai, Yunhai, SLAC
Casagrande, Fabio, FRIB
Charlie Reece, JLab
Cousineau, Sarah, ORNL
Dunham, Bruce, Cornell Univ.
Frisch, Joe, SLAC
Geddes, Cameron, LBNL
Gennady Stupakov, SLAC
Harkay, Katherine, ANL
Hurh, Patrick, FNAL
Marchetto, Marco, TRIUMF
Marques, Sergio, LNLS
Minty, Michiko, BNL
Murokh, Alex, Radiabeam
Neuenschwander, Regis, LNLS
Piot, Philippe, NIU
Prestemon, Søren, LBNL
Russo, Tom, FRIB
Rybarcyk, Larry, LANL
Sagan, David, Cornell Univ.
Seiya, Kiyomi, FNAL
Yakimenko, Vitaly, SLAC

ASIA

Akai, Kazunori, KEK
Chou, Ping-Jung, NSRRC
Dowd, Rohan, Australian Synchrotron
Fukunishi, Nobuhashi, RIKEN Nishina
Hama, Hiroyuki, Tohoku Univ.
Inagaki, Takahiro, RIKEN, SPring-8
Jeon, Dong-O, RISPKR
Joshi, Satish, RRCAT
Kako, Eiji, KEK
Kim, Changbum, PAL
Kim, Eun San, Kyungpook Nat. Univ.
Kim, Yujong, KAERI
Kinsho, Michikazu, JAEA, J-Parc
Klysubun, Prapong, SLRI
Noda, Koji, NIRS
Park, Chawon, KIRAMS
Singh, Pitamber, BARC
Soutome, Kouichi, JASRI
Wang, Chaoen, NSRRC
Wang, Dong, SINAP
Wang, Lin, USTC
Wang, Sheng, IHEP
Yan, Xueqing, Beijing Univ.
Zhang, Linwen, CAEP
Zhao, Hongwei, IMP

EUROPE

Aiba, Masamitsu, PSI
Bousson, Sébastien, IPN
Malyshev, Oleg, STFC
McIntosh, Peter, STFC/DL/ASTeC

IOP Conf. Series: Journal of Physics: Conf. Series **874** (2017) 011002 doi:10.1088/1742-6596/874/1/011002

STUDENTS WITH GRANTS

AMERICAS

Cahill, Alexander (University of California) USA
Deitrick, Kirsten (Old Dominion University Center for Accelerator Science) USA
Hall, Daniel (Cornell University (CLASSE)) USA
Hwang, Yoonwoo (University of California Irvine) USA
Lueangaramwong, Anusorn (Northern Illinois University) USA
MacArthur, James (SLAC National Accelerator Laboratory) USA
Malin, Lucas (Arizona State University) USA
Mihara, Kentaro (Brookhaven National Laboratory) USA
Neveu, Nicole (Argonne National Laboratory) USA

IOP Conf. Series: Journal of Physics: Conf. Series **874** (2017) 011002 doi:10.1088/1742-6596/874/1/011002

Petrushina, Irina (National Research Nuclear University) USA
Porter, Ryan (Cornell University (CLASSE)) USA
Storey, Douglas (TRIUMF) Canada
Tuggle, James (Virginia Polytechnic Institute and State University) USA
Wu, Wanwei (University of Mississippi-Oxford) USA

ASIA
Bahng, Jungbae (Kyungpook National University) South Korea
Bian, Tianjian (Institute of High Energy Physics) China
Charles, Tessa (Monash University) Australia
Gan, Pingping (Peking University) China
Hsieh, Chia-Ying (National Central University) Taiwan
Hu, Nan (University of Science and Technology of China National Synchrotron Radiation Laboratory) China
Huang, Xiyang (Chinese Academy of Sciences Institute of High Energy Physics) China
Jia, Yongzhi (Institute of Modern Physics, Chinese Academy of Sciences) China
Karmakar, Mithun (Saha Institute of Nuclear Physics) India
Kosaentor, Kittiya (Chiang Mai University) Thailand
Lee, Yumi (Korea University Sejong Campus) South Korea
Li, Zongbin (Shanghai Institute of Applied Physics) China
Lin, Zhenyang (Tsinghua University in Beijing Accelerator Laboratory) China
Liu, Zhicong (Chinese Academy of Sciences Institute of High Energy Physics) China
Liu, Wei (University of Science and Technology of China National Synchrotron Radiation Laboratory) China
Luo, Hao-Wen (National Tsing-Hua University) Taiwan
Mei, Zhiyuan (Huazhong University of Science and Technology) China
Meng, Xiangcong (Tsinghua University in Beijing Accelerator Laboratory) China
Nakazato, Yusuke (Waseda University) Japan
Ono, Hiroya (Waseda University) Japan
Qiao, Yusi (Institute of High Energy Physics) China
Shim, Chi Hyun (Pohang University of Science and Technology) South Korea
Shou, Yinren (Peking University) China
Song, Yifan (University of Science and Technology of China National Synchrotron Radiation Laboratory) China
Su, Xiaolu (Tsinghua University in Beijing Accelerator Laboratory) China
Tao, Yue (Institute of Modern Physics, Chinese Academy of Sciences) China
Thongpakdi, Wuttiphong (Chiang Mai University) Thailand
Wang, Zihao (Institute of High Energy Physics) China
Wang, Wei (Tsinghua University in Beijing Accelerator Laboratory) China
Wang, Zhizhuo (Institute of High Energy Physics) China
Wang, Qian (University of Science and Technology of China National Synchrotron Radiation Laboratory) China
Wibowo, Sigit (Sokendai, the Graduate University for Advanced Studies Department of Accelerator Science) Japan

IOP Conf. Series: Journal of Physics: Conf. Series **874** (2017) 011002 doi:10.1088/1742-6596/874/1/011002

Wu, Yipeng (Tsinghua University in Beijing) China
Wu, Xiaowei (CERN) Switzerland
Yamashita, Hiroki (Kyoto University) Japan
Yang, Lei (Institute of Modern Physics, Chinese Academy of Sciences) China
Yang, Li (Chinese Academy of Sciences Institute of High Energy Physics) China
Yang, Zi-Hui (University of Science and Technology of China National Synchrotron Radiation Laboratory) China
Yuan, Haowei (Chinese Academy of Sciences Institute of High Energy Physics) China
Zhang, Lige (Huazhong University of Science and Technology) China
Zhou, Kai (Shanghai Institute of Applied Physics) China

EUROPE
Ahmadiannamin, Sasan (Iranian Light Source Facility Institute for Research in Fundamental Sciences (IPM)) Iran
Alexopoulos, Andreas (CERN) Switzerland
Andre, Thomas (Synchrotron Soleil) France
Baschke, Markus (Goethe Universität Frankfurt) Germany
Bazyl, Dmitry (Technische Universität Darmstadt) Germany
Bergamaschi, Michele (CERN) Switzerland
Bisesto, Fabrizio (Istituto Nazionale di Fisica Nucleare Laboratori Nazionali di Frascati) Italy
Björklund Svensson, Jonas (Lund University) Sweden
Brosi, Miriam (Karlsruhe Institute of Technology) Germany
Esposito, Raffaele (CERN) Switzerland
Ferran Pousa, Ángel (Deutsches Elektronen-Synchrotron) Germany
Filippi, Francesco (Istituto Nazionale di Fisica Nucleare Laboratori Nazionali di Frascati) Italy
Gamelin, Alexis (Laboratoire de l'Accélérateur Linéaire (LAL) CNRS/IN2P3 Université Paris-Saclay) France
Ghaith, Amin (Synchrotron Soleil) France
Ghazaryan, Nvard (CANDLE Synchrotron Research Institute) Armenia
Haghtalab, Saeed (Shahid Beheshti University) Iran
Hakobyan, Laura (CANDLE Synchrotron Research Institute) Armenia
Herrod, Alexander (Cockcroft Institute) UK
Hostettler, Michael (CERN) Switzerland
Hunt, James (Cockcroft Institute) UK
Hähnel, Hendrik (Goethe Universität Frankfurt) Germany
Kazantseva, Erika (Technische Universität Darmstadt Institut für Theorie Elektromagnetischer Felder) Germany
Lobach, Ihar (Belarus State University) Belarus
Medina Medrano, Luis (CERN) Switzerland
Moisieienko, Maksim (National Science Centre Kharkov Institute of Physics and Technology) Ukraine
Ogur, Salim (CERN) Switzerland
Pacey, Thomas (Cockcroft Institute The University of Manchester) UK

IOP Conf. Series: Journal of Physics: Conf. Series **874** (2017) 011002 doi:10.1088/1742-6596/874/1/011002

Pang, Chengguo (Laboratoire de l'Accélérateur Linéaire (LAL) CNRS/IN2P3 Université Paris-Saclay) France
Panofski, Eva (Helmholtz-Zentrum Berlin für Materialien und Energie GmbH Elektronen-Speicherring BESSY II) Germany
Parrella, Alessandro (Instituto Superior Técnico) Portugal
Plassard, Fabien (CERN) Switzerland
Quartullo, Danilo (CERN) Switzerland
Romano, Annalisa (CERN) Switzerland
Rossi, Roberto (CERN) Switzerland
Schenk, Michael (Ecole Polytechnique Federale de Lausanne) Switzerland
Turner, Marlene (CERN) Switzerland
Vaughan, Tom (John Adams Institute) UK
Weikum, Maria (Deutsches Elektronen-Synchrotron) Germany
Wolfenden, Joseph (Cockcroft Institute) UK
Yang, Renjun (Laboratoire de l'Accélérateur Linéaire (LAL) CNRS/IN2P3 Université Paris-Saclay) France
Zhadan, Alina (National Science Centre Kharkov Institute of Physics and Technology) Ukraine

JUAS 2017 Best Student
Kostoglou, Sophia (CERN) Switzerland

Peer review statement

All papers published in this volume of *Journal of Physics: Conference Series* have been peer reviewed through processes administered by the proceedings Editors. Reviews were conducted by expert referees to the professional and scientific standards expected of a proceedings journal published by IOP Publishing.

The "Multimat" experiment at CERN HiRadMat facility: advanced testing of novel materials and instrumentation for HL-LHC collimators

F Carra[1,2,5], A Bertarelli[1], E Berthomé[1], C Fichera[1], T Furness[3], M Guinchard[1], L K Mettler[1], M Portelli[1,4], S Redaelli[1], O Sacristan de Frutos[1]

[1] CERN, Route de Meyrin 385, 1217 Meyrin, Switzerland
[2] Politecnico di Torino, Corso Duca degli Abruzzi, 24, 10129 Turin, Italy
[3] University of Huddersfield, Queensgate, Huddersfield HD1 3DH, UK
[4] University of Malta, Msida MSD 2080, Malta

Abstract. The increase of the stored beam energy in future particle accelerators, such as the HL-LHC and the FCC, calls for a radical upgrade in the design, materials and instrumentation of Beam Intercepting Devices (BID), such as collimators Following successful tests in 2015 that validated new composite materials and a novel jaw design conceived for the HL-LHC collimators, a new HiRadMat experiment, named "HRMT36-MultiMat", is scheduled for autumn 2017. Its objective is to determine the behaviour under high intensity proton beams of a broad range of materials relevant for collimators and beam intercepting devices, thin-film coatings and advanced equipment. The test bench features 16 separate target stations, each hosting various specimens, allowing the exploration of complex phenomena such as dynamic strength, internal damping, nonlinearities due to anisotropic inelasticity and inhomogeneity, effects of energy deposition and radiation on coatings. This paper details the main technical solutions and engineering calculations for the design of the test bench and of the specimens, the candidate target materials and the instrumentation system.

1. Introduction to the experiment

In the next years, the HL-LHC upgrade [1] will increase the energy stored in LHC circulating beams by almost a factor of two (from 360 to 680 MJ). In the case of future proposed accelerators such as the FCC [2], the energy stored will be even higher, in the order of 8500 MJ. Among the components interacting with the particle beam, the collimators, which are used for beam cleaning and machine protection, are exposed to risks of accidental beam impacts, for example in the case of asynchronous dump or injection error. In the case of HL-LHC and FCC, the thermal load expected to be discharged on the collimator jaws will far exceed energy densities of 10 kJ/cm^3 [3], involving potential melting or vaporization of the material in the impacted volume, as well as intense shockwaves propagating towards the surroundings, with risk of plasticity or spallation to the near components. The numerical codes adopted for the simulation of beam impact strongly rely on material models which have to be experimentally benchmarked against the extreme conditions induced by the event. This can take place only in dedicated tests in facilities such as HiRadMat [4].

In 2004 and 2006, carbon collimator jaws were tested and survived the impact of the intense SPS beam [5]. In 2012, an experiment run in the HiRadMat facility allowed deriving the threshold of

[5] federico.carra@cern.ch

damage for tungsten tertiary collimators [6]. Finally, in 2015, tests in HiRadMat validated the design of the novel metal-diamond and ceramic-graphite jaws designed for HL-LHC collimators, for beam brightness values beyond LIU [7]. In parallel with the tests on full-scale collimator components, an experiment, known as *HRMT14*, was performed in 2012 [8] on specimens of simple geometry, fully instrumented, to benchmark the numerical results obtained with codes such as ANSYS [9]0, Autodyn [10] and LS-Dyna [11]. The test confirmed the precision of the numerical results, when the material models, for the extreme conditions produced by the particle beam impact, are well mastered [12],[13].
A new experiment of this kind, named *Multimat*, is under preparation for tests in autumn 2017 in HiRadMat. The experiment profits of the experience of HRMT14, aiming at testing at HL-LHC energy densities new collimator materials and coatings developed in recent years [14], in addition to novel instrumentation devices for the monitoring and correction of jaw distortion, under design for HL-LHC and FCC. With respect to past tests, the new experiment will focus on the derivation of less known material properties such as damping, porosity, anisotropic wave propagation, dynamic behaviour of coatings.
The test bench (Figure 1) features a leak-tight aluminium container (hidden in the figure) hosting 16 target stations, each 1 m long and supported on a rotatable sample holder, separated by protective metal sheets. The target stations will be extensively instrumented (Table 1), with strain gauges, pressure sensors and thermal probes, as well as remote instrumentation such as rad-hard camera and Laser Doppler Vibrometer (LDV) placed in a shielded bunker.

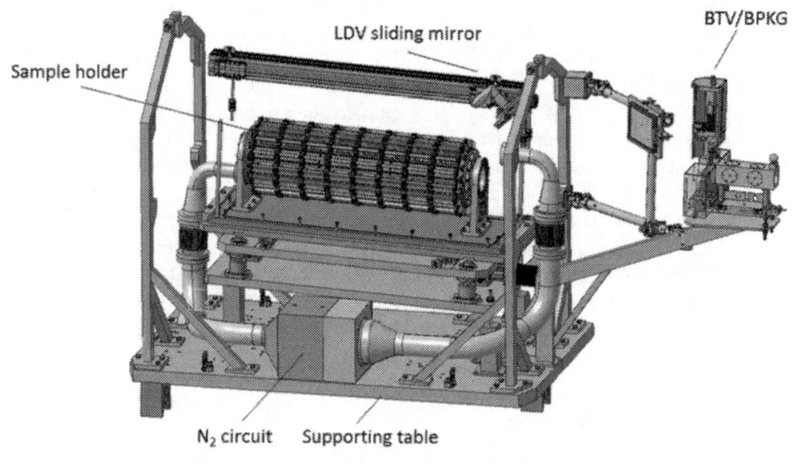

Figure 1. Inner view of the Multimat test bench. The tank is hidden in this view. The beam comes from the right.

2. Experiment design

The sample holder is enclosed in a leak-tight aluminium vessel, maintained under inert atmosphere, preventing oxidation of the samples. A forced flow of nitrogen will decrease the cooling time of specimens after each impact. Large windows guarantee the optical access to the specimens for the LDV and the camera. The beamline is equipped with two beryllium windows, reinforced with CFC, designed to withstand the load induced by the beam passage.

Table 1: DAQ overview.

Physical Effect	Amplitude Threshold	Frequency Range	Sampling Frequency	Sensors selected	Acquisition Selected
Axial waves	20 μm/m	50 kHz	1 MHz	Strain gauges	Real time PXI
Bending waves	20 μm/m	2 kHz	1 MHz	Strain gauges	Real time PXI
Temperature	20°C	1Hz	100 Hz	Pt100	Real time PXI
Transverse oscillation	0.05 m/s	2 kHz	1MHz	LDV	Real time PXI
Permanent deformation	0.1mm	Static	Few Hz	FOS strain sensors	Optical Interrogator
Surface damage	-	-	-	Rad-hard camera	-

An actuation system guarantees the rotation of the sample holder and the vertical/horizontal adjustment of the specimen position in the plane perpendicular to the beam line. Stepper motors an leadscrews are used for the horizontal and vertical movement of the whole tank, with an available stroke of ± 15 mm. The accuracy of the actuator is ±10 μm.

Figure 2. Mock-up of the Geneva mechanism.

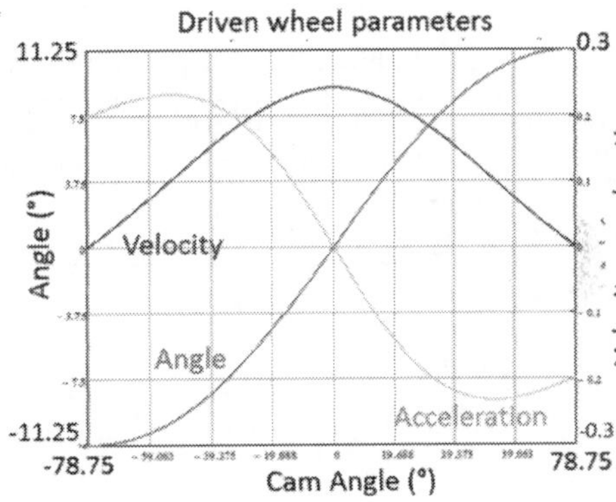

Figure 3. Driven wheel parameters as function of the cam angle.

The rotation of the sample holder (+/– 180°) is assured by a Geneva mechanism, consisting of a stainless steel driven wheel and a brass cam. Extensive calculations were performed to optimize the geometry and verify the stresses and accelerations on the elements in operation. Moreover, due to the criticality of the mechanism, a mock-up was built and tested, to verify the precision and repeatability of the system after a high number of cycles (Figure 2). The sample holder is an aluminium rotatable barrel, with 16 positions for material samples and instrumentation (Figure 3), featuring elliptical holes for the passage of the instrumentation wires.

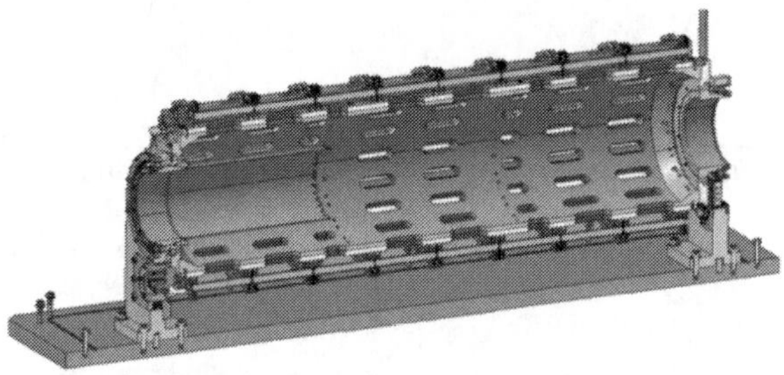

Figure 4. Section view of the sample holder.

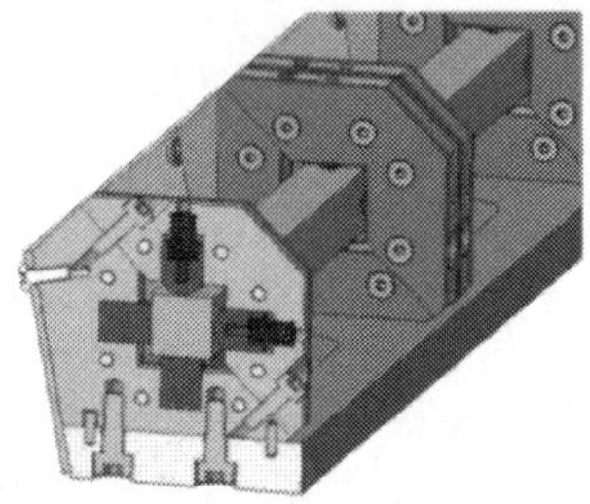

Figure 5. Specimen supporting system.

The supporting system hosts targets of variable length, for a total of 1 m per line, and a section with typical dimension comprised between 8 and 12 mm. Supports are adjustable by means of pins, to match the required specific length for each material (Figure 4). The targets can be grouped in six categories:

- *Pure carbon materials*, such as CFC, isotropic and pyrolytic graphite, and carbon foams.
- *Metal carbide – graphite composites,* with a graphitic matrix reinforced by molybdenum and/or titanium carbides, with the possible addition of carbon fibres [15],[16]. Coated samples will also be tested.
- *Copper – diamond*, of different grades, exploring in particular the nature of binders enhancing the resistance of the interface matrix/particle [17].
- *Heavy alloys*, mostly based on molybdenum, tantalum and tungsten [18],[19]. The upgraded compositions do not present a low melting phase, to enhance the robustness to beam impact.
- *Silicon carbide*, possessing high thermo-mechanical robustness, also appealing for other applications outside particle accelerators (aerospace, nuclear plants, braking elements).
- *Monitoring and actuating devices* to correct beam-induced jaw distortion in HL-LHC collimators.

Table 2: First flexural frequency of five target materials. Results are in Hz.

	Carbon foam	Metal Graphite	Silicon carbide	Pure Mo	Tungsten alloy
Fixed ends	84.2	170.7	1485.0	754.9	592.7
Simply supported	38.3	77.7	676.2	344.4	271.0
Elastic supports	41.4	83.9	729.8	375.7	295.2

The specimens are placed on graphitic elastic supports at the two extremities, with a contact pressure granted by springs with a spring ratio of 1 N·mm^{-1} (Figure 5). A spacer is adopted to change the spring pre-compression adapting the specimen to the desired size. Graphite was chosen because of the transparency to the beam, for its shock damping properties and low shock impedance [20]0. Modal analyses were performed to calculate the flexural frequency of each sample, which will be excited through off-centre beam impacts, and the results show that the system is close to an ideal configuration of simple supports at each end. Table 2 shows the modal results for five different materials.

One target station is dedicated to the testing of a monitoring and actuation system under development at the University of Huddersfield and CERN. Aim of the system is to dynamically correct distortions in the collimator jaws, caused by in-operation events. The distortion detection is performed via Dispersed Reference Interferometry (DRI) by means of optical fibres. The actuation is currently based on piezo-actuators with a rise time in the order of µs, which is the period of flexural oscillations induced on the collimator jaw in case of beam impact. The system is also capable of a static correction of shape errors due to machining tolerances. The principle will be tested on a dedicated target station, correcting the flexural oscillations which will be generated on an aluminium sample (Figure 6).

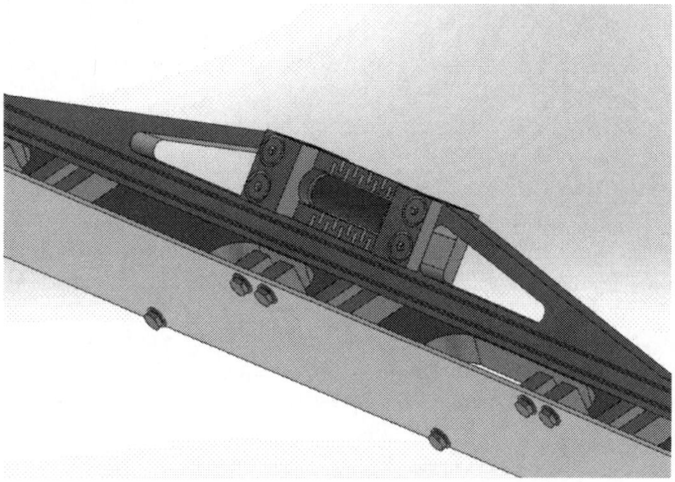

Figure 6. Piezo-actuator for bending correction.

3. Sample dynamic response

Simulations and analytical methods can well predict the behaviour of the material under beam impact [21]. However, these calculations typically include a number of simplifications and assumptions, usually neglecting the dispersive effects due to geometry, impulse shape, material inelasticity, internal friction and porosities. In order to better investigate such mechanisms, the target geometry was optimized to generate uniaxial strains easily detectable by the instrumentation. In HRMT14, measurements on the low-density materials were weak and sometimes in the noise range, as the setup was studied for exciting radial waves, decaying in amplitude with the radius. In Multimat, axial waves will be excited; in this sense, it is important to minimize the dispersion effects due to radial inertia, which affects the quality of the acquired signal and the velocity profile of the wave. This, for a mechanical impulse, can be verified with the *Pochhammer equation* [22]:

$$\frac{c_p}{c_0} = 1 - 3v^2\pi^2\left(\frac{r}{\Lambda}\right)^2 \tag{1}$$

where c_p is the longitudinal wave velocity, c_0 is the sound speed in uniaxial stress condition, v is the Poisson's ratio, r is the bar radius and Λ the length of the wave. As similar studies for thermal-induced stress waves are scarce in literature, an analytical model not accounting for radial inertia and a numerical one, including radial dispersive components, material nonlinearities, boundary effects, as well as thermal energy distribution variable in z, were built to evaluate the target response. As seen in Figure 7 and Figure 8, for the optimized sample geometry the two models are in good agreement, and the radial inertia contributions are negligible, with the wave travelling at a velocity c_p~c_0. Interestingly, due to the variable energy distribution over z, the two planar waves have a different amplitude.

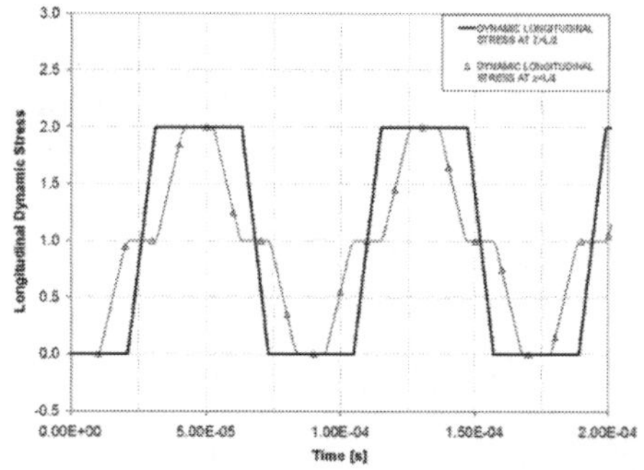

Figure 7. Axial waves in graphitic target impacted by particle beam, analytical model.

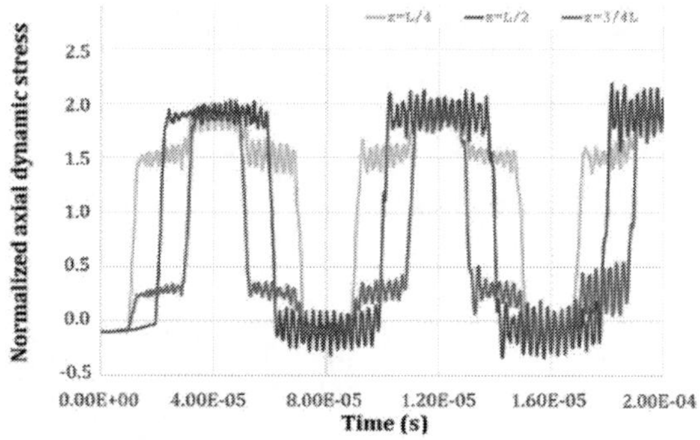

Figure 8. Axial waves in graphitic target impacted by particle beam, numerical model.

The frequency of the axial wave depends on the material, and ranges from 5 to 50 kHz. An off-centre beam impact will induce a higher amplitude, lower frequency (0.1÷2 kHz) flexural oscillation, which superposes with the axial wave (Figure 9, Figure 10).

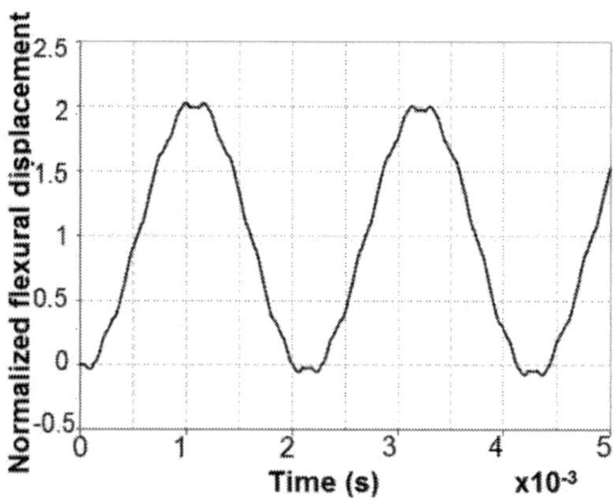

Figure 9. Flexural waves under off-centre impact on graphite targets, analytical model.

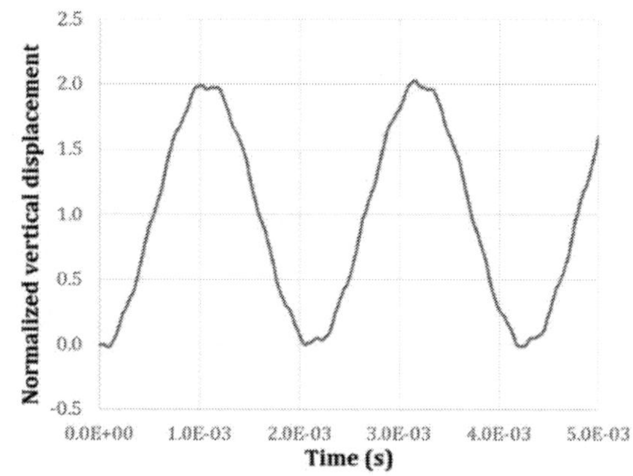

Figure 10. Flexural waves under off-centre impact on graphite targets, numerical model.

As mentioned, the numerical model is able to take into account a number of complexities to the problem; however, its validity must be benchmarked against a real scenario, in particular when dealing with novel materials for which strain, temperature and time dependent properties are scarce. Moreover, to the authors' knowledge, no dynamic model exists to predict the behaviour under beam impact of thin coatings, here adopted to increase the material electrical conductivity.

4. Conclusions

The energy increase in future particle accelerators calls for the development of unconventional materials and technologies. Although numerical codes can predict the response of a material under particle beam impact, the precision of the results is strongly related to the detailed understanding of the material models, such as the equation of state and the constitutive response. Furthermore, the dynamic behaviour of coatings applied on advanced composites, in the case of quasi-instantaneous heating, is not easy to explore with alternative methods. An experiment is foreseen in the HiRadMat facility at CERN at the end of 2017, to explore unusual properties of beam intercepting device materials, shedding light on the less explored mechanisms of the material behaviour, validating the material solution for HL-LHC.

5. Acknowledgements
Part of the work described in this thesis was developed in the scope of the EuCARD-2 Project, WP11 "ColMat-HDED", co-funded by the partners and the European Commission under Capacities 7th Framework Programme, Grant Agreement n. 312453. Research supported by the HL-LHC project.

6. References

[1] Apollinari G, Béjar Alonso I, Brüning O, Lamont M, Rossi L 2015 High-Luminosity Large Hadron Collider (HL-LHC): Preliminary Design Report (CERN) *Rep. CERN CERN-2015-005.*
[2] Benedikt M, Zimmerman F 2015 Future circular colliders *Rep. CERN CERN-ACC-2015-164.*
[3] Bertarelli A 2016 Beam-induced damage mechanisms and their calculation. *CERN Yellow Reports* **2** p 159.
[4] Efthymiopoulos I *et al.* 2011 HiRadMat: a New Irradiation Facility for Material Testing at CERN *Proc. IPAC'11 (San Sebastián)* pp 1665–1667.
[5] Bertarelli A *et al.* 2016 Permanent deformation of the LHC collimator jaws induced by shock beam impact: an analytical and numerical interpretation *Proc. EPAC'06 (Edinburgh)* pp 1801–1803.
[6] Cauchi M *et al.* 2014 High energy beam impact tests on a LHC tertiary collimator at the CERN high-radiation to materials facility *Phys. Rev. ST Accel. Beams* **17** 021004.
[7] Carra F *et al.* 2014 Mechanical engineering and design of novel collimators for HL-LHC *Proc. IPAC'14 (Dresden)* pp 369–372.
[8] Bertarelli A *et al.* 2012 High Energy Tests of Advanced Materials for Beam Intercepting Devices at CERN HiRadMat Facility *Proc. HB2012 (Beijing)* MOP240.
[9] Ansys Mechanical User's Guide – Release 15.0, ANSYS, Inc. 2013.
[10] Ansys AUTODYN User's Manual – Release 15.0, ANSYS, Inc. 2013.
[11] Gladman B *et al.* 2007 LS-DYNA® Keyword User's Manual – Volume I – Version971 *LSTC.*
[12] Bertarelli A *et al.* 2013 Behaviour of advanced materials impacted by high energy particle beams *Journal of Physics: Conference Series* **451** conference 1.
[13] Bertarelli A *et al.* 2013 An experiment to test advanced materials impacted by intense proton pulses at CERN HiRadMat facility *Nuclear Instruments and Methods in Physics Research B* **308** pp 88–99.
[14] Quaranta E *et al.* 2016 Towards optimum material choices for HL-LHC collimator upgrade *Proc. IPAC16 (Busan)* pp 2498–2501.
[15] Bertarelli A, Carra F, Mariani N and Bizzaro S 2014 Development and testing of novel advanced materials with very high thermal shock resistance *Proc. Tungsten, Refractory and Hardmetals Conference (Orlando).*
[16] Bertarelli A *et al.* 2015 Innovative MoC – graphite composite for thermal management and thermal shock applications *Thermal Measurement, Modeling & Management Symposium (SEMI-THERM)* ISSN 10652221.
[17] Bertarelli A *et al.* 2011 Research and development of novel advanced materials for next-generation collimators *Proc. IPAC'11* pp 2888–2890.
[18] Scapin M, Fichera C, Carra F, Peroni L 2015 Experimental investigation of the behaviour of tungsten and molybdenum alloys at high strain-rate and temperature *EPJ Web of Conferences* **94** 01021.
[19] Scapin M, Peroni L, Carra F 2016 Investigation and Mechanical Modelling of Pure Molybdenum at High Strain-Rate and Temperature J. *dynamic behavior mater.* **2**: 460.
[20] Peroni L, Scapin M, Carra F, Mariani N 2013 Investigation of Dynamic Fracture Behavior of Graphite *Key Engineering Materials* **569-570** pp 103–110.
[21] Bertarelli A, Dallocchio A, Kurtyka T Dynamic Response of Rapidly Heated Cylindrical Rods: Longitudinal and Flexural Behavior *J. Appl. Mech.* **75** 031010.
[22] Pochhammer L 1876 On the propagation velocities of small oscillations in an unlimited isotropic circular cylinder *J. Reine Angew. Math.*, **81**, 324.

Anomaly Detection for Beam Loss Maps in the Large Hadron Collider

Gianluca Valentino[1,2], Roderik Bruce[1], Stefano Redaelli[1], Roberto Rossi[1], Panagiotis Theodoropoulos[1], Sonja Jaster-Merz[3]

[1] CERN, Geneva 23, CH1211, Switzerland
[2] University of Malta, Msida, MSD2080, Malta
[3] University of Hamburg, 20146 Hamburg, Germany

E-mail: gianluca.valentino@cern.ch

Abstract. In the LHC, beam loss maps are used to validate collimator settings for cleaning and machine protection. This is done by monitoring the loss distribution in the ring during infrequent controlled loss map campaigns, as well as in standard operation. Due to the complexity of the system, consisting of more than 50 collimators per beam, it is difficult to identify small changes in the collimation hierarchy, which may be due to setting errors or beam orbit drifts with such methods. A technique based on Principal Component Analysis and Local Outlier Factor is presented to detect anomalies in the loss maps and therefore provide an automatic check of the collimation hierarchy.

1. Introduction

The Large Hadron Collider (LHC) at CERN accelerates and collides two counter-rotating beams [1]. Proton beams are injected at an energy of 450 GeV, before being ramped to flat top (presently 6.5 TeV). The beams are then squeezed by reducing the β-function at the collision points, before bringing them into collisions. Throughout the machine cycle, the LHC machine is protected by a beam collimation system [2], which ensures that beam losses are absorbed before they can reach the machine aperture and possibly quench the superconducting magnets. The 108 LHC collimators form a multi-stage hierarchy, with the primary collimators (TCP) positioned closest to the beam, followed by the secondary collimators (TCSG), tertiary collimators (TCTP) and absorbers (TCLA). Most of the collimators are located in the betatron cleaning Insertion Region (IR7), with the remainder concentrated in IR3 for off-momentum particle cleaning and in front of the experiments.

In order to monitor beam losses, the LHC is equipped with a Beam Loss Monitoring (BLM) system [3], consisting of around 3600 ionization chambers (ICs). The beam orbit around the ring in the horizontal and vertical planes can be measured using approximately 1100 beam position monitors (BPMs) [4]. Although the TCTPs and IR6 secondary collimators are equipped with embedded BPMs following a recent upgrade [5], the rest of the collimators need to be aligned using a beam-based technique, in which each jaw is moved in steps of 5-20 μm towards the beam until a spike is observed in the BLM signal, indicating that the beam halo has been reached [6]. The beam center is computed as the average of the two aligned jaw positions.

The collimator jaw positions are then qualified by creating high losses and observing the resulting loss map, a snapshot of the BLM signals as a function of their longitudinal position in the ring. The IR7 betatron cleaning system is qualified by adding white noise to the beam using the transverse damper (ADT), to enhance diffusion [7]. The collimation hierarchy is deemed to be correctly set up when the highest losses in the ring occur at the IR7 TCPs, followed by the TCSGs and TCLAs, such that the collimation system cleaning inefficiency, defined as the ratio of the leakage of primary particles into the IR7 dispersion suppressor superconducting magnets to the losses at the primary collimator, is below $\sim 10^{-4}$ at 6.5 TeV.

Beam-based alignment and the qualification of the determined operational positions using loss maps needs to be performed at injection, flat top, at the end of the squeeze and in collisions, with functions to move the collimators during dynamic phases [8]. Thanks to the good reproducibility of the beam orbit, a yearly alignment campaign has been sufficient, however the system is re-qualified periodically, and the loss maps are qualitatively analysed. In this paper, an anomaly detection technique is proposed to detect minor changes in the loss maps over time due to collimator settings errors or orbit variations.

2. Loss Map Outlier Analysis

The IR7 collimation hierarchy in betatron loss maps performed in 2015 and 2016 for proton beams was analyzed. A total of 12 loss maps were performed at injection (450 GeV), while 40 were performed at top energy (6.5 TeV), for both beams (B1 and B2) in the horizontal (H) and vertical (V) planes. The loss maps at top energy were done to qualify various configurations of the machine cycle (flat top, squeezed separated and colliding beams), however they can be accumulated into a single dataset as the hierarchy in IR7 is independent of the different stages which only affect the experimental regions.

A 'background' consisting of the BLM signals with beam in the machine without any excitation is first subtracted from the readings at the loss map timestamp, and the resulting values are normalized to the highest loss. From the 260 BLMs in the IR7 collimation region (19400 - 20600 m), 77 BLMs at the collimator locations in both beams, as well as empty collimator slots, were used as features for anomaly detection. In addition, all BPMs in the same region were considered. Principal Component Analysis (PCA) uses an orthogonal transformation to reduce the dimensions of a dataset by projecting it to a lower dimensional subspace. As the largest two eigenvalues account for 95% of all information in the loss map datasets [9], PCA was used to map the data to two dimensions to visualize the outliers. Local Outlier Factor (LOF) [10] is an unsupervised learning technique which finds anomalous data points by measuring the local deviation of a given data point with respect to its neighbors. Therefore, LOF was used to compute a score for each loss map utilizing all features, with larger scores indicating a larger likelihood that the loss map is an outlier with respect to others.

3. Results

3.1. Collimator Displacement Tests at 450 GeV

A dedicated test was performed in May 2015, in which a B1V loss map was performed at 450 GeV with the center of the TCLA.A6R7.B1 (vertical collimator) shifted by 200 μm with respect to the beam-based alignment center measured in beam commissioning a month earlier. The loss maps for these two cases are shown in Fig. 1, where the red and blue lines indicate BLMs in the warm and cold regions of the machine respectively, while the black lines indicate BLMs at collimators. A slight increase in the BLM signals of the loss map with center displacement is apparent at $\sim$20100 m (indicated by the green marker). This loss map appears as an outlier with respect to other 450 GeV B1V loss maps in 2015 after performing PCA, shown in Fig. 2 (index 2 marked in red). The loss map indices are sorted chronologically. The BPM readings in IR7 B1V during the 8 loss maps performed at 450 GeV in 2015 are reported in Fig. 3, indicating a good reproducibility of the orbit throughout the period when the loss maps were done.

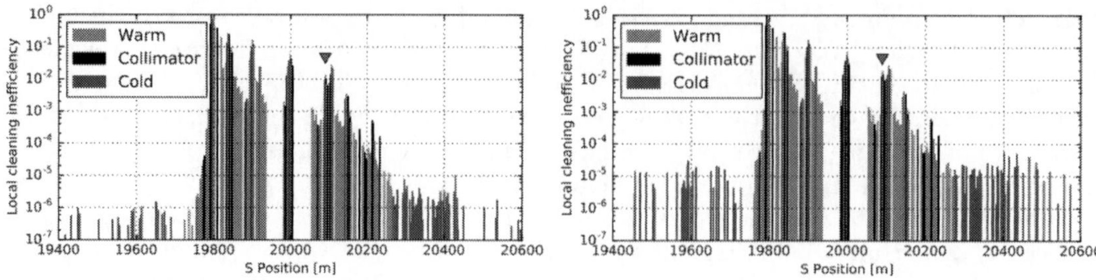

Figure 1. B1V loss maps at 450 GeV (IR7 zoom) for nominal collimator settings (left) and a 200 μm displacement in the TCLA.A6R7.B1 center (right).

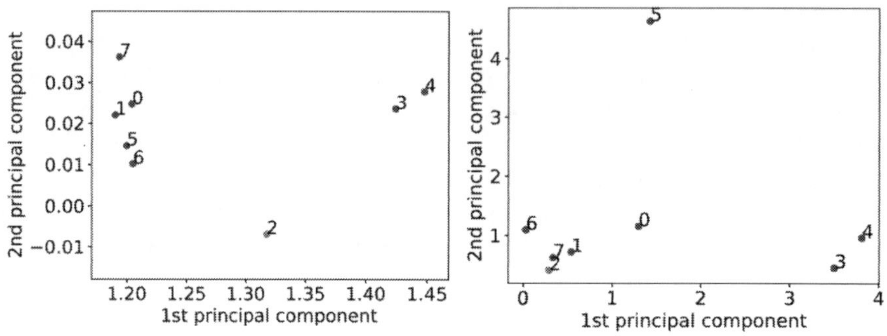

Figure 2. Principal Components Analysis of the BLM (left) and BPM (right) datasets for 450 GeV B1V loss maps.

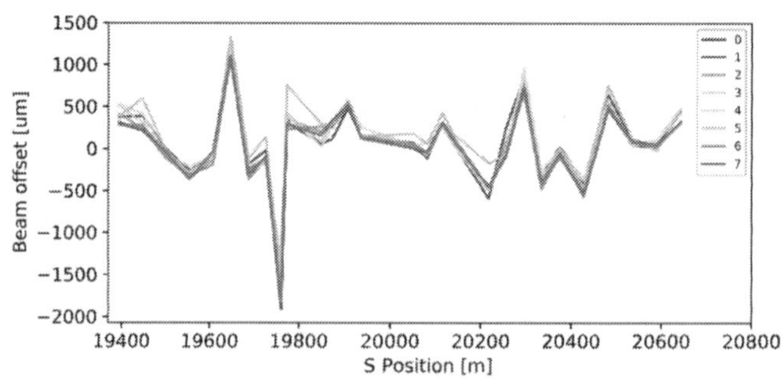

Figure 3. B1V BPM values in IR7 for 450 GeV loss maps.

PCA analysis of the BPM readings shows that index 2 forms part of a cluster, therefore confirming that this particular loss map is different than the rest due to a collimator setting change rather than an orbit shift. Loss maps 3 and 4 were performed a few months afterwards, and from the BPM PCA an orbit shift is noted which is also reflected in the BLM PCA. Loss map 5 appears to be an outlier in the BPM PCA plot, however the deviation with respect to the other points is along the second principal component, which only represents ~10% of the information, and no corresponding difference is seen in the BLM PCA plot.

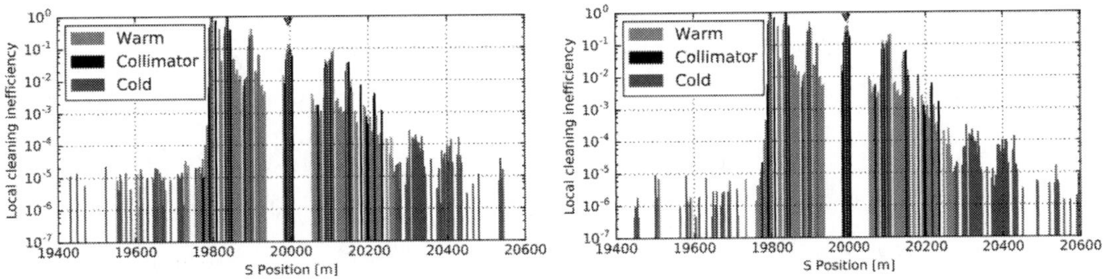

Figure 4. B1H loss maps at 6.5 TeV flat top (zoom in IR7) for standard collimator settings (left) and tight collimator settings during the hierarchy limits tests (right).

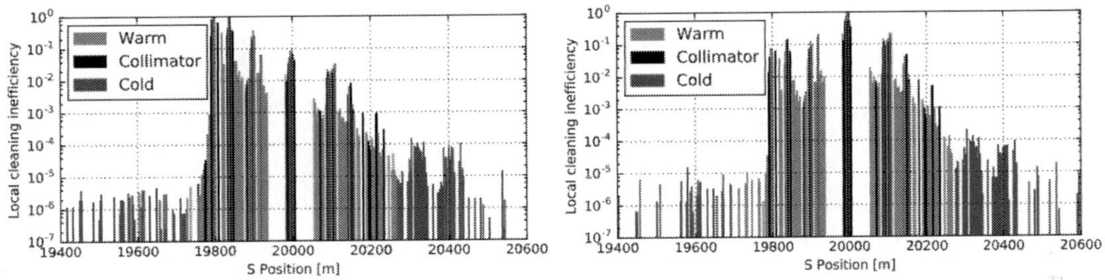

Figure 5. B1V loss maps at 6.5 TeV flat top (zoom in IR7) for nominal collimator settings (left) and tight collimator settings during the hierarchy limits tests (right).

3.2. Hierarchy Limits Tests at 6.5 TeV

The objective of the hierarchy limits tests, held in July 2016, was to evaluate the IR7 collimation hierarchy with a tighter retraction of 1 σ between the TCPs (which are placed at 5.5 σ) and TCSGs, as opposed to the usual 2 σ, to be able to achieve a lower β^* [11]. First, loss maps were performed with the standard collimator settings. Then, the IR7 TCSGs were moved from 7.5 σ to 6.5 σ and another set of loss maps was taken, and following a beam-based alignment of all IR7 collimators, another two sets of loss maps were performed, first with the collimator gaps calculated using measured beam sizes determine from the procedure in [6], and then using nominal beam sizes ($1\sigma = \sqrt{\beta\epsilon}$).

Loss maps for B1H and B1V, performed with standard 2016 collimator settings, and with tighter settings during the test before beam-based alignment, are shown in Fig. 4 and Fig. 5. For B1H, the hierarchy appears to be maintained, however on closer inspection there is a slight increase in losses at ~20000 m (see green marker), which is expected as the TCSGs are now closer to the TCPs. This is reflected in the PCA analysis in Fig. 6, in which indices 10 and 12 marked in red correspond to the loss maps done with the TCSGs at 6.5 nominal σ before alignment, and with the TCSGs at the same position but with settings calculated using nominal beam sizes, respectively.

On the other hand, a clear hierarchy breakage is evident for B1V, which is also visible in the PCA analysis in Fig. 7 (index 10). Indices 10 and 12 appear as outliers in both cases due to a tilt in the collimator tank of the TCSG.D4L7.B1, causing its aperture to be tighter than foreseen using the nominal σ. However, this issue disappears if the measured beam sizes are used (index 11), as the tilt error is reflected in the beam size and the resulting collimator gaps are now wider. The BPM readings during the B1V loss maps are reported in Fig. 8 for reference,

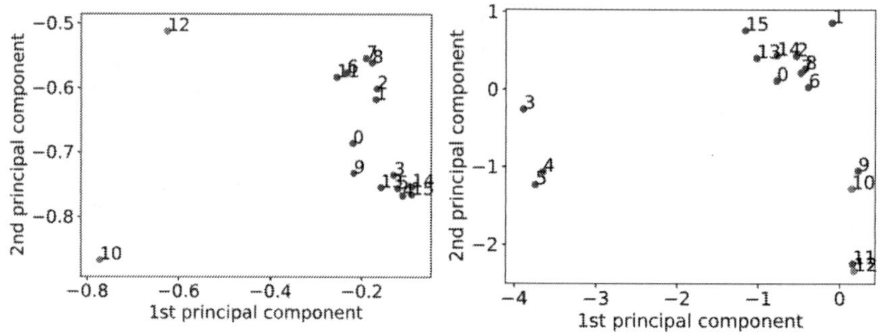

Figure 6. Principal Components Analysis of the BLM (left) and BPM (right) datasets for B1H loss maps at 6.5 TeV.

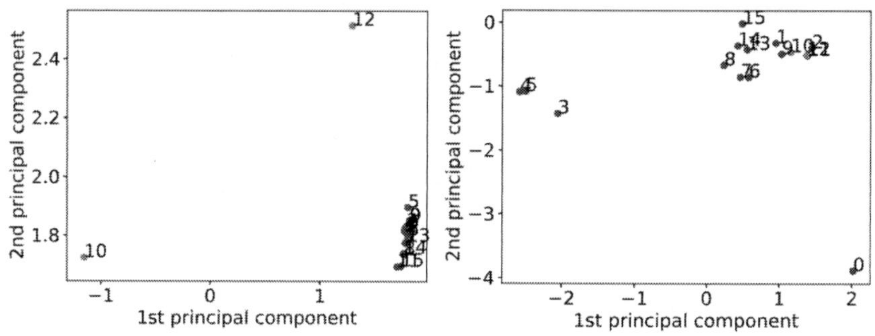

Figure 7. Principal Components Analysis of the BLM (left) and BPM (right) datasets for B1V loss maps at 6.5 TeV.

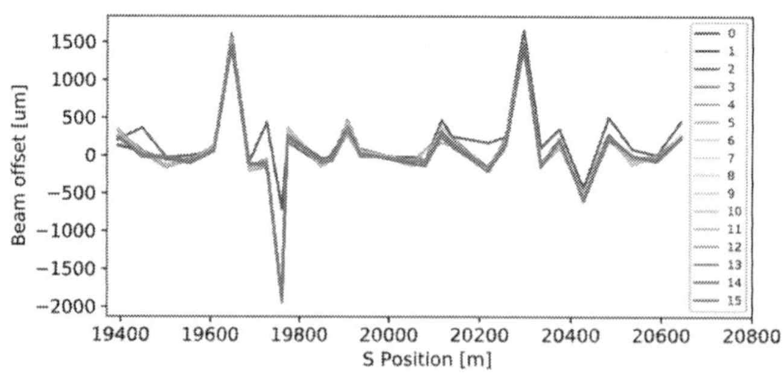

Figure 8. B1V BPM values in IR7 for loss maps at 6.5 TeV.

where it can be seen that the readings in the first loss map are different from the rest. The LOF scores obtained for the BLM and BPM datasets for B1V loss maps at 6.5 TeV are shown in Fig. 9, where it can be seen that the same outliers as in the PCA analysis are obtained.

4. Conclusions and Outlook

The LHC collimation system is qualified for various points of the machine cycle at regular intervals via beam loss maps, and visual checks of the hierarchy of beam losses in IR7 is

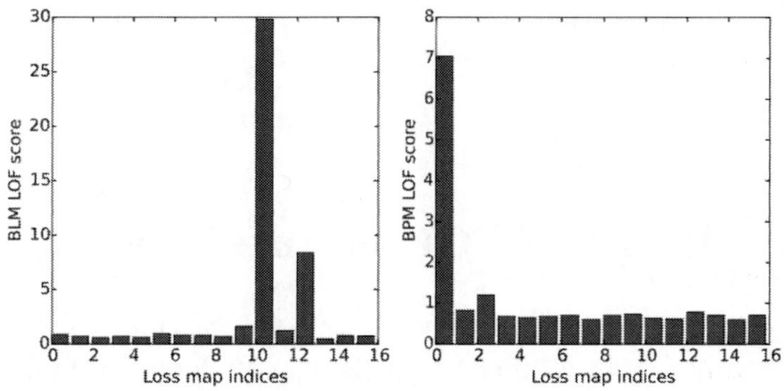

Figure 9. LOF scores obtained for BLM (left) and BPM (right) datasets for B1V loss maps at 6.5 TeV.

a good qualitative indicator of the correctness of the set up. In this paper, a quantitative method was proposed based on principal component analysis to reduce the problem to a two-dimensional space. In addition, a LOF model based on unsupervised learning was used in order to automatically identify anomalous loss maps based on the BLM and BPM measurements. This technique could also be extended to off-momentum loss maps, and to monitor losses during standard physics fills.

References

[1] Bruning O *et al.*, "LHC Design Report: the LHC Main Ring", CERN Report No. CERN-2004-003-V1, 2004.
[2] Assmann R *et al.*, *Proc. European Particle Accelerator Conference*, Edinburgh, UK, pp. 1538-1540, 2006.
[3] Dehning B *et al.*, *Proc. Particle Accelerator Conference*, Albuquerque, USA, pp. 4192-4194, 2007.
[4] Jones O, *Proc. Particle Accelerator Conference*, Albuquerque, USA, pp. 2630-2634, 2007.
[5] Valentino G *et al.*, *Proc. 6th Evian Workshop on LHC Beam Operation*, Evian, France, pp. 149-154, 2015.
[6] Valentino G *et al.*, *Phys. Rev. ST. Accel. Beams* **15**, 051002 (2012).
[7] Hofle W *et al.*, *Proc. International Particle Accelerator Conference*, New Orleans, USA, pp. 4059-4061, 2012.
[8] Valentino G, Bruce R and Redaelli S, *Proc. Machine Protection Panel Workshop*, pp. 87-92, Annecy, France, 2013.
[9] Theodoropoulos P, M.Sc. Thesis, Univ. College London, London, England, CERN-THESIS-2015-128, 2015.
[10] Breunig M, Kriegel H, Ng R and Sander J, *Proc. ACM SIGMOD*, pp. 93-104, 2000.
[11] Mereghetti A *et al.*, "β^*-reach - IR7 Collimation Hierarchy Limit and Impedance", to be published, 2017.

Preliminary design of FCC-ee pre-injector complex

S Ogur[1,2], A Barnyakov[3], K Furukawa[4], N Iida[4], A Levichev[3], F Miyahara[4], D Nikiforov[3], K Oide[4], Y Papaphilippou[2], F Zimmermann[2]

[1] Bogazici University, Bebek 34342, Istanbul, Turkey
[2] CERN, CH-1211, Geneva 23, Switzerland
[3] BINP RAS, Novosibirsk, Novosibirskaya oblast' 630090, Russia
[4] High Energy Accelerator Research Organization (KEK), Tsukuba 305-0801, Japan

E-mail: salim.ogur@cern.ch

Abstract. The design of a 100 km circular e^+e^- collider with extremely high luminosity is an important component of the global Future Circular Collider (FCC) study hosted by CERN. FCC-ee is being designed to serve as Z, W, H and top factory, covering beam energies from 45.6 to 175 GeV. For the injectors, the Z-operation is the most challenging mode, due to the high total charge and low equilibrium emittance in the collider at this energy. Thus, fulfilling the Z-mode will also meet the demands for all other modes of FCC-ee. This goal can be achieved by using a 6 GeV NC linac with an S-band RF frequency of 2.856 GHz and a repetition rate of 100 Hz. This linac will accelerate two bunches per RF pulse, each with a charge of 6.5 nC. Positrons will be generated by sending 4.46 GeV e^- onto a hybrid target so that the e^+ created can still be accelerated to 1.54 GeV in the remaining part of the same linac. The emittance of the e^+ beam will then shrink to the nm level in a 1.54 GeV damping ring. After damping, the e^+ will be reinjected into the linac and accelerated to 6 GeV. The e^- and e^+ will then be accelerated alternately to 45.6 GeV in the booster, before they are injected into the collider.

1. Introduction

In a 100 km circular tunnel, the FCC will host a lepton-lepton, a hadron-hadron, and a lepton-hadron colliders, respectively as the project evolves. Among those, the FCC-ee will provide e^+e^- collisions at high luminosity in order to enable precision studies of several fundamental elementary particles, such as the Z and W bosons, the Higgs boson and the top quark. The FCC-ee baseline parameters depend on the particle type to be produced and the associated beam energy [1].

Table 1. FCC-ee baseline parameters on the Z pole.

parameter	value
beam energy	45.6 GeV
luminosity	0.9×10^{36} cm^{-2}s^{-1}
no. bunches / beam	91500
bunch population	3.3×10^{10}
horiz., vert. emittance	90, 1 pm

The injector design has focused on supplying the beam required for operation on the Z pole, where the luminosity and beam current are the highest, and the emittance the lowest. These parameters are shown in Table 1. Producing the heavier particles (W, H and $t\bar{t}$) requires a higher extraction energy of the booster, but lower beam intensity from the injector.

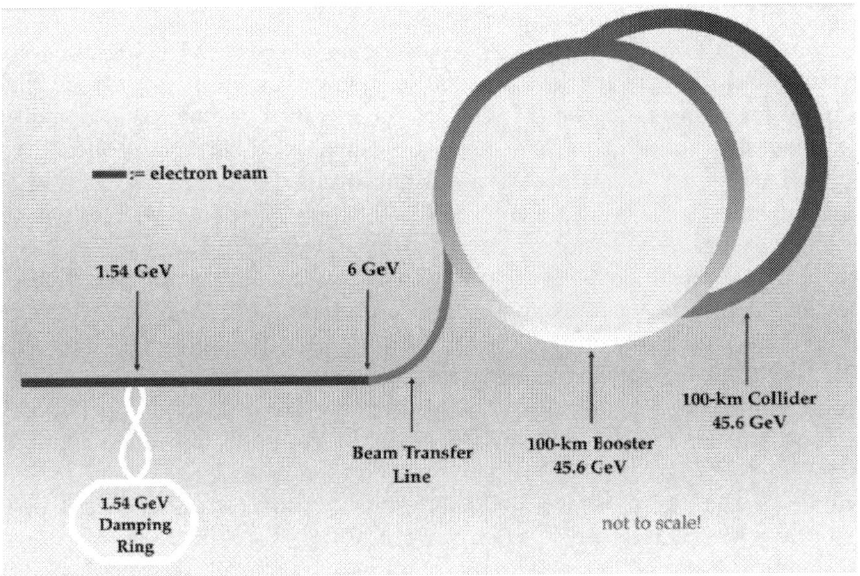

Figure 1. FCC-ee electron flow scheme for Z-pole operation.

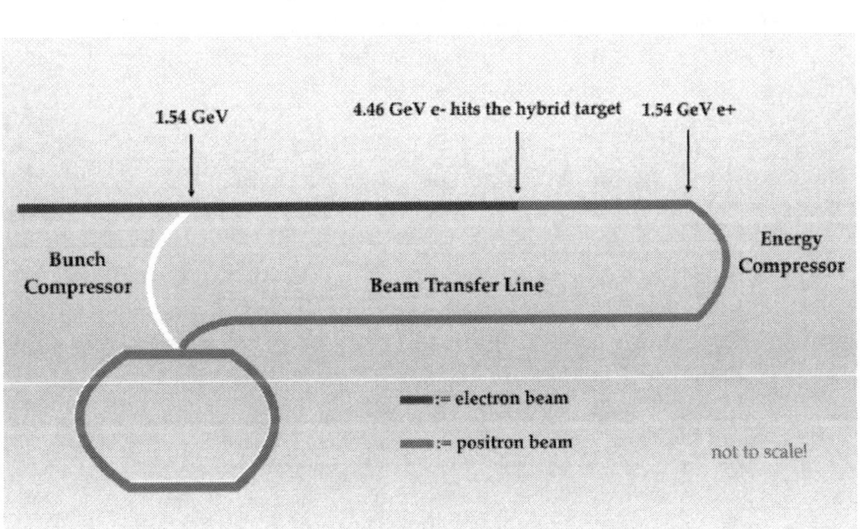

Figure 2. FCC-ee positron flow scheme for the Z-pole operation.

In this paper we do not discuss the detailed design of the booster [2] or of a possible pre-booster [3], but we focus on the upstream linac and damping ring (DR), whose energies will stay fixed for all modes of operation. The performance of linac and DR has been simulated and optimized using the code SAD [4].

The design of the FCC-ee pre-injector is based on available accelerator technology and aims at achieving target performance with minimum cost. Alternatively, CLIC injector parameters can also be considered [5]. A normal-conducting linac operating up to 6 GeV can meet the objectives. An S-band frequency of 2.856 GHz has been chosen for the linac radiofrequency cavities, in line with existing large S-band linacs (such as SLAC, Pohang, KEK), so as to profit from the available infrastructure and experience.

Recent design effort has considered that the electrons, accelerated in the linac up to 6 GeV, will be directly injected into the 100 km top-up booster, which will further raise their energy to 45.6 GeV, suitable for transfer to the collider. If the electron emittance is acceptable without DR, the latter may stay idle during the e^- beam delivery. The corresponding electron flow scheme is depicted in Figure 1. Alternative options would be extending the linac to 10 GeV or adding a pre-booster [3] covering an intermediate range from 6 to 20 GeV. The same linac and the same DR (if employed) as for e^- are used for e^+. Hence, electrons at 4.46 GeV in the linac hit a hybrid e^+ target [6], generating positrons. The remaining portion of the 6 GeV linac will accelerate the captured e^+ to 1.54 GeV. These positrons are then sent through an energy compressor and transferred to the 1.54 GeV DR. After radiation damping, the positrons will be sent back to the linac, via a bunch compressor section, so as to finally also reach 6 GeV, when they can be injected into the top-up booster. This paper will consider the option of 6 GeV linac, the creation of positrons at 4.46 GeV inside the same linac, and a damping ring of 1.54 GeV, as shown in Figure 2.

FCC-ee will accumulate up to about 90,000 e^- and e^+ bunches in less than 1 hour [7]. Using the same accelerators alternately for e^+ and e^- results in a tight time schedule for the injector. Still, for the schedule developed, each positron would spend 50 milliseconds inside the DR.

2. 6 GeV S-Band Linac

We have designed the optics of a normal conducting linac as shown in Figure 3 and listed the design parameters in Table 2. This linac comprises two types of accelerating structures: 1) large aperture cavities (diameter 40 mm) with a module length of 1.50 meter; 2) narrow aperture cavities (20 mm) with a module length of 2.97 meter.

Table 2. Linac Design Parameters.

Parameter	Value
injection-extraction energy	12 MeV-6 GeV
initial emittance (x/y)	0.35/0.5 μm
final emittance w/o blowup (x/y)	0.7/1.0 nm
linac repetition	100 Hz
bunch charge, bunches/RF pulse	6.5 nC, 2
unloaded cavity gradient	$\leq$ 26 MV

Tracking simulations were set up including both transverse and longitudinal wakefields [8]. Quadrupoles and cavities are randomly displaced according to a Gaussian distribution with an rms misalignment of 0.1 mm. Also a random injection error of 0.1 mm and 1 mrad rms is considered. Beam position monitors are taken to be attached to the entrances and the exits of the accelerating structures. Simulations were performed for 10^6 macro-particles with a Gaussian distribution in all planes, and an rms length of σ_z =1 mm. Although at most 3.3×10^{10} e^+ and e^- per bunch are required if a full collider bunch is to be injected in a single shot, the initial bunch

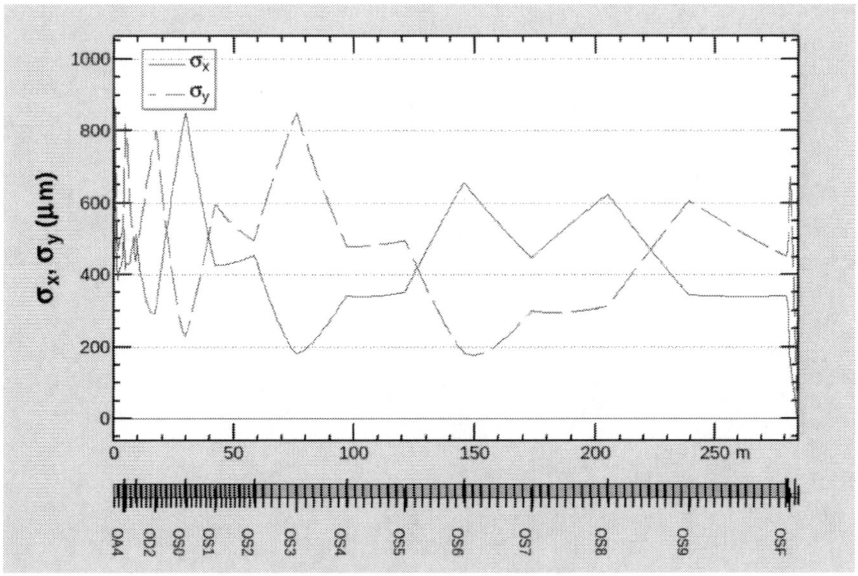

Figure 3. Linac optics including pairs of steerers and larger-aperture cavities.

population is taken to be even higher, namely 4.0×10^{10} (i.e. 6.5 nC) per bunch, to introduce a margin for downstream losses.

Introducing misalignment and injection errors dramatically changes the optics and the transmission [9]. Emittance blow up and beam loss are mitigated by steering magnets and by the larger-aperture cavities. Pairs of adjacent steering correctors are used to execute an automatic orbit correction, so as to steer the beam to the cavity center, controlling both position and angle. Up to 1.1 GeV 28 larger-aperture cavities are deployed to reduce the impact of the wakefields. These are followed by 64 smaller-iris 3 m long accelerating structures. The tracking was carried out for many random seeds. The emittance blow up is always smaller than a factor of ten, and the transmission above 95%. As an illustration, the beam profile for one random seed is displayed in Figure 4. In this example, the transmission is 100% at 6 GeV, the transverse (x/y) emittances are 2.3/5.0 nm, i.e. a few times higher than the emittances expected without any blow up, see Table 2. However, this blow up can be damped in the booster in 1 sec or, possibly, in the DR in few ms.

3. 1.54 GeV damping ring

The positron creation by high-energy electrons hitting a target inevitably results in a huge energy spread and large transverse emittance. A damping ring is required to reduce the e^+ emittance by about 3 orders of magnitude.

The positron production and positron transport from the target up to the damping ring has been modelled using KEK e^+ software [10], and assuming the relevant FCC-ee parameters.

Table 3 shows the main DR parameters. The design goal was to minimize natural emittance and damping time. The DR can host up to 6 bunches, or 3 trains of 2 bunches, with 121 nanoseconds of separation between any two bunches. This bunch separation enables a kicker to ramp up, flatten and ramp down, with a total duration of less than 363 ns. The kicker pulse should have a plateau lasting more than 121 ns such that trains consisting of two bunches can be injected or extracted. The injection and extraction will happen in the same straight section of the DR, possibly sharing a kicker. Each positron bunch experiences 50 ms of damping[7].

The DR consists of 2 straight sections housing in total 4 wigglers of 6.64 m length each. One

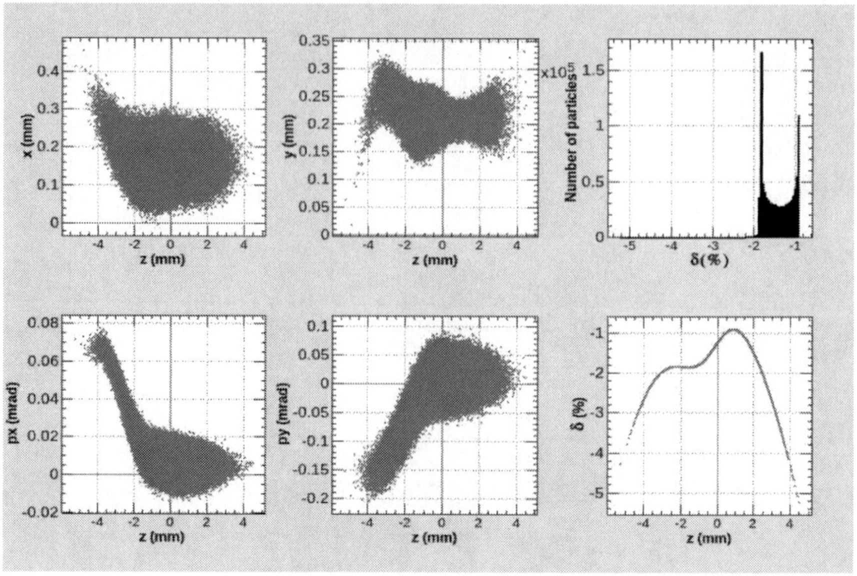

Figure 4. Linac beam profile at 6 GeV, after orbit steering.

Table 3. Damping ring design parameters.

parameter	value
energy	1.54 GeV
circumference	217.6 m
no. trains, bunches/train	3, 2
bunch charge, spacing	6.5 nC, 121 ns
no. of cells in arc, cell length	49, 1.56 m
FODO cell phase advance (x/y)	69.5/66.1 deg
betatron tune (x/y)	21.19/20.14
natural emittance (x/y)	1.00/- nm
damping time (x/y)	10.4/10.7 ms
bending radius, wiggler field	7.1 m , 1.66 T
RF voltage, frequency	5 MV, 400 MHz

of the straight sections also contains a 8.56 m of drift space reserved for injection/extraction and the opposite section hosts two LHC-type 400 MHz superconducting cavities of 1.5 meter length (3.5 m with cryostat). The DR optics is displayed in Figure 5.

The dynamic acceptance of the DR simulated over two damping times, $2 \times \tau_x = 20.8$ ms (i.e. 28900 turns), and including synchrotron radiation, is presented in Figure 6.

The DR performance is summarized in Table 4. The positron emittance can be damped from about 1 μm down to approximately 1 nm in 50 ms (i.e. about 5 damping times), that is indeed better than required. In other words, even if the emittance of the e^+ injected into the DR were a few times larger, one could still obtain the target emittance; see Table 4.

IOP Conf. Series: Journal of Physics: Conf. Series **874** (2017) 012003 doi:10.1088/1742-6596/874/1/012003

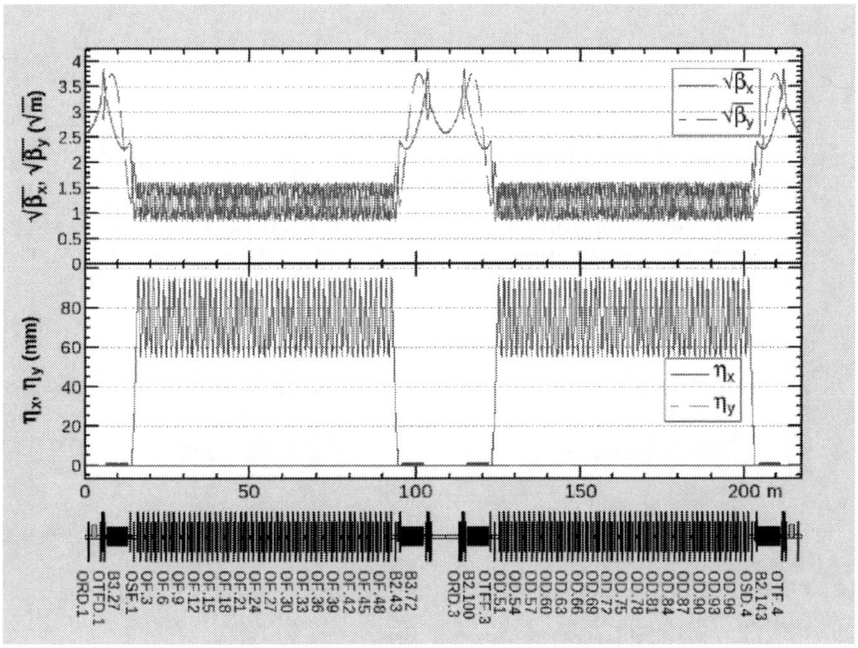

Figure 5. Damping ring optics.

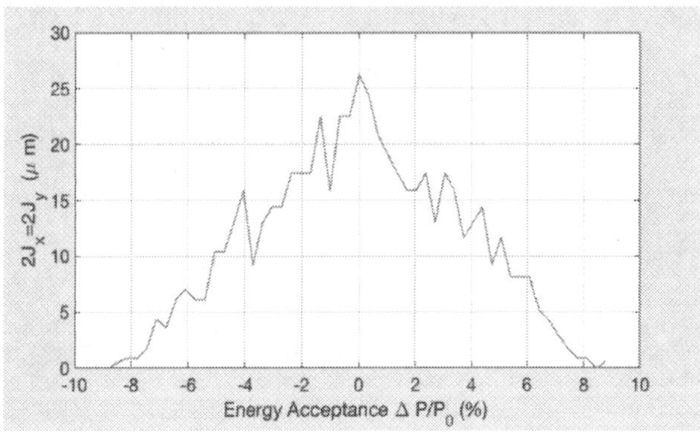

Figure 6. Simulated dynamic acceptance of the DR calculated over two damping times.

4. Conclusions

The 6 GeV S-band linac promises a nearly perfect transmission. Transverse wakefields increase the normalized emittance by a factor of a few, which seems acceptable. The linac is 285.5 meters long; the cavity gradient $\leq$26 MV/m, for all cavities. Positrons must be cooled in the 1.54 GeV damping ring. Optionally, the electron beam can also be passed through this DR. Without errors the DR as designed has a dynamic aperture of $\geq 100\sigma$ in both transverse and longitudinal directions.

Near-term plans include acceleration of more bunches per RF pulse to reduce the first fill time, linac simulations with BPM errors, optics models of all transfer lines and compressors, a baseline configuration for the e^+ source and the design of an e^- RF gun with 0.35/0.50 μm (x/y) emittance at 12 MeV.

Table 4. Damping ring performance without errors.

parameter	value
natural emittance (x)	1.00 nm
dynamic aperture	162 σ_x
longitudinal natural emittance	1.29 μm
longitudinal dynamic aperture	$\pm$ 125 σ
energy acceptance	$\pm$8.1 %
bucket height	8.3 %
energy spread	6.76 $\times 10^{-4}$
injected emittance (x/y)	0.76/0.72 μm
extracted emittance (x/y)	1.07/0.07 nm
target emittance (x/y)	2.66/3.90 nm

Acknowledgments

This work was supported in part by the European Commission under the HORIZON2020 Integrating Activity project ARIES, grant agreement 730871.

Authors are thankful to S. Polozov and colleagues at MEPhI for their contributions to the RF Gun study. Meanwhile the positron creation and capture study are initiated by the collaboration with I. Chaikovska and R. Chehab of LAL and P. Martyshkin of BINP.

References

[1] Oide K *et al.* 2016 Design of Beam Optics for the FCC-ee Collider Ring *Proceedings of IPAC2016* Busan S. Korea 3821
[2] Harer B *et al.* 2017 Challenges and Status of the Rapid Cycling Top-up Booster for FCC-ee *Proceedings of IPAC2017* Copenhagen Denmark WEPIK031
[3] Etisken O *et al.* 2017 Conceptual Design of a Pre-Booster Ring for the FCC e^+e^- Injector *Proceedings of IPAC2017* Copenhagen Denmark MOPVA029
[4] SAD (Strategical Accelerator Design) Program, `http://acc-physics.kek.jp/SAD/index.html`
[5] Papaphilippou Y *et al.* 2016 Design Guidelines for the Injector Complex of the FCC-ee *Proceedings of IPAC2016* Busan S. Korea 3488
[6] Chaikovska I *et al.* 2017 Experimental Activities on High Intensity Positron Sources Using Channeling *Proceedings of IPAC2017* Copenhagen Denmark WEPIK002
[7] Ogur S *et al.* FCC-ee Pre-booster Accelerators *Proceedings of CERN-BINP Workshop* Geneva Switzerland
[8] Yokoya K 1998 Short-Range Wake Formulas for Infinite Periodic Pill-Box (unpublished)
[9] Ogur S *et al.* 2016 Towards a Preliminary FCC-ee Injector Design *Proceedings of eeFACT2016* Manchester United Kingdom TUT2H5
[10] Iida N *et al.* 2011 Beam Dynamics in Positron Injector Systems for the Next Generation b-Factories *Proceedings of IPAC 2011* San Sebastian Spain 2857

Synchrotron radiation backgrounds for the FCC-hh experiments

F Collamati[1], M Boscolo[2], H Burkhardt[3] and R Kersevan[3]

[1] Istituto Nazionale di Fisica Nucleare, INFN, Rome, IT;
[2] Istituto Nazionale di Fisica Nucleare, INFN, Frascati, IT;
[3] CERN, Geneve, CH

E-mail: francesco.collamati@roma1.infn.it

Abstract. In this paper we present a detailed analysis of the Synchrotron Radiation emitted by the 50 TeV protons of the FCC-hh in the last bending and quadrupole magnets upstream of the interaction region. We discuss the characteristics of this radiation in terms of power, flux, photon spectrum and fans with and without crossing angle for comparison. We mainly focus our study on the fraction of photons that may hit the detector, with a full tracking in GEANT4 that simulates the interaction within the central beam pipe.

1. Introduction

The international Future Circular Collider (FCC) study [1] aims at a design of p-p, e^+e^-, e-p colliders to be built on a new 100 km tunnel in the Geneva region. The final goal is a hadron collider at centre-of-mass energies of the order of 100 TeV, the intermediate steps being a e^+e^- collider for the 90 to 400 GeV centre of mass energy range. To reach such unprecedented energies, in addition to the requirement for ultra high luminosities, a careful study of all the background sources both in the machine and in the experimental area must be performed, including the evaluation of Synchrotron Radiation emitted by hadrons.

2. Motivation and Scope

The amount of power radiated by Synchrotron Radiation (SR) strongly depends on the relativistic γ factor of the particle, and thus on its mass: $P \propto \gamma^4 \rightarrow P \propto m^{-4}$.

For this reason, SR power emitted by protons is several orders of magnitude lower than the one emitted by electrons, and has been so far an almost negligible contribution to the background also for high energy hadron colliders like LHC.

However, as we aim at unprecedented centre-of-mass energy as high as 100 TeV, the amount of SR power should be also evaluated. In fact, even if the increase of FCC-hh in the centre-of-mass energy with respect to LHC is of about a factor 7, the increase in SR power is of about a factor 170. On the other hand, the *critical energy* of the photons emitted by the proton beam goes with γ^3/ρ, and thus for FCC-hh turns out to be about a factor 100 greater than the one in LHC. This shifts the spectrum from hard ultraviolet for LHC which is easily absorbed into the X-ray range for FCC, for which the Beryllium of the inner beam pipe starts to become transparent. In this paper we present the study of the SR produced in the bending magnets upstream of the Interaction Region and we investigate the SR that may reach the detector

Table 1. Summary of s position along the lattice, length, strength, magnetic field, SR critical energy and total power for the last 4 bending magnets

Element	S [m]	l [m]	B [T]	E_{crit} [keV]	P [W]
$D1_A$	231	12.5	-4.3	1.15	32
$D1_B$	245	12.5	-4.3	1.15	32
$D2_A$	427	15	3.6	0.96	27
$D2_B$	443	15	3.6	0.96	27

passing through the main absorber that protects the insertion quadrupoles from collision debris, called TAS as for the LHC, which stands for *Target Absorber Secondaries.* The goal of the study is to understand the impact in the experiments in terms of photon energy or flux. In Figure 1 we show the present design of the FCC-hh Interaction Region [2]. To perform this study, we used two different software tools: *MDISim* [3] and *SYNRAD+* [4], quite different in their approach. *MDISim* is a toolkit that combines existing standard tools (MAD-X, ROOT and GEANT4). It reads the MAD-X optics files, and uses its twiss output file to generate the geometry and the magnetic field information in a format which can be directly imported in Geant4 to perform full tracking, including the generation of secondaries and detailed modeling of the relevant absorption processes. Current limitations in GEANT4 are that the generation of the angular distribution is approximate ($1/\gamma$ cone) and that hard photon optical mirror reflection is not included [5].

SYNRAD+ takes as input a geometry of arbitrary complexity either from CAD or STL format, and magnetic fields, and uses them to simulate the beam, whose starting characteristics at a certain point must be provided, generating and tracking SR photons (no $1/\gamma$ approximation is performed). *SYNRAD+* is able to perform a full simulation of the optical interaction (reflection, absorption) of the incident radiation with the pipe material, and has been benchmarked against peer-reviewed literature [6].
We find that results are consistent with each other, as discussed in the next paragraph.

3. Simulation Studies

3.1. *MDISim*

MDISim simulation study has been performed on the FCC-hh optics named LATTICE_V8 [7]. This optics foresees a $\beta^* = 0.3\,\text{m}$ and $L^* = 45\,\text{m}$. After running MAD-X on the lattice file, MDISim creates the ROOT file for the geometry of the beam pipe around the IP, as shown in Figure 1, where the vertical dimension has been zoomed by a factor 1000 in order to gain visibility. The apertures shown are those present inside the optics file, with the exception of the ones in proximity to the IP (i.e. behind the TAS), that were not provided in the optics description and instead added by hand to match the current design of the pipe (Beryllium pipe radius of 2 cm). The trajectory of the beam generated by MAD-X within the twiss file is also visible in Figure 1 as a continuous line inside the pipe.

Its displacement from the centre is due to the fact that the picture represents the case with the 89 μrad crossing angle between the two beams turned on. At this first level of the analysis, we calculate with MDISim the total amount of SR power produced by these magnets, from magnet strength, length and particle energy. The results are shown in table 1, which shows that the total SR power radiated towards the IP is about 100 W.

The contribution of a possible last spectrometer magnet has been estimated to be of the order of 1 W.

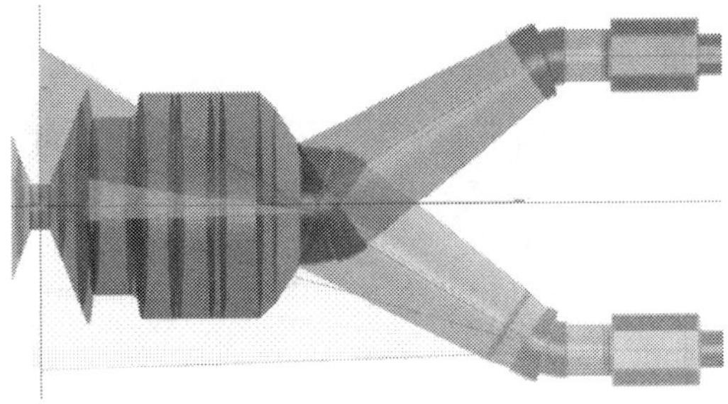

Figure 1. Top view of the geometry of the pipe around the IP as generated by MDISim and shown in ROOT. The transverse dimension has been zoomed by a factor 1000 in order to gain readability. The red area shows the angular distribution of SR photons emitted by the bending elements, the yellow the fraction reaching the TAS. The ratio of the two angles gives the fraction of produced power that actually enters the TAS

Table 2. Summary of solid angle for SR photons emitted by the last 4 bending magnets and total power that enters the TAS, with (Cr) and without ($\not{Cr}$) the crossing angle. P is the total power emitted, while f is the fraction of it entering the TAS.

El.	**P [W]**	**$f_{\not{Cr}}$ [%]**	**$P_{\not{Cr}}$ [W]**	**f_{Cr} [%]**	**P_{Cr} [W]**
$D1_A$	32	40	13	77	25
$D1_B$	32	0	0	0	0
$D2_A$	27	15	4	17	5
$D2_B$	27	0	0	0	0
TOT	-	-	17	-	30

As a second step of the analysis, we calculate the SR radiation cones for each magnetic element and illustrate them using the visualization provided by MDISim-ROOT TEvent interface, shown in Figure 1. We can see that only a fraction of the radiation emitted towards the IP will impact on the TAS absorber.

The numbers are summarized in table 2, evaluated both with and without crossing angle scheme. The power load on the TAS corresponds to about 20 W.

As third and final step, we perform the full tracking simulation for the last ± 700 m around the IP using the magnet geometry and fields imported into GEANT4. This is illustrated in Figure 2 for 10 protons starting on axis at a distance of 700 m from the IP.

The Geant4 simulation has then been used to evaluate the amount of power entering the TAS. The spectra of SR photons entering the TAS region for both cases (with and without Crossing Angle) are shown in Figure 3, while the summary of the power load is reported in table 3.

A similar approach has been used to evaluate the particles reaching the 16 m long Beryllium pipe surrounding the IP, and thus the power load on it, also shown in table 3. In summary, the SR power hitting the Be pipe is less than 1 W with crossing angle, a safe value in term of power. However, also the abundant flux of these low energy photons may play a role.

To evaluate the effect of the interaction of these photons in the Be pipe, a devoted simpler simulation has been set up, in order to avoid the unneeded complexity of the full one and obtain

Figure 2. Output of a simple 10 protons simulation, shown again in ROOT with the usual 1000x enlargement on the vertical scale.

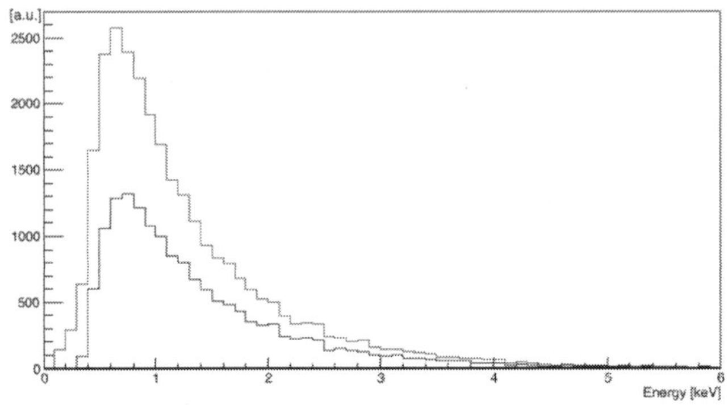

Figure 3. Spectrum of photons entering the TAS region with (in red) and without the Crossing Angle scheme (in blue).

Table 3. Summary of SR power emitted by the last 500 m from the IP that enters the TAS (P_{TAS}) or hits the Be pipe (P_{Be}), coming from the full Geant4 simulation, with or without Crossing Angle. Values are per bunch.

CrAn.	$\mathbf{N}_{\gamma TAS}$	$\bar{E}$ **[keV]**	$\mathbf{P}_{TAS}$ **[W]**	$\mathbf{P}_{Be}$**[W]**
Yes	2.9×10^9	1.28	14.6	0.8
No	1.6×10^9	1.38	8.6	0.5

a much quicker response for a high number of primaries. The geometry is simply represented by a Be tube of 2 *cm* radius and 1.2 *mm* thickness on which photons generated with the right SR spectrum (taken from the previous full simulation) impinge with a 70 μrad grazing angle.

The main conclusion from the GEANT4 simulations is that the energy spectrum of the photons hitting the Beryllium pipe remains in the few keV range where only a tiny fraction, of order 10^{-9} corresponding to one photon per bunch crossing would traverse the beam pipe and enter the detector region.

3.2. SYNRAD+

SYNRAD+ generates and traces SR photons emitted by a given beam in a given magnetic field. The user has thus to provide the magnetic fields (dipoles, quadrupoles and so on), the beam

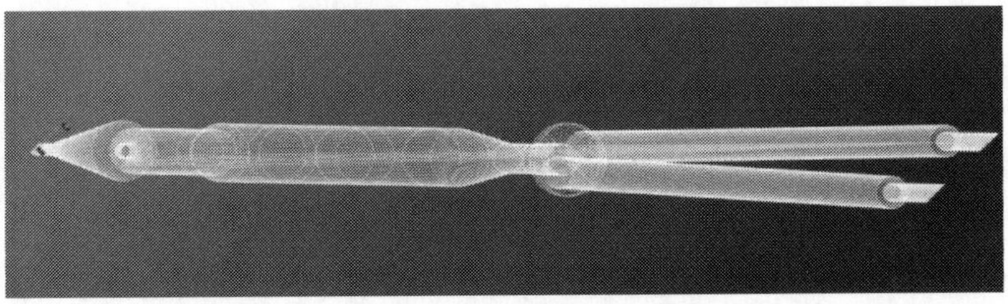

Figure 4. SYNRAD+ simulation of the last 700m towards the IP. In the middle it is visible the "LHC-like" recombination chamber, discussed in the text.

Table 4. Summary of SR power emitted in the last 500 m that enters the TAS , coming from the SYNRAD+ simulation.

Element	$P_{\not{Cr}}$ [W]	P_{Cr} [W]
$Q1$	0.01	10^{-6}
$Q2_A$	0	10^{-4}
$Q2_B$	0.1	2.2
$Q3$	0	1.2
$D1_A$	5.0	5.8
$D1_B$	0	4×10^{-5}
$D2_A$	0.1	10^{-3}
$D2_B$	0	0
TOT	**5.3**	**9.2**

parameters (emittance, energy spread and so on) at the entrance of each magnetic element, since the program does not track beam particles outside magnetic elements (i.e. in drifts), while it performs a full track of the emitted photons. Also the actual beampipe geometry has to be provided, since the program is able to perform a full simulation of the optical interaction (reflection, absorption) of the incident radiation with the pipe material, based on the the appropriate optical properties provided to SYNRAD+. For this part of the study, the optics version *fcc_hh_v6_45* was used [8]. In particular, the twiss file generated by MAD-X was used to input the characteristics of the beam at each point, while geometry and magnetic fields for the last $\sim 700\ m$ of pipe were manually created from optics apertures. All the materials were assumed to be perfect absorbers of the incident radiation, to be coherent with the MDISim part of the study.

There is a significant difference between the SYNRAD+ geometry and the MDISim one, given by the fact that some unphysical situations arise when simply importing optics apertures, in particular in the recombination chamber between the two beams, as shown in Figure 4. In the SYNRAD+ simulation, a "LHC-like" geometry for the recombination chamber has been used, as visible in Figure 4, where SR photons are also shown.

Once the 3D model is built, SYNRAD+ allows us to calculate power and spectrum of radiation hitting a certain surface by simply selecting it. The study was performed independently for the different magnetic elements (dipoles and quadrupoles), and for both cases with or without crossing angle. The results are summarised in table 4.

IOP Conf. Series: Journal of Physics: Conf. Series **874** (2017) 012004 doi:10.1088/1742-6596/874/1/012004

4. Summary

In the development of a Future Circular Collider with unprecedented energies and luminosities a careful study of all the background sources is crucial, also regarding Synchrotron Radiation entering the Interaction Area. In this paper, a systematic study has been presented, following two different and independent programs, MDISim and SYNRAD+. Both models predict that the power load inside the TAS will be close to 10 W, with small differences between the case with crossing angle and without it. Concerning the load on the inner Be pipe, about 0.5 W are expected, with a very low number of photons traversing the beam pipe.

Both programs allow for a fast update of the SR estimate to be performed after optics changes. However, we do not expect significant lattice modifications that could change our conclusions on the SR in the experiments.

The difference between the two approaches for the power entering the TAS can be attributed to differences in the vacuum chamber geometry implemented in the two codes. While MDISim uses the horizontal and vertical apertures provided in the optics files, with circular/elliptical apertures, SYNRAD+ implements more detailed vacuum chamber geometries similar to the ones used in the LHC and foreseen for HL-LHC, in particular the recombination chamber between the D1 and D2 dipoles. This discrepancy is believed not to affect the conclusions.

It has to be noted that for both calculations no real geometry for the Inner Triplets' (IT) beam screens (BS) has been considered. At this point in time a geometry envisaging an internal distributed tungsten liner from Q1a to at least D1, similar to what is being done for HL-LHC, is probable, but lack of details at time of writing this paper have not allowed us to formulate a detailed geometry of the FCC-hh IT BS and analyse it.

References

[1] FCC web site http://cern.ch/fcc

[2] Seryi A *et al.* 2017 Overview of design development of FCC-hh experimental interaction regions *Proc. of IPAC17* Copenhagen TUPVA040

[3] Burkhardt H and Boscolo M 2015 Tools for flexible optimisation of IR designs with application to FCC *Proc. of IPAC15* Richmond USA TUPTY031

[4] Kersevan R 1993 SYNRAD: a Monte Carlo synchrotron radiation ray-tracing program *Conf.Proc. C930517* Washington vol **5*** 3848-3850

[5] Allison J *et al.* 2016 Recent developments in Geant42 *Nucl. Instr. Meth.* A **835** 186–225.

[6] Ady M Monte Carlo simulations of ultra high vacuum and synchrotron radiation for particle accelerators *CERN/EPFL PhD thesis CERN-THESIS-2016-047* http://cds.cern.ch/record/2157666?ln=fr

[7] Chance A *et al.* 2017 Updates on the optics of the future hadron-hadron collider FCC-hh *Proc. of IPAC17* Copenhagen TUPVA002

[8] Martin R *et al.* 2016 Beta* reach studies *talk at FCC Week 2016, Rome*

Assessment of thermal loads in the CERN SPS crab cavities cryomodule[1]

F Carra[2,4], J Apeland[2], R Calaga[2], O Capatina[2], T Capelli[2], S Verdú-Andrés[3] and C Zanoni[2]

[2] CERN, Route de Meyrin 385, 1217 Meyrin, Switzerland
[3] BNL, Upton, NY 11973, USA

Abstract. As a part of the HL-LHC upgrade, a cryomodule is designed to host two crab cavities for a first test with protons in the SPS machine. The evaluation of the cryomodule heat loads is essential to dimension the cryogenic infrastructure of the system. The current design features two cryogenic circuits. The first circuit adopts superfluid helium at 2 K to maintain the cavities in the superconducting state. The second circuit, based on helium gas at a temperature between 50 K and 70 K, is connected to the thermal screen, also serving as heat intercept for all the interfaces between the cold mass and the external environment. An overview of the heat loads to both circuits, and the combined numerical and analytical estimations, is presented. The heat load of each element is detailed for the static and dynamic scenarios, with considerations on the design choices for the thermal optimization of the most critical components.

1. Introduction and heat budget

One key element of the next LHC upgrade, known as HL-LHC, is represented by the adoption of a crab-crossing scheme, provided by a set of compact SRF crab cavities [1]. Prior to installation in the LHC, a prototype two-cavity cryomodule (Figure 1) was designed and is under manufacturing, in view of beam tests in the SPS in 2018 [2],[3].

The cavities are cooled with superfluid helium at 2 K, and a second circuit with helium gas, at a temperature between 50 K and 70 K, is connected to the thermal screen, also serving as heat intercept for all the elements connecting the cold mass and the external environment. Detailed calculations were performed in order to evaluate the thermal losses to the two cooling circuits, a necessary input for the design of the cryogenic infrastructures. The thermal losses breakdown is shown in Table 1, referring to both circuits in static (without RF power) and dynamic (with RF power, cavity delivering 3.4 MV deflecting voltage) nominal HL-LHC working conditions.

[1] Work supported by the European Union HL-LHC Project and by US DOE through Brookhaven Science Associates LLC under contract No. DE-AC02-98CH10886 and the US LHC Accelerator Research Program (LARP). Research supported by the HL-LHC project.
[4] federico.carra@cern.ch

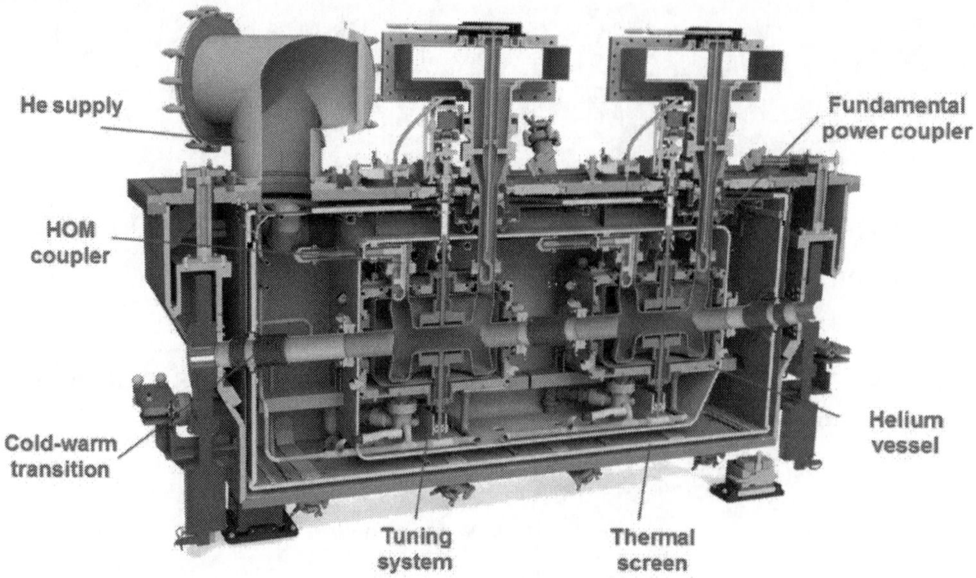

Figure 1. Cavity and interfaces.

The main items contributing to the total losses and the thermal design choices are summarized in the next sections. While the He gas circuit is at a temperature comprised between 50 K and 70 K, the connection to the intercepted components is done by means of copper flexures at a temperature conservatively assumed equal to 80 K.

Table 1: Cryomodule Heat Loads. Values are in Watts

	2 K	**80 K**
Static		
Radiation	3.4	30
Cold/Warm transitions	0.2	10
Supports	2	40
FPC	4	100
Instrumentation	2.3	10
HOM/Pickup	3.9	40
Tuner	1	10
Total static	**16.8**	**240**
Dynamic		
Cavity deflecting mode	11	0
FPC	5	10
HOM/Pickup	4.9	10
Beam current	0.5	0
Total Dynamic	**21.4**	**20**
TOTAL	**38.2**	**260**

2. Supports, tuning system and cold/warm transitions

These elements are not traversed by electrical currents and contribute therefore only to the static losses of the system. Their thermal design keeps into account the heat flowing by conduction from the thermal screen circuit to the helium tank (80 K – 2 K), and exchanged between the external ambient and the thermal screen (300 K – 80 K). The contribution of thermal radiation was also investigated and found to be negligible; therefore the problem is solved analytically, for each element, with the *Fourier's equation*:

IOP Conf. Series: Journal of Physics: Conf. Series **874** (2017) 012005 doi:10.1088/1742-6596/874/1/012005

$$Q_{2K}=\frac{A}{L_{int}}\int_{2K}^{80K} k(T)\mathrm{d}T$$
$$Q_{80K}=\frac{A}{L\text{-}L_{int}}\int_{80K}^{300K} k(T)\mathrm{d}T \quad (1)$$

where k is the thermal conductivity, A and L are section and length of the connection, L_{int} is the distance between the cold mass and the intercept on the 80 K circuit. This is valid for geometries of constant thickness; for more complex shapes or 3D problems, the calculation was refined with a finite element analysis (FEA). Good thermal performances have to be reached without endangering the structural resistance of the component, and the technical solutions aim at the best thermal/mechanical compromise.

2.1. Supporting System
As shown in Figure 2, the dressed cavities are mechanically supported via the external tube Fundamental Power Coupler (FPC) and two 316LN blades [4]. The helium biphase line is supported on 316LN and Ti-6Al elements connected to the vacuum tank. All these components are quite massive; however, stainless steel and titanium are bad thermal conductors and the power losses to the cold mass, calculated with Equation (1), are acceptable.

Figure 2. Cavity supports.

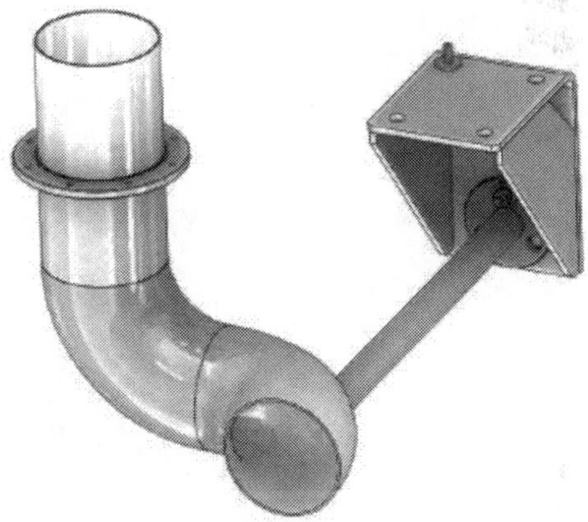
Figure 3. Helium line support.

2.2. Cold/warm Transitions (CWT)
The CWT are elements connecting the cavity to the warm beam line, and are constituted of a stainless steel beam pipe and bellows of the same material (see Figure 1). The system can be represented by two resistances in series, and since the bellows thickness is extremely low (0.15 mm), the thermal losses are negligible.

2.3. Tuning System
The tuner changes quasi symmetrically the distance between the two capacitive parts of the cavity. This is done by an actuator outside the cryostat that, acting on a titanium pushrod, creates a displacement between two vertical stainless steel concentric tubes [5].

Figure 4. Tuning system with detail of the connection frame/helium vessel.

The thermal losses to the cold mass are generated through the pushrod, the tubes and a rigid titanium frame surrounding the helium tank. The frame is in contact with the helium tank in four points, designed to maximize the thermal resistance while sustaining the mechanical stress given by the tuning actuation, avoiding the risks of buckling (Figure 4).

3. Instrumentation

The cryomodule is extensively instrumented, by means of temperature probes and magnetic flux sensors. Moreover, heaters are included at the end of each cold/warm line, at the vacuum tank level, to avoid the formation of ice (Figure 5). The wires for the powering of these systems act as resistances in parallel. Although each wire is very resistive, due to the high number of wires the total loss to the cold mass is quite relevant (above 2 W).

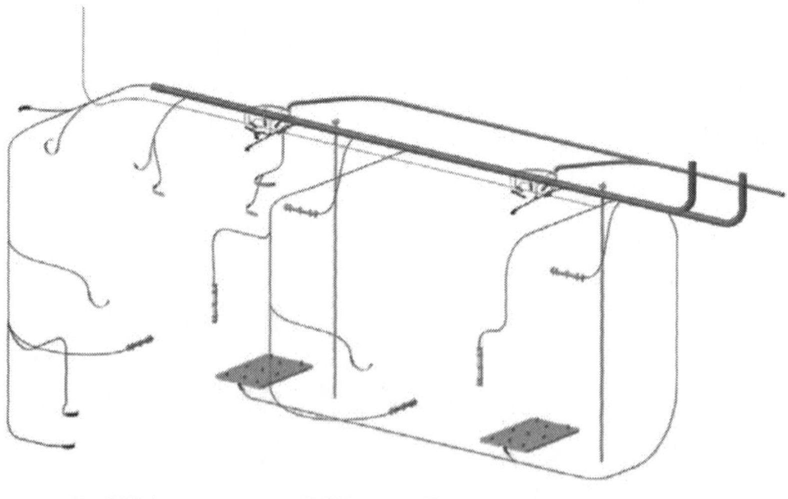

Figure 5. Wiring system. Different lines highlighted: thermal probes (purple), magnetic flux sensors (blue) and heaters (red).

4. Thermal screen

The thermal screen is a copper structure wrapped with multilayer insulation (MLI) and directly thermalized by the 80 K helium gas circuit (Figure 6). Its role is to minimise the thermal losses by radiation to the cold mass at 2 K [6]. In principle, the contribution of thermal radiation to the heat balance is not difficult to estimate for two grey bodies at a known emissivity by means of the

IOP Conf. Series: Journal of Physics: Conf. Series **874** (2017) 012005 doi:10.1088/1742-6596/874/1/012005

Stefan-Boltzmann law. However, the addition of MLI complicates the problem, and the losses on the two circuits were estimated based on experimental measurements performed on LHC cryostats [7]. The thermal screen also presents openings required by the alignment instrumentation, allowing an optical access to the targets. At these openings, the cold mass exchanges heat by radiation directly with the external ambient. These additional losses were estimated with FEA.

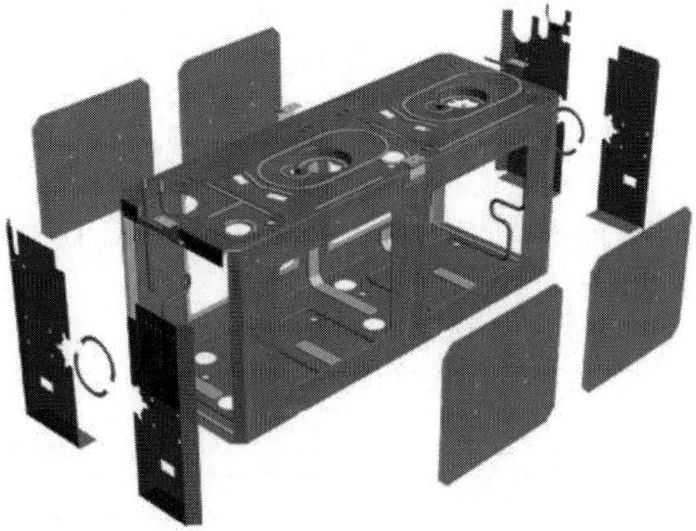

Figure 6. Exploded view of the thermal screen.

5. Fundamental Power Coupler

The FPC is one of the main contributor to the losses on the cold mass. It is made of a copper-coated 316LN tube, 3 mm thick, and a warm copper internal antenna (Figure 7). At nominal operation the antenna dissipates 100 W by Joule's effect. This could lead to high temperatures on the antenna, with possible issues of creep, outgassing, and radiation losses to the cold mass. Iterative analyses with HFSS [8] and ANSYS were performed to find the optimal design of the component, which is actively cooled by a room-temperature water flow of 1.5 $m \cdot s^{-1}$ circulating in an internal channel dug in the antenna straight section. The component reaches 100 °C (Figure 8) at the tip of the antenna, and emits heat by radiation towards the cold mass. These losses in dynamic conditions were also estimated with FEA.

Figure 7. FPC antenna and can.

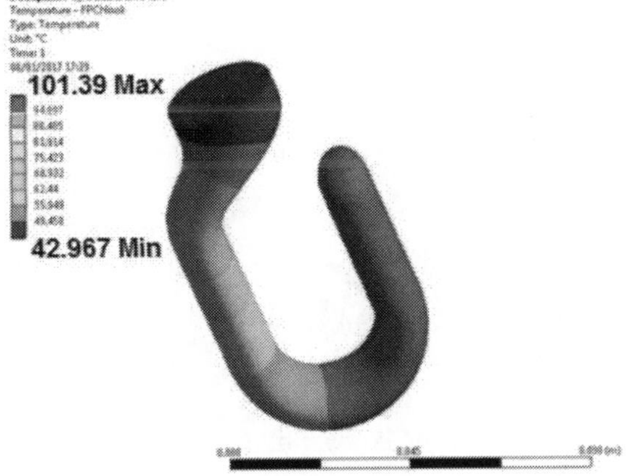

Figure 8. Thermal FEA of the tip of the cooled antenna.

The stainless steel tube, is subjected mostly to conduction heat transfer. In this case, however, Equation (1) can describe only the losses in static conditions. When the RF power is active, additional losses by Joule effect have to be accounted. An analytical solution of the problem is extremely complex, as the electrical resistivity of the component depends on the temperature field and vice versa. A semi-analytical approach was adopted, discretizing the system in several nodes and applying to each node the energy conservation law. The temperature of each discrete element of length *dx* is defined by three equations:

$$Q=-\int_{T_{(x_i-\mathrm{d}x/2)}}^{T_{x_i+\mathrm{d}x/2}} k(T)\frac{A}{dx}\mathrm{d}T$$

$$P_i=\frac{1}{2}\int_{x_i-\mathrm{d}x/2}^{x_i+\mathrm{d}x/2} I^2(x)\cdot R(x)\mathrm{d}X \qquad (2)$$

$$I(x)=I_0\cdot 2sin\left(\frac{\omega}{c}x\right)$$

where the index i is referred to the i-th discrete element, T_a and T_b are the temperature at the inlet and outlet of the element, I is the current intensity and R the electrical resistance, P is the power, ω is the wave frequency and c the speed of light in vacuum. As it can be seen in Table 1, the Joule effect generated on the FPC tubes for a nominal duty cycle ($P = 40$ kW) is relevant, even with a copper coating of the 316LN walls, representing the 25% of the total dynamic losses of the cryomodule.

6. HOM filters and pickup

The HOM and pickup lines feature a special design [9] with thin coaxial stainless steel tubes, copper coated (Figure 9). Commercial coaxial cables were also investigated, but the power loss at the duty cycle (1 kW on each) would provoke melting of the cable in operation. The calculation of the heat load to the cold mass in static and dynamic conditions was performed semi-analytically, with the method detailed in the previous section.

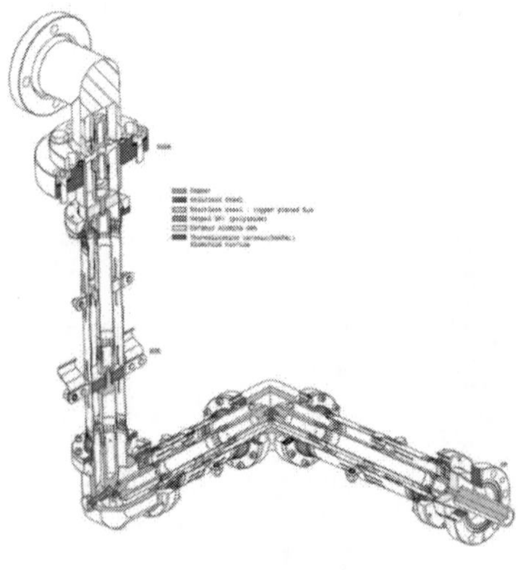

Figure 9. HOM coaxial lines.

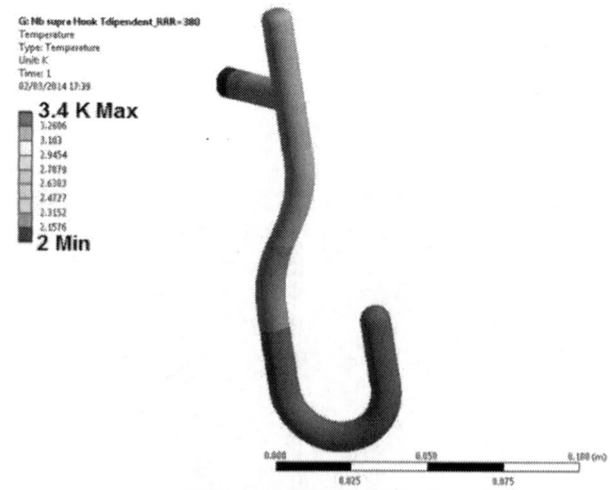

Figure 10. HOM antenna.

The HOM antenna is in superconducting niobium, as initial simulations showed that the losses on a copper component would be too high. The maximum temperature on the antenna without cooling is 3.4 K. As the niobium critical temperature is 9.2 K, this presents enough margin against quench, with negligible losses to the cold mass (see Figure 10, FEA performed on the preliminary version of the antenna). As an additional precaution, the antenna is cooled with superfluid helium. The pickup antenna, which is less loaded, does not require active cooling. For each line, additional 0.4 W of dynamic losses were included in the estimation, to account for a 50 μm shape tolerance.

7. Beam and cavity losses

The beam passage inside the cavities generates surface losses due to resistive-wall impedance. These losses have been estimated to be 0.5 W based on previous experience on superconductive cavities at CERN [10]. The thermal losses on the cavity in operation depend on the thermal efficiency, expressed by the Q factor, degrading at increasing gradients. Recent crab cavity cold tests conducted at CERN for the SPS cavities showed a power loss of about 5 W/cavity at the nominal gradient of 3.4 MV [11]. In the estimation of the total heat budget, 5.5 W per cavity were assumed, to keep into account possible deviations in operation with respect to the measured values.

8. Conclusions

The thermal performance of the SPS crab cavity cryomodule was optimized intervening on the design of the cooling circuits and the elements connected to the cold mass or radiating towards it. The design solutions detailed in this paper are the best compromise between the thermal and mechanical requirements of the different components, minimising the heat loss to the cold mass without endangering the structural resistance. The developed analytical and numerical calculation methods were also discussed.

References

[1] Apollinari G, Béjar Alonso I, Brüning O, Lamont M and Rossi L 2015 High-Luminosity Large Hadron Collider (HL-LHC): Preliminary Design Report (CERN) *Rep. CERN CERN-2015-005*.

[2] Carra F *et al.* 2015 Crab cavity and cryomodule development for HL-LHC *Proc. SRF2015 (Whistler)* pp 1160–1466.

[3] Zanoni C *et al.* 2017 The crab cavities cryomodule for SPS test *presented at IPAC'17 (Copenhagen) this conference* MOPVA096.

[4] Zanoni *et al.* 2015 Design of dressed crab cavities for the HL-LHC upgrade *Proc. SRF2015 (Whistler)* pp 1284–1288.

[5] Artoos K *et al.* 2015 Development of SRF Cavity Tuners for CERN *Proc. SRF2015 (Whistler)* pp 1247–1251.

[6] Templeton N *et al.* 2015 Design of the Thermal and magnetic Shielding for the LHC High Luminosity Crab Cavity Upgrade *Proc. SRF2015 (Whistler)* pp 852–856.

[7] Parma V 2013 Cryostat design *Lecture at CERN Accelerator School (Erice)* https://indico.cern.ch/event/194284/contributions/1472773/attachments/281469/393535/Cryostat_design.pdf

[8] Verdú-Andrés S *et al.* Thermal Losses in Couplers and Ports of a SPS Double-Quarter Wave Crab Cavity *Proc. SRF2015 (Whistler)* pp 1219–1221.

[9] Zanoni C *et al* Engineering Design and Prototype Fabrication of HOM Couplers for HL-LHC Crab Cavities *Proc. SRF2015 (Whistler)* pp 1279–1283.

[10] Capatina O 2013 private communication.

[11] Castilla Loeza *et al.* 2017 CERN cavity results *International review of the crab cavity performance for HiLumi (CERN)* https://indico.cern.ch/event/590988/contributions/2489635/

IOP Conf. Series: Journal of Physics: Conf. Series **874** (2017) 012006 doi:10.1088/1742-6596/874/1/012006

Multiparametric response of the LHC Dynamic Aperture in presence of beam-beam effects

D Pellegrini[1], F Antoniou[1], S Fartoukh[1], G Iadarola[1] and Y Papaphilippou[1]
[1] CERN, Geneva, Switzerland
E-mail: dario.pellegrini@cern.ch

Abstract. We performed extended simulations of LHC dynamic aperture (DA) in the presence of beam-beam effects in the weak-strong approximation, evaluating the contributions of parameters such as: tunes, optics, bunch intensity, crossing angle, emittance, chromaticity and current in the Landau octupoles. Here we present a summary of these studies, giving an overview of the amplitude of the LHC operational space and pointing out the remaining margins for mitigation of instabilities. These studies supported the actions deployed during the 2016 run of the LHC, which aimed at maximising its performances. Examples of such actions are the switch to lower emittance beams, the reduction of crossing angle and tune trims. More recently, DA scans have been used to help the definition of the operational scenarios for the 2017 run. Additional room for improvements, for instance by deploying crossing angle levelling, will be explained.

1. Introduction

The performance optimisation of the Large Hadron Collider (LHC) relies on a careful tuning of a number of operational parameters, balancing many conflicting constraints. Tunes, optics, bunch intensity, crossing angle, emittance, chromaticity and current in the Landau octupoles, they all play a role in the maximisation of the luminosity, the mitigation of non-burnoff losses and the prevention of beam instabilities.

The duration of the LHC cycle, which typically extends up to several hours, together with a number of constraints from machine protection, does not allow to probe such a vast parameter space in an experimental manner. It is therefore essential levering on detailed computer simulations aimed at identifying the possible steps for improvements, to be validated in Machine Development sessions or directly applied to the machine.

The 2016 run of the LHC has been closely followed up by means of dynamic aperture (DA) investigations. These studies helped to steer performance of the collider towards the reach of the design luminosity and beyond. Although the many uncertainties involved and the fact that a precise relation between beam lifetime and DA is not available, the latter has been proven a valuable estimator of margins and boundaries. Excellent response from the machine has been obtained; confidence has been gained in the model and in the technique, allowing for reliable predictions.

2. Simulation Framework and Settings

The simulations are performed in a weak-strong approximation in which only a single beam is tracked and the beam-beam lenses (both for the head-on and long-range interactions) are static. This simplification allows for a substantial computational speed up and applies well to the particles with an action of some r.m.s. beam size (σ), whose dynamics is relevant for the determination of the DA and is not much influenced by the coherent motion of the beam core.

The model relies on the MADX [1], SixTrack [2], SixDesk [3] environment and includes all the IPs. The tolerances on alignment and multipolar errors are such that they can be effectively corrected and, on average, no significant impact is observed in presence of beam-beam effect.

We consider the minimum value of DA determined over 1×10^6 turns for 5 angles equally spaced in the positive quadrant of the configuration space. Although the statistics may appear limited, the fine granularity of the parametric scans compensates for it, proving to be adequate in most of the cases. The studies focussed on Beam 1, which has shown the weakest lifetime all along the run.

3. Predictions and Follow Up of the 2016 Run

3.1. Emittance reduction

After the start-up with nominal beams, the availability of a new method for production of physics beams (BCMS [4]), resulted in a progressive reduction of the emittance from 3.75 µm to about 2.3 µm during the first part of the 2016 run.

Smaller emittances increase the strength of the head-on beam-beam, but at the same time the sampling of multipolar components is reduced. It has been found that although the physical value of DA (measured in mm) decreases for smaller emittances (indicating that the head-on component has a visible impact on DA), in units of beam σ, the DA improves. This is illustrated in Figure 1 for a region around the emittance change. The divergent colour scale is used to mark the areas of good DA (blue) and of DA leading to potentially unacceptable loss rates (red). The machine, indeed, responded to the smaller emittance with improved beam lifetime [5].

3.2. Reduction of Crossing

The bunch intensity along 2016 has been limited by the heating of the injection kicker [6]. This, together with the increased margin from the reduced emittance, allowed for a consistent reduction of the crossing angle. The decision for a switch from 185 µrad to 140 µrad, was based both on previous experience [7] and new simulations [8].

The new crossing angle caused a minor increase of losses during the first hour in Stable Beams (see Figure 2), whose impact on the luminosity production was superseded by the enhancement

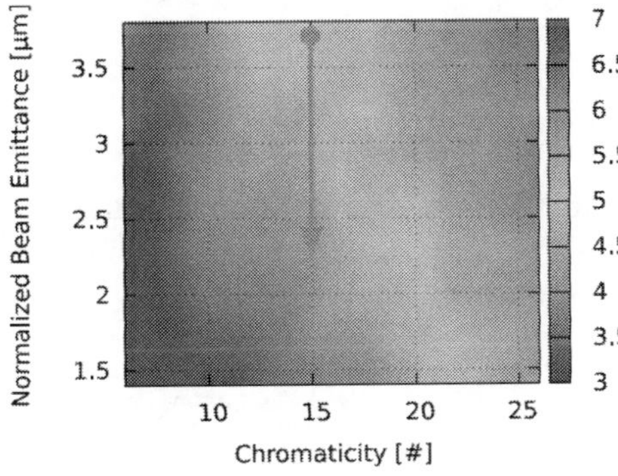

Figure 1. The additional DA gained for small emittances allows to run with higher chromaticity and/or with smaller crossing angle (not shown). The arrow indicates the reduction performed in 2016.

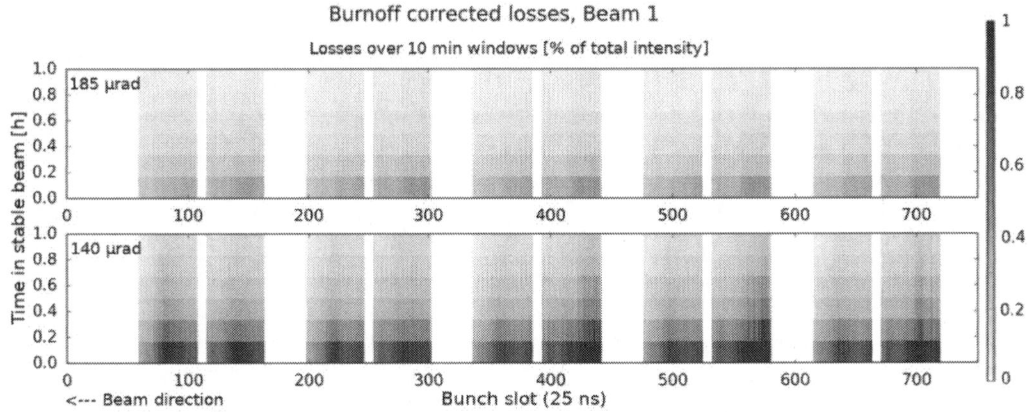

Figure 2. Losses observed for some trains of bunches of Beam 1 during the first hour in stable beam as percent of the total intensity, computed over 10 min windows and averaged over several fills before (top) and after (bottom) the reduction of the crossing angle from 185 to 140 µrad. The progressive increase of the losses along the trains indicates some e-cloud build-up, while the signature of LR interactions is visible as the enhanced losses in the centre of the trains, where bunches suffer more LR interactions. Beam 2 presents a similar behaviour although the losses are milder.

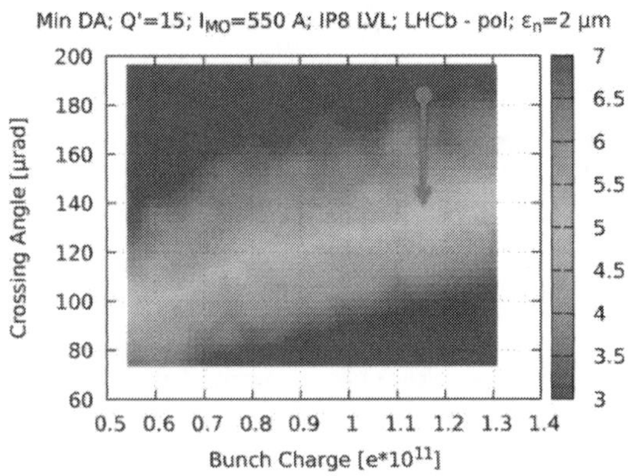

Figure 3. DA response to bunch intensity and crossing angle. The arrow indicates the crossing angle reduction performed during 2016. The DA at the beginning of the fill has been reduced to 5 σ, the intensity decay gives more margin later in the fill.

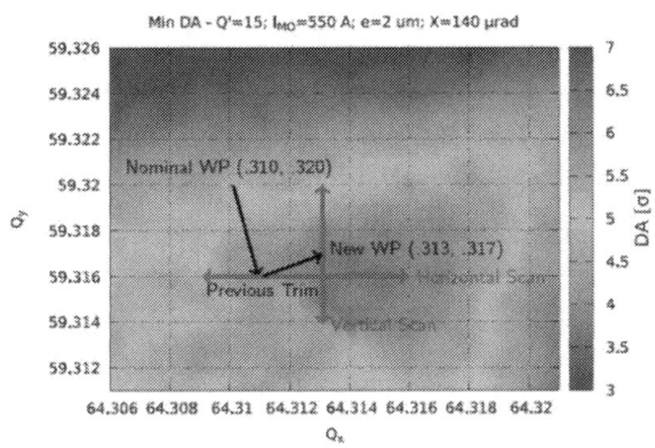

Figure 4. A tune scan showing that the performed adjustment to the tune. The arrows represent the area explored experimentally, carefully monitoring the beam lifetime.

of the geometrical reduction factor [9].

Figure 3 shows the prediction from simulations indicating a DA of 5 σ at the beginning of the fill. This justifies the fact that a DA around of 6 σ is going to be used as a starting point also in 2017, with a possible reduction to 5 σ as a second step during the run. More experience will be necessary in order to better qualify these values.

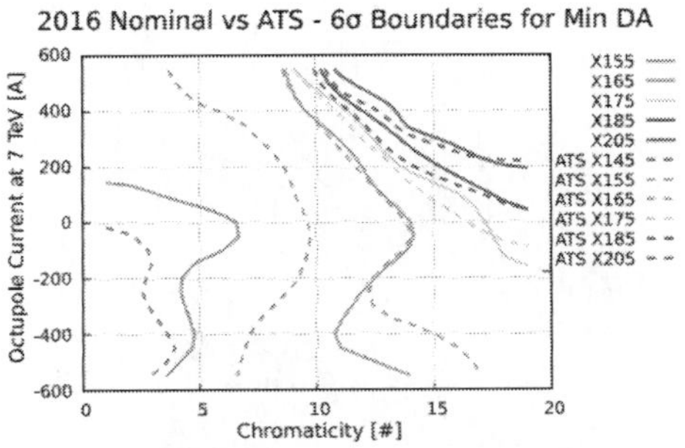

Figure 5. 6σ DA boundaries in the chromaticity-octupole space, comparing the performances of Nominal and ATS optics for several crossing angles in IP1 and IP5. The legend indicates the half crossing angle values in µrad.

3.3. Tune Adjustment
During the 2016 Run the measured lifetime of Beam 1 has consistently been shorter than the one of Beam 2. Variations of this asymmetry were noted for changes of the polarity of the LHCb spectrometer, whose main effect is the introduction of a tune shift of about 1×10^{-3} for the bunches colliding there (the majority) [10].

DA tune scans were performed for Beam 1 (see Figure 4) and confirmed that the nominal working point (0.31, 0.32) is suboptimal, and, for well corrected linear coupling, areas of better DA are observed closer to the diagonal. The response of Beam 1 lifetime to small tune trims was monitored at the beginning of a fill; the working point selected by this procedure (0.313, 0.317) was in remarkable agreement with the predictions from simulations and resulted in a complete recovery of Beam 1 lifetime in the next fills [10].

4. ATS Optics
The achromatic telescopic squeeze (ATS) is the baseline optics for the HL-LHC, allowing to reach lower values of β^* [11]. During 2016 the ATS concepts have been successfully demonstrated in MDs [12] and it has been chosen to use it already in 2017 with moderate settings.

Figure 4 shows a comparison of the performances of Nominal and ATS optics. It can be noted that ATS is predicted to perform remarkably better with small chromaticity allowing for tighter crossings, in particular with negative octupoles which allow for LR compensation [13].

5. Preparation for the 2017 Run
Several studies have been performed in order to help the definition of the 2017 settings [14, 15]. Figure 6 summarises the DA sensitivity to the tune. With the 2017 settings, the nominal tune (0.31, 0.32) is particularly suboptimal. We propose an optimised tune of (0.313, 0.317), which was already demonstrated during 2016 and proved to be effective even in presence of small residual coupling. It should also be noted that if the good DA area in the tune-space becomes too small, it might be impossible to accommodate all the different classes bunches, e.g. pacman bunches [16], bunches without collision in IP8...

Figure 7 shows an intensity-crossing scan for the optimised tune. The starting point at 150 µrad and 1.25×10^{11} proton per bunch presents significant margin for a further crossing reduction and/or crossing anti-levelling, which consists in following the iso-DA curve during the intensity decay. The predicted gain of integrated luminosity with crossing angle anti-levelling reaches up to 3 %, with a continuous adjustment of the crossing angle along the fill [17].

6. Conclusions
Multi-parametric DA scans have been applied to the LHC allowing to gain more confidence in the sensitivity of its parameter space. The 2016 Run has been closely followed up. Input has been provided for the main decisions taken along the year and the predictions turned out to be extremely accurate in spite of the many uncertainties present. The studies played a central role

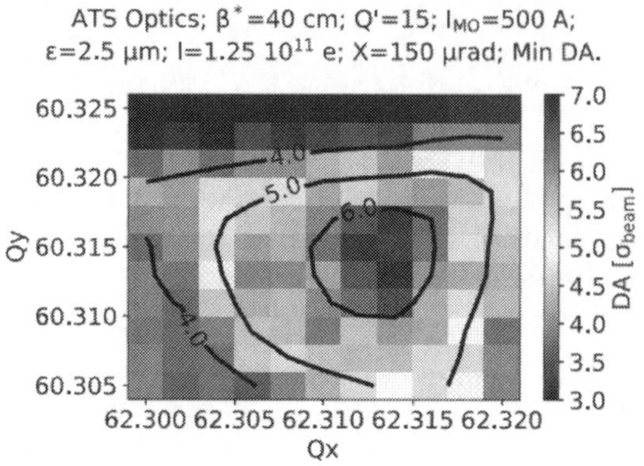

Figure 6. Tune scan for the proposed 2017 settings at collision. Good DA is obtained moving closer to the diagonal, provided that linear coupling is properly corrected.

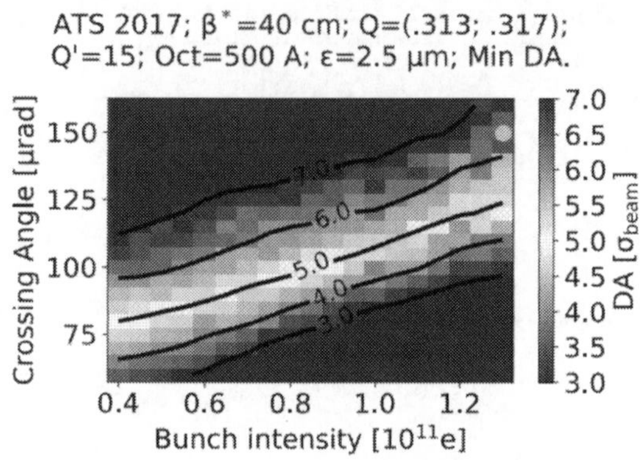

Figure 7. A Crossing-Intensity scan for the 2017 settings and optimised tune. The yellow dot is the proposed starting point for 2017, which is expected to give substantial margin.

also in the definition of the parameters for 2017. Although the start up will be more aggressive, margins have already been identified and they may result in a substantial performance increase along the year. In particular a careful tune control at the level of 1×10^{-3} allows for a substantial gain of DA, which together with a potential reduction of chromaticity and octupoles, may allow to further push the crossing angle and/or β^*.

References

[1] Grote H and Schmidt F 2016 *MAD-X User's Guide* URL `http://madx.web.cern.ch/madx/`
[2] Schmidt F 2012 *SixTrack User Manual* URL `http://sixtrack.web.cern.ch/SixTrack/`
[3] McIntosh E and De Maria R 2013 *SixDesk User Manual* URL `http://sixtrack.web.cern.ch/SixTrack/`
[4] Steerenberg R, Arduini G, Argyropoulos T, Bartosik H, Bohl T, Cornelis K, Damerau H, Findlay A, Garoby R, Gilardoni S *et al.* 2012 *Proc. LHC Beam Operation Workshop, Evian*
[5] Hostettler M *et al.* 2016 *Update on analysis of lumi lifetime with BCMS beams* LHC Machine Committee meeting, CERN URL `https://indico.cern.ch/event/559439/`
[6] Day H, Barnes M, Ducimetière L, Vega Cid L and Weterings W 2016 *Proceedings of IPAC2016* (Busan, Korea) THPMW031 URL `http://cds.cern.ch/record/2207460`
[7] Pieloni T *et al.* 2016 *BBLR studies and possible reduction of the crossing angle: machine results* LBOC Meeting, CERN URL `https://indico.cern.ch/event/561910/`
[8] Pellegrini D *et al.* 2016 *BBLR studies and possible reduction of the crossing angle: simulations* LBOC Meeting, CERN URL `https://indico.cern.ch/event/561910/`
[9] Antoniou F *et al.* 2016 *7th Evian Workshop* (France) URL `https://indico.cern.ch/event/578001/`
[10] Pellegrini D *et al.* 2016 *AOB: beam 1 lifetime recovery* LHC Machine Committee Meeting, CERN URL `https://indico.cern.ch/event/578380/`
[11] Fartoukh S 2013 *Phys. Rev. ST Accel. Beams* **16**(11) 111002 URL `https://link.aps.org/doi/10.1103/PhysRevSTAB.16.111002`
[12] Fartoukh S, Albert M, Le Borgne Y, Bracco C, Bruce R, Carlier F S, Coello De Portugal Martinez Vazquez J M, Garcia-Tabares Valdivieso A, Fuchsberger K, Giachino R, Maclean E H, Malina L, Mereghetti A, Mirarchi D, Nisbet D, Normann L, Papotti G, Persson T H B, Pojer M, Ponce L, Redaelli S, Salvachua Ferrando B M, Skowronski P K, Solfaroli Camillocci M, Suykerbuyk R, Tomas Garcia R, Valuch D, Wegscheider A and Wenninger J 2016 *ATS MDs in 2016* Tech. Rep. CERN-ACC-2017-0003 CERN Geneva URL `https://cds.cern.ch/record/2242513`
[13] Fartoukh S, Valishev A, Papaphilippou Y and Shatilov D 2015 *Phys. Rev. ST Accel. Beams* **18**(12) 121001 URL `https://link.aps.org/doi/10.1103/PhysRevSTAB.18.121001`
[14] Papaphilippou Y 2016 *7th Evian Workshop* (France) URL `https://indico.cern.ch/event/578001`

[15] Bordry F 2017 *LHC Performance Workshop* Chamonix, France URL `https://indico.cern.ch/event/580313/`
[16] Herr W 1996 *Effects of PACMAN bunches in the LHC* Tech. Rep. LHC-Project-Report-39. CERN-LHC-Project-Report-39 CERN Geneva URL `https://cds.cern.ch/record/314386`
[17] Wenninger J 2017 *Crossing Angle Plans* LHC Machine Committee Meeting, CERN URL `http://indico.cern.ch/event/627059/`

IOP Conf. Series: Journal of Physics: Conf. Series **874** (2017) 012007 doi:10.1088/1742-6596/874/1/012007

Multiparametric response of the HL-LHC Dynamic Aperture in presence of beam-beam effects

D Pellegrini[1], S Fartoukh[1], N Karastathis[1] and Y Papaphilippou[1]

[1] CERN, Geneva, Switzerland

E-mail: dario.pellegrini@cern.ch

Abstract. We performed extended simulations of HL-LHC dynamic aperture in the presence of beam-beam effects in the weak-strong approximation, evaluating the contributions of parameters such as: bunch intensity, crossing angle, chromaticity, current in the Landau octupoles and multipole errors.

From the beam dynamics point of view, the main difference between the LHC (until 2017) and the HL-LHC is the deployment of the achromatic telescopic squeezing (ATS) optics, allowing not only for a smaller β^* reach, but also modifying the phase advances between the lattice correctors (sextupoles, octupoles) and the main IPs, and increasing the peak β functions in the arcs. These correctors become therefore more efficient for the chromatic correction, but also a mitigation of the beam-beam long range interactions using the Landau octupoles is enabled, resulting in a possible reduction of the normalised crossing angle.

The limits have been investigated in a tracking simulation campaign aimed at exploring the operational space for the HL-LHC and two possible options for luminosity levelling.

1. Introduction

The High Luminosity LHC (HL-LHC) is an approved upgrade of the LHC, aiming at the increase of the integrated luminosity to $250\,\mathrm{fb}^{-1}$ per year, enabling $3000\,\mathrm{fb}^{-1}$ over its lifetime [1]. This goal will be achieved through an upgrade of the high luminosity insertion regions allowing to reduce β^* and to implement crab crossing to counteract the reduction of the luminosity due to the large crossing angle, required to minimize long range beam-beam effects. In addition, the brightness of the beams delivered by the injectors will be increased by almost a factor of 3. The β^* levelling will be used to maintain the instantaneous luminosity at $5 \times 10^{34}\,\mathrm{Hz\,cm^{-2}}$, so to keep the event pile-up to acceptable levels for the experiments.

Precise indications of the dynamic aperture (DA) evolution can be used to define the operational scenarios and the achievable performance. Multi-parametric DA scans have been proven to be a valuable technique for the prediction of the LHC margins for performance tuning [2]. The studies have therefore been extended to the HL-LHC in an exploratory campaign aimed at qualifying better margins and the potential performance reach of the machine. These estimations follow from our current knowledge of the LHC, extrapolated to the HL-LHC, and aim at identifying possible means to improve the performance of the baseline scenario, whose latest review can be found in [3].

In this paper, we present the impact of the settings of chromaticity and Landau octupoles, a detailed proposal for the simultaneous evolution of β^* and crossing angle during the luminosity levelling and the first estimate of the impact of magnetic errors in presence of beam-beam effects.

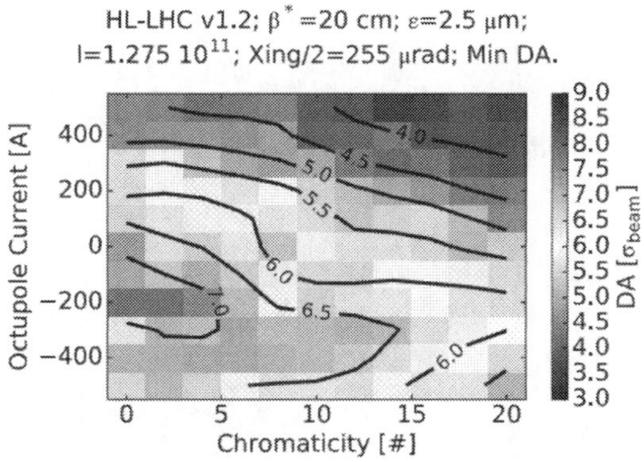

Figure 1. DA response to octupoles and chromaticity for the nominal crossing angle of 510 µrad and end-of-levelling settings with head-on collisions in LHCb and 5 σ separated beams in Alice. The negative setting for octupoles improves the DA, allowing for a larger chromaticity and/or reduced crossing angle.

2. Simulation Framework and Settings

The simulations are performed in a weak-strong approximation in which only a single beam is tracked and the beam-beam lenses (both for the head-on and long-range interactions) are static. This simplification allows for a relevant speed up and applies well to the particles with an action of some sigmas, whose dynamics is relevant for the determination of the DA and is not much influenced by the coherent motion of the beam core. We rely on the MADX [4], SixTrack [5], SixDesk [6] environment commonly used at CERN.

We consider the minimum value of DA determined over 1×10^6 turns for 5 angles equally spaced in the positive quadrant of the configuration space. Although the statistics may appear limited, the fine granularity of the parametric scans compensate for it, proving to be adequate in most of the cases. The studies are performed with the HL-LHC optics v1.2 [7] and are focussed on Beam1.

In order to perform the estimations of the intensity decay, we adopt an inelastic cross section of 81 mb at 7 TeV, in contrast with the total cross section of 111 mb used in [3]. Indeed, the current observations in the LHC indicate that this could be achieved [8], provided that other sources of beam losses are understood and controlled in HL-LHC. The 4 σ bunch-length is fixed at 1.2 ns. Concerning the emittance we consider a "round-beam" scenario with constant normalised emittances of 2.5 µrad, and a second scenario in which synchrotron radiation, intra-beam scattering and elastic scattering at the IPs, result in a reduction of emittance mainly in the vertical plane [8]. As additional sources of emittance blowup are observed in the LHC, we consider the "round-beam" scenario as more realistic.

3. Chromaticity-Octupole Scans

The settings of chromaticity and Landau octupoles play a crucial role in guaranteeing the stability of the beam in presence of collective effects such as e-cloud and wakefields. However these settings have also a severe impact on the DA, potentially reducing the beam lifetime.

Chromaticity-octupole scans allow to identify sharp DA transitions and help in the selection of the working point. Figure 1 presents an example for the beam parameters at the end of levelling, including all the interaction regions. The divergent colour scale is used to mark the areas of good DA (blue) and of potentially unacceptable DA (red), with the transition taking place at the target DA of 6 σ. It can be noted that negative octupoles are much more favourable for the HL-LHC. In addition the detrimental effect of chromaticity is highlighted: in case it will be possible to reduce it (by means of enhanced feedback systems or suppression of the e-cloud), some additional margin can be obtained [9].

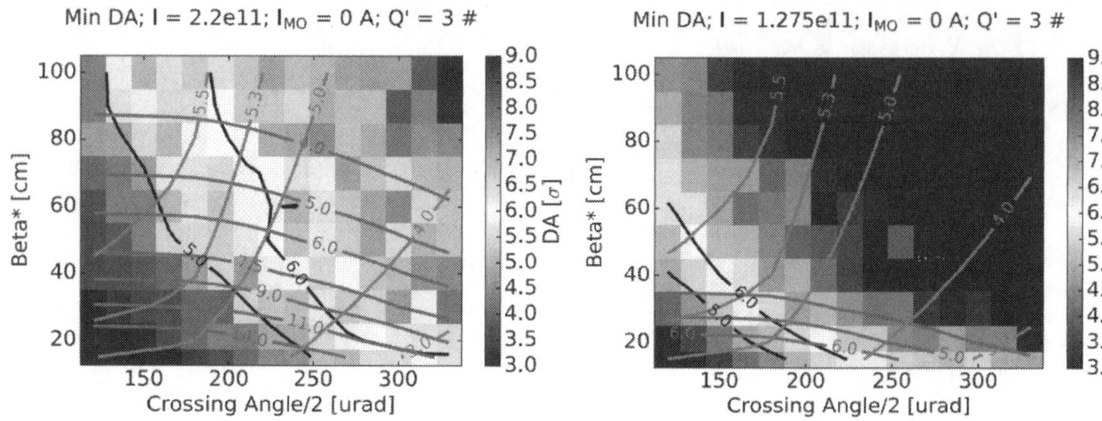

Figure 2. Multi-parametric scans of minimum DA in beam sigma for various values of β^* and half crossing angle for the start (left) and the end (right) of the levelling process. The superimposed curves correspond to the various contours of Dynamic Aperture (black), Luminosity (red) and r.m.s. length of the luminous region (gold). The study was performed with 3 units of chromaticity and the octupoles off.

4. Beam Dynamics in the levelling process

We have investigated both the nominal levelling scenario, relying on the progressive reduction of β^* in IP1 and IP5, and an alternative based on beam separation. DA-related studies show that both the levelling techniques are possible [10], but we chose to focus here on the levelling with β^* which is the current baseline.

The possibility to adjust the crossing angle during the fill is also considered in order to contain the requirements on the maximum crabbing angle, now limited to ~350 µrad, which is the maximum achievable with two crab cavities per beam per IP side without specific optics optimisations [7].

With two degrees of freedom: β^* and crossing angle; two constraints are needed: the target luminosity and the minimum DA; the first is obtained from numerical integration based on the techniques presented in [11], while the latter is obtained from tracking simulations. The intersections of the iso-luminosity with the iso-DA curves, computed for decreasing beam intensities, allow reconstructing the time evolution of the operational parameters during the levelling process. In addition, other observables such as the peak pile-up density and the length of the luminous region, which benefit from operating with large β^* and small crossing angle, are automatically optimised.

In terms of DA requirements, two scenarios can be formulated: a "relaxed" one, in which the DA is chosen at 6 σ_{beam}, and an "aggressive" one, which allows for 5 σ_{beam} DA. The benefits of pushing for a smaller DA are related to the pileup density and the length of the luminous region. Due to the sinusoidal shape of the crabbing voltage, these are still improved even for crossing angles smaller than the maximum crabbing angle. For crossing angles above the maximum crabbing angle, marginal gain in term of the duration of the levelling process can also be achieved.

It should be noted that the aggressive DA settings also limit the tune space available for accommodating the different classes of bunches whose dynamics is especially affected by the missing collision in IP8 and by other coherent effects. Furthermore operation at high chromaticity might be required, although octupole currents in the range of −200 to −300 A are expected to recover some DA.

Figure 2 shows the DA simulation results for the study of the β^* levelling process. The

IOP Conf. Series: Journal of Physics: Conf. Series **874** (2017) 012007 doi:10.1088/1742-6596/874/1/012007

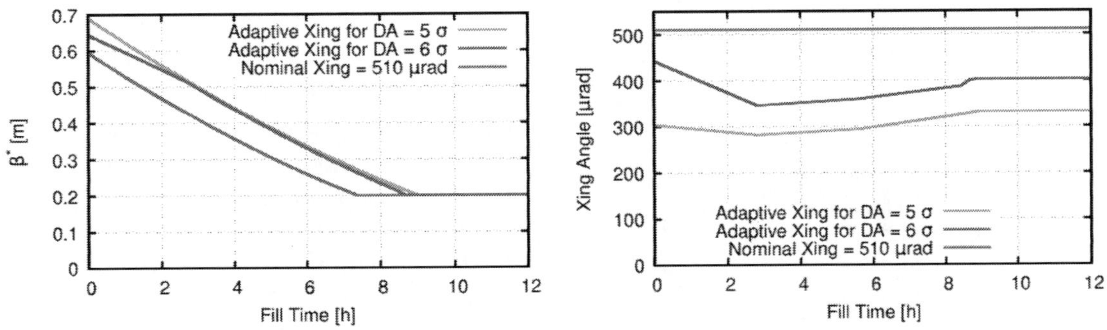

Figure 3. The evolution of β^* (left) and crossing angle (right) as a function of time for the relaxed and the aggressive scenarios, with round emittance. The nominal case with the crossing angle fixed at 510 μrad is also shown.

selected plots refer to 2.2×10^{11} protons per bunch (the beginning of the fill) and 1.275×10^{11} protons per bunch (the nominal end of levelling). The black contours represent the DA levels in units of σ_{beam} and the red contours are the levels of luminosity over the same range of parameters in units of $1 \times 10^{34}\,\text{Hz}\,\text{cm}^{-2}$. Finally, the gold contours represent the r.m.s. length of the luminous region in units of cm.

The resulting time evolution of the β^* and the crossing angle is shown in Figure 3. Both the relaxed and the aggressive scenarios are shown, for the more conservative round emittance case. The RF curvature of the crab cavities and the hourglass effect are taken into account according to [11]. It is interesting to note that a crabbing angle of 350 μrad still allows to substantially compensate the crossing angle even for the relaxed scenario.

5. Impact of Magnetic Field Errors

We chose the bunch intensity of 1.275×10^{11} protons, close to the end of the levelling process, to perform a statistical analysis in order to estimate the impact of magnetic errors. In addition to the unperturbed machine, DA simulations for 60 different realisations have been performed with the last version (v5) of the field error table without fringe fields, in an area spanning over ± 4 cm in β^* and ± 20 μrad in crossing angle around the point resulting in the nominal luminosity.

Overall, the average DA over the full statistical sample is 6.4 σ_{beam}, while the average standard deviation is found to be 0.56 σ_{beam}. The global minimum DA for the statistical population over all configurations is 4.8 σ_{beam}, and, as expected, it is found at the furthest configuration from the nominal point corresponding to a lower iso-DA curve. Furthermore, the dependence of the statistical results on the intensity, and thus time during the levelling process, is found to be well within the simulation uncertainties.

Summary results for a selected crossing angle as a function of β^* are collected in the violin plot shown in Figure 4. For each seed the minimum DA value is determined over the angles in the configureation space. The density of the observed minimum DA values is represented by the shaded areas, while the horizontal bars show the maximum, average and minimum DA value of each configuration. The dots superimposed to each violin show the minimum DA for the case without errors.

These results indicate that the field quality optimised according to the specifications resulting from the DA simulations without beam-beam effect appear adequate also in presence of these effects. Nevertheless, the final determination of the machine performance will require to take into account several additional aspects such as: operation with high chromaticity and octupoles, bunch-by-bunch tune spread, sources of emittance blow-up, noise and effect of fringe fields.

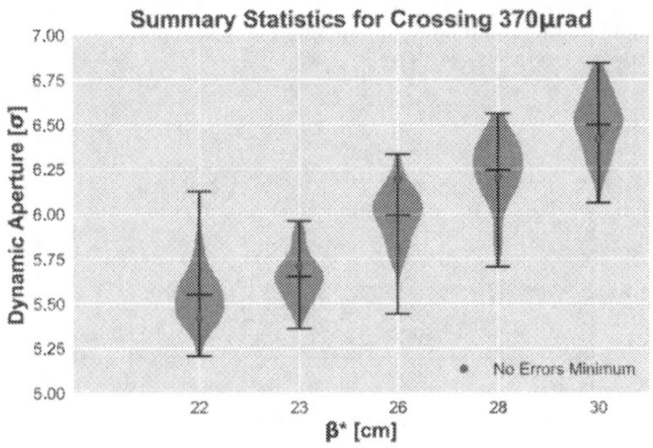

Figure 4. The DA as a function of the β^* is illustrated for 370 μrad crossing angle. The three bar points for each violin correspond to the minimum, average and maximum value of the statistical population. The shaded area represents the density of seeds presenting the same DA. The red bullets on top of each violin correspond to the minimum DA of the case without errors.

6. Conclusions

By applying several multi-parametric DA scans, we started to produce an overview of the parameter space of the HL-LHC. The simultaneous operation of β^* and crossing angle during the levelling process has been determined with a rigorous approach based on DA estimations. In addition, the impact of magnetic imperfections has been evaluated in the presence of beam-beam effects. The current tolerances and correction strategies appear to be adequate in order to guarantee the operation of the machine.

7. Acknowledgement

We thank F. Antoniou and G. Iadarola for their support and tools; R. De Maria for providing the squeeze; G. Arduini and R. Tomas for the comments and discussion.
Research supported by the HL-LHC project.

References

[1] Apollinari G, Béjar Alonso I, Brüning O, Lamont M and Rossi L 2015 *High-Luminosity Large Hadron Collider (HL-LHC): Preliminary Design Report* Geneva URL `https://cds.cern.ch/record/2116337`
[2] Pellegrini D *et al.* 2017 *Proceedings of IPAC2017* (Copenhagen, Denmark) TUPVA009
[3] Medina L *et al.* 2017 *Proceedings of IPAC2017* (Copenhagen, Denmark) TUPIK089
[4] Grote H and Schmidt F 2016 *MAD-X User's Guide* URL `http://madx.web.cern.ch/madx/`
[5] Schmidt F 2012 *SixTrack User Manual* URL `http://sixtrack.web.cern.ch/SixTrack/`
[6] McIntosh E and De Maria R 2013 *SixDesk User Manual* URL `http://sixtrack.web.cern.ch/SixTrack/`
[7] *High-Luminosity Large Hadron Collider (HL-LHC) technical design report V0* edited by Apollinari G, Béjar Alonso I, Brüning O, Lamont M, and Rossi L; unpublished URL `https://edms.cern.ch/ui/doc/1558149`
[8] Antoniou F *et al.* 2016 *7th Evian Workshop* (France) URL `https://indico.cern.ch/event/578001/`
[9] Métral E *et al.* 2017 *Operational scenario update (stability vs DA constraints)* HI-LUMI WP2 Meeting, CERN URL `https://indico.cern.ch/event/623917/`
[10] Papaphilippou Y *et al.* 2017 *LHC Performance Workshop* (Chamonix, France) URL `https://indico.cern.ch/event/580313/`
[11] Fartoukh S 2014 *Phys. Rev. ST Accel. Beams* **17**(11) 111001 URL `https://link.aps.org/doi/10.1103/PhysRevSTAB.17.111001`

Modelling and measurements of bunch profiles at the LHC

S Papadopoulou[1,2], F Antoniou[1,3], T Argyropoulos[1], M Fitterer[4], M Hostettler[1], Y Papaphilippou[1]

[1] CERN, Geneva, Switzerland
[2] Institute of Theoretical and Computational Physics (ITCP), Crete, Greece
[3] Department of Physics, University of Liverpool, United Kingdom
[4] Fermi National Accelerator Laboratory, Batavia, IL 60510, USA

E-mail: stefania-parthena.papadopoulou@cern.ch

Abstract. The bunch profiles in the LHC are often observed to be non-Gaussian, both at Flat Bottom (FB) and Flat Top (FT) energies. Especially at FT, an evolution of the tail population in time is observed. In this respect, the Monte-Carlo Software for IBS and Radiation effects (SIRE) is used to track different types of beam distributions. The impact of the distribution shape on the evolution of bunch characteristics is studied. The results are compared with observations from the LHC Run 2 data.

1. Introduction

For the LHC luminosity studies, a model including the effects of intrabeam scattering (IBS), synchrotron radiation (SR) and luminosity burn-off is used [1]. The model was constructed based on analytical models which assume Gaussian beam distributions. A comparison of the evolution of the bunch characteristics as predicted by this model with the measured ones revealed an extra (on top of IBS and SR) transverse emittance blow up in the measured data. One of the attempts to explain this blow up concerns the bunch profiles that appear to have non-Gaussian shapes both at flat bottom (FB) and flat top (FT) energies, i.e. 450 GeV and 6.5 TeV respectively. The aim of this study is to quantify the impact of the distribution's shape on the emittance and luminosity evolution, extending the usual approach of employing the analytical formulas for modelling IBS, which are based on 3D Gaussian beam assumptions [2]. For this, the Monte Carlo multiparticle simulation code for IBS and Radiation Effects (SIRE) is being used [3]. The comparison of the code output with analytical formulas has already been studied for the FB [4] and the nominal collision energy (7 TeV) [5]. In this paper, examples of measured bunch profiles at FT are presented for both planes. The evolution of these profiles is compared to the ones calculated by this code. A comparison between the SIRE code, the Bjorken-Mtingwa analytical formalism [2] and experimental data, is presented for the bunch length evolution at the FT energy.

2. Bunch profiles at the FT energy

It has been generally observed that the bunch profiles in the LHC, both at FT and FB energies, appear to have tails that differ from the ones of a normal distribution. In order to describe more accurately the bunch shape, a generalized Gaussian function, called the q-Gaussian [6], can be used. This distribution has a probability density function given by:

$$f(x) = \frac{\sqrt{\beta}}{C_q}\left[1-(1-q)\beta x^2\right]^{\frac{1}{1-q}} . \tag{1}$$

The parameter q describes the weight of the tails. The heavy tail domain corresponds to $1 < q < 3$, in the limit of $q \to 1$ the distribution becomes a normal distribution and when $q < 1$ the tails are lighter compared to the ones of the Gaussian. The normalization factor C_q differs for specific ranges of the q parameter. The parameter β is always a positive number and for a specific q value the probability density function grows with larger β.

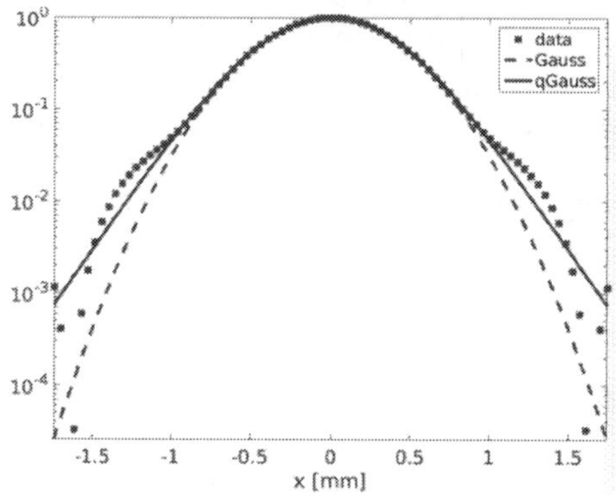

Figure 1. A horizontal bunch profile at the FT energy, as measured by the BSRT, is denoted by blue stars. It is fitted with the Gaussian (dashed line) and the q-Gaussian (solid line) functions.

Table 1. Fit parameters for a horizontal bunch profile, as measured by the BSRT, at FT energy.

Horizontal distribution	Fit parameters
Gaussian (RMSE=0.010)	$\sigma_{rms} = 0.380 \pm 0.002$ mm
q-Gaussian (RMSE=0.004)	$\sigma_{rms} = 0.403 \pm 0.005$ mm $q = 1.12 \pm 0.01$

An example of a horizontal bunch profile is presented in Figure 1 in logarithmic scale. Two fitting methods were applied; a Gaussian fit represented by a dashed line and a q-Gaussian fit represented by a solid line. The results of the fits are summarized in Table 1. Since the tails of the distribution are non-negligible, the q-Gaussian approaches the profile much better than the Gaussian function, with a reduction of the RMSE (root mean squared error) by a factor of 2. The measurements were performed with the Beam Synchrotron Radiation Telescope (BSRT) [7] which is a diffraction-limited instrument [8]. Even after using a fast Fourier transform (FFT) filter and a cut at 3 σ, the significant diffraction patterns at the tails of the bunch profiles still persist, implying that with the present instrument, it is difficult to have conclusive estimations of the tails. When $q \neq 1$ the standard deviation of the q-Gaussian fit differs from the Gaussian one and so do the resulted beam sizes (σ_{rms} values in Table 1).

During the energy ramp, the bunches in the LHC are blown up longitudinally in order to avoid instabilities due to the loss of Landau damping [9]. This results in non-Gaussian longitudinal distributions at the start of collisions [10]. The longitudinal distribution is measured by the wall current monitors [11] and the longitudinal synchrotron light monitor (BSRL) [12]. An example of the initial (at the start of collisions) and the final (after 11.5 h) longitudinal bunch profiles [13] are presented in logarithmic scale in Figure 2. Both profiles are fitted with the

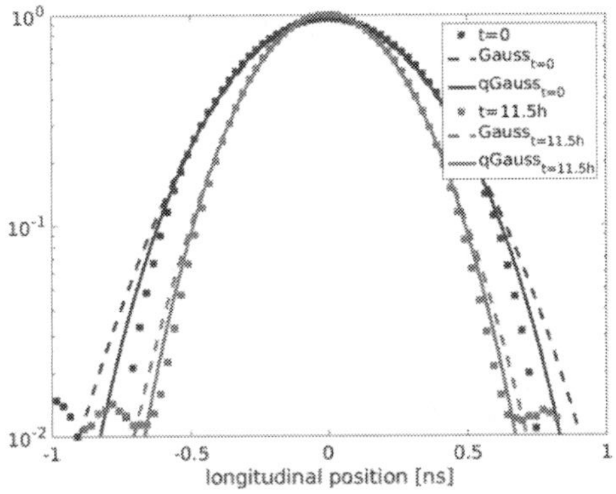

Figure 2. The initial (at the start of collisions) and the final (after 11.5 h) longitudinal bunch profiles, as measured at the LHC, are denoted by blue and red stars respectively. They are fitted with the Gaussian (dashed line) and the q-Gaussian (solid line) functions.

Table 2. Fit parameters for a longitudinal bunch profile coming from experimental data at FT energy.

Longitudinal distribution	Initial fit parameters
Gaussian (RMSE=0.020)	$\sigma_{rms} = 0.299 \pm 0.003$ ns
q-Gaussian (RMSE=0.011)	$\sigma_{rms} = 0.286 \pm 0.004$ ns $q = 0.88 \pm 0.03$
Longitudinal distribution	**Final fit parameters**
Gaussian (RMSE=0.013)	$\sigma_{rms} = 0.233 \pm 0.002$ ns
q-Gaussian (RMSE=0.009)	$\sigma_{rms} = 0.227 \pm 0.002$ ns $q = 0.93 \pm 0.03$

Gaussian (dashed line) and the q-Gaussian (solid line) functions. The results of the fits are summarized in Table 2, with the RMSE showing again that the q-Gaussian fit is better than the Gaussian one. As expected from the strong effect of SR at FT, the rms beam size decreases with time. During these 11.5 hours in collisions the q parameter remains smaller than 1. The fact that it slightly increases (around 2%/h) indicates that the distribution becomes slowly Gaussian.

3. Comparison with simulations

In the case of non-Gaussian beam distributions no analytical IBS models exist. In order to study the impact of the distribution shape on the emittance and distribution evolution, the Monte-Carlo code SIRE, inspired by MOCAC (MOnte CArlo Code) [14], has been developed [3]. In addition to IBS it also takes into account the effects of synchrotron radiation damping and quantum excitation. After specifying the beam distribution and the optics along a lattice, SIRE iteratively computes intrabeam collisions between pairs of macro-particles. The beam distribution is updated and the rms beam emittances are recomputed, giving finally as output the emittance evolution with time.

The SIRE code accepts any type of distribution as an input. In order to compare the experimental observations with the results of the code, a particle distribution generated from

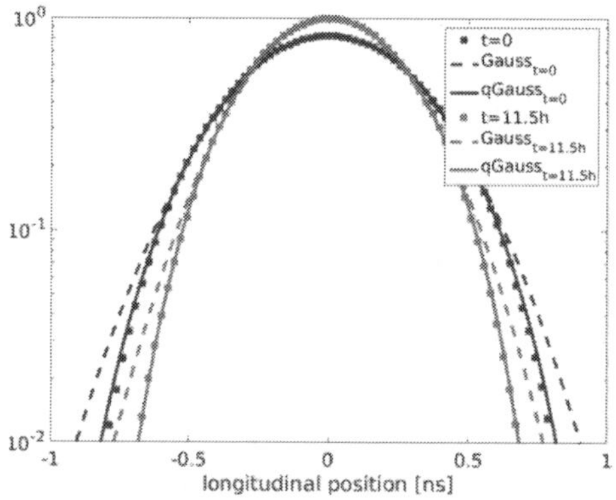

Figure 3. The initial ($t = 0$) and the final ($t = 11.5$ h) longitudinal bunch profiles, as calculated by SIRE, are denoted by blue and red stars respectively They are fitted with the Gaussian (dashed line) and the q-Gaussian (solid line) functions.

Table 3. Fit parameters for a longitudinal bunch profile, as calculated by SIRE, at FT energy.

Longitudinal distribution	**Initial fit parameters**
Gaussian (RMSE=0.014)	$\sigma_{rms} = 0.307 \pm 0.002$ ns
q-Gaussian (RMSE=0.003)	$\sigma_{rms} = 0.290 \pm 0.001$ ns $q = 0.85 \pm 0.01$
Longitudinal distribution	**Final fit parameters**
Gaussian (RMSE=0.014)	$\sigma_{rms} = 0.254 \pm 0.002$ ns
q-Gaussian (RMSE=0.003)	$\sigma_{rms} = 0.240 \pm 0.001$ ns $q = 0.84 \pm 0.01$

a q-Gaussian function with parameters given in Table 2 was tracked for 11.5 h, taking into account the bunch population decrease with time. The transverse distributions are assumed to be Gaussian since at FT the shape of their tails is not clear due to diffraction. In Figure 3, the input (t=0) and output (t=11.5 h) distributions, as calculated by the code, are denoted by blue and red stars, respectively. The fit results for the Gaussian (dashed line) and the q-Gaussian (solid line) functions are presented in Table 3. There is no change at the tails of the tracked distribution, while in reality the profiles become more Gaussian. The q-Gaussian fit shows that within this 11.5 h the rms beam size gets 21% smaller. For the measured bunch profile (in Figure 2) this reduction is around 26%. This difference can be explained by the fact that in the measurements there is an extra (on top of IBS and SR) transverse emittance blow up that is not included in the simulations yet.

The 4σ-bunch length evolution during several hours in stable beams is shown in Figure 4 and 5, when assuming Gaussian and q-Gaussian distributions, respectively. The results given by SIRE are denoted by blue lines. The red line corresponds to the evolution calculated by the IBS module of MADX [15], which is based on the analytical formulation of B-M and always assumes Gaussian distributions. The bunch length together with the two standard deviation error-bars, when fitting the data with the Gaussian and the q-Gaussian functions is represented by a black and a grey line, respectively. The initial (at the start of collisions) and final (after 11.5 h) bunch parameters are summarized in Table 4. The 4 σ bunch length when assuming Gaussian and q-Gaussian distributions, in the SIRE code and for the fitted data, is denoted by σ_{l_G} and $\sigma_{l_{qG}}$. The bunch length values differ for the two distribution functions used. Generally, for a light tailed distribution the rms value is overestimated by fitting a Gaussian, whereas the opposite is true for a heavy tailed distribution.

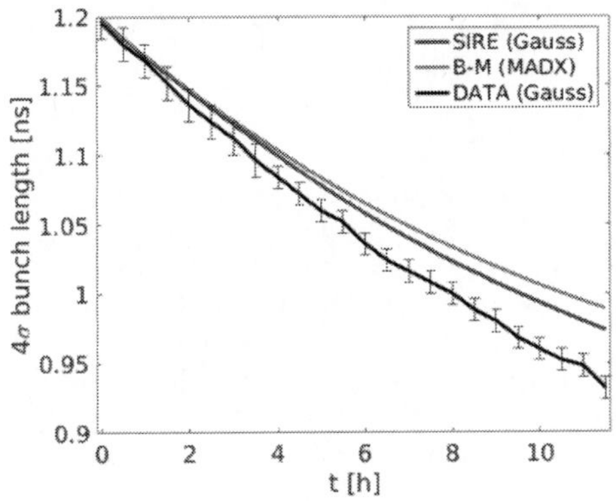

Figure 4. The bunch length (4σ) evolution during several hours in stable beams, as computed by the SIRE code (blue), the B-M analytical formalism (red) and as measured by the longitudinal profile monitors (black) when assuming a Gaussian distribution.

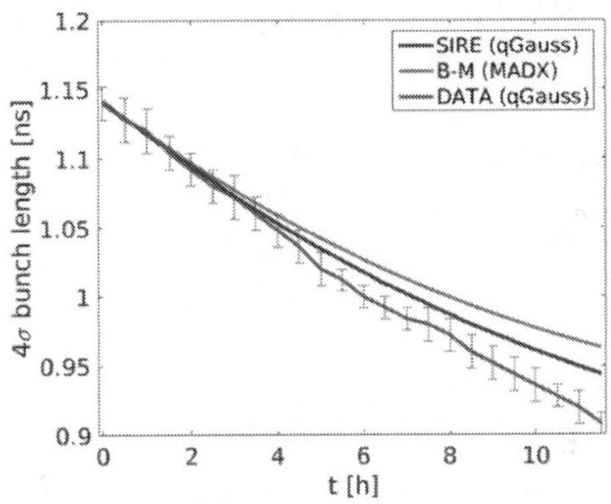

Figure 5. The bunch length (4σ) evolution during several hours in stable beams, as computed by the SIRE code (blue), the B-M analytical formalism (red) and as measured by the longitudinal profile monitors (grey) when assuming a q-Gaussian distribution.

Table 4. Initial and final bunch parameters at FT energy.

Parameters	Initial	Final
Bunch population [10^{11}]	1.05	0.86
ϵ_x, ϵ_y [μm.rad]	3.5, 3.5	2.85, 2.45
σ_{l_G} (DATA, SIRE, B-M) [ns]	1.2	(0.93, 0.97, 0.99)
$\sigma_{l_{qG}}$ (DATA, SIRE, B-M) [ns]	1.14	(0.91, 0.94, 0.97)

In Figure 4, the divergence between the SIRE and MADX [16] for longer time-spans is probably due to the fact that the two codes make use of different approaches to calculate the IBS effect [1]. The bunch length evolution as calculated by the code is closer to the measured data for the q-Gaussian than for the Gaussian case. In order to make the simulations more realistic the extra transverse emittance blow up should be taken into account in the code.

4. Summary and next steps

In the LHC, the interplay between a series of effects can lead to non-Gaussian distributions. In this paper, the q-Gaussian function, which is able to describe much more accurately the bunch

[1] SIRE uses the classical Rutherford cross section which is closer to the Piwinski formalism [17]

profiles in the case of non-Gaussian tail populations, is employed. Examples of non-Gaussian bunch profiles at FT were presented. The measured transverse bunch profiles can not be used for conclusive estimations because of the diffraction that affects their tails. However, in the longitudinal plane the tails are clearly underpopulated compared to a normal distribution. For specific bunch parameters, a Gaussian and a q-Gaussian longitudinal distributions tracked for several hours by the SIRE code are compared to the measured profiles from the machine. The bunch length evolution calculated by the code is close to the real data during the first hours at stable beams but start diverging after some hours. This difference can be explained by the extra emittance blow up observed in the transverse plane in the LHC data, while the simulation includes only IBS and radiation effects. The comparison with the analytical model of B-M for Gaussian bunch distributions is also presented. In order to understand the impact on the evolution of the bunch characteristics and finally on the luminosity evolution, various cases of non-Gaussian distributions, also in the transverse plane, are currently being tracked.

References

[1] Antoniou F *et al.* 2016 Can we predict luminosity? *Proc. of the 7th Evian Workshop*
[2] Bjorken J C and Mtingwa S K 1983 Intrabeam scattering *Part. Accel.* **13** pp 115-143
[3] Martini M and Vivoli A 2010 Intra-Beam scattering in the CLIC damping rings *CERN-ATS-2010-094*
[4] Papadopoulou S *et al.* 2016 Modelling and measurements of bunch profiles at the LHC flat bottom *Proc. of IPAC'16* TUPMW022
[5] Martini M and Vivoli A 2010 Effect of intrabeam scattering and synchrotron radiation damping when reducing transverse emittances to augment the LHC luminosity *sLHC Project Report* 0032
[6] Curado E M F and Tsallis C 1992 Generalized statistical mechanics: connection with thermodynamics *J. Phys.* A: Math. Gen. **25** 1019
[7] Jung R *et al.* 2002 The LHC 450 GeV to 7 TeV Synchrotron radiation profile monitor using a superconducting undulator *Proc. of the Beam Instrumentation Workshop* p 220
[8] Trad G 2013 Development and optimisation of the SPS and LHC beam diagnostics based on synchrotron radiation monitors PhD Thesis
[9] Baudrenghien P *et al.* 2013 Longitudinal emittance blowup in the Large Hadron Collider *Nuclear Instruments and Methods in Physics Research* A: Accelerators, Spectrometers, Detectors and Associated Equipment **726** pp 181-190
[10] Timko H *et al.* 2015 Lessons learned in LHC operation in 2015 *Proc. of the 6th Evian Workshop*
[11] Bohl T and Malo J F 2009 The APWL wideband wall current monitor *CERN-BE-2009-006*
[12] Jeff A *et al.* 2012 Longitudinal density monitor for the LHC *Phys. Rev. ST Accel.Beams* **15** 032803
[13] Hostettler M *et al.* 2016 How well do we know our beams? *Proc. of the 7th Evian Workshop*
[14] Zenkevich P, Boine-Frankenheim O and Bolshakov A 2007 Last advances in analysis of intrabeam scattering in the hadron storage rings *Nucl. Instr. and Meth.* A **577** pp 110-116
[15] Antoniou F and Zimmermann F 2012 Revision of intrabeam scattering with non-ultrarelativistic corrections and vertical dispersion for MAD-X *CERN-ATS-2012-066*
[16] MAD-X homepage, http://mad.web.cern.ch/mad
[17] Piwinski A 1974 *Proc. of 9th Int. Conf. on High Energy Accelerators*

100 km CEPC parameters and lattice design*

D Wang†, J Gao, C H Yu, Y Zhang, Y W Wang, F Su, J Y Zhai, S Bai, H P Geng, T J Bian, N Wang, X H Cui, C Zhang and Q Qin

IHEP, Beijing 100049, China

wangd93@ihep.ac.cn

Abstract. The 100km double ring configuration with shared superconducting RF system has been defined as baseline by the circular electron positron collider (CEPC) steering committee. Based on this new scheme, we will get higher luminosity for Higgs (+170%) keeping the beam power in preliminary conceptual design report (Pre-CDR) or to reduce the beam power (19 MW) while keeping same luminosity. CEPC will be compatible with W and Z experiment. The luminosity for Z is designed at the level of 10^{35} $cm^{-2}s^{-1}$. The requirement for the energy acceptance of Higgs has been reduced to 1.5% by enlarging the ring to 100 km. The optics of arc and final focus system (FFS) with crab sextupoles has been designed, and also some primary dynamic aperture (DA) results were introduced.

1. Introduction

According to the physics goal of CEPC at the energy of Higgs and Z-pole, it is required that the CEPC provides e+e- collisions at the center-of-mass energy of 240 Gev and delivers a peak luminosity of 2×10^{34} $cm^{-2}s^{-1}$ at each interaction point. At Z-pole the luminosity is required to be larger than 1×10^{34} $cm^{-2}s^{-1}$ per IP. The schematic layout of CEPC is shown in Figure 1.

In the beginning of 2015, Pre-CDR of CEPC-SppC [1] has been completed with 54km circumference and single ring scheme. After that the size and the collision scheme of CEPC-SPPC was reconsidered. The 100 km double ring configuration with shared SCRF which is similar as FCC-ee [2-5] has been defined as baseline by CEPC steering committee on Jan. 14th of 2016. Meanwhile the advanced partial double ring scheme with 8 partial double ring regions was defined as CEPC alternative scheme with the aim of reducing the construction cost which may need long efforts to study the possible solution to the sawtooth and beamloading effects. The conceptual design report (CDR) with 2mm vertical β^* and 32 MW beam power will be finished by the end of 2017.

CEPC was proposed as a compatible machine which will allow stringent tests of the Standard Model (SM) with precision measurements at the Z pole and WW thresholds. At the energy of Higgs, all bunches will distribute in the half ring due to the shared RF system. While for W and Z, bunches will distribute in the whole ring thanks to the independent RF system so that more bunches are possible. The scheme of CEPC bunch distribution is shown in Figure 2.

* Work supported by the National Key Programme for S&T Research and Development (Grant NO. 2016YFA0400400) and the National Natural Science Foundation of China (11505198, 11575218, 11605210 and 11605211).

† corresponding author

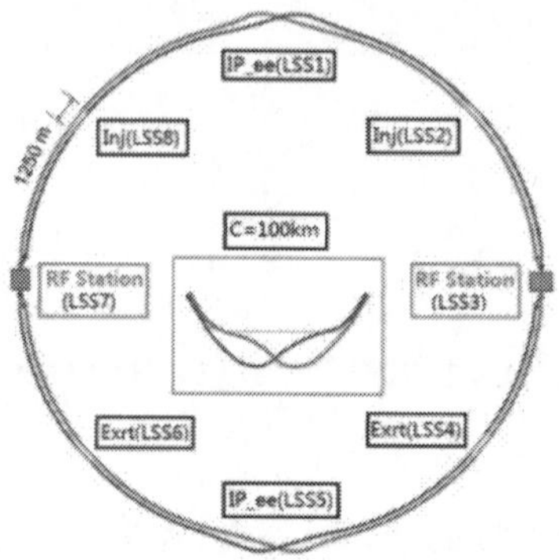

Figure 1. Layout of CEPC double ring scheme.

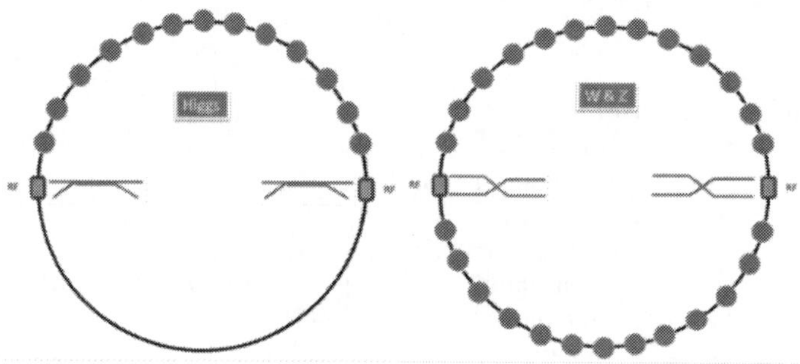

Figure 2. CEPC bunch distribution scheme (left: Higgs, right: W&Z).

2. CEPC parameter choice for 100 km double ring

To make an optimization for a collider, started from the goals, such as energy, luminosity/IP, number of IPs, etc, one has to consider very key beam physics limitations, such as beam-beam effects and beamstrahlung effect, and also take into account of economical and technical limitations, such as synchrotron radiation power and high order mode power in each superconducting RF cavity. By taking into account all these limitations in an analytical way, an analytical electron positron circular collider optimized design method has been developed both for head-on collision [6] and crab-waist collision [7].

2.1. *Constraints for parameter choice*

- Limit of beam-beam tune shift

$$\xi_y = \frac{2845}{2\pi}\sqrt{\frac{U_0}{2\gamma E_0 N_{IP}}} \times F_l \tag{1}$$

The real beam-beam tune shift should not larger than the above limit. The first part of this formula is the beam-beam limit without crossing angle [8]. We introduced a factor F_l here which is the luminosity enhancement factor by crab waist scheme and so far we assume it is 1.5 for Higgs, 1.7 for W and 2.6 for Z.

- Beam lifetime due to beamstrahlung

$$\frac{N_e}{\sigma_x \sigma_z} \leq 0.1\eta \frac{\alpha}{3\gamma r_e^2} \tag{2}$$

We hope to make sure the beam life time is not shorter than half hour and we have used Telnov's formula [9].

- Extra energy spread due to beamstrahlung

$$A = \delta_0 / \delta_{BS} \ (A \geq 5) \tag{3}$$

The energy spread due to beamstrahlung should be much smaller than the natural energy spread in order to control the beam quality after collisions.

- Higher order mode (HOM) power per cavity

$$P_{HOM} = k(\sigma_z)eN_e \cdot 2I_b \leq 2 \quad kw \tag{4}$$

where the HOM loss factor is $k(\sigma_z) = \frac{1.8}{\sqrt{\sigma_z / 0.00265}} \quad V/pC$.

*2.2. CEPC parameter potential with 1 mm β_y**

Using the method in reference 7, we got the parameter design for CEPC double ring scheme with 1mm vertical β*. At the energy of Higgs, with 50 MW synchrotron radiation (SR) power, the luminosity will be 5.4×10^{34} $cm^{-2}s^{-1}$ which is 2.7 times of Pre-CDR luminosity. If we reduce SR power to 19 MW, still we can reach our luminosity goal 2×10^{34} $cm^{-2}s^{-1}$. Luminosity potential of 100 km CEPC with 50 MW SR power is shown in Figure 3.

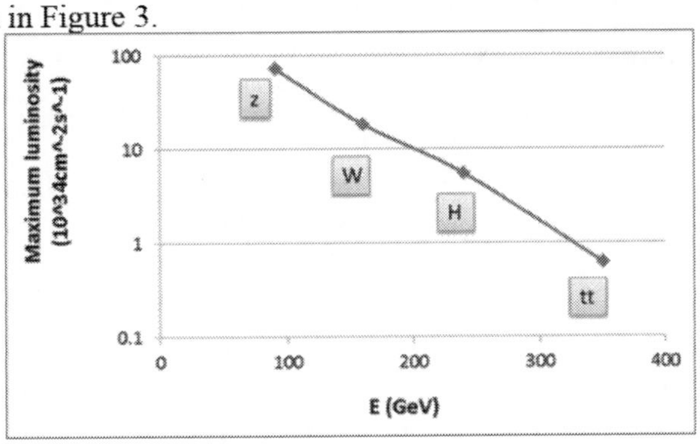

Figure 3. Luminosity potential of 100 km CEPC with 1mm vertical β* and 50 MW SR power/beam.

2.3. CEPC intermediate parameters toward CDR

Since the FFS design and DA study is the bottleneck for CEPC physics design, as the first step, we decided to choose 2 mm vertical β* and lower SR power (32 MW) for CDR report which will be published by the end of 2017. Using the method in reference 7, we got the parameter table for 2 mm vertical β* as table 1. For Higgs, we will need 32 MW SR power to reach the luminosity goal 2×10^{34} $cm^{-2}s^{-1}$. For W, with 32 MW SR power, the luminosity will be 5×10^{34} $cm^{-2}s^{-1}$. At Z pole, we plan to use 21300 bunches and the luminosity will be $1.2\times10^{35}cm^{-2}s^{-1}$.

3. Main ring lattice design

We chose 90°/90° FODO cell and non-interleave sextupole scheme for the arc. The emittance of the ring is 1.3 nm. The twiss parameters of arc are shown in Figure 4.

Figure 5 shows the design for FFS. We need a triplet following the final doublet to produce 90° phase difference between horizontal plane and vertical plane. Here IP beta function is 0.171 m/0.002 m as CDR parameter in table 1. Both main sextupoles can work as the crab sextupoles (in Figure 5). According to the issue of detector background, we have controlled the critical energy of dipoles under 100 kev within 560 m distance from IP and under 200 keV within 800 m within this FFS design.

Here we need a pair of weak sextupoles with about -10% of the main sextupoles' strength to correct the nonlinearity of main sextupole length effect [10]. This approach is essential otherwise the on-momentum DA will be very small.

Table 1. Parameters for 100 km CEPC double ring with 2 mm vertical β*.

	Pre-CDR	Higgs	W	Z
Number of IPs	2	2	2	2
Energy (Gev)	120	120	80	45.5
Circumference (km)	54	100	100	100
SR loss/turn (Gev)	3.1	1.67	0.33	0.034
Half crossing angle (mrad)	0	16.5	16.5	16.5
Piwinski angle Φ	0	3.19	5.69	4.29
N_e/bunch (10^{11})	3.79	0.968	0.365	0.455
Bunch number	50	412	5534	21300
Beam current (mA)	16.6	19.2	97.1	465.8
SR power /beam (MW)	51.7	32	32	16.1
Bending radius (km)	6.1	11	11	11
Momentum compaction (10^{-5})	3.4	1.14	1.14	4.49
$\beta_{IP\ x/y}$ (m)	0.8/0.0012	0.171/0.002	0.171 /0.002	0.16/0.002
Emittance x/y (nm)	6.12/0.018	1.31/0.004	0.57/0.0017	1.48/0.0078
Transverse σ_{IP} (um)	69.97/0.15	15.0/0.089	9.9/0.059	15.4/0.125
ξ_x/ξ_y/IP	0.118/0.083	0.013/0.083	0.0055/0.062	0.008/0.054
V_{RF} (GV)	6.87	2.1	0.41	0.14
f_{RF} (MHz)	650	650	650	650
Nature σ_z /Total σ_z (mm)	2.14/2.65	2.72/2.9	3.37/3.4	3.97/4.0
HOM power/cavity (kw)	3.6 (5cell)	0.41(2cell)	0.36(2cell)	1.99(2cell)
Energy spread (%)	0.13	0.098	0.065	0.037
Energy acceptance requirement (%)	2	1.5		
Energy acceptance by RF (%)	6	2.1	1.1	1.1
n_γ	0.23	0.26	0.15	0.12
Life time due to beamstrahlung_cal (minute)	47	52		
L_{max}/IP ($10^{34}cm^{-2}s^{-1}$)	2.04	2.0	5.15	11.9

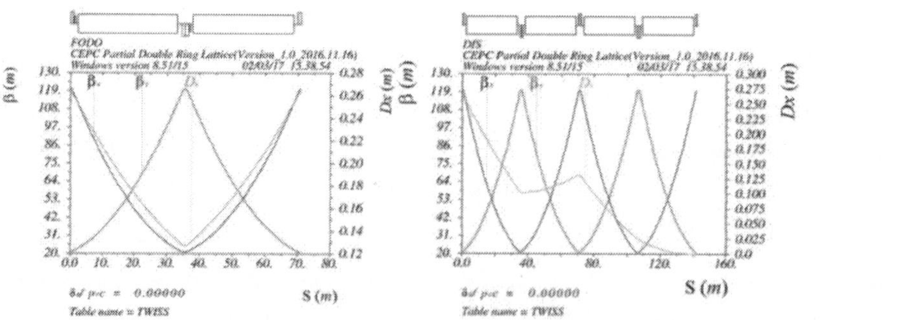

Figure 4. Twiss parameter of arc with 90°/90° phase advance per cell (left: FODO, right: dispersion suppressor).

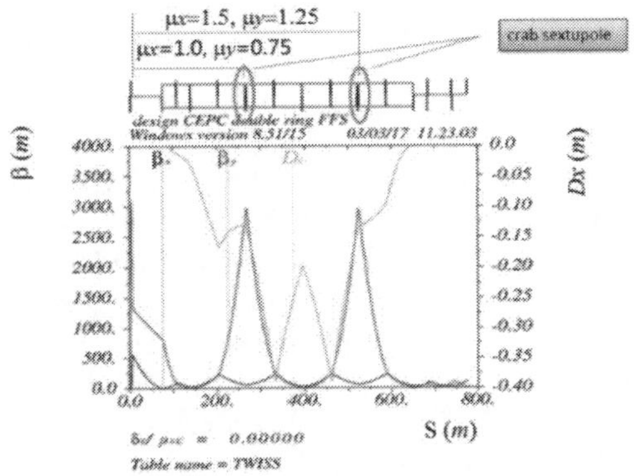

Figure 5. FFS optics with crab sextupoles.

4. Dynamic aperture

We made a whole ring by connecting the arc and FFS. The detail of RF sections and crossing collision was not included by this ideal lattice however the main nonlinearities for CEPC were there. The DA of the ideal ring with only two families sextupoles in the arc and without damping effect was shown in Figure 6. The on-momentum DA is good enough while the energy acceptance is not satisfying (0.5%). The further optimization of DA bandwidth with multi-sextupoles is undergoing.

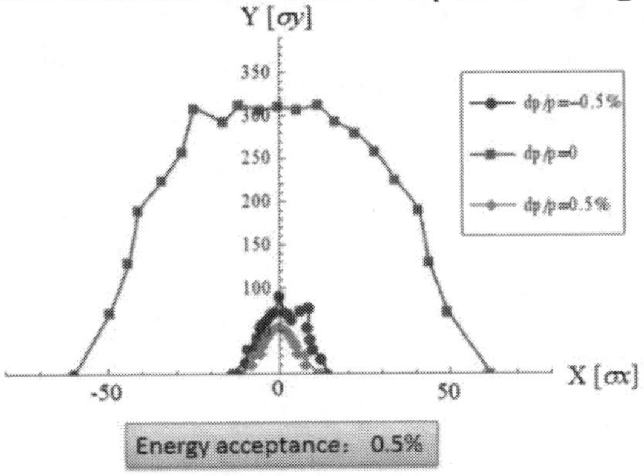

Figure 6. DA before optimization (radiation damping is close.).

5. Conclusion

We have developed a consistent method for CEPC parameter choice with crab waist scheme by using analytical expression of maximum beam-beam tune shift and beamstrahlung beam lifetime started from given IP vertical beta, beam power and other technical limitations. Based on 100 km double ring scheme, we hope to get higher luminosity for Higgs keeping the Pre-CDR beam power (52 MW) or to reduce the beam power by 40% keeping the Pre-CDR luminosity. The luminosity can be crosschecked by beam-beam simulations.

We chose 2 mm vertical β^* and 32 MW SR power for CDR. The optics of arc and IR with crab waist has been designed. The dynamic aperture of the whole ring is good enough for on momentum particle while the off momentum DA is still need to been optimized.

Acknowledgments

The authors thank Professors Katsunobu Oide, Yunhai Cai, Frank Zimmermann, Michael Koratzinos, and Valery Telnov for their helpful discussions.

References

[1] The CEPC-SPPC Study Group 2015 CEPC-SPPC preliminary Conceptual Design Report *Institute of High Energy Physics Report (Vol 2 Accelerator)* IHEP-CEPC-DR-2015-01
[2] Blondel A *et al.* 2012 LEP3: A High Luminosity e+ e- Collider in the LHC Tunnel to Study the Higgs Boson *Int. Conf. of Particle Accelerator (New Orleans)* TUPPR078
[3] Koratzinos M *et al.* 2013 TLEP: A High Performance Circular e+e- Collider to Study the Higgs Boson *Int. Conf. of Particle Accelerator (Shanghai)* TUPME040
[4] Koratzinos M *et al.* 2015 The FCC-ee Study: Progress and Challenges *Proc. Int. Conf. of Particle Accelerator (Richmond)* TUPTY060
[5] Oide K *et al.* 2016 Design of beam optics for the FCC-ee collider rings *Phys. Rev. ST Accel. Beams* **19** 111005
[6] Wang D *et al.* 2013 Optimization parameter design of a circular e+e- Higgs factory *Chinese Phys.* C **37** 9
[7] Wang D *et al.* 2016 CEPC partial double ring scheme and crab-waist parameters *International Journal of Modern Physics* A **31** 1644016.
[8] Gao J 2004 Eittance growth and beam lifetime limitations due to beam-beam effects in e+e- storage rings *Nucl. Instr. and methods* A **533** 270-4
[9] Telnov V 2012 Restriction on the energy and luminosity of e+e- storage rings due to beamstrahlung *Preprint* arXiv:1203.6563
[10] Bogomyagkov A *et al.* 2009 Effect of the sextupole finite length on dynamic aperture in the collider final focus *Preprint* arXiv:0909.4872

Experimental validation of the Achromatic Telescopic Squeezing (ATS) scheme at the LHC

S Fartoukh, R Bruce, F Carlier, J Coello De Portugal, A Garcia-Tabares, E Maclean, L Malina, A Mereghetti, D Mirarchi, T Persson, M Pojer, L Ponce, S Redaelli, B Salvachua, P Skowronski, M Solfaroli, R Tomas, D Valuch, A Wegscheider, J Wenninger

CERN, Geneva, Switzerland

E-mail: stephane.fartoukh@cern.ch

Abstract. The Achromatic Telescopic Squeezing scheme offers new techniques to deliver unprecedentedly small beam spot size at the interaction points of the ATLAS and CMS experiments of the LHC, while perfectly controlling the chromatic properties of the corresponding optics (linear and non-linear chromaticities, off-momentum beta-beating, spurious dispersion induced by the crossing bumps). The first series of beam tests with ATS optics were achieved during the LHC Run I (2011/2012) for a first validation of the basics of the scheme at small intensity. In 2016, a new generation of more performing ATS optics was developed and more extensively tested in the machine, still with probe beams for optics measurement and correction at $\beta^* = 10$ cm, but also with a few nominal bunches to establish first collisions at nominal β^* (40 cm) and beyond (33 cm), and to analysis the robustness of these optics in terms of collimation and machine protection. The paper will highlight the most relevant and conclusive results which were obtained during this second series of ATS tests.

1. Introduction

The Achromatic Telescopic Squeezing (ATS) scheme is a novel optics concept enabling the matching of ultra-low β^* in the LHC (and other hadron circular colliders), while correcting the chromatic aberrations induced by the inner triplet [1]. This scheme is essentially based on a two-stage telescopic squeeze. In a first phase, a so-called pre-squeeze is achieved by using exclusively, as usual, the matching quadrupoles of the high luminosity insertions IR1 and IR5. In a second phase, the squeeze continues by acting only on the insertions located on either side of IR1 and IR5 (i.e. IR8/2 for the telescopic squeeze of IR1, and IR4/6 for IR5). As a result, sizable β-beating bumps are induced in the four sectors on either side of IP1 and IP5. These waves of β-beating are then also necessary in order to boost the efficiency of the lattice sextupoles for the chromatic correction. The complete validation of the ATS scheme at high intensity is a very important milestone in the overall upgrade plan of the LHC. The first series of ATS machine developments (MD) took place in Run I (2011 and 2012), where most of the ATS principles were demonstrated, but only with pilot beams [2]-[5]. All LHC and HL-LHC ATS optics versions developed so far however showed very unfavorable phase advances, nearly equal to 90 degrees in the horizontal plane, between the extraction kickers (MKD) in the dump insertion (IR6) and some tertiary collimators (TCT) in IR1 and IR5, in particular the most exposed one in case of asynchronous dump (TCT.R5B2). When discussing the possibility to

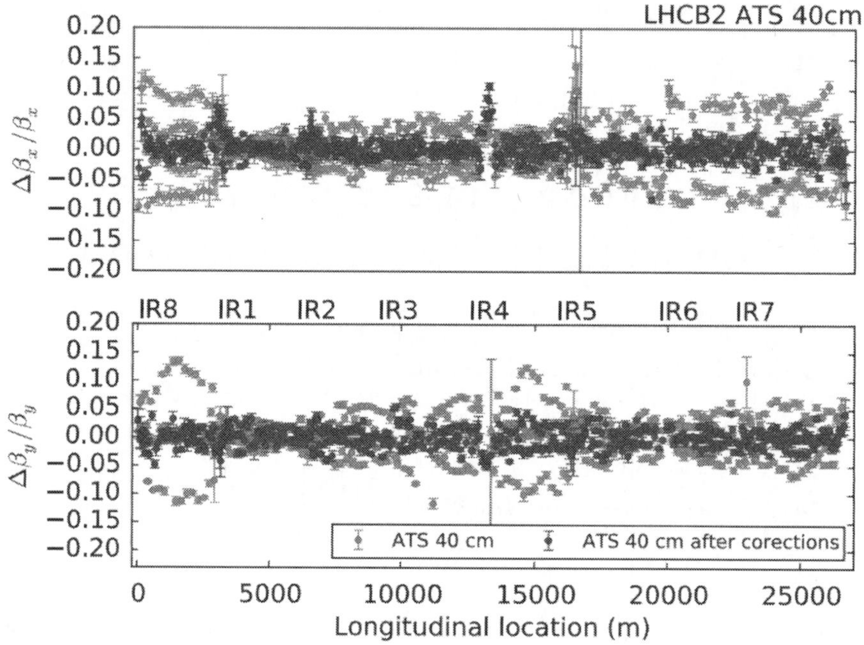

Figure 1. Hor. (H) and vert. (V) β-beating measured at $\beta^* = 40$ cm (example for beam2) before and after (global) correction. A local correction knob (triplet quadrupole trims) was preset based on the 40 cm nominal optics.

directly use ATS optics in order to restart the LHC for 6.5 TeV operation in 2015, this feature was shown to be a clear weakness of ATS optics for the LHC, which rapidly discarded this option [6]. Very recently, a new generation of ATS optics was then deployed in order to bring a definite cure to the above mentioned limitation, offering phase advances very close to optimal (within $20 - 30°$) between the MKDs and TCTs, for both beams and both IR1 and IR5 [7]. This opened the possibility to test these new optics with higher intensity beams. Basically, two types of ATS MDs took place in 2016, namely: (i) MDs for optics measurements and corrections achieved with probe beams to re-validate the pre-squeezed and fully telescopic optics down to $\beta^* = 40$ cm and 10 cm, respectively, (ii) MDs with a major component related to collimation, conducted with (quasi-)nominal collimator settings, and a filling scheme containing two colliding nominal bunches per beam and/or sparse non-colliding pilot bunches (for loss maps or aperture measurement).

2. ATS pre-squeezed optics with $\beta^* = 40$ cm

The first ATS MD commissioned the new ATS injection optics, its ramp up to 6.5 TeV, and the pre-squeeze down to $\beta^* = 40$ cm at IP1 and IP5 (3 m at IP8), using low intensity beams and a flat machine (crossing bumps switched off). The optics was measured, then successfully corrected at the 5-10% level in terms of β-beating, at injection, flat top, and at $\beta^* = 40$ cm (see Figure 1). Dedicated chromatic measurements were conducted at $\beta^* = 40$ cm, showing an as-expected off-momentum β-beating pattern (shown in Figure 5 at $\beta^* = 21$ cm) and a vanishing non-linear chromaticity, which is one key feature of the ATS scheme. Finally, a complete fill was dedicated to the first tests of the crossing bumps from injection to $\beta^* = 40$ cm. The second ATS MD demonstrated a full ATS cycle from injection to collision with two nominal bunches, and nominal or quasi-nominal collimation and machine protection settings. In particular, a

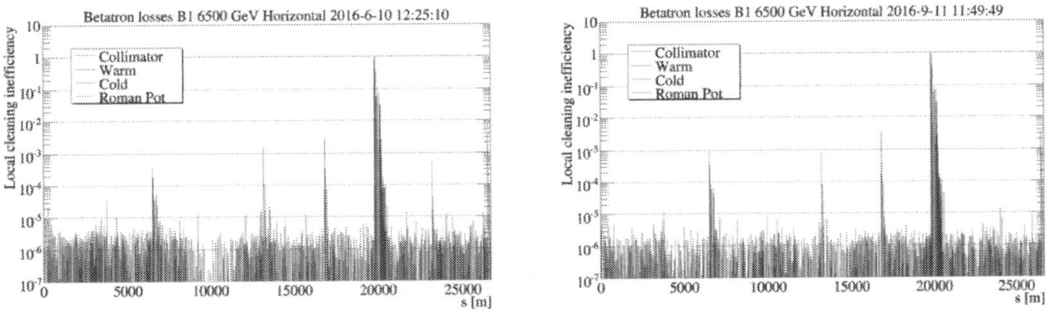

Figure 2. Loss maps measured for B1H at $\beta^* = 40$ cm: nominal (left) and ATS pre-squeezed (right) optics.

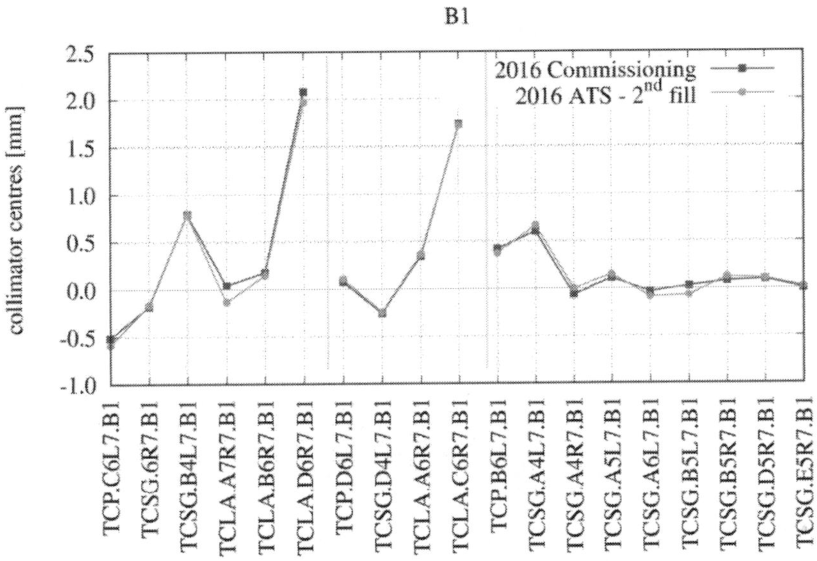

Figure 3. IR7 collimator centers measured for beam1: nominal vs. ATS optics at $\beta^* = 40$ cm.

new reference orbit was established, collisions at all four IPs were rapidly found and optimized (with a typical luminosity of 5E30 $cm^{-2}s^{-1}$ in ATLAS and CMS). Using the nominal collimator and dump protection settings in IR3, IR7 and IR6 (thank to the very small changes of ATS optics w.r.t. the nominal optics in these 3 insertions), and after a beam-based re-alignment of the TCTs, betatron loss maps were successfully conducted at injection, showing no particular anomalies. Loss maps done at $\beta^* = 40$ cm also did not show any problems in terms of collimation hierarchy and inefficiency, even without need of re-aligning the IR7 collimators (see Figure 2). Then, a beam-based collimator realignment campaign was conducted at $\beta^* = 40$ cm, leading to a marginal re-centering of the collimator jaws compared to the nominal optics (see Figure 3). A few fills were needed to cover these activities. Two of them ended up with a (programmed) asynchronous dump with the TCT settings at $9\,\sigma$ and $8\,\sigma$ respectively, in order to validate the new MKD-TCT phase advances. The TCT losses measured for beam1 were rather conclusive, with beam2 in the right ballpark compared to expectations.

3. Telescopic optics with $\beta^* = 10$ cm

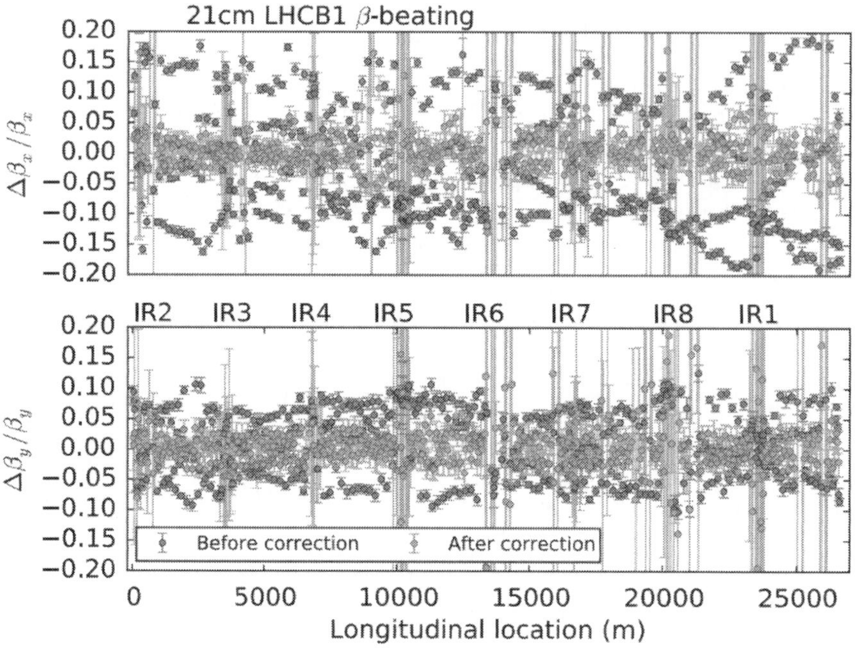

Figure 4. H and V β-beating measured at $\beta^* = 21$ cm (example of beam1) before and after (global) correction.

Considering the several validation steps already done with the 40 cm pre-squeezed optics, the aim of the third ATS MD was to (re-)validate the telescopic techniques of the scheme. The target was fixed to be the HL-LHC ultimate β^* of 10 cm at IP1/5, and passing through a moderately telescopic optics with $\beta^* = 33$ cm, as one potential candidate for pushing the LHC performance in the end of the 2017 Run. Probe beams were injected, ramped and pre-squeezed down to 40 cm, where the crossing bumps were switched off, and the collimator and machine protection settings relaxed in order to liberate enough aperture to reach $\beta^* = 10$ cm. The mechanics of the telescopic squeeze was successfully demonstrated down to $\beta^* = 10$ cm. First optics measurements took place at $\beta^* = 33$ cm, showing not more than 20% peak β-beating, which was deemed small enough in order to continue the telescopic squeeze without applying any correction yet. The first global optics corrections (below $\beta^* = 40$ cm) were calculated and successfully implemented at $\beta^* = 21$ cm, bringing the $\beta-$beating level back to the range of 5-10% (see Figure 4). The Montague functions also showed as-expected behaviour, with off-momentum $\beta-$beating waves induced by a dedicated powering of the lattice sextupole families in the sectors 81/12/45/56 adjacent to the high luminosity insertions, and arriving exactly in phase in order to compensate for the chromatic betatron kicks induced by the inner triplets (see Figure 5). Optics and dispersion measurements also took place at $\beta^* = 14$ cm and 10 cm, showing a β-beating in the 20-25% range (without any further corrections).

4. Telescopic optics with $\beta^* = 33$ cm

The last ATS MD focused on a (moderately) telescopic collision optics with $\beta^* = 33$ cm and a half-crossing angle of $\pm 140\,\mu$rad at IP1 and IP5 (i.e $9.0\,\sigma$ for an emittance of $\gamma\epsilon = 2.2\,\mu$m at 13 TeV c.m. energy). The aim was (i) to measure the triplet aperture in the end of the squeeze, (ii) to establish and optimize the collisions at all four IP's, and (iii) to preliminary assess the

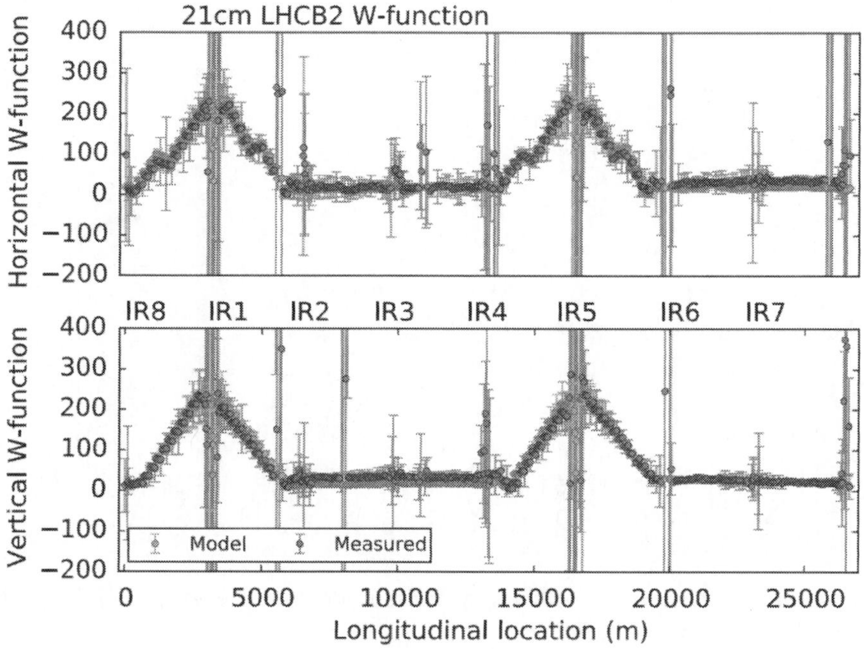

Figure 5. H and V Montague functions measured at $\beta^* = 21$ cm (example of beam2). $W \equiv 100$ units corresponds to an off-momentum peak β-beating of 20% at $\delta_p = 10^{-3}$.

collimation system via a series of loss maps (on- and off-momentum, with the beams separated or in collision at $\beta^* = 33$ cm). Two consecutive fills were needed to meet these objectives. The first one was dedicated to triplet aperture measurement, filling each ring with 8 pilot bunches and using techniques described in [8]. First, all collimators were opened at 33 cm, and a pilot bunch blown up in a given beam and a given plane. The triplet quadrupole corresponding to the aperture bottleneck was then easily found by looking at the BLM response (spikes) during the excitation, and the aperture finally determined via a beam-based alignment of the TCT in front of the triplet under consideration. A normalised aperture larger than or equal to $9.7\,\sigma$ was measured for both beams and both planes. In the second fill, two nominal bunches were injected per beam. Collisions were successfully established and optimized in all 4 IPs (with a luminosity of about 8E30 $\text{cm}^{-2}\text{s}^{-1}$ recorded by CMS). Before and after putting the beams into collisions at $\beta^* = 33$ cm, the TCT centers were realigned based on BPM data, and loss maps were conducted, both on and off-momentum in collision (applying an RF trim of $\Delta f = \pm 30$ Hz corresponding to a momentum shift of $\delta_p \sim \pm 2.5 \times 10^{-4}$). These measurements again did not show any unexpected features. Figure 6 shows a condensed summary of the collimation inefficiency measurement results which took place over the full 2016 LHC run, which is a clear illustration of the robust behaviour of the collimation system for ATS optics.

5. Summary and outlook

Using the latest ATS optics solution, the fundamental principles of the scheme were re-demonstrated with probe beams, in particular the telescopic squeeze down to $\beta^* = 10$ cm at constant sextupole strength beyond the 40 cm pre-squeezed optics. But also, state-of-the-art optics and coupling measurement and correction techniques, which were developed for the LHC nominal optics [10]-[13], such as β^*/α^* measurement with K-modulation, segment by segment

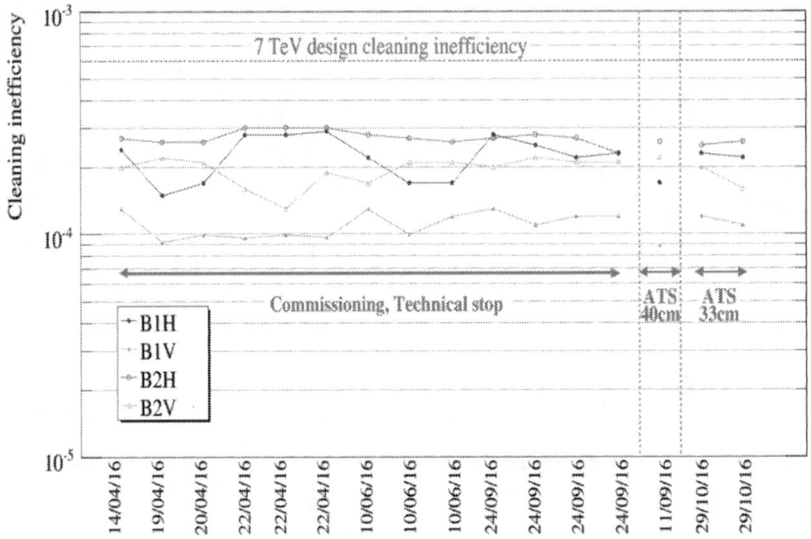

Figure 6. Collimation inefficiency measured over the 2016 LHC run in various optics and machine conditions [9].

local corrections and weighted global corrections, were successfully applied for the first time to telescopic optics, demonstrating their universality but also robustness at unprecedentedly small β^*. ATS pre-squeezed optics or moderately telescopic optics were also tested for the first time with a few nominal bunches, to establish and optimize collisions in all experimental insertions, and to re-assess the LHC collimation system with ATS optics. Completing these beam tests with dedicated studies, in particular beam-beam studies [14], the decision was taken to switch to ATS optics for the 2017 LHC luminosity run.

Acknowledgments
The successful beam tests with ATS optics obviously required the joint and coordinated effort of many different LHC teams, in particular the LHC Operation team, but also the Optics Measurement and Correction, the Collimation, the ADT and the LHC Beam Dump teams, without forgetting the constant support and advices from the Machine Protection Panel, the LHC MD Coordination and the management of the CERN Beams Department.

References
[1] Fartoukh S 2013 Achromatic telescopic squeezing scheme and its application to the LHC and its luminosity upgrade *Phys. Rev. ST Accel. Beams* **16** 111002
[2] Fartoukh S *et al.* 2011 The Achromatic Telescopic Squeezing (ATS) MD Part I *CERN-ATS-Note* **2011-033**
[3] Fartoukh S *et al.* 2011 The Achromatic Telescopic Squeezing (ATS) MD part II *CERN-ATS-Note* **2011-060**
[4] Fartoukh S *et al.* 2011 The Achromatic Telescopic Squeezing (ATS) MD part III *CERN-ATS-Note* **2011-132**
[5] Fartoukh S *et al.* 2013 The 10 cm beta* ATS MD, *CERN-ATS-Note* **2013-004**
[6] Bruce R *et al.* 2014 ATS optics validation, *LHC Machine Committee* **188** `http://indico.cern.ch/event/337929/`
[7] Fartoukh S 2015 Towards an asynchronous dump free ATS optics for LHC and HL-LHC *HiLumi WP2 Meeting* 61 `https://indico.cern.ch/event/402184/`
[8] Hermes P *et al.* 2016 Improved Aperture Measurements at the LHC and Results from their Application in 2015 *Proc. 7th International Particle Accelerator Conference (Busan: Korea)* TUPMW014
[9] Mirarchi D *et al.* 2016 Collimation: Experience and Performance *7th LHC Operation Workshop (Evian: France)* `https://indico.cern.ch/event/578001/`

[10] Aiba M *et al.* 2009 First β-beating measurement and optics analysis for the CERN Large Hadron Collider *Phys. Rev. ST Accel. Beams* **12** 081002

[11] Tomás R, Brüning O, Giovannozzi M, Hagen P, Lamont M, Schmidt F, Vanbavinckhove G, Aiba M, Calaga R and Miyamoto R 2010 CERN Large Hadron Collider optics model, measurements, and corrections *Phys. Rev. ST Accel. Beams* **13** 121004

[12] Tomás R *et al.* 2012 Record low beta-beating in the LHC *Phys. Rev. ST Accel. Beams* **15** 091001

[13] Persson T and Tomás R 2014 Improved control of the betatron coupling in the Large Hadron Collider *Phys. Rev. ST Accel. Beams* **17** 051004

[14] Pellegrini D, Antoniou F, Fartoukh S, Iadarola G and Papaphilippou Y 2017 Multiparametric Response of the LHC Dynamic Aperture in Presence of Beam-Beam Effects *Proc. 8th International Particle Accelerator Conference Copenhagen: Denmark)* TUPVA009

Acceleration of polarized protons and deuterons in the ion collider ring of JLEIC

A M Kondratenko[1], M A Kondratenko[1], Yu N Filatov[2], Ya S Derbenev[3], F Lin[3], V S Morozov[3], and Y Zhang[3]

[1] Science & Technique Laboratory Zaryad, Novosibirsk, Russia
[2] Moscow Institute of Physics and Technology, Dolgoprudny, Moscow Region, Russia
[3] Jefferson Lab, Newport News, VA, USA

E-mail: morozov@jlab.org

Abstract. The figure-8-shaped ion collider ring of Jefferson Lab Electron-Ion Collider (JLEIC) is transparent to the spin. It allows one to preserve proton and deuteron polarizations using weak stabilizing solenoids when accelerating the beam up to 100 GeV/c. When the stabilizing solenoids are introduced into the collider's lattice, the particle spins precess about a spin field, which consists of the field induced by the stabilizing solenoids and the zero-integer spin resonance strength. During acceleration of the beam, the induced spin field is maintained constant while the resonance strength experiences significant changes in the regions of "interference peaks". The beam polarization depends on the field ramp rate of the arc magnets. Its component along the spin field is preserved if acceleration is adiabatic. We present the results of our theoretical analysis and numerical modeling of the spin dynamics during acceleration of protons and deuterons in the JLEIC ion collider ring. We demonstrate high stability of the deuteron polarization in figure-8 accelerators. We analyze a change in the beam polarization when crossing the transition energy.

1. Preservation of ion polarization in figure-8 accelerators

A characteristic feature of JLEIC [1] is its figure-8-shaped rings [2]. Such a ring topology is transparent to the spin: the combined effect of arc fields on the spin in an ideal collider lattice reduces to zero after one particle turn on the design orbit, i.e. any orientation of the particle spin at any orbital location repeats from turn to turn. To preserve the polarizations of the proton and deuteron beams during acceleration from 8 GeV/c to 100 GeV/c in the ion collider ring, it is sufficient to use a weak solenoid with a field integral of 1.2 T·m, which does not perturb the design orbit and has practically no effect on the beam's orbital parameters [3-10]. The solenoid then stabilizes longitudinal spin polarization at its location. A solenoid with the indicated field integral allows one to induce a spin tune ν of 10^{-2} for protons and $3\cdot10^{-3}$ for deuterons, i.e., when a particle with a vertical spin makes one turn on the design orbit, its spin tilts by an angle of $2\pi\nu$ from its initial orientation.

For polarization stability, one must ensure that the spin tune ν induced by the solenoid significantly exceeds [3-6] the strength of the zero-integer spin resonance ω: $\nu \gg \omega$. The resonance strength is the average spin field $\vec{\omega}$ (the zero-integer Fourier harmonic of the spin perturbation without a stabilizing solenoid) determined by deviation of the trajectory from the design orbit due to machine element errors and beam emittances. In the absence of a solenoid, the spin precesses by an angle of $2\pi\omega$ about the $\vec{\omega}$ direction in one particle turn. The resonance strength consists of two parts: a coherent part arising due to additional transverse and longitudinal fields on a trajectory deviating from the design orbit and an incoherent part associated with the particles' betatron and synchrotron oscillations (beam emittances) [8, 9]

$$\vec{\omega} = \vec{\omega}_{coh} + \vec{\omega}_{emitt}\,,\ \omega_{coh} \gg \omega_{emitt}$$

In practice, the coherent part ω_{coh} significantly exceeds the incoherent one ω_{emitt}. The coherent part does not cause beam depolarization and only results in a simultaneous rotation of the polarization about the field determined by the strength and alignment errors of collider elements. In principle, the direction and size of the coherent part of the resonance strength can be measured and taken into account for polarization control. To preserve the polarization, it is then sufficient to satisfy a weaker condition: $\nu \gg \omega_{emitt}$.

2. Calculation of zero-integer spin resonance strength in JLEIC

Figure 1 shows β functions of the JLEIC collider lattice in the acceleration mode [11] used in our spin dynamics calculations. The origin of the coordinate frame is located at the collider's IP. Figure 1 also indicates the location of the solenoid stabilizing the spin during acceleration. The difference from the collision mode [12] where β functions in the IP region reach 2.5 km is that, in the acceleration mode, the β functions in the detector section do not exceed 150 m.

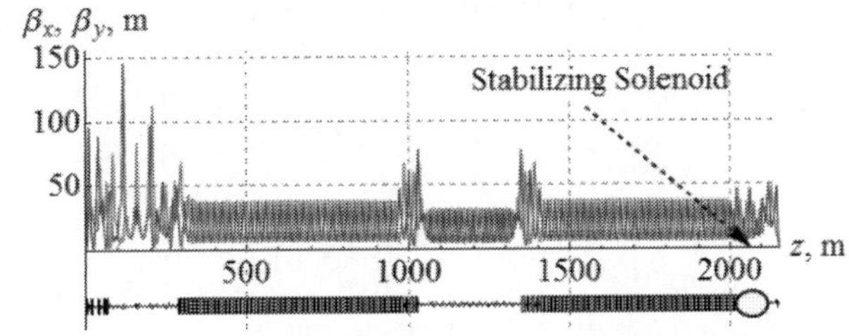

Figure 1. β functions of the ion collider ring.

2.1. Incoherent part of the spin resonance strength

Figure 2 shows the incoherent part of the proton resonance strength for normalized emittance values of 1 mm·mrad in both radial and vertical directions. As we can see, the value of the incoherent part does not exceed $2{\cdot}10^{-4}$ practically in the whole momentum range of the collider with the exception of narrow "interference" peaks where spin perturbations add up coherently in the arc magnets. The presented calculation confirms that the spin tune value of 10^{-2} induced by the solenoid field is sufficient to stabilize the spin with normalized emittances of the betatron motion equal to 1 mm·mrad.

Figure 3 shows a similar graph of the dependence of the incoherent part of the deuteron resonance strength on momentum. Our calculations assumed that the transverse size of the deuteron beam was equal to the proton beam size, i.e. the transverse beam emittances were 0.5 μm.

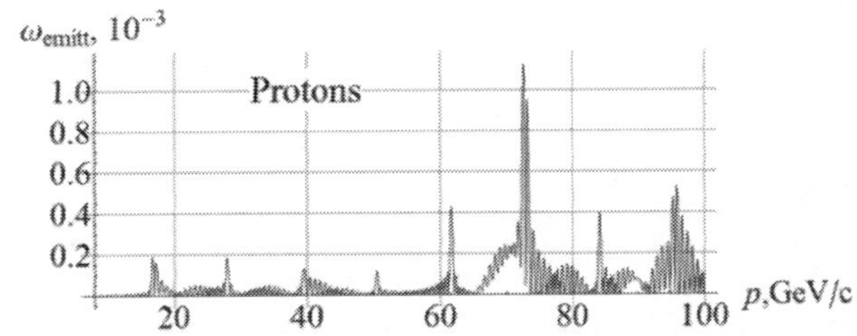

Figure 2. Incoherent part of the proton resonance strength in the JLEIC ion collider ring.

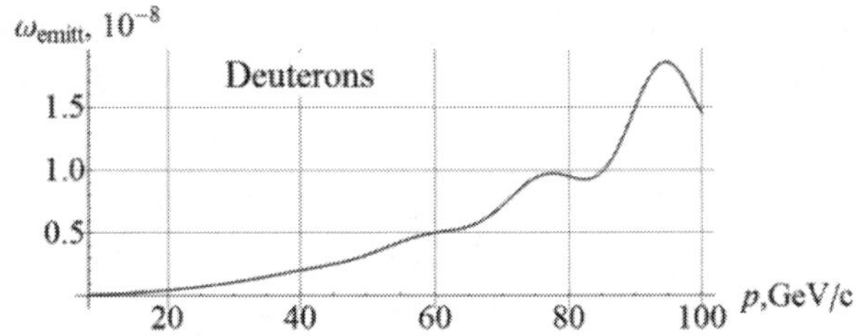

Figure 3. Incoherent part of the deuteron resonance strength in the JLEIC ion collider ring.

In contrast to protons, due to the small value of the deuteron anomalous magnetic moment, deuterons have only one interference peak at the momentum of 93 GeV/c whose value does not exceed $2{\cdot}10^{-8}$. Thus, a deuteron spin tune of $3{\cdot}10^{-3}$ induced by a solenoid significantly exceeds the incoherent part of the resonance strength.

2.2. Coherent part of the spin resonance strength

Let us calculate the resonance strength for a proton beam using a model with random shifts of all quadrupoles in the transverse directions. Figure 4 shows the coherent part of the proton resonance strength in the ion collider ring with random quadrupole misalignments resulting in a transverse closed orbit distortion of about 100 μm rms.

The statistical model calculates the most probable magnitude of the coherent part of the resonance strength not specifying its direction, which lies in the collider's plane. As in the case of the incoherent part, the coherent part of the resonance strength has interference peaks whose maximum values do not exceed $1.5{\cdot}10^{-2}$, which has an order of magnitude comparable to the field induced by the stabilizing solenoid. Figure 5 shows a graph of the coherent part of the deuteron resonance strength calculated using the statistical model of random quadrupole misalignments.

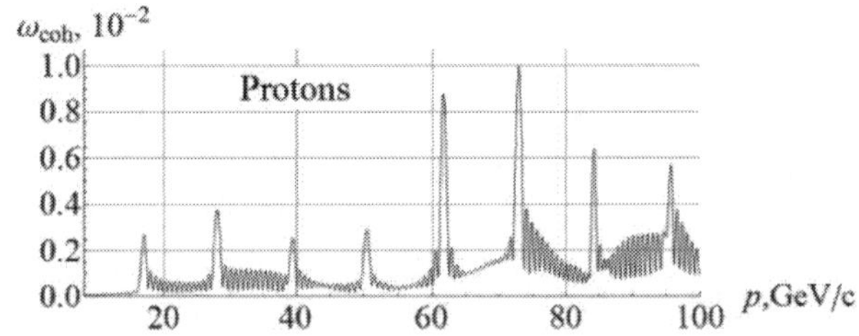

Figure 4. Coherent part of the proton resonance strength in the ion collider ring.

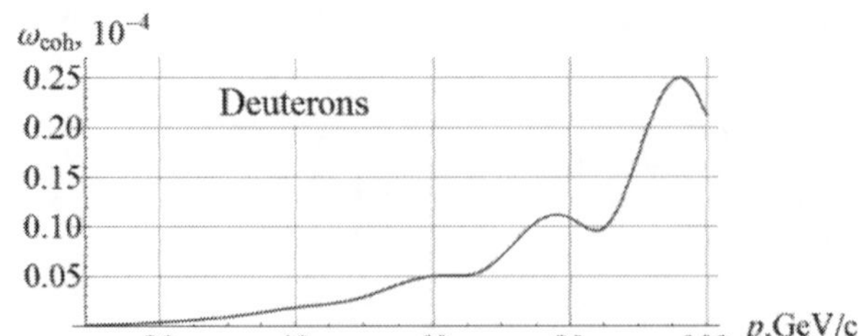

Figure 5. Coherent part of the deuteron resonance strength in the ion collider ring.

3. Acceleration of polarized ions in the JLEIC collider

The spin dynamics during acceleration in the ion collider ring is a precession about the spin field $\vec{h}$, which consists of the field $\vec{h}_{\text{sol}}$ induced by the stabilizing solenoid and the resonance strength $\vec{\omega}$: $\vec{h} = \vec{h}_{\text{sol}} + \vec{\omega}$ [8]. During acceleration the field $\vec{h}_{\text{sol}}$ is maintained constant while the resonance strength $\vec{\omega}(t)$ experiences significant changes in the regions of interference peaks.

The beam polarization substantially depends on the field ramp rate in the arc magnets. When using superconducting magnets with a field ramp rate of ~3 T/min, acceleration happens adiabatically, which means that, in a characteristic time of change in the spin field, the spin makes a large number of turns. During adiabatic acceleration, the spin follows the $\vec{h}$ field direction, which can significantly deviate from the longitudinal direction at the locations of the interference peaks of the coherent part of the resonance strength. However, this does not signify polarization loss, the beam polarization restores its longitudinal direction in places where $h_{\text{sol}} \gg \omega_{coh}$.

Let us present calculations of the spin dynamics during acceleration of protons and deuterons in the JLEIC ion collider ring made using a spin tracking code Zgoubi [13]. Figure 6 shows the longitudinal spin components in the ion collider ring during acceleration of 3 protons with $\Delta p/p = 0$ (green line), $\Delta p/p = 10^{-3}$ (red line) and $\Delta p/p = -10^{-3}$ (blue line). As we can see, the graphs of the longitudinal spin components practically do not differ from each other (the red line covers up the blue and green lines), i.e. synchrotron energy modulation does not give a noticeable contribution to the ion spin motion when stabilizing the polarization by a weak solenoid in the JLEIC ion collider ring. All particles were launched with the same initial conditions: $S_{z0} = 1$, $x_0 = 0.61$ mm, $x_0' = 0$ rad, $y_0 = 0.27$ mm, $y_0' = 0$ rad. The field ramp rate was ~3 T/min (the particles were accelerated in 8.3 million turns). During acceleration, the spin preserves its component along the spin field, which lies in the orbit plane and noticeably deviates from the longitudinal direction in the regions of the "interference" peaks of the coherent part of the resonance strength at momenta of about 60 GeV/c and 75 GeV/c, where the resonance strength becomes approximately equal to the size of the solenoid spin field. The spin tune induced by the solenoid during acceleration is 10^{-2}.

The simulation in Figure 6 is done with a closed orbit excursion of 100 μm rms. If needed tolerances to alignment of the lattice elements can be relaxed. The strength of the stabilizing solenoid can be increased. One then has to account for the fact that the solenoid itself gives a contribution to the spin resonance strength due to an angle between the distorted closed orbit and solenoid axis. This results in a transverse magnetic field component, which has practically no effect on the orbital motion but can have a strong effect on the spin motion especially for a proton beam at high energies ($\gamma G \gg 1$). This contribution can be minimized either by a more precise alignment of the solenoid axis or by choosing such a collider lattice, which has a sufficiently small value of the spin response function at the solenoid location. The response function is the spin Green's function determined and controlled by the linear lattice. It describes the effect of the ring as a whole on the spin due to a δ-function-like radial field [8]. Another option is to compensate the coherent part of the spin resonance strength at the experimental energy using a 3D spin rotator [4-8, 10].

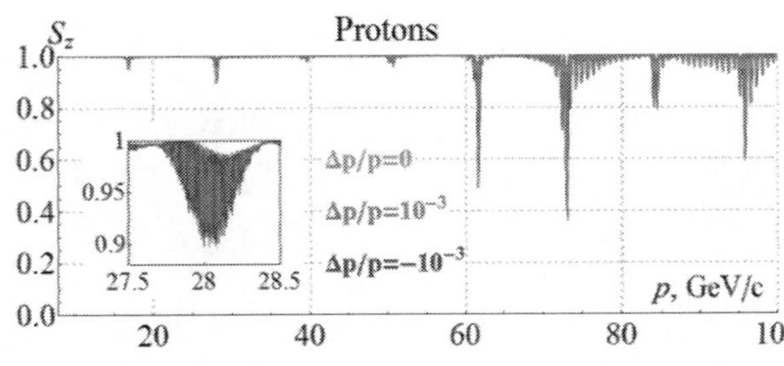

Figure 6. Longitudinal spin component during acceleration of three protons in the ion collider ring.

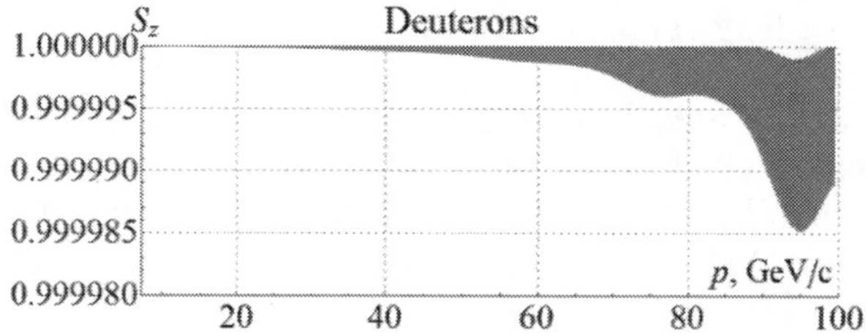

Figure 7. Longitudinal spin component during acceleration of three deuterons in the ion collider ring.

Similar graphs for the longitudinal components of the deuteron spin are shown in Figure 7. The initial conditions and solenoid field strength during acceleration were chosen the same as in the proton case. In contrast to protons, the change in the deuteron longitudinal polarization during acceleration does not exceed $2 \cdot 10^{-5}$ even in the interference peak. This example demonstrates a high stability of the deuteron polarization in figure-8 rings, which can be used for high-precision experiments. To the contrary, in conventional accelerators with preferred periodic spin orientation, control of the deuteron polarization and its preservation during acceleration to 100 GeV/c is a practically unrealistic task.

3.1. Crossing the transition energy

When accelerating polarized protons and deuterons in the JLEIC ion collider ring in the momentum range from 8 GeV/c to 100 GeV/c, one has to consider the question of preserving the beam polarization during transition energy crossing. The relativistic Lorentz factor of the transition energy in the ion collider ring equals 12.453, which corresponds to a momentum of 11.65 GeV/c for protons and 23.3 GeV/c for deuterons.

For our calculations, we chose a conventional model, in which crossing of the transition energy is done by a fast jump of the RF cavity phase at the exact moment of the crossing from the value φ_s to the value $\varphi_s^* = \pi - \varphi_s$ at a constant field ramp rate.

The phase space trajectories of two protons with the initial momentum offsets of $\Delta p/p = 10^{-3}$ (red line) and $\Delta p/p = -10^3$ (blue line) are shown in Figure 8. As the energy approaches transition, the amplitude of the synchrotron phase deviation from the equilibrium value of $\varphi_s \approx 0.3$ rad reduces while the amplitude of the momentum deviation grows. After crossing the transition energy, the particles are captured inside a new separatrix and undergo oscillations about a new equilibrium phase of $\varphi_s^* \approx 2.84$ rad. The amplitude of the momentum deviations damps as the energy gets further away from transition. Our calculations indicate that, if transition energy crossing is organized without significant excitation of the emittances, its effect on the polarization is negligible. Further studies with a more representative particle distribution and a more realistic model of transition energy crossing are needed.

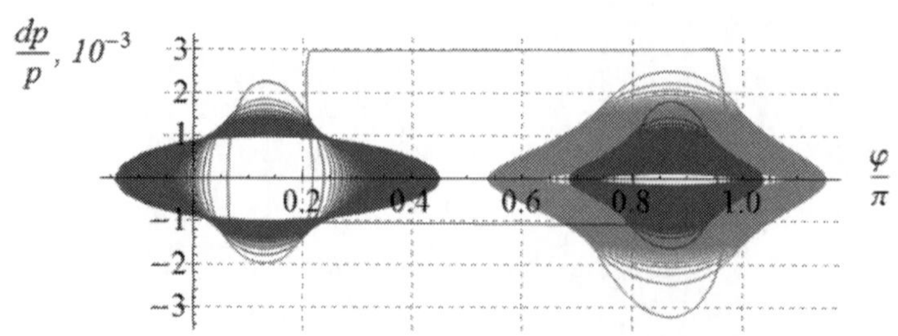

Figure 8. Proton phase space trajectories.

4. Conclusion

Figure-8 rings with weak solenoids provide an elegant solution to preservation of the polarization when accelerating particles of any kind. Our calculations made using a spin tracking code Zgoubi verify the validity of our scheme for preserving the polarization during acceleration of protons and deuterons in the JLEIC ion collider ring with transition energy crossing.

Acknowledgments

This material is based upon work supported by the U.S. Department of Energy, Office of Science, Office of Nuclear Physics under contracts DE-AC05-06OR23177 and DE-AC02-06CH11357.

References

[1] Abeyratne S *et al.* 2015 MEIC Design Summary *arXiv*:1504.07961 [physics.acc-ph]
[2] Derbenev Ya S 1996 The Twisted Spin Synchrotron *University of Michigan report* UM HE 96-05
[3] Derbenev Ya S *et al.* 2013 Ion Polarization Control in MEIC Rings Using Small Magnetic Field Integrals *PoS (PSTP 2013)* 026
[4] Derbenev Ya S *et al.* 2016 Polarization Preservation and Control in a Figure-8 Ring *Int. J. Mod. Phys. Conf. Ser.* **40** 1660090
[5] Derbenev Ya S *et al.* 2014 Polarized Ion Beams in Figure-8 Rings of JLab's MEIC *Proc. Int. Part. Accel. Conf. 14 (Dresden)* paper MOPRO004 pp 68-70
[6] Kondratenko A M *et al.* 2015 Baseline Scheme for Polarization Preservation and Control in the MEIC Ion Complex *Proc. Int. Part. Accel. Conf. 2015 (Richmond)* paper TUPWI029 pp 2301-2303
[7] Kondratenko A M *et al.* 2015 Numerical Calculation of the Ion Polarization in MEIC *Proc. Int. Part. Accel. Conf. 2015 (Richmond)* paper TUPWI030 pp 2304-2306
[8] Kondratenko A M *et al.* 2016 Ion Polarization Scheme for MEIC *arXiv*:1604.05632 [physics.acc-ph]
[9] Kondratenko A M *et al.* 2016 Ion Beam Polarization Dynamics in the 8 GeV Booster of the JLEIC Project at JLab *Proc. Int. Part. Accel. Conf. 2016 (Busan)* paper WEPMW017 pp 2460-2462
[10] Kondratenko A M *et al.* 2016 Spin Flipping System in the JLEIC Collider Ring *Proc. North-American Part. Accel. Conf. 2016 (Chicago)* paper TUPOB30
[11] Morozov V S *et al.* 2016 Injection Optics for the JLEIC Ion Collider Ring *Proc. Int. Part. Accel. Conf. 2016 (Busan)* paper WEPMW012 pp 2445-2447
[12] Nosochkov Y *et al.* 2015 Progress on Optimization of the Nonlinear Beam Dynamics in the MEIC Collider Rings *Proc. Int. Part. Accel. Conf. 2015 (Richmond)* paper TUPWI032 pp 2311-2314
[13] Méot F 1999 The Ray-Tracing Code Zgoubi *Nucl. Instr. Meth.* A **427** 353

IOP Conf. Series: Journal of Physics: Conf. Series **874** (2017) 012012 doi:10.1088/1742-6596/874/1/012012

Injector linac stability requirements for high precision experiments at MESA

F Hug and R Heine

Institute for Nuclear Physics, Johannes Gutenberg-Universität Mainz, Johann-Joachim-Becher Weg 45, 55128 Mainz, Germany

flohug@uni-mainz.de

Abstract. MESA is a recirculating superconducting accelerator under construction at Johannes Gutenberg-Universität Mainz. It will be used for high precision particle physics experiments in two different operation modes: external beam (EB) mode and energy recovery (ERL) mode. The operating beam current and energy in EB mode is 0.15 mA with polarized electrons at 155 MeV. In ERL mode an unpolarized beam of 1 mA at 105 MeV will be available. In a later construction stage of MESA the beam current in ERL-mode shall be upgraded to 10 mA. In order to achieve high beam stability and low energy spread in recirculating operation for external beam the acceleration in the main linac sections will be done on a certain phase with respect to the maximum of the accelerating field (off crest) while the return arcs provide longitudinal dispersion. On specific longitudinal working points this can result in a setting where any RF phase or magnitude jitters from main linac do not contribute to the resulting energy spread of the final beam at all. Then the resulting energy spread of the beam at the experiment is mostly determined by the beam properties provided by the injector linac. On the other hand the acceleration in ERL operation mode most likely needs to be done on crest of the accelerating field aiming for the highest efficiency in the energy recovering process albeit we are currently investigating different recirculation schemes for the ERL mode as well. Using on crest acceleration the achievable energy spread is determined by the longitudinal phase space properties behind the injector linac again but mostly by the bunch length of the beam injected to the main linac. Within this contribution we will investigate the requirements on the stability of the MESA injector linac MAMBO for achieving the experimental goals under both operating conditions.

1. Introduction

The design work on the MESA accelerator has undergone several changes since the accelerator has been proposed the first time in approx. 2009 [1] Since that time the layout of accelerator as well as requirements of the experiments have been refined continuously [2,3]. The latest layout, which is now defined as the final one, is given in figure 1. MESA will run in two completely different operation modes, an external beam mode and a multi-turn ERL mode, which implies different beam dynamics for running these operation modes.

Using the external beam mode the polarized electrons can be accelerated to a maximum energy of 155 MeV and to a maximum beam current of 150 μA. The P2 experiment served in external mode aims on a precise determination of the electroweak mixing angle (Weinberg angle), using the method of a parity violation asymmetry measurement [4,5]. To do so the polarization of the beam is flipped

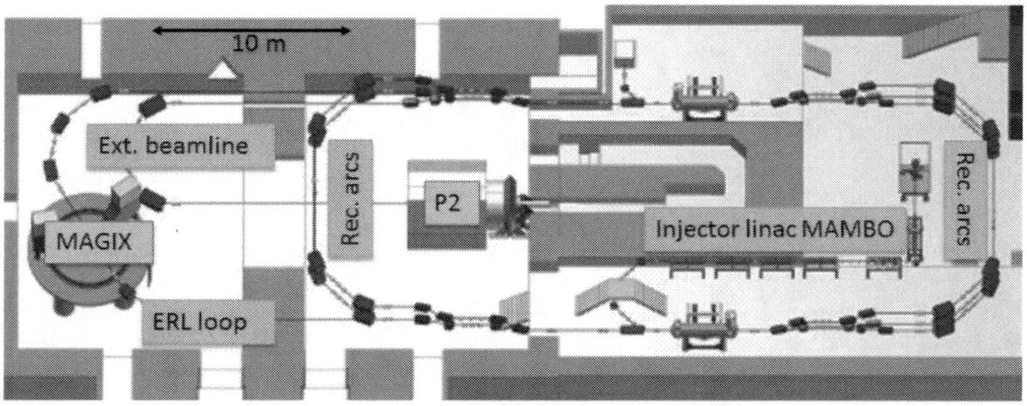

Figure 1. Layout of the MESA accelerator and planned main experiments MAGIX and P2.

with a frequency of 1 kHz using a pockels cell [6]. In order to achieve the precision needed for that experiment highest beam quality in energy spread, beam current and beam position is crucial. Concerning the latest an active stabilization of beam position is being developed [7]. In particular the errors correlated to polarization flips need to be suppressed strongly. Otherwise the experiment will measure false asymmetries and overestimate the parity violation effect. In this contribution we are focussing on the error contributions at the experiment resulting from the injector linac MAMBO.

In ERL mode maximum beam energy of 105 MeV and maximum beam current of 1 mA are planned, while in a later construction stage of MESA this beam current shall be increased to a maximum of 10 mA. For such high beam currents no polarization of the beam is planned so far due to very short lifetimes of polarized photocathodes at such high currents. The experimental setup running with ERL beam is the high resolution spectrometer facility named MAGIX [8]. Experimental goals of MAGIX amongst others are searching for dark photons and measurements of the proton radius. In electron scattering experiments the achievable resolution is determined by the statistical errors resulting from the beam energy spread and the resolution of the spectrometer itself. At MAGIX a relative energy spread of the beam of approx. 10^{-4} (RMS) is needed for not being the main source of error. Again we will focus on the contribution resulting from injector beam properties.

2. MESA injector chain

The electrons for the MESA accelerator will be extracted from a polarized inverted DC photo gun [9]. The gun section is followed by a low energy beam transport containing the required spin manipulation for P2 as well as a chopper-buncher section for longitudinal matching into the accelerating structures of the injector. The normal conducting injector linac MAMBO consists of four accelerating modules. The first one is a graded-β cavity followed by three fixed-β cavities (one for $\beta = 0.977$ and two for $\beta = 1$). It is capable for beam currents of up to 10 mA [10,11]. MESA will be operated in a continuous wave mode with each longitudinal bucket filled by a bunch (duty cycle 100%). The spacing of two bunches equals one RF wavelength or 796 ps. After leaving the injector the beam is transferred into the main linac following a 180° arc. This arc can be used for further bunch compression when needed.

2.1. Longitudinal phase space of the injector beam

The longitudinal phase space provided by the MAMBO injector strongly depends on the beam current and on the chosen accelerating phases of the four linac sections. When operating in EB mode with 150 μA beam current (corresponding bunch charge 0.12 pC) the injector will be tuned for minimum energy spread of the extracted beam. The beam is then transferred to the main accelerator and accelerated to final energy using a non-isochronous recirculation scheme like presented in [12-14]. In such a setting the energy spread of the injected beam determines the final energy spread and therefore

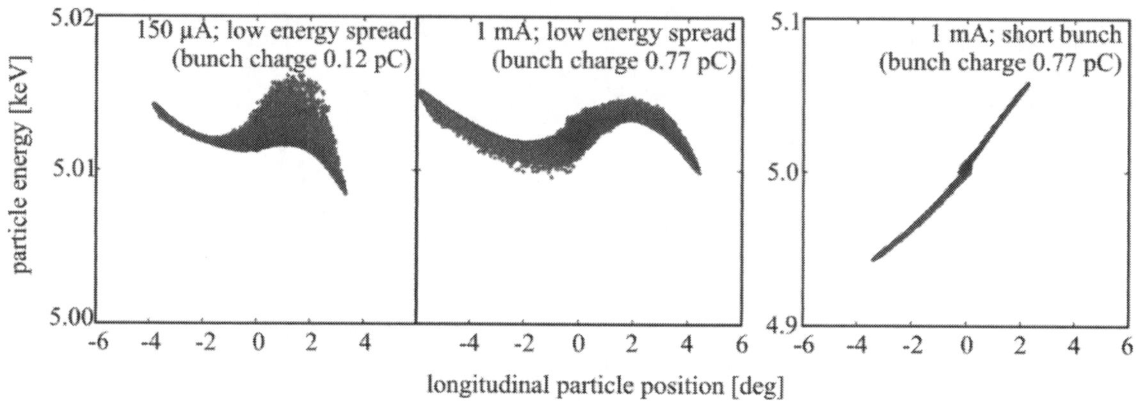

Figure 2. Simulated longitudinal particle distributions at the end of the MAMBO injector for different beam currents and optimized for different operating conditions. The RMS bunch length and energy spread is listed in table 1.

needs to be minimized. The longitudinal phase space of the 150 μA beam is plotted in figure 2. The corresponding longitudinal beam properties are given in table 1.

Table 1. Longitudinal Properties of the MAMBO Beam.

Beam current (bunch charge)	σE [keV]	σφ [deg]	ε_{rms} (deg·keV)
150 μA (0.12 pC)	0.72	1.36	0.978
1 mA (0.77 pC)	0.87	1.91	1.59
1 mA, short bunch	20.4	0.95	2.76

When operating in ERL mode bunches should be optimized for short length rather than for best energy spread as motivated in the following section. In figure 2 the 1 mA beam (bunch charge 0.77 pC) is plotted once optimized for best energy spread and once for shortest bunch-length. The increasing non-linear space charge forces are deforming the particle distribution in phase space. This results in an increase of the RMS emittance of the beam which appears in an increase of RMS bunch length and RMS energy spread. Nevertheless the beam optimized for short bunches is capable to be further compressed within the 180° injection arc.

2.2. Injector stability

The operational stability of the injector affects the beam properties of the electrons provided to the main accelerator as well. Phase or magnitude jitters in the first section affect bunch-length and time of flight through the injector mostly whereas errors in the sections 2-4 have larger impact on the mean energy of the MAMBO beam.

3. Beam properties at experiments

The simulated bunches from the injector have been taken as input for beam dynamics simulations of the recirculating main linac. The simulation method is discussed in more detail in [15]. Here we need to distinguish again between EB and ERL operation modes.

3.1. Investigations on external beam stability

In EB mode non-isochronous recirculation with acceleration on edge of the accelerating field is used. In this operation mode the energy spread is reduced significantly if the longitudinal phase advance over the complete acceleration process reaches a half- or full-integer number of synchrotron oscillations in phase space. The optimized working point for EB mode yields to a relative energy spread of $\Delta E_{rms}/E = 5.5 \cdot 10^{-5}$ (isochronous: $\Delta E_{rms}/E = 3.4 \cdot 10^{-4}$) [16]. The achieved energy spread fulfils the experimental requirements. In subsequent calculations we investigated the stability of the optimized longitudinal working point against variations of the injected beam in mean phase and energy. The results are plotted in figure 3. As it can be seen the mean energy of the beam at the experimental setup varies with respect to the fluctuations in the injected beam. These results are of importance for the P2 experiment. E.g. if the helicity flip of the spin-polarization causes a time of flight difference of the particles before entering the injector this can be seen as a phase offset of the first accelerating MAMBO cavity. As a consequence the energy error at the experimental setup would be helicity correlated and can cause false asymmetries.

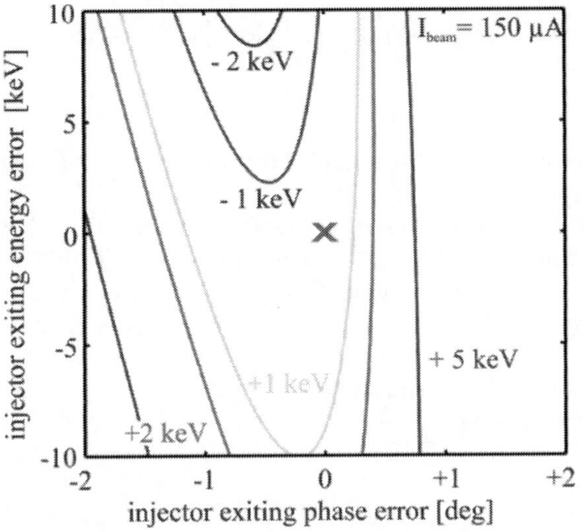

Figure 3. Simulated deviation of mean energy (contours of same levels) to design energy at the P2 experiment under variation of injected mean energy and mean phase. X marks the design energy (0 keV deviation).

3.2. Investigations on ERL beam stability

In ERL mode the acceleration of the beam most likely needs be done on crest of the accelerating field. Therefore the injector beam affects the final energy spread mostly by bunch length as simulations given in figure 4 show. On the contrary in ERL mode the energy spread of the MAMBO beam will not be crucial for resulting beam quality. Relaxing the demands on energy spread allows for much shorter bunch lengths and for an improvement in energy spread. Doing so the energy spread can be reduced from $\Delta E_{rms}/E = 7.2 \cdot 10^{-4}$ to $\Delta E_{rms}/E = 1.99 \cdot 10^{-4}$ using the short bunch setting.

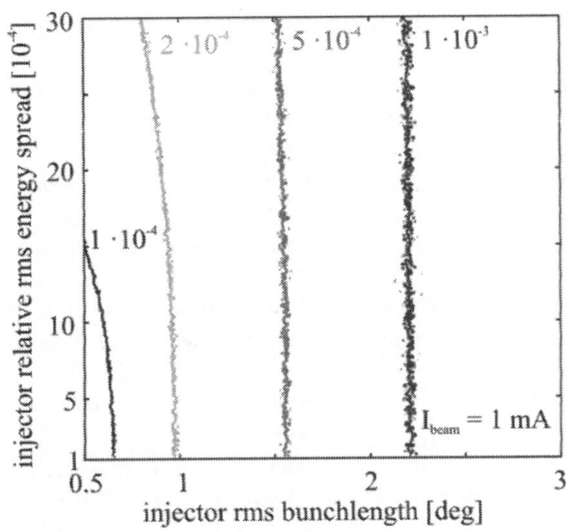

Figure 4. Hill plot of simulated energy spread at the position of the MAGIX experiment in dependency of injected energy spread and bunch length.

4. Summary and outlook

The MESA accelerator can operate either as ERL or recirculating linac with external beam and will serve two different experiments with high demands on beam quality. Within this work we presented the influence of injector beam properties on the resulting beam stability and quality at the experimental setups. In order to achieve best experimental conditions the injector needs to be well understood and the requirements on RF-control of the injector need to be chosen according to the required beam stability. Work on this topic is going on. Furthermore, investigations on the benefit of an additional feedback system to stabilize the time of arrival of bunches at the main linac are planned in the future.

Acknowledgments

This work has been supported by DFG through the PRISMA cluster of excellence EXC 1098/2014 and Research Training Group GRK 2128.

References

[1] Aulenbacher K and Jankowiak A 2009 Polarized Electrons and Positrons at the MESA Accelerator *Proc. of PST 2009 (Ferrara, Italy)* 49
[2] Heine R, Aulenbacher K and Eichhorn R 2012 MESA-sketch of an energy recovery linac for nuclear physics experiments at Mainz *Proc. of IPAC '12 (New Orleans, LA, USA)* pp 1993-5
[3] Simon D, Aulenbacher K, Heine R and Schlander F 2015 Lattice and beam dynamics of the energy recovery mode of the Mainz energy recovering superconducting accelerator MESA *Proc. of IPAC '15 (Richmond, VA, USA)* 220-2
[4] Aulenbacher K 2011 Opportunities for parity violating electron scattering experiments at the planned MESA facility *Hyperfine Interactions* **200** pp 3-7
[5] Becker D, Baunack S and Maas F 2013 P2 - a new measurement of the weak charge of the proton *Hyperfine Interactions* **214** pp 141-8
[6] Aulenbacher K, Alexander I and Tioukine V 2012 The polarimetry chain for the P2 experiment *Nuovo Cim.* **4** pp 186-91
[7] Dehn M, Aulenbacher K, Diefenbach J, Fichtner F, Herbertz R and Klag W 2015 Testing a digital beam position stabilization for the P2-experiment at MESA *Proc. of IPAC '15 (Richmond, VA, USA)* pp 888-90
[8] Aulenbacher S 2014 Design and Simulation of the Internal Gas-Target for MAGIX *diploma thesis, Phys. Dept., Johannes Gutenberg-Universität Mainz (Mainz, Germany)*
[9] Friederich F and Aulenbacher K 2015 Test electron source for increased brightness emission by near band gap photoemission *Proc. of IPAC '15 (Richmond, VA, USA)* pp 1512-4
[10] Heine R and Aulenbacher K 2013 Injector for the MESA facility *Proc. of IPAC'13 (Shanghai, China)* pp 2150-2
[11] Heine R, Aulenbacher K, Hein L and Matejcek C 2016 Current status of the milliampere booster for the Mainz energy-recovering superconducting accelerator *Proc. of IPAC '16 (Busan, Korea)* pp 1743-5.
[12] Herminghaus H 1991 The polytron as a cw electron accelerator in the 10 GeV range *Nucl. Instr. and Meth.* A **305** pp 1-9
[13] Herminghaus H 1992 On the inherent stability of non-isochronous recirculating accelerators *Nucl. Instr. and Meth.* A **314** pp 209-11
[14] Hug F, Burandt C, Eichhorn R, Konrad M. and Pietralla N 2012 Measurements of a reduced energy spread of a recirculating linac by non-isochronous beam dynamics *Proc. of LINAC '12 (Tel Aviv, Israel)* pp 22-4
[15] Hug F, Araz A, Eichhorn R and Pietralla N, Reducing energy spread of the beam by non-isochronous recirculation at the S-DALINAC *Proc. of IPAC '10 (Kyoto, Japan)* pp 4470-2
[16] Hug F 2017 Application of non-isochronous beam dynamics in ERLs for improving energy spread and beam stability *Proc. IPAC'17 (Copenhagen, Denmark) Preprint* MOPVA013

S2E simulation of an ERL-based high-power EUV-FEL source for lithography

N Nakamura[1,2], R Kato[1,2], T Miyajima[1,2], M Shimada[1,2], T Hotei[2], R Hajima[3]

[1] High Energy Accelerator Research Organization (KEK), Tsukuba, Japan
[2] The Graduate University for Advanced Studies (SOKENDAI), Tsukuba, Japan
[3] National Institute for Quantum and Radiological Science and Technology (QST), Tokai, Japan

E-mail : norio.nakamura@kek.jp

Abstract. Energy recovery linac (ERL) based extreme ultraviolet (EUV) free electron lasers (FELs) are candidates of a next-generation high-power EUV source for lithography. An ERL-based EUV FEL source has been designed in order to demonstrate the feasibility of generating a 10-kW class EUV power. Start-to-End (S2E) simulation including the injection beam optimization, bunch compression, FEL lasing and bunch decompression is performed for the designed EUV source. As a result it is demonstrated that the EUV FEL can produce high power more than 10 kW at 10 mA and that the electron beam can be well transported throughout the EUV source without beam loss.

1. Introduction

The technologies on EUV lithography are progressing based on laser-produced plasma (LPP) source [1], which is expected to produce the EUV power of 250 W or more. However it is important to develop a new-type EUV source to meet future demand for higher power than 1 kW [2]. ERL-based FELs are possible candidates of a high-power EUV source that can distribute 1 kW power to multiple scanners simultaneously. The ERLs can provide high-current and high-quality electron beams for the high-power FELs and also greatly reduce the dumped beam power and activation compared to ordinary linacs.

An EUV-FEL source based on an 800-MeV ERL has been designed using available technologies and resources to produce 10-kW class EUV power for lithography [3]. S2E simulation from the electron gun to the exit of the decelerating main linac is performed for this design to track the electron beam distribution and parameters and to evaluate the FEL performance. This paper will present results of the S2E simulation for the designed EUV-FEL source.

2. Overview of EUV source

Figure 1 shows the image of the designed ERL-based EUV-FEL source [3]. In the injector design, the electron beam is generated by a 500-kV photocathode DC gun with the same structure as the 2nd gun developed for the cERL [4] and accelerated to 10.5 MeV or more by six 2-cell superconducting (SC) cavities in two cryomodules, which have the same structure as the cERL injector cavities [5]. The merger system that guides the injection beam into the recirculation loop is slightly modified from the cERL merger.

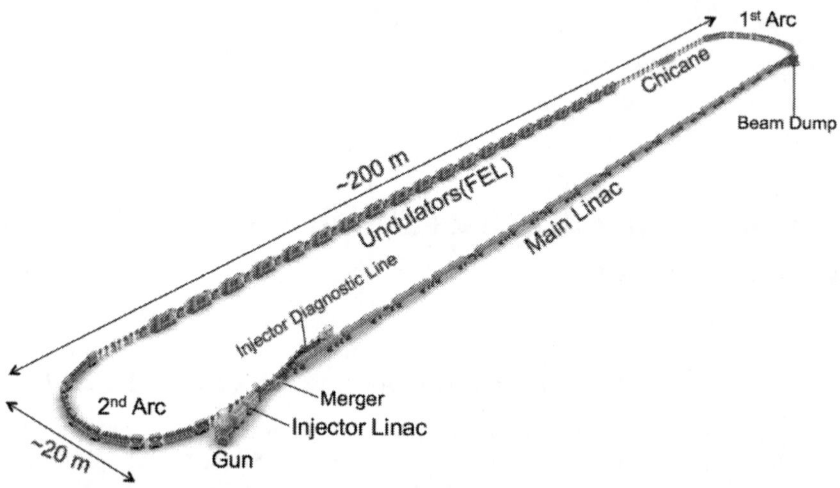

Figure 1. Image of the designed ERL-based EUV-FEL.

The main-linac SC cavity is a tesla-type 9-cell cavity with a large-aperture HOM-damped (higher-order-mode-damped) beam pipe. This cavity has a reduced ratio of the peak to acceleration electric fields by a factor of 1.5 compared to the cERL main-linac cavity, which is stably operated at ~8.5MV/m against the field emission [6]. Therefore stable operation at 12.5 MV/m is promising. The main linac consists of sixty-four main-linac cavities in sixteen cryomodules to accelerate the beam to 800 MeV. Quadrupole triplets are placed at every two cryomodules for beam focusing. The betatron function is optimized against beam breakup caused by the cavity HOMs and it is symmetric for acceleration and deceleration phases. After the off-crest acceleration by the main linac, the electron bunches are compressed by the 1st arc and/or a chicane down to several ten fs to increase the peak current for obtaining high FEL power and, after emitting the EUV FEL light, decompressed before the decelerating main linac by the 2nd arc. Finally they are decelerated by the main-linac cavities and dumped.

3. Injection beam optimization and bunch compression

The injection beam parameters from the gun to the entrance of the main linac are optimized by the tracking code *GPT* [7] and genetic algorithm. The bunch charge of 60 pC and the injection beam energy of 10.5 MeV are assumed. In addition, the bunch length at the entrance of the main linac is set to ~1 ps so that high peak current and low energy spread should be ensured after the bunch compression. After optimizing the injector and merger parameters and performing the optics matching between the merger and the main linac, the vertical and horizontal normalized emittances are obtained to be 0.65 and 0.73 mm·mrad for the energy spread of 0.347 % at the entrance of the main linac.

The injection beam is accelerated off-crest by the main linac up to 800 MeV and then compressed by the 1st arc and/or a chicane to increase the peak current for obtaining high FEL power. In this S2E simulation, we use three DBA (Double-Bend Achromat) cells as the 1st arc for the bunch compression in place of two TBA (Triple-Bend Achromat) cells so far used [3]. DBA lattice has been studied to suppress coherent synchrotron radiation (CSR) effects on beams in bending sections even for a changeable bunch length [8]. The DBA optics is designed so that the betatron phase between the two bending magnets in the DBA cell is 180 degrees and two Twiss parameters at the center of the 2nd bending magnet, β_{x2} and α_{x2}, can be adjusted to optimize the horizontal emittance or performance of the bunch compression. Figure 2 shows the 1st arc optics with β_{x2}=1.0 m and α_{x2}=1.6. The bending radius and angle are 2.2 m and 30 degrees and the longitudinal dispersion of the 1st arc, R_{56}, is about 0.31 m.

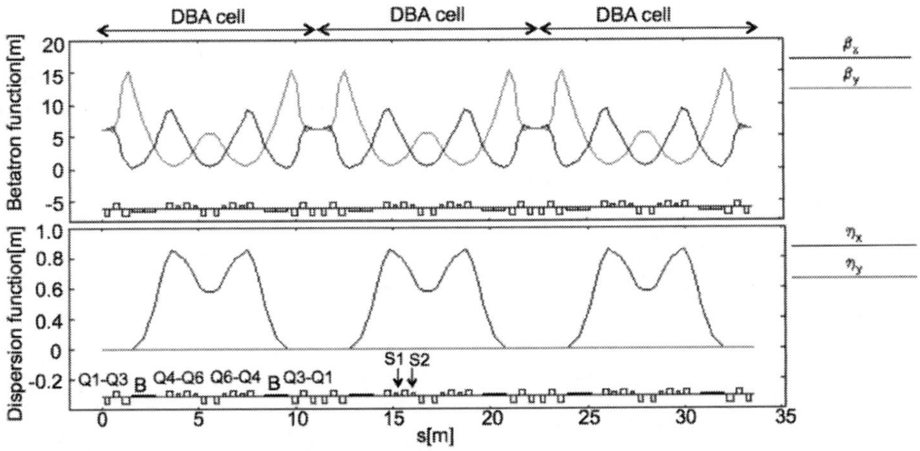

Figure 2. Betatron and dispersion functions of the 1st arc with three DBA cells having β_{x2}=1.0 m and α_{x2}=1.6.

Figure 3 shows the simulation result of the bunch compression by the simulation code *elegant* [9] for the injection beam described above. The RF phase of the main linac and the field strengths of two-family sextupole magnets (S1 and S2) are adjusted to optimize the bunch compression in the 1st arc optics. As shown in figure 3, the bunch is compressed to 47 fs with the energy spread of 0.112 % and the horizontal and vertical normalized emittances of 2.02 and 0.72 mm·mrad. Current and normalized horizontal and vertical slice emittances at the exit of the 1st arc are shown in figure 4. Thanks to the DBA optics, the peak current exceeds 700 A and the horizontal slice emittance is kept within 1.5 mm·mrad around the peak current and significantly smaller than the projected one.

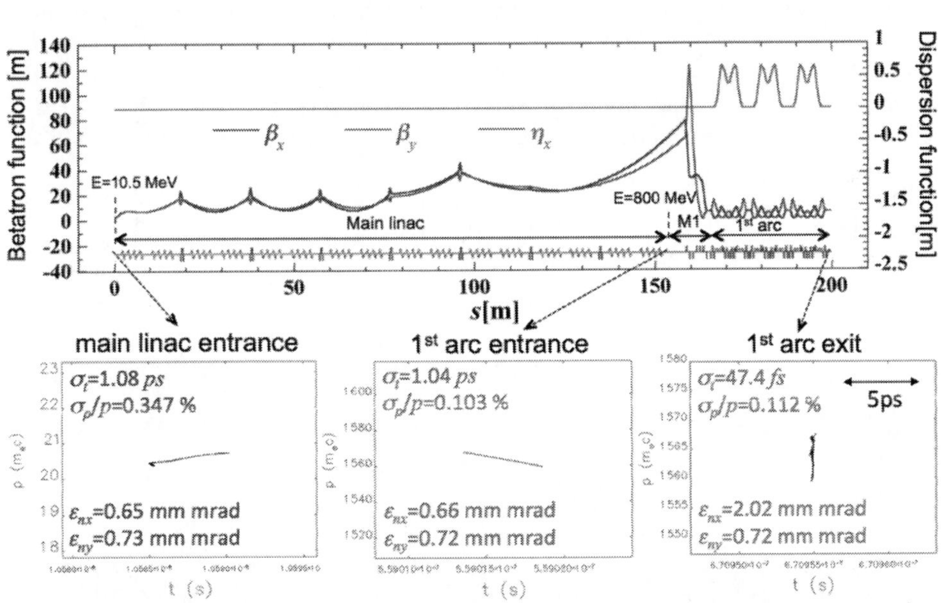

Figure 3. Betatron and dispersion functions from the main-linac entrance to the 1st arc exit (upper) and the time-momentum distributions with the basic beam parameters at the main-linac entrance, the 1st arc entrance and exit (lower).

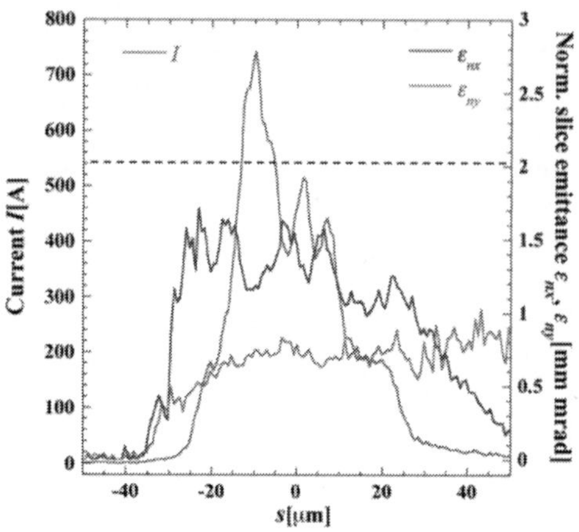

Figure 4. Current and horizontal and vertical normalized slice emittances in the bunch at the exit of the 1st arc. The broken line indicates the projected horizontal normalized emittance.

4. SASE FEL

The self-amplified spontaneous emission (SASE) FEL simulation is performed with the FEL simulation code *Genesis* (version 3) [10] and the electron distribution after the bunch compression. The FEL system consists of eighteen 4.9-m circular-polarizing undulator segments with the undulator period of 28 mm and the undulator K-value of 1.656 for the wavelength of 13.5 nm (the photon energy of 91.84 eV). Seventeen focusing quadrupole magnets are placed in 1.12-m gaps between each two segments and two quadrupole magnets with the half length are 5.6 m before the 1st segment and after the last segment. Quadrupole magnets are also placed in the matching section in front of the FEL system to maximize the FEL power by adjusting the Twiss parameters at the FEL entrance.

Figures 5a to 5c show results of the SASE-FEL simulation. The Twiss parameters are adjusted to β_x=14 m, β_y=14 m, α_x=α_y=0 at the FEL entrance. The FEL pulse energies without undulator tapering and with the optimum linear tapering of 2 % are shown as a function of the undulator section length in figure 5a. For bunch frequency of 162.5/325 MHz, the FEL power with the 2% undulator tapering reaches 14.4/28.8 kW at the average beam current of 9.75/19.5 mA. The FEL temporal profile with the electron beam current profile and the FEL spectrum are shown in figures 5b and 5c. The FEL spectrum is found to be narrow compared to the reflectivity bandwidth of the Mo/Si multi-layer mirror.

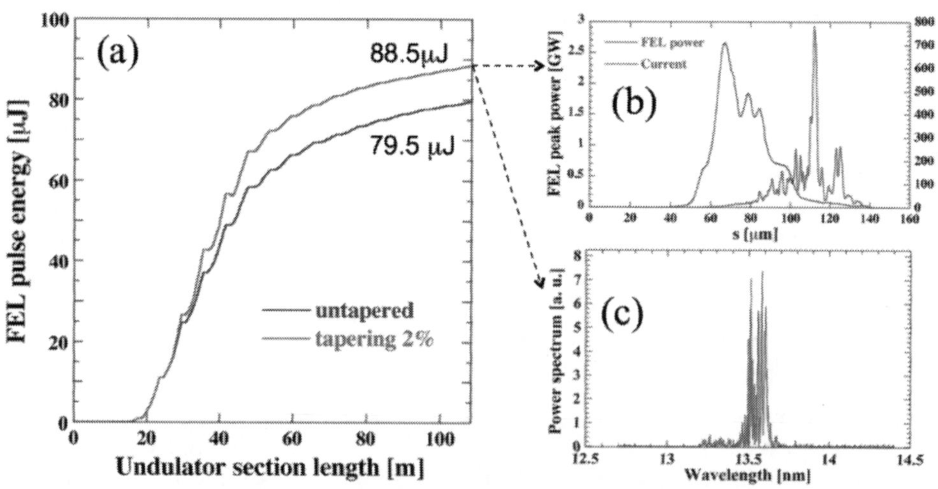

Figure 5. Results of the FEL simulation: (a) the FEL pulse energy without and with tapering as function of the undulator section length, (b) the FEL temporal profile and (c) the FEL power spectrum for 2% tapering at the FEL exit.

IOP Conf. Series: Journal of Physics: Conf. Series **874** (2017) 012013 doi:10.1088/1742-6596/874/1/012013

The six-dimensional electron distribution just after the FEL is also produced by *Genesis* in order to proceed with the S2E simulation. Figures 6a and 6b shows the time to momentum (*t-p*) distributions at the FEL entrance and exit. As shown by comparing these figures, electrons at high currents in the bunch lose part of the energies or momenta because of the EUV-FEL lasing and as a result the averaged beam energy is slightly decreased and the energy spread is significantly increased up to ~0.3 %, while the horizontal and vertical emittances are almost unchanged.

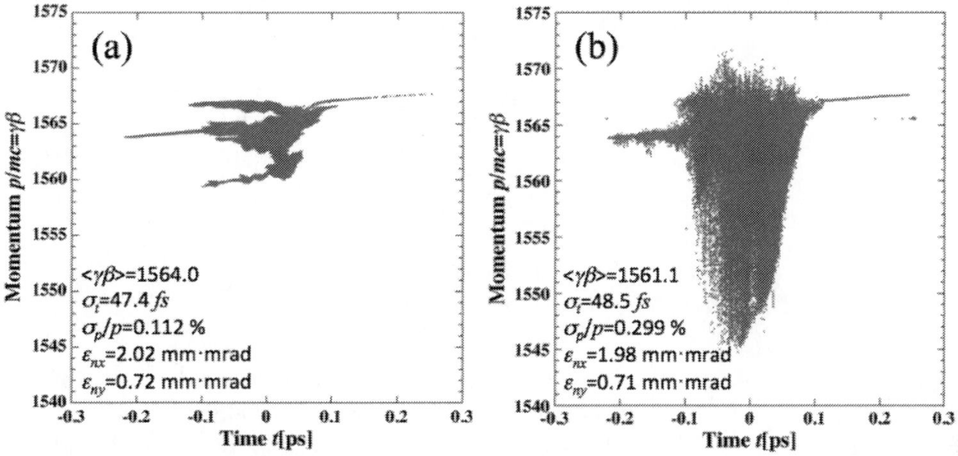

Figure 6. Time-momentum (*t-p*) distributions in the electron bunch at (a) the entrance and (b) the exit of the FEL system.

5. Bunch decompression and energy recovery

The electron bunch is decompressed in the 2nd arc after the FEL and then decelerated by the main linac in order to achieve efficient energy recovery without significant beam loss. Two TBA cells with R_{56}=-0.25 m are used as the 2nd arc for the bunch decompression. Simulation from the FEL exit to the exit of the decelerating main linac is performed by *elegant* with the six-dimensional electron distribution at the FEL exit. Figure 7 shows the betatron and dispersion functions from the FEL exit to the exit of the decelerating main linac and the time-momentum distributions with the basic beam parameters at the FEL exit, the 2nd arc exit and the exit of the decelerating main linac. The decelerating RF phase of the main linac and the field strengths of two-family setupole magnets of the 2nd arc are adjusted to optimize the bunch decompression and the energy recovery. The beam is decelerated by the main linac to about 10.5 MeV and the momentum spread is 1.55 % for the beam energy. The bunch is finally decompressed to about 2.5 ps at the exit of the main linac and the bunch length is 2.5-times as long as that of the injection beam mainly due to the FEL lasing. The horizontal maximum beam size is 3.1 mm in the dispersion section of the 2nd arc and small enough for the assumed horizontal half-aperture of the beam pipe (35 mm). As a result the electron beam is transported to the exit of the decelerating main linac without any beam loss.

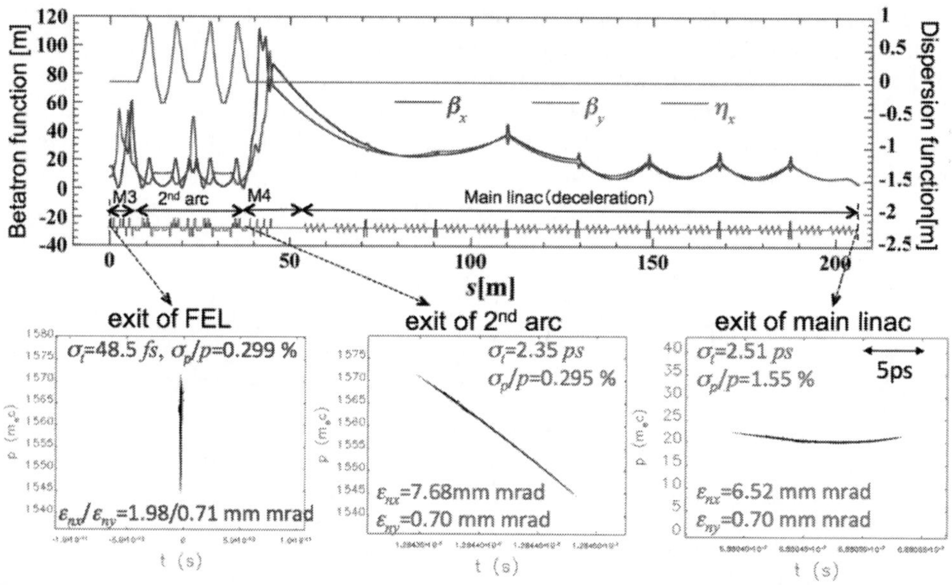

Figure 7. Betatron and dispersion functions from the FEL exit to the exit of the decelerating main linac (upper) and the time-momentum distributions with the basic beam parameters at the FEL exit, the 2nd arc exit and the exit of the decelerating main linac (lower).

6. Conclusions

In the S2E simulation, FEL power of higher than 10 kW is generated at the average beam current of about 10 mA in the ERL-based EUV-FEL source designed with available technologies and resources. It is also shown that the electron beam is successfully transported throughout the EUV-FEL source without any beam loss even after the FEL lasing. This result greatly enhances the feasibility of the high-power ERL-based EUV source for lithography.

References

[1] Mizoguchi H *et al.* 2017 *Proc. SPIE* **10143** ed E M Panning and K A Goldberg 101431J
[2] Hosler E R *et al.* 2015 *Proc. SPIE* **9422** ed O R Wood and E M Panning 94220D
[3] Nakamura N *et al.* 2015 *Proc. ERL2015* (Stony Brook USA) pp 4-9
[4] Yamamoto Y *et al.* 2015 *Proc. ERL2015* (Stony Brook USA) paper TUIBLH1020
[5] Kako E *et al.* 2012 *Proc. IPAC2012* (New Orleans USA) pp 2239-2241
[6] Sakai H *et al.* 2015 *Proc. SRF2015* (Whistler Canada) pp 592-596
[7] *GPT* http://www.pulsar.nl/gpt/index.html
[8] di Mitri S *et al.* 2016 *Nucl. Instrum. and Methods* A **806** 184
[9] Borland M 2000 *Advanced Photon Source LS-287*
[10] *Genesis* http://genesis.web.psi.ch

Conceptual design of a pre-booster ring for FCC e^+e^- injector

O Etisken[1*], Y Papaphilippou[2], A K Ciftci[3]

[1]Ankara University, Ankara, Turkey
[2]CERN, Geneva, Switzerland
[3]Izmir University of Economics, Izmir, Turkey

*ozgur.etisken@cern.ch

Abstract. The FCC-e^+e^- injector complex needs to produce and to transport a high-intensity e^+/e^- beam at a fast repetition rate of about 0.1 Hz for topping up the collider at its collision energy. A basic parameter set exists for all the collider energies, assuming a 10 GeV linac operating with a large number of bunches being accumulated in the existing SPS, which serves as pre-accelerator and damping ring before the bunches are transferred to the high-energy booster. The purpose of this study is to provide the conceptual design of an alternative accelerator ring, replacing the SPS in the present scheme. This ring will have injection energy of 6 GeV and extraction energy of 20 GeV. In this study, the basic parameters of the ring are established, including the optics design and layout. Preliminary consideration for non-linear dynamics optimization and the impact of intra beam scattering are also presented.

1. Introduction

The initial basic parameters for FCC e^+e^- pre-injector were established in order to satisfy the demanding collider flux requirements, especially for the full machine filling of the lowest energy collider flavour (45.6 GeV). This design is based, on a CLIC (Compact Linear Collider) based linac and using the CERN Super Proton Synchrotron (SPS) as a damping and pre-booster Ring. Since there may be issues for using the SPS as pre-injector, such as machine availability, synchrotron radiation and new RF system, just to name a few, a "green field" alternative pre-booster damping ring (PBDR) design was mandatory [2]. In parallel, another alternative linac study has also started [3], based on the design approaches of the SUPERKEKB pre-injectors. Additionally, a booster design study is also continuing [4]. Figure 1 shows the FCC e^+e^- injector complex schematic layout.

The extraction energy of the PBDR has to be as high as possible, in order to raise as much as possible the field of the bending magnets of the main booster at injection. Based on parameter scaling and basic calculations, the extraction energy was established at 20 GeV, for having the lowest circumference within a reasonable energy loss per turn limits of around 50 MeV [5].

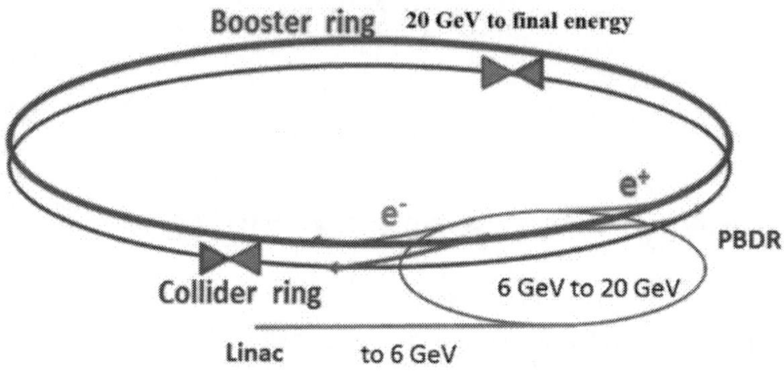

Figure 1. Layout of FCC ee injector complex.

2. Emittance at extraction

The emittance at extraction of the PBDR has to satisfy the requirement for the injected emittance in the main booster ring, which in turn is established by the demands of the collider. The lowest injection emittance target of the collider ring is 0.09 nm.rad [6]. Assuming no extra blow-up effects, the emittance $\varepsilon(t)$ at a certain instant t can be given by

$$\varepsilon(t) = \varepsilon_{inj}\, e^{-\frac{2t}{\tau}} + \varepsilon_{equ}(1 - e^{-\frac{2t}{\tau}}) \tag{1}$$

where ε_{inj} and ε_{equ} are the injected and equilibrium emittances, respectively and τ the damping time. Taking into account that the different parameters are energy dependent, a parametric relation can be established connecting the injected emittance as a function of a linear ramping time t_r, for a given emittance at extraction. Figure 2 uses this relation to establish the injected emittance in the booster ring (or extraction in the PBDR assuming again no transfer blow-up). Taking into account that the cycle time of the main booster ring is around 12 seconds [1], and assuming equal ramp-up and down times and negligible injection and extraction plateaux, the emittance at the entrance of the booster ring should be about 1 nm.rad, corresponding to an accelerating time of 6 seconds.

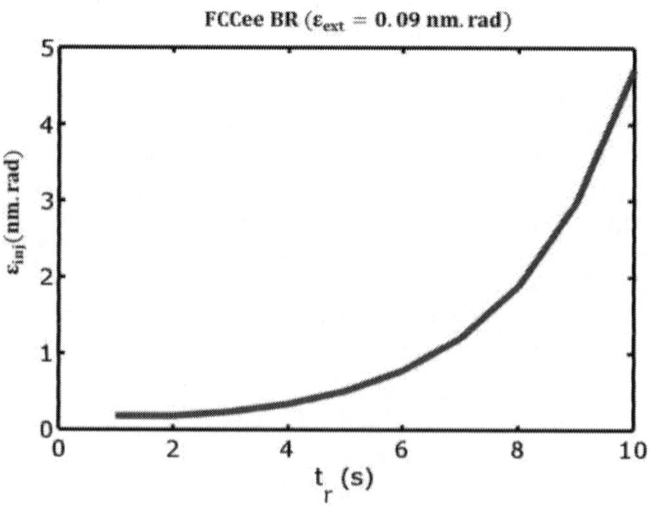

Figure 2. Injection emittance of booster ring with ramping time.

All the following calculations are made by targeting an extraction emittance of the PBDR of around 1 nm.rad. The same approach can be used for establishing the injection emittance in the PBDR. For a cycle of 1.2 s, the injected emittance obtained is 1.5 nm.rad. This may already point to the direction of needing an extra damping stage with a dedicated damping ring for both species.

3. Parameter scaling and lattice design

After fixing the extraction emittance of the PBDR as around 1 nm.rad, the scaling of parameters such as energy loss per turn, damping time, energy spread and emittance, as a function of the circumference C and ring filling factor FF (total dipoles' length over circumference) can be established, while constraining the energy loss to below 50 MeV/turn [5]. This is an important step to further determine the layout of the ring and provide the main lattice cell.

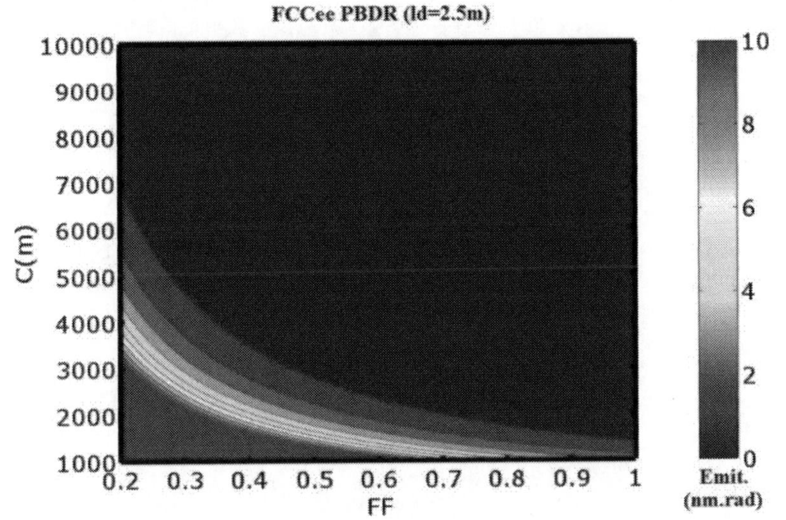

Figure 3. Scaling of emittance with filling factor and circumference.

Figure 3 shows an example of this parametrisation, between ring circumference and the filling factor color-coded with the emittance at extraction energy of 20 GeV, considering a reasonable dipole length of 2.5 m and FODO cells. As expected, the higher filling factor allows lower emittance for shorter ring circumference. For a filling factor of 0.7, the PBDR should have a circumference of around 3 km for obtaining around 1 nm.rad emittance at extraction. This choice is optimal, as it provides the minimum circumference for a limited RF power, while satisfying the transverse emittance requirements.

Additionally, there is a strict relation amongst emittance, phase advance and chromaticity for any given cell. Thus, the relations between these parameters are also checked. The basic idea is to find an adequate phase advance that satisfies the emittance requirement at extraction, while keeping the chromaticity low.

After analytic calculations and simulations, the basic structure of PBDR has been achieved. The ring is of racetrack shape consisting of 2 arcs and 2 straight sections; each arc has 166 FODO lattice with sextupole magnets in each main cell, whereas each straight section has two matching cells. The total circumference is 2736 m. Table 1 shows the main parameters of the magnets in PBDR.

The main cell is shown schematically in figure 4. The cell consists of two 2.5 m-long dipoles sandwiched between quadrupoles with 30 cm length. The dipoles have a 70 Gauss field at injection.

IOP Conf. Series: Journal of Physics: Conf. Series **874** (2017) 012014 doi:10.1088/1742-6596/874/1/012014

The chromaticity is controlled by two families of 10 cm-long sextupoles. The horizontal (red) and vertical (green) beta functions and horizontal dispersion (blue) of the cell are presented in figure 5. The present cell employs separate function magnets, although combined function ones may be considered if found to be more cost-effective.

Table 1. Magnets in PBDR.

Magnets	Num. of Magnets	Length	Field or Strength
Bending	700	2.5 m	0.07T/0.239T
FQ	362	0.3 m	-0.94 m^{-2}
DQ	362	0.3 m	1.59 m^{-2}
FS	332	0.1 m	109 m^{-3}
DS	332	0.1 m	-138 m^{-3}

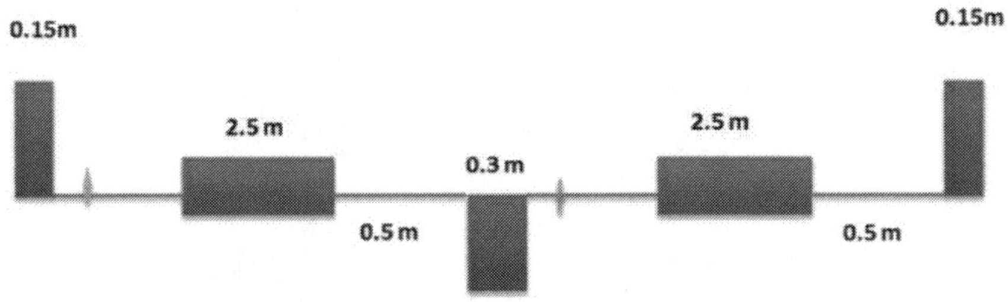

Figure 4. Main FODO cell of PBDR.

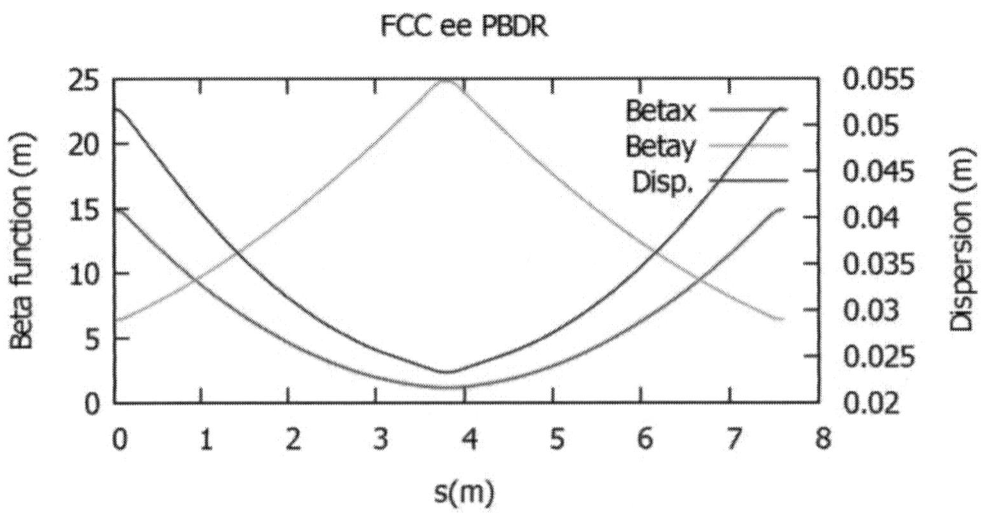

Figure 5. Beta functions and dispersion of the main cell.

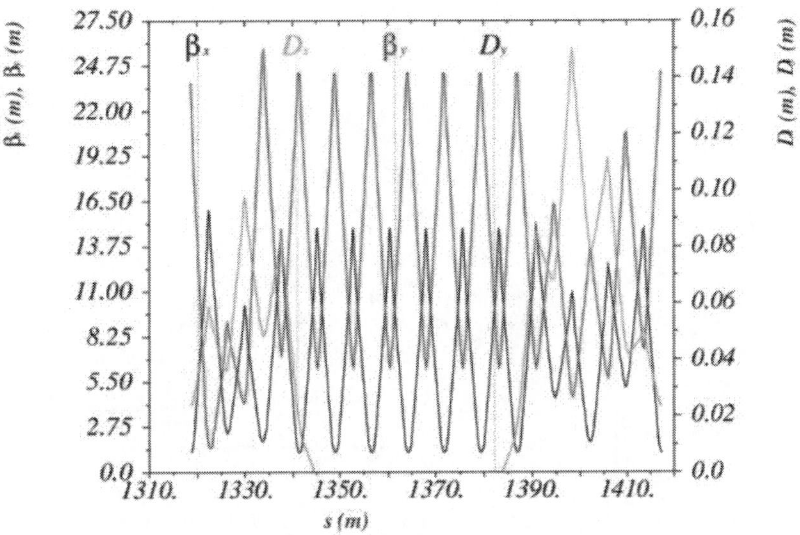

Figure 6. Straight section optics of PBDR.

Zero-dispersion straight sections with adequately allocated space for the RF, the injection and the extraction elements, are built with identical bending-free cells. They are connected to the arc with the help of matching cells. The optics functions of one such straight section including the matching cells is shown in figure 6.

During the design process, three different codes, namely OPA [7], MADX [8] and ELEGANT [9] were used. A comparison of general ring parameters is presented in table 2. All three codes show very good agreement. For these design parameters, the emittance at 20 GeV is around 0.7 nm rad, which provides some margin with respect to the 1 nm rad target requirement. Note that a remarkably low equilibrium emittance of around 65 pm.rad is achieved at injection energy of 6 GeV. The energy loss per turn is 51 MeV and the damping times are quite short (a few ms).

Table 2. Main Parameters of PBDR.

	OPA	MADX	ELEGANT
Energy [GeV]		20	
Circumference [m]		2736	
Emittance [pm.rad]	710	718	712
Energy loss / turn [MeV]	50.8	50.8	50.8
Natural chromaticity	-189/-110	-189/-110	-189/-108
Dx max [m]	0.149	0.156	0.149
Betax max [m]	16.0	16.0	16.0
Betay max [m]	25.7	24.6	25.8
Damping times [ms]	7.2/7.2/3.6	7.2/7.2/3.6	7.2/7.2/3.6

3.1. Chromaticity correction and dynamic aperture

The high chromaticity of the ring has to be controlled with a pair of sextupoles, which inevitably impact the rings dynamics aperture (DA). Figure 7 shows the horizontal versus vertical DA for different momentum deviations using the ELEGANT code [8].

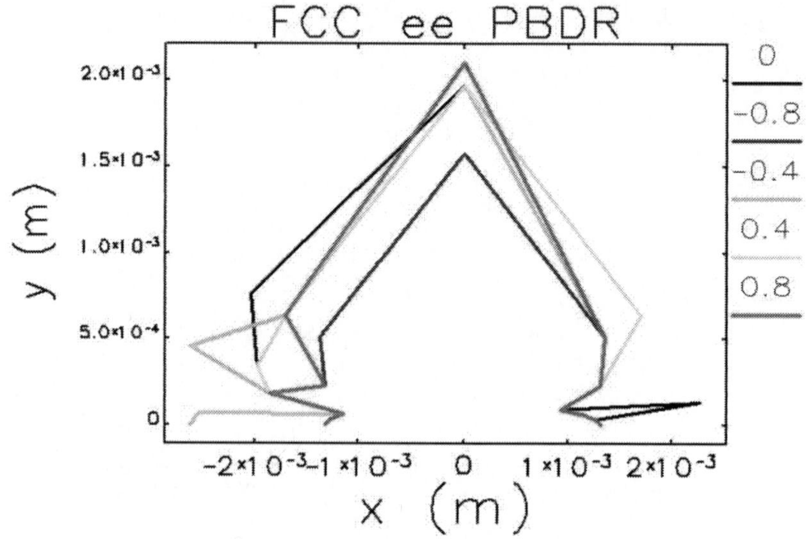

Figure 7. Dynamic aperture for different momentum deviations (%±0.4, ±0.8).

For the injected emittance of 1.5 nm.rad, the beam size is equal to 0.3 mm, so the minimum DA obtained is around 5 σ. This is a preliminary study that needs further optimization (e.g. of the working point); this can be indeed improved by a better working point choice, which is not presently optimised.

3.2. Intrabeam scattering effect

The Intrabeam scattering (IBS) effect causes emittance growth in all three dimensions. It basically depends on the optics and beam properties and in particular with the beam energy. The scaling of the horizontal emittance including the IBS effect using ELEGANT [9] is presented in Figure 8 (assuming no coupling or residual vertical dispersion and a bunch population is 8.3 x 10^9), along with the "zero-current" equilibrium emittance. The horizontal steady state emittance is IBS dominated at injection energy, with an almost 10-fold increase. At high energy, the effect of IBS is indeed negligible.

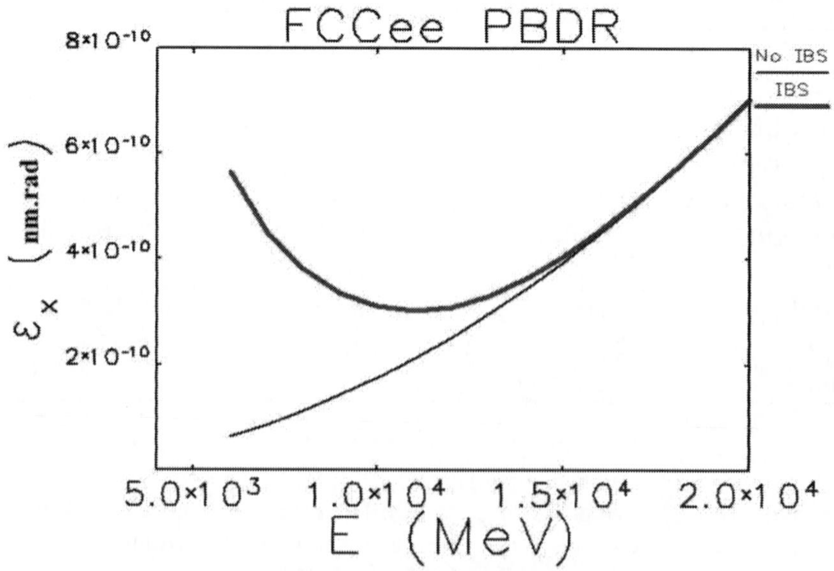

Figure 8. Emittance evaluation with energy changes by IBS effect.

IOP Conf. Series: Journal of Physics: Conf. Series **874** (2017) 012014 doi:10.1088/1742-6596/874/1/012014

4. Conclusion

In this study, a preliminary design of the pre-booster damping ring of the FCC e^+e^- injector complex is undertaken. Further detailed studies are in progress, in particular, the establishment of magnet and alignment tolerances impacting the vertical emittance and the dynamic aperture and the study of other collective effects and synchrotron radiation handling.

References

[1] Papaphilippou Y *et al.* 2016 Design guidelines for the injector complex of the FCC-ee *Proceeding of IPAC* Busan (Korea) THPMR042

[2] Etisken O *et al.* 2016 Preliminary design study of a pre-booster damping ring for the FCC-ee injector Rome Italy *FCC week*

[3] Ogur S *et al.* 2017 Preliminary design of FCC-ee pre-injector complex *Proceeding of IPAC* Kopenhag (Denmark) TUPAB014

[4] Haerer B *et al.* 2017 Challenges and status of the rapid cycling top-up booster for FCC-ee *Proceeding of IPAC* Kopenhag (Denmark) WEPIK031

[5] Etisken O *et al.* 2016 Preliminary design study of a pre-booster damping ring for the FCC ee injector *Proceeding of CERN BINP Workshop* Geneva (Switzerland) ID 378

[6] Ogur S *et al.* 2016 Towards a preliminary FCC-ee injector design Geneva Switzerland *Proceeding of CERN BINP Workshop*

[7] CERN-BE/ABP Accelerator Beam Physics Group MAD (Methodical accelerator design)

[8] SLS Lattice design code *OPA*

[9] Borland M Advance photon source *Elegant* program

Field quality of 1.5 m long conduction cooled superconducting undulator coils with 20 mm period length

S Casalbuoni[1], N Glamann[1], A W Grau[1], T Holubek[1], D Saez de Jauregui[1], C Boffo[2], Th A Gerhard[2], M Turenne[2], W Walter[2]

[1]Institute for Beam Physics and Technology, Karlsruhe Institute of Technology, Hermann-von-Helmholtz-Platz 1, D-76344 Eggenstein-Leopoldshafen, Germany
[2]Babcock Noell GmbH, Würzburg 97080, Germany

E-mail: sara.casalbuoni@kit.edu

Abstract. The Institute for Beam Physics and Technology (IBPT) at the Karlsruhe Institute of Technology (KIT) and the industrial partner Babcock Noell GmbH (BNG) are collaborating since 2007 on the development of superconducting undulators both for ANKA and low emittance light sources. The first full length device with 15 mm period length has been successfully tested in the ANKA storage ring for one year. The next superconducting undulator has 20 mm period length (SCU20) and is also planned to be installed in the accelerator test facility and synchrotron light source ANKA. The SCU20 1.5 m long coils have been characterized in a conduction cooled horizontal test facility developed at KIT IBPT. Here we present the local magnetic field and field integral measurements, as well as their analysis including the expected photon spectrum.

1. Introduction

Superconducting undulators can produce, with respect to permanent magnet ones, a higher peak field on axis for the same vacuum gap and period length, allowing to increase the brilliance and the tunability of the photon spectrum. The collaboration between the Institute for Beam Physics and Technology (IBPT) at the Karlsruhe Institute of Technology (KIT) and the industrial partner Babcock Noell GmbH (BNG) is developing superconducting undulators for ANKA and low emittance light sources. The first full length device with 15 mm period length has been successfully tested in the ANKA storage ring for one year [1]. Presently the collaboration is focused on the development of a full scale superconducting undulator with 20 mm period length with 75.5 full periods (SCU20). The device is conduction cooled and wound with a round NbTi wire with a diameter of 0.7 mm; each full winding package consists of 113 single turns and the magnetic gap is 8 mm. More details on the layout are provided in References [2,3].

The coils have been measured in a unique test facility developed allowing precise measurements in vacuum and at low temperatures of the field profile and field integrals of superconducting undulator coils. The coils are tested in the KIT IBPT test facility CASPER II [3] in a horizontal arrangement and conduction cooled, in a similar configuration to the one that they will have in the final cryostat. The results of the training, the stability and the thermal behavior of the coils are described in Reference [4]. The field integrals are measured using a stretched wire, while the local field profile along the coils is measured using a Hall probe mounted on a sledge which moves along the coils and is kept below 50 K.

2. Field integrals

With the aim of minimizing the horizontal first and second field integrals the end field design shown in Figure 1 has been chosen. Two auxiliary coil pairs and Helmholtz coils are wound with a NbTi round wire with 0.25 mm diameter. The first groove of the main coil has 23 turns and the second one 53 turns. Similar configurations are at all 3 other ends.

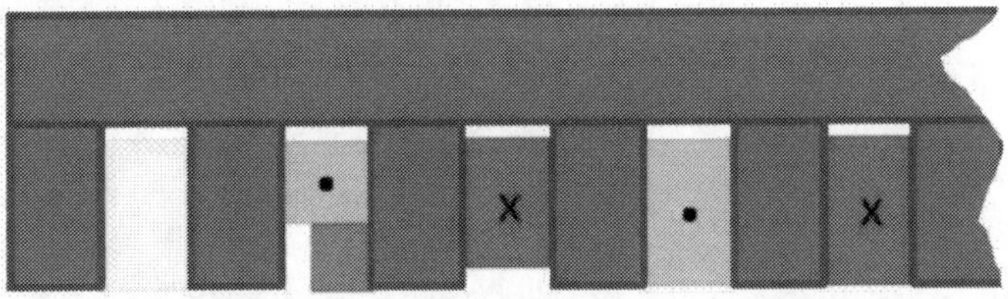

Figure 1. End field configuration at the beginning of the upper coil. In yellow are indicated auxiliary coils 1 (AUX1), in magenta auxiliary coils 2 (AUX2), and in red and green the main coil, wound in opposite direction to create the alternating magnetic field. Similar configurations are at all 3 other ends.

To keep the first and second vertical field integrals (I_{1v} and I_{2v}) within the specified values, only auxiliary coils 1 (AUX1) and the Helmholtz coil downstream (HH DS) are needed. The AUX1 upstream and downstream are connected to the same power supply and are used to compensate the second vertical field integral, while the HH DS is used to compensate the first vertical field integral. All other correction coils are not needed. The measured values of the first and second vertical and horizontal field integrals, presented in Table 1 for five different currents in the main coil in the range from 50 to 395 A, are all within the specified values: $|I_{1v}| < 3\ 10^{-5}$ T m, and $|I_{2v}| < 4\ 10^{-4}$ T m^2.

The horizontal field integrals are specified: $|I_{1h}| < 3\ 10^{-6}$ T m, and $|I_{2h}| < 10^{-5}$ T m^2. In order to reach these values correctors will be added outside the cryostat.

Table 1. First and second vertical field integrals minimized by powering the AUX1 and HH DS coils to values up to 5.6 A and 0.82 A, respectively.

Current (A)	I_{1v} (10^{-5} T m)	I_{2v} (10^{-4} T m^2)
50	0.6	-0.29
100	0.5	0.64
200	0.6	0.09
300	0.9	0.28
395	1.3	-0.8

3. Integral field multipoles

The first vertical field integral has been measured at different positions of the transverse coordinate x at the five different currents in the main coils indicated in Table I, and powering the AUX1 and HH DS coils to the same currents minimizing the first and second vertical field integrals on the magnetic axis (x = y = 0 mm, being y the coordinate along the magnetic gap).

The vertical integrated field multipoles have been calculated from the measurements of I_{1v} shown in Figure 2.

Only the vertical integrated field multipoles have been measured. The quadrupole Q, sextupole S and octupole component O are for all measured currents: |Q|<0.005 T, |S|<5 T/m and |O|<15 T/m^2. With these values of the integrated field multipoles the dynamic aperture for the 2.5 GeV optics is not affected.

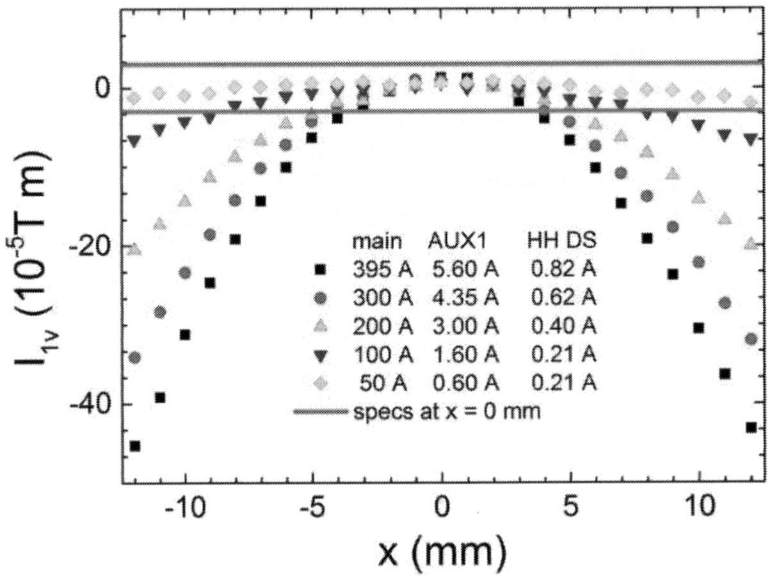

Figure 2. Measurements of the vertical first field integral as a function of the transverse coordinate x at different currents in the main coils. These measurements are usedto calculate the multipole integral components.

4. Longitudinal field profile

With an operating current of 395 A and a magnetic gap of 8 mm an average peak field of 1.187 T is reached. The measured absolute values on axis (x = 0 mm) of the peaks of the magnetic field profile are plotted and compared to the simulated one with Radia [5] with a current in the main coils of 395 A and in the AUX1 coils of 5.6 A, see in Figure 3 (lower plot). The simulation is performed considering the deviations measured at room temperature of the half period length before winding < 20 μm and of the pole height < 50 μm and winding height < 100 μm measured after impregnation. Figure 3 (upper plot) shows also the half period length obtained from the measured and simulated field profiles. The larger deviation of the period length observed at room temperature is at the connection of the building blocks of the coils. The smaller average period length observed in cold conditions is due to the thermal contraction of the yoke. The larger deviations observed in the half period length in cold conditions with respect to the ones at room temperature might be due to the loosening of the connections between the blocks as well as to an increase of the winding height deviations. This could explain a larger variation also in absolute value of the peak field on axis of the measured peaks with respect to the ones simulated.

The roll off has been obtained at each pole by measuring the longitudinal field profile with the same Hall probe positioned at x = 0 mm, -10 mm, +10 mm. In Figure 4 are reported the absolute values of the maxima and minima measured with 200 A in the main coils, 3 A in the AUX1, and 0.4 A in the HH DS coils.

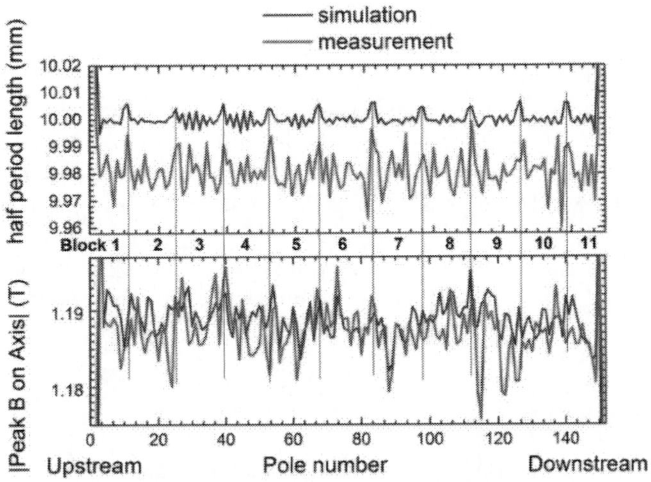

Figure 3. Comparison of the half period length and of the peak field on axis as a function of the pole number (counting from upstream to downstream) of the measured and simulated magnetic field. The dashed line shows the peak field of the field profile simulated with Radia considering the mechanical accuracies (pole height, half period length and winding height deviations from the ideal values) measured at room temperature.

The calculated roll off is defined as follows:

$$\text{Roll off} = \frac{|B(x = \pm 10\text{ mm}) - B(x = 0\text{ mm})|}{|B(x = 0\text{ mm})|}.$$

The roll off obtained from the local magnetic field measurements at different currents in the main coils and in the correction coils to minimize the vertical field integrals is below 0.5% at higher currents and it increases up to 0.7-0.8 % at lower currents. The roll off measured for the five currents in the main coils as indicated in Table I induces a negligible dynamic kick [6], and does not constitute a problem for the operation of the SCU20.

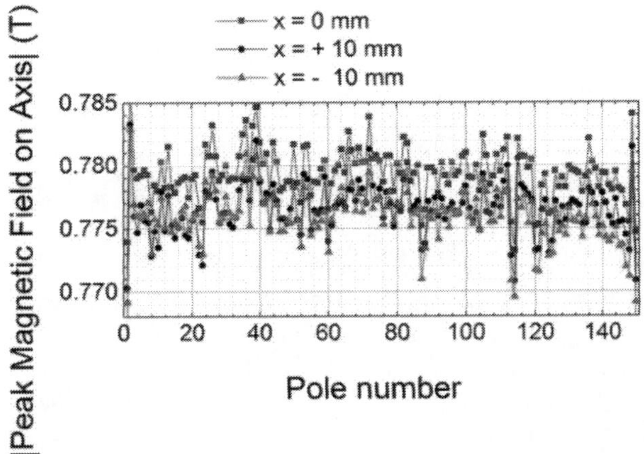

Figure 4. Peak magnetic field measured along the SCU20 coils powered with a current of 200 A at x = 0 mm, +10 mm and –10 mm.

5. Calculated spectrum

The advantage of using the SCU20 with respect to an ideal (without mechanical errors and perfect end fields) cryogenic permanent magnet undulator (CPMU) with the same parameters as the one built at SOLEIL [8] for the same vacuum gap of 7 mm can be appreciated in Figure 5. The CMPU has a period length of 18 mm (CPMU18), a peak field of 0.82 T (highest field reachable with PrFeB) and 2 m magnetic length.

The flux calculated with B2E [7] from the measured field at maximum current of the SCU20 is compared in Figure 5 with the flux of a CPMU18 with an ideal field, that is without mechanical errors and perfect end fields.

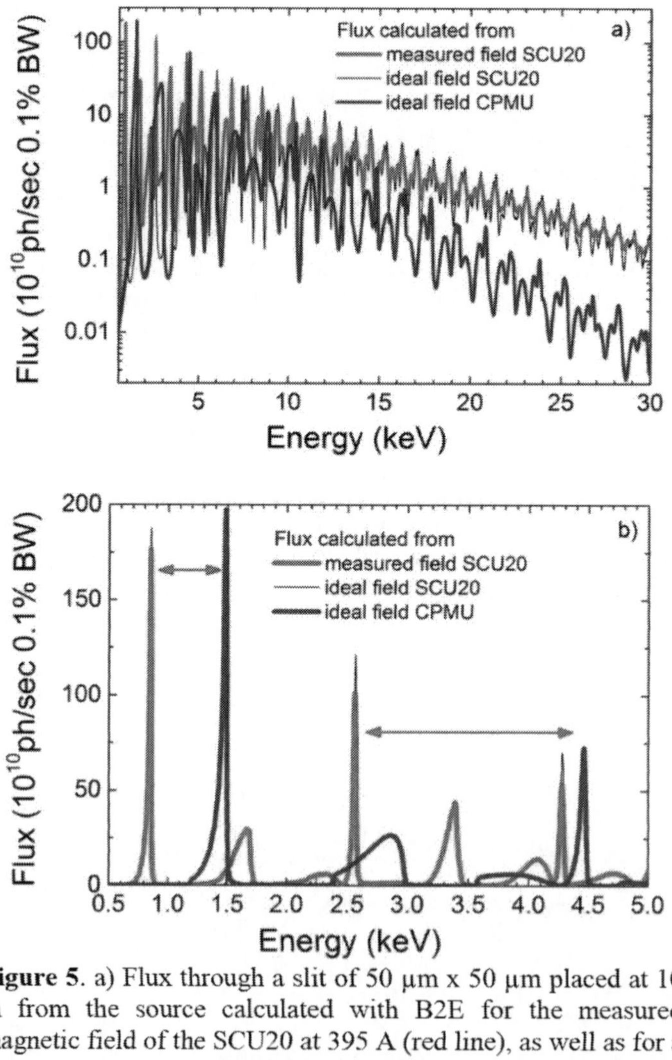

Figure 5. a) Flux through a slit of 50 µm x 50 µm placed at 10 m from the source calculated with B2E for the measured magnetic field of the SCU20 at 395 A (red line), as well as for a SCU20 (black line) and a CPMU18 (blue line) with ideal field profile (without mechanical errors and perfect end fields), respectively with 1.5 m and 2 m magnetic length. b) Zoom of a) in the low energy range and with a linear scale: the red arrows indicate the extended energy region available with the SCU20 with respect to the CPMU18.

Please note that the magnetic length considered is 1.5 m for SCU20 and 2 m for the CPMU18.

To give a flavor of a comparison in terms of brilliance, the flux is calculated through a pinhole of 50 μm x 50 μm at 10 m from the source. The spectra presented have been simulated using the ANKA beam parameters: beam energy 2.478 GeV, beam current 100 mA, energy spread 0.001, horizontal emittance 41 nm rad, vertical emittance 0.3 nm rad, horizontal beta function 19 m, and vertical beta function 1.7 m.

6. Discussion

The flux calculated from the measured field of the SCU20 is compared also to the flux of SCU20 with an ideal field. As expected, a slight reduction (less than 28% up to 30 keV) in flux for the odd harmonics, due to the mechanical errors and to the non-ideal end field configuration, is observed.
The comparison with the ideal CPMU18 with the same parameters as the one in use at SOLEIL shows the larger flux/brilliance of the SCU20 at high energies up to factor of five (see Figure 5a) and at low photon energies the energy regions allowed with the SCU20, and not reachable with the CPMU18 (see Figure 5b).

7. Conclusion

The field integrals, the longitudinal magnetic field profile and the roll off of the 1.5 m long coils of the SCU20 have been measured in the test facility CASPER II.

The magnetic field profile has been used to simulate the expected photon spectrum at ANKA. The advantages of the spectrum produced by the measured field profile of the SCU20 with respect to an ideal PrFeB CPMU18 (without mechanical errors and perfect end fields) with the same parameters as the one built at SOLEIL and same beam stay clear have been demonstrated (see Figure 5).

References

[1] Casalbuoni S, Cecilia A, Gerstl S, Glamann N, Grau A W, Holubek T, Meuter C, Saez De Jauregui D, Voutta R, Boffo C, Gerhard Th, Turenne M and Walter W 2016 Characterization and long term operation of a novel superconducting undulator with 15 mm period length in a synchrotron light source *Phys. Rev. Accel. Beams* **19** 110702

[2] Casalbuoni S, Gerstl S, Glamann N, Grau A, Holubek T, Saez De Jauregui D, Boffo C, Turenne M and Walter W 2014 Test of short mockups for optimization of superconducting undulator coils *IEEE Trans. Appl. Supercond.* **24-3** 4101905

[3] Grau A, Casalbuoni S, Gerstl S, Glamann N, Holubek T, Saez De Jauregui D, Voutta R, Boffo C, Gerhard Th, Turenne M and Walter W 2016 Characterization of 30-cm-Long Superconducting Undulator Coils With the Magnetic Measurement System CASPER II *IEEE Trans. Appl. Supercond.* **24-4** 4100804

[4] Grau A, Casalbuoni S, Glamann N, Holubek T, Saez De Jauregui D, Boffo C, Gerhard Th, Turenne M and Walter W 2017 Training and Characterization of 1.5m Long Conduction Cooled Superconducting Undulator Coils with 20 mm Period Length *Proceedings of the 8th Int. Particle Accelerator Conf. (IPAC'17)* Copenhagen, Denmark TUPAB036

[5] Chubar O, Elleaume P, Chavanne 1998 A 3D Magnetostatics Computer Code for Insertion devices *J. Synchrotron Rad.* **5** 481-484

[6] Safranek J, Limborg C, Terebilo A, Blomqvist K I, Ellaume P, Nosochkov Y 2002 Nonlinear dynamics in a SPEAR wiggler *Phys. Rev. ST Accel. Beams* **5** 010701

[7] Elleaume P, Marechal X 1991 B2E. A software to compute Synchrotron Radiation from Magnetic field data Report ESRF-R/ID-9154

[8] Benabderrahmane C, Valleau M, Ghaith A, Berteaud P, Chapuis L, Marteau F, Briquez F, Marcouille O, Marlats J L, Tavakoli K, Mary A, Zerbib D, Lestrade A, Louvet M, Brunelle P, Medjoubi K, Nagaoka R, Herbeaux C, Bechu N, Rommeluere P, Somogyi A, Chubar O, Kitegi C, Couprie M E 2017 Development and operation of a $Pr_2Fe_{14}B$ based

IOP Conf. Series: Journal of Physics: Conf. Series **874** (2017) 012015 doi:10.1088/1742-6596/874/1/012015

cryogenic permanent magnet undulator for a high spatial resolution x-ray beam line *Phys. Rev. Accel. Beams* **20** 033201

Installation management for the European XFEL main accelerator

M Bousonville, F Eints and S Choroba

DESY Deutsches Elektronen-Synchrotron, Hamburg, Germany

michael.bousonville@desy.de

Abstract. By end of 2016, the main accelerator of the European XFEL was completed. To build this complex machine in a minimum of time, certain management methods were introduced in mid 2015, which accelerated the installation process substantially. In the following 64 weeks additional 84 % of the main accelerator were set up. This was possible due to an improved planning, the reinforcement of two teams as well as a permanent controlling and optimizing of the installation process. In this paper, the installation process from July 2015 to end 2016 and the measures, which speeded up the workflow, are described.

1. Introduction

In the 2 km tunnel of the European XFEL, 96 superconducting accelerator modules were installed. Typically, 12 accelerator modules form a so-called cryo-string (CS), which was identical to one installation section. It is particular that no supply tunnel exists. Therefore, the accelerator modules hang from the ceiling while the RF stations, electronic racks, power and cooling water supply etc. are located below them on the ground (Figure 1, details in [1]).

The installation of all these components in one single tunnel was a complex undertaking, which was done by several expert groups of DESY and some external companies. The main tasks were split into 13 work packages (WPs):

WP 01 – RF System
WP 02 – LLRF System
WP 03 – Accelerator Modules
WP 05 – Power Couplers
WP 08 – Cold Vacuum
WP 13 – Cryogenics
WP 28 – Accelerator Controls
WP 32 – Survey & Alignment
WP 33 – Tunnel Installation
WP 34 – Utilities
WP 36 – General Safety
MDI – Cabling for some WPs
IT – Fibre Optics and Ethernet

IOP Conf. Series: Journal of Physics: Conf. Series **874** (2017) 012016 doi:10.1088/1742-6596/874/1/012016

At the beginning of 2015, the interaction between these WPs was not ideal. For this reason, the installation speed was not sufficient to ensure the completion in 2016. So a new installation management approach was introduced in mid 2015.

Figure 1. Completed main accelerator in one single tunnel. On top accelerator modules and cable trays. Below RF stations, electronic racks, power and water supply etc.

2. Installation Management

The installation management methods used were based on [2, 3] and included three key elements: planning, controlling, and continual improvement.

2.1. Planning

A new process plan was worked out in cooperation with all involved WP leaders in order to optimise the interaction of the WPs (scheme described in [4]). In broad outline the installation was effected as follows. First of all, 12 modules were fixed to the tunnel ceiling and their cryogenic pipes as well as the beam pipe were interconnected (Figure 2 Block „Installation of 12 Modules"). Apart from a few exceptions (e.g. positioning of klystrons and pulse transformers), all other process steps, which had to be carried out underneath the modules, were not allowed to start until the connection of the pipes and their final checks were completed. This was necessary to enable the fastest possible welding of the cryogenic pipes, which was essential for completing the whole installation process in time.

After the module installation was done, the electronic racks were brought into position and the installation of waveguides, power and water supplies started (Figure 2 „RF Stations"). Simultaneously, the "Optical Fibres" were laid. The installation of cable trays started a few days later and afterwards, cables for most of the WPs were laid ("Cable"). The "Ethernet" cable laying started three weeks later because some of the waveguides had to be installed before. After the electronic racks had been supplied with power, cooling water, Ethernet etc., some subsystems, which are necessary to operate the RF stations, were commissioned („Preparation Personnel Interlock"). Afterwards, the functionality of the personnel interlock was tested ("TÜV"). Subsequently, the remaining process steps were done and the conditioning of the RF couplers could start afterwards.

In order to meet the overall deadline in 2016, it was necessary to complete one "Installation of 12 Modules" every 8 weeks. The process analysis showed, however, that this would not be possible with the welders available at that time. Therefore it was decided to commission a second welding team. Plan # 1 came into force in July 2015.

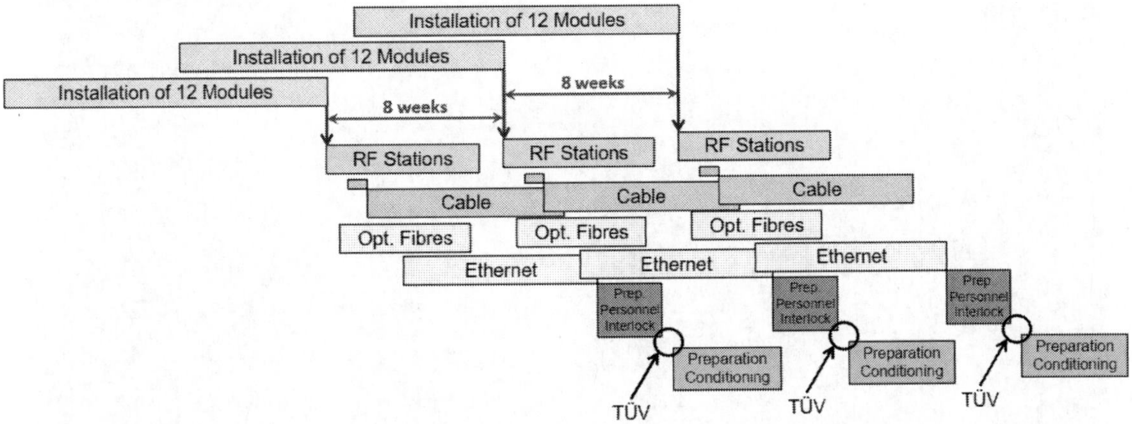

Figure 2. Original planning in mid 2015, each block represents 4 to 24 process steps.

2.2. *Controlling*

To control the installation process, the following four tools were used [4]:

- *Timetables* – four weeks in advance, each WP leader got a timetable when to start which process step
- *Weekly tunnel meeting* (Figure 3)
- *Daily tunnel inspections*
- *Monitoring* of the installation status

The installation status was recorded once a week. It included all installation steps with three possible states: not started, started or finished. The total number of process steps started and finished per week was used as figure of merit to indicate the installation activity and was displayed in a graph (Figure 4). The fluctuations are due to the fact that sometimes many process steps were started in the same week, e.g. after the installation of modules in one cryo-string was finished. For that reason, the trend line is a better indicator. It is correlated to the actual installation speed.

2.3. *Continual Improvement*

In contrast to other approaches it was assumed that the process could be speeded up permanently because of the growth of experience in the interaction of the WPs. So in the on-going installation each opportunity was caught which allowed to start a process step earlier than originally planned. In addition, the initial plan # 1 was optimized three times, from one cryo-string installation to the next.

Figure 3. Weekly meeting in the accelerator tunnel at the site of installation.

3. Actual Installation Process

On 28 July 2015, plan # 1 came into force with CS 3 (Figure 4). Despite certain deviations, plan # 1 worked well and due to the continual improvement, the personnel interlock test (Figure 4 "TÜV") was brought forward by 15 days. So the installation management prove to be effective for CS 3. The problem at this time was that the interval between completion of "Installation of 12 Modules" in CS 3 and CS 4 was 12 instead of 8 weeks because the second welding team was not trained enough at this time for their difficult task. On the basis of the processing time measured for "RF stations + Utilities", a decision was made to assign a second waveguide installation team in order to reduce this processing time. The success of this action can be seen in the further process. Before the "Installation of 12 Modules" in CS 4 was finished, the optimised plan # 2 became effective, which lead to a precision landing of the CS 4 personnel interlock test. However, now the duration of the cabling ("Cable") became time-critical. As countermeasure, some cables were preassembled outside the tunnel and the work on different cryo-strings could be partly parallelized. A drawback was the increasing gap between the finalization of "Installation of 12 Modules" in CS 4 and CS 5 from 12 to 13 weeks, but afterwards it shortened to 6 weeks between CS 5 and CS 6. This lead to the breakthrough in speeding up the installation process because now it was possible to work on up to four cryo-strings at the same time.

As a result of all these measures the installation speed was increased by a factor of 10 (graph in Figure 4) and 84 % of the main accelerator were built in the period from July 2015 to October 2016.

On 11 October 2016, a pressure test of the Helium exhaust line was carried out. During this test, the line fell down on a length of 200 m and damaged the installation of CS 8 and 9. Despite the necessary repair work, the main accelerator was completed on 6 December 2016. The cool down of the superconducting accelerator modules started at 10 December 2016 and the first beam through the complete electron accelerator was reported on 23 February 2017 [5].

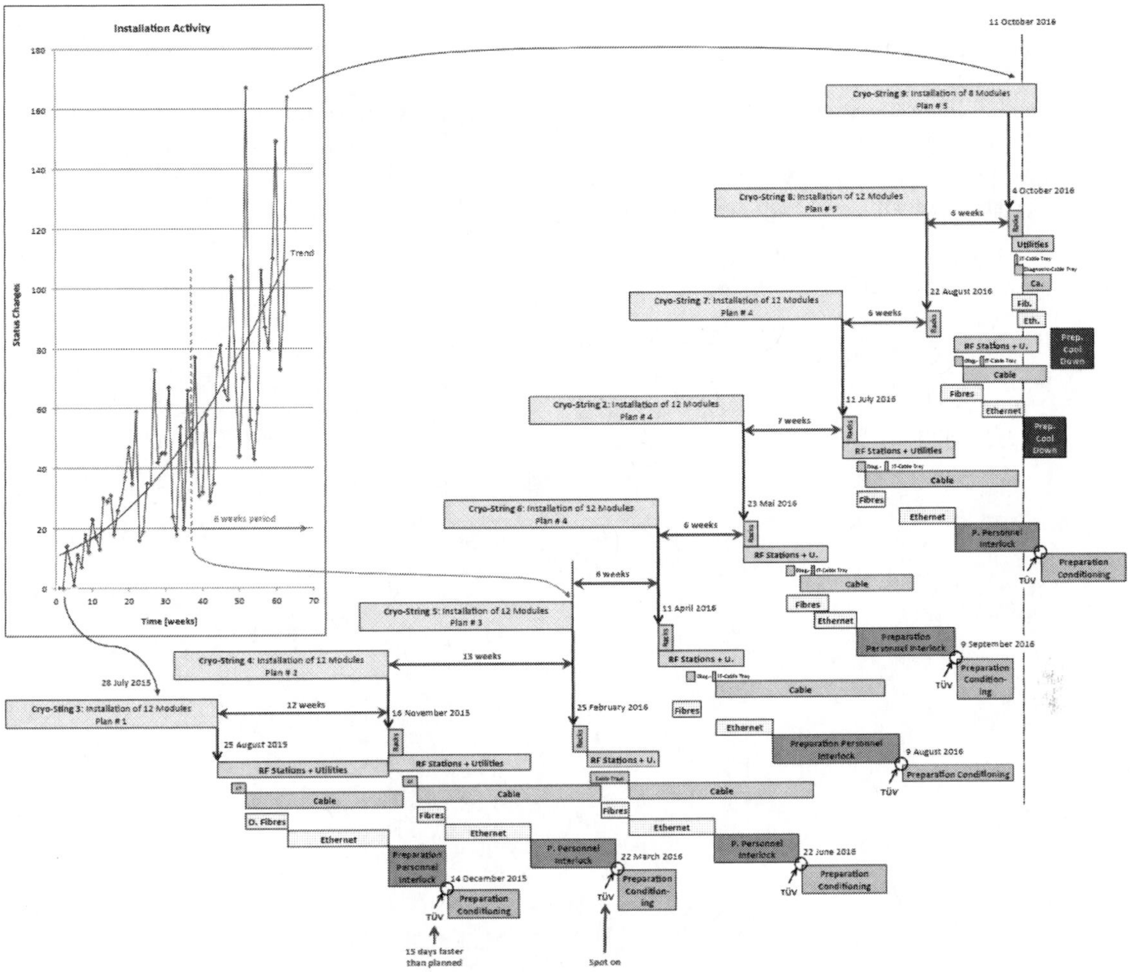

Figure 4. Installation process from July 2015, when the installation management took effect, to October 2016. In this period of just over 15 months, 84 % of the main accelerator was installed. The trend line indicates the increase of installation speed.

4. Summary

The implementation of the installation management led to an increase of the installation speed by a factor of 10. Thus the main accelerator of the European XFEL could be completed in 2016. In relation to future projects, it is fair to say that with adequate process management, the installation of such a machine can take place in 2 years.

Acknowledgments

Thanks to all co-workers involved and the XFEL linear accelerator project management (DESY) for target-driven cooperation.

References

[1] Altarelli M, Brinkmann R *et al.* 2007 *The European X-Ray Free-Electron Laser Technical design report* (Hamburg: DESY XFEL Project Group, European XFEL Project Team and Deutsches Elektronen-Synchrotron)

[2] 2005 *Prozessmanagement für Praktiker – Leitfaden für das Erkennen, Beschreiben, Bewerten, Umsetzen und Verbessern von Prozessen* (Frankfurt: Deutsche Gesellschaft für Qualität e.V.

IOP Conf. Series: Journal of Physics: Conf. Series **874** (2017) 012016 doi:10.1088/1742-6596/874/1/012016

vol 14-26)
[3] Malik F 2006 *Managing Performing Living: Effective Management for a New Era* (Frankfurt: Campus)
[4] Bousonville M, Eints F and Choroba S 2016 Planning and Controlling of the Cold Accelerator Sections Installation in XFEL *Proc. of IPAC 2016* Busan (Korea) MOPOW006 pp. 716–8.
[5] Decking W *et al.* 2017 Commissioning of the European XFEL *Proc. of IPAC 2017* Copenhagen (Denmark) MOXAA1 pp. 0–5.

Long beam pulses with SLED compression in DAΦNE LINAC

Paolo Valente[1,*], Maurizio Belli[2], Bruno Buonomo[2], Riccardo Ceccarelli[2], Alberto Cecchinelli[2], Renato Clementi[2], Claudio Di Giulio[2], Luca Gennaro Foggetta[2], Graziano Piermarini[2], Luis Antonio Rossi[2], Serena Strabioli[2] and Raffaele Zarlenga[2]

[1] INFN – Sezione di Roma, P.le Aldo Moro 2, I-00185 Rome, Italy.
[2] INFN – Laboratori Nazionali di Frascati, Via Enrico Fermi 40, I-00044 Frascati (Rome), Italy.

*paolo.valente@roma1.infn.it

Abstract. The DAΦNE LINAC is a ~60 m long, S-band (2856 MHz) linear accelerator, made up by four 45 MW klystrons with SLED compression, and by 15 travelling-wave, $2/3\pi$, SLAC-type, 3 m long accelerating sections. It serves as injector of the DAΦNE e+ e– collider, providing 510 MeV, 10 ns long, electron and positron pulses of ≈1 nC, and to the Beam-Test Facility extraction line, with variable beam energy and intensity pulses, of length in the range 1.5 to 40 ns. A new pulsing system for the gun allows longer beam pulses, but the shape of the accelerating field in the sections due to the SLED compression has to be taken into account. We describe the tuning of the RF power, phase and delays in the pre-buncher, buncher and following accelerating sections, and the results of the tests performed in order to reach >200 ns, 500 MeV electron pulses and the characterization of the quality of the beam in terms of energy spread, time distribution, etc.

1. The DAΦNE LINAC

1.1. General Layout

The DAΦNE LINAC is a ~60 m long, S-band (2856 MHz) linear accelerator, made up by four 45 MW klystrons (Thales TH-2128C) with SLED (SLAC Energy Doubler) compression, and by 15 travelling-wave, $2/3\pi$ phase advance, 3 m long accelerating sections of SLAC type.

The RF distribution scheme is shown in Figure 1: three of the klystrons have exactly the same configuration, consisting of an evacuated rectangular waveguide network with three 3 dB splitters arranged in order to divide the power into four equal parts, feeding each one an accelerating section. The configuration of the fourth klystron is different: half the power is sent to the capture section (CS), the first section downstream of the positron converter (PC), while the second half is equally divided between two branches, the first one feeding the accelerating section P1, the second one feeding the pre-buncher, the buncher and the accelerating section E1.

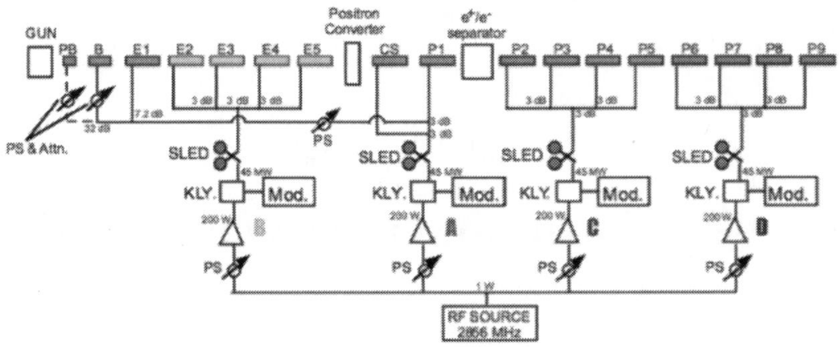

Figure 1. RF distribution scheme of the LINAC: four klystrons with SLED compression, powered by line modulators, feed a total of 15 accelerating sections.

The phase between the sections (including pre-buncher and buncher) can be adjusted by means of low-power 360° phase shifters upstream of the RF amplifiers of each klystron, and by a high-power 360° phase shifter uncoupling the CS from E1.

The four modulators produce a pulse of 4.5 μs flat top with a repetition rate of 50 Hz: a HV power supply with resonant circuit charges the pulse forming network (PFN), composed by 9 LC cells up to 50 kV, and a switching thyratron.

The RF pulse from the klystron is then compressed by the SLAC system [1], in order to reach a higher energy with respect to the flat power: the RF power from the klystron is first stored in two resonant cavities, coupled by means of a 3 dB hybrid coupler, and then is discharged towards the accelerator by applying a phase inversion to the input power from the klystron.

This transforms the flat power from the klystron, 4.5 μs long, into the typical peaked shape at the output of the SLED, shown in Figure 2.

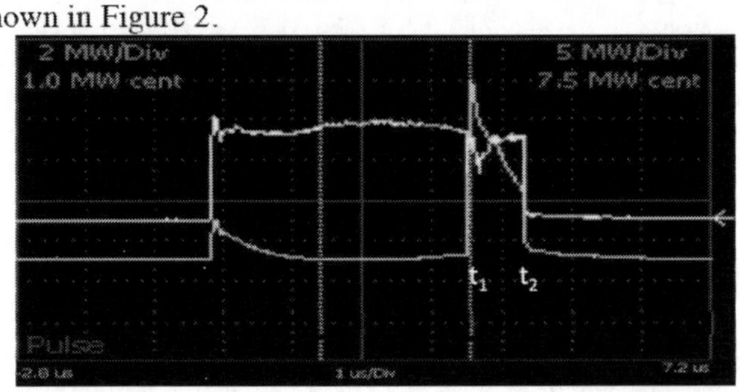

Figure 2. Klystron power (yellow trace) and SLED output power (cyan trace) for one of the DAΦNE four RF stations, measured by a power-meter.

The corresponding accelerating voltage will be approximately double with respect to the uncompressed case, but with a sharply peaked shape from the time of the phase inversion t_1, followed by an exponential decrease until the klystron pulse end time t_2. One effect of the SLED compression is the fact that longer pulses accelerated close to the peak at t_1 will exhibit an energy spread due to the variation of the accelerating voltage during the pulse length.

Electrons are produced by a gridded electron gun with replaceable cathode, high-voltage deck with up to 150 KV power supply (operational 120 KV), isolation transformer and corona shielding, allowing to reach full voltage in air. The gun is pulsed at 50 Hz with a rectangular waveform of 10 ns length during DAΦNE injections. The system used up to Summer 2016 can adjust the beam pulse height in the 300-750 V range and the pulse length between 1.5 and 40 ns.

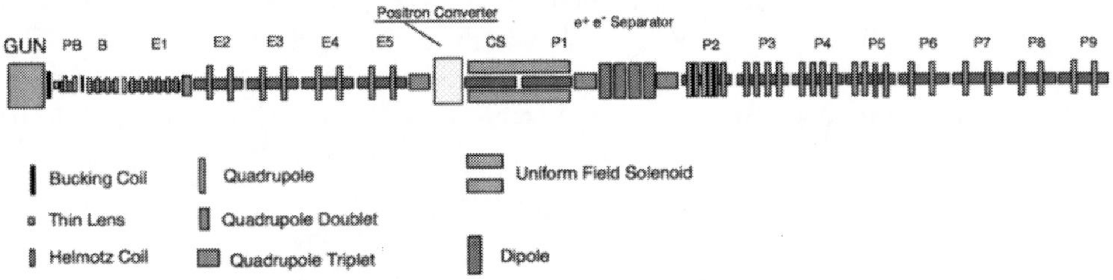

Figure 3. LINAC focusing system scheme.

The focussing system, schematically shown in Figure 3, varies its configuration along the length of the LINAC: in the first part the bucking coil reduces the field in the cathode region, while the so-called thin-lens solenoidal field focalize the electron current in the pre-buncher and buncher region; a FODO in the first accelerating sections transports the beam up to the converter region (matched with the up-stream solenoidal focussing by means of quadrupole doublet); downstream of the positron converter with the strong solenoids, a four-dipoles achromatic bump separates the electrons from the positrons; in the remaining part of the LINAC a FODO completes the focussing.

1.2. Accelerating voltage and beam loading

When the pulse is shorter with respect to the structure filling time, the effect of beam loading can be neglected and the beam is accelerated near the peak of the accelerating voltage at the output of the SLED (the small "flat top" close to the time t_2, as shown in Figure 4, adapted from [2]).

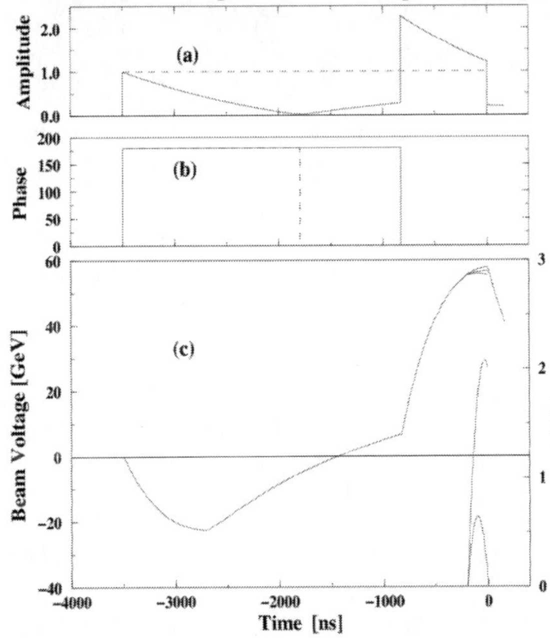

Figure 4. From the top: input (dashed) and output (solid) amplitude (a) and phase (b) of the SLED and (c) unloaded and loaded beam voltage and energy spread (lower curves), adapted from [2].

On the contrary, the beam loading can be used in order to compensate for the varying accelerating voltage when longer beam pulses are fed to the structures, by turning on the beam prior to the peak: the energy loss due to loading will be – at least partially – compensated by the still increasing energy gain curve, thus flattening the beam voltage and reducing the energy spread.

In addition to the beam loading effect, any shift between different RF phases or delays in the four stations feeding the four different portions of the DAΦNE LINAC can further increase the flat top of the beam voltage.

2. Long beam pulse results

2.1. New gun pulser

In order to produce electron pulses from the LINAC gun longer than 40 ns (for dedicated, low pile-up, high energy physics experiments [3]), a new pulsed power supply was necessary: the detailed specifications, the installation and commissioning is described in [4]: the new system is now routinely used since September 2016 in the standard DAΦNE operations with 10 ns long pulses.

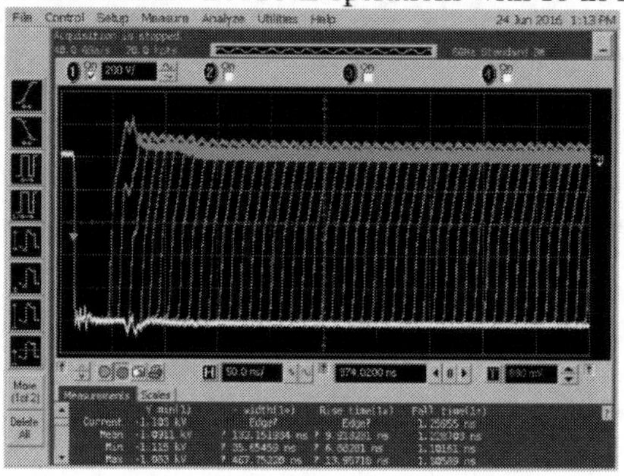

Figure 5. Test pulses of different width from the gun supply system.

Commissioning with beam and acceleration tests were performed on September 15-17, 2016. First of all the pulse height for the square waveform to be applied to the gun has been optimized vs. the counter-potential of the stopping grid, in order to have a clean shape of the time profile of the electron current emitted by the cathode: increasing both the gun pulse and the grid potential has been useful for reducing the effect of the ringing of the output of the pulser, clearly visible in Figure 5. While the pulse at the gun can be easily extended up to 300 ns, the time profile of accelerated electrons along the LINAC gets instead a triangular shape, with a FWHM of few tens of ns, shorter duration, due to the head-tail effect during acceleration, since the RF power stations (including the SLED's) were not modified.

2.2. Buncher power

Referring to the scheme of Figure 1, the distribution of RF power from the modulator "A" is the following: half of the power is fed to the "capture section" (CS), immediately following the positron converter (thus operating at a higher gradient of 26.5 MV/m, instead that 17 MV/m of all standard sections). The remaining half is further split: 25% of the modulator A power is fed to the P1 section (the first after the CS), while the last 25% can be phase-shifted and is distributed to the pre-buncher (PB), buncher (B) and first section (E1).

In order to get a significantly longer pulse width of the fully accelerated beam, the first parameter that we had to change was the relative power in the buncher with respect to the following first section E1: we have decreased the pre-buncher fraction from 49% to 32%, while increasing the overall power of modulator A from 64% to 76%, obtaining a triangular beam time distribution at the end of the LINAC with ≈150 ns base and 70 ns FWHM, when applying a 200 ns pulse to the gun (see Figure 6).

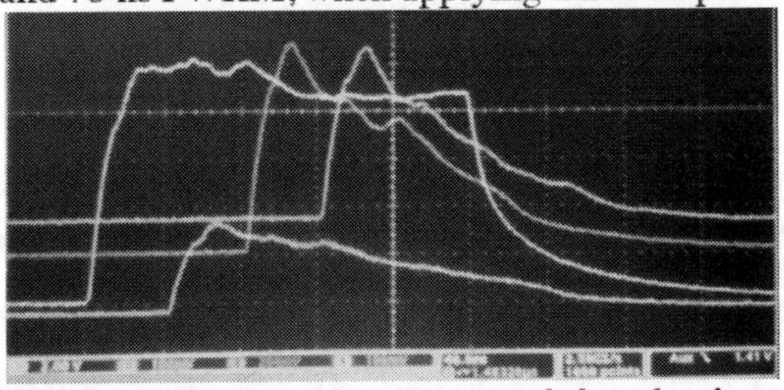

Figure 6. Beam pulse measured by the beam current monitors along the LINAC, from the exit of the gun (yellow trace) to the last accelerating section (green trace), for a gun pulse width set to 200 ns, buncher power at 32%.

2.3. RF delay
In order to get a long flat top in the accelerated beam, the second important parameter that has been changed is the timing of the pulse with respect to the accelerating field inside the structures: if we move from the maximum of the field, in particular moving the head of the macro-bunch prior to it, towards the still increasing part of the accelerating potential (negative times in the reference of Figure 4), we can get a smaller variation of the accelerating field of micro-bunches in the head and the tail of the pulse. This was done by changing the relative delay between the RF and the gun: 100 ns was the optimal value we found.

2.4. RF power and phase
In order to smooth the distribution of electrons at the end of the LINAC the other parameters that have been optimized were of course the power of the four modulators and the phase shifters. Of course changing the power and the phases of the modulators also affects the energy gain and thus the final energy at the end of the LINAC. The best result, in terms of a long and flat beam pulse, is shown in Figure 7: a 250 ns long distribution with a relative variation at the level of 25% (peak to peak) has been obtained, at a final energy of 550 MeV.

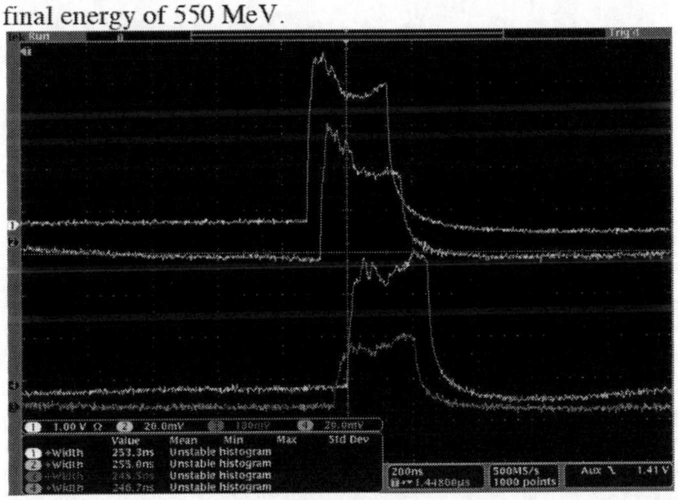

Figure 7. Beam current monitor signals after the gun (yellow trace), and at different positions along the LINAC (the green trace is right at the end of the accelerating sections): optimized 250 ns almost flat electron beam pulse, at 550 MeV final energy.

IOP Conf. Series: Journal of Physics: Conf. Series **874** (2017) 012017 doi:10.1088/1742-6596/874/1/012017

3. Conclusions and future plans

At the end of 1990's at the SLAC LINAC – which indeed has been the model for the realization of the Frascati DAΦNE one – some studies for producing the longest possible, high-energy electron beam has been performed, in the framework to the experiment E-155.

By applying multiple phase inversion to the klystron power, a longer flat top in the SLED output can be obtained, in particular with a couple of additional phase inversions, a nearly constant, no-load accelerating voltage as long as 500 ns has been obtained, although at the price of a reduced beam charge (from 10^{11} down to $2.9 \cdot 10^9$ particles) and slightly higher energy spread [5-6].

Another possibility for getting even longer pulses is to modulate the low-level RF that is amplified by the klystron and fed to the SLED and from there distributed to the accelerating structure, in such a way to compensate the variation of accelerating voltage over a longer time. For instance, moving the beam pulse with even larger advance with respect to the flat top, a decreasing input RF power would compensate for the (larger) increase of voltage at the output of the SLED, of course at the price of a reduced energy gain.

References

[1] Farkas Z D, Hogg H A, Loew G A and Wilson P B 1974, *SLAC-PUB-1453*
[2] Decker F-J, Farkas Z D, Rinolfi L and Truher J 1996, *SLAC-PUB-7214*
[3] Valente P et al 2016, Linear Accelerator Test Facility at LNF. Conceptual Design Report *Preprint* arXiv:1603.05651 [physics.acc-ph] (2016); INFN-16-04-LNF
[4] Buonomo B 2015, *IPAC 2015* TUPWA056
[5] Decker F-J, Farkas Z D and Minty M G 1999, *SLAC-PUB-8111*
[6] Decker F-J, Farkas Z D and Turner J 1999, *SLAC-PUB-8113*

Acknowledgements

This work is supported by the H2020 project AIDA-2020, GA no. 654168.

Research of L-band disk-loaded waveguides travelling wave accelerating structures for a high power Linac

Y M Zhang[1,2], Y J Pei[2,3], L S Sheng[2] and Y F Song[2]

[1] School of Nuclear Science and Technology, University of Science and Technology of China, Hefei 230029, P. R. China
[2] National Synchrotron Radiation Laboratory, University of Science and Technology of China, Hefei 230029, P. R. China

Abstract. L-band Electron Accelerator has been widely utilized for industrial irradiation. In this paper, we designed a constant-impedance, disk-loaded structure which operates on 2π/3 mode. CST and SUPERFISH code were used for the design of bunching and accelerating cavities respectively. The geometrical parameters of the cavities were studied, and optimized RF parameters were obtained. We calculated the beam dynamics which presented that the electrons can be accelerated to 50 MeV. The model cavities have been fabricated and tested. Some valuable experimental results were obtained, which can provide a beneficial datum for the design and manufacture of L-band travelling-wave accelerating structures of 50 MeV LINAC.

1. Introduction

Electron linear accelerators have been widely used for industrial applications, such as cargo inspection, irradiation processing, food preservation sterilization and so on. [1] In order to meet the needs of the production of radioactive drugs via photonuclear reaction, it usually requires the accelerators to provide electron beams with higher energy and better stability. In order to get higher beam energy, L band travelling-wave accelerating structure is a good candidate which can also simplify the accelerating system.

In this paper, an L-band disk-loaded waveguides travelling wave accelerating structure for a high energy LINAC of 50 MeV was developed. SUPERFISH, a well-known RF design code [2], was used for the design of cavities and beam dynamics was simulated with MATLAB code. At the same time, model cavities have been fabricated and the experimental results were displayed.

2. Design of the accelerating structure

We designed a constant-impedance structure which operates on 2π/3 mode and the operating frequency is 1.3 GHz. This structure consists of 3 bunching cavities and 63 uniform accelerating cavities. (Shown in figure. 3) In order to accelerate the particles continuously, the length of each cavity must equal $\beta\lambda/3$ (λ refers the RF wave-length and β is electron traveling speed over light speed). The schematic diagram of the 1/4 section of the uniform accelerating cavity is shown in figure 1. In this figure, D refers the cycle length, 2a is beams aperture, t is the disk thickness, b is waveguide radius, d is disk interval and ρ is the disk hole edge radius. Subtle changes of the above dimension would affect the basic parameter of the accelerating structure. So, it's of great importance to get an accurate cavity dimension, which was calculated with two-dimensional SUPERFISH code.

[3] yjpei@ustc.edu.cn

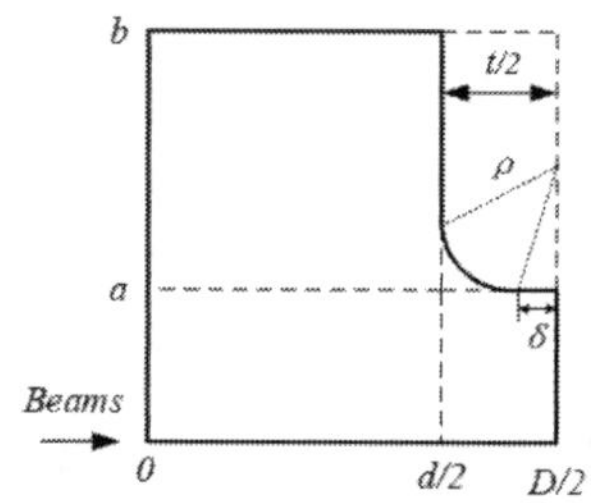

Figure 1. Schematic diagram of cavity dimension.

The dimensions of the uniform cavity was optimized in order to get better parameters of accelerating structure. The optimized dimensions are shown in table 1 and the relevant basic parameters of the accelerating structure are listed in table 2.

Table 1. The accelerating cavity parameters.

parameter	value(mm)
D	76.8699
d	69.8699
2a	40
2b	179.2242
t	7
ρ	3.6

Table 2. Basic parameters of accelerating structure.

parameter	value
Frequency（f_{rf}/GHz）	1.3
Quality factor (Q)	21333.0
Attenuation constant (α)	0.07802
Shunt impedance (R_s/MΩ·m^{-1})	51.58
Group velocity (v_g)	0.008216

In this paper, SUPERFISH code was used to calculated the effect of small changes in size of accelerating cavity dimension on basic parameters. It can be seen that the RF frequency changes linearly with the cavity dimension as the figure 2 shows and the changing rate were listed in Table 3. It can be seen that the waveguide radius b have the great influence on RF frequency and the effect of cycle length D and δ on f is relatively small in Table 3.

Table 3. RF frequency changing rate with different cavity dimensions.

Cavity dimension derivative	Changing rate (kHz/μm)	Standard error
$\partial f / \partial t$	0.69205	0.00142
$\partial f / \partial r$	-1.27936	0.00391
$\partial f / \partial a$	2.89212	0.00271
$\partial f / \partial \delta$	-0.33764	0.00801
$\partial f / \partial b$	-14.92903	0.00334
$\partial f / \partial D$	-0.27264	0.000302

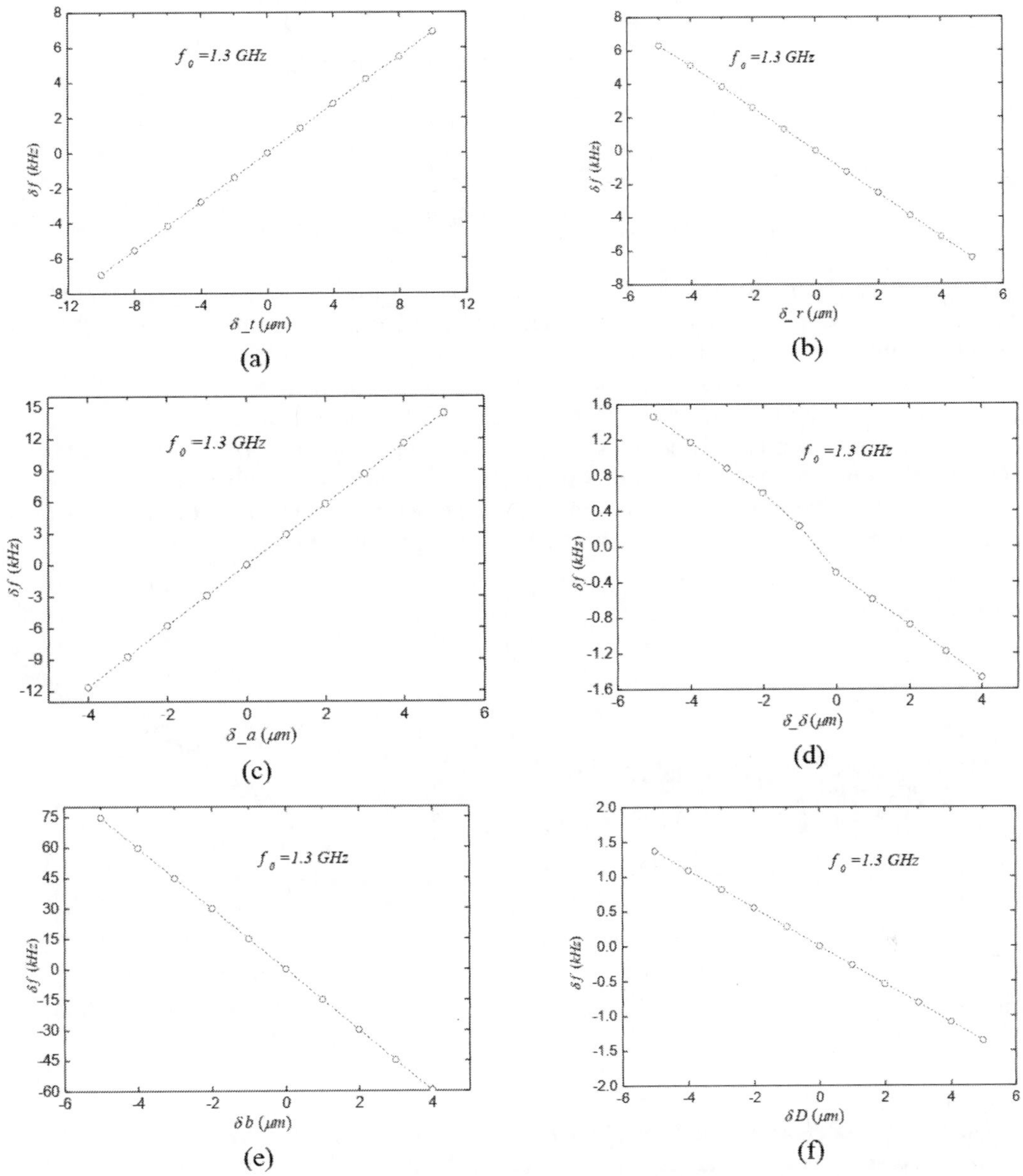

Figure 2. RF frequency changing rates with different cavity dimensions.

3. Beam dynamics

CST code was used to calculate the bunching cavity size, shunt impedance and quality factor etc. under different phase velocities, which can be used to simulate the bunching cavity parameters.

figure 3 shows the schematic diagram of the accelerating tube. The electron beam generated by the electron gun is accelerated in the tube. Different combination of β_φ in the bunching cavities can result in different parameters of beam dynamics. The beam loading effect was taken into account in the simulation of beam dynamics.

IOP Conf. Series: Journal of Physics: Conf. Series **874** (2017) 012018 doi:10.1088/1742-6596/874/1/012018

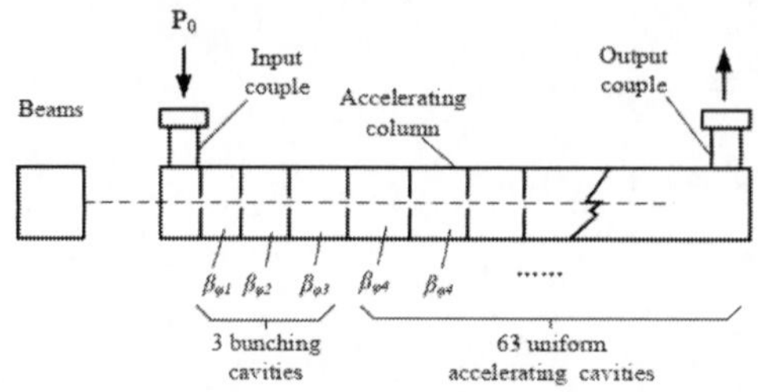

Figure 3. Schematic diagram of the accelerating tube.

The voltage of the electron gun is 50 kV and the input Power from the coupler is 10 MW. The beam current is design to be 0.3 A. In this paper, the parameters of the three bunching cavities and uniform accelerating cavities were set as follows:

$\beta_{\varphi 1}$=0.56 (bunching 1), $\beta_{\varphi 2}$=0.68 (bunching 2)

$\beta_{\varphi 3}$=0.95 (bunching 3), $\beta_{\varphi 4}$=0.9999 (63 uniform cavities)

The beams dynamics was calculated with a self-developed program on MATLAB platform and the results are shown in figures. 4 and 5. Figure 4 refers the phase orbits of electrons with different initial phase. Figure 5 refers the energy gain of electrons with different initial phase and it could reach 50 MeV and the relative energy spread is small, which meets our design requirements.

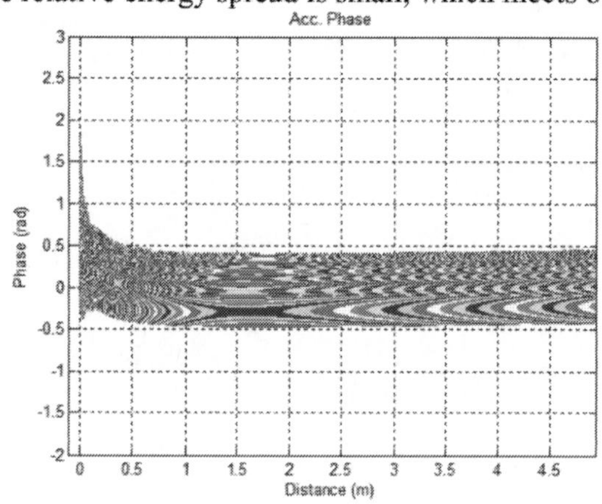

Figure 4. Phase along the accelerating cavity.

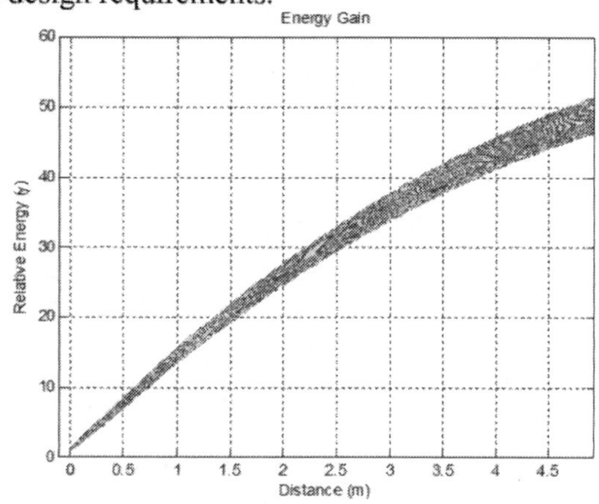

Figure 5. Energy gain along the accelerating cavity.

4. RF cold test

Four complete cells model cavities and two half-cells model cavities were fabricated in NSRL workshop and both were made up with Aluminum. The two half-cells model cavities were placed at the left and right sides of test cavity (as 1 and 4 in figure. 6). RF signal was fed into the cavity with a monopole antenna and the signal was analyzed with the network analyzer Agilent E5071C meanwhile. The monopole antenna and network analyzer was connected with N-type connector in this experiment. Figure 6 gives a brief drawing of the whole experiment platform and the physical map are showed in figure. 7.

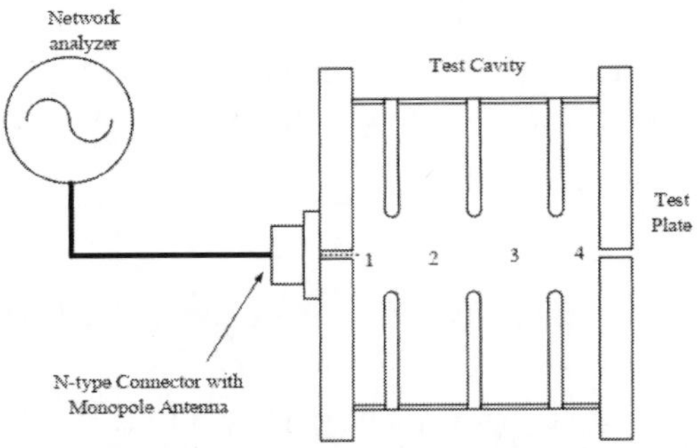

Figure 6. The experiment schematic drawing.

Figure 7. Physical drawing of the testing platform.

There exist $\pi/2$ model for one whole cavity and two half, $\pi/3$ and $2\pi/3$ models for two whole cavities and two half cavities. At the same time, there are $\pi/4$, $2\pi/4$, $3\pi/4$ for three and two half cavities. $\pi/5$, $2\pi/5$, $3\pi/5$ and $4\pi/5$ for four and two half cavities. Experimental test was performed under temperature 20°C and humidity 66% and the RF frequency results of different testing cavities (with different numbers of whole cavities) under this condition were showed in figure. 8.

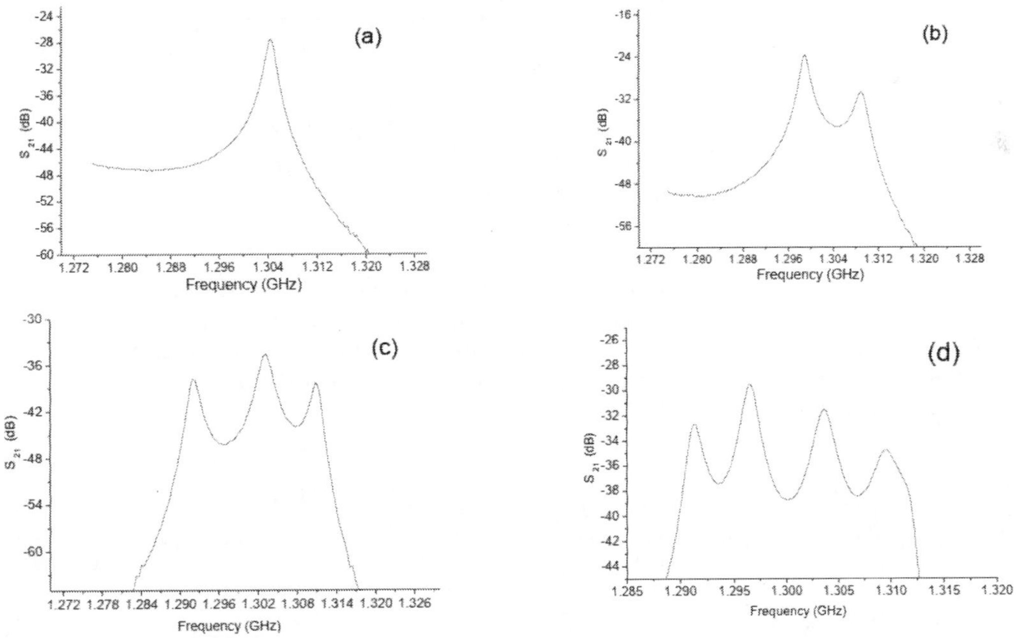

Figure 8. Testing result of cavities. (a) refers S21 curves of one and two half cavities; (b) refers two and two half cavities; (c) is three and two half cavities; (d) is four and two half cavities.

IOP Conf. Series: Journal of Physics: Conf. Series **874** (2017) 012018 doi:10.1088/1742-6596/874/1/012018

The cool test was conducted under atmosphere condition while the simulation was under the vacuum assumptions. The testing resonant frequency f_r and the vacuum resonant frequency f_0 was related in the following formula:

$$f_r = \frac{f_0 + \Delta f_T}{\sqrt{\varepsilon_r}} \tag{1}$$

The ε_r stands for the air relative permittivity which was defined as follows:

$$\varepsilon_r = 1 + 210\times10^{-6}\frac{P_d}{T} + 180\times10^{-6}\left(1 + \frac{3580}{T}\right)\frac{P_W}{T} \tag{2}$$

P_d and P_w were the pressure (the unit was Torr) that the dry air and the water vapor provide. $P_w=P_s\times H$, $P_d=760-P_w$ and the Ps was the humidity related saturated vapor pressure. The Δf_T is the frequency difference resulting from the difference between testing and working temperature, which was defined as:

$$\Delta f_T = \frac{\partial f}{\partial T}\bullet\Delta T \tag{3}$$

The temperature affected coefficient measured was -40 kHz/°C. On this condition, the frequency difference Δf_P out of the pressure difference is -0.413 MHz. the testing frequency of the network analyzer is 20 MHz and the related measured resolution Δf_S is ±0.0125 MHz.

The directly testing frequency f_t is 1.29905 GHz. Considering the above errors, the experimental RF frequency equal:

$$f = f_t - (f_P + f_s) = 1299.05 - (-0.413 + 0.0125) = 1299.4505 \text{ MHz} \tag{4}$$

Table 4. Comparison of calculated and experimental RF frequency results.

Model	Calculated results f_1/GHz	Testing results f_2/GHz	(f_1-f_2) /MHz	Relative deviation
2π/ 3	1.299 999 704	1.299 450 5	0.549 204	0.042%

In table 4, it can been seen that the difference between the calculated and experimental RF frequency results is very little (about 0.042%). The present work demonstrates numerical simulation can provide a beneficial datum for the design and manufacture of L-band travelling-wave accelerating structures.

5. Conclusions

In this paper, the bunching and accelerating cavities was designed, some experimental model cavities were fabricated and tested. At the same time, beam dynamics was calculated and the simulation have a great agreement with the experiment. Designing the structures systematically, fabricating the model cavities and making tests can provide an integrated and systematic method for the research and design of an L-band accelerating structure and fabrication of the L-band LINAC.

References

[1] Todd A M M 1998 *Proc. Intl. LINAC Conf.* (Chicago: USA) pp 23–28
[2] Halbach K and Holsinger R F 1976 *Proton Linac Conf.* (Ontario: Canada/Springfield) p 122.

IOP Conf. Series: Journal of Physics: Conf. Series **874** (2017) 012019 doi:10.1088/1742-6596/874/1/012019

Optimized undulator to generate low energy photons from medium to high energy accelerators

Ting-Yi Chung[1,a)], Mau-Sen Chiu[1], Hao-Wen Luo[2], Chin-Kang Yang[1], Jui-Che Huang[1], Jyh-Chyuan Jan[1] and Ching-Shiang Hwang[1]

[1]National Synchrotron Radiation Research Center, Hsinchu, Taiwan
[2]Department of Engineering and System Science, National Tsing Hua University, Hsinchu, Taiwan

[a)]Corresponding author: Chung.albert@nsrrc.org.tw

Abstract. While emitting low energy photons from a medium or high energy storage ring, the on-axis heat load on the beam line optics can become a critical issue. In addition, the heat load in the bending magnet chamber, especially in the vertical and circular polarization mode of operation may cause some concern. In this work, we compare the heat loads for the APPLE-II and the Knot-APPLE, both optimized to emit 10 eV photons from the 3 GeV TPS. Under this constraint the heat load analysis, synchrotron radiation performance and features in various polarization modes are presented. Additional consideration is given to beam dynamics effect.

1. Introduction

To pursue a high brightness hard x-ray source, it is customary to increase the electron beam energy and decrease the period length of an undulator. For maximum research flexibility broad band, high brightness synchrotron radiation (SR) from VUV to soft X-ray is desired as well in high and medium energy storage rings. To serve as a low energy photon source, an undulator must have a high strength parameter, which in turn results in a high radiation power made up of mostly undesired photons. The radiation power causes a heat load on the beam line optics while the wiggling of the electron beam increases the impact on beam dynamics. These are two critical challenges which we will address in this paper.

To solve the heat load issue one could increase the period length with a corresponding decrease of the magnetic field. The concept was implemented in the designs [1-3] with the additional advantage of a rapid change of the polarization. The other solution is to offset the peak radiation power from the axis, leading to a decrease of the heat load on the beam line optics [4]. This scheme transfers the heat load from the beam line optics to high heat load components in the front end while keeping a reasonable heat load on the ring vacuum chamber.

In addition to the requirement of low photon energy, variation of SR polarization is desired. An APPLE-II with a long period length is a feasible design choice at the expense of on-axis flux density. To be an adjustable polarization source with low on-axis radiation power and a high on-axis flux density an undulator, called Helical-8, was proposed and developed at Spring-8 [5]. The undulator requires magnetization of magnets at different angles and therefore relies on available manufacturing technology for permanent magnets. An alternative design, called Knot-APPLE [6], is based on the APPLE-II design and has been constructed for use at the SSRF [7]. The undulator consists of four Knot arrays outside and four APPLE arrays inside. Similar to the APPLE-II design, the polarization change of the Knot-APPLE also relies on mobile magnet arrays.

In this paper, we compare the performance of an optimized APPLE-II and Knot-APPLE to emit low energy photons at TPS. We organize this paper as follows. In Section 2, we discuss the heat load on the beam line optics for 10 eV photons with horizontal polarization. In Section 3, the heat load on the ring vacuum chamber will be discussed. The performance of SR with circular and vertical polarization is discussed in some detail in this section. In Section 4, the effect on beam dynamics will be discussed and a summary is given in Section 5.

2. Heat load on the beam line optics

The fundamental photon energy E_p, radiated from an undulator, is for K >>1, given by

$$E_p \propto \frac{E_e^2}{\lambda_u K^2} \tag{1}$$

E_e is the electron beam energy, λ_u is the period length of an undulator, K is the undulator strength parameter. E_p is proportional to the square of E_e, which means that an increased E_e is suitable for high photon energies, E_p but requires large values for K and λ_u to deliver low energy photons.

Additionally, a high photon flux density is desired which pushes the number of periods to higher and the period length λ_u, to lower values. In other words, pursuing a low energy photon beam with high flux density, the undulator should have a short period λ_u but a high strength parameter. This condition causes a heat load issue. The on-axis angular power density is given by

$$\frac{dP}{d\Omega}(0,0) \propto \frac{N \cdot E_e^2 \cdot K \cdot G(K)}{\lambda_u} \tag{2}$$

Where N is the number of periods, the function $G(K)$ is close to 1 for $K > 1$. In the case of the 3 GeV TPS, an APPLE-II with $\lambda_u = 124\ mm$ serving as a 10 eV photon source, the fractional power passing through a slit aperture of ±2sigma of photon beam divergence is up to 1.74 kW. The allowable heat load for the first optics elements in the beam line should not exceed 1 kW utilizing liquid nitrogen cooling. This means the heat load from this APPLE-II cannot be absorbed by the beam line optics.

A straightforward remedy would be to reduce N, but the angular flux density would then drop. For our example, the period number N must be reduced from 32 to 12 to reduce the power to 1 kW, but at the same time the angular flux density would drop by 84%. The second method is to increase the period length λ_u. Figure 1 shows that for λ_u =155 mm (APPLE155), the heat load becomes acceptable, while the angular flux density is only reduced by 16%.

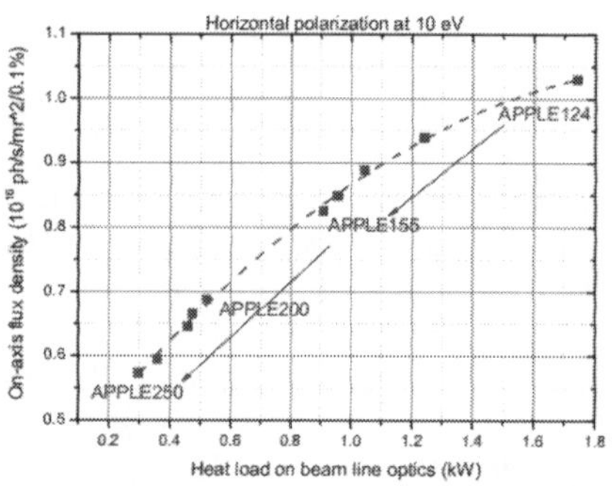

Figure 1. On-axis angular flux density versus heat load on beam line optics for different APPLE-II period lengths.

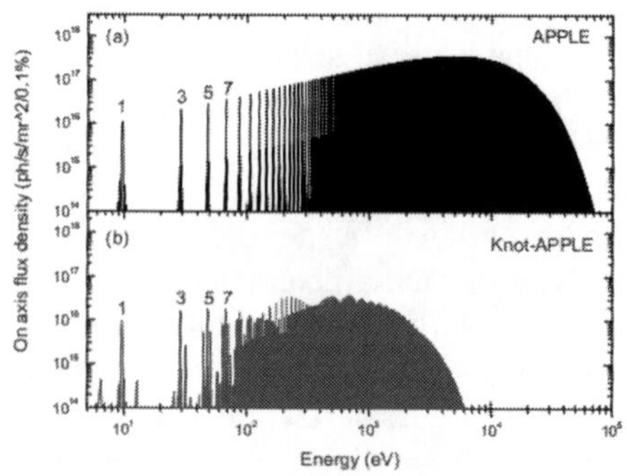

Figure 2. Comparison of the on-axis angular flux density spectra for the APPLE124 (a) and the Knot-APPLE (b), both with a period length of 124 mm. Indices show the harmonic number.

To understand the source of the heat load, Figure 2(a) shows the on-axis flux density of APPLE124. The desired photon energies can be obtained from the first and third harmonics. As seen from Figure 2(a) most of the radiation power comes from higher harmonics and must be absorbed by beam line optics elements. In the Knot-APPLE design, a horizontal field exists even in the horizontal polarization mode. This field deflects the electron beam vertically and with it the radiation power from the mid plane. Compared to the APPLE-II, Figure 2(b) shows that the on-axis angular flux density of the Knot-APPLE becomes significantly smaller above 1 keV, which explains why the Knot-APPLE results in a smaller heat load than a conventional APPLE-II. In the calculation, the period ratio of the Knot and APPLE arrays is 1.5. The period length of the APPLE array is still 124 mm.

We compare the characteristics of the APPLE124 and Knot-APPLE in Figure 3. Each of them produces a total power in excess of 10 kW, which, however, still can be handled by the front end. While the APPLE-II produces excessive power to the beam line, the power from the Knot-APPLE is reduced to 0.38 kW. The on-axis flux density of the first harmonic is also suppressed but only by 6%. Figure 4 shows the tuning curve of four different IDs serving as 10 eV photon sources. The Knot-APPLE not only produces the lowest heat load on the beam line optics, but also the highest angular flux density to the end user over the full operating range.

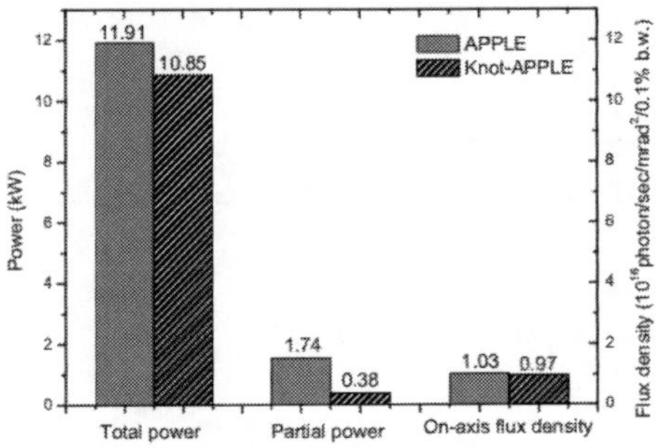

Figure 3. Comparison of APPLE124 and Knot-APPLE parameters. The bar chart shows the total power, the partial power and the on-axis flux density.

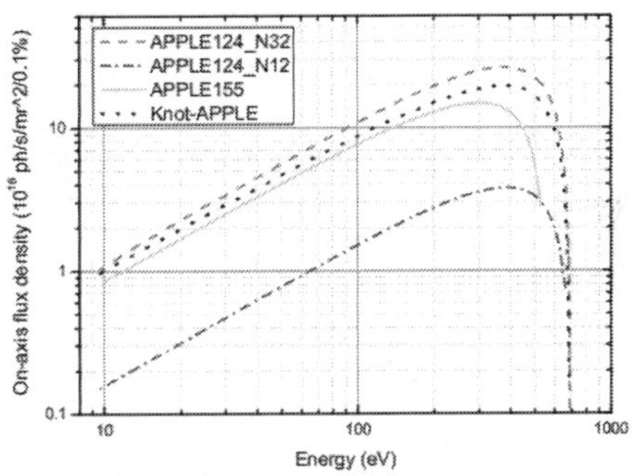

Figure 4. Tuning curve for four different IDs serving as 10 eV photon sources with horizontal polarization.

3. Heat load in the ring vacuum chamber

To decrease the partial power, the Knot-APPLE deflects the peak power from the axis due to the horizontal field generated in the magnet array. However, the limited vertical aperture in the vacuum chamber from undulator to the front end limits the horizontal deflection parameter K_x, because the radiation deflection is related to K. This limitation also applies to the APPLE-II and Knot-APPLE while operated in any non-horizontal polarization mode. In the case of TPS, a conservative value for K_x is 3.5. Under this condition, the allowable minimum photon energy at vertical and circular polarization modes for the APPLE-II and Knot-APPLE are determined. Figure 5 shows the gap-dependent of the fundamental energy and K for the Knot-APPLE and APPLE124. Both have a similar gap-dependent of K_x but the Knot-APPLE has a large effective field contribution from the Knot array leading to lower photon energies than the APPLE124. The effective undulator strength K_{eff} of APPLE124 is the same as K_x. Both sources have a nearly ideal degree of polarization at the fundamental energy. It is worth noting that the degree of polarization changes with energy except for the fundamental energy. If a user would like to use, for example, the third harmonic, the radiation turns out not to be purely polarized in the vertical plane. For circular polarization, the angular flux density of the Knot-APPLE at the fundamental energy is much lower than that of APPLE-II which has a longer period length 155 mm, as seen in Figure 6. This can be understood by the magnetic field distribution as shown in the insert of Figure 6. The magnetic field is still periodic but consists of multi-harmonics.

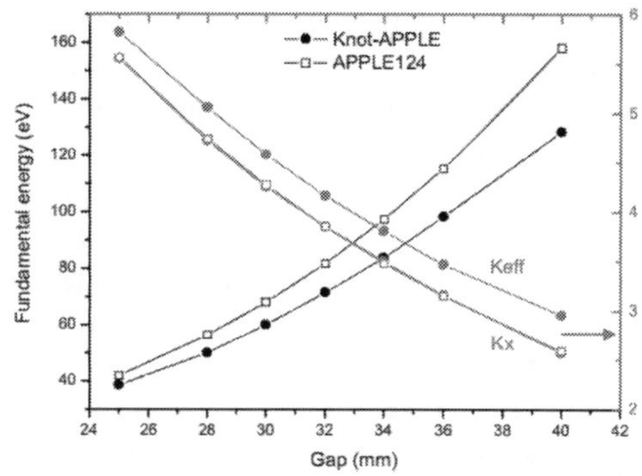

Figure 5. The gap-dependent of the fundamental energy and K. The Knot-APPLE (solid circle) and APPLE124 (open square).

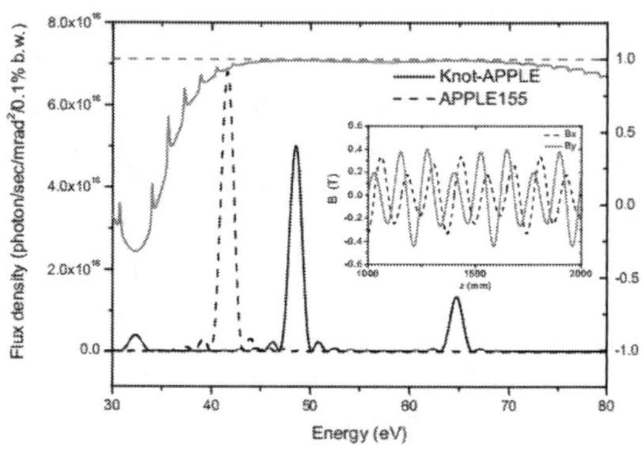

Figure 6. Angular flux density and degree of circular polarization for the Knot-APPLE (solid line) and APPLE155 (dashed line). The insert shows the magnetic field distribution in the Knot-APPLE.

4. Beam dynamics effect

The good field region should now also include the space needed for the wiggling of the electron beam. It is known that the APPLE-II has an inherently small uniform region especially for vertical polarization, resulting in an obviously second-order kick [8]:

$$\theta_{x,y}^{2nd}(x,y) = -\frac{1}{2(B\rho)^2}\frac{\partial}{\partial x,y}\int_0^L[\left(\int_0^{\lambda_u} B_x(x,y,s')ds'\right)^2 + \left(\int_0^{\lambda_u} B_y(x,y,s')ds'\right)^2]ds \quad (3)$$

$B\rho$ is the magnetic rigidity proportional to the beam energy, L is the insertion device (ID) length. The second-order kick becomes significant for a long ID, a large magnetic field, and a small electron beam energy. Due to the partial differential in Equation (3), a small uniform magnetic field gives rise to large higher order multipole errors, causing tune shifts and a reduction of the dynamic aperture. For the Knot-APPLE, the second-order kick was analysed using Equation (3) as shown in Figure 7. Compared to an APPLE-II of the same period length, the Knot-APPLE exhibits a similar magnitude for the second-order kick which contributes the most influence to the vertical polarization. To compensate the adverse effects, an active [9] or passive method [10] was demonstrated for the APPLE-II and is proposed for the Knot-APPLE as well.

Another part of the study aimed to evaluate the impact on the electron beam energy spread and emittance due to the synchrotron radiation integrals. For the TPS operating in a low emittance mode (non-achromatic mode), the Knot-APPLE decreases the energy spread by 0.4% but increases at the same time the emittance by 0.2%.

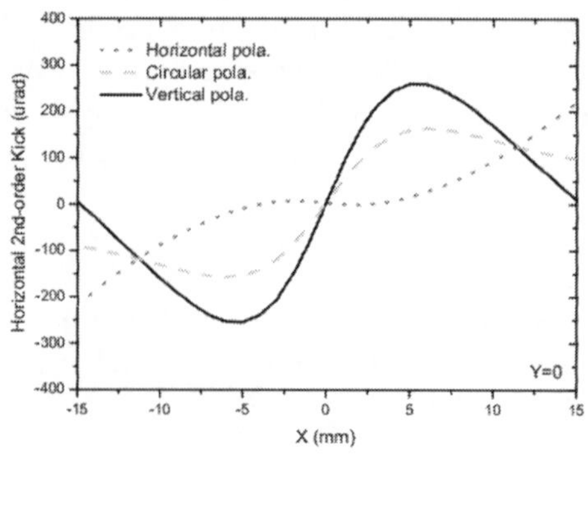

Figure 7. Horizontal second-order kick (Equation (3)) for the Knot-APPLE in the mid plane for horizontal (dotted), circular (dashed) and vertical (solid) polarization.

5. Discussion

To produce low energy photons from a medium or high energy storage ring, this study addresses the major issues for an optimized undulator, including heat load on the beam line optics, on the ring vacuum chamber and beam dynamics effects. The characteristics of SR of various polarizations are compared for the APPLE-II and Knot-APPLE. The Knot-APPLE generates the highest angular flux density with the lowest heat load on beam line optics. Even so, two somewhat negative characteristics of the Knot-APPLE must be noted. The first is that poor polarization can be obtained only within a small energy range. Second, the Knot-APPLE produces a non-uniform distribution of the heat load, which increases the risk of non-uniform mechanical distortion of beam line optics.

References

[1] Schmidt T *et al. Proc. EPAC2002* Paris (France) 2631-33.
[2] Marcouille O *et al.* 2007 *AIP Conf. Proc.* **879** 311-14; A. Batrakov *et al.* 2007 *AIP Conf. Proc.* **879** 396-99.
[3] Jaski M *et al. Proc. PAC2013* Pasadena (USA) 1064-66.
[4] Tanaka T and Kitamura H, 1995 *Nucl. Instrum. Methods Phys. Res.* A **364** 368-73.
[5] Tanaka T and Kitamura H, 2011 *Nucl. Instrum. Methods Phys. Res.* A **659** 537-42.
[6] Sasaki S *et al. Proc. PAC2013* Pasadena (USA) 1043-45.
[7] Ji F *et al.* 2015 *J. Synchrotron Rad.* **22** 901-07.
[8] Elleaume P, *Proc. EPAC1992* Berlin (Germany) 661-63.
[9] Bahrdt J *et al. Proc. EPAC2008* Genoa (Italy) 2222-24.
[10] Chung T Y *et al.* 2016 *Nucl. Instrum. Methods Phys. Res.* A **826** 48-54.

IOP Conf. Series: Journal of Physics: Conf. Series **874** (2017) 012020 doi:10.1088/1742-6596/874/1/012020

The experience of Taiwan photon source commissioning and operation

Y C Liu, C H Chen, J Y Chen, M S Chiu, P J Chou, C S Huang, Sam Fann, C C Kuo, T Y Lee, C C Liang, G H Luo, H J Tsai, F H Tseng
Synchrotron Radiation Research Center, Hsinchu 30076, Taiwan

E-mail: yichihl@nsrrc.org.tw

Abstract. TPS commissioning occurred between August 2014 and March 2016. The experience of phase I (bare lattice 2014.8~2015.3) and phase II (SRF and insertion devices 2015.9~2016.3) commissioning will be discussed. The Taiwan Photon Source (TPS) started user operation in March 2016 and delivery of user time has reached 3,211 hours in 2016. Continuous improvements of integrated accelerator performance and future developments are described and discussed.

1. Introduction

The TPS is a 3 GeV third generation light source. To diminish the difficulty and uncertainty from an extremely tight installation schedule, the TPS commissioning was divided into two phases, i.e., phase I [1] and phase II [2, 3]. In phase I, the main commissioning goal was to store beam, measure beam parameters and pursue vacuum cleaning to become ready for installation of two SRF cavities and gain enough experiences for TPS phase II commissioning. In phase II, 10 insertion devices, two SRF cavities and seven beamlines were installed and commissioned for tests of their performance. In order to ready beamline hutches for experimentations, a liquid nitrogen pipe, as well as beamline end stations were installed, and the regulation radiation dose limit was set to 2μSv per 4 hour interval to guarantee radiation safety for all people in the experimental hall during phase I and phase II beam commissioning.

The TPS started with 150 mA in top-up operation with a 1.3 % relative beam current stability on March 24. The stored beam was limited by a radiation hot spot in the area of the TPS-09 experimental hutch. After reinforcing radiation shielding in the beam collimator, radiation safety requirements of 2μSv per 4 hours were met. Then the stored beam current was gradually raised to 300 mA on May 26, 2016 in 50 mA increments per week since May 12. In order to minimize the impact of injection transients, we chose 0.66% beam current stability and 4 to 4.5 minutes filling interval while the beam lifetime was 9 to 11 hours [4, 5]. The top-up refilling process is triggered whenever the stored beam current has decayed from 302 mA to 300 mA. Due to radiation safety regulations, the stored beam current was raised eventually to 500 mA in three stages. The first stage occurred at 300 mA for meeting the existing operation license, later, when a new operating license was approved, the maximum current was raised to 400 mA and the final stage was 500 mA. Currently, the maximum current is set to 400 mA in top-up operation for the time being in 2017.

2. TPS commissioning experience

Booster beam commissioning started on August 12, 2014, but got delayed for a week due to a fire in the dipole power supply during AC mode testing. The resistor of the leakage current detector in the booster dipole PS was designed for 100 Ohm without sufficient safety margin, such that it heated up to more than 1000 °C and caused nearby circuits to burn during AC mode testing. The resistance of resistor was increased to 1 M Ohm and extra cooling fans were added to carry the heat away from the dipole power supply cabinet.

In the first step of booster commissioning it took three months without getting a stored beam at 150 MeV. Several improvements were implemented, including modification of the injection kicker field uniformity and post pulse residual from 2% to 0.4%, realignment of the booster vacuum chamber from ±6 mm to less than ±1 mm in the vertical plane, improvement of the chamber alignment by installing fixtures at both sides of dipole magnets and quadrupoles to ensure an accurate gap between chamber and magnets. Roughly 30000 turns of injected electron beam could be achieved as shown. Figure 1 shows the radiation dose distribution as measured at every booster dipole to possibly detect the most likely beam loss location.

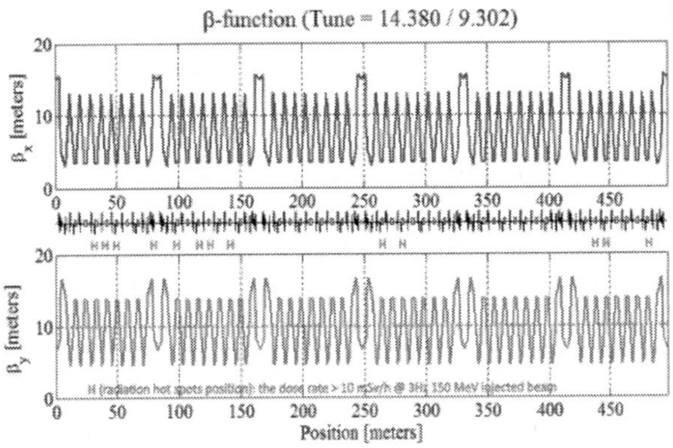

Figure 1. Booster beta functions in DC mode with the red H indicating the distribution of high radiation dose rates (> 10 mSv/h), large horizontal orbit (± 4 mm) was measured. Simulation with distorted magnetic fields indicated that linear optics were severely changed and dynamical aperture was very poor.

The booster vacuum chamber was found to be magnetic by touching the vacuum chamber with strong magnets on November 12, 2014. It took one month to uninstall all vacuum chambers, to anneal them at 1050 °C for demagnetization, reinstallation and final pump down. The injected beam survived then 50 ms on December 11, and after turn on of the RF system stored beam was achieved on December 12. The beam injected at 150 MeV was then successfully accelerated to 3 GeV on December 16 [6]. After replacing several silicon controlled pulser rectifiers for the injection septum, 3 GeV storage ring commissioning started on December 31, and a stored beam could be accumulated up to 5 mA only a few hours later.

In phase I commissioning, the maximum stored beam current reached 100 mA, limited by the maximum safe working power of the ceramic RF windows for the two 5 cell PETRA cavities. A beam dose of 35 A·hr was accumulated and the vacuum conditions improved well enough to consider replacement of the PETRA cavities with two SRF cavities.

To protect IDs and vacuum chamber from upstream synchrotron radiation due to miss-steering of the electron beam, an orbit interlock is required [7]. Checking orbit interlock criteria, it became clear that the 9 mm internal height of the B1 chamber was too small. Some of the synchrotron radiation from an upstream EPU48 (Elliptical Polarized Undulators) in vertical polarization mode was hitting the chamber and deposited a very high heat load (2.4kW on each side) on the B1 vacuum chamber at 500 mA. Four B1 chambers needed to be modified, and the modifications call for the vertical aperture to be increased from 9 mm to 18 mm. The new chambers have been replaced in Jan. 2016 [8]. The impedance of TPS storage ring was investigated that found the threshold of microwave instability was about 3 mA bunch current [9].

A local abnormal vacuum pressure problem occurred on October 4, when in the storage ring cell 2 the pressure increased exponentially once the stored beam current exceeded 180 mA. A vacuum leak check had been carried out several times before, including a helium leak check even with stored beam and crotch absorber replacement. However, no vacuum leak was found. Following an innovative plan to replace the cell 2 B1 chamber in late November 2015, the chamber was cut and welded in situ. A PVC pad and a small screw were found inside the B1 chamber. This in situ replacement successfully demonstrated a well-controlled dust contamination and a state of the art vacuum cleaning approach without requiring bake out. The stored beam current was raised to 520 mA on December 12, proving the performance of the two RF cavities and two SRF modules fully satisfying fulfil the power expectations for 500 mA stored beam current operation. Meanwhile, we observed a limit for high current operation that was caused by vacuum bursts inside the horizontal stripline kicker due to its large loss factor and a redesigned kicker was installed in January 2017. We pushed high current operation during the early stages of commissioning to get early signs of problems and giving time to plan for their solutions.

To compensate for linear optics perturbations caused by insertion devices (IDs) we employed orbit feedforward tables for all IDs as well as tune and coupling feedforward tables for three elliptical polarization undulator (EPU). The tune shifts introduced by EPUs are minimized by adjustment of local nearby quadrupole pairs, while the coupling introduced by EPUs is minimized by a pair of nearby skew quadrupoles. Figure 2 shows the horizontal and vertical orb bit shifts due to gap changes of seven In Vacuum Undulators (IVU) from 40 mm to 7 mm with and without feedforward and FOFB.

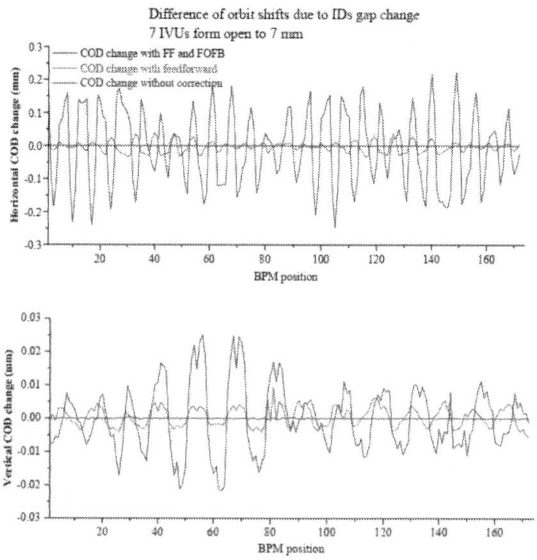

Figure 2. Difference of the orbit shift for 40 mm and 7 mm gaps of seven IVUs with and without feed forward and FOFB.

3. TPS user operation

Optimization of the TPS was pursued toward good reliability and high performance for both experimental use and beamline commissioning on March 24, 2016, just seven months after installation of the two SRF cavities and ten insertion devices.

Four standard 7 meter and three 12 meter long straight sections have been assigned to ten insertion devices reducing the six fold symmetry to a threefold symmetric lattice. A double minimum βy (DMB) lattice could preserve the original low emittance design to provide extremely high brilliance and high flux beamlines. The seven IVUs will allow a magnetic gap as small as 7 mm (3 IVUs located in 7 meter straight section have the potential to allow a magnetic gap of only 5 mm), while the two EPUs with 48 mm period length and one EPU with 46 mm period length have variable magnetic gaps down to 13 mm and 14 mm, respectively. The optical functions and horizontal emittance of the electron beam at the insertion device positions determine photon beam sizes and divergences. Lower horizontal emittance can, for example, provide a smaller source size, especially suited for nano focusing applications.

To achieve top-up operations with submicron electron orbit stability (1/10 beam size criteria), requirements of radiation safety must be fulfilled first, followed by completeness of the machine protection system, accelerator system capabilities and related reliability and finally integrated performance of the accelerator complex.

The electron orbit interlock has been merged with the machine protection system to protect IDs and vacuum chambers from radiation damage caused by miss-steered electron beams, at the cost of a reduction in accelerator availability. Any BPM failure can lead to a false action of the orbit feedback, or faulty BPMs close to IDs can activate the orbit interlock and dump the beam. It is crucial to mitigate false signals from critical components such as SRF modules, in vacuum undulators and BPMs, by delaying the response time or by reduction of the EMI noise. In this respect, the reliability of accelerator components have been tested and improved during beam commissioning.

There are two filling modes in top-up operation, 650 bucket (75 %) filling and 75 % filling with a 2 mA isolated single bunch in the center of the beam gap (hybrid mode). In hybrid mode, the bunch current of the single bunch is over 4 times higher than multibunch currents depending on the vertical chromaticity.

The injection efficiency is defined as the ratio of charge increase in the storage ring to the delivered charge from the booster. The injection efficiency gradually improved by matching optics of the transfer line to its design values at the storage ring entrance, fine tuning of launch conditions and 45 seconds running of the booster extraction pulser power supplies. With these efforts, the injection efficiency is between 80 % and 100 % due to small losses within a bunch train into different buckets.

A number of procedures are now in use or under test to ensure reliable performance of beamlines. A diagnostic beamline, including a visible light synchrotron radiation interferometer and an X-ray pinhole camera, is used to measure the beam size from a bending magnet. The electron beam emittance can be determined from optical functions at the source point and measured beam sizes. The horizontal and vertical emittances are measured to be 1.64 nm·rad and 15.7 pm·rad, respectively, while the emittance coupling ratio is measured to be 0.96 %. This result is in good agreement with the expected natural emittance of 1.6 nm·rad and 1 % coupling ratio. The linear optics of the storage ring is corrected by the widely used algorithm LOCO, "linear optics from closed orbit". LOCO is used to calibrate the linear optics and to maintain the symmetry of the lattice and working tunes. Figure 3 shows that the LOCO procedure should be performed every month to maintain reliable daily operation.

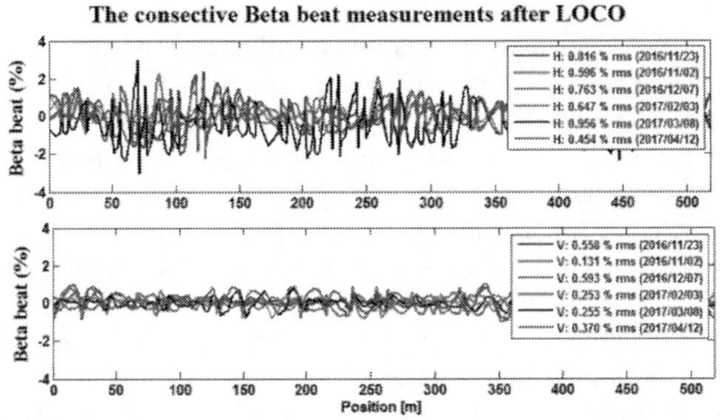

Figure 3. The consecutive beta beat measurements show that the LOCO procedure should be performed every month.

To understand the performance of the accelerator facility, it is essential to measure three key parameters of machine operation, like user beam availability, Mean Time between Failures (MTBF) and Mean Time to Recovery (MTTR). The scheduled user beam time for TPS was 3351 hours in 2016, and the delivered beam time was 3211 hours resulting in 95.82 % beam availability for the first year of user operation, which just meets a typically expected user beam availability. The MTBF is 52.35 hours, based on a total of 64 faults and 2.15 hours MTTR. Monthly statistics of user operation are shown in Figure 4.

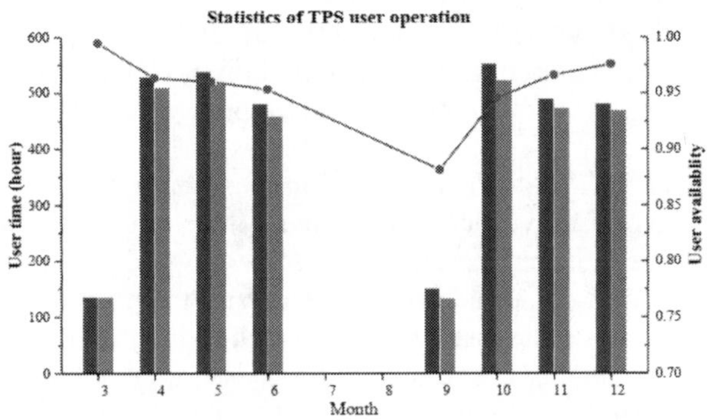

Figure 4. Statistics of monthly user beam availability.

4. Discussion

Incomplete system integration can cause many other problems. For example, a missing demagnetization step in the production of the booster vacuum chamber, foil damage incident of IU22B-09, synchrotron light from a miss-steered electron beam in the upstream IU22A-09, located in the 12 meter long DBM straight section. Many damage incidents can be prevented by more sophisticated quality control. After the foil damage event, we modified the operating procedures from floating target to fixed orbit interlock target, and activate first the orbit interlock before turning on the FOFB. Reducing the injection transients by Mu-metal shielding in the upstream vacuum chamber of the kicker magnet 2 and breaking the circuit for induced eddy currents in the injection section vacuum chambers, the horizontal and vertical injection transient oscillations could be reduced to 0.5 mm/0.1 mm from 5 mm/0.7 mm peak to peak value, respectively. Every problem that we found during TPS commissioning and operation drives us to improve the performance of TPS, and it can provide higher availability and better beam quality to user operation.

Acknowledgement
It is a pleasure to thank Professor S Y Lee for useful discussions, and Dr. Hitoshi Tanaka and the Spring-8 expert team for useful technical suggestions to reduce injection transients.

References

[1] Kuo C C *et al.* 2015 *Proc. International Particle Accelerator Conference (Richmond)* pp 1314-8
[2] Chiu M S *et al.* 2016 *Proc. International Particle Accelerator Conference (Busan)* pp 3360-2
[3] Tseng F H *et al.* 2016 *Proc. International Particle Accelerator Conference (Busan)* pp 3357-9
[4] Shoji Y *et al.* 2004 *The 14th Symposium on Accelerator Science and Technology (Tsukuba)*
[5] Lebasque P *et al.* 2008 *Proc. European Particle Accelerator Conference (Genoa)* pp 2183-5
[6] Tsai H J *et al.* 2015 *Proc. International Particle Accelerator Conference (Richmond)* pp 1741-3
[7] Safranek J *et al. 2004 Proc. Asian Particle Accelerator Conference (Gyeongju)* pp 39-43
[8] Sheng I C *et al.* 2016 *Proc. International Particle Accelerator Conference (Busan)* pp 3676-8
[9] Kuo C C *et al.* 2016 *Proc. International Particle Accelerator Conference (Busan)* pp 630-3

IOP Conf. Series: Journal of Physics: Conf. Series **874** (2017) 012021 doi:10.1088/1742-6596/874/1/012021

Parallel operation of SASE1 and SASE3 undulator sections of European XFEL

A A Sargsyan[1,3], V V Sahakyan[1] and W Decking[2]

[1] CANDLE Synchrotron Research Institute, Acharyan 31, 0040 Yerevan, Armenia
[2] Deutsches Elektronen-Synchrotron (DESY), Notkestraße 85, D-22607 Hamburg, Germany

[3] asargsyan@asls.candle.am

Abstract. In the current paper the numerical simulation results for parallel (decoupled) operation of SASE1 and SASE3 undulator sections of European XFEL are presented. The study was based on the idea of betatron switcher implementation. It was shown that it is possible to avoid energy spread growth in SASE1 and to reach the saturation in SASE3 in desirable range of radiation wavelengths by a trajectory kick before SASE1 and its correction before SASE3.

1. Introduction

At the current stage three undulator lines are installed in European XFEL: SASE1, SASE2 and SASE3. SASE3 undulator line is placed after SASE1 in the same electron beamline. All undulators are 5 m long and the spacing between the modules is 1.1 m. Undulators of SASE1 and SASE2 are identical. Each line consists of 35 modules, each having 40 mm period length and a gap variating in the range of 10-20 mm. SASE3 undulator line (with total length of 128.1 m) consists of 21 modules with a period length of 68 mm and a gap tunability range of 10-25 mm. The schematic layout of European XFEL undulator sections is presented in figure 1.

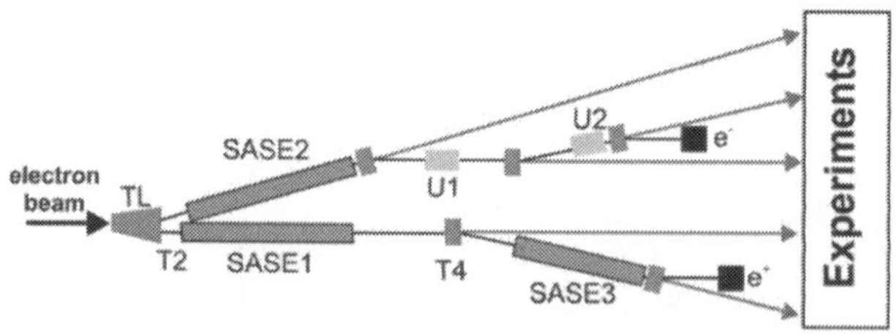

Figure 1. The schematic layout of European XFEL undulator sections.

For European XFEL operation four different energy modes are foreseen: 17.5 GeV, 14 GeV, 12 GeV and 8.5 GeV. The operating range for bunch charge is from 20 pC to 1 nC, the peak current is 5 kA, and the normalized rms emittance is between 0.3 and 1 mm-mrad, depending on bunch charge. The tunability range of undulator parameter is 1.65 - 3.9 and 4 - 9 for SASE1/2 and SASE3, respectively. Thus, SASE1 and SASE2 will cover the range of radiation wavelengths from 0.04 to 0.62 nm, and SASE3 - from 0.27 to 5.17 nm.

The usage of the electron beam, coming from SASE1, puts limitations on SASE3 operation. The general problem is that the lasing process in SASE1 causes high-induced energy spread in the electron beam which leads to degradation of SASE3 performance. Thus, the operation of SASE3 as an afterburner of SASE1 is possible with a reduced range of accessible wavelengths and radiation energy. Decoupling of SASE1 and SASE3 operation can be performed by the application of betatron switcher proposed by R. Brinkmann, E. Schneidmiller and M. Yurkov [1-2]. The idea is to use a fast kicker in front of SASE1 undulator line and give angular kicks to a part of bunches. Trajectories of kicked bunches strongly deviate from the straight path and as a result these bunches do not lase in SASE1, thus, avoiding the increase of energy spread. The kick is compensated before SASE3 by using two corrector magnets. After this bunches go straight and lase to saturation in SASE3.

In this paper the results of numerical study for SASE1 and SASE3 parallel operation, based on betatron switcher implementation, are presented.

2. Numerical simulations

In the numerical study ASTRA [3] output beam files before TL section (figure 1) were used. First of all, they were converted to ELEGANT [4] input files. Then particles were tracked through TL and T2 sections, where they are kicked by one of two available fast kickers (K1 or K2). Before entering the first undulator of SASE1 the files of the kicked beams were converted from ELEGANT output into GENESIS [5] input. Further, after GENESIS time-dependent FEL simulations for SASE1, they were converted back into ELEGANT files and tracked through T4 section, where the corrections of trajectories were made. Before entering the first undulator of SASE3, the beam files once again were converted into GENESIS input files and then were used for FEL simulations in SASE3.

In the study electron beams with 1 nC bunch charge were used. In table 1 the considered values of beam energy and radiation wavelengths are given.

Table 1. Beam energy and radiation wavelength.

Beam energy (GeV)	λ_{SA1} (nm)	λ_{SA3} (nm)
17.5	0.1	0.4
14	0.23	0.4
8.5	0.62	1.1

In simulations, the boundary values of wavelengths were considered according to undulator K parameter's possible values in SASE1 and SASE3, since on the one hand, in longer wavelengths the lasing goes faster and it is harder to disturb it by an induced betatron motion in SASE1, and on the other hand in shorter wavelengths it becomes harder to reach saturation in SASE3. In other words, the most difficult cases for realizing SASE1 and SASE3 decoupled operation were considered.

2.1. Conversion Scripts

During simulations beam files were converted several times between simulation programs. For these conversions SDDS toolkit was used and several MATLAB scripts were developed. For the conversion from GENESIS output into ELEGANT input the written MATLAB script takes into account different treatments of macro-particles in these programs. In GENESIS output file the beam is presented in the form of slices, and all slices have the same number of macro-particles, but different currents, which is not the case for ELEGANT file, where all macro particles are identical. Taking all macro-particles from GENESIS output would lead to the loss of beam current profile, and it would become impossible to perform correct FEL simulations for SASE3. For solving this disagreement and retaining the current profile after conversion, the written script takes not all macro-particles from *i*-th slice, but randomly chooses (since distributions of particle transvers coordinates within slices are uniform) an amount according to

$$N_i^T = N_i \cdot \frac{I_i}{I_{\max}}$$

IOP Conf. Series: Journal of Physics: Conf. Series **874** (2017) 012021 doi:10.1088/1742-6596/874/1/012021

where N_i is the number of macro particles in *i*-th slice, I_i is the current of *i*-th slice, and I_{max} is the maximal value of slice currents.

3. Simulation results

In this section the results of numerical simulations for 17.5 GeV, 14 GeV and 8.5 GeV operating energy cases are presented.

For 17.5 GeV beam energy case the kick value equal to 2, 4 and 6 μrad at K1 fast kicker and 6 μrad at K2 were considered. In figure 2 GENESIS time dependent simulation results for SASE1 are presented. It can be seen that for 2 μrad kick value at K1 the radiation energy is reduced but the saturation remains. For 4 and 6 μrad kick values the radiation energy becomes significantly lower and the lasing process in SASE1 is destroyed. In these considered cases the maximum deviation of beam centroid is less than 280 μm. Note that if 6 μrad kick is given at K2 fast kicker, the destructive effect on lasing process in SASE1 is as strong as when the same kick is given at K1, but the maximum deviation of beam centroid is 21% less, which makes the second kicker more preferable in point of view of resistive wall wakefields.

In figure 3 the beam energy spread profile after SASE1 is shown where one can see that when the kick value increases, the energy spread becomes lower. In particular, when the kick value is 6 μrad at K1 it has almost the same profile as for input beam, which points out a complete destruction of lasing in SASE1.

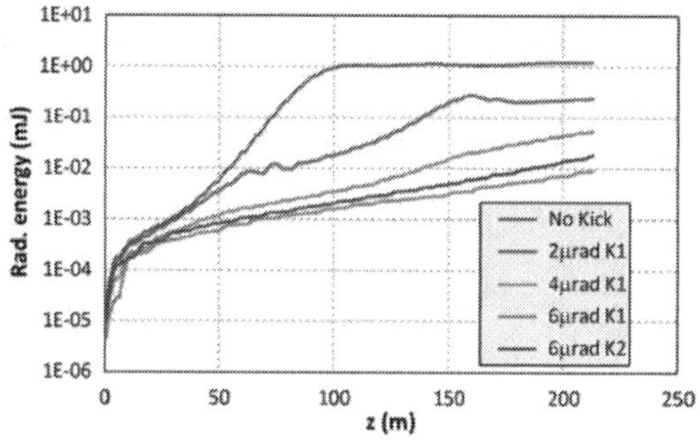

Figure 2. Radiation energy along SASE1.

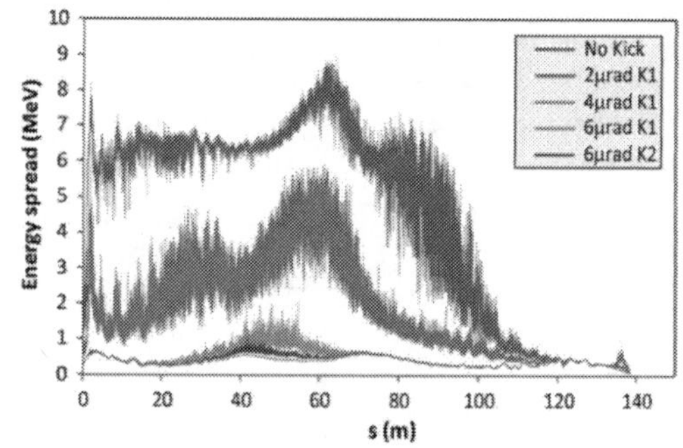

Figure 3. Beam energy spread profiles after SASE1.

In figure 4 the radiation energy along SASE3 is presented. It can be seen from the plot that in case of undisturbed bunches the energy spread growth in SASE1 led to a significant reduction of radiation energy in SASE3. On the other hand, the realization of the described scheme for 4 and 6 μrad kicks at K1 and for 6 μrad kick at K2 resulted in the growth of radiation energy almost in the same way as if SASE1 would be off for an undisturbed beam.

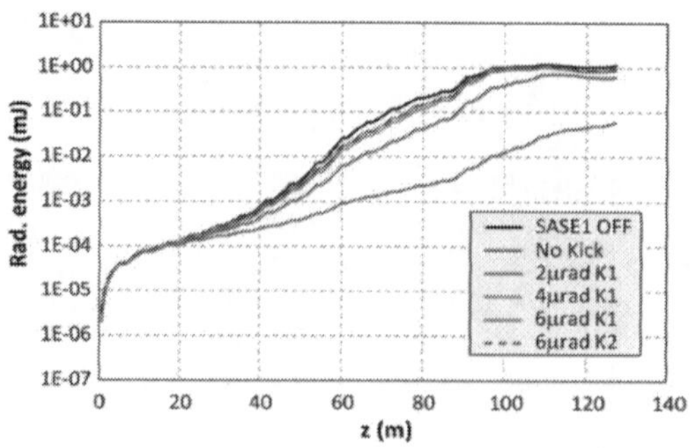

Figure 4. Radiation energy along SASE3.

For 14 GeV beam energy case 4, 6 and 8 μrad kick values at K1 fast kicker and 8 μrad at K2 were considered. Here, except for 4 μrad kick case at K1, the radiation energy becomes significantly lower and the lasing process in SASE1 is destroyed. For the considered kick values the maximum deviation of beam centroid was less than 360 μm. As simulations showed the growth of radiation energy along SASE3 in case of 8 μrad kick value at K1 is almost the same way as for SASE1 off case.

In 8.5 GeV beam energy operation mode, when the radiation wavelength in SASE1 is longer, as it was expected, it was harder to destroy the lasing process by inducing betatron oscillations. 8 μrad kick value at K1 and K2 were considered. In both cases the saturation is maintained and the energy at saturation is not significantly changed (see figure 5).

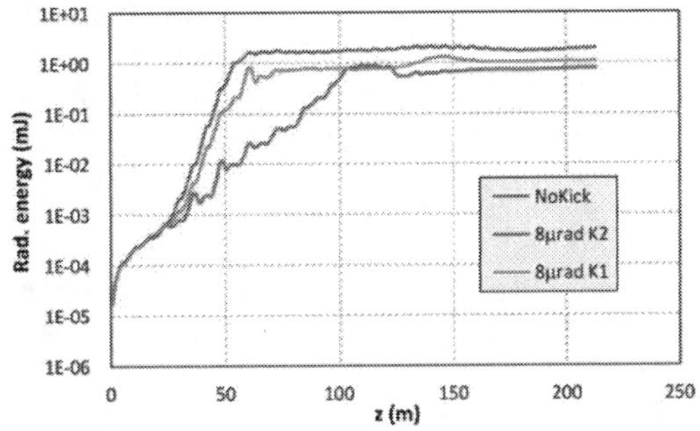

Figure 5. Radiation energy along SASE1.

Bigger kick values were not considered because of undesirable beam centroid large deviations they would cause. Hence, in this mode reasonable kick values are not able to make the scheme work. Two solutions were considered: 1) turn off the part of undulators which are located after the saturation point of undisturbed beam, 2) increase the number of betatron oscillations by decreasing the average beta function in SASE1. In case of the first approach it turned out that 8 μrad kick by the second fast kicker is enough to get satisfactory result in SASE3. As a second solution, the average value of beta function in SASE1 was changed from 33 m to 17 m, which doubled betatron oscillations that the beam centroid performs. As compared to 33 m average beta case, here, in the same range of kick values, the radiation energy in SASE3 is higher. However, these results are worse compared with the results given by the first solution.

Simulation results for the considered preferable kick values are summarized in table 2, where trajectory maximum deviations in SASE1 and relative changes of saturation length and radiation energy at saturation point in SASE3 are given.

Table 2. Trajectory maximum deviation in SASE1 and relative change of saturation length and radiation energy at saturation for the considered preferable kick values.

Beam energy (GeV)	λ_{SA1} (nm)	λ_{SA3} (nm)	Pref. kick value	Traj. max. dev. in SASE1 (μm)	SASE3	
					$L_{sat}/L_{sat}^{(SA1off)}$	$E_{sat}/E_{sat}^{(SA1off)}$
17.5	0.1	0.4	4μrad at K1	173	1.04	0.91
14	0.23	0.4	8μrad at K1	352	1.02	0.78
8.5 (β_{av}= 33m) short SASE1	0.62	1.1	8μrad at K2	300	1.01	0.98
8.5 (β_{av}= 17m)			8μrad at K2	227	1.78	0.45

4. Summary

In the current paper the parallel (decoupled) operation of SASE1 and SASE3 undulator sections of European XFEL has been studied. It was shown, that by inducing betatron oscillations in SASE1 undulator line, by using fast kickers of section TL and correcting beam trajectory before SASE3 with two corrector magnets of section T4, it is possible to avoid energy spread growth in SASE1 and to reach saturation in SASE3 in the desirable range of radiation wavelengths. The cases of 17.5, 14 and 8.5 GeV beam energy were considered.

For 17.5 GeV beam energy case the radiation wavelength was assumed to be 0.1 nm in SASE1 and 0.4 nm in SASE3. Different kick values at the positions of two fast kickers, located in section TL, were examined. Numerical simulation results showed that kick values bigger than 4 μrad are enough to destroy the lasing process in SASE1 and provide saturation in SASE3.

The same investigation was done for 14 GeV energy case, considering the longest possible radiation wavelength in SASE1 (0.23 nm) and the shortest wavelength in SASE3 (0.4 nm). Here it was shown that 8 μrad kick by the first fast kicker brings almost to the same result for radiation energy in SASE3 as in the case when SASE1 is off for an undisturbed beam.

In 8.5 GeV energy mode 0.62 nm and 1.1 nm boundary values of radiation wavelengths in SASE1 and SASE3, respectively, were considered. The results showed that in this case reasonable kick values are not able to make the scheme work, necessitating the implementation of additional solutions such as shortening SASE1 undulator line (opening undulators after the saturation point of undisturbed beam) or lowering the average beta function.

Acknowledgments

Authors would like to express their gratitude to Dr. G. Feng for helpful discussions and provided ASTRA beam files.

References

[1] Brinkmann R, Schneidmiller E and Yurkov M 2010 *Nucl. Instr. Meth. In Phys. Res.* A **616** 81-7
[2] Schneidmiller E and Yurkov M 2014 *Proc. of FEL'14* Basel Switzerland **MOP066** 204-9
[3] Floettmann K *ASTRA* http://www.desy.de/~mpyflo/
[4] Borland M *ELEGANT: a flexible SDDS-compliant code for accelerator simulation* 2000 Advanced Photon Source LS-287
[5] Reiche S 1999 *Nucl. Instr. Meth. In Phys. Res.* A **429** 243-8

Beam loss simulations for the implementation of the Hard X-Ray Self-Seeding system at European XFEL

Shan Liu, Winfried Decking, Lars Fröhlich

Deutsches Elektronen-Synchrotron (DESY), Notkestraße 85, 22607, Hamburg

shan.liu@desy.de

Abstract. The European XFEL is designed to be operated with a nominal beam energy of 17.5 GeV at a maximum repetition rate of 27000 bunches/second. The high repetition rate together with the high loss sensitivity of the undulators raises serious radiation damage concern, especially for the implementation of the Hard X-ray Self-Seeding (HXRSS) system, where a 100 μm thick diamond crystal will be inserted close to the beam in the undulator section. Since the seeding power level highly depends on the delay of the electron beam with respect to the photon beam, it is crucial to define the minimum electron beam offset to the edge of the crystal in the HXRSS chicane. At European XFEL a ~200 m long post-linac collimation section has been designed to protect the undulators. In the HXRSS scheme, however, beam halo particles hitting the crystal can generate additional radiation. Particle tracking simulations have been performed using GEANT4 and BDSIM for the undulator and the collimation section, respectively. The critical number of electrons allowed to hit the crystal is estimated for a certain operation mode and the efficiency of beam halo collimation is investigated to predict the minimum HXRSS chicane delay.

1. Introduction

The Hard X-ray Self-Seeding (HXRSS) is a well-known scheme to increase the X-ray longitudinal coherence [1]. At European XFEL [2], the HXRSS will be first equipped at the SASE2 (3-25 keV) photon beam line. In order to reduce the heat load on the 100 μm thick diamond crystal a 2 cascades system will be implemented (see Figure 1). This scheme can also significantly increase the signal to noise ratio of the photon spectrum, thus generating Fourier transform-limited X-ray pulses [3].

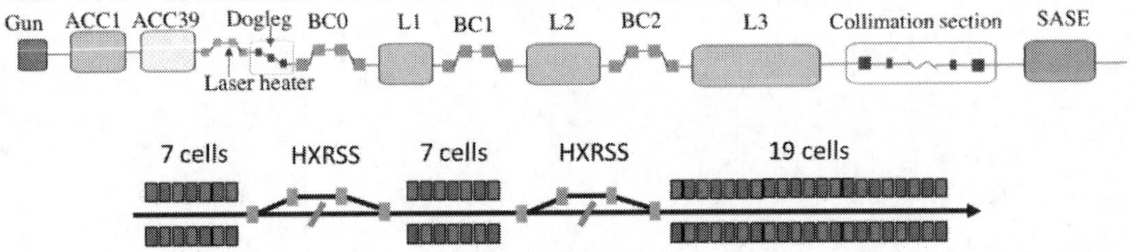

Figure 1. Schematic layout of European XFEL beam line (top) and HXRSS in SASE2 (bottom).

One of the most important parameters, which determine the performance of HXRSS, is the seeding power level. The seeding power level highly depends on the delay of the electron beam with respect to the photon beam (i.e. the horizontal offset of the e- beam in the chicane). With a delay increase of 10 fs (~0.5 mm increase of offset), the seeding power can decrease by a factor 2 [4]. Since the crystal is inserted to "filter" the frequency of the photon beam, the electron beam offset also refers to the distance of the electron beam center to the crystal edge, which is critical for the radiation protection (beam halo particles hitting the crystal can generate additional radiation). At LCLS and SACLA, where the HXRSS has already been implemented, the minimum distances from the crystal edge to the beam are set to be ~2.5 mm, which permits a seeding with ~20 fs delay [5, 6]. For European XFEL, however, due to the much higher repetition rate[1], there is more concern about radiation damage. Therefore, it is crucial to study the crystal insertion position limit.

At European XFEL, a ~200 m long post-linac collimation section has been designed to protect the undulators. This section is designed as a second order achromatic and first-order isochronous section to satisfy the requirement of an energy acceptance of ±1.5% of the nominal energy. Particle tracking simulations through the collimation section had been performed during its design [7]. The simulation results showed that the R=3 mm aperture in the collimation section is enough to protect an undulator vacuum chamber with 3 mm radius. However, in these simulations, secondary particles generation and scattering were not included, and the collimators were considered as "black absorber" (i.e. any particle touching them is considered as lost). In the HXRSS scheme, however, beam halo particles, which lost only a small fraction (<1.5%) of their energy in the collimation section (e.g. by scattering), may still be transported to the undulator section and hit the crystal.

Therefore, for the HXRSS implementation, we first performed GEANT4 [8] simulation with simplified geometry in the undulator section to estimate the critical number of beam halo particles allowed to hit the crystal. And then we used BDSIM[2] [9] to track a uniformly distributed beam halo particle distribution through the collimation section to study the maximum number of beam halo particles that are able to reach the crystal.

2. Estimation of critical number of beam halo particles

High energy electrons hitting the diamond crystal can generate high energy photons which can in turn generate photonuclear process in different materials downstream of the HXRSS system. Fast neutrons generated in photonuclear processes are considered to be the main source of demagnetization of permanent Nd-Fe-B magnets [10]. GEANT4 simulations have been performed to estimate the maximum neutron fluence generated by the 17.5 GeV e- beam hitting the 100 μm diamond crystal.

A simplified geometry is applied in GEANT4, as shown in Figure 2, with the aperture of each component shown on the right plot. Quadrupoles, chicane magnets, absorbers, undulator magnet and poles, vacuum chambers and the diamond crystal are included in the geometry with the same longitudinal dimensions as in the design of the HXRSS and undulator sections. The quadrupoles and bending magnets are approximated as iron boxes with a cylindrical hole in the center.

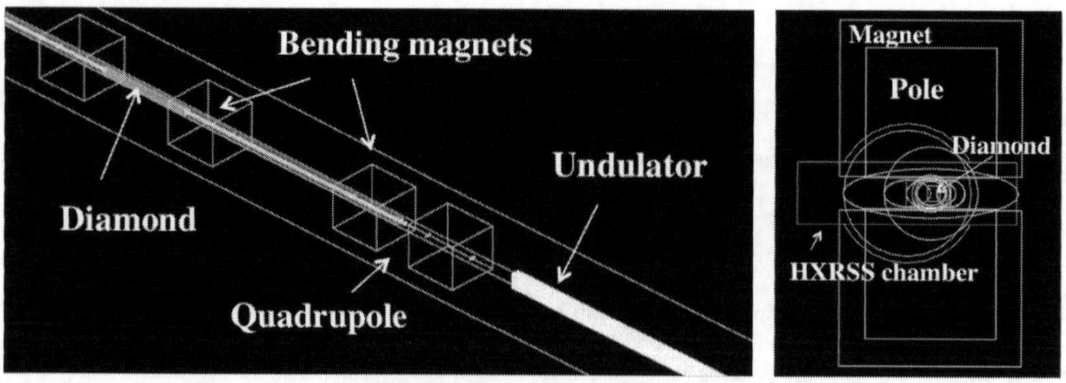

Figure 2. GEANT4 geometry layout (left) and apertures of different components (right).

Physics list QGSP_BERT_HP is used in the simulation. This list includes a high precision neutron package that can transport the neutrons below 20 MeV down to thermal energies. A biasing factor of 50 is applied to the Bremsstrahlung production inside the diamond crystal and to the photonuclear cross section in all the materials[3].

The simulation starts at the 1st HXRSS section (which is located in the 8th undulator section) followed by 7 undulators plus a 2nd HXRSS section and another 19 undulators (see Figure 1). For the first simulation, 10^7 electrons are generated, which hit the crystal with normal incidence. The simulation results are shown in Figure 3.

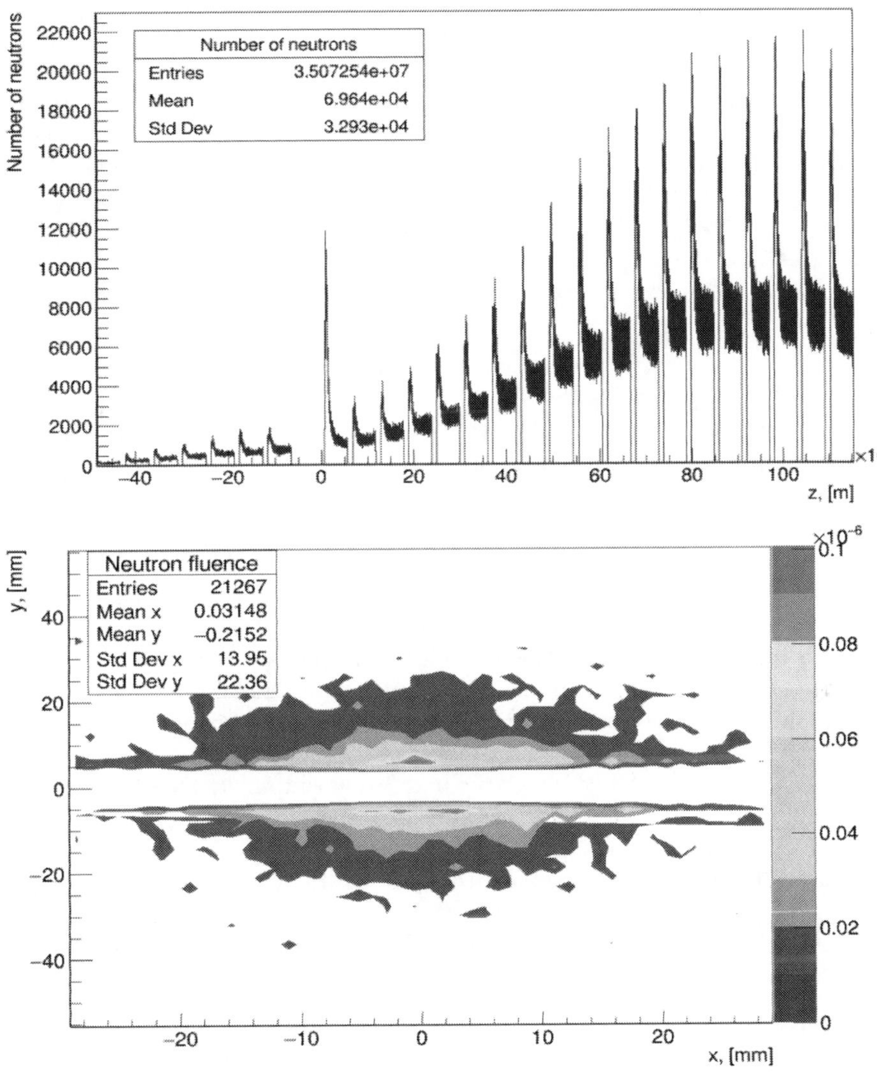

Figure 3. Number of neutrons along the undulators starting from the first HXRSS section (top) and transverse distribution of the neutron fluence at the location with maximum number of neutrons (bottom).

[1] The maximum repetition rate is 4.5 MHz at European XFEL, 120 Hz at LCLS and 60 Hz at SACLA.
[2] BDSIM is a toolkit of GEANT4 and it extends the ability of GEANT4 to allow for fast geometry building and fast tracking with design optics.

The top plot in Figure 3 shows the histogram of the number of neutrons ($N_{neutrons}$) along the beam line. Note that the 0 position is set to the beginning of the undulator section right after the 2nd HXRSS section, where a large jump of $N_{neutrons}$ can be observed. Similar spikes can be found at the beginning of each undulator section, since at this location the undulator magnets and poles are exposed directly to the radiation shower generated by the quadrupoles in the inter-sections between two undulators. The following magnets and poles are, however, more and more protected by the "self-shielding". Thus, we observe a decrease of $N_{neutrons}$ after the spikes. $N_{neutrons}$ reaches maximum at around 134 m (~22nd undulator sections) downstream of the 1st HXRSS section. This value is quite close to the estimated typical distance of bremsstrahlung impact (L) based on the characteristic angle (θ_e) of emission for thin-target bremsstrahlung: $L = R/\tan\theta_e = \frac{R}{\tan\left(\frac{m_e}{E_e}\right)} \approx 137$ m, where $R = 25$ mm is the radius of the vacuum chamber, and E_e is the electron beam energy [10].

Figure 3 bottom plot shows the transverse distribution of the neutron fluence at the location with maximum $N_{neutrons}$ along the undulator beam line (beginning of the 22nd undulator). It can be seen that the maximum neutron fluence is located in the center of the magnet with the value of ~10^{-7} n/cm^2/e$^-$. The maximum neutron flux allowed for 0.01 % demagnetization[4] of Nd-Fe-B magnets is measured experimentally in [11] to be ~$1\text{x}10^{11}$ n/cm^2. Therefore, if we assume 0.01% demagnetization of the magnets in 20 years with 10 shifts (8 hours each) /month for HXRSS operation (i.e. $6.912\text{x}10^7$ seconds), the maximum allowed number of e$^-$/bunch (with 27000 bunches/s) can be calculated as:

$$N_{critical} = \frac{10^{11}}{6.912\times10^{7}\times27000\times10^{-7}} \approx 5\times10^{5}\ \text{e}^{-}/\text{bunch} \quad (1)$$

Comparing with the maximum charge per bunch: N_{total} =1nC ≈ $6.25\text{x}10^9$ e-/bunch, the maximum allowed number of e-/bunch is around 10^{-4} of N_{total}[5]. This number will be further compared with the maximum number of beam halo particles that may hit the crystal estimated in the following section.

3. Tracking simulation in collimation section

Tracking simulations with ±50 sigma beam halo particles uniformly distributed in the (x,xp,y,yp) 4D phase space [13] have been performed in BDSIM[3]. The beam halo particles distribution is matched to the design optics at the beginning of the collimation section and tracked through the section using BDSIM. Four main collimators made of titanium with inner radius of 2 mm and three supplementary collimators made of aluminium with inner radius of 10 mm are used in the simulation[6].

The energy loss map along the collimation section with the design optics and the locations of the collimators are shown in Figure 4. Since the collimation section is designed as a "dogleg", the secondary photons and neutrons can not reach the undulator section. On the other hand, the electrons that lost only a small fraction of their energy (e.g. due to scattering on the collimator or chamber wall) can still be transported up to the undulator section.

Figure 5 (a) and (b) shows the histogram of energy distribution of the primary and secondary electrons at the end of the four main collimators with 10^5 input electrons. One can see that at COLM#1 and COLM#3 more low energy secondary electrons are produced than at COLM#2 and COLM#4.

[3] In GEANT4 the production cuts for the e$^-$, e$^+$ and γ are set as 0.1 μm in the crystal, 1 μm in the undulator region and 0.7 mm in the other regions. In BDSIM cut of 0.7 mm is used for all particles in all regions.
[4] This is the maximum allowed demagnetization to preserve FEL performance without retuning the undulators [11].
[5] This is a rough estimation, since no transverse position or angular distributions of the e$^-$ are taken into account and the electrons are considered as monochromatic. Besides, note that GEANT4 can underestimate the photon neutron production by ~30% according to Ref. [12].
[6] The main collimators can be moved vertically to change the apertures (2 mm, 3 mm, 4 mm and 10 mm are available), while the apertures of the supplementary collimators are fixed as 10 mm. For tracking convenience, the horizontal and vertical coordinates are exchanged.

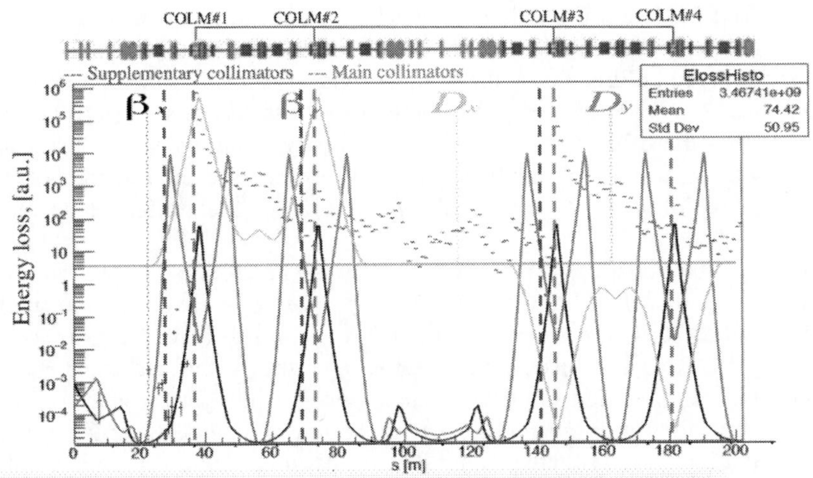

Figure 4. Energy loss map and the design optics (not to scale) along the collimation section.

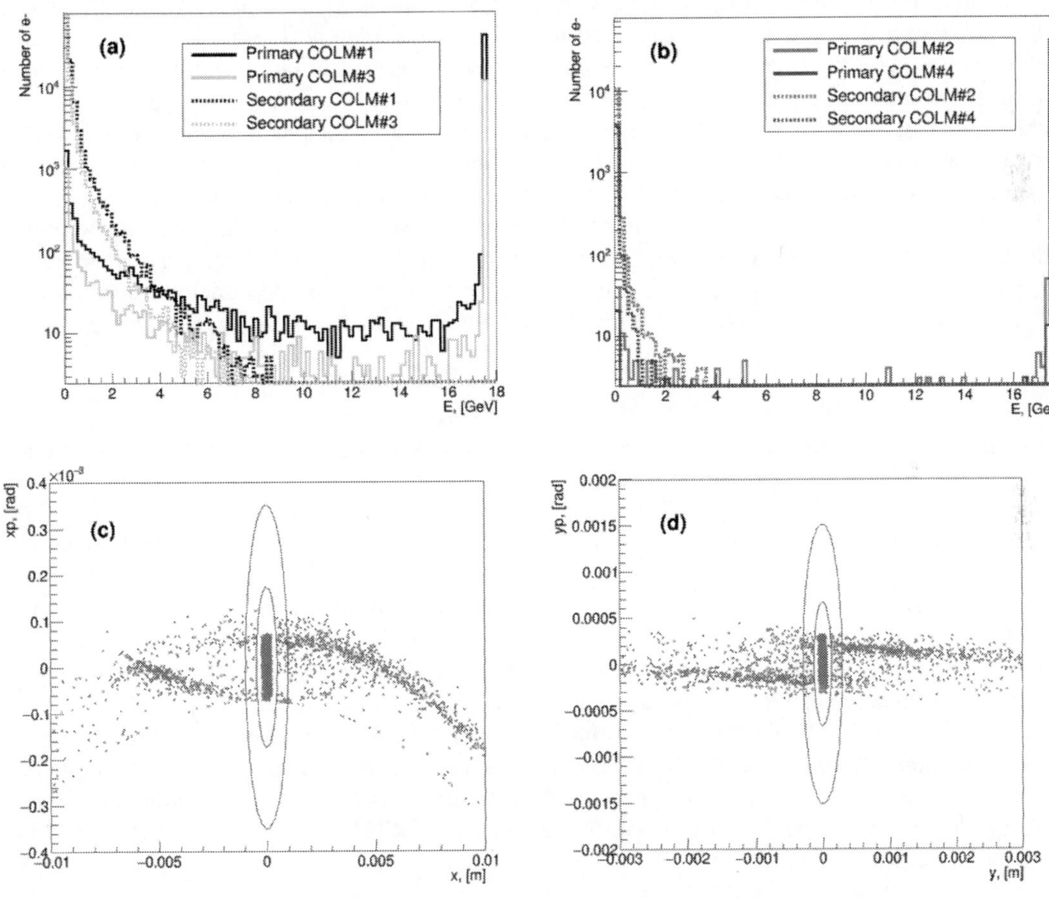

Figure 5. Energy distribution of the primary and secondary beam halo particles at the end of COLM#1 and COLM#3 (a) and at end of COLM#2 and COLM#4 (b) with 10^5 input electrons; Phase space distributions at the end of the collimation section for the horizontal (c) and vertical (d) plane with 10^7 input electrons.

This is due to the fact that COLM#1 and COLM#3 are the first main collimators in the first and second arcs (see Figure 4), and the phase space is rotated by 90° between them, so most primary betatron and angular halo particles are collimated by them. In the end of COLM#4 there are only few secondary electrons with less than 2 GeV energy left. We assume that those secondary electrons would get lost in the bending magnet after the COLM#4 and only those primary electrons, which lost a small fraction of their energy (<1.5%), can reach the undulators. In order to increase the statistics of these electrons, we repeated the simulation with 10^7 input electrons without recording the secondary particles.

Figure 5 (c) and (d) shows the phase space distribution at the end of the collimation section for the primary particles with energy higher than 17.2375 GeV with 10^7 input electrons. The dynamic aperture in the undulator section with radius R=4 mm (minimum undulator aperture) and R=2 mm are also shown in Figure 5 as two ellipses. Particles outside the dynamic aperture of the undulator chamber will be stopped at the undulator entrance and only the particles inside the aperture may reach the crystal location, which is 7 undulators downstream. The center part are the particles which can pass through the collimator freely, and the particles between the R=2 mm and R=4 mm apertures are those which may hit the crystal (assuming that the crystal is 2 mm away from the beam center).

By estimating the number of electrons between the two ellipses (N_{hits}) and comparing this number with the maximum allowed number of electrons ($N_{critical}$) calculated in Eq. (1), one can set the minimum distance that the crystal can approach to the beam center (i.e. the minimum HXRSS chicane delay). In the case that the crystal is 2 mm away from the center, N_{hits} is estimated to be 27±6 out of the total number of electrons $N_{total}=10^6$. This is 3 times lower than $N_{critical}$, which is estimated as 10^{-4} of N_{total}. Note that in our simulation the input transverse beam distribution is uniform, while the normal beam should be Gaussian-like, which can reduce N_{hits} dramatically. Meanwhile, the crystal will be inserted horizontally from the side of the beam, which means only half of the horizontal halo particles can hit the crystal. Therefore, with the conditions mentioned before, and assuming that there is no additional beam halo particles generation after collimation section, we can conclude that the crystal can be inserted to a minimum distance of 2 mm from the beam center, which means a chicane delay of ~13 fs.

4. Conclusions and prospects

We have performed beam halo particles tracking simulations for the implementation of HXRSS at European XFEL. Using a simplified geometry in GEANT4 we first studied the maximum number of beam halo particles allowed to hit the diamond crystal ($N_{critical}$). Then, in order to estimate the number of beam halo particles that may hit the crystal (N_{hit}), we tracked a uniformly distributed beam halo particles through the collimation section (with 2 mm apertures) using BDSIM. Finally, by extracting the number of electrons which can pass through the aperture of the undulators and can hit the crystal inserted to 2 mm from the beam center, we conclude that, even in this worst scenario, N_{hit} is ~3 times smaller than $N_{critical}$. Therefore, we can safely insert the crystal down to 2 mm from the beam center (i.e. a minimum chicane delay of ~13 fs) with the 2 mm collimator apertures.

Further simulations can be performed with more realistic beam halo particles distribution and with different collimator apertures. Similar beam loss studies can also be carried out for other applications with very tight aperture requirement (e.g. corrugated structure [14], afterburner undulator). It is also possible to include the GEANT4 model with undulators in BDSIM and perform a start to end beam halo particles tracking.

Acknowledgments

The authors would like to thank colleagues at DESY and European XFEL: I. Agapov, V. Balandin, G. Feng, G. Geloni, N. Golubeva, J. Pflueger and many colleagues from LCLS, especially J. Welch, H.-D. Nuhn and M. Santana-Leitner, for helpful discussions. And thanks to S. Boogert and L. Nevay from RHUL for their support on BDSIM.

References

[1] Geloni G, Kocharyan V and Saldin E 2011 A novel self-seeding scheme for hard X-ray FELs *Journal of Modern Optics* **58** (16) pp.1391-403
[2] Altarelli M, Brinkmann R *et al.* 2006 XFEL: The European X-Ray Free-Electron Laser Technical Design Report *DESY Report* 2006-097
[3] Geloni G, Kocharyan V and Saldin E 2010 A Cascade self-seeding scheme with wake monochromator for narrow-bandwidth X-ray FELs *DESY Report* 10-080
[4] Liu S and Decking W 2016 Self-Seeding implementation at European XFEL *8th Hard X-ray FEL Collaboration Meeting* Pohang South Korea
[5] Amann J *et al.* 2012 Demonstration of self-seeding in a hard-X-ray free-electron laser *Nature photonics* **6** (10) pp. 693-8
[6] Inagaki T *et al.* 2014 Hard x-ray self-seeding set-up and results at SACLA in *Proc. FEL' 14*, Basel Switzerland paper TUC01 pp. 603-8
[7] Balandin V, Brinkmann R, Decking W, and Golubeva N 2007 Optics Solution for the XFEL Post-Linac Collimation Section *TESLA-FEL Report* 2007-05
[8] Agostinelli S *et al.* 2003 GEANT4—a simulation toolkit *Nucl. Instr. Meth.* A **506** (3) pp. 250-303
[9] Agapov I *et al.* 2009 BDSIM: A particle tracking code for accelerator beam-line simulations including particle–matter interactions *Nucl. Instr. Meth.* A **606** (3) pp. 708-12
[10] Fasso A 2005 Dose Absorbed in LCLS Undulator Magnets I. Effect of a 100 μm Diamond Profile Monitor Radiation *Physics Note RP*-05-05
[11] Santana-Leitner M *et al.* 2010 Radiation protection studies for LCLS tune up dump SLAC *SLAC-PUB*-14020
[12] Quintieri L *et al.* 2016 Comparison of validations of GEANT4 and FLUKA codes on photo-nuclear predictions in the high-energy range *13th Meeting on Shielding Aspects of Accelerators Targets and Irradiation Facilities (SATIF-13)* Dresden Germany
[13] Yang R *et al.* 2016 Modeling and Experimental Studies of Beam Halo at ATF2 in *Proc. 7th Int. Particle Accelerator Conf. (IPAC'16)* Busan South Korea paper MOPMB008 pp. 88-91
[14] Zagorodnov I, Feng G and Limberg T 2016 Corrugated structure insertion for extending the SASE bandwidth up to 3% at the European XFEL *Nucl. Instr. Meth.* A **837** pp. 69-79

Experience with Multi-Beam and Multi-Beamline FEL-Operation

J Rönsch-Schulenburg, B Faatz, K Honkavaara, M Kuhlmann, S Schreiber, R Treusch, M Vogt
Deutsches Elektronen-Synchrotron DESY, 22607 Hamburg, Germany
E-mail: juliane.roensch@desy.de

Abstract. DESY's free-electron laser FLASH provides soft X-ray pulses for scientific users at wavelengths down to 4 nm simultaneously in two undulator beamlines. They are driven by a common linear superconducting accelerator with a beam energy of up to 1.25 GeV. The superconducting technology allows the acceleration of electron bunch trains of several hundred bunches with a spacing of 1 microsecond or more and a repetition rate of 10 Hz. A fast kicker-septum system directs one part of the bunch train to FLASH1 and the other part to FLASH2 keeping the full 10 Hz repetition rate for both. The unique setup of FLASH allows independent FEL pulse parameters for both beamlines. In April 2016, simultaneous operation of FLASH1 and FLASH2 for external users started. This paper reports on our operating experience with this type of multi-beam, multi-beamline set-up.

1. Introduction

The superconducting soft X-ray free-electron laser FLASH [1, 2, 3] delivers FEL pulses in the wavelength range from 4 to 90 nm for user experiments. The FEL radiation is generated by the self-amplified spontaneous emission (SASE) process. The extension of the FLASH facility by a second undulator beamline [4, 5] allows multi-user operation. Both undulator beamlines are driven by the same linear accelerator. Figure 1 shows a schematic layout of the FLASH facility with its common accelerator part, the two undulator beamlines and the third FLASHForward beamline which is currently under construction for a pioneering plasma-wakefield acceleration experiment [6]. The common accelerator part consists of a laser driven rf gun with high QE cathode, seven superconducting accelerating modules, a third-harmonic module and two bunch compressors. FLASH is able to generate pulse trains with a duration of up to 800 μs. The first part of the bunch trains enters into the FLASH1 beamline which contains 6 fix gap undulators each with a length of 4.5 m. A fast kicker system and a Lambertson septum are used to extract the second part of the bunch train into the FLASH2 beamline. FLASH2 has 12 variable gap undulators each with a length of 2.5 m, which allow choosing the wavelength to a certain extent independently from FLASH1. In order to operate FLASH2 almost independently of FLASH1, three injector laser systems are available. They allow a certain flexibility to tailor the laser parameters (bunch charge, bunch pattern, laser pulse length) to the need of both beamlines independently. A flexible low level RF system enables steps in the high voltage of each accelerating module of more than 10 MV in amplitude and more than 5° in phase between the FLASH1 and FLASH2 part of the bunch train. Applying different RF phases in the acceleration

IOP Conf. Series: Journal of Physics: Conf. Series **874** (2017) 012023 doi:10.1088/1742-6596/874/1/012023

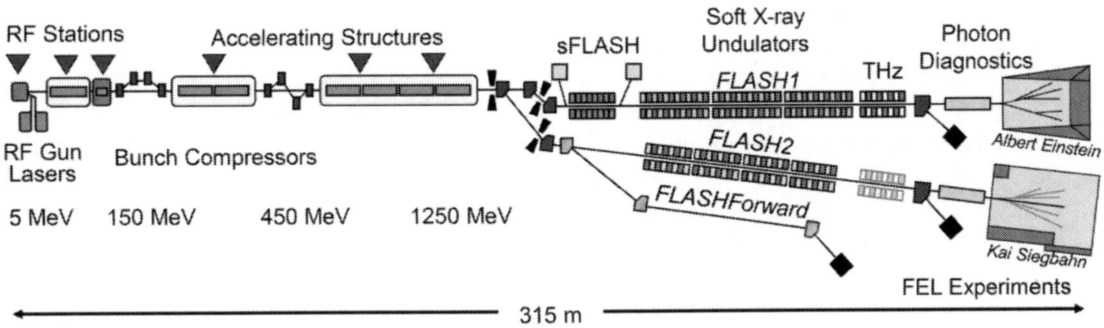

Figure 1. Layout of the FLASH facility.

modules before the two bunch compressors allows an individual setup of the compression for both beamlines.

First lasing of FLASH2 was achieved in August 2014 simultaneous to FLASH1 user operation [7]. On 8th of April 2016 two user experiments were running in parallel for the first time. The operation and capabilities of two user experiments in parallel with individually selected photon beam and pulse train characteristics is discussed.

2. Challenges of Operating two FEL-Beamlines

The simultaneous user operation entails some challenges.

- The setup of the common part of the beamline and the extraction region requires a beam in both beamlines FLASH1 and FLASH2. A later setup of one of the beamlines may cause beam losses in that beamline or changes in the other one and thus also a new setup.
- For the operation with long pulse trains, the available RF flat-top pulse duration needs to be shared between the FLASH1 and FLASH2 experiments but the bunch pattern can be chosen individually for both beamlines.
- Since FLASH1 is still equipped with fixed gap undulators all wavelength changes in FLASH1 require a change of the electron beam energy. In consequence, each change of wavelength in FLASH1 also requires a new set-up and SASE-tuning for FLASH2 as well. For FLASH2, a wavelength change is much easier. The variable gap undulators allow tuning the wavelength between 1 and 3 times the FLASH1 wavelength.
- The requirements on the FEL beam depend on the experiments. Experiments requests for short FEL pulses are typically set up with a low bunch charge of 100 to 200 pC. Contrariwise experiments requiring high FEL pulse energy are typically generated with a high bunch charge of up to 1 nC. At FLASH1 a planar electromagnetic undulator is installed downstream of the SASE undulators to produce —on request— THz radiation. The produced THz pulses are naturally synchronized with the SASE pulses. Also the set up of THz radiation is typically performed with a bunch charge of about 1 nC. Setting up such different settings in the two beamlines in parallel requires a careful setup of common accelerator beamline, since although the compression can be chosen independently, the beam optics in the common part is the same for both beams.

2.1. Operation of FLASH1 and FLASH2 with Different Bunch Charges

Two standard photo-injector lasers are available with an rms pulse duration of about 4.5 and 6.5 ps. A third one provides shorter, adjustable pulse length from 0.8 to 1.6 ps. Typically the injector laser with the slightly longer pulse duration is used at FLASH1 and the other at FLASH2 but an exchange is possible and has been demonstrated. Transverse shaping of the laser beams is obtained with a hard edge aperture imaged onto the photocathode of the RF gun. With this, a truncated Gaussian transverse beam shape is realized. The laser spot size on the cathode is determined by the diameter of this aperture. Various diameters can be selected. Unfortunately, two lasers share the same aperture, only the third laser has an own aperture. The optimal choice of the laser spot size depends on the bunch charge required.

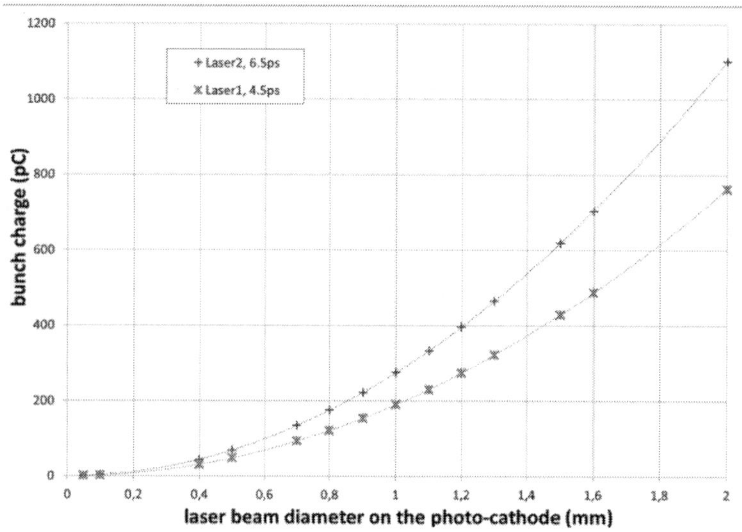

Figure 2. The beam size at the photo-cathode is selected depending on the emitted bunch charge in order to keep the charge density constant for the two standard laser at FLASH. The bunch charge is plotted as a function of the transverse laser beam diameter for the optimized setup.

In order to allow a simple setup of the common beamline and its optics, the beam size at the photo-cathode is selected depending on the bunch charge to keep the charge density on the cathode constant. The control of the space charge forces acting on the bunch in the RF gun at linear order allows to minimize the transverse beam emittance. Figure 2 shows the transverse laser beam size diameter at the photo-cathode and the corresponding optimized bunch charge. The markers represent the available apertures at FLASH. When operating with two completely different charges a compromise has to be found. Also the focussing with the solenoid around the RF-gun is charge dependent.

The following example illustrates the potential difficulties related to an injector set up for two apparently incompatible charges and our operational solution. A user experiment in FLASH1 was mainly interested in high THz-radiation pulse energy, while at the same time a FLASH2 user experiment requested short pulses. Thus a large difference in bunch charge is required to set up both experiments properly. For the specific user experiment, FLASH1 has been set-up with a charge of 680 pC and FLASH2 with 140 pC. The optimal transverse laser beam diameter at the photo-cathode would have been 1.6 mm in the first case and 0.9 mm in the second case. As a compromise, a diameter of 1.2 mm has been chosen which works well for both cases. The beam properties have been optimized for both cases independently using different RF-gun launch phases and different compression schemes.

Figure 3 shows an example of photon single pulse energy measured with an absolute calibrated gas monitor detector (GMD) [8] during 54 hours of the experiments discussed above. The plot illustrates, that while beam parameters at FLASH1 have been changed frequently, FLASH2 was operating stable and untouched by these changes.

3. Capabilities of FLASH2 operation

The FLASH2 wavelength is tuneable thanks to the variable gap undulators. In total, 12 undulators with a period of 31.4 mm and a length of 2.5 m each are installed [4]. A larger gap

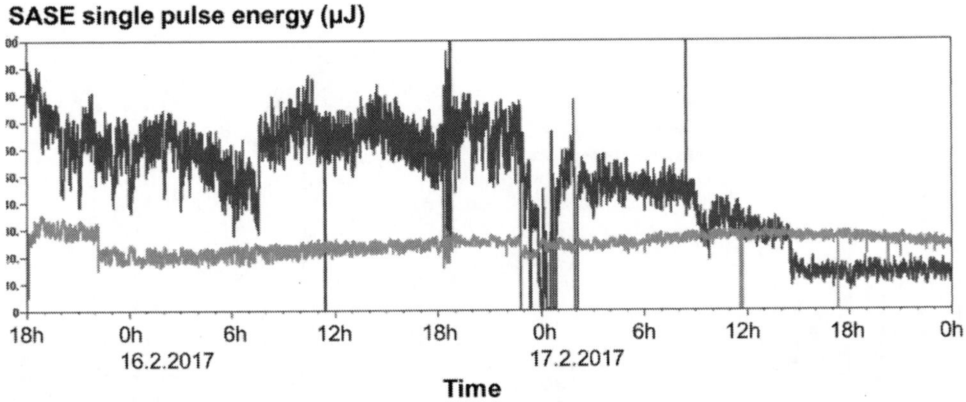

Figure 3. SASE single pulse energy at FLASH1 (blue) and FLASH2 (purple) during 54 hours of the experiments in parallel operation. The wavelength was 20.8 nm, single bunch with 680 pC, and 53 nm, 30 bunches with 10 µs spacing (100 kHz), and 140 pC resp. At FLASH1, also THz-radiation was produced.

width decreases the radiation wavelength and increases the saturation length. Figure 4 shows

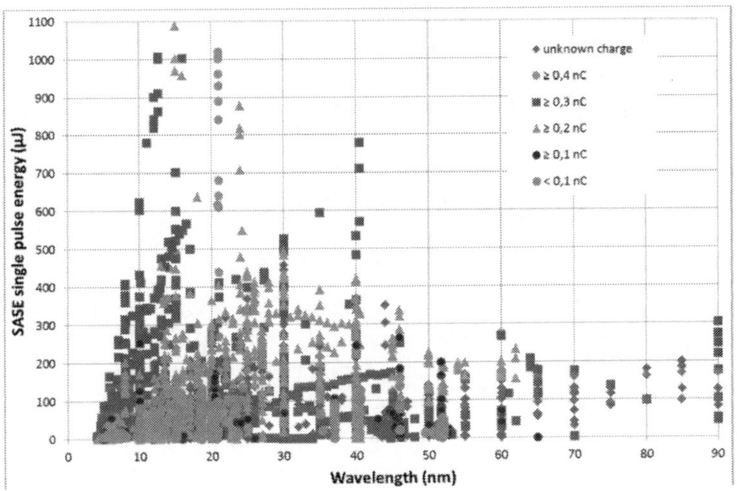

Figure 4. Single pulse SASE energies achieved in FLASH2 for the different wavelength.

the SASE single pulse energies reached in FLASH2 for different wavelengths. As a highlight, a single pulse energy of 1 mJ has been demonstrated at a wavelength of 21 nm. This corresponds to 10^{14} photons, never achieved before at this wavelength in a single pulse. Similar results have been achieved at other wavelengths as well. Single pulse energies above 200 µJ have been demonstrated for many wavelengths between 7 nm and 90 nm. The different colors in Figure 4 represent different charge ranges. Typically FLASH2 is running with bunch charges between 20 pC and 350 pC but higher charges are possible as well. Figure 5 shows which wavelengths have been achieved at FLASH2 as a function of the beam energy. All these measurements have been performed in simultaneous operation with FLASH1. The black line represents the FLASH1 wavelength, while the gray line shows the upper limit with closed gap. The two lines limit the range of the design wavelength. The high SASE energies are reached for the small undulator gap width, where the saturation length is small. Some results in the plots of Figure 4 and 5 have been obtained using tapering techniques which allow a doubling of the SASE energy [9].

IOP Conf. Series: Journal of Physics: Conf. Series **874** (2017) 012023 doi:10.1088/1742-6596/874/1/012023

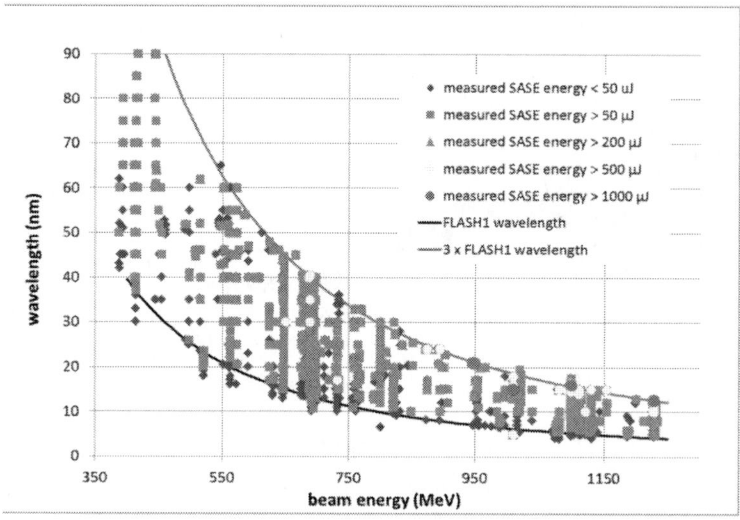

Figure 5. SASE single pulse energies at FLASH2 depending on the beam energy and wavelength defined by the undulator gap width.

Another important parameter required for a large class of user experiments is the FEL-pulse duration. Many experiments need pulses well below 50 fs. As mentioned above the pulse duration is controlled by changing the compression factor and adjusting the bunch charge accordingly. The FEL pulse duration, however, via slippage also depends on the effective undulator length relative to the gain length. In FLASH2 with its variable gap undulators it is possible to choose the effective undulator length by closing only those undulators required to just reach saturation. The last undulator closed defines the source point. Some experiments require high SASE energies for alignment purposes where the pulse duration does not play a role. This can be quickly achieved by closing upstream undulators without having to change any other parameter. Thus, in an experimental situation, one can easily switch between these two operation modes.

4. Summary

Parallel operation of FLASH1 and FLASH2 for different machine settings has been demonstrated successfully. Even with a factor 5 difference in charge between FLASH1 and FLASH2 an independent setup of both beamlines is possible. Such a setup is challenging since the injector laser beam size at the photo-cathode can currently not be chosen independently but in future it is planed to modify the laser beamline to allow different beam sizes. This will further increase the flexibility of the setup.

The variable gap undulators at FLASH2 offers a high degree of freedom. Single pulse energies above 1 mJ at 21 nm have been achieved for the first time corresponding to 10^{14} photon per pulse. Energies above 200 µJ are easy to achieve and are demonstrated over almost the whole wavelength range.

References

[1] Ackermann W *et al* 2007 *Nature Photonics*, **vol. 1** pp 336-342

[2] Schreiber S and Faatz B, 2015 The free-electron Laser FLASH In:*High Power Laser Science and Engineering* **vol. 3** e20

[3] Vogt M *et al* 2016 The superconducting soft X-ray free-electron laser user facility FLASH *Proc. 7th Int. Particle Accelerator Conf. (IPAC'16)* (Busan, Korea) pp 729-731

[4] Faatz B, Plönjes E *et al* 2016 Simultaneous operation of two soft x-ray free-electron lasers driven by one linear accelerator *New J. Phys.*, **vol. 18** 062002

[5] Scholz M, Faatz B, Schreiber S, Zemella J 2015 First Simultaneous Operation of Two Sase Beamlines in FLASH *Proc. 37th Int. Free Electron Laser Conf.(FEL'15)* (Daejeon, Korea) pp 297-200

[6] Aschikhin A *et al* 2016 The FLASHForward Facility at DESY *Nucl. Instrum. & Methods in Phys. Research A* **vol. 806** pp 175-183
[7] Schreiber S and Faatz B 2014 First Lasing at FLASH2 *Proc. 36th International Free Electron Laser Conference (FEL'14)* (Basel, Switzerland) pp 7-8
[8] Tiedtke K *et al* 2009 The soft x-ray free-electron laser FLASH at DESY: beamlines, diagnostics and end-stations *New J. Phys.* **vol .11** 023029
[9] Schneidmiller E and Yurkov M 2016 Application of Statistical Methods for Measurements of Coherence Properties of the Radiation from SASE FEL *Proc. 7th Int. Particle Accelerator Conf. (IPAC'16)* (Busan, Korea) pp 738-740

Construction and commissioning of direct beam transport line for PF-AR

Nao Higashi, Seiji Asaoka, Kazuro Furukawa, Kaiichi Haga, Kentaro Harada, Toshiyasu Higo, Tohru Honda, Hiroyuki Honma, Naoko Iida, Hiroshi Iwase, Kazuhisa Kakihara, Takuya Kamitani, Mitsuo Kikuchi, Yuji Kishimoto, Yukinori Kobayashi, Kota Kodama, Kikuo Kudo, Tatsuya Kume, Katsuhiko Mikawa, Toshihiro Mimashi, Fusashi Miyahara, Hiroshi Miyauchi, Shinya Nagahashi, Hajime Nakamura, Norio Nakamura, Takuya Natsui, Kazuyuki Nigorikawa, Yasuhiro Niwa, Takashi Nogami, Takashi Obina, Yujiro Ogawa, Masaaki Ono, Toshiyuki Ozaki, Hidenori Sagehashi, Toshiya Sanami, Masayuki Sato, Masanori Satoh, Tsuyoshi Suwada, Mikito Tadano, Toshihiro Tahara, Ryota Takai, Hiroyuki Takaki, Seiji Takasaki, Madoka Tanaka, Yasunori Tanimoto, Masafumi Tawada, Nobu Toge, Takashi Uchiyama, Akira Ueda, Yusuke Yamada, Masahiro Yamamoto and Mitsuhiro Yoshida

High Energy Accelerator Research Organization, KEK, 1-1, Oho, Tsukuba, Ibaraki, Japan

E-mail: nao.higashi@kek.jp

Abstract. PF-AR was constructed as an accumulator ring for TRISTAN, and in the KEKB era it has been revitalized as a 6.5 GeV synchrotron radiation source. The injection energy was 3 GeV and the beam was accelerated to 6.5 GeV prior to the user run. The original beam transport line (BT) from the LINAC to the PF-AR shared its upstream part with the the BT line of KEKB High Energy Ring (HER). The injection-mode change from PF-AR to HER or vice versa needs about 10 minutes for the magnet cycling procedure of the shared part. In SuperKEKB, the upgrade of KEKB, the lifetime of HER is about 10 minutes. The mode-switch operation of the BT is, therefore, not allowed for maintaining the highest luminosity of the SuperKEKB. In order to avoid this problem, a new 6.5 GeV BT line dedicated to PF-AR has been constructed. This also enables the top-up injection for the user run. The commissioning of the new BT line has been completed in this March, and now the first user run has been operated successfully.

1. Introduction

The Photon Factory Advanced Ring (PF-AR) is a storage ring light source of Hard X ray region dedicated to the single bunch operation. With the old beam transport (BT), 3 GeV electrons were injected and accelerated to 6.5 GeV before the user run. KEK LINAC provides injection beams for four rings; 7 GeV electron for SuperKEKB HER (High Energy Ring), 4 GeV positron for LER (Low Energy Ring) [1], 2.5 GeV electron for PF ring, and 6.5 GeV electron for PF-AR now. Since the pulse-to-pulse switching of LINAC and the beam switching yard had been made

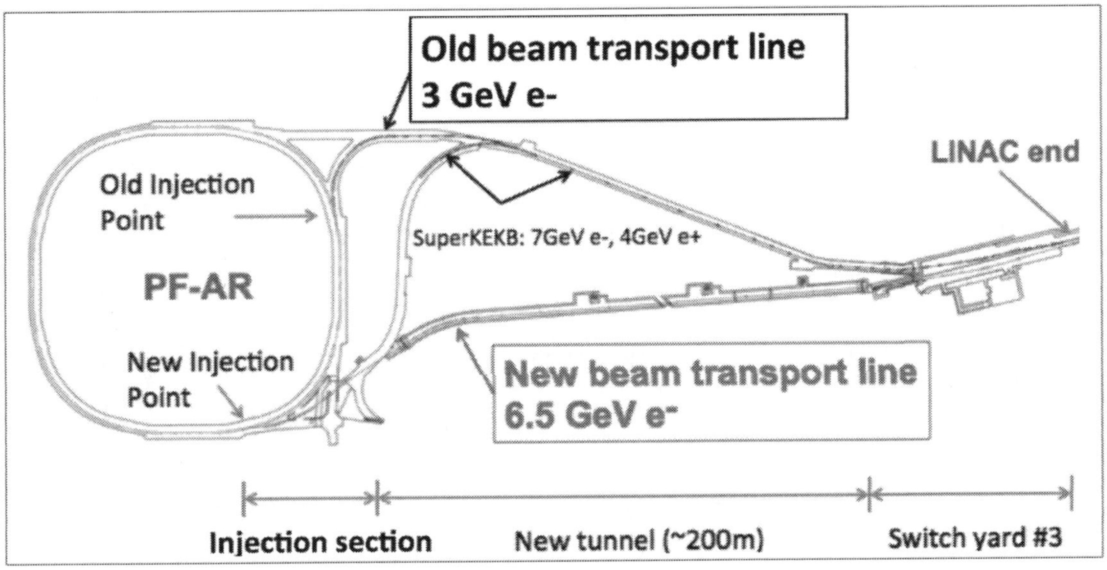

Figure 1. The schematic view of the old and the new direct beam transport line of PF-AR.

for three rings of PF, HER and LER for old KEKB, PF-AR injection interrupted it for about 10 minutes at least due to the magnet cycling procedure for the upstream part of the old BT that is used not only for PF-AR but also for HER. Since SuperKEKB requires a continuous injection of 50 Hz for HER and LER because of their short Touschek lifetime of about 10 minutes [2], the 10 minutes interruption is very critical for SuperKEKB. Thus a new independent BT for PF-AR that enables the pulse-to-pulse injection for all four rings has been constructed [3, 4]. The new BT is designed for 6.5 GeV and enables the top-up operation that will begin in autumn 2018 after the completion of Phase 2 operation of SuperKEKB. The construction of a new tunnel for the independent BT began in 2012 and finished in 2015 with the infrastructure. The installation of accelerator components of magnets and vacuum ducts finished in 2016. In order to reuse all bending magnets, the removal and the installation were conducted in parallel. Finally the commissioning began in February 2017. Here, we report the construction and commissioning of the new BT for PF-AR.

2. Configuration of PF-AR direct beam transport line

The new BT transport is about 320 m long from the exit of the pulsed bending magnet at the end of LINAC to the new injection point of PF-AR (Figure 1). The optical functions are shown in Figure 2. We describe the configuration of the new BT in three parts; LINAC Switch Yard #3 (SY3), new tunnel, and ring injection section.

2.1. Switch Yard #3

As shown in Figure 3, SY3 hall is located at the exit of LINAC to switch electron and positron beams for HER (purple line), LER (blue), PF (green) and PF-AR (red). The beam is kicked by the pulsed bending magnet with the maximum repetition rate of 25 Hz for PF and PF-AR. With the same magnetic field of the magnet for different beam energies, BTs for PF and PF-AR can be separated. This section is the most crowded and complicated area in this construction. The geometrical design of the BT, the hardware design and alignments have overcome many difficulties.

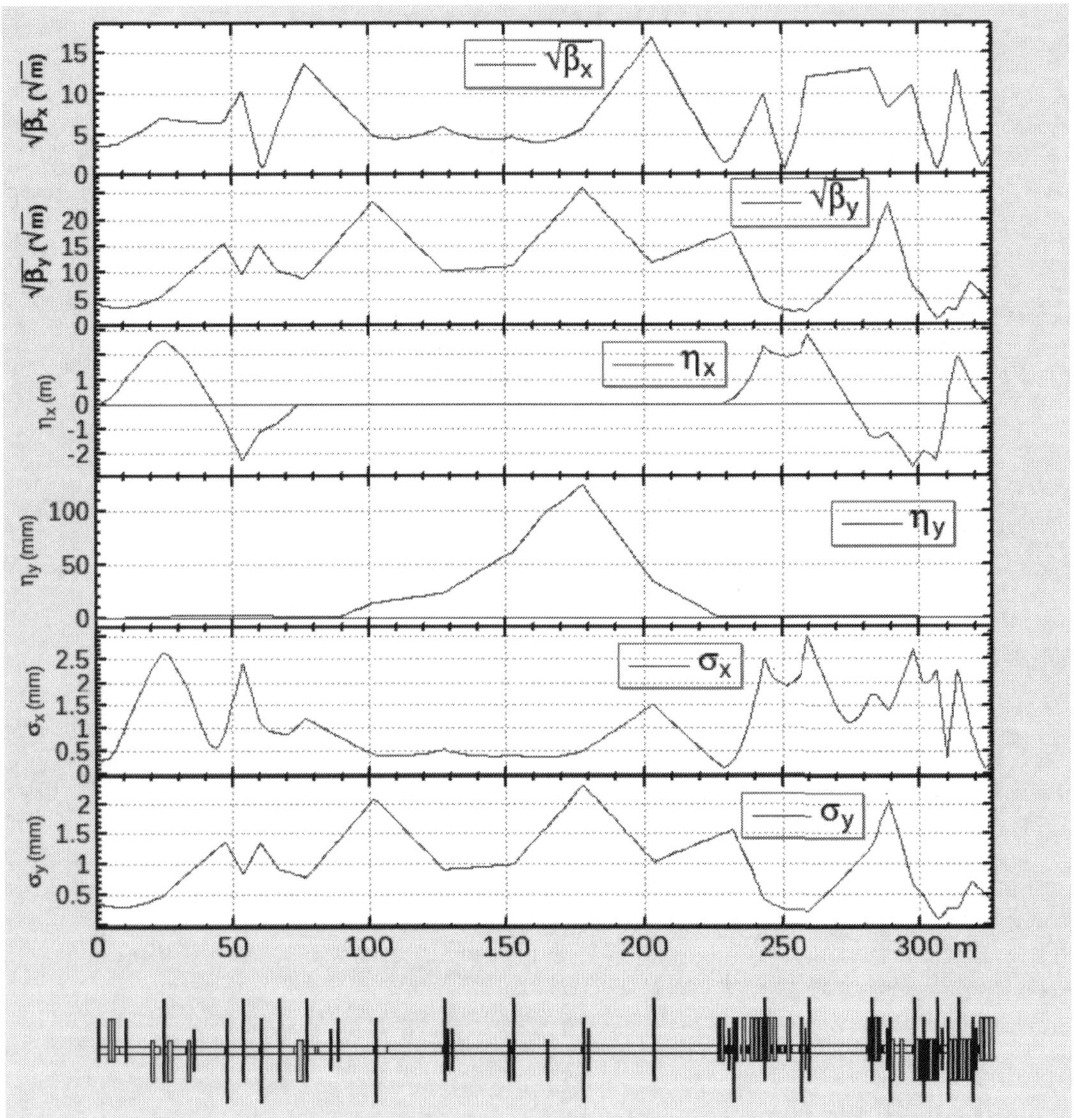

Figure 2. Optical functions from the end of the injector linac to the injection point of the PF-AR. In order to calculate the beam size, we supposed that the normalized emittance is 100×10^{-6} mrad, and the energy spread is 1.0×10^{-3} as the 1 σ of Gaussian distribution at the initial point of the BT.

2.2. New tunnel

The new BT tunnel is about 200 m straight line divided by an existing utility conduit (Figure 4). The upstream and downstream tunnels are connected to each other to prevent geometrical fluctuation without fixing the utility conduit. In order to make the vacuum duct pass through the conduit without interference with existing utility racks, the beam trajectory is bumped to +60 mm using four vertical bending magnets. The level adjustment between LINAC and PF-AR is made with this bump. The height of the trajectory at the exit of the bump is -28 mm. The optics is designed as achromatic for this vertical asymmetrical bump.

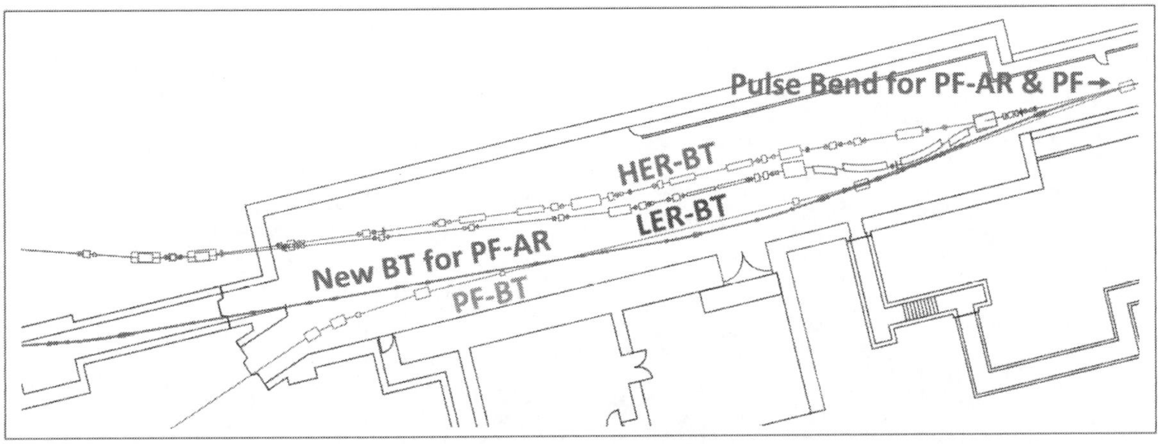

Figure 3. Four beam transport lines in LINAC Switch Yard #3.

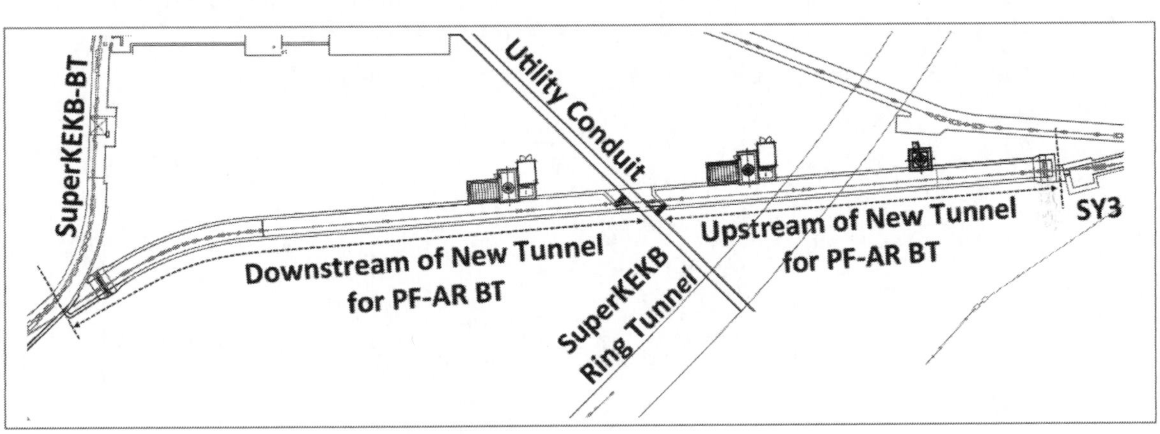

Figure 4. The new tunnel section of the direct beam transport line for PF-AR.

2.3. Injection section

Behind the new tunnel, the new BT crosses the existing SuperKEKB BTs that is 0.6 m below (Figure 5). The BT for PF-AR is 2.4 m high from the floor of the SuperKEKB BTs and thus the three new 4 m vacuum ducts are hanged on the ceiling of the SuperKEKB BT tunnel.

For the injection system of PF-AR, three pulsed kicker magnets and two pulsed septum magnets are newly manufactured for 6.5 GeV [5–7]. In order to install these new kickers, vacuum ducts are renewed in the range of 25 m.

3. Commissioning

The beam commissioning began on February 13, 2017. The beam passed through the new BT and reached the injection point on the first day. The beam storage was observed on the second day. However, a large loss occurred in the accumulation, and the injection efficiency was suppressed below only 1 %. This problem was resolved with the precise measurement and adjustment of the timing of the injection kickers, and the beam injection efficiency was drastically improved to about 80 % on the fifth day. This was calculated as the ratio of the increase of the accumulated beam current detected by the DCCT of PF-AR ring to the charge amount detected by the CT at the end of LINAC (Figure 6). After this adjustment, the injection efficiency got stable around 84 - 88 % through the commissioning.

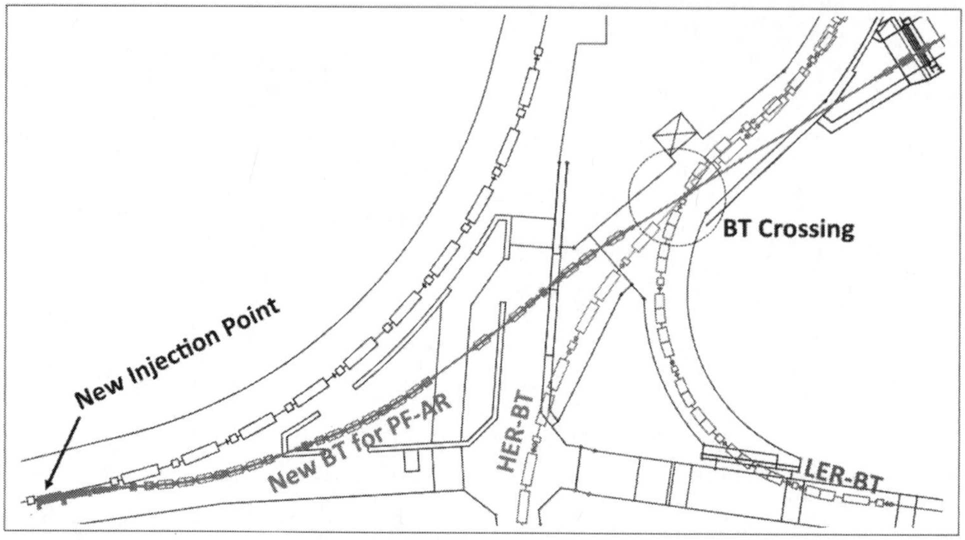

Figure 5. The injection section of the direct beam transport line for PF-AT.

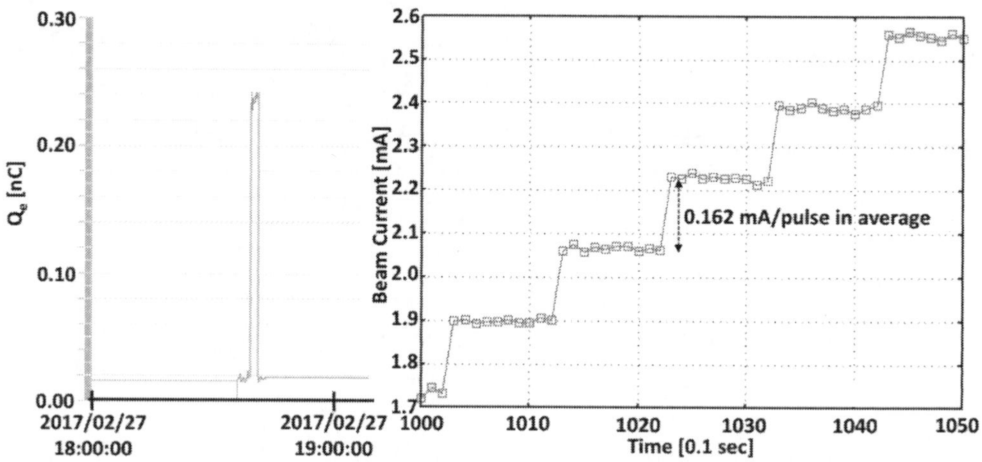

Figure 6. (Left) Charge amount detected at the end of Linac. (Right) Accumulated beam current of PF-AR ring.

Figure 7 shows the BPM measurement results for the new BT with the very rough first calibration. Figure 8 shows the history of the maximum stored current of the day and typical lifetime during the commissioning period. The stored current already reached 60 mA, the design goal [1]. The vacuum pressure and beam lifetime is almost recovered as shown in Figure 9. On March 1, the facility inspection by Nuclear Safety Technology Center was implemented and on March 16, the notification of the acceptance was made. From April 2017, PF-AR successfully operates the user run with the new BT and injection system.

[1] After confirming the accumulation of 60 mA, the current was lowered to 50 mA for the device protection.

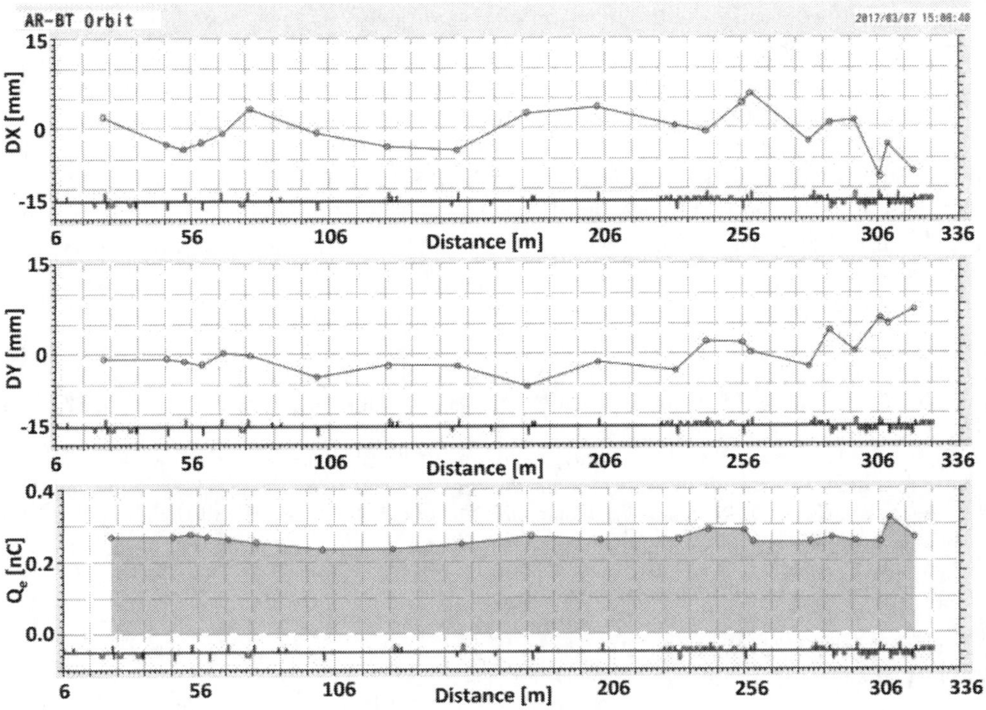

Figure 7. A example of the beam position and charge amount through the new direct beam transport line, measured by BPMs with the very rough first calibration. The starting point of "Distance" is the pulsed bending magnet in SY3, indicated in Figure 3.

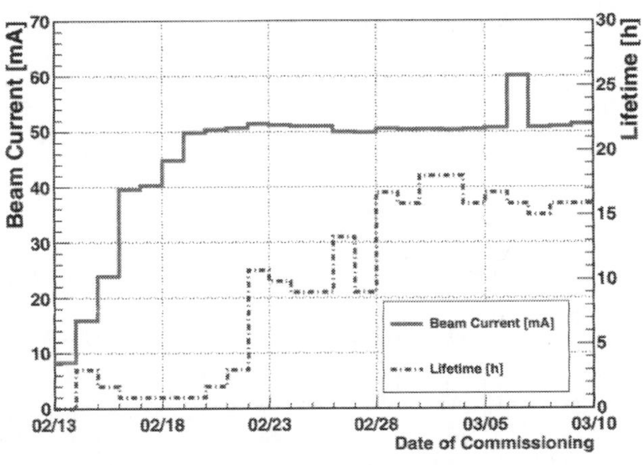

Figure 8. The history of the beam current and the lifetime. The displayed values of the beam current are the maximum ones, and those of the lifetime are the typical ones recorded for each day.

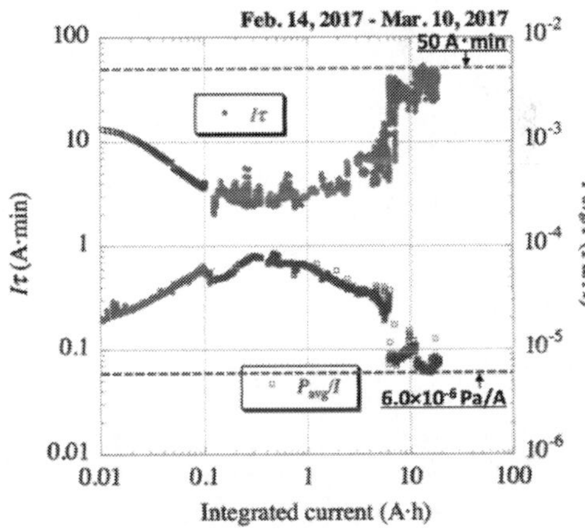

Figure 9. The state of the vacuum scrubbing of the PF-AR ring. Each dashed line indicates the goal estimated from the past results before the construction of the new BT.

IOP Conf. Series: Journal of Physics: Conf. Series **874** (2017) 012024 doi:10.1088/1742-6596/874/1/012024

4. Future work

The upgrade of LINAC for Phase 2 operation of the SuperKEKB is scheduled from this June for five months [8]. LINAC finally enables the continuous injection for four rings with the positron damping ring in autumn 2017. After the commissioning for SuperKEKB and beam study for PF and PF-AR, the user run with the top-up injection for PF and PF-AR will begin in autumn 2018.

Also, the full energy injection resolves the problem of the very severe beam instability effect at 3 GeV which occurs due to the strong HOMs excited by the comparatively short bunch. We will begin the machine study for the lower emittance optics with the careful hardware protection in the next step.

References

[1] Kamitani T *et al.* 2016 *Proc. of the 13th Annual Meeting of PASJ* (Chiba) pp 24-8
[2] Ohnishi Y, Koiso H, Morita A, Ohmi K, Oide K, Sugimoto H and Zhou D 2014 *Proc. of HF2014* (Beijing) pp 73-8
[3] Takaki H *et al.* 2014 *Proc. of the 11th Annual Meeting of PASJ* (Aomori) pp 990-4
[4] Nagahashi S *et al.* 2016 *Proc. of the 13th Annual Meeting of PASJ* (Chiba) pp 567-70
[5] Ueda A, Asaoka S, Honda T, Nagahashi S, Nakamura N, Nogami T, Takaki H and Uchiyama T 2016 *Proc. of the 13th Annual Meeting of PASJ* (Chiba) pp 1222-6
[6] Ueda A, Asaoka S, Honda T, Nagahashi S, Nakamura N, Nogami T, Takaki H and Uchiyama T 2017 *Proc. of IPAC2017* (Copenhagen) WEPVA060
[7] Ueda A, Asaoka S, Honda T, Nagahashi S, Nakamura N, Nogami T, Takaki H and Uchiyama T 2017 *Proc. of IPAC2017* (Copenhagen) WEPVA059
[8] Funakoshi Y *et al.* 2016 *Proc. of IPAC2016* (Busan) pp 1019-21

New HMBA lattice for PF-AR

N Higashi, K Harada, Y Kobayashi, S Nagahashi, N Nakamura and A Ueda

High Energy Accelerator Research Organization, KEK, 1-1 Oho, Tsukuba, Ibaraki, Japan

E-mail: nao.higashi@kek.jp

Abstract. Photon Factory Advanced Ring (PF-AR) has been operated for users about 30 years from 1987. The lattice and optics are almost not changed from the original one as the TRISTAN booster ring constructed in 1984. The lattice employs FODO structure and the horizontal emittance for the 6.5 GeV user run is about 300 nmrad. In order to improve the performance of PF-AR dramatically, the full replacement of the accelerator to the ESRF type HMBA (Hybrid multi bend achromat) lattice is examined. In order to geometrically fit the new lattice to the present PF-AR tunnel, the new ring consists of 12 cells with four long straight sections. The emittance is improved to about 500 pmrad at 3 GeV. With the present user experimental hall at the northern half of the ring, at least eight undulator beam lines can be constructed. The simulated dynamic aperture is about 1.5 cm at the long straight section with reasonable magnetic errors and COD correction. The Touschek lifetime is about 6 hours. The beam injection with conventional injection system is no problem and the beam lifetime is long enough.

1. Introduction

HMBA lattice is developed for the ESRF EBS project [1]. The design concept of ESRF type HMBA, the normal cell has two dispersion bumps by the bending magnets with small curvature and longitudinal gradients. The chromatic sextupoles are only installed to this section where the effective chromaticity correction can be made by large dispersion function. The combined function bending magnet with defocusing field at the centre of the cell can keep emittance small. In order to suppress the non-linear effect of the sextupoles, the tune advance between two dispersion bumps is fixed at the half integer. The original lattice has very small amplitude and momentum dependent tune shifts and the large dynamic apertures. In addition, HMBA has great flexibility. For example, the 3 GeV EBS lattice of 20 cells [2] was the starting point for KEK-LS and the short straight section of 1.2 m length was added to the original lattice [3]. If we add the quadrupoles for the straight section, the performance is improved in the aperture and lifetime [4,5]. Since KEK-LS is fully new facility project, the replacement of existing ring is also possible by HMBA type lattice. In this presentation, we show the design and performance of the lattice replacement study by HMBA type for PF-AR.

2. Present status of PF-AR

PF-AR is 6.5 GeV synchrotron radiation (SR) light source dedicated to the single bunch operation. The ring originally constructed as an accumulation ring (AR) for TRISTAN project in 1984. Because KEKB adopts full energy injection from LINAC, AR becomes PF-AR, the full-time synchrotron radiation facility. The circumference of PF-AR is 377 m with four long straight sections. East and west long straight sections are used for RF cavities. North and south straights are originally used for detector development for TRISTAN. Presently, the insertion device is installed in the north straight section and just small accelerator components in south. Because the original injection and extraction systems are installed to the south half of the ring, the experimental hall for SR users are concentrated only in the north half of the ring.

The original beam transport line (BT) is designed for 2.5 GeV and 3 GeV injection began in 2003. Because of the hardware limitation, full energy injection is impossible. For the SuperKEKB, continuous injection is essentially required for HER and LER due to the short Touschek lifetime with low emittance and large current. In order not to prevent SuperKEKB injection, a new BT was constructed in 2017 [6]. The injection point moved from the south-east straight section to the south-west section. If a new insertion device is installed at the old injection point, a new beamline can be constructed in existing south experimental hall. The consideration for the new beamline just began. The emittance of the present FODO structure lattice is about 300 nmrad for 6.5 GeV that is about two orders of magnitude worse than those of present advanced SR facilities.

The available beamline numbers and extremely old infrastructures are the disadvantage of the PF-AR. Comparing with PF ring that is oval shape ring of 187 m circumference, the lattice upgrade is easier for PF-AR and potential improved performance is much better because of its longer circumference and tunnel shape.

3. Replacement lattice design

In order to fit the circumference, the cell number is 12. We started the design from the 12 cells symmetrical ring consisting of identical twelve normal cells with the circumference of about 330 m and the emittance of about 0.5 nmrad for 3 GeV. It is 2 nmrad for 6 GeV. The lattice and optics of the normal cell of this starting point are shown in Figure 1. We add the short straight section to double the numbers of available insertion devices.

In order to fit the ring for the existing accelerator tunnel, we adjusted the length of the straight sections of symmetrical 12 cells ring and made it asymmetrical. The required geometrical adjustments are shown in Figure 2. There are four long straight sections more than 20 m in the ring where the horizontal tune advance of the cell cannot be kept to be the same as that of the arc normal cells. The tune survey and the tune balance survey between arc cells and 20 m straight cells show that the operating point for the best dynamic aperture with errors is the same as the case with the complete symmetry and the tune difference should be as small as possible. To expand long straight section from 5.4 m to more than 20 m, the horizontal tune of the cell increased by 0.175 for one cell as shown in Table 1. The horizontal tune advance of LL cell is $2.3833 + 0.175 = 2.5583$. To fix the total tune of the ring, the horizontal tune of the arc cell is $(28.6 + 2.5583 \times 4)/8 = 2.2958$. The optics of 1/8 part of the ring is shown in Figure 3. The parameters are shown in Table 2. The emittance without IBS is about 350 pmrad for 2.5 GeV, 500 pmrad for 3 GeV and 2 nmrad for 6 GeV. The potential performance seems to be great.

Assuming the magnetic errors similar to present PF ring as the Gaussian random alignment errors of 50 μm, field fluctuation of 0.05 %, and rotation of 0.1 mrad, the simulated dynamic aperture after COD correction is show in Figure 4. With errors, the momentum aperture is about 3.5 % and the horizontal amplitude 150 σ_x. Here, σ_x is the amplitude normalized with the horizontal beam size. The dynamic aperture is large enough to adopt the conventional injection system and the estimated Touscheck lifetime is about 5.6 hours.

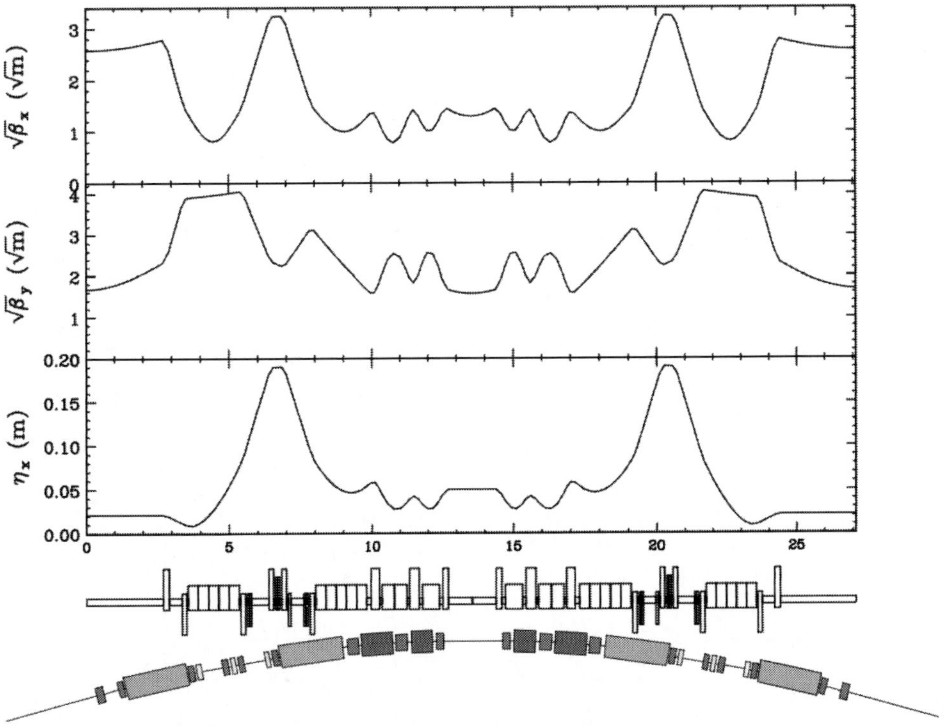

Figure 1. Lattice and optics of the normal cell of 12 cells symmetrical HMBA ring as a starting point of the design. Light blue show the bending magnet with longitudinal gradient, deep blue the combined function bending magnet, red the quadrupole magnets, and yellow the sextupole magnets. The short straight section middle in the figure is 1.6 m and the long one both sides 5.4 m.

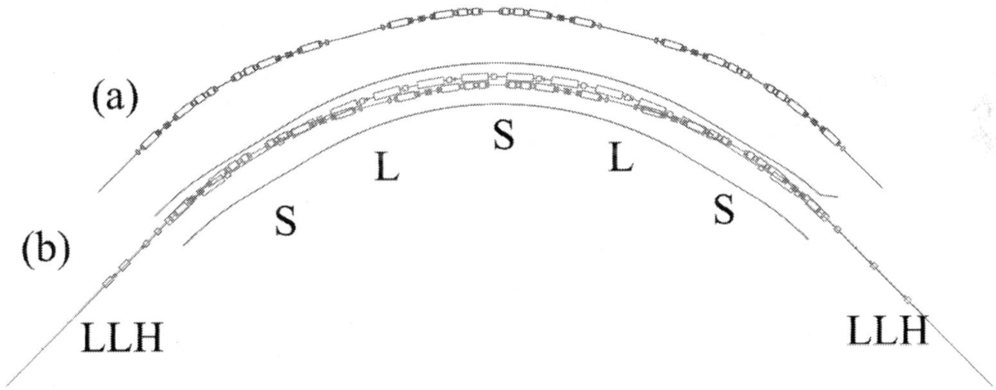

Figure 2. Geometrical adjustment of the long straight section to fit the existing tunnel arc. Figures show the quarter of the ring. The blue figure (a) is the symmetrical ring with twelve identical normal cells for the starting point of the design. The red figure in (b) shows the existing PF-AR lattice of FODO structure and the tunnel wall of the arc section. The green figure shows the new lattice for the replacement. S is the short straight section of 1.68 m, L the arc long straight of 2.7 m. LLH is the very-long straight section that has about 23 m length between two bending magnets at the both sides.

Table 1. Distribution of the betatron tune for geometrical matching.

		Num.	Hor.	Ver.
Symmetrical ring	Arc cell	12	2.3833	0.8417
	Total	-	28.6000	10.1000
AR fit	LL cell	4	2.5583	0.8417
	Arc cell	8	2.2958	0.8417
	Total	-	28.6000	10.1000

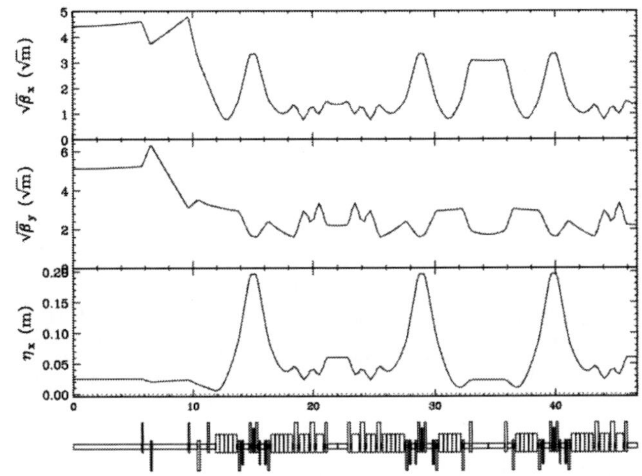

Figure 3. Optics of the 1/8 part of the ring.

Table 2. Parameters of the ring.

	Symbol [Unit]	SuperAR 3.0	
Energy	E [GeV]	3	
Cell number	N_s	12	
Circumference	C [m]	374.28	
RF freq.	f_{RF} [MHz]	500	
Harmonic Number	h	544	
Energy loss	[MeV/rev]	0.4259744	
Momentum compaction	α	5.6715×10^{-4}	
Damping time (x, y, z)	[ms]	8.365, 17.58, 19.59	
RF voltage	V_{RF} [MV]	2.5	
Bucket height	[%]	3.58	
Betatron tune (x, y)	ν	28.7, 10.2	
Beam current	[mA]	0	500
Horizontal emittance	[pmrad]	481.15	520.23
Vertical emittance	[pmrad]	-	7.8
y/x coupling	[%]	-	1.5
Touschek lifetime (3.5 % $\sigma_{\Delta p}$, 150 σ_x)	[hour]	-	5.6
Energy spread	-	9.79×10^{-4}	1.01×10^{-3}
Bunch length	[mm]	5.21	5.37

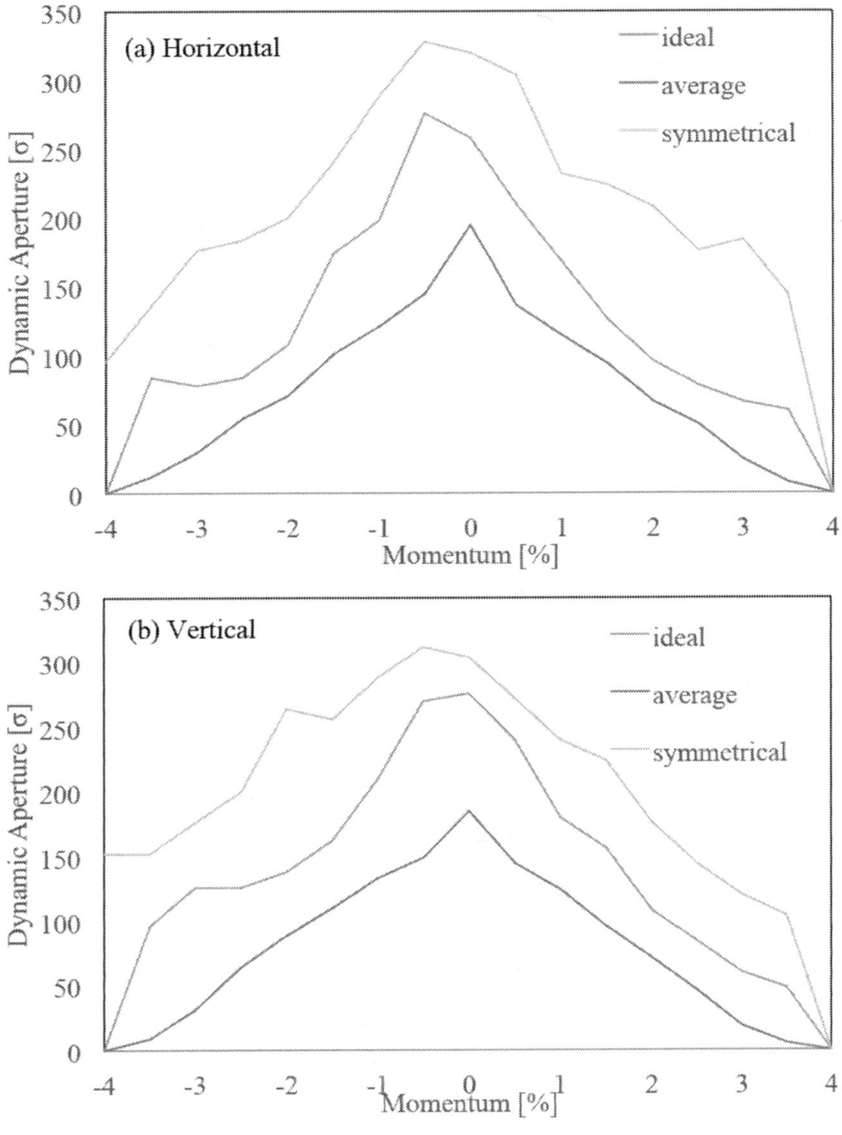

Figure 4. Dynamic aperture with reasonable magnetic errors after COD correction. The line "average" shows the averaged dynamic aperture for 100 random error seeds. The line "ideal" shows the aperture without errors and "symmetrical" for the symmetrical 12 cells case without errors.

4. Beamline configuration

Because of existing KEKB BT tunnel, the experimental hall and user beamlines are presently concentrated in only north half of the PF-AR. The estimated numbers of available beam lines are eight as shown in Figure 5. Because the injection point was moved from east to west with new direct BT, the another insertion device can be installed to the old injection point and the new beam line can be constructed in the existing south experimental hall. If we can use old beam transport tunnel as the new experimental hall, another few new beamlines can be constructed.

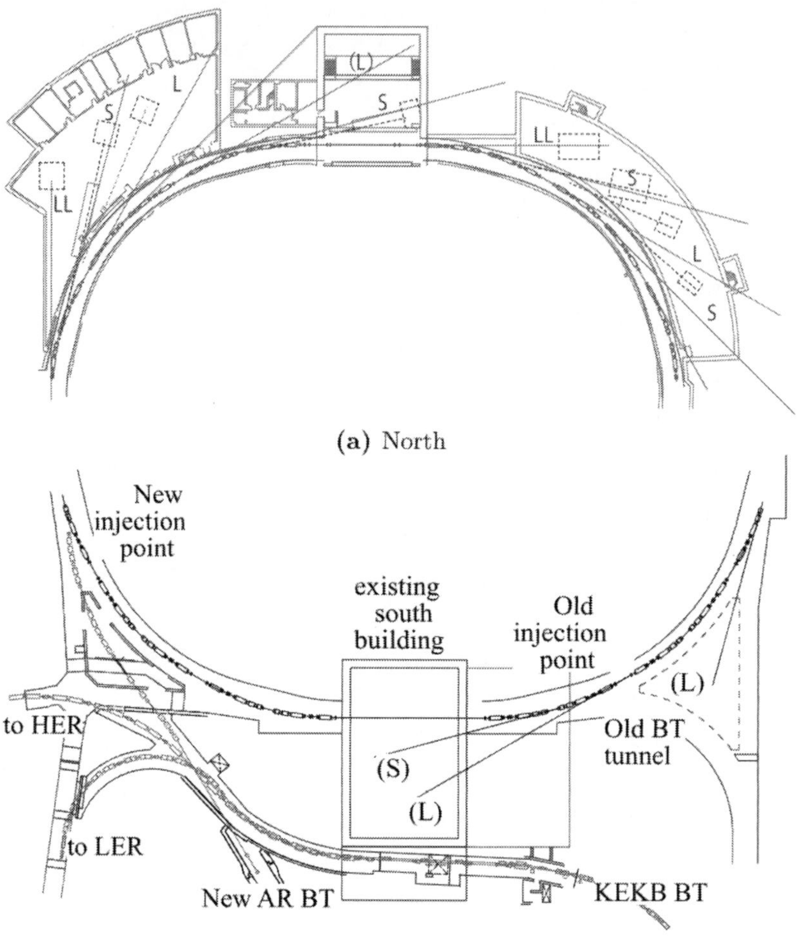

(a) North

(b) South

Figure 5. Beamline configuration for the existing experimental hall. "LL" shows the straight line from the about 20 m (B-B) straight section at the symmetrical point of the ring, "L" 2.7 m straight section at the arc and "S" 1.68 m. (a) shows the north side of the ring and (b) south.

5. Summary

PF-AR has double circumference of the PF ring and the shape is almost the circle. The potential performance of replacement lattice seems to be great. The old infrastructures and the small space for the experimental beam line are the problem. By making the best use of new space at the east-west side of the ring where the old BT and injection system were installed, the number of the beamlines can be increased (Figure 5 (b)).

References

[1] *ESRF Orange Book* 2014 `http://www.esrf.eu/Apache_files/Upgrade/ESRF-orange-book.pdf`
[2] Raimondi P 2013 private communications
[3] *KEK-LS CDR* 2016 `http://kekls.kek.jp/`
[4] Liuzzo S 2016 private communications
[5] Harada K, Higashi N, Nagahashi S, Nakamura N, Sakanaka S, Ueda A and Liuzzo S M 2017 *Proc. of IPAC2017* (Copenhagen) WEPAB043
[6] Higashi N *et al.* 2017 *Proc. of IPAC2017* (Copenhagen) WEPAB044

Characterization of cold model cavity of cryocooled C-band 2.6-cell RF gun at 20 K

T Tanaka[1], T Sakai[1], K Nogami[1], K Hayakawa[1], Y Hayakawa[1], K Nakao[1], K Takatsuka[1], M Fukuda[2], D Satoh[2], T Takatomi[2], N Terunuma[2], J Urakawa[2] and M Yoshida[2]

[1] Laboratory for Electron Beam Research and Application, Institute of Quantum Science, Nihon University, 7-24-1 Narashinodai, Funabashi, Japan
[2] High Energy Accelerator Research Organization, 1-1 Oho, Tsukuba, Japan

E-mail: tanaka@lebra.nihon-u.ac.jp

Abstract. A cryogenic C-band 2.6-cell photocathode RF electron gun has been studied at Nihon University in collaboration with KEK for future use in a compact electron-accelerator based monochromatic X-ray source. The cold model cavity with an input coupler was fabricated at KEK in 2016 by ultraprecision machining and diffusion bonding techniques. The RF characteristics of the cavity obtained at cryogenic measurements have been in agreement with the result of the CST Studio simulation. The unloaded quality factor of ~73000 has been obtained at the cavity temperature of 20 K, which is 5.5 times higher than that at room temperature and consistent with the simulation taking into account the anomalous skin effect of the cavity surface resistance. The input coupling coefficient of approximately 20 at 20 K has well reproduced the design value. The shift in the accelerating π-mode resonant frequency due to the cavity temperature change from 296.65 to 20 K has been 19.02 MHz, being 0.12 MHz greater than the value based on the linear expansion coefficients for copper by NIST.

1. Introduction

An RF cavity made of high-purity copper operating at low temperatures is known to be low power loss due to low surface resistance compared to room temperature operation [1, 2], though the effect is not as drastic as a superconducting cavity. In addition, significantly high thermal conductivity and low linear expansion coefficient at low temperatures are favorable properties for stable operation of the cavity at high RF input power [3]. Motivated by these advantages, the development of a cryogenic C-band 2.6-cell photocathode electron gun has begun in 2013 at Nihon University in collaboration with KEK intending to use as the injector of a compact X-ray generating electron linac [4]. The cavity temperature of 20 K has been chosen by considering that the lowest surface resistance and the highest thermal conductivity are reached at around and below 20 K, and that a refrigerator with refrigeration capacity required for high power operation at 20 K is readily available.

From the result of the simulations for improvement of the power efficiency and the coupler property in the previous test cavity, a final cold model was fabricated in 2016 [5, 6]. The RF properties obtained at room temperature have been in good agreement with the CST Studio simulations [7]. The cryogenic experiments of the cavity carried out afterward have shown excellent results as expected. This paper reports the characterization of the cold model cavity at 20 K measured by supplying the test RF power through a thermal insulating waveguide.

2. Structure of the 2.6-cell cavity

The cryogenic C-band RF gun cavity consists of 5712 MHz $TM_{01\pi}$-mode 2.6 accelerating cells. The combination of a mode converter from rectangular TE_{10} to circular TM_{01} mode and a short circular waveguide, in a manner similar to an X-band RF gun [8], works as an input coupler that feeds the accelerating cells with the RF power along the cavity axis as shown in figure 1. The cavity dimensions were optimized to give high efficiency acceleration and low insertion loss in the converter by the simulations using Superfish and CST Studio [5, 6, 9].

The cavity, of high purity 6N8 copper, was fabricated at KEK with ultraprecision machining and diffusion bonding techniques. The diffusion bonding is considered to be an excellent method of fabrication to avoid the degradation of the surface resistance at low temperatures. Due to need for the measurement of the field on the cavity axis as a cold model, a hole with a diameter of 3.6 mm has been made in the center of the left-side end plate. Each cell is equipped with 4 tuners placed every 90° around the axis.

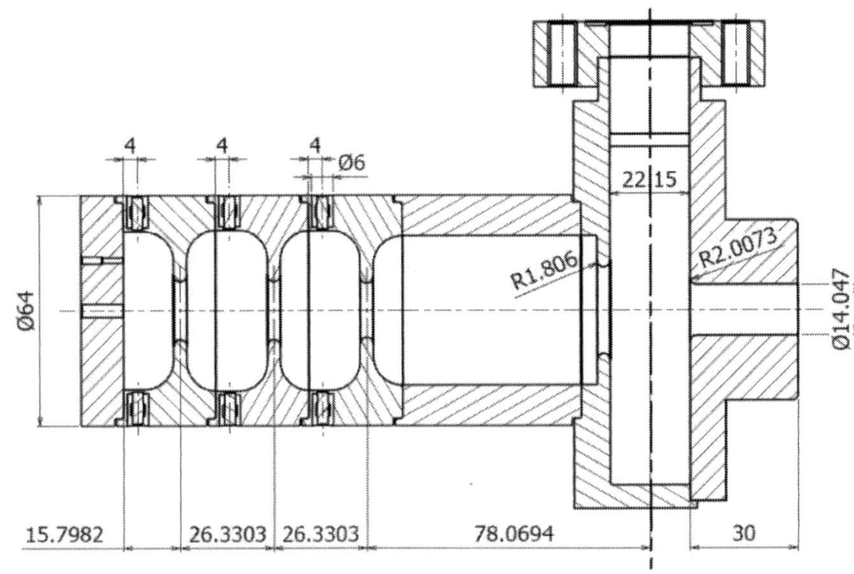

Figure 1. Cross-sectional view of the cold model of a C-band 2.6-cell RF gun cavity. The dimensions shown in the drawing are those for machining at 23.5 °C.

3. Shunt impedance at room temperature

Figure 2 shows the behavior of the bead-pull perturbation measured on the cavity axis over the entire structure length, where the frequency shift is plotted at 0.5 mm step as an averaged result of 10 runs of the computer controlled automatic sweep. The radius r_b of the metal bead used in the measurement was 1.0 mm.

The frequency shift Δf of the $TM_{01\pi}$-mode resonant peak by the presence of the bead in the cavity is expressed as

$$\Delta f = -\frac{\pi \varepsilon_0 E_z^{\,2} r_b^{\,3}}{U} f_0, \tag{1}$$

where ε_0 is the dielectric constant of vacuum, f_0 the unperturbed frequency, E_z the axial electric field amplitude at the position of the bead and U the stored energy in the cavity [10]. The shunt impedance to quality factor ratio R/Q_0 is given by the relation

$$\frac{R}{Q_0} = \left(\int |E_z| ds\right)^2 (4\pi f_0 U)^{-1}, \qquad (2)$$

where $|E_z|$ is integrated along the axis over the 2.6 cells. Thus the ratio can be deduced from the frequency shift data by the perturbation method as

$$\frac{R}{Q_0} = \left(\int \sqrt{|\Delta f|} ds\right)^2 \left(4\pi^2 \varepsilon_0 f_0^{\,2} r_b^{\,3}\right)^{-1}. \qquad (3)$$

The experimental ratio deduced from equation (3) has been $R/Q_0 = 293\ \Omega$ in agreement with the result estimated by equation (2) from the axial field by the CST Studio simulation. In figure 2 the thin orange curve shows the frequency shift calculated by equation (1).

The effective maximum accelerating voltage V_{acc} in the 2.6-cell cavity without beam loading is given by

$$V_{acc} = \frac{2\sqrt{\beta}}{1+\beta}\sqrt{PZT^2L}, \qquad (4)$$

where β is the input coupling coefficient of the cavity, P the source RF power, L the cavity length, Z (= $2R/L$) the shunt impedance per unit length, and T (= 0.763) the transit time factor. Appling the unloaded quality factor Q_0 = 13200 at room temperature that has been reported elsewhere [5] and 16 % higher than the case in the previous cold model [11], the shunt impedance Z of the cavity is estimated to be 113 MΩ/m.

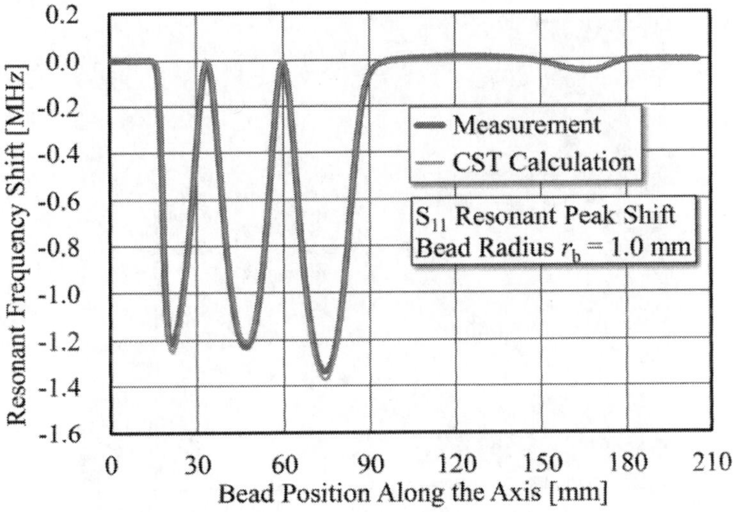

Figure 2. Result of the on-axis field measurement of the accelerating field by the bead-pull perturbation method.

4. Cryogenic system assembling for cold test

Figure 3 shows the top view of the cryogenic system assembled for low-temperature cavity measurements. The cold model cavity is mounted on the cold head of the refrigerator (Suzuki Shokan RF273S) attached to the bottom of the vacuum chamber. The chamber is pumped with a 250 l/sec turbo-molecular pump. The refrigeration capacity of the refrigerator is 5 W at 20 K. The test RF power from the network analyzer (Agilent E5071C) is supplied to the cavity through the R48-type waveguide inserted into the chamber.

The waveguide is made of stainless steel with 1.5 mm thick wall over a partial length of 275 mm [12]. The inner wall was plated with 20 μm thick copper. Figure 4 shows the picture of the cavity connected to the waveguide. The other end of the waveguide is connected to a coaxial-to-waveguide converter outside the chamber. The heat flow from the room-temperature end of the waveguide has been estimated to be 4 W. The network analyzer is calibrated at the waveguide flange of the converter before the vacuum chamber is pumped.

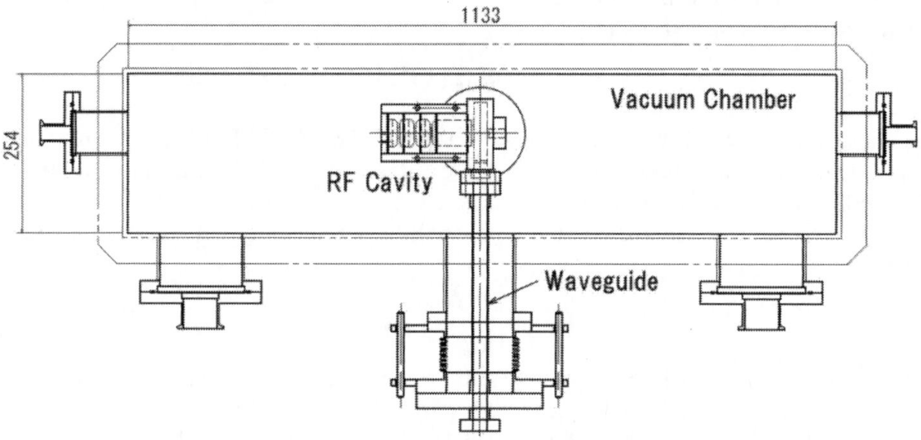

Figure 3. Top view of the cryostat assembled for the cold test of the cavity.

Figure 4. Picture of the cavity connected to the stainless steel waveguide.

5. Experiments and analyses

The changes in the temperature and the resonant frequency of the cavity measured during a cooling experiment down to around 20 K are plotted in figure 5 as a function of the elapsed time since several minutes before the refrigerator was turned on. The cavity temperature was always monitored using 3 semiconductor thermometers attached to the different positions on the surface of the structure, i.e. (1) the 0.6-cell end plate, (2) near the cooling heatsink and (3) near the waveguide flange, respectively. In the course of the experiment the temperature was lowest at (2) and highest at (3) as expected. However, the difference was not greater than 0.5 K, and less than 0.1 K at the temperatures below 35 K. The temperature data plotted in figure 5 is the result measured at the 0.6-cell end plate. The lowest cavity temperature reached by this cryogenic system was 19.61 K. No temperature stabilization has been done in the system.

The Smith Chart data and the temperatures of the cavity were taken every 2 minutes and stored in a PC simultaneously. The resonant frequency of the cavity at each temperature in figure 5 has been obtained by the least square fitting of the RF reflection coefficients in the vicinity of the resonance crest to a parabolic function. The frequency shift at the temperatures below 25 K has been less than 13 kHz. The average shift rate of 2.4 kHz/K over this region is approximately 1/40 of that in the room temperature region due to the small linear expansion coefficients of copper.

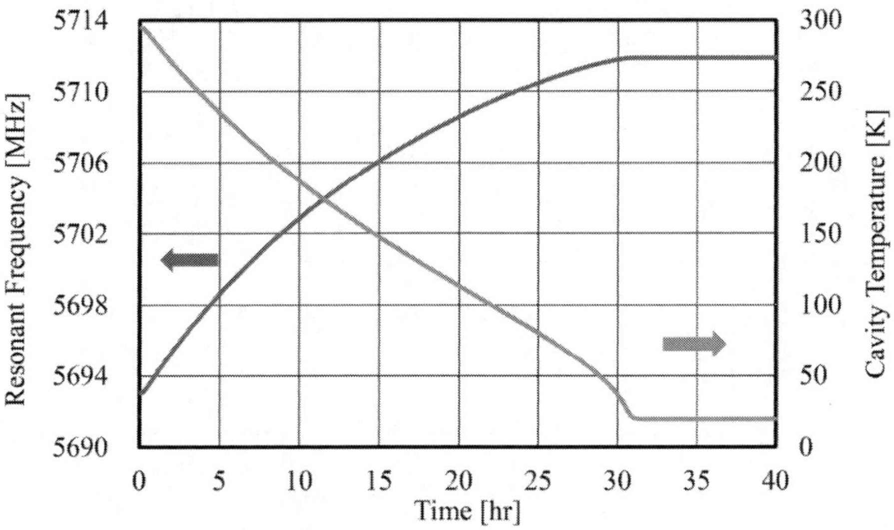

Figure 5. The temperature and the resonant frequency of the cold model cavity during a cooling experiment over 40 hours.

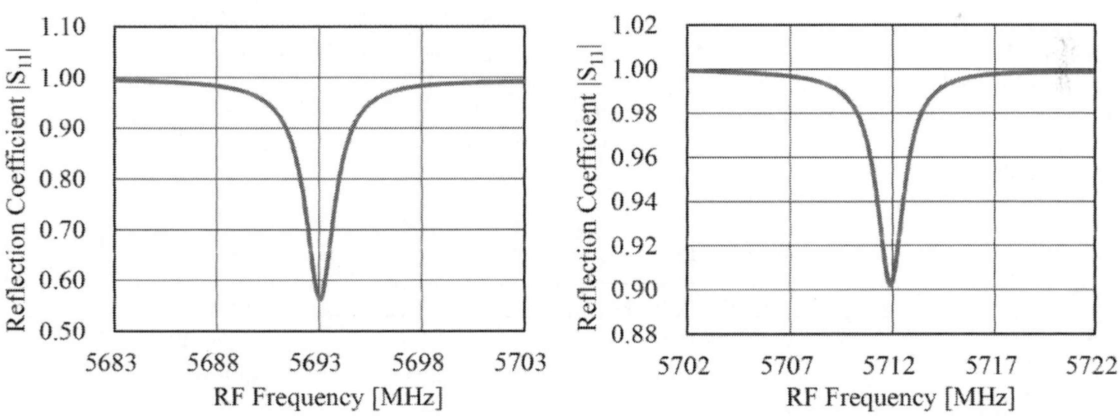

Figure 6. Examples of the RF reflection coefficients $|S_{11}|$ around the $TM_{01\pi}$ mode resonance measured at cavity temperatures of (a) 294 K and (b) 20 K.

Examples of the RF reflection coefficients, $|S_{11}|$, around the resonant frequency measured at cavity temperatures of 294 K and 20 K are shown in figure 6. In both cases the vacuum chamber was pumped with a turbo-molecular pump. Each $|S_{11}|$ spectrum over ±10 MHz around the peak frequency obtained in a series of experiments has been fitted to a Lorentzian curve by the least square fitting method to estimate the coupling coefficient and the loaded quality factor of the cavity.

The cooling experiments on this cold model were carried out 5 times in 3 months. The behavior of the Q_0 values evaluated from the analyses of the Smith Chart data is shown in figure 7 as a function of the cavity temperature for each measurement run. The different final Q_0 values resulted after cooled down to around 20 K can be related to the RF power loss in the coaxial cable which is very sensitive to ambient temperature.

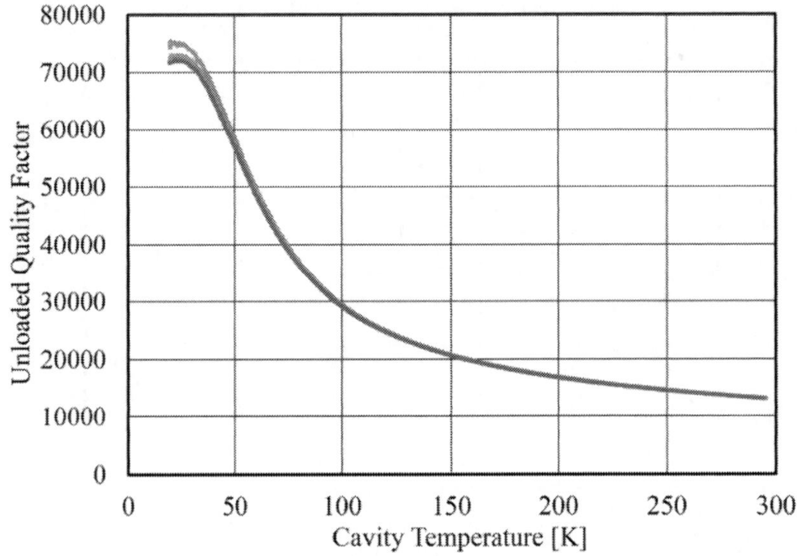

Figure 7. The behavior of the Q_0 values measured during 5 runs of the cooling experiment.

The average values of the properties obtained from the experiments are listed in Table 1, where the values in parentheses are the results of the CST Studio simulations using the surface resistances based on the theory of the anomalous skin effect [2, 13]. Though the difference is not large, the resonant frequency shift caused by the cooling has been 0.12 MHz greater than that was estimated from the cavity contraction. The factor of increase in the frequency deduced from the NIST data [3] is denoted by an asterisk. The Q_0 values and the coupling coefficients are in agreement with the prediction by the CST Studio simulation.

Table 1. The properties obtained by the experiments.

	at 296.65 K	at 20 K	ratio
Q_0	13176 ±20 (13310)	72775 ±1168 (73029)	5.52 ±0.09 (5.48)
Coupling coefficient	3.566 ±0.008 (3.52)	19.79 ±0.26 (19.30)	5.55 ±0.07 (5.48)
Resonant frequency [MHz]	5692.863±0.006 (5692.791)	5711.883±0.009 (5712.015)	1.0033409 *1.0033203

6. Conclusion

The cold model of a C-band 2.6-cell photocathode RF gun cavity has been characterized by the cryogenic cooling experiments. The unloaded quality factor of approximately 73000 at 20 K, with an enhancement factor of 5.2 from the room temperature operation, is in agreement with the result of the CST Studio calculation using the surface resistance based on the theory of the anomalous skin effect. The shunt impedance of 113 MΩ/m measured at room temperature corresponds to 624 MΩ/m at 20 K, which offers possibilities of high efficiency and high accelerating gradient RF guns at a high peak RF input power of the order of 10 MW: a peak field of 180 MV/m on the cathode surface is expected at 10 MW. Fabrication of a high power model cavity is under consideration for future high gradient experiments.

Acknowledgments
This work was supported by Photon and Quantum Basic Research Coordinated Development Program of the Ministry of Education, Culture, Sports, Science, and Technology (MEXT), Japan.

References

[1] Septier A 1970 The surface impedance of superconducting metals at high frequencies *Linear Accelerators* ed P M Lapostolle and A L Septier (Amsterdam: North-Holland) pp 1090–2
[2] Reuter G E H and Sondheimer E H 1948 The theory of the anomalous skin effect in metals *Proc. Royal Soc. (London)* A **195** pp 336–64
[3] http://cryogenics.nist.gov/MPropsMAY/OFHC%20Copper/OFHC_Copper_rev.htm
[4] Fukuda M, Araki S, Aryshev A, Honda Y, Terunuma N, Urakawa J, Sakaue K and Washio M 2013 Status and future plan of the development of a compact x-ray source based on ICS at Laser Undulator Compact Xray (LUCX) *Proc. 25th North American Particle Accelerator Conf.* (September 2013, Pasadena) pp 589–91
[5] Sakai T *et al.* 2016 Cold model cavity for 20-K cryocooled C-band photocathode RF gun *Proc. 7th Int. Particle Accelerator Conf.* (May 2016 Busan) pp 2635–7
[6] Takatsuka K *et al.* 2016 Optimization of C-band RF input coupler as a mode converter for 20-K cryocooled photocathode RF gun *Proc. 7th Int. Particle Accelerator Conf.* (May 2016 Busan) pp 2638–40
[7] *CST STUDIO SUITE*, Computer Simulation Technology AG, Germany, http://www.cst.com
[8] Taniguchi Y, Sakamoto F, Natsui T, Hashimoto T, Hashimoto E, Uesaka M, Yoshida M, Higo T, Fukuda S and Akemoto M 2008 Development of X-band thermionic RF electron gun using choke structure *Proc. 5th Annual Meeting of Particle Accelerator Society of Japan* (August 2008 Higashihiroshima) pp 954–6
[9] Billen J H and Young L M 2006 LA-UR-96-1834 *Los Alamos National Laboratory* http://laacg.lanl.gov/laacg/services/download_sf.phtml
[10] Ginzton E L 1957 *Microwave Measurements* (New York: McGraw-Hill) p 439
[11] Tanaka T *et al.* 2015 RF input coupler for 20 K cooled C-band 2.6-cell photocathode RF gun *Proc. 6th Int. Particle Accelerator Conf.* (May 2015 Richmond) pp 2522–5
[12] Iino A *et al.* 2016 High-power test of the C-band accelerating structure with a high-Q factor at 20 K *Proc. 13th Annual Meeting of Particle Accelerator Society of Japan* (August 2016 Chiba) pp 863–6
[13] Tanaka T *et al.* 2014 Basic design of a 20K C-band 2.6-cell photocathode RF gun *Proc. 5th Int. Particle Accelerator Conf.* (June 2014 Dresden) pp 658–60

IOP Conf. Series: Journal of Physics: Conf. Series **874** (2017) 012027 doi:10.1088/1742-6596/874/1/012027

Simultaneous generation of drive and witness beam for collinear wakefield acceleration

G Ha[1], J G Power[2], M Conde[2], D S Doran[2], and W Gai[2]
[1]Pohang Accelerator Laboratory, Pohang, Gyeongbuk, 790-784, KOREA
[2]Argonne National Laboratory, Argonne, IL 60439, USA
E-mail: homehe@postech.ac.kr

Abstract. Generating the drive and witness bunch for collinear wakefield acceleration (CWFA) requires precise control of the longitudinal bunch shape for each bunch as well as the controlling their separation. The emittance exchange (EEX) beamline and a transverse mask can be used to achieve all of these requirements. First, this EEX-based method can independently control the longitudinal bunch shape of each bunches so that the drive bunch is shaped to generate a high transformer ratio while witness bunch is shaped to suppress its energy spread. Second, the timing jitter between the drive and witness bunch poses a serious limitation to the CWFA scheme but the EEX-based method eliminates this since both bunches are generated at the same time and share the exactly same beamline so there are no relative errors. In this paper, we confirm the feasibility of this EEX-based method for simultaneous generation with simulation for CWFA in a dielectric structure.

1. Introduction

Generation of the drive and witness bunch for collinear wakefield acceleration (CWFA) requires precise control of the longitudinal profile of each bunch and control of their separation. The longitudinal bunch shape of the drive bunch needs to be controlled to generate a high transformer ratio [1] and the witness bunch must be shaped to suppress its energy spread for efficient acceleration [2]. Timing jitter between the drive and witness bunch in CWFA poses a serious limitation to the scheme. This is because a jitter in the high frequency CWFA scheme can significantly change the energy gain of the witness bunch.

All three of these requirements can be met with an emittance exchange (EEX) based method [3] that simultaneously generates the drive and witness bunch. The method uses a transverse mask in the EEX beamline to independently control the longitudinal distribution of each bunch [4, 5] as well as controlling the separation between the bunches. This method uses a single mask, but with two shapes, to carve a single input beam into two beamlets with different transverse profiles.These beamlets pass through the EEX beamline which exchanges the properties of the transverse and longitudinal phase spaces. Therefore, the two transverse beamlets that enter the EEX beamline become two longitudinally separated bunches (i.e. drive and witness bunches). This method allows control over temporal profiles of both bunches via the shapes on the mask. Further, control over the final bunch lengths and their separation can be controlled from a combination of the separation of the shapes on the mask and the quadrupoles in front of EEX beamline [5]. Finally, this method eliminates the timing jitter between the bunches since they

IOP Conf. Series: Journal of Physics: Conf. Series **874** (2017) 012027 doi:10.1088/1742-6596/874/1/012027

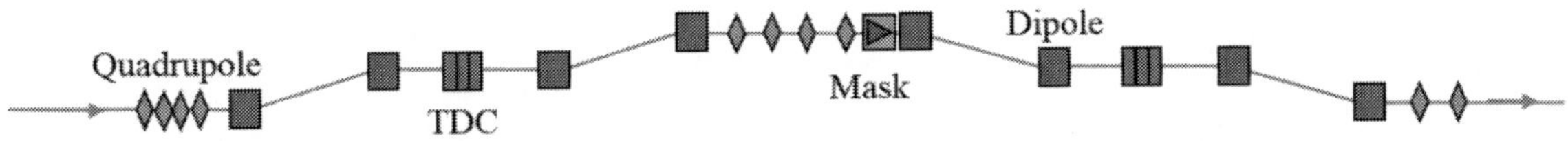

Figure 1. Configuration of a double EEX bemaline. The beam is moving from left to the right in this figure.

were carved out of the same bunch and propagate through the same beamline so that the bunches experience the same jitters along the beamline.

In summary, the EEX-based method provides distinct advantages for CWFA.

- High transformer ratio from drive beam shaping
- High efficiency from witness beam shaping
- Controlled drive-witness separation
- Relative jitter free acceleration

In this paper, we explored the simultaneous generation and shaping of drive and witness bunches. Also, bunches travel through a rectangular CWFA structure to demonstrate a high transformer ratio and an efficient acceleration. Particle tracking code GPT [6] is used for the simulation. This code includes 3D space charge effect and coherent synchrotron radiation.

2. Generation and tailoring of drive and witness beams

The simulation is performed with a double EEX beamline (Figure 1) and 6D Gaussian beam artificially generated at the entrance to the beamline. The bending angle of the dogleg is 15°, and the spacing between the dipoles is 1.5 m. This angle is too large to transport a high charge beam without significant emittance growth from CSR but we used 15° in order to install the experiment in the space available at the Argonne Wakefield Accelerator facility [5]. The dispersion from the dogleg is 0.5 m, and the corresponding transverse deflecting cavity (TDC) kick strength is $\kappa = 2\ m^{-1}$. The input beam parameters are given in Table 1. Quadrupoles in the middle of the beamline are set to satisfy the transverse parameters (with the subscript 2) in Table 1. The mask is located in between the last quadrupole in the middle and the first dipole of the second EEX beamline. The shape of the mask is determined from Equation (2) in Reference [7].

Table 1. Beam parameters for 6D Gaussian input. Subscript 1 and 2 stands for the entrance to the first and second EEX beamline respectively.

Input beam parameters	Value	Unit
Energy	40	MeV
Charge	12	nC
σ_{x1}, σ_{y1}, σ_{z1}	5.0, 3.0, 1.5	mm
S_{x1}, S_{y1}, S_{z1}	-0.2, 0.0, 1.0	m^{-1}
σ_{x2}, σ_{y2}	8.0, 3.0	mm
S_{x2}, S_{y2}	0.0, 0.0	m^{-1}

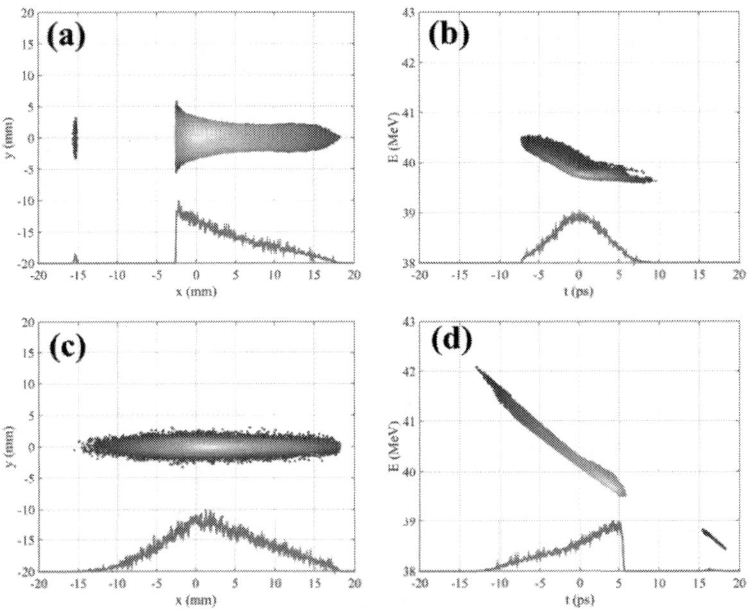

Figure 2. (a) x-y projection and (b) longitudinal phase space after the mask. (c) x-y projection and (d) longitudinal phase space after the second EEX beamline. Red curves are projetions to the x-axis (i.e. horizontal profile or temporal profile)

Figure 2 shows the transverse projection and the longitudinal phase space of the beam before and after the second EEX. The mask separates 12 nC beam to two beamlets whose charges are 4.14 nC (drive) and 22.5 pC (witness). The mask size is designed to make the drive bunch length 2λ, the witness bunch length 10° of , and the separation $5/4\lambda$, where λ is the wavelength of the wakefield. Since the R51 of the second EEX is 0.25, the dimensions on the mask are 20 mm, 0.3 mm, and 13 mm, respectively (Figure 2a). Separation and bunch length control are accomplished by adjusting the horizontal mask or changing R51 of the EEX beamline with the quadrupoles in front of the second EEX beamline.

Horizontally shaped profiles of the beamlets become a single horizontally symmetric profile after the second EEX beamline (Figure 2c). Similarly, the initially symmetric longitudinal profile (Figure 2b) becomes two beamlets (Figure 2d) at the end. The drive beam has a clear triangle profile and its bunch length is 17 ps which is close to 2λ. Unfortunately, the profile of the witness beam is collapsed. The ideal bunch length is 0.23 ps, but the result is 2 ps. The lengthening occurs due to the thick-lens effect (dominant) and second order effects [3]. The thick-lens effect term, $L_c/4\kappa^2\xi$, is 0.06, so an initial rms bunch length of 1.5 mm becomes at least 1.1 ps long. This can be overcome by introducing a fundamental mode cavity to the beamline to suppress the thick-lens effect [8].

3. Wakefield acceleration with longitudinally tailored beams

To simulate the wakefield acceleration, we used CST to generate the wake function of a 120 GHz rectangular CWFA structure [9]. Since the horizontal beam size easily blows up due to CSR along this beamline (due to its large bending angle), rectangular structures are preferred.

IOP Conf. Series: Journal of Physics: Conf. Series **874** (2017) 012027 doi:10.1088/1742-6596/874/1/012027

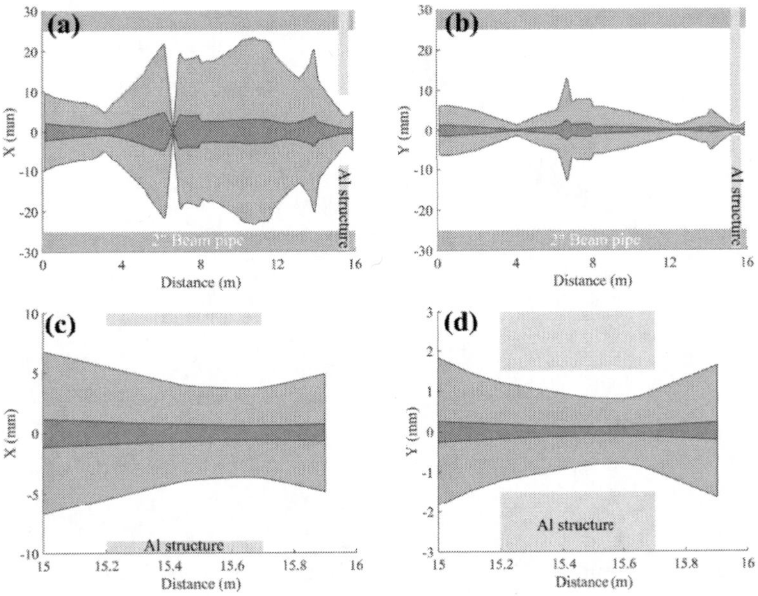

Figure 3. Horizontal (a) and vertical beam (b) envelope along the double EEX beamline. The red envelopes show the simga for x and y, and the blue envelopes shows the full size of the beam for x and y. (c) and (d) shows the beam envelopes near the structure.

Dimensions in Table 2 were chosen to generate a measurable energy loss on the drive and energy gain (~1 MeV) on the witness beam during the planned experiment.

Since the structure is only 50 cm long, three quadrupoles are used to focus the beam to minimize its size near the center of the structure. Horizontal and vertical beam envelops along the beamline are shown in Figure 3. All particles pass through the structure without any loss.

Table 2. Dimensions of the CWFA structure generating a wakefield. The shape of the structure is a rectangle.

Dimensions	Value	Unit
Height (vacuum)	3	mm
Width (vacuum)	18	mm
Length	500	mm
Thickness	130	mm

The wakefield generated by the drive bunch in the structure is calculated by taking the convolution of the wake function and the current profile (Figure 2). Figure 4 shows the temporal profiles of the drive and the witness bunches and the corresponding wakefield. Two oscillations of the wakefield inside the drive bunch show that the bunch length is 2λ. Although the witness bunch is lengthened by the second order and thick-lens effects, the separation is well-controlled.

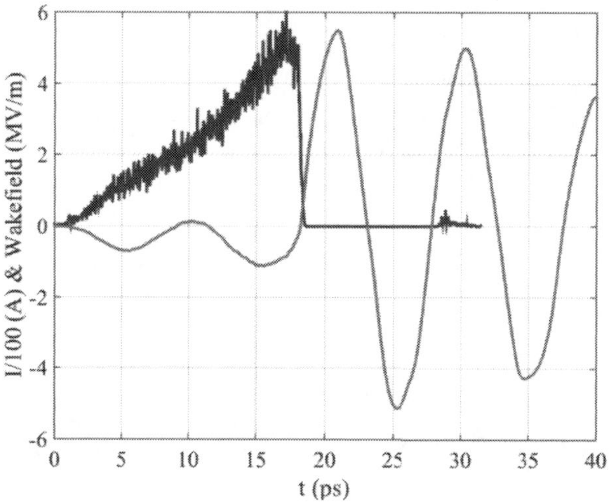

Figure 4. Current profiles at the entrance to the structure (blue). Wakefield calculated by the convolution of the current profile and the wake function (red).

The witness beam is $\sim 5/4\lambda$ away from the end of drive beam where the accelerating wakefield becomes maximum.

The ideal transformer ratio due to a bunch length of 2λ is 6.3 [10]. The simulated transformer ratio, however, is 4.5 which is 71% of the ideal due to the curvature in the triangle shape. This curvature creates an imbalance between the two peaks of the wakefield inside the drive bunch. It results in a higher second peak (1.12 MV/m), so the ratio of maximum accelerating wakefield (5 MV/m) to the maximum decelerating wakefield (1.12 MV/m) is lower than the ideal value. This curvature originates from the initial horizontal profile (Figure 2a) (i.e. initial horizontal profile is not a perfect triangle).

Figure 5 shows the particle tracking simulation results. A homemade wake module in GPT calculates the wakefield by convoluting the wake function and the current profile. The module also calculates the effect of the wake function on the bunch. Figure 5a shows the longitudinal phase space before the CWFA structure and 5b shows it after. The longitudinal phase space of the drive beam also has an oscillating shape due to the oscillating decelerating wakefield (Figure 4) while the witness beam shows a typical RF curvature with an acceleration. The maximum energy loss of the drive beam is about 0.4 MeV (39.5→39.1 MeV), and the maximum energy gain of the witness beam is about 2.5 MeV (38.5→41.0 MeV). Therefore the transformer ratio should be in between 4 and 6 which is larger than 2 (transformer ratio limit for symmetric bunches) [11]. This result agrees with the estimated number.

4. CSR and emittance growth

Although the shaping and a high transformer ratio is feasible, one of the major limitation of present EEX method is the emittance growth from CSR. This method uses eight dipole magnets which makes the beam has more chances to interact with CSR. Also, it requires a high charge beam due to the low transmission (30 – 40%). When the 12 nC beam passes through the first EEX, the four dipoles contribute CSR that dilutes the emittance significantly (30→200 μm). More seriously, CSR directly changes both the initial horizontal and longitudinal phase spaces due to the exchange process. This effect increases both projected and slice emittances. We are

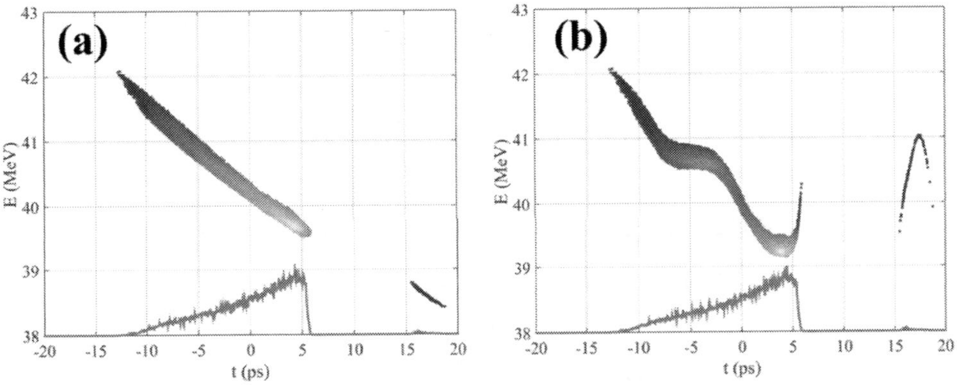

Figure 5. Longitudinal phase space before (a) and after (b) the CWFA structure. Red curves are temporal profiles.

still investigating methods for suppressing CSR and its effect on the emittance in double EEX beamline.

Acknowledgement

This work is supported by Department of Energy, Office of Science, under Contract No. DE-AC02-06CH11357.

References

[1] Jiang B, Jing C, Schoessow P, Power J and Gai W 2012 *Phys. Rev. Accel. Beams* **15** 011301
[2] Zholents A *et al.* 2016 *Nucl. Instrum. Methods Phys. Res. Sect.* A **829** 190
[3] Emma P, Huang Z, Kim K J and Piot P 2006 *Phys. Rev. Accel. Beams* **9** 100702
[4] Sun Y E, Piot P, Johnson A, Lumpkin A H, Maxwell T J, Ruan J and Thurman-Keup R 2010 *Phys. Rev. Lett.* **105** 234801
[5] Ha G *et al.* 2017 *Phys. Rev. Lett.* **118** 104801
[6] www.pulsar.nl/gpt
[7] Power J G, Zholents A, Kim K J, Conde M, Jing C, Cho M H, Namkung W and Ha G 2014 *Proc. International Partical Accelerator Conference 2014* (Geneva: JACoW) p 1506
[8] Zholents A and Zolotorev M S 2011 A new type of bunch compressor and seeding of a short wave length coherent radiation (*Preprint* ANL-APS-LS-327)
[9] Barber B private communication
[10] Lemery F and Piot P 2015 *Phys. Rev. Accel. Beams* **18** 081301
[11] Bane K L F, Chen P and Wilson P B 1985 On collinear wakefield acceleration (*Preprint* SLAC-PUB-3662)

IOP Conf. Series: Journal of Physics: Conf. Series **874** (2017) 012028 doi:10.1088/1742-6596/874/1/012028

First results with the novel petawatt laser acceleration facility in Dresden

U Schramm, M Bussmann, A Irman, M Siebold, K Zeil, D Albach, C Bernert, S Bock, F Brack, J Branco, JP Couperus, TE Cowan, A Debus, C Eisenmann, M Garten, R Gebhardt, S Grams, U Helbig, A Huebl, T Kluge, A Köhler, JM Krämer, S Kraft, F Kroll, M Kuntzsch, U Lehnert, M Loeser, J Metzkes, P Michel, L Obst, R Pausch, M Rehwald, R Sauerbrey, HP Schlenvoigt, K Steiniger, O Zarini

Helmholtz-Zentrum Dresden-Rossendorf, 01328 Dresden, Germany
Technische Universität Dresden, 01069 Dresden, Germany

E-mail: u.schramm@hzdr.de

Abstract. We report on first commissioning results of the DRACO Petawatt ultra-short pulse laser system implemented at the ELBE center for high power radiation sources of Helmholtz-Zentrum Dresden-Rossendorf. Key parameters of the laser system essential for efficient and reproducible performance of plasma accelerators are presented and discussed with the demonstration of 40 MeV proton acceleration under TNSA conditions as well as peaked electron spectra with unprecedented bunch charge in the 0.5 nC range.

1. Introduction

Plasma based concepts providing compact high gradient acceleration of electron and ion bunches receive growing interest in the particle accelerator community. Highest peak currents at ultra-short pulse duration, high initial energies, and point-like sources represent beam parameters that may open numerous applications from medical therapy to compact and bright light sources. On the other hand, a number of parameters like energy spread, average power, reproducibility, or matched diagnostics remains to be understood and controlled before demanding applications can be served in routine operation, standard to conventional rf accelerator technology.

For this reason the German Helmholtz Association has recently launched the dedicated program "Accelerator Research and Development (ARD)" building on and focusing the unique combination of research in superconducting rf accelerator physics and relativistic plasma physics at DESY Hamburg, KIT Karlsruhe, GSI Darmstadt (together with HI-Jena) and HZDR Dresden. In this article we introduce the Dresden "ELBE center for high power radiation sources" where in the extended building (Figure 1) of the superconducting cw linear electron accelerator ELBE two conceptually different Petawatt laser systems are currently under commissioning.

IOP Conf. Series: Journal of Physics: Conf. Series **874** (2017) 012028 doi:10.1088/1742-6596/874/1/012028

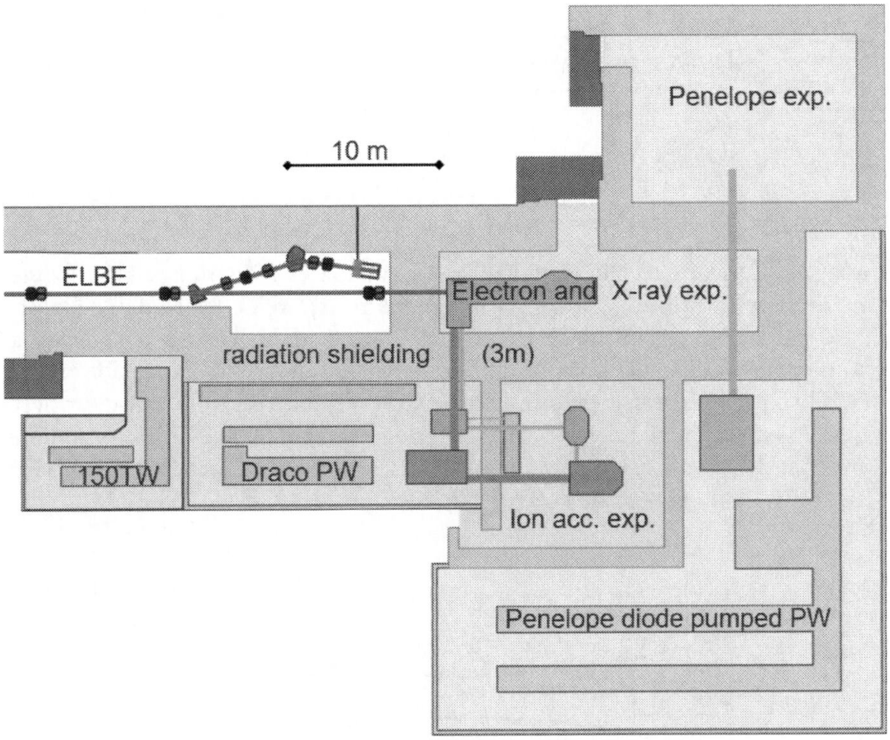

Figure 1. Floor plan of the upgraded HZDR ELBE center for high power radiation sources, illustrating the placement of the two PW lasers as close as possible to the dedicated target areas presently under operation or development (green: DRACO (Dresden Laser Acceleration Source), orange: PEnELOPE (Petawatt energy efficient laser for plasma experiments)) and the connection to the beamline of the linear accelerator (red) with its user facility radiation sources (not shown) extending to the left.

2. PW laboratory infrastructure

On the one hand, energy efficient laser technology supporting highest peak and average power simultaneously in a presently unaccessible range is developed in the framework of the PEnELOPE laser project [1]. It relies on direct diode laser pumping of the broad-band gain medium Yb:CaF_2 [2, 3, 4], expected to support pulse durations of 100-150 fs and energies of up to 150 J on target. PEnELOPE serves as a prototype for the development of 1 Hz pulse repetition rate compatible pumping and cooling techniques and as a driver for plasma proton acceleration with the focus on studying the applicability for radiation tumor therapy [6, 5]. The project, having reached the 10 J energy level in the lasers second to last major amplifier is closely linked to the ongoing development of the POLARIS laser system operated at the HI-Jena on target at a 250 TW level with highest available pulse intensity contrast in this pulse duration regime [7].

On the other hand, and complementing lased development with plasma accelerator research, the ultra-short pulse 150 TW Ti:Sapphire laser DRACO [8] has been upgraded (Amplitude Technologies) to a dual beam double chirped pulse amplification (CPA) system providing full Petawatt (30 J in 30 fs) and 150 TW (4.5 J in 30 fs) on target with optimized temporal pulse contrast and beam quality. The concept of the laser system, incorporating active spectral phase

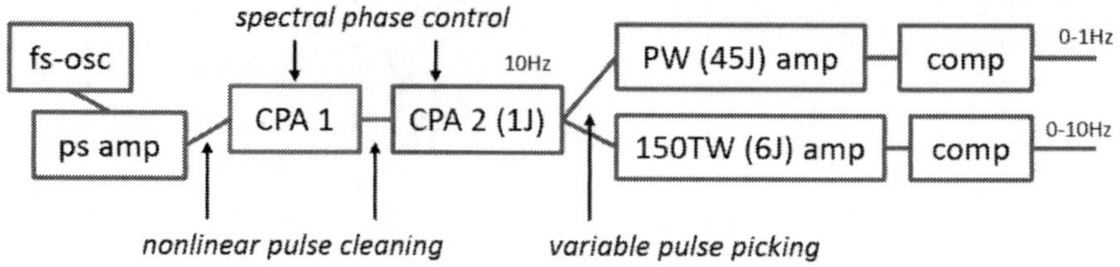

Figure 2. Concept of the double CPA (chirped pulse amplification) dual beam Ti:Sapphire 150 TW / PW laser chain DRACO at HZDR. Colors indicate stages, the fs front-end (red), a first 20 mJ CPA stage consisting of regenerative and multi-pass amplifiers with full active spectral phase and amplitude control (green), pulse cleaning (XPW), a second CPA stage where the beam is split on the J-level to seed final (partially cryogenic) amplifiers and independently operating compressors. Beam quality on target is fed back to the active elements of the second CPA stage.

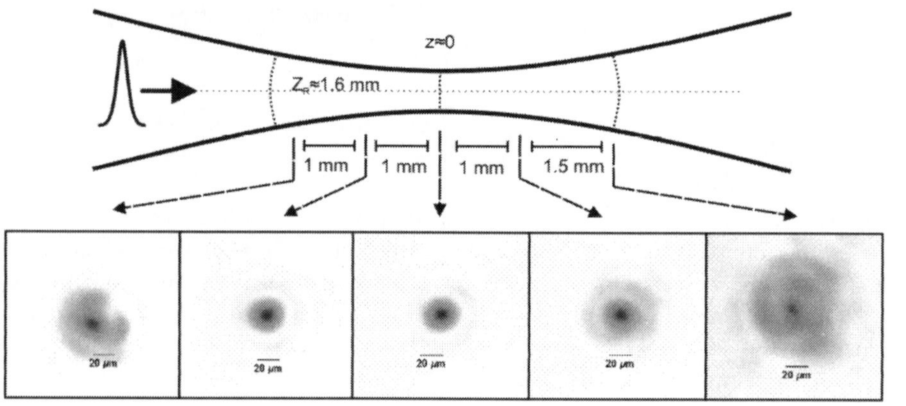

Figure 3. Development of the transverse intensity profile of the laser pulse in the vicinity of the focal spot generated with an F/20 off-axis parabolic mirror.

control for optimum temporal pulse profiles, booster amplifiers and cross-polarized wave (XPW) pulse cleaning for optimum temporal contrast, pump beam profile homogenization for smooth near field profiles, active beam pointing stabilization, full active back-reflection protection, and multiple adaptive optics for wavefront correction is sketched in Figure 2.

An illustration of the spatial beam quality is given with Figure 3 where the propagation of the beam at full energy through the focus of a plasma wakefield accelerator is investigated [9]. Beam quality on target thus corresponds to a Strehl-ratio of better than 0.9 ensuring sufficient energy content in the focal spot as well as controlled propagation even under non-linear plasma conditions. Improvement of the temporal pulse contrast, as mandatory for the interaction with down to skin-length (few nm) thick solid density targets, is provided for both chains close to the

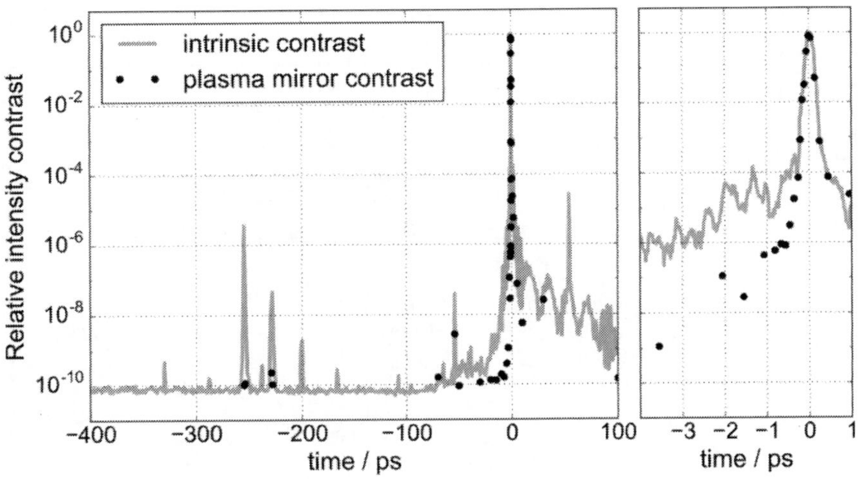

Figure 4. Intensity pulse contrast, measured with a scanning third order autocorrelator for the standard configuration (intrinsic) and behind a single stage plasma mirror (dots), highlighting the contrast improvement on the ps-scale.

target by single-stage plasma mirrors [10]. The actual temporal contrast representing conditions at full laser energy on target is displayed in Figure 4. A sharp central peak (width 30 fs, here measurement resolution limited) can be provided within the dynamic range of active spectral phase measurement and feedback (10^{-4}) and is preceded by a pedestal slowly raising between -20 to -10 ps from the ASE background level of few 10^{-11} (monitored up to -2 ns). Few discrete pre- and post-pulses still appear, which can however be controlled by the plasma mirror.

Figure 1 summarizes the laboratory infrastructure built at HZDR for advanced accelerator research. Connected and synchronized to the intense ps long bunches of the ELBE accelerator, pulses of both laser systems can be guided to permanently installed and equipped target areas with dedicated topical focus and corresponding radiation protection measures. Ion acceleration experiments require short focal length irradiation of solid targets and, ideally, options for reference irradiation under well-defined conditions for the direct comparison of the performance of novel target concepts to reference targets. Electron acceleration in laser driven wakefields requires long focal length for matching laser beam propagation with the guiding properties of the generated long plasma channel. X-ray generation requires long forward observation geometries [11] and potentially colliding pulse scenarios as for the case of two beam Thomson-scattering experiments. These can be exploited as ultra-short synchronized probes of plasma experiments or for the unique phase space diagnostics of acceleration processes [12].

3. Ion acceleration

The realization of ion acceleration with high power lasers [13, 14] holds the promise of providing compact sources for applications where, due to the limited availability of pulse repetition rate, only low average currents are required. The most prominently discussed application is medical, namely particle beam cancer therapy [5] and related radio-biological studies [6]. The most established and robust acceleration mechanism is the so-called Target Normal Sheath Acceleration (TNSA), where ions are accelerated in the sheath field set by laser-energized MeV electrons at the target surfaces. TNSA beams are used in applications, which exploit their

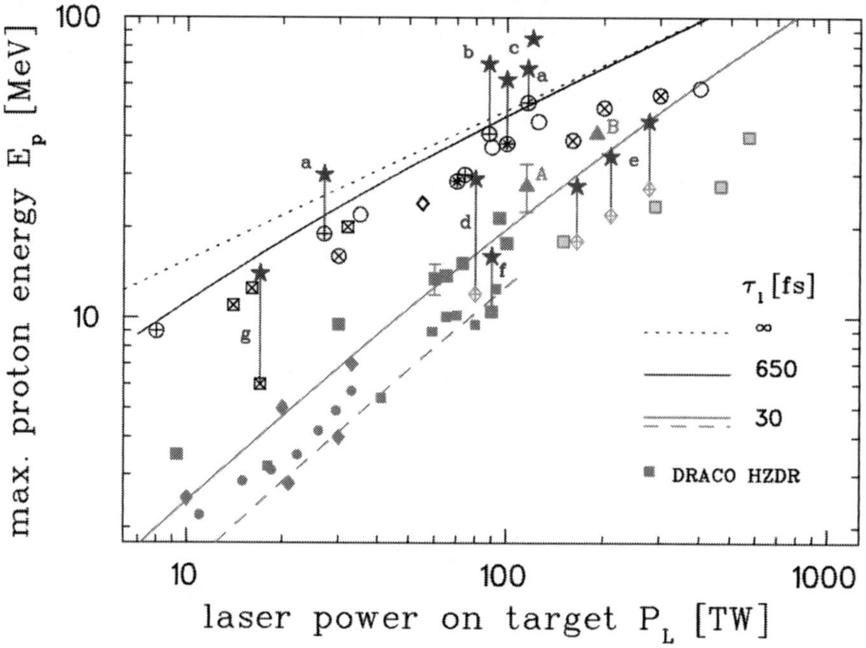

Figure 5. Overview of maximum proton energies achieved with linear polarization as a function of laser power P_L on target based on [15]. The plot is updated with recently published work and emphasizes energy enhancing mechanisms (blue stars, connected to individual TNSA references). Open black symbols as in [8] represent data from single-shot glass laser systems where record energies were reached with a) hollow cone targets [16] and b,c) in regimes where relativistic transparency sets in [17, 18, 19]. Filled symbols depict results obtained with state-of-the-art Ti:Sapphire laser ultra-short pulse systems (as in [8]), squares representing DRACO (HZDR) data [8, 20], circles UHI100 (CEA Saclay) data [21] and A) and B) highlighting recent TNSA improvements [22, 23] on the multi 100 TW laser scale. Energies could be further increased by using e) ultra-thin targets (comparison between thinnest and thickest targets still assumed to be dominated by TNSA) [24], d) carbon nanotube foam coated thin targets [25], or f,g) flat targets of finite size (reduced mass targets) [26, 27]. Promising enhancement factors for protons are typically observed to reach values between 1.5 and 2.5. Recent DRACO PW commissioning data from $1-2\,\mu$m thick Ti-foils is highlighted in green.

advantageous properties such as short burst duration and low emittance (point source). A number of alternative acceleration mechanisms have emerged which promise higher efficiency, and, in some cases, enhanced spectral properties beyond the state-of-the-art exponential spectra characterized by a distinct cut-off energy. The ongoing improvement of laser parameters (pulse energy and contrast, see Figure 4) and diagnostics tools, jointly with technological innovation in target manufacturing are all key factors driving the field. A summary of published data on proton acceleration is presented in Figure 5 illustrating the continuous increase in maximum energy with laser performance and highlighting (stars) the potential for a variety of advanced target concepts, including (for ps-class pulses and ultra-thin foils) the onset of volumetric interaction in the relativistic transparency regime or geometric, field enhancing effects.

Recent commissioning data, where the DRACO-PW beam was focused with an F/2.5 off-axis paraboloid to a preliminary intensity of $I \sim 6 \cdot 10^{21}$ W/cm^2 on 1-2 μm thick Ti-foils is depicted as green squares. Proton energies follow the general trend [14] up to $\sim$ 40 MeV, although absolute values of maximum energies still remain below the expectation set by scaling data from the 150 TW laser arm. Here, and for different target geometries and types up to 10 MeV proton cut-off energy per Joule of laser energy could be achieved up to a maximum value of about 25 MeV for ultra-thin targets.

4. Electron acceleration

Laser-plasma wakefield acceleration (LWFA) has demonstrated ultra-short (fs) electron bunches with peaked energy distribution up to the multi GeV-range on a centimeter scale. Various injection mechanisms into linear and non-linear wakes have been established in order to control initial phase space and thus energy and energy spread [28], yet, scaling the bunch charge to the nC-level and thus to unprecedented peak current in the hundreds of kA as originally predicted [29] for the non-linear bubble-regime has not yet been achieved.

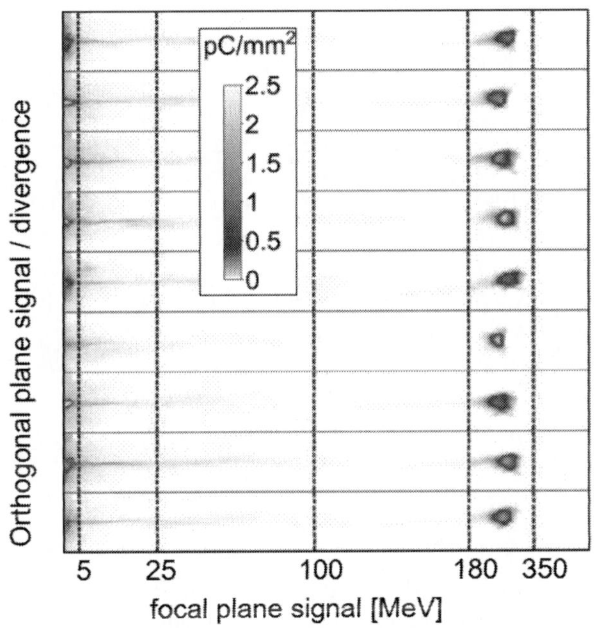

Figure 6. Spectrometer raw-data for 9 consecutive shots illustrating the stability of the peaked energy distribution at low medium energy backround level. Typical charge per bunch is 250 pC (FWHM) in the peak at a relative bandwidth of 15 %.

Schemes relying on ionization injection, where the helium-gas based acceleration medium is doped with small (percent level) fractions of nitrogen, allow for a controlled injection of nitrogen K-shell electrons near the intensity peak of the driving laser pulse. These schemes are known to provide high bunch charge but rather large energy spread. Based on the high quality of the focal spot profile (see Figure3) we recently demonstrated [9] that carefully controlled laser and wakefield evolution enables the restriction of the injection time and thus of the final energy spread. Previous reports on this self truncated ionization injection scheme [30] already demonstrated improved energy bandwidth, yet, at the cost of limited bunch charge.

Well reproducible high charge performance of the plasma wakefield accelerator is summarized in Figure 6, showing a small fraction of a shot sequence in spectrometer raw-data format. Peaked distributions exhibiting total charges of up to 0.4 nC were observed with low background. Only

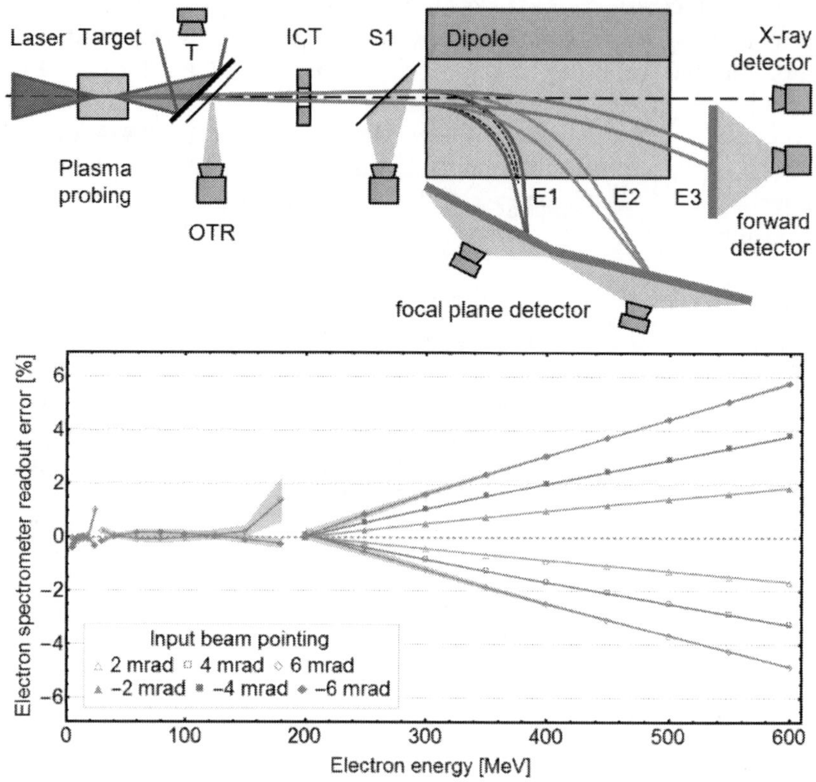

Figure 7. Schematic LWFA electron beam diagnotics setup. Electron bunches are decoupled from co-propagating laser light, pass diagnostics screens for the generation of transition radiation (OTR) or profile and pointing monitoring (S1), are globally charge analyzed by integrating current transformers (ICTs), are energy analyzed in a dispersive dipole magnet and finally dumped. Spectrometers are used in imaging (E1,E2) and non-imaging (E3) configuration with linear read-out planes covering a wide energy range and absolutely calibrated charge response. Different levels of energy resolution are achieved as presented in the graph based on ray-tracing calculations in the mapped field. Shaded areas indicate the influence of vertical position errors of the incoming beam (±0.2 mm). X-rays emitted by the electron bunch are measured after separation from the electron beam behind the magnet.

at the lowest energies (below few MeV) significant background charge is present, presumably due to down-ramp injection at the end of the plasma source.

Characterization of not only a wide range of electron energies in single-shot but also of pulse duration is mandatory for the characterization of LWFA pulses for peak current sensitive applications in radiation sources, in particular in compact undulator based [31] FELs and optical undulators [32, 33]. Figure 7 shows a typical reference setup for the longitudinal characterization of electron bunches. Measurement of differential charge relies on the optical read-out of Lanex-type phosphorescing screens, which for the study presented here have been systematically cross-calibrated against well-defined bunches of the ELBE accelerator. Energy resolution well below the typical percent-level energy spread (lower Figure7) can easily be reached by placing sets of linear (Lanex-) screens in the imaging and non-imaging planes of the spectrometer after modelling.

Laboratory generated high peak current bunches as presented above with estimated peak currents of $\sim 50\,\text{kA}$ [9] and a spectral charge density of around $10\,\text{pC/MeV}$ are not only of interest for driving compact light sources ranging from the far infrared THz to the X-ray region. They further open a path to the investigation of fundamental aspects [34] of beam driven plasma accelerators [35] independent from large scale infrastructure.

5. Acknowledgement

J.P.C. & J.M.K. acknowledge LA3NET (No. 289191) from the EU 7th Framework Programme. The work has been partially supported by Horizon 2020 LASERLAB-EUROPE / LEPP (Contract No. 654148) and EUCARD2 (No. 312453). M.G. and A.H. acknowledge support from the European Cluster of Advanced Laser Light Sources EUCALL (No. 65422).

6. References

[1] Siebold M *et al.* 2013 PEnELOPE: a high peak-power diode-pumped laser system for laser-plasma experiments *Proceeding SPIE* **8780** 878005
[2] Petit V *et al.* 2014 CW and tunable laser operation of Yb^{3+} doped CaF_2 *Applied Physics* B **78** 681
[3] Siebold M *et al.* 2009 Yb:CaF_2 a new old laser crystal *Applied Physics* B **97** 327
[4] Körner J *et al.* 2014 Spectroscopic characterization of Yb^{3+} doped laser materials at cryogenic temperatures *Applied Physics* B **75** 116
[5] Ledingham K *et al.* 2014 Towards Laser Driven Hadron Cancer Radiotherapy: A Review of Progress *Applied Sciences* **4** 402
[6] Zeil K *et al.* 2013 Dose-controlled irradiation of cancer cells with laser-accelerated proton pulses *Applied Physics* B **110** 437
[7] Hornung M *et al.* 2016 54J pulses with 18nm bandwidth from a diode-pumped chirped-pulse amplification laser system *Optics Letters* **41** 5413
[8] Zeil K *et al.* 2010 Scaling of proton energies in ultra-short pulse laser plasma acceleration *New Journal of Physics* **12** 045015
[9] Couperus JP *et al.* 2017 Demonstration of a beam loaded nanocoulomb-class laser wakefield accelerator (unpublished)
[10] Roedel C *et al.* 2011 High repetition rate plasma mirror for temporal contrast enhancement of terawatt femtosecond laser pulses by three orders of magnitude *Applied Physics* B **103** 295
[11] Köhler A *et al.* 2016 Single-shot betatron source size measurement from a laser-wakefield accelerator *Nuclear Instruments and Methods* A **829** 265
[12] Jochmann A *et al.* 2013 High Resolution Energy-Angle Correlation Measurement of Hard X Rays from Laser-Thomson Backscattering *Physical Review Letters* **111** 114803
[13] Macchi A, Borghesi M and Passoni M 2013 Ion acceleration by superintense laser-plasma interaction *New Journal of Physics* **85** 751
[14] Schreiber J, Bolton PR and Parodi K 2016 Hands-on laser-driven ion acceleration: A primer for laser-driven source development and potential applications *Review of Scientific Instruments* **87** 071101
[15] Borghesi M and Schramm U 2016 Summary of Working Group 2: Ion beams from plasmas *Nuclear Instruments and Methods in Physics Research* A **829** 137
[16] Kluge T *et al.* 2012 High proton energies from cone targets: electron acceleration mechanisms *New Journal of Physics* **14** 023038
[17] Jung D *et al.* 2013 Beam profiles of proton and carbon ions in the relativistic transparency regime *New Journal of Physics* **15** 123035
[18] Wagner F *et al.* 2015 Simultaneous observation of angularly separated laser-driven proton beams accelerated via two different mechanisms *Physics of Plasmas* **22** 063110
[19] Wagner F *et al.* 2016 Maximum Proton Energy above 85 MeV from the Relativistic Interaction of Laser Pulses with Micrometer Thick CH2 Targets *Physical Review Letters* **116** 205002
[20] Metzkes J *et al.*, 2014 Experimental observation of transverse modulations in laser-driven proton beams *New Journal of Physics* **16** 023008

IOP Conf. Series: Journal of Physics: Conf. Series **874** (2017) 012028 doi:10.1088/1742-6596/874/1/012028

[21] Passoni M *et al.* 2009 Proton Maximum Energy Cutoff Scaling Laws For Bulk Targets *AIP Conference Proceedings* **1153** 159
[22] Green JS *et al.* 2014 High efficiency proton beam generation through target thickness control in femtosecond laser-plasma interactions *Applied Physics Letters* **104** 214101
[23] Ogura K *et al.* 2012 Proton acceleration to 40 MeV using a high intensity, high contrast optical parametric chirped-pulse amplification Ti:Sapphire hybrid laser system *Optics Letters* **37** 2868
[24] Kim IJ *et al.* 2013 Transition of proton energy scaling using an ultrathin target irradiated by linearly polarized femtosecond laser pulses *Physical Review Letters* **111** 165003
[25] Bin JH *et al.* 2015 Ion acceleration using relativistic pulse shaping in near-critical-density plasmas *Physical Review Letters* **115** 064801
[26] Buffechoux S *et al.* 2010 Hot electrons transverse refluxing in ultraintense laser-solid interactions *Physical Review Letters* **105** 015005
[27] Zeil K *et al.* 2014 Robust energy enhancement of ultrashort pulse laser accelerated protons from reduced mass targets *Plasma Physics and Controlled Fusion* **56** 084004
[28] Esarey E, Schroeder CB and Leemans WP 2009 Physics of laser-driven plasma-based electron accelerators *Review of Modern Physics* **81** 1229
[29] Pukhov A and Meyer-ter-Vehn J 2002 Laser wake field acceleration: the highly non-linear broken-wave regime *Applied Physics* B **74** 355
[30] Mirzaie M *et al.* 2015 Demonstration of self-truncated ionization injection for GeV electron beams *Scientific Reports* **5** 14659
[31] Eichner T *et al.* 2007 Miniature magnetic devices for laser-based, table-top free-electron lasers *Physical Review ST Accelerators and Beams* **10** 082401
[32] Debus A *et al.* 2010 Traveling-wave Thomson scattering and optical undulators for high-yield EUV and X-ray sources *Applied Physics* B **100** 61
[33] Steiniger K *et al.* 2014 Optical free-electron lasers with Traveling-Wave Thomson-Scattering *Journal of Physics* B **47** 234011
[34] Chou S *et al.* 2016 Collective Deceleration of Laser-Driven Electron Bunches *Physical Review Letters* 117 144801
[35] Heinemann T *et al.* 2017 Investigating the key parameters of a staged laser- and particle driven plasma wakefield accelerator experiment *Proc. 8th Int. Particle Accelerator Conf. (IPAC17) Copenhagen Denmark* TUPIK010

Horizon 2020 EuPRAXIA design study

P A Walker[1], P D Alesini[2], A S Alexandrova[3,4], M P Anania[2], N E Andreev[5,6], I Andriyash[7], A Aschikhin[1], R W Assmann[1], T Audet[8], A Bacci[9], I F Barna[10], A Beaton[11], A Beck[12], A Beluze[13], A Bernhard[14], S Bielawski[15], F G Bisesto[2], J Boedewadt[1], F Brandi[16], O Bringer[17], R Brinkmann[1], E Bründermann[14], M Büscher[18], M Bussmann[19], G C Bussolino[16], A Chance[17], J C Chanteloup[13], M Chen[20], E Chiadroni[2], A Cianchi[21], J Clarke[3,22], J Cole[23], M E Couprie[7], M Croia[2], B Cros[8], J Dale[1], G Dattoli[24], N Delerue[25], O Delferriere[17], P Delinikolas[11], J Dias[26], U Dorda[1], K Ertel[27], A Ferran Pousa[1,28], M Ferrario[2], F Filippi[2], J Fils[17,29], R Fiorito[3,4], R A Fonseca[26], M Galimberti[27], A Gallo[2], D Garzella[17], P Gastinel[17], D Giove[9], A Giribono[2], L A Gizzi[16,30], F J Grüner[28,31], A F Habib[1], L C Haefner[32], T Heinemann[1,11,28], B Hidding[3,11], B J Holzer[33], S M Hooker[34,35], T Hosokai[36], A Irman[19], D A Jaroszynski[11], S Jaster-Merz[1,28], C Joshi[37], M C Kaluza[38,39], M Kando[40], O S Karger[28], S Karsch[41], E Khazanov[42], D Khikhlukha[43], A Knetsch[28], D Kocon[43], P Koester[16], O Kononenko[1], G Korn[43], I Kostyukov[42], L Labate[16,30], C Lechner[1], W P Leemans[44], A Lehrach[18], F Y Li[11], X Li[17], V Libov[28], A Lifschitz[45], V Litvinenko[46,47], W Lu[48], A R Maier[28,31], V Malka[45], G G Manahan[11], S P D Mangles[23], B Marchetti[1], A Marocchino[49], A Martinez de la Ossa[28], J L Martins[26], F Massimo[45], F Mathieu[13], G Maynard[8], T J Mehrling[1], A Y Molodozhentsev[43], A Mosnier[17], A Mostacci[2,49], A S Mueller[14], Z Najmudin[23], P A P Nghiem[17], F Nguyen[24], P Niknejadi[1], J Osterhoff[1], D Papadopoulos[13], B Patrizi[50], R Pattathil[27], V Petrillo[9], M A Pocsai[10], K Poder[1], R Pompili[2], L Pribyl[43], D Pugacheva[5,6], S Romeo[2], A R Rossi[9], E Roussel[7], A A Sahai[23], P Scherkl[11], U Schramm[19], C B Schroeder[44], J Schwindling[17], J Scifo[2], L Serafini[9], Z M Sheng[11,20], L O Silva[26], T Silva[26], C Simon[17], U Sinha[26], A Specka[12], M J V Streeter[51], E N Svystun[1], D Symes[27], C Szwaj[15], G Tauscher[1], A G R Thomas[51], N Thompson[3,22], G Toci[50], P Tomassini[16], C Vaccarezza[2], M Vannini[50], J M Vieira[26], F Villa[2], C-G Wahlström[52], R Walczak[34,35], M K Weikum[1,11], C P Welsch[3,4], C Wiemann[18], J Wolfenden[3,4], G Xia[3,53], M Yabashi[54], L Yu[20], J Zhu[1], A Zigler[55]

1 Deutsches Elektronen-Synchrotron, 22607 Hamburg, Germany
2 INFN, Laboratori Nazionali di Frascati, 00044 Frascati, Rome, Italy
3 Cockcroft Institute, Warrington WA4 4AD, UK
4 University of Liverpool, Liverpool L69 7ZE, UK
5 JIHT, Moscow, 125412, Russia
6 Moscow Institute of Physics and Technology, Dolgoprudny, 141701, Russia
7 Synchrotron SOLEIL, Gif-sur-Yvette 91192, France
8 LPGP, CNRS, Univ. Paris-Sud, Université Paris-Saclay, 91405 Orsay, France
9 INFN, Sezione di Milano, Milan, Italy
10 Wigner Research Centre for Physics, Budapest, Hungary
11 SUPA, Department of Physics, University of Strathclyde, Glasgow G4 0NG, UK
12 LLR, CNRS, Ecole Polytechnique, Palaiseau and Université Paris Saclay, France
13 LULI, Ecole Polytechnique, CNRS, CEA, UPMC, 91128 Palaiseau, France

14 Karlsruhe Institute of Technology, 76131 Karlsruhe, Germany
15 PHLAM, UMR-CNRS 8523, Université de Lille, France
16 CNR Istituto Nazionale di Ottica, 56124 Pisa, Italy
17 CEA, IRFU, SACM, Université Paris Saclay, F-91191 Gif-sur-Yvette, France
18 Forschungszentrum Jülich, 52428 Jülich, Germany
19 Helmholtz Institute Dresden-Rossendorf, 01328 Dresden, Germany
20 Shanghai Jiao Tong University, Shanghai 200240, P. R. China
21 University of Rome Tor Vergata, 00173 Rome, Italy
22 STFC Daresbury Laboratory, Sci-Tech Daresbury, Warrington, U.K.
23 John Adams Institute, Blackett Laboratory, Imperial College London, UK
24 ENEA, Centro Ricerche Frascati, 00044 Frascati, Rome, Italy
25 LAL, CNRS/IN2P3 Univ. Paris Sud, Orsay, and Université Paris Saclay, France
26 GoLP/Instituto de Plasmas e Fusão Nuclear, Instituto Superior Técnico, Universidade de Lisboa, Lisbon, Portugal
27 Central Laser Facility, RAL, Didcot, Oxfordshire OX11 0QX, UK
28 Universität Hamburg, 22761 Hamburg, Germany
29 GSI Helmholtzzentrum für Schwerionenforschung, 64291 Darmstadt, Germany
30 INFN, Sezione di Pisa, Italy
31 Center for Free Electron Laser Science, 22607 Hamburg, Germany
32 Lawrence Livermore National Laboratory, Livermore, CA 94550, United States
33 CERN, 1211 Geneva 23, Switzerland
34 John Adams Institute, Oxford University, UK
35 University of Oxford, Oxford OX1 2JD, UK
36 Osaka University, Osaka Prefecture, 565-0871, Japan
37 University of California Los Angeles, Los Angeles, CA 90095, USA
38 Helmholtz Institute Jena, 07743 Jena, Germany
39 Institut für Optik und Quantenelektronik, 07743 Jena, Germany
40 KPSI- QST, Kyoto 619-0215, Japan
41 Ludwig-Maximilians-Universität München, 80802 Munich, Germany
42 IAP RAS, Nizhnij Novgorod, 603155, Russia
43 ELI-Beamlines, Dolni Brezany, Czech Republic
44 Lawrence Berkeley National Laboratory, Berkeley, CA 94720, USA
45 LOA, ENSTA-CNRS-École Polytechnique UMR 7639, Palaiseau F-91761, France
46 Brookhaven National Laboratory, Upton, NY 11973, USA
47 Stony Brook University, Stony Brook, NY 11794, USA
48 Tsinghua University, Beijing, 100084, P. R. China
49 Sapienza, University of Rome, 00161, Rome, Italy
50 CNR Istituto Nazionale di Ottica, I-50019 Sesto Fiorentino, Italy
51 Lancaster University, Lancaster LA1 4YB, UK
52 Lund University, 223 62 Lund, Sweden
53 University of Manchester, Manchester M13 9PL, UK
54 RIKEN SPring-8 Center, Hyogo, 679-5148, Japan
55 Hebrew University of Jerusalem, Jerusalem, Israel

E-mail: andreas.walker@desy.de

Abstract. The Horizon 2020 Project EuPRAXIA ("European Plasma Research Accelerator with eXcellence In Applications") is preparing a conceptual design report of a highly compact and cost-effective European facility with multi-GeV electron beams using plasma as the acceleration medium. The accelerator facility will be based on a laser and/or a beam driven plasma acceleration approach and will be used for photon science, high-energy physics (HEP) detector tests, and other applications such as compact X-ray sources for medical imaging or material processing. EuPRAXIA started in November 2015 and will deliver the design report in October 2019. EuPRAXIA aims to be included on the ESFRI roadmap in 2020.

1. Introduction

The EuPRAXIA collaboration is the first plasma accelerator collaboration on this scale bringing together 16 European partner laboratories and additional 22 associated partners from the EU, Israel, China, Japan, Russia and the USA [1]. EuPRAXIA is structured into 14 working packages each headed by two work package leaders from different institutions. Eight work packages receive EU funding and their topics include: plasma and laser simulations (WP2), plasma accelerator structures (WP3), laser design (WP4), conventional beam physics (WP5), FEL radiation (WP6), and a table-top test beam for HEP and other applications (WP7). Two further EU work packages work on the management of the collaboration (WP1) and the outreach to the public (WP8). In-kind work packages (WP9 - WP14) include additional approaches: beam driven plasma acceleration PWFA (WP9), hybrid acceleration schemes (WP14), alternative radiation generation (WP13) and alternative laser sources such as fiber lasers (WP10). WP11 and WP12 connect to prototyping on plasma-based FEL's and facility access for experiments until 2019. Industry partners Amplitude Technologies, Thales and TRUMPF Scientific take part in the scientific advisory board and contribute their experience towards a successful completion of the design report.

2. Plasma acceleration

Scientists, medical doctors and engineers have used radio-frequency (RF) based particle accelerator beams for ninety years to probe nature, to produce new particles, to generate light of exquisite quality or to irradiate tumors. The accelerators are of outstanding quality, but have grown in size and cost due to the materials used for construction, which can only sustain accelerating fields of around 100 MV/m before electrical breakdown occurs. Plasma accelerators are not subject to these electrical breakdown limits and the accelerating field reaches 100 GV/m, three orders of magnitude larger than in an RF accelerator. As a consequence, the size of plasma accelerators can potentially be quite small, reducing kilometer scale machines to the meter scale. A new generation of cost-efficient and compact accelerators could open completely new usages of particle accelerators, for example in hospitals and universities. This requires suitable stability and repetition rates.

The great potential of plasma waves for particle acceleration was first recognized by Veksler [2] and Tajima and Dawson [3]. The longitudinal plasma waves can be excited by both electron beams (plasma wakefield acceleration, PWFA) or intense laser pulses (laser wakefield acceleration, LWFA) and are well suited for accelerating charged particles to relativistic energies [4]. Electron beams that are accelerated inside a plasma accelerator structure can originate from the background plasma within the plasma accelerator structure itself ("internal injection") or from an accelerator that is situated in front of the plasma accelerator structure ("external injection"). Within the last two decades, the beam quality of LWFA accelerators has significantly improved [5-13] and the current peak energy lies at 4.2 GeV [14]. Using these beams, various types of X-ray radiation such as betatron, synchrotron, and undulator radiation down to the water-window wavelengths were produced [15-21]. While several tens of laboratories use laser systems to accelerate electrons, few laboratories have the electron beam needed for beam-driven plasma acceleration [22-28]. FACET at SLAC achieved energy doubling within a single electron beam in 2007 [24] and energy was transferred successfully from a drive beam to a witness bunch in 2014 [25].

In the EuPRAXIA study, both laser driven and beam driven approaches as well as combined plasma acceleration schemes - using LWFA-produced beams as drivers of PWFA stages [29, 30] - are taken into consideration. The final EuPRAXIA design report in 2019 will include various configurations of a possible EuPRAXIA facility. Depending on available budget and the targeted science case, one of these options, or a combination of options, might be the best choice. The design report will compare size, cost, and performance on a common basis. The first iteration of the design goals were defined in

October 2016 [31] and from these, the initial goal parameters for the 5 GeV electron beam at the entrance of the undulator are shown in Table 1. The agreed possible configurations are:

Configuration 1: LWFA with internal injection;
Configuration 2: LWFA with external injection from an RF accelerator;
Configuration 3: LWFA with external injection from a laser plasma injector;
Configuration 4: PWFA with an RF electron beam; and
Configuration 5: PWFA with LWFA produced electron beam (hybrid schemes).

In addition to the 5 GeV electron beam, the facility aims to provide a medical imaging X-ray source as well as FEL radiation ultimately concentrating in the range between 1 nm and 0.1 nm. TW laser pulses synchronized to the electron and X-ray radiation will be available in the user areas. Parameter tables for medical imaging and a table-top test beam for HEP and other applications are currently being finalized.

3. Laser and electron beam drivers

The laser used in the LWFA cases is being studied in work package 4 (WP4) with colleagues from Thales and Amplitude industry. WP4 reviewed current laser systems in 2016 [32] and proposed preliminary specifications of the EuPRAXIA laser, the so-called "100 cube" laser challenge (an energy of 100 Joule, a pulse length of 100 fs (FWHM), and a repetition rate of 100 Hz, with a contrast of 10^{10} at 10 ps). The present work towards this challenging goal disfavors a complete Ti:Sa laser system and is considering a diode-pumped solid-state laser pumping scheme. A second laser system, used for the plasma injector [33], will operate at lower energy and shorter pulse length.

Design work on the drive beam for the PWFA case is being performed in WP5. One option under discussion is that both configuration 2 (LWFA) and 4 (PWFA) use at low energy the same S-band injector and RF linac [34]. The simulated transverse phase space of a possible electron drive beam for a PWFA application is shown in Figure 1. This electron beam has an energy of 548 MeV, a peak beam current of 1 kA, transverse normalized emittances of 1 µrad m and an energy spread of below 0.07%. After acceleration through S-band and X-band structures, the beam is focused by both conventional, electro-magnets, and permanent quadrupole magnets before entering the plasma.

Table 1. Target values for the 5 GeV electron beam parameters at the entrance of the undulators [31].

Quantity	Symbol	Value
Particle type	e	Electrons
Energy	E	5 GeV
Charge	Q	30 pC
Bunch length (FWHM)	τ	10 fs
Peak current	I	3 kA
Repetition rate	f	10 Hz
Number of bunches	N	1
Total energy spread (RMS)	σ_E/E	1%
Slice energy spread (RMS)	$\sigma_{E,S}/E$	0.1%
Trans. Norm. emittance	$\varepsilon_{N,x}$, $\varepsilon_{N,y}$	1 µrad m
Alpha function	α_x, α_y	0
Beta function	β_x, β_y	5 m
Trans. beam size (RMS)	σ_x, σ_y	22 µm
Trans. divergence (RMS)	$\sigma_{x'}$, $\sigma_{y'}$	4.5 µrad

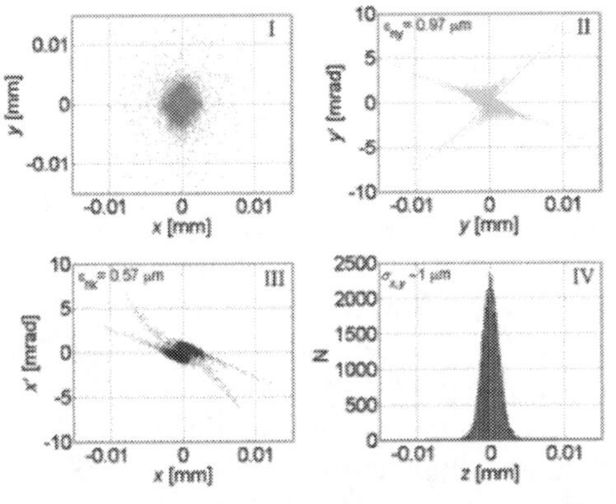

Figure 1. Preliminary simulation results for the transverse phase space of a possible PWFA drive beam (configuration 4) with an energy of 548 MeV, a peak beam current of 1 kA, transverse normalized emittances of 1 μrad m and an energy spread of below 0.07%.

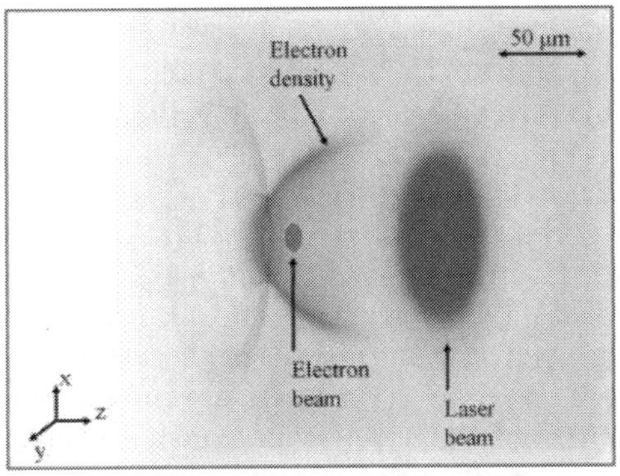

Figure 2. PIC simulation [36] of a LWFA case (configuration 2). The laser pulse (red) propagates the plasma (electron density shown in blue) from left to right and excites a wakefield, which accelerates electrons (shown in green) from 0.1 to 1 GeV in 2.5 cm.

4. Plasma accelerator structure

Components necessary for the design of the plasma accelerator structure were reviewed by WP3 in 2017 [35] in which published experimental results were examined and compared not only in terms of achieved electron properties, but also regarding their reliability, stability, or scalability to larger electron energy, or repetition rate. The proposed criteria from [35] for selecting a specific plasma accelerator structure will be used to decide which types of plasma accelerator will ultimately be incorporated into the design report.

Figure 2 shows a particle-in-cell (PIC) simulation [36, 37] performed with the OSIRIS code [38] in which a 1 PW laser traverses a plasma accelerator structure of 1.2 10^{17}cm^{-3} electron density. The externally injected electron beam (initially: energy E = 100 MeV; relative energy spread $\sigma_E/E = 0.1\%$; transverse emittance $\varepsilon_{N,x}$ = 1 μrad m) exits the plasma after 2.5 cm with an energy of 1 GeV (σ_E/E = 1.5%; $\varepsilon_{N,x}$ = 1 μrad m). While emittance is well preserved, the energy spread is significantly increased due to the sizable variation of the accelerating field along the injected bunch. Beam loading techniques

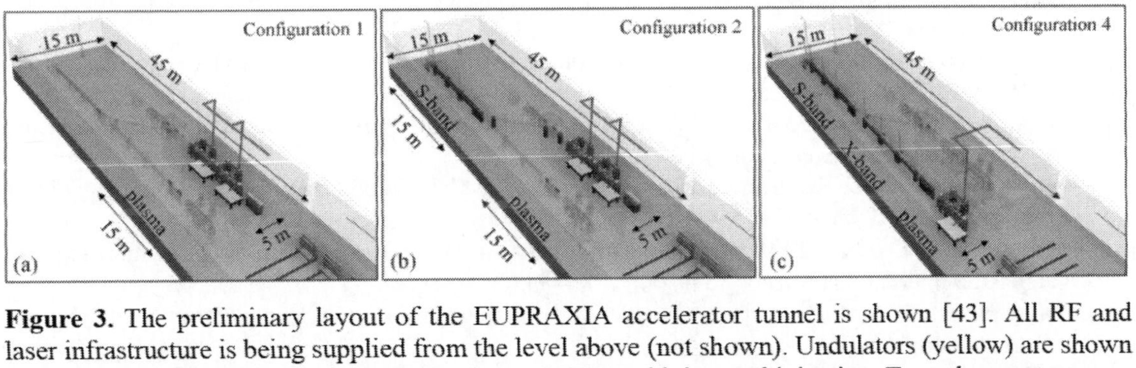

Figure 3. The preliminary layout of the EUPRAXIA accelerator tunnel is shown [43]. All RF and laser infrastructure is being supplied from the level above (not shown). Undulators (yellow) are shown in the bottom right corners. (a) Configuration 1: LWFA with internal injection. Two plasma stages are included which are supplied with two laser beams (red). (b) Configuration 2: LWFA with external injection from an RF accelerator. The RF gun and three S-band structures are shown in front of a dogleg which transports the electrons to the two plasma stages. (c) Configuration 4: PWFA. Using the same infrastructure of RF gun and S-band structure, the PWFA case uses additional X-band structures to accelerate beams to several hundred MeV before using it inside a single plasma accelerator stage.

will be used in order to compensate this gradient on the accelerating field and minimize the induced energy spread [39-42]. After completion of 1 GeV simulations with conservation of all beam qualities, simulations of the 5 GeV beam will continue.

5. Layout considerations

The preliminary layout of the EuPRAXIA accelerator tunnel [43] is shown in Figure 3, excluding user areas. Configurations 1, 2, and 4 are visualized. In the current layout, laser and RF infrastructure are situated on the level above the accelerator level floor (not shown). If individual configurations were built separately, the area for the accelerator tunnel for configuration 1, 2, 3, and 4 are 75 m^2, 175 m^2, 150 m^2, and 225 m^2, respectively and configuration 1 to 4 can incorporate configuration 5. Hence the footprint of the accelerator tunnel can be up to 5 times smaller than in conventional accelerator facilities. EuPRAXIA is a site-independent design study. Potential sites will be included in the design report and EuSPARC (Frascati, Italy), SINBAD (Hamburg, Germany), CILEX (Paris, France), CLF (Didcot, UK) and ELI (Prague, Czech Republic) have been discussed as potential sites.

6. Summary

The EuPRAXIA collaboration is preparing a conceptual design report for a multi-GeV plasma-based accelerator with outstanding beam quality. The facility design aims to include FEL radiation in the soft (to hard) X-ray range, a table-top test beam for HEP detectors and industry, and a compact X-ray source for medical imaging. Synchronized TW laser beams will be available in the user areas. Both laser and electron beams are considered as power sources for the plasma accelerator. EuPRAXIA will prepare a proposal to be included on the ESFRI roadmap in 2020 as an innovative European research infrastructure. Ultimately, EuPRAXIA will: use the world-wide leading high power lasers from European industry, drive laser innovation in the connected companies, provide for the first time usable electron beam quality from a plasma accelerator, and serve pilot users from science, engineering, medicine and industry.

References

[1] European Plasma Research Accelerator with eXcellence In Applications http://www.eupraxia-project.eu
[2] Veksler V 1956 Coherent principle of acceleration of charged particles *Proc. of the CERN Symposium on High Energy Accelerators and Pion Physics* Geneva Switzerland
[3] Tajima T and Dawson J M 1979 Laser electron-accelerator *Phys. Rev. Lett.* **43**(4):267–270
[4] Esarey E *et al.* 2009 Physics of laser-driven plasma based electron accelerators *Rev. Mod. Phys.* **81**:1229–1285
[5] Modena A *et al.* 1995 Electron acceleration from the breaking of relativistic plasma waves *Nature* **377**:606 – 608
[6] Umstadter D *et al.* 1996 Laser injection of ultrashort electron pulses into wakefield plasma waves *Phys. Rev. Lett.* **76**(12):2073–2076
[7] Malka V *et al.* 2002 Electron acceleration by a wake field forced by an intense ultrashort laser pulse *Science* **298**(5598):1596–1600
[8] Mangles S P D et al. 2004 Monoenergetic beams of relativistic electrons from intense laser plasma interactions *Nature* **431**(7008):535–538
[9] Faure J *et al.* 2004 A laser-plasma accelerator producing monoenergetic electron beams *Nature* **431**(7008):541–544
[10] Geddes C G R *et al.* 2004 High-quality electron beams from a laser wakefield accelerator using plasma-channel guiding *Nature* **431**(7008):538–541
[11] Leemans W P *et al.* 2006 GeV electron beams from a centimetrescale accelerator *Nat. Phys.* **2**(10):696–699
[12] Wang X *et al.* 2012 Petawatt-laser-driven wakefield acceleration of electrons to 2 GeV in 1017cm3 plasma *Proc. AIP* 1507(1):341–344

[13] Assmann R W *et al.* 2014 Accelerator Physics Challenges towards a Plasma Accelerator with Usable Beam Quality *Proc. of 5th International Particle Accelerator Conference* Dresden Germany
[14] Leemans W P *et al.* 2014 Multi-GeV Electron Beams from Capillary-Discharge-Guided Sub petawatt Laser Pulses in the Self-Trapping Regime *Phys. Rev. Lett.* **113** 245002
[15] Kneip S *et al.* 2009 Near-GeV acceleration of electrons by a nonlinear plasma wave driven by a self-guided laser pulse *Phys. Rev. Lett.* **103**:035002
[16] Fuchs M *et al.* 2009 Laser-driven soft-X-ray undulator source *Nature Physics* **5** 826 – 829
[17] Maier A R 2012 Stabilized Water-Window X-Ray Pulses from a Laser-Plasma Driven Undulator *PhD thesis* Ludwig-Maximilians-Universität München
[18] Lambert G *et al.* 2012 Progress on the generation of undulator radiation in the UV from a plasma-based electron beam *Proc. FEL Conf.* Nara Japan
[19] Cipiccia S *et al.* 2012 A tuneable ultra-compact high-power, ultra-short pulsed, bright gamma-ray source based on bremsstrahlung radiation from laser-plasma accelerated electrons *Journal of Applied Physics* **111**(6):063302
[20] Anania M P *et al.* 2014 An ultrashort pulse ultra-violet radiation undulator source driven by a laser plasma wakefield accelerator *Applied Physics Letters* **104** 264102
[21] Khrennikov K *et al.* 2015 Tunable all-optical quasimonochromatic Thomson X-ray source in the nonlinear regime *Phys. Rev. Lett.* **114**(19):195003
[22] Chen P 1985 Acceleration of Electrons by the Interaction of a Bunched Electron Beam with a Plasma *Phys. Rev. Lett.* **54** 693
[23] Wang S *et al.* 2001 Observation of spontaneous emitted X-ray betatron radiation in beam-plasma interactions *Proc. of 19th IEEE PAC* Chicago IL USA
[24] Blumenfeld I *et al.* 2007 Energy doubling of 42 GeV electrons in a metre-scale plasma wakefield accelerator *Nature* **445** 741
[25] Litos M *et al.* 2014 High-efficiency acceleration of an electron beam in a plasma wakefield accelerator *Nature* **515** 7525:92-5
[26] Ferrario M *et al.* 2013 SPARC_LAB present and future *Nuclear Instruments and Methods in Physics Research B* **309** 183–188
[27] Aschikhin A *et al.* 2016 The FLASHForward facility at DESY *Nucl. Instrum. Meth.* Vol. **806** 175-183
[28] Kasilnikov M *et al.* 2012 Experimentally minimized beam emittance from an L-band photoinjector *Phys. Rev. ST Accel. Beams* **15** 100701
[29] Hidding B *et al.* 2010 Monoenergetic Energy Doubling in a Hybrid Laser-Plasma Wakefield Accelerator *Phys. Rev. Lett.* **104** 195002
[30] Martinez de la Ossa A *et al.* 2013 High-Quality Electron Beams from Beam-Driven Plasma Accelerators by Wakefield-Induced Ionization Injection *Phys. Rev. Lett.* **111** 245003
[31] Walker P A *et al.* 2016 *Report defining preliminary study concept* EuPRAXIA Deliverable Report 1.2
[32] Gizzi L A *et al.* 2016 *Benchmarking of existing technology and comparison with the requirements* EuPRAXIA Deliverable Report 4.1
[33] Cros B *et al.* 2017 Electron injector for multi-stage laser-driven plasma accelerators *presented at the 8th Int. Particle Accelerator Conf.* Copenhagen Denmark WEPVA001
[34] Chiadroni E *et al.* 2016 *Preliminary RF accelerator specifications* EuPRAXIA Milestone Report 5.2
[35] Cros B *et al.* 2017 *Design for an electron injector and a laser plasma stage proposed* EuPRAXIA Milestone Report 3.1
[36] Ferran Pousa A *et al.* 2017 VisualPIC: A New Data Visualizer and Post-Processor for Particle-in-Cell Codes *presented at the 8th Int. Particle Accelerator Conf.* Copenhagen Denmark TUPIK007
[37] Mosnier A *et al.* 2017 *Report designing baseline designs* EuPRAXIA Deliverable Report 2.1

[38] Fonseca R A *et al.* 2002 *OSIRIS: A Three-Dimensional, Fully Relativistic Particle in Cell Code for Modeling Plasma Based Accelerators* Sloot P.M.A., Hoekstra A.G., Tan C.J.K., Dongarra J.J. (eds) Computational Science — ICCS 2002. ICCS 2002. Lecture Notes in Computer Science vol 2331 Springer Berlin Heidelberg
[39] Katsouleas T *et al.* 1987 Beam loading in plasma accelerators *Particle Accelerators* **22**:81–99
[40] Tzoufras M *et al.* 2008 Beam loading in the nonlinear regime of plasma-based acceleration *Phys. Rev. Lett.* **101**:145002
[41] Rechatin C *et al.* 2009 Observation of Beam Loading in a Laser-Plasma Accelerator *Phys. Rev. Lett.* **103**:194804
[42] Schroeder C B *et al.* 2013 Beam loading in a laser-plasma accelerator using a near-hollow plasma channel *Physics of Plasmas* **20** 123115
[43] Walker P A *et al.* 2017 Space considerations and possible layouts for future plasma accelerator facilities in the context of EuPRAXIA *presented at the 8th Int. Particle Accelerator Conf.* Copenhagen Denmark TUPIK012

Acknowledgments
This work was supported by the European Union's Horizon 2020 research and innovation programme under grant agreement No. 653782.

Possible limits of plasma linear colliders

F Zimmermann
CERN, Geneva, Switzerland
E-mail: frank.zimmermann@cern.ch

Abstract. Plasma linear colliders have been proposed as next or next-next generation energy-frontier machines for high-energy physics. I investigate possible fundamental limits on energy and luminosity of such type of colliders, considering acceleration, multiple scattering off plasma ions, intrabeam scattering, bremsstrahlung, and betatron radiation. The question of energy efficiency is also addressed.

1. Introduction

Plasma acceleration [1] is one of the proposed technologies for realizing a highest energy e^+e^- collider, e.g. [2, 3]. A plasma linear collider would accelerate electrons and positrons in the strong fields achievable in a plasma wake, and collide them at an interaction point (IP). The plasma wake itself can be excited in a variety of ways, e.g. by short high-power laser pulses [4], by driving electron bunches [5], or by a self-modulated proton beam [6]. The maximum energy reach, luminosity and electrical power efficiency are crucial figures of merit, which might be limited by various physical effects in such novel type of accelerator.

2. Acceleration and focusing

In the blow-out [7, 8] or bubble regime [9], the maximum accelerating electric field is [1] $E_s \approx m_e c^2 \sqrt{4\pi n_e r_e}/e$, where $m_e c^2$ denotes the electron rest mass, r_e the classical electron radius, and n_e the plasma (electron) density. In practical units we have E_s [V/m] $\approx 0.1 \left(n_e[\mathrm{m}^{-3}]\right)^{1/2}$. Typical plasma densities proposed for future plasma linear colliders are of order $10^{22} - 10^{24}$ m^{-3} [2], corresponding to gradients $E_s \approx 10 - 100$ GV/m. Introducing $a \equiv eE_s/(m_e c^2)$, for constant acceleration the beam energy E_b grows as $E_b(s) = (as + \gamma_0)m_e c^2$, where γ_0 is the initial value of the Lorentz factor, and

$$a \approx \sqrt{4\pi n_e r_e} \,, \tag{1}$$

the acceleration in units of 1/metre. The Lorentz factor changes as $\gamma(s) = as + \gamma_0$. An approximately linear focusing force acts on the electrons [7],

$$\frac{d^2 r}{ds^2} \approx -\frac{2\pi r_e n_e r}{\gamma(s)} \,, \tag{2}$$

where r refers to the radial distance from the axis.

Assuming the beam focusing is provided only by the plasma, the plasma-matched beta function $\beta \equiv \beta_x = \beta_y$ is energy dependent and equals

$$\beta(s) \approx \sqrt{\frac{\gamma(s)}{2\pi n_e r_e}} \,. \tag{3}$$

To avoid blow-out and to stay in the so-called "quasilinear" regime, the beam density would need to be lower than the plasma density and the beta function greater than [10]

$$\beta_{\rm ql}(s) \geq \frac{N\gamma(s)}{(2\pi)^{3/2} n_e \varepsilon_N \sigma_z} , \tag{4}$$

where σ_z denotes the rms bunch length, N the bunch population, and ε_N the normalized emittance.

3. Emittance evolution

Acceleration reduces the geometric emittance ε as

$$\left.\frac{d\varepsilon}{ds}\right|_{\rm ad} \approx -a\frac{\varepsilon}{\gamma} . \tag{5}$$

Multiple small angle elastic scattering of the beam particles off the background ions (if present) leads to emittance growth, which we approximately estimate as [11]

$$\left.\frac{d\varepsilon}{ds}\right|_{\rm ms} \approx \beta(s)\left(\frac{13.6\,{\rm MeV}}{E_b}\right)^2 \frac{1}{X_0} , \tag{6}$$

where $\beta(s)$ denotes the beta function at location s, and X_0 the radiation length of the plasma ions in units of m. In the expression for the average squared scattering angle leading to (6), we have dropped a logarithmic term $(1 + 0.038\ln(s/X_0))^2$ [11]. In the blow-out regime, a lower bound on the inverse radiation length is [12]

$$\frac{1}{X_0} = 4\alpha r_e^2 n_e \left(Z^2[L_{\rm rad} - f(Z)] + (Z-1)L'_{\rm rad}\right) , \tag{7}$$

where we have assumed singly ionized plasma ions, Z designates the atomic number, α the fine-structure constant, and the second term in the parentheses corresponds to the contribution from the remaining electrons. Up to uranium, the function $f(Z)$ can be represented as [13]

$$f(Z) \approx \frac{f^2}{1+f^2} + 0.2f^2 - 0.037f^4 + 0.008f^6 - 0.002f^8 , \tag{8}$$

where $f \equiv \alpha Z$. The quantities $L_{\rm rad}$ and $L'_{\rm rad}$ were calculated by Tsai [12]. For $Z > 4$ they can be approximated as $L_{\rm rad} \approx \ln(184.15\, Z^{-1/3})$ and $L'_{\rm rad} \approx \ln(1194\, Z^{-2/3})$, using the Thomas-Fermi-Moliere model [12]. Table 1 shows some example values for commonly used alkali plasmas, considering a typical plasma electron density of $n_e = 10^{23}$ m^{-3}, corresponding to a peak field of about 30 GV/m.

Table 1. Scattering and bremsstrahlung parameters for alkali plasmas at a plasma electron density of $n_e = 10^{23}$ m^{-3}, assuming single ionization.

el.	Z	$f(Z)$	$L_{\rm rad}$	$L'_{\rm rad}$	X_0 [km]
Li	3	0.0006	4.74	5.81	1600
Rb	37	0.083	4.01	4.68	16
Cs	55	0.17	3.88	4.41	7.6

In addition to scattering off the plasma elements, the beam particles can scatter off each other. Consider a round Gaussian bunch of N electrons with transverse normalized rms emittances

$\varepsilon_N \equiv \gamma\varepsilon_x = \gamma\varepsilon_y$, and normalized longitudinal rms emittance $\varepsilon_{z,N} \equiv \gamma\sigma_\delta\sigma_z$, where σ_δ denotes the rms relative momentum spread. Assuming that the dispersion function is zero everywhere, and that $\gamma/\sigma_\delta \gg \sqrt{\beta/\varepsilon}$ (generally fulfilled for relativistic beams), the analytical formula for the emittance growth due to intrabeam scattering (IBS) [14, 15, 16] reduces to the closed expression

$$\left.\frac{d\varepsilon}{ds}\right|_{\rm IBS} \approx -\frac{r_e^2 N \beta^{1/2} \log}{16 \varepsilon_N^{3/2} \sigma_z \gamma^3} , \tag{9}$$

where log $\approx 15-20$ denotes the Coulomb logarithm, and we have used $\int_0^\infty d\lambda(\lambda+b)/(\lambda^3 + x\lambda^2 + 2bx\lambda + b^2x)^{3/2} \approx \pi/(2\sqrt{b}x^{3/2})$ for $x \gg b$. The transverse emittance shrinks (!), but more weakly the higher the energy.

Inserting the matched beta function (3), (9) becomes

$$\left.\frac{d\varepsilon}{ds}\right|_{\rm IBS} \approx -\frac{r_e^{7/4} N \log}{16(2\pi)^{1/4} n_e^{1/4} \varepsilon_N^{3/2} \sigma_z \gamma^{11/4}} , \tag{10}$$

which still decreases strongly with energy.

The emittance at location s is determined by

$$\left.\frac{d\varepsilon}{ds}\right|_{\rm tot} = \left.\frac{d\varepsilon}{ds}\right|_{\rm ad} + \left.\frac{d\varepsilon}{ds}\right|_{\rm ms} + \left.\frac{d\varepsilon}{ds}\right|_{\rm IBS} \tag{11}$$

At high energy $\gamma \gg \gamma_0$ the local "equilbrium emittance" follows from $d\varepsilon/ds|_{\rm tot} = 0$. Again assuming the beta function (3), and introducing the two auxiliary constants

$$B_{\rm ms} \equiv \frac{1}{2\pi\sqrt{2r_e}} \left(\frac{13.6\,{\rm MeV}}{m_e c^2}\right)^2 \frac{1}{X_0 n_e} , \tag{12}$$

$$C_{\rm IBS} \equiv \frac{r_e^{5/4} \log}{32\,(2)^{1/4}\,\pi^{3/4}} , \tag{13}$$

the equilibrium condition becomes

$$\varepsilon_{N,{\rm eq}} = B_{\rm ms}\gamma^{1/2} - C_{\rm IBS} \frac{N}{n_e^{3/4} \varepsilon_N^{3/2} \sigma_z \gamma^{7/4}} . \tag{14}$$

In the limit of zero bunch charge ($N = 0$), the normalized equilibrium emittance is given by $\varepsilon_N = B_{\rm ms}\gamma^{1/2}$. For a Li plasma $B_{\rm ms} \approx 0.2$ μm, for Cs $B_{\rm ms} \approx 40$ μm; at 1 TeV, $B_{\rm ms}\gamma^{1/2}$ becomes 0.25 nm and 50 nm, respectively – negligible for round-beam operation.

The emittance growth due to multiple scattering has been examined previously, e.g., in [17] (for crystals) and [18, 19, 2, 10] (for plasmas), in each case invoking somewhat different assumptions and approximations. Our estimate with emittance values around 1 nm is of the same order of magnitude as those in [18, 2]. Reference [10] obtained a larger value of 10 μm, by considering the beta function (4) for the quasilinear regime, in place of the plasma-matched beta function (3).

Concerning the IBS contribution, $C_{\rm IBS} \approx 1.3 \times 10^{-19}$ m$^{5/4}$ (with log ≈ 17.5). Then, for example, using $n_e \approx 10^{23}$ m^{-3}, $\varepsilon_N \approx 1$ μm, $N \approx 10^9$, $\sigma_z \approx 1$ μm, $\sigma_\delta \approx 10^{-3}$ at a beam energy of 1 TeV, we obtain $C_{\rm IBS}N/(n_e^{3/4}\varepsilon_N^{3/2}\sigma_z\gamma^{7/4}) \approx 10^{-23}$ m, an insignificant effect.

IBS also changes the beam energy spread, namely as

$$\left.\frac{d\sigma_\delta^2}{ds}\right|_{\rm IBS} \approx 2\frac{r_e^2 N \log}{16\beta^{1/2}\varepsilon_N^{3/2}\sigma_z\gamma} . \tag{15}$$

IOP Conf. Series: Journal of Physics: Conf. Series **874** (2017) 012030 doi:10.1088/1742-6596/874/1/012030

Comparing with (9), we note that

$$\frac{d}{ds}\left[\frac{\sigma_\delta^2}{\gamma^2}+2\frac{\varepsilon}{\beta}\right]_{\rm IBS} \equiv \frac{d}{ds}\left[\frac{\sigma_\delta^2}{\gamma^2}+\frac{\varepsilon_x}{\beta_x}+\frac{\varepsilon_y}{\beta_y}\right]_{\rm IBS} = 0\,, \tag{16}$$

a relation already derived by Piwinski [20].

Using (9) and (3), we find for the plasma-focused case

$$\left.\frac{d\sigma_\delta}{ds}\right|_{\rm IBS} = -\frac{\gamma^2}{\sigma_\delta \beta}\left.\frac{d\varepsilon}{ds}\right|_{\rm IBS} \approx \frac{r_e^{9/4}(2\pi)^{1/4}\log N n_e^{1/4}}{16\varepsilon_N^{3/2}\varepsilon_{z,N}\gamma^{1/4}}\,. \tag{17}$$

or, for large γ, $\Delta\sigma_\delta \to 4\, d\sigma_\delta/ds|_{\rm IBS}\,\gamma/(3a)$. Assuming the same parameters as above, this amounts to $d\sigma_\delta/ds|_{\rm IBS} \approx 2\times 10^{-8}$/m, and $\Delta\sigma_\delta \approx 10^{-6}$ — negligible.

4. Energy loss

At very high energies and except at the high-energy tip of the bremsstrahlung spectrum, in the "complete screening case" (though in our case atoms are partially ionized and nuclei not fully shielded) the cross section can be approximated as [12]

$$\begin{aligned}\left.\frac{d\sigma}{dk}\right|_{\rm bs} \approx \frac{1}{k}4\alpha r_e^2 \Big\{ \left(\frac{4}{3}-\frac{4}{3}y+y^2\right)\Big(Z^2[L_{\rm rad}-f(Z)] \\ +(Z-1)L'_{\rm rad}\Big) - \frac{1}{9}(1-y)(Z^2+Z-1)\Big\}\,,\end{aligned} \tag{18}$$

where k is the photon energy, and $y = k/E_b$ the fraction of the electron energy transferred to the radiated photon. If the second term (of order of 2% of the total) is ignored the above can be expressed through the radiation length as

$$\left.\frac{d\sigma}{dk}\right|_{\rm bs} \approx \frac{1}{X_0 n_e k}\left(\frac{4}{3}-\frac{4}{3}y+y^2\right)\,. \tag{19}$$

The average energy loss per unit length becomes

$$\left.\frac{dE_b}{ds}\right|_{\rm bs} = n_e\int_{E_{\rm min}}^{E_b} dk\,\left.\frac{d\sigma}{dk}\right|_{\rm bs} k \approx \frac{E_b}{X_0}\,. \tag{20}$$

This can be a noticeable effect for plasma atoms of higher atomic number (e.g., Rb, Cs) — see Table 1.

The energy loss due to scattering off thermal photons [21, 22] might need to be taken into acount if the plasma temperature is high. The additional contribution from scattering off atomic electrons [22] is negligible.

More important is the following effect. A nonzero betatron amplitude together with the linear focusing force of the plasma leads to the emission of a special type of synchrotron radiation called betatron radiation (br) [23]. The resulting energy loss is given by

$$\left.\frac{dE_b}{ds}\right|_{\rm br} = -\frac{2}{3}r_e m_e c^2\gamma^2(2\pi r_e n_e)^2 r^2\,. \tag{21}$$

Averaging over a Gaussian bunch we have $< r^2 >= 2\beta(s)\varepsilon()$, and with the beta function of (3)

$$\left\langle \left.\frac{dE_b}{ds}\right|_{\rm br}\right\rangle = -\frac{4}{3}\left(r_e^{5/2}(2\pi)^{3/2}n_e^{3/2}m_e c^2\right)\gamma(s)^{5/2}\varepsilon(s)\,. \tag{22}$$

The acceleration comes to a standstill if

$$a m_e c^2 + \left.\frac{dE_b}{ds}\right|_{\rm bs} + \left.\frac{dE_b}{ds}\right|_{\rm br} = 0 . \tag{23}$$

For simplicity, dropping the bremsstrahlung and retaining only the betatron-radiation loss, this condition becomes

$$\frac{1}{\gamma_{\rm max,br}(s)\varepsilon(s)} = \frac{4\sqrt{2}\pi}{3} r_e^2 n_e \gamma_{\rm max,br}(s)^{3/2} . \tag{24}$$

The right side of the equation increases strongly with γ. A lower emittance boosts the energy reach:

$$\gamma_{\rm max,br} = \left(\frac{3}{4\sqrt{2}\pi r_e^2 \varepsilon_N n_e}\right)^{2/3} . \tag{25}$$

Let us consider an example: With $\varepsilon_N \equiv \gamma\varepsilon = 10$ μm, $n_e = 10^{23}$ m^{-3}, the inverse of (25) yields $\gamma_{\rm max} \approx 7.7 \times 10^6$, or a maximum energy of 4 TeV, where betatron radiation loss outweighs the acceleration.

Combining the relations (1) and (25) allows us to estimate the minimum length of a plasma accelerator. Assuming a constant (maximum possible) plasma density determined by the final energy, this minimum length becomes

$$l_{\rm min} = \frac{\gamma}{a_{\rm max}} = \frac{2^{1/4} r_e^{1/2}}{3^{1/2}} \gamma^{7/4} {\varepsilon_N}^{1/2} . \tag{26}$$

This length increases roughly quadratically with energy, also taking into account (14). The relation (26), for fixed ε_N, is illustrated in Figure 1. Weaker focusing (quasi-linar regime) yields a more favorable conclusion on the accelerator length [2], e.g., by considering the beta function of (4).

5. Efficiency

In the classical beamstrahlung regime (i.e. critical photon energy E_c smaller than E_b, or $\Upsilon \equiv (2/3)E_c/E_b \ll 1$), the luminosity of a linear e^+e^- collider may be written as [24]

$$L \approx \left(\frac{1}{16\pi\alpha r_e}\right) \frac{P_{\rm wall}}{E_b} N_\gamma \frac{\eta_{\rm tot}}{\sigma_y^*} , \tag{27}$$

where $P_{\rm wall}$ is the wall plug power, and η the conversion efficiency of wall-plug (w.p.) power into total average beam power, $P_{\rm beam} = 2 f_{\rm rep} N E_b$, and $f_{\rm rep}$ the (average) bunch repetition rate. N_γ denotes the average number of beamstrahlung photons per colliding particle, and should not much exceed 1, in order to preserve a well-defined luminosity spectrum. Since cross sections decrease as $1/\gamma^2$ the luminosity should increase with energy as γ^2. The vertical beam size at the collision point, σ_y^*, is limited by various effects. Values of σ_y^* below 1 nm have rarely been proposed.

Equation (27) reveals the general feature that the luminosity is proportional to the overall conversion efficiency $\eta_{\rm tot}$. For linear colliders like CLIC and ILC the value of $\eta_{\rm tot}$ varies between 4 and 6%. For a plasma collider the efficiency is the product of the efficiency of generating the driver, and the transfer efficiencies from the driver to the plasma wake, and from the wake to the main beam, i.e.:

$$\eta_{\rm tot} = \eta_{\rm w.p.\to driver}\, \eta_{\rm driver\to wake}\, \eta_{\rm wake\to beam} . \tag{28}$$

The driver can be a proton beam, an electron beam, or a laser pulse. As examples for possible proton drivers, the typical power consumption of the LHC accelerator is 80 MW, and its pre-injector, the SPS, consumes about 60 MW [25]. Reference [26] discusses how the LHC cycle

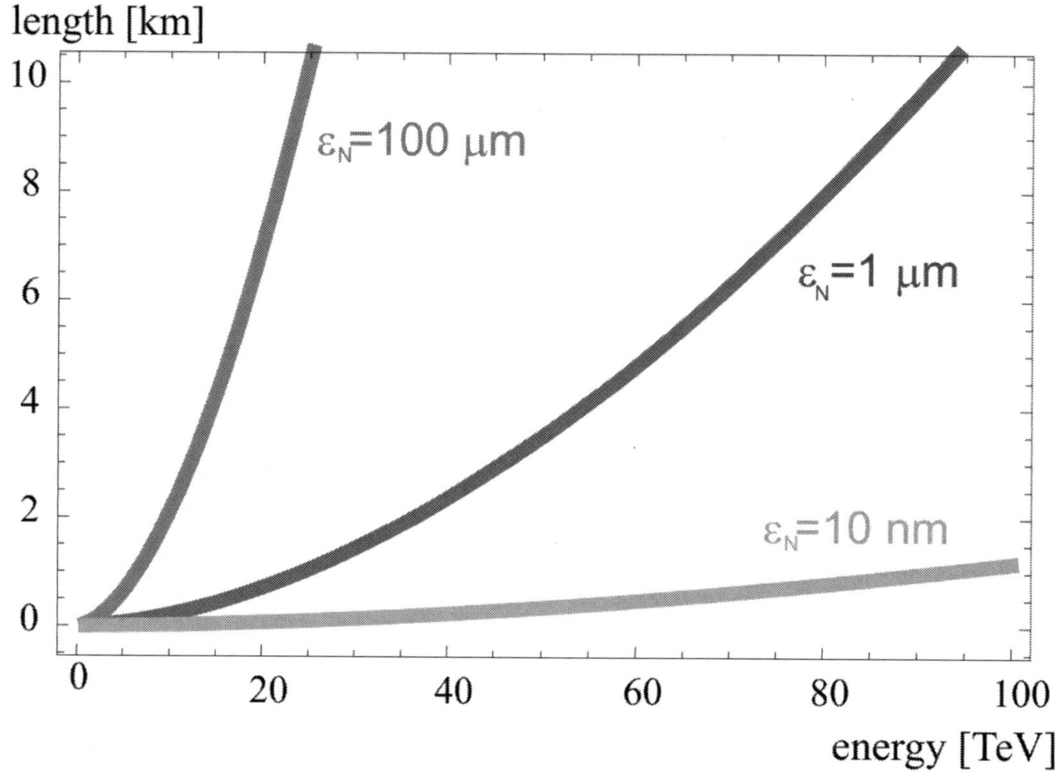

Figure 1. Minimum plasma accelerator length versus beam energy, according to (26); the various curves correspond to different values of the transverse emittance as indicated.

time with a top energy of 3.3 TeV could be shortened to about 30 minutes (incl. 15 minutes injection plateau); with 3000 bunches of 10^{11} protons each, per beam, the average total beam power which could then be delivered at 3.3 TeV is 0.2 MW, i.e. less than 0.3% of the electrical power (80 MW) spent on the LHC alone. On the other hand, a highly efficient electron beam driver was proposed in Reference [3], and for fibre lasers record efficiencies around 48% and 50% were reported in Reference [27], not including the power consumption of the lasers' chillers. From these considerations and references, we estimate that

$$\eta_{\mathrm{w.p.}\to\mathrm{driver}} \approx \left\{ \begin{array}{ccc} < 0.5\% & \text{protons} & (5 \times \text{faster LHC}) \\ \approx 40\% & e^- & (\text{cw SC linac}) \\ \approx 50\% & \text{laser} & (\text{future fibre lasers}) \end{array} \right. \tag{29}$$

According to simulations with Gaussian bunches a realistic transfer efficiency from driver to wake is $\eta_{\mathrm{driver}\to\mathrm{wake}} \approx 77\%$ [3]. For $\eta_{\mathrm{wake}\to\mathrm{beam}}$ a peak value of about 50% was demonstrated at FACET [5], and 65% or above appears attainable [3, 28]. However, transverse stability constraints may yet reduce these optimistic efficiency figures [29].

6. Conclusions

Equations (14), (17) and (25) impose constraints on high-energy plasma-based accelerators. The scattering off plasma nuclei and plasma electrons is almost negligible in the blow-out regime, but may become significant in the opposite, quasilinear regime. The inverse dependence is true for betatron radiation. Both effects might be mitigated by accelerating in a hollow plasma channel

[30, 31], at the expense of a lower accelerating gradient and, perhaps, additional complication [31]. The betatron radiation would then be determined by the electromagnetic fields in this channel [32].

It appears that if driven by fibre lasers or electron beams plasma accelerators might achieve overall efficiencies approaching, or exceeding, those of conventional accelerators: $\eta_{\rm tot} \approx 50\% \times 77\% \times 65\% \approx 25\%$. Considering only the efficiency constraint, from (27) a 3 TeV e^+e^- collider with this optimistic value of $\eta_{\rm tot} \approx 25\%$, along with $E_b = 1.5$ TeV, $\sigma_y^* \approx 1$ nm, $N_\gamma \approx 1$ and a wall-plug power of 500 MW could reach a luminosity of 5×10^{34} cm^{-2}s^{-1} (or possibly 2× more if operating at $\Upsilon \gg 1$ [3]), which still is 6× (or 3×) lower than the luminosity of a 100 TeV circular *pp* collider, FCC-hh [33], at comparable electric power.

Acknowledgements
I thank M. Benedikt for suggesting this study, G. Arduini and M. Giovannozzi for several helpful comments, and the two IPAC'17 referees for their excellent reviews.

This work was supported in part by the European Commission under the HORIZON2020 Integrating Activity project ARIES, grant agreement 730871.

References

[1] Tajima T and Dawson J M 1979 Laser Electron Accelerator *Phys. Rev. Lett.* **43** 267
[2] C.B. Schroeder C, Esarey E, Geddes C G R, Benedetti C, and Leemans W P 2010 Physics Considerations for Laser-Plasma Linear Colliders *Phys. Rev. ST Accel. Beams* **13** 101301
[3] Adli E, Delahaye J-P, Gessner S J, Hogan M J, Raubenheimer T, An W, Joshi C and Mori W 2013 A Beam-Driven Plasma-Wakefield Linear Collider: From Higgs Factory to Multi-TeV *Preprint* arXiv:1308.1145
[4] Leemans W P *et al.* 2014 Multi-GeV Electron Beams from Capillary Discharge-Guided Subpetawatt Laser Pulses in the Self-Trapping Regime *Phys. Rev. Lett.* **113** 245002
[5] Litos M *et al.* 2014 High Efficiency Acceleration of an Electron Beam in a Plasma Wakefield Accelerator *Nature* **515** 92
[6] Gschwendtner E *et al.* 2016 AWAKE, The Advanced Proton-Driven Plasma Wakefield Acceleration Experiment at CERN *Nucl. Instrum. Methods A* **829** 76
[7] Rosenzweig J B, Breizman B, Katsouleas T and Su J J 1991 Acceleration and Focusing of Electrons in Two-Dimensional Nonlinear Plasma Wake Fields *Phys. Rev. A* **44** R6189
[8] Lotov K B 2004 Blowout Regimes of Plasma Wakefield Acceleration *Phys. Rev. E* **69** 046405
[9] Pukhov A and Meyer-ter-Vehn J 2002 Laser Wake Field Acceleration: the highly Non-Linear Broken-Wave Regime *J. Appl. Phys. B* **74** 355
[10] Lebedev V and Nagaitsev S, Comment on 'Beamstrahlung considerations in laser-plasma-accelerator-based linear colliders' *Phys. Rev. ST Accel. Beams* **16** 108001
[11] The Particle Data Group 2014 Review of Particle Physics (32. Passage of Particles through Matter (rev.)) *Chinese Physics C* **38** 090001
[12] Tsai Y S 1974 Pair Production and Bremsstrahlung of Charged Leptons *Rev. Mod. Phys.* **46** 815
[13] Davies H, Bethe H A, and Maximon L C 1954 Theory of Bremsstrahlung and Pair Production. II. Integral Cross Section for Pair Production *Phys. Rev. 93* **788**
[14] Bjorken J D and Mtingwa S K 1983 Intrabeam Scattering *Particle Accelerators* **13** 115
[15] Conte M and Martini M 1985 Intrabeam Scattering in the CERN Antiproton Accumulator *Particle Accelerators* **17** 1
[16] Antoniou F and Zimmermann F 2012 Revision of Intrabeam Scattering with Non-Ultrarelativistic Corrections and Vertical Dispersion in MAD-X *European Organization for Nuclear Research (CERN) Report* CERN-ATS-2012-066
[17] Chen P and Noble R 1997 Crystal Channel Collider: Ultrahigh-Energy and Luminosity in the Next Century, *AIP Conf. Proc.* **398** 273
[18] Kirby N A, Berry M K, Blumenfeld I, Hogan M J, Ischebeck R and Siemann R 2007 Emittance Growth From Multiple Coulomb Scattering in a Plasma Wakefield Accelerator, *Proc.* 2007 IEEE Particle Accelerator Conference 25–29 June 2007 Albuquerque USA
[19] Dodin I Y and Fisch N J 2008 Charged Particle Acceleration in Dense Plasma Channels *Physics of Plasmas* **15** 103105
[20] Piwinski A 1985 Intra-beam Scattering *European Organization for Nuclear Research (CERN) Report*

Proc. CERN Accelerator School: Accelerator Physics 16–27 September 1985 Oxford UK CERN-1987-003-V-1 402
[21] Telnov V I 1987 Scattering of Electrons on Thermal Radiation Photons in Electron-Positron Storage Rings *Nucl. Instrum. Methods A* **260** 304
[22] Reichel I, Zimmermann F, Raubenheimer T O and Tenenbaum P 1998 Thermal Photon and Residual Gas Scattering in the NLC Beam Delivery *Stanford Linear Accelerator Center (SLAC) Report* SLAC-PUB-8012
[23] Kiselev S, Pukhov A, and Kostyukov I 2004 X-ray Generation in Strongly Nonlinear Plasma Waves *Phys. Rev. Lett.* **93** 135004
[24] Zimmermann F 2000 Tutorial on Linear Colliders *AIP Conf. Proc.* **592** 494
[25] 2011 Powering CERN and the LHC *European Organization for Nuclear Research (CERN) Report* CERN Bulletin 06/2011
[26] Milanese A, Goddard B and Solfaroli Camillocci M 2015 Faster Ramp of LHC for Use as an FCC High Energy Hadron Booster *European Organization for Nuclear Research (CERN) Report* CERN-ACC-2015-0133
[27] Hatcher M 2015 Laser 2015: Efficiencies on the Up as Fiber Laser Competition Grows *Preprint* optics.org/news/6/6/43
[28] Tzoufras M, Lu W, Tsung F S, Huang C, Mori W B 2008 Beam Loading by Electrons in Nonlinear Plasma Wakes *Phys. Plasmas* **16** 056705
[29] Lebedev V, Burov A and Nagaitsev S 2017 Efficiency versus Instability in Plasma Accelerators *Preprint* arXiv:1608.02498
[30] Sprangle P, Esarey E, Krall J and Joyce G 1992 Propagation and Guiding of Intense Laser Pulses in Plasmas *Phys. Rev. Lett.* **69** 2200
[31] Gessner S *et al.* 2016 Demonstration of a Positron-Beam-Driven Hollow Channel Plasma Wakefield Accelerator *Nature Communications* **7** 11785
[32] Schroeder C B, Whittum D H and Wurtele J S 1999 Multimode Analysis of the Hollow Plasma Channel Wakefield Accelerator *Phys. Rev. Lett.* **82** 117
[33] Benedikt M, Schulte D and Zimmermann F 2015 Optimizing Integrated Luminosity of Future Hadron Colliders *Phys. Rev. ST – Accel. Beams* **18** 101002

IOP Conf. Series: Journal of Physics: Conf. Series **874** (2017) 012031 doi:10.1088/1742-6596/874/1/012031

Upgrade of the two-screen measurement setup in the AWAKE experiment

Marlene Turner[1], Vincent Clerc, Ishkhan Gorgisyan, Edda Gschwendtner, Stefano Mazzoni and Alexey Petrenko

CERN, Geneva, Switzerland

[1] also at Technical University of Graz, Graz, Austria

E-mail: marlene.turner@cern.ch

Abstract. The AWAKE project at CERN uses a self-modulated 400 GeV/c proton bunch to drive GV/m wakefields in a 10 m long plasma with an electron density of $n_{pe} = 7 \times 10^{14}$ electrons/cm^3. We present the upgrade of a proton beam diagnostic to indirectly prove that the bunch self-modulated by imaging defocused protons with two screens downstream the end of the plasma. The two-screen diagnostic has been installed, commissioned and tested in autumn 2016 and limitations were identified. We plan to install an upgraded diagnostics to overcome these limitations.

1. Introduction

The Advanced Proton Driven Plasma Wakefield Acceleration Experiment (AWAKE)[1] is a proof-of-principle R&D project at CERN. The goal of AWAKE is to accelerate electrons using plasma wakefields with GV/m amplitude created by a self-modulated proton bunch. The proton bunch provided by the CERN SPS has a momentum of 400 GeV/c, a longitudinal size σ_z of 12 cm and a radial beam rms radius of $\sigma_r = 0.2$ mm. A 100 fs long, 450 mJ laser pulse ionizes 10 m of rubidium vapour and creates a plasma channel with a radius of 1 mm and a plasma density ranging from $10^{14} - 10^{15}$ electrons/cm^3.

The optimum proton bunch length for a given plasma density is given by $k_{pe}\sigma_z \cong \sqrt{2}$, where $k_{pe} = \omega_{pe}/c$ is the plasma wavenumber, ω_{pe} the angular plasma frequency and c the speed of light. For a plasma density of 7×10^{14} electrons/cm^3 (AWAKE baseline) the optimum bunch length is in the mm-range, much shorther than the $\sigma_z = 12$ cm of the proton bunch. Hence AWAKE relies on the development of the self-modulation-instability (SMI) [2] to modulate the long proton bunch into micro-bunches spaced at the plasma wavelength $\lambda_{pe} = 2\pi/k_{pe}$. The self-modulation instability is seeded by a sharp turn-on of the plasma, realized by placing the ionizing laser pulse in the center (temporally) of the proton bunch.

Seed wakefields periodically focus and defocus protons. As the instability develops, micro-bunches are formed due to the proton density modulation near the axis. These micro-bunches can then resonantly drive the plasma wave. AWAKE studies of the SMI are scheduled for 2016 and 2017, and electron acceleration of externally injected electrons is planned for 2018.

In this article we discuss the upgrade of a proton beam diagnostics to indirectly demonstrate that the self-modulation instability develops within the 10 m of plasma. The basic version of this setup is described in [3] and was successfully operated during a first AWAKE run in December

IOP Conf. Series: Journal of Physics: Conf. Series **874** (2017) 012031 doi:10.1088/1742-6596/874/1/012031

2016. We explain the limitations we observed with this setup and discuss a major upgrade on the system, to be operational before the next beam time in June 2017.

2. The goal of the indirect SMI two-screen measurement system

The basic idea of the two-screen diagnostic is to image the protons that are defocused by the transverse plasma wakefields and to measure their maximum defocusing angle. As indicated in Figure 1, we want to measure maximum radial displacement of the protons by inserting scintillating screens 2 and 10 m downstream the plasma. As the protons traverse the screen, the screen emits scintillation light. The radial, time-integrated proton beam profile is reconstructed by imaging the light onto a camera.

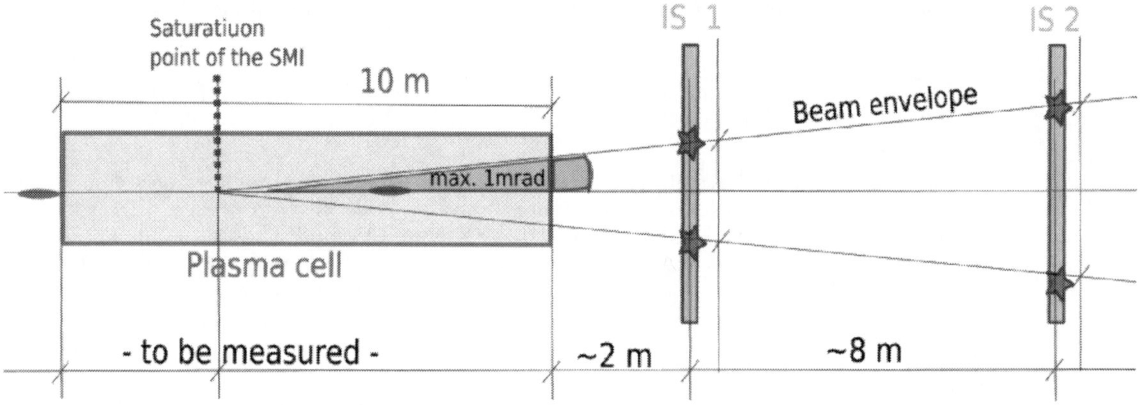

Figure 1. Schematic drawing of the principle behind the two screen measurement setup.

Figure 1 also suggests that we can calculate the maximum proton defocusing angle from the two beam images by connecting the measured maximum radial displacements. If SMI develops successfully we expect to measure a maximum defocusing angle around 1 mrad instead of the ≈ 0.05 mrad without plasma and SMI. Defocusing the proton beam from its natural divergence up to $\approx$1 mrad over a distance of ≈ 0.5 m (as expected from LCODE [4, 5] simulations) requires GV/m defocusing plasma wakefields.

Figure 2 shows the expected radial proton beam profiles at the two screen locations in case of no plasma and no SMI (blue line) and with SMI in plasma with a density of 7×10^{14} electrons/cm^3 (red line).

We want to detect the outermost proton beam edges and one major challenge for measuring these edges is that the beam density on the screen ranges over 4-5 orders of magnitude, while the dynamic range of the cameras is limited to 2-3 orders of magnitudes. We previously found that the Chromox screen light yield is directly (linearly) proportional to the proton beam density traversing the screen [6]. We do not expect any light yield degradation over time because of the low number of protons per bunch N_p ($N_p = 3\times10^{11}$) and the low repetition rate (1 bunch every 30 s).

3. The two screen measurement setup

Previously, we installed two imaging stations (IS1, IS2) 2 and 10 m downstream the end of the plasma (see Figure 1). Each imaging station consisted of a vacuum tank, a scintillating screen (Chromox:Al_2O_3:CrO_2), a filter wheel, a camera lens, a camera and an image readout system.

Since the dynamic range of the camera is limited to 2-3 orders of magnitude, we cut a hole with a radius of 3 and 5 mm into the first and the second screen, respectively (see vertical

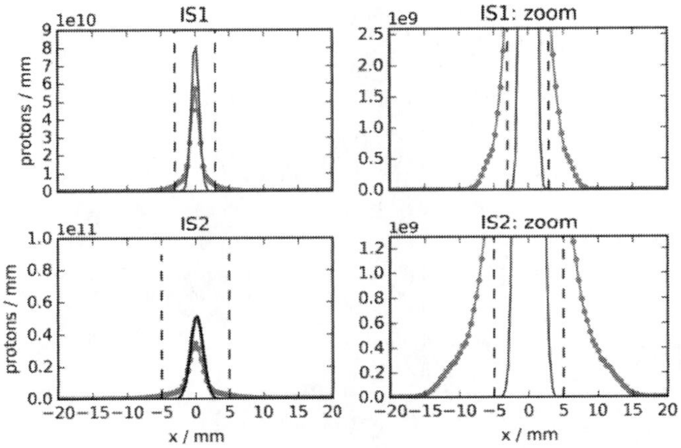

Figure 2. From LCODE [4, 5] simlations expected transverse profile of the self-modulated proton bunch for the AWAKE baseline parameters [3] (red) and no plasma (blue). Dashed lines indicate the size of the holes that were cut into the imaging screens (see the following section).

dashed lines in Figure 2). For one screen installed in IS2 we filled the hole with a low light yield material (aluminum) to create a combined screen. The two imaging stations were commissioned and successfully operated in December 2016, but we identified several limitations of the system:

- If the experimental observations differ from the expected simulation results shown in Figure 2 or we change for example the plasma density, we need to change the size of the hole in the screens. Changing screens is a major intervention that takes several days because it involves breaking the beam-line vacuum.
- The proton beam has to pass through the center of the hole in the screen with an accuracy of 0.1 mm. If the experiment decides to change the proton beam trajectory, the screens have to be realigned. Realignment can be done by precisely moving the vacuum tank, but due to safety reasons can imply breaking the vacuum.
- The imaging stations use analogue video cameras (standard system at CERN working well in radiation environment). The cameras are triggered every 20 ms and integrate over 20 ms, independently of the arrival time of the proton bunch. Hence we are unable to compare measurements in terms of signal intensity.
- Information regarding the beam core is either lost (if the screen has a hole) or limited (combined screen), due to the fixed light yield ratio between the screen materials in the combined screen.

We propose an upgrade to overcome the observed limitations and to make the system more flexible for the experiment.

4. The upgraded two screen measurement system

A schematic of the upgraded optics system for IS1 is shown in Figure 3.

We replaced the scintillating Chromox screen with a hole by a full Chromox screen. Instead of imaging the light emitted by the scintillator screen directly onto a camera, we split the light using a 3 inch pellicle beam splitter that reflects 8% and transmits 92% of the emitted light.

The reflected light is directly imaged onto the camera C1 by the lens L1. The image of camera C1 will provide information about the size and position of the beam core. The transmitted light is imaged by lens L2 (3 inch, f=200 mm) onto the plane M, where the mask is placed. The mask blocks the intense beam core by an opaque region in mask. A technical drawing of the mask setup is shown in Figure 4. The mask is reimaged by lens L3 onto camera C2.

The mask system consists of a 1 mm thick fused silica window with different sized opaque coatings, a glass holder, two stepper motors for horizontal and vertical alignment and a manual

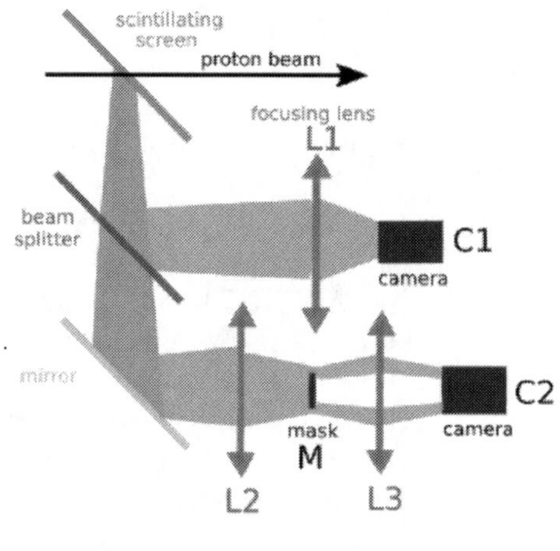

Figure 3. Schematic drawing of the optics setup of the upgraded imaging system of IS1.

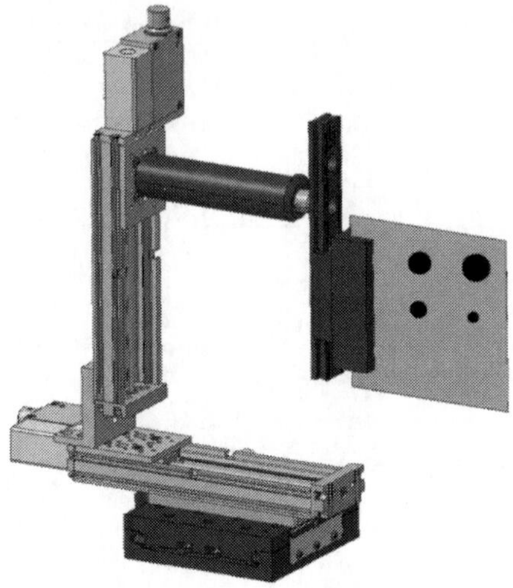

Figure 4. Schematic drawing of the light-core blocking mask system.

translation stage to faciliate imaging by longitudinal positioning on the table. The distances are chosen to have an image magnification of 0.5 in the mask plane. The light passing around the mask is then reimaged onto an externally triggered 12 bit camera (Basler ac1600-60gm).

The optics for IS2 are very similar to the one from IS1 (see Figure 3), but due to space restrictions the beam splitter comes after the mirror.

We have several different mask sizes available and we can remotely adjust and align the mask for various cases (i.e. plasma densities, changing proton beam trajectory, etc.) by varying the radius to which we want to block the beam core. We use a camera, that is synchronized to the proton beam -to capture all the emitted light- so that we can directly compare different events and explore the radial proton beam profile in details. We can now optimally observe the beam core and the defocused protons in terms of light intensity and mask size.

The procedure will be as follows: We locate the beam without mask, align the mask onto the beam and then adjust filters in the optical line to get the right exposure of the camera.

We previously found [3] that the main background contribution comes from secondary particles directly impacting onto the camera. The laser dump ≈ 30 cm upstream IS1 shields the imaging screen from low-energy electrons from the plasma. Since the plasma wakefield amplitude is low in this experiment, trapping and accelerating electrons is unlikely.

5. Summary

We discussed an upgrade of the indirect SMI two-screen measurement system to be installed in the AWAKE experiment and ready for operation in June 2017. The idea of the two screen measurement setup is to image protons that got defocused by the transverse plasma wakefields to indirectly show that the proton bunch self-modulated and wakefields were created in the 10 m of plasma. The upgraded system offers different mask sizes to block the light from the beam core, facilitates alignment of the masks with respect to the proton beam, enables comparability between measurements and images also the proton beam core.

In the upgraded system we split the light emitted by the scintillator screen and image one part directly onto a camera to obtain information about the beam core. The other part blocks the beam core with a mask with several mask sizes available, to detect the maximum defocused protons.

References

[1] Caldwell A *et al.* 2016 *Nucl. Instrum. Meth.* A **829** 3-16
[2] Kumar N *et al.* 2010 *Phys. Rev. Lett.* **104** 255003
[3] Turner M *et al.* 2017 *Nucl. Instrum. Meth.* A **854** 100-106
[4] `www.inp.nsk.su/~lotov/lcode/`
[5] Lotov K V 2003 *Phys. Rev. ST Accel. Beams* **6** 061301
[6] Burger S *et al.* 2016 *Proc. Int. Beam Instrumentation Conference* (Barcelona) 268

IOP Conf. Series: Journal of Physics: Conf. Series **874** (2017) 012032 doi:10.1088/1742-6596/874/1/012032

External injection into a laser-driven plasma accelerator with sub-femtosecond timing jitter

A Ferran Pousa[1,2], R Assmann[1], R Brinkmann[1] and A Martinez de la Ossa[1,2]

[1] DESY, 22607 Hamburg, Germany
[2] Universität Hamburg, 22761 Hamburg, Germany

E-mail: angel.ferran.pousa@desy.de

Abstract. The use of external injection in plasma acceleration is attractive due to the high control over the electron beam parameters, which can be tailored to meet the plasma requirements and therefore preserve its quality during acceleration. However, using this technique requires an extremely fine synchronization between the driver and witness beams. In this paper, we present a new scheme for external injection in a laser-driven plasma accelerator that would allow, for the first time, sub-femtosecond timing jitter between laser pulse and electron beam.

1. Introduction

Over the last decades, laser-driven plasma wakefield acceleration (LWFA) [1] has been positioned as a possible alternative to conventional RF technology for future compact accelerators. A series of milestones like the experimental measurement of gradients on the hundreds of GV/m [2, 3], the development of controlled injection techniques [4, 5, 6] or the production of quasi-monoenergetic beams [7, 8, 9] showcase its potential.

Unfortunately, in terms of stability, emittance and energy spread, the quality of these electron beams is still inferior to those produced at classical RF accelerators.

One of the proposed methods for improving the beam quality is to use an external injection scheme, where the electron beam is provided by an external RF linac, rather than produced within the plasma. This allows a better control of the beam properties so that the emittance and energy spread growth during acceleration can be minimized by properly shaping [10] and matching [11, 12] the beam into the plasma.

However, operating a plasma accelerator with external injection brings in new technical challenges. In particular, it requires an extremely low timing jitter between the drive laser and witness beam. Due to the small size of the accelerating plasma buckets (typically on the order of $10 - 100$ μm) and the steep field gradients in them, an accuracy on the femtosecond level is required, smaller than what is available today in state-of-the-art setups.

2. A scheme for timing jitter compensation

In this paper, we propose a new concept for external injection that would allow to achieve unprecedented sub-femtosecond timing jitter between laser and witness beam. The scheme, as seen in Figure 1, relies on adding an intermediate synchronizing stage before the main plasma acceleration module.

In this concept, the laser pulse is initially split in two, with one of them containing only a small fraction of the energy (e.g. $\sim 5\%$). These two pulses are intrinsically synchronized and will serve different purposes: the weak one will be used in the synchronizing stage, while the strong pulse will be used as driver in the main plasma acceleration module.

As seen in Figure 1, the main idea is to use the weak pulse to drive a first plasma stage in which the electron beam will get a small energy change proportional to its arrival time. In the speed-of-light frame, $\xi = z - ct$, beams with an offset $\Delta\xi_i = \xi_i - \xi_0$ with respect to the zero crossing, ξ_0, of the accelerating field, E_z, will gain or lose energy depending on whether they arrived too late ($\Delta\xi_i < 0$) or too soon ($\Delta\xi_i > 0$), therefore correlating arrival time and beam energy.

Due to this correlation, a magnetic chicane will then be able to correct the initial offsets thanks to the differences in path length for each beam energy, virtually removing the timing jitter between electron beam and main laser pulse for injection into the second plasma stage.

Additionally, the scheme contains two quadrupole sections to match the beam into and out of the plasma modules.

The laser parameters should be such that the weak pulse is able to excite a linear wake in the first plasma stage, so powers on the order of 1 – 10 TW should suffice. For the strong one, the necessary power will depend on the desired final energy of the beam after acceleration.

The design energy of the chicane, E_0, should be selected as the energy, after the first plasma, of a beam with $\Delta\xi = 0$. In case of no beam loading, this means that $E_0 = E_i$, where E_i is the initial energy of the beams coming from the linac.

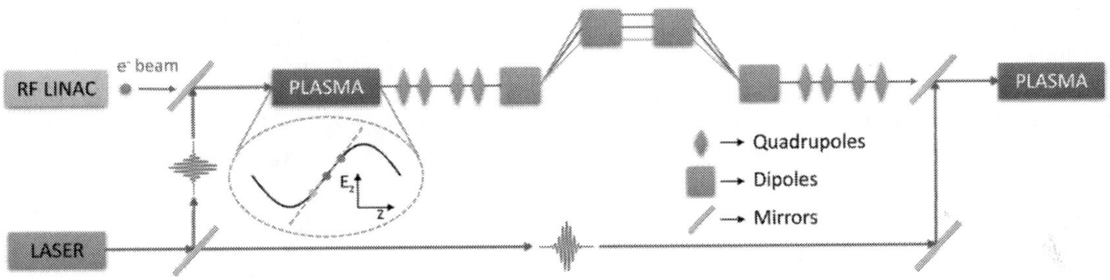

Figure 1. Schematic view of the synchronizing stage.

3. Linear model

In the first plasma stage, beams will get an energy deviation $\delta = \frac{E-E_0}{E_0}$ depending on their longitudinal position. For small offsets this can be approximated as

$$\delta = -\frac{eE_z' L_p}{E_0}(\xi_i - \xi_0) \ , \tag{1}$$

where e is the electron charge, $E_z' = \partial_\xi E_z$ is the slope of the accelerating field around ξ_0 and L_p is the length of the plasma stage.

To first order, the final position of the beam after the chicane will be given by

$$\xi_f = \xi_i - R_{56}\delta \ , \tag{2}$$

where the minus sign comes from the fact that in the comoving coordinate ξ, as commonly used in plasma acceleration, the bunch head is on the right.

Then, in order to correct the initial offsets, $\xi_f = \xi_0$ should be imposed. This, together with Equation (1), yields

$$R_{56} = -\frac{E_0}{eE'_z L_p} \ . \tag{3}$$

For a plasma in the linear regime ($a_0 \ll 1$) [13, 14] and a circularly polarized Gaussian pulse, $a^2 = a_0^2 \exp(-\xi^2/2\sigma_z^2)$, the slope of the field E_z around its zero crossing is

$$E'_z = \frac{mc^2}{e}\sqrt{\frac{\pi}{2}} a_0^2 k_p^3 \sigma_z e^{-\frac{k_p^2\sigma_z^2}{2}} \ , \tag{4}$$

where m is the electron mass, c is the speed of light, a_0 is the peak amplitude of a, the laser's normalized vector potential, k_p is the plasma wavenumber and σ_z is the laser rms length. In case of linear polarization a_0^2 should be replaced by $a_0^2/2$. By substituting this into Equation (3) one finds the general expression

$$R_{56} = -\sqrt{\frac{2}{\pi}} \frac{E_0}{mc^2 a_0^2 k_p^3 \sigma_z L_p} e^{\frac{k_p^2\sigma_z^2}{2}} \ . \tag{5}$$

Now, if the laser pulse length satisfies the resonant condition ($\sigma_z k_p = 1$), one can find a simplified equation which, written in a convenient way, reads

$$R_{56}\,[\mathrm{mm}] = -7.27 \times 10^{13} \frac{E_0\,[\mathrm{MeV}]}{a_0^2\, n_p\,[\mathrm{cm}^{-3}]\, L_p\,[\mathrm{mm}]} \ . \tag{6}$$

In reality, depending on the beam energy, it might be necessary to take into account the path length effect due to the deviation from the speed of light not being negligible. This would decrease the necessary R_{56} at the chicane, but a precise value can only be given once the full geometry of the beamline up to the second plasma has been decided. Also, Equations (5) and (6) do not take into account beam loading so, if this effect becomes significant, R_{56} might have to be determined from the chirp of δ as obtained from simulations.

Another important remark is that the quality of the jitter correction will depend on the stability of the parameters in Equation (5), as well as the pointing jitter of the electron beam after the first plasma. For the case presented below, assuming the pointing jitter does not exceed the natural beam divergence, the resulting maximum timing jitter would be around 0.2 fs per meter of beamline downstream of the plasma, still on the sub-fs range.

Moreover, δ should always be much bigger than the intrinsic energy spread of the beam, otherwise its longitudinal phase space might be perturbed in the chicane.

4. Numerical simulations

In order to test this method, a series of start-to-end simulations going from the injection of the beam into the first plasma until the end of the chicane have been performed.

The simulation of the plasma stage was performed in 3D with the fully relativistic Particle-in-Cell code OSIRIS [15] using the Ponderomotive Guiding Center (PGC) algorithm [16]. WinAGILE [17] and MAD-X [18] were used for the optimization of the quadrupole section, and the tracking of the electron beam from the plasma exit until the end of the chicane was done with ELEGANT [19].

Effects derived from Coherent Synchrotron Radiation (CSR) at the dipoles have been taken into account, but a full treatment including space charge after exiting the plasma has not yet been possible. However, due to the low charge density of the beam after the plasma and its relativistic energy, strong space charge effects are not expected.

4.1. Main Parameters

The electron beam after the linac is Gaussian with a charge of 0.1 pC in 1 fs, an energy of 100 MeV with 0.1% relative energy spread, a normalized emittance $\epsilon_{nx,y}$ of 0.3 μm and initial offsets due to timing jitter between −20 and 20 fs. These parameters were chosen in order to avoid significantly perturbing the plasma wakefield with the beam (see Figure 2) and are within the range of those expected to be produced at the SINBAD facility at DESY [20, 21, 22]. The first plasma stage has a length of 2 mm with a density of 10^{17} cm^{-3}, and the laser parameters are based on those considered for the EuPRAXIA conceptual design [23], using only around 3.5% of its energy for the synchronizing stage. This means an 800 nm linearly polarized pulse of 3.5 J with an a_0 of 0.6, a waist w_0 of 54μm and a length of 93 fs (FWHM in intensity), having a peak power of 35 TW. With these parameters, the matched beam size for emittance preservation [12] is $\sigma_x = 1.3$ μm for the focusing fields around ξ_0.

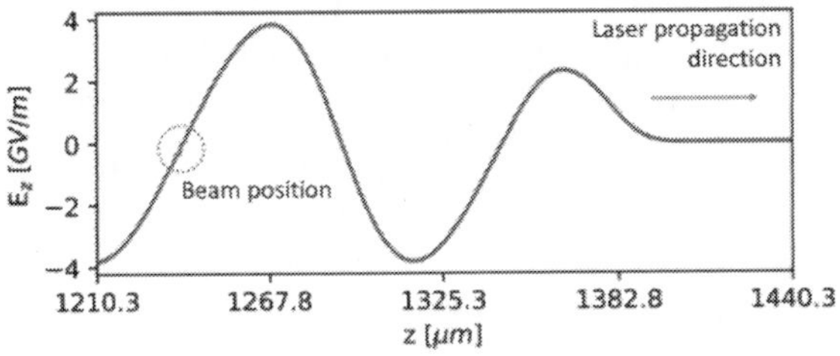

Figure 2. On-axis longitudinal electric field after 1.2 mm of propagation in the plasma with the electron beam placed at the zero crossing (case with $\Delta\xi = 0$).

4.2. Quadrupole Section

The set of magnets is placed 4 cm after the plasma and consists on 4 permanent quadrupoles with a width of 1 cm separated by drifts of 4, 6 and 4 cm. Their strengths were left as free parameters for optimization in order to achieve, as seen in Figure 3, a beam waist at the center of the chicane and a beta function, $\beta_{x,y} = \gamma_r \sigma^2_{x,y}/\epsilon_{nx,y}$, as constant as possible, where γ_r is the Lorentz factor of the beam.

The resulting geometric strengths of the quadrupoles read, from first to last, -878, 1406, -1497 and 823 in units of m^{-2} which, for this beam energy, translate into field gradients of up to 500 T/m, requiring tight apertures.

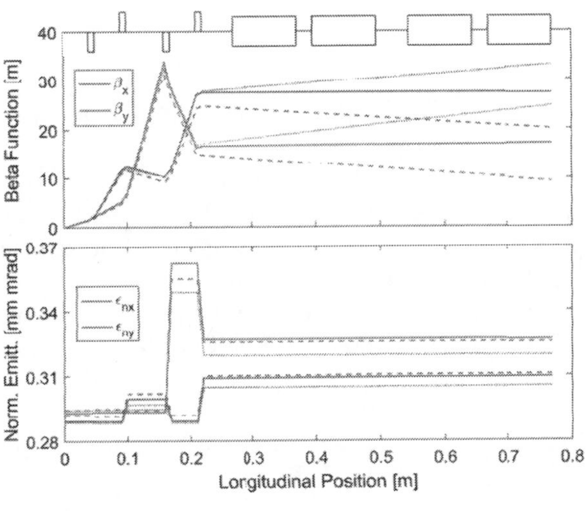

Figure 3. Beta function and emittance evolution for beams with different offsets. The solid line corresponds to $\Delta\xi_i = 0$, while the dashed and dotted lines represent offsets of 20 and -20 fs, respectively.

4.3. Chicane Setup
For this case, Equation (5) predicts an $R_{56} = -0.222$ mm. However, for higher accuracy, the slope of δ was directly measured from the simulation results (see Figure 4), obtaining an $R_{56} = -0.267$ mm. This could be due to the laser a_0 being 0.6, slightly high for the linear theory to hold, since $a_0 \ll 1$ is assumed.

The bending angle θ of the dipoles was determined from

$$R_{56} = -2\theta^2 \left(L_d + \frac{2}{3} L_m \right) \tag{7}$$

by assuming a magnet length $L_m = 10$ cm and a drift length $L_d = 2.5$ cm, obtaining $\theta = 2.19°$.

4.4. Simulation Results
As seen in Figure 4, the plasma stage correctly imprints a linear kick to the beam energy depending on its initial offset, allowing the chicane to compensate the jitter. All beam offsets between ±10 fs have been reduced to sub-fs level, showcasing the potential of this method.

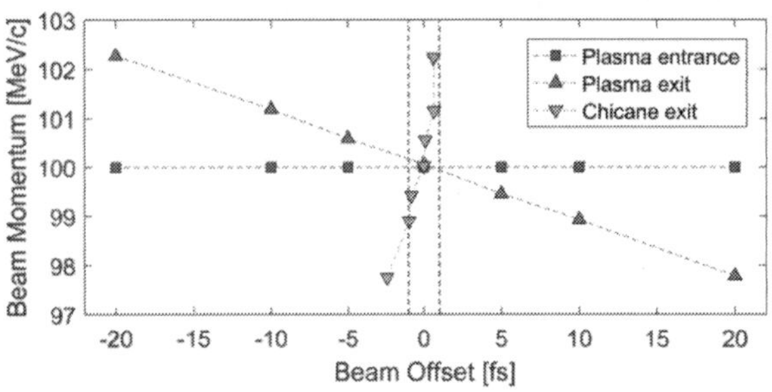

Figure 4. Jitter correction thanks to plasma stage and chicane. The vertical lines delimit a 1 fs jitter.

In terms of quality, as seen in Figure 3, the beam experiences an increase in emittance of around 10 − 20% due to chromatic effects at the quadrupoles. Additionally, due to the strong focusing, the beam develops a slightly curved shape in the transverse plane, increasing its rms length in about 0.5 fs.

Also, as seen in Figure 3, beams with $\Delta\xi_i \neq 0$ will present variations in the evolution of the beta function due to their different energies, so some way of compensating this chromatic effect before the second plasma stage will be required.

5. Conclusion
We have presented a new scheme for arrival time jitter minimization between laser pulse and witness beam for LWFA with external injection. Simulation results show that jitters of 10 fs can be reduced to the sub-femtosecond level with minimal loss of beam quality.

Further studies are required to determine the stability and tolerances of the setup, as well as the influence of space charge effects and finding ways of matching the beam into the second plasma stage taking into account the variations in beta function seen in Figure 3.

Also, although only ultrashort beams have been tested, it is possible that for longer beams ($\sigma_z \sim 10$ fs) this scheme could be used both for synchronization and compression.

Acknowledgments
The authors would like to give special thanks to Jun Zhu and Daniel Marx for fruitful discussions regarding the beamline design and its simulation.

This project has received funding from the European Union's Horizon 2020 research and innovation programme under grant agreement No. 653782.

References

[1] Tajima T and Dawson J M 1979 Laser electron accelerator *Phys. Rev. Lett.* **43** 267
[2] Malka V *et al* 2002 Electron acceleration by a wake field forced by an intense ultrashort laser pulse *Science* **298** 1596–1600
[3] Leemans W P *et al* 2006 GeV electron beams from a centimetre-scale accelerator *Nat. Phys.* **2** 696–9
[4] Faure J *et al* 2006 Controlled injection and acceleration of electrons in plasma wakefields by colliding laser pulses *Nature* **444** 737
[5] Pak A *et al* 2010 Injection and trapping of tunnel-ionized electrons into laser-produced wakes *Phys. Rev. Lett.* **104** 025003
[6] Gonsalves A J 2011 *et al* Tunable laser plasma accelerator based on longitudinal density tailoring *Nat. Phys.* **7** 862–6
[7] Faure J *et al* 2004 A laserplasma accelerator producing monoenergetic electron beams *Nature* **431** 541–4
[8] Geddes C G R *et al* 2004 High-quality electron beams from a laser wakefield accelerator using plasma-channel guiding *Nature* **431** 538-541
[9] Mangles S P D *et al* 2004 Monoenergetic beams of relativistic electrons from intense laserplasma interactions *Nature* **431** 535–8
[10] Tzoufras M *et al* 2008 Beam loading in the nonlinear regime of plasma-based acceleration" *Phys. Rev. Lett.* **101** 145002
[11] Assmann R and Yokoya K 1998 Transverse beam dynamics in plasma-based linacs *Nucl. Instr. and Meth.* A **410** 544–8
[12] Mehrling T, Grebenyuk J., Tsung F S, Floettmann K and Osterhoff J 2012 Transverse emittance growth in staged laser-wakefield acceleration *Phys. Rev. ST Accel. Beams* **15** 111303
[13] Esarey E, Schroeder C B and Leemans W P 2009 Physics of laser-driven plasma-based electron accelerators *Rev. Mod. Phys.* **81** 1229–85
[14] Gorbunov L M and Kirsanov V I 1987 Excitation of plasma waves by an electromagnetic wave packet *Sov. Phys. JETP* **66** 40
[15] Fonseca R *et al* 2002 OSIRIS: a three-dimensional, fully relativistic particle in cell code for modeling plasma based accelerators *Proc. ICCS* (Amsterdam) p 342-351
[16] Gordon D F, Mori W B and Antonsen T M 2000 A ponderomotive guiding center particle-in-cell code for efficient modeling of laser-plasma interactions *IEEE transactions on plasma science* **28** 1135–43
[17] Bryant P J, 2000 AGILE - a tool for interactive lattice design *Proc. 7th European Particle Accelerator Conf.* (Vienna) pp 1357–59
[18] Grote H and Schmidt F 2003 MAD-X-an upgrade from MAD8 *Proc. 20th IEEE Particle Accelerator Conf.* (Portland) p 3497
[19] Borland M 2000 Elegant: a flexible SDDS-compliant code for accelerator simulation *Advanced Photon Source* LS-287.
[20] Assmann R, Behrens C, Brinkmann R *et al* 2014 SINBAD - A proposal for a dedicated accelerator research facility at DESY *Proc. 5th Int. Particle Accelerator Conf.* (Dresden)
[21] Zhu J *et al* 2017 Matching space-charge dominated electron bunches into the plasma accelerator at SINBAD *Proc. 8th Int. Particle Accelerator Conf.* (Copenhagen)
[22] Zhu J, Assmann R, Dohlus M, Dorda U and Marchetti B 2016 Sub-fs electron bunch generation with sub-10-fs bunch arrival-time jitter via bunch slicing in a magnetic chicane *Phys. Rev. Accel. Beams* **19** 054401
[23] Walker P A *et al* 2016 *EuPRAXIA deliverable report D 1.2 report defining preliminary study concept*

IOP Conf. Series: Journal of Physics: Conf. Series **874** (2017) 012033 doi:10.1088/1742-6596/874/1/012033

Application of optical emission spectroscopy to high current proton sources

G Castro[1], M Mazzaglia[1], D Nicolosi[1], D Mascali[1], R Reitano[1,2], B Zaniol[3], L Celona[1], O Leonardi[1], F Leone[1,2,4], E Naselli[1,2], L Neri[1], G Torrisi[1], S Gammino[1]
[1]INFN-LNS, Catania, Italy
[2]University of Catania, Catania, Italy; INFN-LNS, Catania, Italy
[3] consorzio RFX, Corso Stati Uniti 4, 35127 Padova, Italy
[4] Istituto Nazionale di Astrofisica-Osservatorio Astrofisico di Catania Via S. Sofia 78, 95123, Catania Italy

E-mail: Giuseppe.Castro@lns.infn.it

Abstract. Optical Emission Spectroscopy (OES) represents a very reliable technique to carry out non-invasive measurements of plasma density and plasma temperature in the range of tens of eV. With respect to other diagnostics, it also can characterize the different populations of neutrals and ionized particles constituting the plasma. At INFN-LNS, OES techniques have been developed and applied to characterize the plasma generated by the Flexible Plasma Trap, an ion source used as "testbench" of the proton source built for European Spallation Source. This work presents the characterization of the parameters of a hydrogen plasma in different conditions of neutral pressure, microwave power and magnetic field profile, along with perspectives for further upgrades of the OES diagnostics system.

1. Introduction

In recent years, different projects have requested the ability to produce intense proton and H_2^+ beams. A 74 mA proton current is requested by the European Spallation Source [1]), while up to 50 mA H_2^+ by the Daeδalus and ISOdar experiments [2, 3]. The fraction of proton and H_2^+ constituting the extracted beam depends on the source parameters such as magnetic field profile, neutral pressure, microwave power and frequency and also on the dimensions of the plasma chamber [4]. Up to now, the proton fraction of a beam is known only after beam extraction; furthermore, some important information, such as the vibrational states of the H_2^+ molecule, can not be measured in any way by standard beam diagnostics by means (higher H_2^+ vibrational states are detrimental for H_2^+ acceleration [3]).

The limits of the actual diagnostics can be overcome by means of Optical Emission Spectroscopy. It enables direct "in plasma" diagnostics measuring not only the electron density and temperature, but also the relative concentration of different neutral species, vibrational temperature of the molecules, and concentration of the different charges states in case of plasma of multicharged ions. At the INFN-Laboratori Nazionali del Sud (INFN-LNS), OES has been applied to characterizing the hydrogen plasma generated by a plasma trap (described in the next section) as a preliminary test for further development on other ion sources (in particular for the proton source of ESS). Hereinafter the first results obtained will be described.

2. Experimental set-up

OES measurements have been carried out with the Flexible Plasma Trap (FPT), a test bench for plasma diagnostics and development of new sources, installed at INFN-LNS. FPT is fed by microwaves in the range 4-7 GHz generated by a TWT, while the magnetic field is obtained by means of three coils, which permit different magnetic configurations. The most interesting magnetic configurations, presented in figure 1, have large importance for ion sources. Flat magnetic field is the typical magnetic field used in a microwave discharge ion source for proton generation; simple mirror configuration increases ion lifetime for multicharged ion generation; finally, magnetic beach configuration allows exploring new mechanisms of plasma heating such as EBW heating [5]. An overall characterization of the FPT can be found in reference [6].

The OES diagnostics system consists of an ImSpector V8E spectrometer, coupled to an ACA2040 CMOS camera. The spectrometer resolution is 2 nm and it is sensitive in a spectral range of 380 - 1000 nm. The whole system is connected to the FPT by means of a 1500 μm diameter fiberglass that is, in turn, properly connected to a quartz window, which "looks" towards the centre of the FPT plasma chamber. The cone of view of the fiberglass is around 22 degree wide and can integrate the plasma light over a well-determined volume. The whole OES experimental set-up has been calibrated by comparison with a calibrated lamp. Figure 2 shows a view of the FPT and of the experimental set-up for OES. The low resolution of our instrumentation (2 nm) makes impossible any attempt to search for vibrational lines of H_2^+ (a few tens of pm needed); however, the resolution is enough for studying the dependence of the H/H_2 ratio and of electron density on the source parameters.

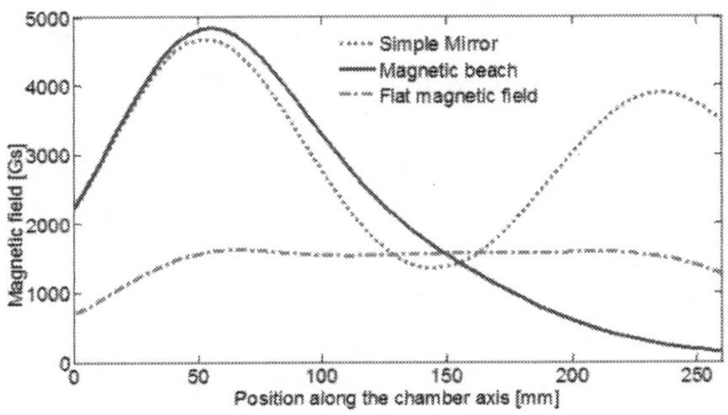

Figure 1. Magnetic field profiles that can be generated by the FPT.

Figure 2. A view of the FPT and the OES set-up.

3. Experimental results

The determination of plasma parameters from emission spectra is still a challenging question. An interesting summary can be found in references [7, 8]. The intensity of emission *I* depends on the particle density in the excited state, which, in turn, depends strongly on plasma parameters, such as electron temperature t_e, electron density n_e and density and temperature of all neutral and ionized particles n_n and t_n. Such a dependence is described by the effective emission rate coefficients X^{eff}, calculated by means of opportune population and depopulation models [7].

The intensity of the emission line, for the transition *l-->m,* is proportional to $X_{lm}^{eff}(n_e, t_e, n_n, t_n)$ and to the particle density in the ground state n_g [7]. The ratio between two emission lines I_{pk}^1 and I_{lm}^2 is therefore:

$$\frac{I_{pk}^1}{I_{lm}^2} = \frac{n_{g1}}{n_{g2}} \frac{X_{pk}^{eff}(n_e, t_e, n_n, t_n)}{X_{lm}^{eff}(n_e, t_e, n_n, t_n)} \quad (1)$$

If one knows the $X_{pk}^{eff}/X_{lm.}^{eff}$ ratio, equation (1) allows calculating n_1/n_2 from I_{pk}^1/I_{lm}^2. This is called the "line ratio method".

We applied this method to calculate the relative abundance n_H/n_{H_2} of atomic and molecular hydrogen in plasma. It can be shown [8], in fact, that the ratio between the H_γ Balmer line (434.1 nm) and the Fulcher band (600-640 nm) is almost independent of electron temperature and assumes a value ~1 for the values of electron density typical of FPT, i.e. $n_H/n_{H_2} \approx I_\gamma/I_{Fulch.}$

The determination of the n_H/n_{H_2} ratio has large importance for proton sources, because it identifies the conditions that favour the H or H_2 presence in plasma. An H-rich plasma favours proton generation, while an H_2-rich plasma favours H_2^+ generation.

Together with the n_H/n_{H_2} ratio, the total intensity of the spectral emission has been calculated. Total intensity gives information about the energy content of the electron population responsible for the optical transitions in plasma, i.e. in the range 1-20 eV.

First, we characterized the n_H/n_{H_2} ratio and total intensity as a function of the RF power and neutral pressure. For experimental measurements, a very stable configuration of FPT has been chosen: the microwave frequency has been fixed at 6.82 GHz and operated in simple mirror magnetic configuration. The hydrogen pressure was fixed at $0.8 \cdot 10^{-4}$ mbar, a value minimizing the reflected power.

Figure 3 shows the dependence of the n_H/n_{H_2} ratio and the total intensity on the microwave power from 20 W to around 120 W. At 20 W, n_H/n_{H_2} is indeterminate, since the experimental error is too high. From 40 to 80 W, n_H/n_{H_2} increases up to 0.5 and then decreases again. A similar behaviour can be found in the total intensity of the signal. Since the total intensity is related to the cold electron population, this means that n_H/n_{H_2} is maximized when the electron energy is high enough for the H_2 molecule break up. Further increase in the electron energy content, however, is detrimental for H generation, probably because the cross section for H_2 break up decreases above 15 eV [9].

This means that either too low or too high RF power could be detrimental for proton generation in a proton source. Furthermore, since higher RF power favours the increase of the emittance, proton beam should be generated by using RF power as low as possible.

Figure 4 shows the dependence of n_H/n_{H_2} ratio and total intensity on neutral pressure. Microwave power has been fixed at 60 W, while the other parameters remained unchanged. The neutral pressure was increased from 1 to $8 \cdot 10^{-4}$ mbar. The highest values are comparable with the typical operating pressure of proton and H_2^+ sources ($5 \cdot 10^{-4}$-$1 \cdot 10^{-3}$ mbar [1]). Total intensity increases strongly with the pressure, indicating an increase of the electron energy content of cold electrons, while the n_H/n_{H_2} ratio decreases because of the continuous addition of H_2 molecules. This means that for higher values of pressure, higher and higher RF power is needed to improve the proton fraction of the extracted beam.

Finally, figure 5 shows the dependence of the n_H/n_{H_2} ratio and total intensity on the mirror ratio of the magnetic field, i.e. on the ratio $B_{max}/B_{min.}$ The mirror ratio is an important parameter because it allows modifying the ion confinement time within the plasma.

To have a stable plasma for any value of the magnetic field profile, microwave' frequency and power have been fixed respectively at 5.63 GHz and 60 W, with neutral pressure at $2 \cdot 10^{-4}$ mbar. In this case, both the n_H/n_{H_2} ratio and the total intensity increase with B_{max}/B_{min} and, consequently, with confinement time. This means that, for larger confinement times, the chance that H_2 molecules are broken into H atoms by electrons is increasing. Furthermore, for lower values of B_{max}/B_{min}, the density of RF power absorbed by plasma decreases, so a strong increase of RF power is needed to guarantee a larger value of the proton fraction.

IOP Conf. Series: Journal of Physics: Conf. Series **874** (2017) 012033 doi:10.1088/1742-6596/874/1/012033

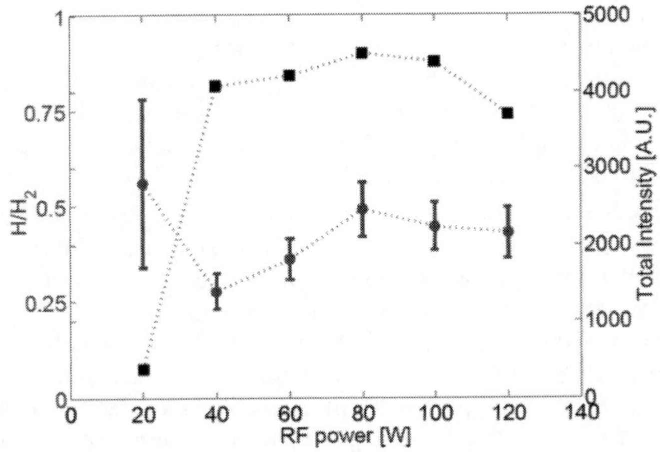

Figure 3. Dependence of the n_H/n_{H_2} ratio on RF power.

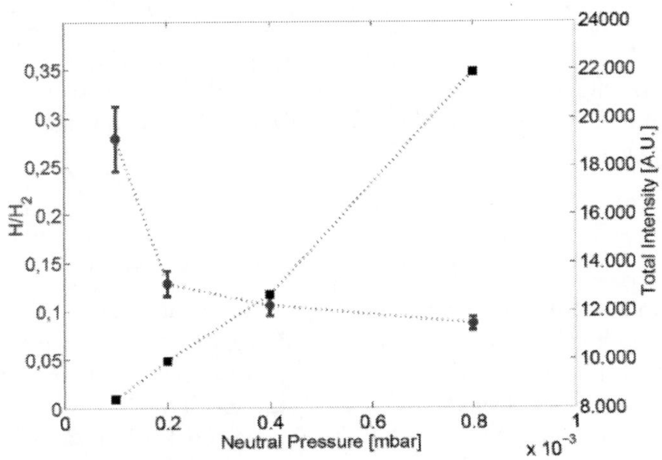

Figure 4. Dependence of the n_H/n_{H_2} ratio on the pressure.

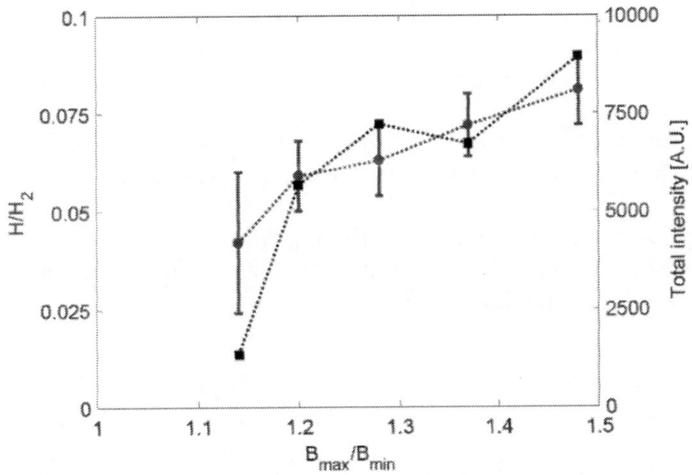

Figure 5. Dependence of the n_H/n_{H_2} ratio on the B_{max}/B_{min} ratio.

4. Conclusions and next steps

Preliminary OES measurements carried out with the FPT permitted the "in plasma" characterization of the relative abundances of the neutral component of a hydrogen plasma and indeed the identification of the source parameters that favour the proton or the H_2^+ generation within the plasma. New studies, based on the radiative collisional model, are expected to permit also the electron density calculation from the ratio between $\mathbf{H}_\beta$ and $\boldsymbol{H}_\gamma$ Balmer lines. The OES technique will be also applied to the PS-ESS source in order to relate, for the first time, the proton fraction to the $\boldsymbol{n_H/n_{H_2}}$ratio and electron density measured by means of the line ratio method. In perspective, we expect that all new proton sources will be equipped with multi-line of sight OES diagnostics, in order to evaluate, during the tuning phase of the source, the best operating conditions for high-stability, high current, proton (or H_2^+) beams.

The low-resolution spectrometer will be supported soon by a high-resolution spectrometer, SARG. SARG is an advanced spectropolarimeter with a resolution of $\lambda/\Delta\lambda$=160000 in the range of 370-900 nm (~3 pm at 400 nm), to be installed at INFN-LNS in late spring in the PANDORA experiment [10]. The new spectrometer will also allow discriminating, "in plasma", the vibrational and rotational states of H_2^+. In the case of plasmas composed of multicharged ions, the determination of the charge state distribution before of the beam extraction will also be possible.

Acknowledgements

The suggestions of U. Fantz has been essential for the present paper. The contributions of E. Zappalà, G. Messina, A. Maugeri, A. Seminara, S. Marletta, G. Pastore, S. Vinciguerra, F. Chines and B. Trovato are warmly acknowledged.

References

[1] Neri L *et al.* Proton Source for the European Spallation Source (PS-ESS) commissioning strategy for high flexibility machine, submitted for publication on *rev. sci. Instrum.*
[2] Bungau A *et al.* 2012 Proposal for an Electron Antineutrino Disappearance Search Using High-Rate ^{8}Li Production and Decay *Phys. Rev. Lett.* **109** 141802
[3] Conrad J M and Shaevitz M H 2010 Multiple Cyclotron Method to Search for CP Violation in the Neutrino Sector *Phys. Rev. Lett.,* **104** 141802
[4] Castro G *et al.* 2016 A new H2+ source: Conceptual study and experimental test of an upgraded version of the VIS-Versatile ion source *Rev. Sci. Instrum.* **87** 083303
[5] Castro G *et al.* 2017 Overdense plasma generation in a compact ion source *Plasma Sources Sci. Technol.* **26** 055019
[6] Gammino *S et al.* The flexible plasma trap for the production of overdense plasmas submitted for publication on *jinst*
[7] Fantz U 2006 Basics of plasma diagnostics, *Plasma Sources Sci. Technol.* **15** S137.
[8] Fantz U *et al. 2006* Spectroscopy-a powerful diagnostic tool in source development *Nucl. Fusion* **46** S297
[9] Zhang H 1999 *Ion sources* (Science Press Beijing and Springer)
[10] Mascali D *et al.* PANDORA a New Facility for Interdisciplinary In-Plasma Physics, accepted for publication on *EPJ A.* [arXiv:1703.00479]

Microwave injection and coupling optimization in ECR and MDIS ion sources

Giuseppe Torrisi[1], Antonio Caruso[1], Giuseppe Castro[1], Luigi Celona[1], Santo Gammino[1], Ornella Leonardi[1], Alberto Longhitano[1], David Mascali[1], Eugenia Naselli[1,2], Lorenzo Neri[1], Gino Sorbello[1,3]

[1] INFN - Laborari Nazionali del Sud, Catania, Italy
[2] Dipartimento di Fisica ed Astronomia, University of Catania, Catania, Italy
[3] Dipartimento di Ingegneria Elettrica Elettronica e Informatica, University of Catania, Catania, Italy

E-mail: giuseppe.torrisi@lns.infn.it

Abstract. The fundamental aspect of coupling between microwave and plasma of the Electron Cyclotron Resonance Ion Source (ECRIS) and Microwave Discharge Ion Source (MDIS) is hereinafter treated together with "ad hoc" microwave-based plasma diagnostics, as a key element for the next progress and variations with respect to the classical ECR heating mechanism. The future challenges for the production of higher-charge states, higher beam intensity, and high absolute ionization efficiency also demand for the exploration of new heating schemes and synergy between experiments and modeling. An overview concerning microwave transport and coupling issues in plasma-based ion sources for particle accelerator will be given in the paper, along with perspectives for the design of next generation sources.

1. Introduction

Electron cyclotron resonance ion sources (ECRISs) and Microwave Discharge Ion Sources (MDISs) are widely used nowadays in many acceleration facilities, for the productions of highly charged ions (HCIs) and of tens of mA of proton beams, respectively.

In this paper we will focus on:

(i) 2.45 GHz Radio frequency (RF) injection system design and measurements for the Proton Source of the European Spallation Source (PS-ESS) [1], currently under commissioning at INFN-LNS.

(ii) Modeling of wave-plasma interaction and test on the Flexible Pasma Trap (FPT)[2] and Plasma Reactor (PR), operating at the INFN-LNS, as test-benches for the investigation of innovative mechanisms of overdense plasmas heating and microwave-based diagnostics measurements [3].

2. RF measurements for PS-ESS

The detailed design and performances of 2.45 GHz Microwave Discharge Ion Sources, based on the principle of the off-resonance microwave discharge, are discussed in many references [4, 5]. Hereinafer we focus on the microwave injection system design and measurements of the PS-ESS source.

IOP Conf. Series: Journal of Physics: Conf. Series **874** (2017) 012034 doi:10.1088/1742-6596/874/1/012034

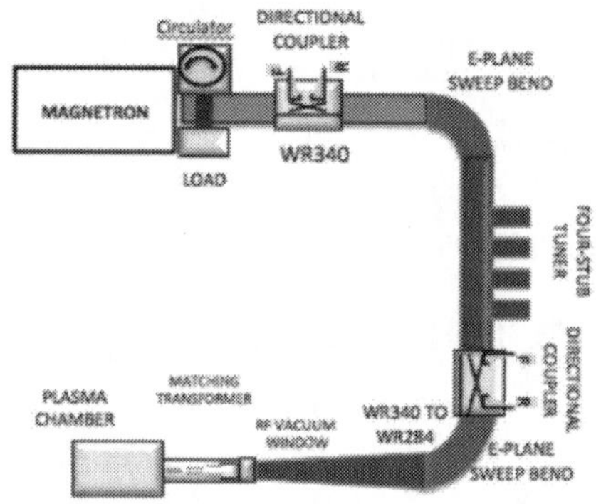

Figure 1. PS-ESS 2.45 GHz microwave injection block scheme.

The microwave line, shown in Figure 1, consists of:

(i) Pulsed 2 kW 2.45 GHz Magnetron generator with continuous power supply on the filament to decrease the ripple, with external remote control of high and low level of the pulses, and On/Off duration time controlled by optical fiber. The emission head of the magnetron is protected from the reflected power by a RF circulator.

(ii) Cross-Guide Directional Coupler for a simultaneous measure of forward (2 kW max) and reflected power at 2.45 GHz (Coupling coefficient= 40 dB and Directivity= 30 dB).

(iii) WR340 Waveguide 4-stub Automatic Tuning Unit (ATU): as soon as the generator delivers microwaves, the auto tuner matches instantaneously the load and maintains permanently the reflected power at a low level [6].

(iv) Broadwall Multi-hole Directional Coupler for a simultaneous measure of forward (2 kW max) and reflected power at 2.45 GHz in a WR340 waveguide (Coupling coefficient= 50 dB and Directivity=40 dB);

(v) WR284 Waveguide quartz pressure window (Max. Power 2 kW Continuous Wave and VSWR<1.15 at 2.45 GHz).

(vi) WR284 waveguide multi-step binomial maximally flat matching transformer which permits a progressive match between the impedance of WR284 waveguide and the impedance of plasma filled chamber.

A waveguide transition WR340 - WR284, two WR340 90° E-plane bends and Diode RF Power probes with a measurement range from -57 dBm to +33 dBm at 10 MHz-18 GHz connected to the ports (N-connectors) of directional couplers, complete the microwave layout.

The use of directional couplers for incident and reflected power measurements placed in two parts of the microwave line (one immediately after the magnetron output and the other one just after the ATU) represents a very important microwave diagnostic tool providing information on the interaction between the electromagnetic waves and the plasma, which depends mainly on the plasma density and absorption efficiency of the energy carried by the microwave [7].

As a first step, we experimentally tested the microwave line in pulsed operation considering a waveguide matched load in place of the plasma chamber and by inserting an impedance tuner between the microwave generator and the matched load. A ripple measurement has been carried out on the output pulse from magnetron when the maximum power of 2 kW is applied on the waveguide matched load. In Figure 2 the signal amplitude is around 200 mV which corresponds to 2 kW value. The signal ripple has an amplitude of about ± 1 mV (± 10 W). Therefore we can conclude that ripple is 1%.

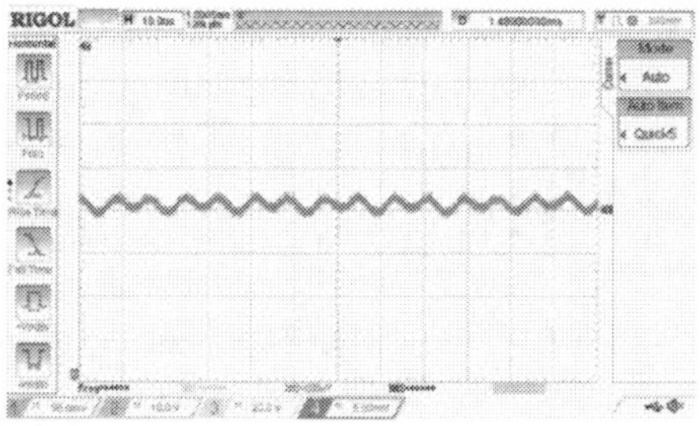

Figure 2. Ripple measurement of ± 0.5% (P_{FW}=2 kW).

During the PS-ESS conditioning, when the real plasma load is present, we acquired reflected and forward RF power from the RF power probes connected to the Broadwall Multi-hole Directional Coupler.

Figure 3 shows the plasma ability to absorb the incoming microwave power that is gradually increased from 20 W to 120 W in CW mode. After this value the plasma density is high enough to absorb more that 99.5 % of the injected power. Results show an optimum matching of power inside the plasma chamber that is very promising for the beam extraction process.

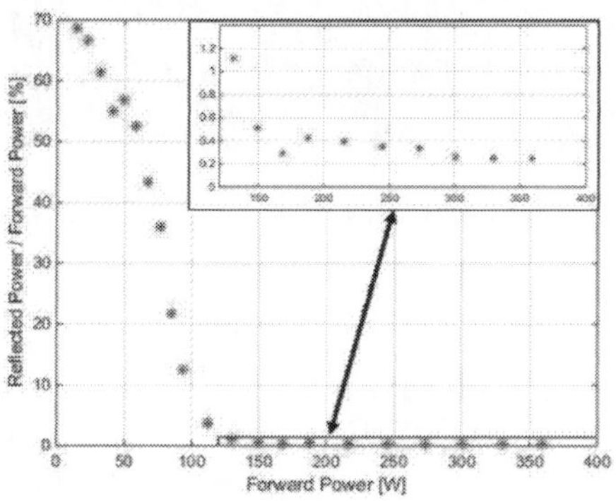

Figure 3. Fraction of reflected microwave power vs injected.

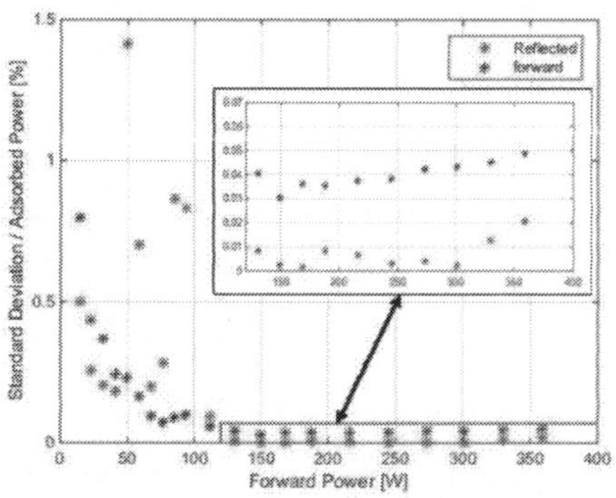

Figure 4. Stability of microwave absorbed power.

The software used for the data acquisition of the RF probes gives us also the possibility to extract the stability of the two measured power values. We select a data acquisition duration of

10 μs and an acquisition rate of 10 Hz; the standard deviation was calculated after 2000 counts. Figure 4 shows the fraction between the standard deviation and the adsorbed power (forward minus reflected) versus the forward power. The result shows that with an injected power greater that 120 W the plasma features a very stable condition with a fluctuation below 0.05%. We estimate that such a value will permit to reach the $\pm 2\%$ of beam stability, as requested by the ESS project.

Finally we acquired forward and reflected power from power probes (which represent an useful RF diagnostic to retrieve plasma condition stability), in typical pulsed condition imposed from the external timing system (in optical fiber) with a pulse width of 6 ms, 14 Hz repetition rate, forward power P_{FW} = 400 W and gas flow rate= 1.41 sccm (Figure 5): it is evident that the overall system is well-matched and that the ATU in automatic mode operation permits to find the optimum working point. If the gas flow rate is too low (under 1.01 sccm), the RF pulse becomes unstable.

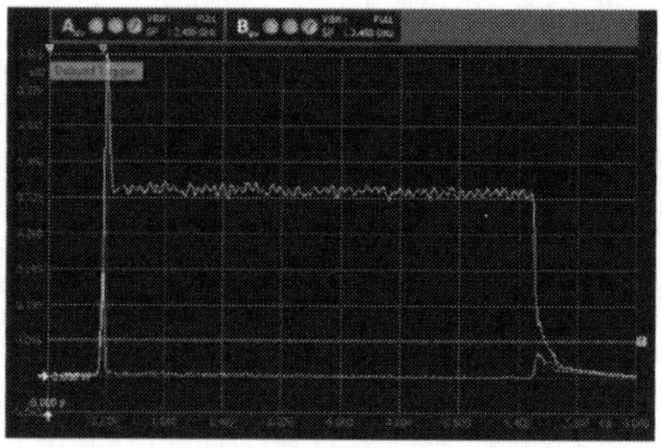

Figure 5. Forward (yellow) and reflected (blue) Power.

3. RF Modeling and experimental results

Up to now in almost all operating ECRIS the increase of the electron density and extracted current is obtained through the increase of microwave frequency and confinement magnetic field. ECRIS devices are density limited, because the electromagnetic waves cannot propagate over the cut-off density. Many experiments, mainly based on slight variations of the microwave frequency (frequency tuning effect) and the use of two or more frequencies for plasma heating, have been carried out to verify the possibility to improve the plasma heating outside the framework traced by the Standard Model. An alternative to the classical ECR interaction is the electrostatic wave heating, driven by Bernstein waves. On the PR operating at INFN-LNS some evidences of Bernstein wave (BW) generation were observed in [8], obtained via the inner plasma Electromagnetic-to-Electrostatic wave conversion. Moreover a novel frequency-sweep microwave interferometry [3] enabled measuring the overall density of the whole plasma. The measurement of the line-integrated density $n_e = 2.1 \pm 1.0 \cdot 10^{18}$ m^{-3} for the PR demonstrates that the plasma density can be measured in a non-intrusive way also in compact ECRIS ans MDIS ion sources with the possibility to improve the tuning of existing sources in terms of magnetic trapping and RF coupling.

The need for a more flexible magnetic field and RF injection system suggested to design and develop a different type of ECRIS-like plasma trap, named FPT[2]. FPT can work with different plasma heating schemes by means of three different microwave systems, one parallel and two perpendicular with respect to the plasma chamber. The axial injection works at frequencies in the range of 4-7 GHz. A directional coupler enables the measure of forward and refected power, while an insulator safeguards the TWT by the reflected power. The innovative perpendicular microwave launcher [9] consists of an array of two waveguides properly phased in order to tilt the injection angle, which plays a key role for modal conversion from the BW generation. The contemporary use of the different microwave inputs on FPT allows operating with a double frequency (first and second harmonic) mode: a fraction of power provided to the plasma by means of the usual ECR-heating process by a standard double-ridge rectangular waveguide, while the additional RF power at the second harmonic is injected along the perpendicular direction with respect to the axis-symmetric magnetic field, in order to excite EBW through O-SX-B conversion mechanism.

4. Conclusion

The future challenges of higher-charge states and beam intensity, high-charge breeding efficiency, and high absolute ionization efficiency need the exploration of new heating schemes and synergy between experiments and modeling as described in this paper. The plasma conditioning phase on the MDIS-type source PS-ESS, started in July 2016: it has shown an excellent RF-plasma coupling, above 99.5%. Reflected power fluctuation below 0.05% was measured providing a great starting point to reach the beam stability requested by the ESS accelerator. The stability of the absorbed microwave power tells us the overall RF design (both in terms of launching and diagnostics) was successful. A deeper study of the mechanisms for overdense plasmas generation and to investigate new strategies for improve the plasma-wave coupling in ECRIS devices is ongoing on the FPT: the use of two perpendicular microwave injection lines at different frequencies enable exploring also the aspects of resonant absorption of waves energy by electrons at the second-harmonic of the cyclotron resonance.

Acknowledgments

The authors acknowledge the valuable support given by F. Chines, G. Pastore, G. Calabrese, S. Vinciguerra, A. Amato, A. Caruso, G. Gallo, S. Passarello, G. Manno.

References

[1] Celona L *et al.* 2015 *in Proc. 6th Int. Particle Accelerator Conf.*, Richmond (Virginia, USA) WEPVA010 pp. 2509-2511.
[2] Gammino S *et al.* 2017 The Flexible Plasma Trap (FPT) for the production of overdense plasmas, submitted for publication on *JINST.*
[3] Mascali D *et al.* 2016 *Rev. Sci. Instrum.* **87** 095109.
[4] Gammino S, Celona L, Ciavola G, Maimone F and Mascali D 2010 *Rev. Sci. Instrum.* **81** 02B313.
[5] Pandit V.S *et al.* 2013 *Nucl. Instr. Meth. Phys. Res.* B **317** 434-437.
[6] Muguira L *et al.* 2014 *Nucl. Instrum. Meth. Phys. Res.* A **741** 95-103.
[7] Cortázar O. D., Komppula J, Tarvainen O, Megía-MacíasA, Vizcaíno-de-Julián A and Koivisto H 2013 *Plasma Sources Sci. Technol.* **22** 015026.
[8] Castro G*et al.* 2017 *Plasma Sources Sci. Technol.* **26** 055019.
[9] Torrisi G Sorbello G Leonardi O Mascali D, Celona L and Gammino S 2016 *Microw. Opt. Technol. Lett.* **58** 2629-34.

Innovative single-shot diagnostics for electrons from laser wakefield acceleration at FLAME

F G Bisesto[1], M P Anania[1], A Cianchi[2], E Chiadroni[1], A Curcio[1], M Ferrario[1], R Pompili[1], A Zigler[3]
[1]INFN-LNF, Via Enrico Fermi 40, 00044 Frascati, Italy
[2]University of Rome Tor Vergata, 00133 Rome, Italy
[3]Racah Institute of Physics, Hebrew University, 91904 Jerusalem, Israel

E-mail: fabrizio.giuseppe.bisesto@lnf.infn.it

Abstract. Plasma wakefield acceleration is the most promising acceleration technique known nowadays, able to provide very high accelerating fields (> 100 GV/m), enabling acceleration of electrons to GeV energy in few centimeters. Here we present all the plasma related activities currently underway at SPARC_LAB exploiting the high power laser FLAME. In particular, we will give an overview of the single shot diagnostics employed: Electro Optic Sampling (EOS) for temporal measurement and Optical Transition Radiation (OTR) for an innovative one shot emittance measurements. In detail, the EOS technique has been employed to measure for the first time the longitudinal profile of electric field of fast electrons escaping from a solid target, driving the ions and protons acceleration, and to study the impact of using different target shapes. Moreover, a novel scheme for one shot emittance measurements based on OTR, developed and tested at SPARC_LAB LINAC, used in an experiment on electrons from laser wakefield acceleration still undergoing, will be shown.

1. INTRODUCTION

The fast development in the laser field during the past decade has provided high power multi-TW system at femtosecond level. The availability of such systems has allowed to study the interaction between ultra intense ($> 10^{19} W/cm^2$) lasers with matter in different fields: astrophysics in laboratory [1], high energy density experiments [2], electromagnetic wave sources [3] and novel schemes for particle acceleration [4, 5, 6, 7, 8]. Particular attention has been given to the latter since the possibility to have a table-top accelerator, exploiting relative small and cheap systems, results very fascinating. Nevertheless, the produced particle (electrons and ions) beams suffer of shot-to-shot instabilities and the physical mechanism is not yet well clear, especially concerning the interaction with solid targets for ion acceleration. In this context, the need for single shot diagnostics to be employed in laser-plasma experiments is crucial in order to achieve a better control. The plasma related activities currently underway at SPARC_LAB exploiting the high power laser FLAME are moving in this direction.

In this work, an overview of the developed single shot diagnostics will be given. In detail, we are focusing on a temporal diagnostics based on Electro Optic Sampling (EOS), to study the emission of fast electrons from thin solid targets responsible of ion and proton acceleration, and on a scheme for single shot emittance measurements based on OTR (Optical Transition Radiation).

2. EOS diagnostics for fs resolution temporal measurements on fast electrons

Ion acceleration from thin foils irradiated by high-intensity short-pulse lasers is one of the most interesting aspect in this research field since it produces a large amount of particles with energies in multi-MeV range [9, 10]. According to the established theoretical models [11], the process starts when some electrons, directly accelerated by the laser, penetrate the target. The majority of them spread and dissipate energy inside of it while the hot component of these electrons is able to reach the target rear side and is released in vacuum [12]. The most energetic electrons escape, leaving behind an electrostatic potential, set up by the unbalanced positive charge left on target, that locks the majority of them near the target [13]. Such potential generates an electric field that ionizes and accelerates surface ions in a process called Target Normal Sheath Acceleration [14] (TNSA).

So far only indirect evidences of the escaping electrons have been detected by measuring the radiated electromagnetic pulses [15, 16] and magnetic fields [17].Here we report direct and temporally resolved measurements of fast electrons by means of a temporal diagnostics based on EOS [18].

2.1. *Experimental Setup and Results*

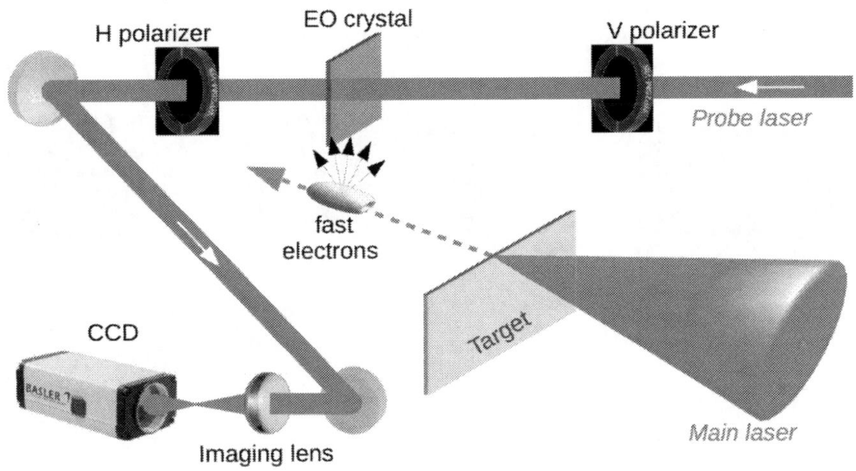

Figure 1. Setup of the experiment. The FLAME laser is focused on a metallic target from where fast electrons can escape. The EOS system, based on a 500μm thick ZnTe crystal placed 1 mm downstream the target, allows to measure the temporal profile of the emitted electrons by means of an ancillary laser beam probing the local birefringence induced by the electric field.

The experiment has been carried out with the FLAME laser at the SPARC_LAB test-facility [19] using the setup in Figure 1. FLAME consists in a 130 TW Ti:Sapphire laser system delivering 35fs (FWHM), up to 4J pulses at 800nm central wavelength and 10Hz repetition rate. The laser beam was focused by a $f/10$ off-axis parabolic mirror with focal length $f = 1$m. The $1/e^2$ diameter of the laser spot on target was approximately 60μm, corresponding to a peak intensity of about $10^{19} W/cm^2$. Exploiting such an intense laser pulse, fast electrons are produced by irradiating the tip (about 10μm thick) of a stainless steel wedged target. The diagnostics is based on a 500μm thick ZnTe EO crystal, placed 1mm downstream the target. The longitudinal profile of the electric field carried by fast electrons is encoded into the transversal plane of a probe laser following the EOS spatial scheme [21]. The probe laser employed to detect the EO signal (35fs FWHM pulse duration) is directly split from the FLAME laser, ensuring a jitter-free synchronization. The synchronization of the main and probe lasers in correspondence of the EOS crystal is obtained by means of an α-cut BBO crystal. The time overlapping, i.e. our reference time, is then retrieved by looking for the light from sum frequency generation (SFG).

Once calibrated for time of flight measurements in order to retrieve the electron energy [20], we studied the influence of target shape by testing planar, blade and tip targets [22]. We changed the position of the laser on the target at each shot. In the case of the planar foil target, the resulting snapshot in Figure 2(a,d) shows the presence of a first emitted bunch with approximately 1.2 nC charge and 7 MeV energy followed by a second broadened structure carrying a larger amount of particles (about 3 nC). For the wedged target, the snapshot in Figure 2(b,e) shows a similar structure. The first bunch now carries a larger amount of electrons (2 nC) at the same energy (7 MeV) while the charge in the second bunch is strongly reduced to 0.3 nC. Electron bunches coming from the tip target are shown in Figure 2(c,f). In this case the interaction with the laser produced a much larger number of released electrons (about 7 nC) at higher energies (about 12 MeV).

In conclusion, we measured for the first time the temporal profile of the electric field carried by fast electrons, related to the ion acceleration in high intensity laser and solid target interactions. Afterwards, we used our diagnostics to test the influence of target shape. In detail, the results provide a direct evidence of charge and energy boost when using sharp tips. Therefore a consequently field enhancement for ion and proton acceleration has been measured.

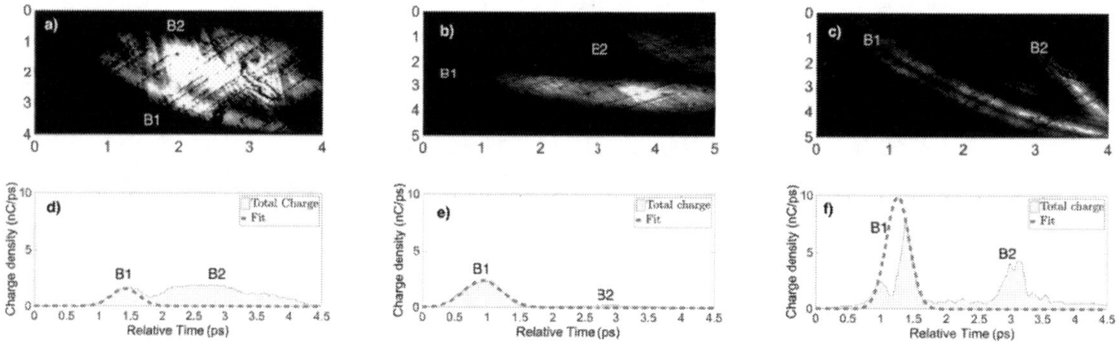

Figure 2. Snapshots with different target shapes. Signatures of the escaping electrons from (a) planar, (b) wedged and (c) tipped targets. The emitted charges are, respectively, (a) 1.2nC (B1) and 3nC (B2); (b) 2nC (B1) and 0.3nC (B2); (c) 7nC (B1) and 3nC (B2). The gaussian envelopes represent the extrapolated charge profiles of each bunch. (d-f) Corresponding longitudinal charge profiles.

3. Single shot emittance measurements based on incoherent OTR

Electron beams from plasma accelerators are characterized by relatively large energy spread (around 5% rms) and suffer shot to shot instabilities. These points make the common diagnostics tools for emittance measurements almost not useful. The quadrupole scan technique [23], for example, is a multi-shot technique and it is very sensitive to energy spread [24].

A solution under studying at SPARC_LAB comes from Optical Transition Radiation (OTR). [25]. The main features of OTR can be found analysing its angular distribution by looking at in the focal plane of a lens. Indeed, it shows a strong dependence from the particle energy: the characteristic lobes (Figure 3) are peaked at $1/\gamma$, where γ is the relativistic Lorenz factor. Furthermore, from [27], it can be found that beam divergence affects the incoherent OTR angular

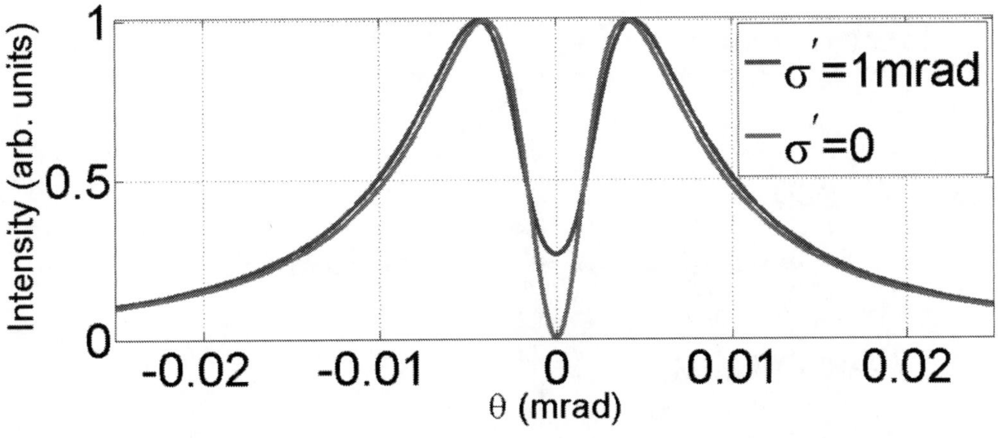

Figure 3. Incoherent OTR angular distribution line profile with (green) and without (blue) angular divergence for a $125MeV$ electron beam. The effect due to $1mrad$ beam divergence is the increase of the central minimum.

distribution. Figure 3 shows the comparison between the line profiles of the angular distributions with and without taking into account the beam angular divergence: the former has a central minimum different from 0. Using the information coming from the beam image and the OTR angular distribution, both values of beam dimension and divergence can be obtained with a single shot. A first scheme based on OTR [26] provided an emittance measurement only in a beam waist where the correlation term between divergence and beam size is zero. However, for its own nature the plasma acceleration has strong shot-to-shot fluctuations and a beam waist cannot be guaranteed. Therefore, in order to fully measure the emittance, it is necessary to measure the correlation term as well. The idea developed in this work is to use a microlens array that allows to correlate the beam divergence with the spatial position by analysing the OTR angular distribution of each microlens.

3.1. Experimental Setup

In Figure 4 the layout of our experimental setup for single shot emittance measurement is shown. The OTR emitted by a silicon aluminated screen is sent to two different arms through a 90:10 beam splitter: the reflected radiation is used for transverse dimension diagnostics, in combination with a high quantum efficiency Hamamatsu Orca II camera, equipped with a Nikon $f = 180mm$ focal length F/2.8; the transmitted part passes through a $400mm$ focal length achromatic doublet which makes a 1:1 replica of OTR source in its imaging plane where a microlens array ($10mm \times 10mm$, $300\mu m$ pitch, $18.7mm$ focal length plano-convex lenses) takes place. Its focal plane is duplicated by a $50mm$ focal length achromatic doublet imaged on an intensified camera (Hamamatsu Orca IV). The whole setup has been extensively simulated by means of Zemax, an optical and illumination design software, in order to evaluate any possible aberrations affecting the measure. In Figure 5 a qualitatively comparison between simulation and experimental results is reported. The experiment has been performed with the SPARC_LAB photoinjector [19], using $200pC$ bunch charge at $125MeV$. Unfortunately, the maximum energy achievable in this run was limited by the use of only two of the three accelerating sections and the minimum detectable angular divergence was equal to $500\mu rad$, while our electron beam had $250\mu rad$ [27]. Even though we did not measure the emittance, this preliminary result demonstrates that it is possible to produce the OTR angular distribution from different part of the beam image.

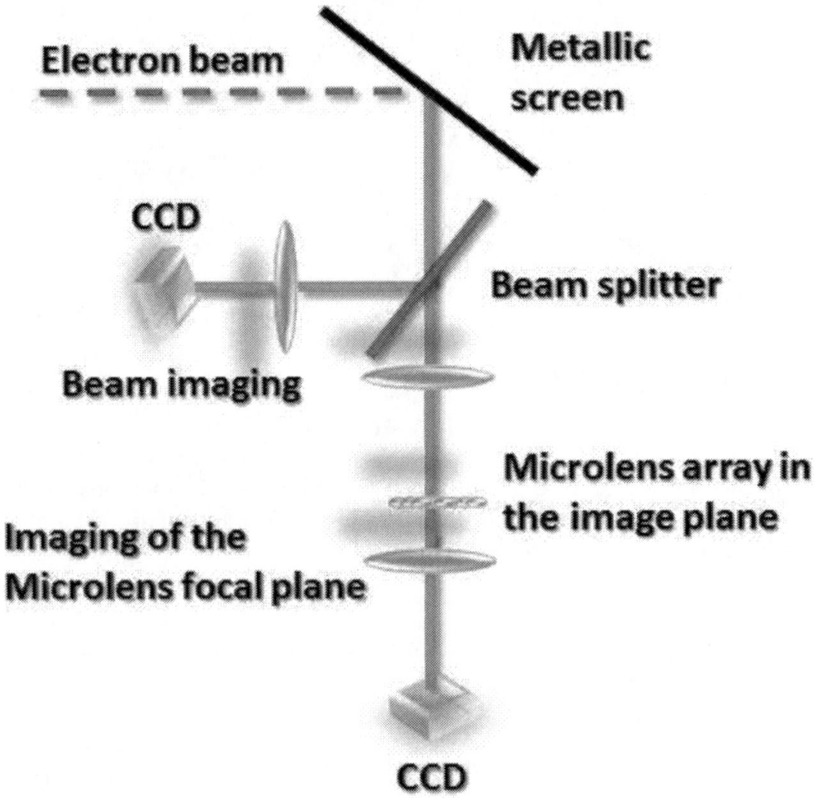

Figure 4. Single shot emittance measurement setup: one arm is for transversal size diagnostics, the other is dedicated to correlation term measurement by means of microlens array. [27]

4. Conclusions

In this work, we presented the development of two single shot diagnostics for ultraintense laser-plasma interaction experiments. A setup based on Electro Optical Sampling has been realized in order to measure for the first time the longitudinal profile of fast electron electric field escaping from a solid target during the interaction with a high intensity laser pulse. With such a diagnostics, we have measured a significant increase in the charge and energy of the escaping electrons (corresponding to an increase in the potential barrier) for sharp structured targets. These results can be used as a guideline in order to achieve higher energies for positively charged ions with respect to what is currently obtained through conventional laser acceleration schemes. Moreover a new scheme for single shot emittance measurements has been proposed. It relies on the characterization of incoherent optical transition radiation. Using a microlens array it is possible to retrieve the correlation term between beam divergence and spatial position. This allows to measure the emittance in one shot. This kind of diagnostics represents a very useful tool to fully characterize an electron beam from LWFA despite its typical shot-to-shot instabilities and relative large energy spread. Some preliminary results obtained with the SPARC_LAB LINAC have been reported, while a measurement on a plasma-accelerated electron beam at FLAME Facility is still under design, both with a self-injection and external injection [28] scheme.

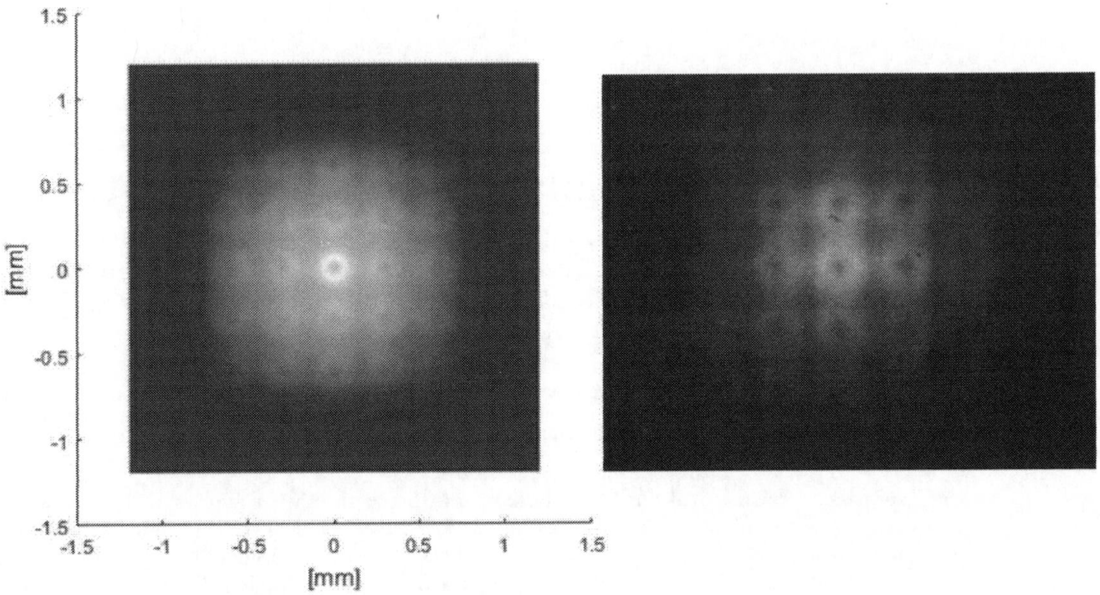

Figure 5. Comparison between a simulation made by means of Zemax (on the left) and a typical experimental measurement (on the right).[27]

References

[1] Remington B A, Arnett D, Paul R, Drake and Takabe H 1999 *Science* **284** 5419 1488-1493
[2] Roth M *et al.* 2001 *Phys. Rev. Lett.* **86** 3
[3] Rousse A *et al.* 2004 *Phys. Rev. Lett.* **93** 13
[4] Tajima T and Dawson J M 1979 *Phys. Rev. Lett.* **43** 4
[5] Geddes C G R *et al.* 2004 *Nature* **431** 7008
[6] Leemans W P *et al.* 2006 *Nat. Phys.* **2** 10
[7] Maksimchuk A, Gu S, Flippo K, Umstadter D and Bychenkov V Yu 2000 *Phys. Rev. Lett.* **84** 18
[8] Macchi A, Borghesi M and Passoni M 2013 *Rev. Mod. Phys.* **85** 2
[9] Clark E L *et al.* 2000 *Phys. Rev. Lett.* **85** 8
[10] Snavely R A *et al.* 2000 *Phys. Rev. Lett.* **85** 14
[11] Dubois J L *et al.* 2014 *Phys. Rev.* E **89** 1
[12] Kumar S P *et al.* 2013 *Phys. Plasma (1994-present)* **20** 11
[13] Poyé A *et al.* 2015 *Phys. Rev.* E **91** 4
[14] Wilks S C *et al.* 2001 *Phys. Plasma (1994-present)* **8** 2
[15] Pfotenhauer S M *et al.* 2010 *New J. Phys* **12** 10
[16] Nilson P M *et al.* 2012 *Phys. Rev. Lett.* **108** 8
[17] Sandhu A S *et al.* 2002 *Phys. Rev. Lett.* **89** 22
[18] Wilke I *et al.* 2002 *Phys. Rev. Lett.* **88** 12
[19] Ferrario M *et al.* 2013 *Nucl. Instr. Meth. Phys. Res.* B **309**
[20] Pompili R *et al.* 2016 *Opt. Exp.* **24** 26
[21] Cavalieri A L *et al.* 2005 *Phys. Rev. Lett.* **94** 11
[22] Pompili R *et al.* 2016 *Sci. Rep.* **6**
[23] Löhl F *et al.* 2006 *Phys. Rev. ST Accel. Beams* **9** 9
[24] Mostacci A *et al.* 2012 *Phys. Rev. ST Accel. Beams* **15** 8
[25] Ter-Mikaelian 1972 *High energy electromagnetic processes in condensed media* (New York, Wiley-Interscience)
[26] Feldman R, Lumpkin A, Rule D and Fiorito R, 1990 *Nucl. Instr. Meth. Phys. Res.* A **296** 1 193198
[27] Cianchi A *et al.* 2016 *Nucl. Instr. Meth. Phys. Res.* A (only available online https://doi.org/10.1016/j.nima.2016.11.063)
[28] Bisesto F G *et al.* 2016 *Nucl. Instr. Meth. Phys. Res.* A **829** 309-313

IOP Conf. Series: Journal of Physics: Conf. Series **874** (2017) 012036 doi:10.1088/1742-6596/874/1/012036

Gas-filled capillaries for plasma-based accelerators

F Filippi[1], M P Anania[1], E Brentegani[1], A Biagioni[1], A Cianchi[2], E Chiadroni[1], M Ferrario[1], R Pompili[1], S Romeo[1] and A Zigler[3]

[1]Laboratori Nazionali di Frascati, INFN, Frascati, Italia
[2]Dipartimento di Fisica, Universit di Roma Tor Vergata, Roma, Italia
[3]Hebrew University of Jerusalem, Jerusalem, Israel

E-mail: `francesco.filippi@lnf.infn.it`

Abstract. Plasma Wakefield Accelerators are based on the excitation of large amplitude plasma waves excited by either a laser or a particle driver beam. The amplitude of the waves, as well as their spatial dimensions and the consequent accelerating gradient depend strongly on the background electron density along the path of the accelerated particles. The process needs stable and reliable plasma sources, whose density profile must be controlled and properly engineered to ensure the appropriate accelerating mechanism. Plasma confinement inside gas filled capillaries have been studied in the past since this technique allows to control the evolution of the plasma, ensuring a stable and repeatable plasma density distribution during the interaction with the drivers. Moreover, in a gas filled capillary plasma can be pre-ionized by a current discharge to avoid ionization losses. Different capillary geometries have been studied to allow the proper temporal and spatial evolution of the plasma along the acceleration length. Results of this analysis obtained by varying the length and the number of gas inlets will be presented.

1. Introduction

Modern high energy particle accelerators require several hundreds of meter to transfer energy from an electro-magnetic field to particles. The easiest way to reduce their dimensions (and cost) is to increase the acceleration gradient, nevertheless its peak value is limited by the wall breakdown that rules the maximum accelerating field that can be sent into a radio-frequency cavity. On the opposite, plasma does not suffer of breakdown and at the same time shows collective effects which are used in plasma-based acceleration techniques to transfer energy to an accelerating particle [1, 2]. In these techniques one or more driver beams, either a laser or particle bunch(es), excite large amplitude plasma waves able to cede energy to a bunch of accelerating particles which may be extracted by the plasma itself (self-injection) or generated by an external source and then injected into the plasma (external injection) [3].

The quality of the accelerated beam depends strongly on the properties of the plasma, especially on its local density, defined as the number of free plasma electrons contained in a cube centimeter (cm^{-3}). Indeed, the amplitude of plasma waves, as well as their spatial dimensions and the consequent accelerating gradient felt by particles, depend on the background electron density. Variation in plasma density along the interaction length results in a variation of the accelerating structures, then it must be controlled and properly tailored to preserve the quality of the accelerated beam. Gas-filled capillaries are well-suited devices to control plasma density up to centimeter scale length. This technique has been studied in the past [5, 4] since it allows to control the evolution of the plasma, ensuring a stable and repeatable plasma density distribution

during the interaction with the drivers. Moreover, in a gas filled capillary plasma is pre-ionized avoiding ionization losses, crucial for low energy drivers [7, 6].
Here, we will show the temporal evolution of the longitudinal plasma density measured in two different capillaries filled with 300 mbar hydrogen and ionized by a current discharge of few hundreds of amperes. The first one is a 3 cm long capillary with circular cross section of 1 mm diameter fed by two gas inlets of 0.5 mm diameter placed at 7.5 mm from each edge of the capillary [8]. The other one is a 1 cm long capillary with 1 mm diameter fed by a single inlet of 0.5 mm diameter placed at the half of the capillary length.

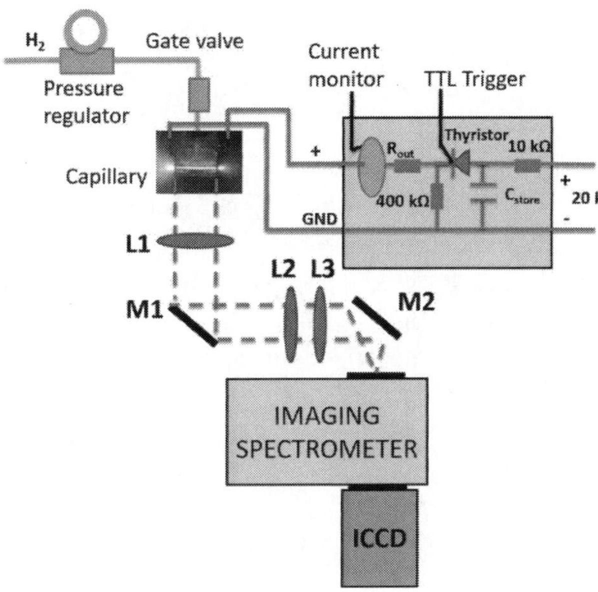

Figure 1. Experimental setup.

The measurements have been performed by using the Stark broadening analysis of the self-emitted Balmer beta line of the hydrogen, as described in our previous works [9, 10]. The setup of the experiment is schematically shown in Figure 1. The hydrogen backing pressure set at the regulator is of 300 mbar, a fast gate valve let it flows into the capillary for 3 ms then a discharge circuit let a current flowing into the capillary ionizing the gas. The plasma density along the capillary length has been measured every 100 ns from the discharge trigger with a temporal resolution given by the gate time of the ICCD camera of 100 ns.

2. Double-inlets capillary

To let the driver(s) interact with the desired plasma density it is possible to wait after the discharge until the expected plasma profile is reached.

A typical trend of the averaged density compared with the discharge evolution in the 3 cm capillary is shown in Figure 2. The experimental data (cross) have been obtained by averaging the measured plasma density along the entire capillary length. As can be seen, the highest density is reached around 500 ns after the discharge trigger, then it decays. The density shows an exponential decay from its maximum to the complete recombination, as expected by the theoretical models [4]. The decay constant is defined as the time needed by the plasma density to reach its maximum value multiplied by e^{-1}. It can be measured by an exponential fit of the data. The fit sowed in Figure 2 presents a decay constant of 340 ns.
To examine more in detail the plasma distribution along the interaction length, we have also

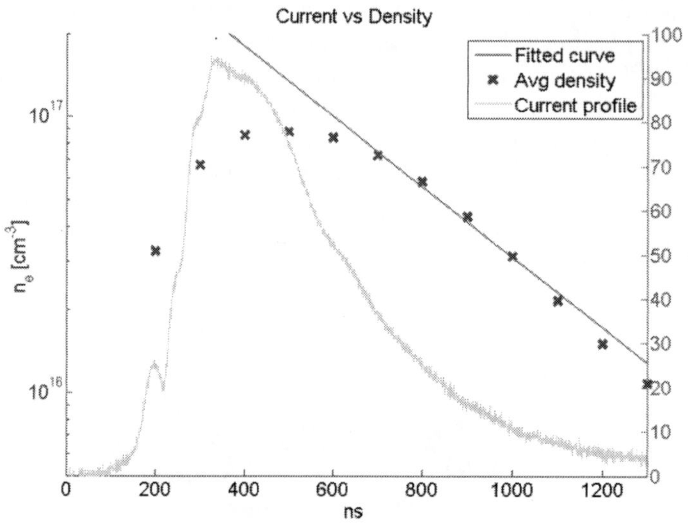

Figure 2. Comparison between profile of the current discharge and the average plasma density along the capillary. The density data have been fitted with an exponential curve (red line).

measured the local density with a spatial resolution of 0.125 mm. The density profile along the entire capillary length is shown in Figure 2. The spatial distribution of the plasma density shows

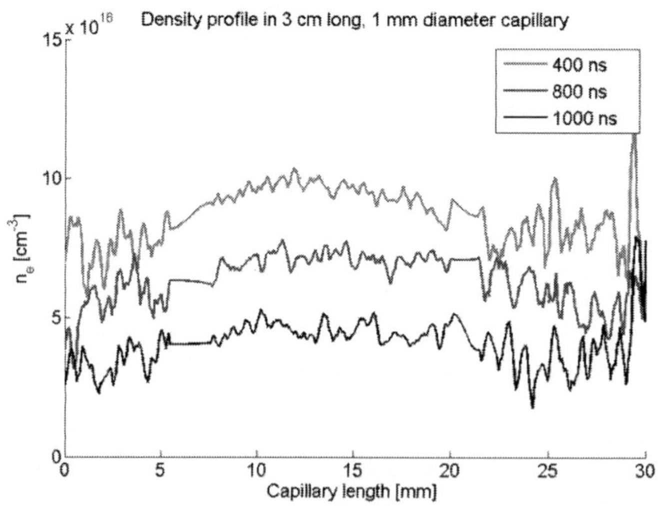

Figure 3. Longitudinal density distribution along a 3 cm capillary with 20 kV of applied voltage

a local maximum that decreases close to the electrodes. In the early instants of the discharge the density appears to be not exactly symmetric probably due to an uneven background density of the gas caused by some geometrical differences between the inlets. These differences have been probably introduced during the manufacturing of the capillary. This unwanted inhomogeneity can be problematic for the propagation of the beam during the acceleration process which may be dephased respect to the accelerating structure. Nevertheless, after few hundreds of nanoseconds the plasma density decreases flattening its profile. At 1000 ns after the discharge, the density is almost flat along the entire capillary.

3. Single-inlet capillary

To experimentally test the effect of a different inlet configuration we have used a 1 cm long capillary fed by a single inlet placed in the middle of the capillary. In this configuration the gas filling is symmetric. The shortest length of the capillary is dictated by the fact that a single inlet can hardly filled homogeneously bigger volumes. For these measurements we also increased the current up to 230 A in order to completely ionize the gas contained into the capillary. The decay constant of the plasma is weakly affected by these variations since it was 380 ns as measured with the same technique exposed before.

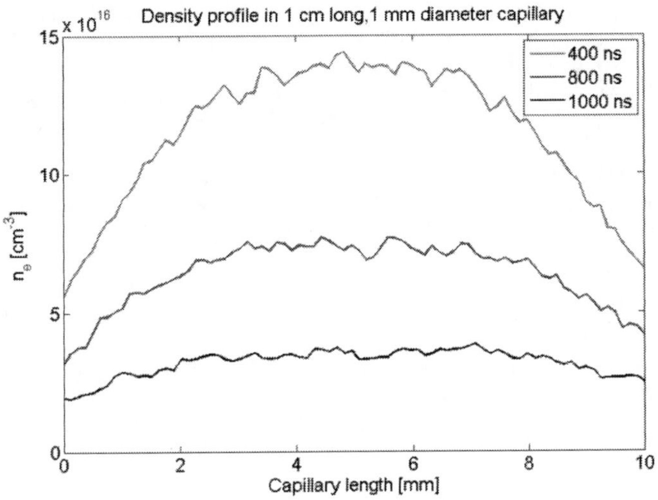

Figure 4. Longitudinal density distribution along a 1 cm capillary with 20 kV of applied voltage.

As expected, the symmetric feeding of the single inlet ensures a more symmetric, then predictable, density, which is evident in Figure 3. The density overcome the value of $1 \cdot 10^{17}$ cm^{-3} at 400 ns after the discharge trigger, then it decreases being flatter with almost the same density shown in previous capillary configuration at 1000 ns after the discharge when the density is almost flat. The good homogeneity of the plasma profile in the latter moments of the discharge is particularly suitable for most of the acceleration process which requires centimeter scale constant plasma density.

4. Conclusion

We have shown the different plasma density distribution achieved by two different capillary geometries, one fed by two inlets placed at 7.5 mm from each edge of the capillary and the other one fed by a single inlet injecting gas in the middle of a 1 cm long capillary. Multiple inlets presented an unexpected density distribution probably caused by small differences between the inlets due to the manufacturing process. This induced to an unwanted density variation along the acceleration length. Nevertheless after some hundreds of nanoseconds after the discharge the density tends to become homogeneous and almost flat. On the contrary, the single-inlet capillary showed a remarkably symmetric plasma distribution and its homogeneity after 1000 ns from the trigger discharge is comparable with what obtained with the longer capillary. The

homogeneous plasma distribution reached with these devices is well suitable for most of the acceleration schemes which requires centimeter scale homogeneous density profile.
Further measurements will investigate the possibility to modify the capillary geometry in order to taper the plasma density along the interaction path.

5. Acknowledgement

This work has been supported by the EU Commission in the Seventh Framework Program, Grant Agreement 312453-EuCARD-2, H2020 research and innovation program under grant agreement n. 653782, and the Italian Minister of Research in the framework of FIRB Fondo per gli Investimenti della Ricerca di Base, Project n. RBFR12NK5K.

References

[1] Tajima T and Dawson J 1979 Laser electron accelerator *Phys. Rev. Lett.* **43**.4 267.
[2] Esarey E, Schroeder C and Leemans W P 2009 Physics of laser-driven plasma-based electron accelerators *Rev. Mod. Phys* vol. **81** no. 3 pp. 12291285.
[3] Ferrario M *et al.* 2013 SPARC_LAB present and future *Rev. Mod. Phys* **309** pp. 183188.
[4] Bobrova N A *et al.* 2001 Simulations of a hydrogen-filled capillary discharge waveguide *Physical Review* E **65**.1 016407.
[5] Spence D and Hooker S 2000 Investigation of a hydrogen plasma waveguide *Physical Review* E **63** 15401.
[6] Kallos E 2008 Plasma wakefield accelerators using multiple electron bunches *Ph.D. thesis*, Electrical Engineering Dept. University of Southern California USA.
[7] Butler A, Spence D and Hooker S 2002 Guiding of High-Intensity Laser Pulses with a Hydrogen-Filled Capillary Discharge Waveguide *Phys. Rev. Lett.* **89** 185003.
[8] Pompili R *et al.* 2017 Experimental characterization of active plasma lensing for electron beams *Applied Physics Letters* **110**.10 104101.
[9] Filippi F *et al.* 2016 Plasma density profile characterization for resonant Plasma Wakefield Acceleration experiment at SPARC_LAB in *Proc. IPAC'16* Busan (Korea) pp. 2554–2556.
[10] Filippi F *et al.* 2016 Spectroscopic measurements of plasma emission light for plasma-based acceleration experiments *Journal of Instrumentation* **11**.09 C09015.

IOP Conf. Series: Journal of Physics: Conf. Series **874** (2017) 012037 doi:10.1088/1742-6596/874/1/012037

Beam commission of the high intensity proton source developed at INFN-LNS for the European Spallation Source

L Neri, L Celona, S Gammino, A Miraglia, O Leonardi, G Castro, G Torrisi, D Mascali, M Mazzaglia, L Allegra, A Amato, G Calabrese, A Caruso, F Chines, G Gallo, A Longhitano, G Manno, S Marletta, A Maugeri, S Passarello, G Pastore, A Seminara, A Spartà, S Vinciguerra

Istituto Nazionale di Fisica Nucleare – Laboratori Nazionali del Sud, Catania, Italy

neri@lns.infn.it

Abstract. At the Istituto Nazionale di Fisica Nucleare – Laboratori Nazionali del Sud (INFN-LNS) the beam commissioning of the high intensity Proton Source for the European Spallation Source (PS-ESS) started in November 2016. Beam stability at high current intensity is one of the most important parameter for the first steps of the ongoing commissioning. Promising results were obtained since the first source start with a 6 mm diameter extraction hole. The increase of the extraction hole to 8 mm allowed improving PS-ESS performances and obtaining the values required by the ESS accelerator. In this work, extracted beam current characteristics together with Doppler shift and emittance measurements are presented, as well as the description of the next phases before the installation at ESS in Lund.

1. Status of commissioning

The design of the PS-ESS source [1-2] started in 2012 taking into account the experience gained at INFN-LNS with the sources TRIPS [3] and VIS [4]. The source construction started in November 2015 and the first plasma was produced in June 2016 while the beam commissioning started in November 2016. During the overall project [5] the interaction with ESS accelerator division and CEA for the beam diagnostics and the control system was intense. The whole design was driven by the future installation at ESS. A major performance upgrade was accepted in 2014 with 25% of increase of the proton current in the linear accelerator (from 50 mA to 62.5 mA), leading to a required ion source proton current of 74 mA. The commissioning was divided in two main stages. The first consisted of the characterization of the source performance with a Faraday Cup (FC), one Emittance Measurement Unit (EMU) and a Doppler shift measurement unit. To increase the reliability of the source measurements, these instruments were inserted as close as possible to the source plasma, 626 mm after the plasma hole. Several thousands of ion source configurations were tested by changing the magnetic profile, microwave power, the hydrogen (H_2) gas flow, and by using two different designs of the extraction electrodes. The last step will characterize the fully assembled Low Energy Beam Transport Line (LEBT) [6-7] with an additional diagnostic box to measure the beam parameters at the LEBT-RFQ interface. This stage will be useful for understanding the effects of different amount and type of gas addition in the LEBT for the space charge compensation of the beam. The magnetic beam transport optimization, the optimum parameter of the chopper and the effects of the Iris will be tested.

IOP Conf. Series: Journal of Physics: Conf. Series **874** (2017) 012037 doi:10.1088/1742-6596/874/1/012037

2. Source commissioning results

The characterization of the source was optimized using a custom code able to test several configurations, one every 10 seconds, practically unmanned. The post analysis of the collected data is ongoing. Table 1 shows a comparison between the required values, and collected data from the commissioning.

Table 1. ESS requirements at source-LEBT interface and already collected data during the commissioning.

Requirement	Value	Measurement
Total beam current	>90 mA	95.6 mA
Nominal proton beam current	74 mA	74.8 mA
Proton beam current range	67-74 mA	60-80 mA
Proton fraction	>75%	78.3%
Pulse length	6 ms	6 ms
Pulse flat top	3 ms	3 ms
Flat top stability	±2 %	±1.5 %
Pulse to pulse stability	±3.5 %	±3 %
Repetition rate	14 Hz	14 Hz
Beam energy	75±5 keV	75 keV
Energy adjustment	±0.01 keV	±0.01 keV
Transverse emittance (99%)	1.8 pi.mm.mrad	1.06 pi.mm.mrad @ 82 mA
Beam divergence (99%)	<80 mrad	50 mrad @ 82 mA
Start-up after source maintenance	32 hours	32 hours

2.1. Beam current verification

The total current produced out of the source was measured with two devices. A large bandwidth (from 3 Hz to 1 MHz) AC Current Transformer (ACCT) inserted on the high voltage cable measured the current delivered to the high voltage platform. An 80 mm diameter FC measured the beam current at a location as close as possible to the ion source. We observed that the divergence of the extracted current increases linearly with the current density. With extracted current larger than 80 mA the beam size became larger than the FC diameter and only the ACCT could measure the beam current value and ripple. The ACCT voltage droop was taken into account and recovered with a custom code. Figure 1 shows the current in a beam pulse for an ion source configuration providing high beam current. The performances of the source need to be verified between 2.9 and 5.9 ms of the beam pulse. The other part of the beam pulse will be removed by the LEBT electrostatic chopper and not delivered to the RFQ. The mean current measured is about 100 mA, 5% of this amount is due to back streaming electrons, so that the remaining net beam current is 95.6 mA. By changing the magnetic configuration of the source and the microwave power we are able to choose a stable current within the 60 to 80 mA range.

2.2. Beam stability verification

The beam stability is one of the crucial parameters of the whole accelerator. It is defined by measuring the range of current fluctuation inside the single pulse and between pulses. The pulse shape acquisition was done with 1 Ms/s card, while the current fluctuations at which the accelerator is sensitive is 50 μs period. The acquired current value, measured for the FC and the ACCT was averaged with 50 μs period and the maxi-mum excursion inside the pulse and between pulses were taken into account for the verification of the requirement. For beam currents above 80 mA, the cur-rent can only be measured by the ACCT as explained before. Figure 1 shows in blue the acquired signal, after the droop voltage correction, and in red the signal averaged with 50 μs period. The pink lines identify the limit in time of the transmitted beam to the RFQ. The stability measured in the shown configuration is ±1.5 % inside the beam flat top and ±3 % between pulses.

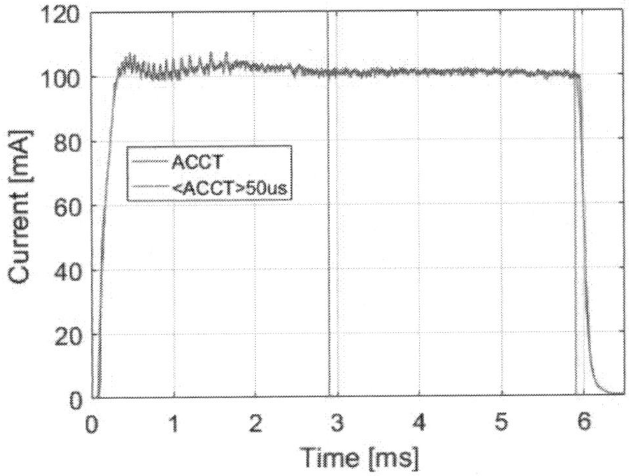

Figure 1. Beam current acquisition using an ACCT on the HV power supply cable.

2.3. Beam emittance verification

The beam emittance was measured using an Allison scanner in the same position of the FC. The same limitation was observed also for the EMU and the measurement was only possible for beam sizes smaller than 95 mm diameter. Figure 2 shows a measurement made with 82 mA total beam current. The beam emittance can be calculated with the beam fraction defined by a multiple of the rms value, or by a percentage. The ESS specifies a beam emittance calculated on the 99% of the beam. The estimated Twiss parameters of the beam are shown in the figure 2: Emitt. [99%] = 1.0559 pi.mm.mrad, Beta = 25.6 mm/pi.mrad, Alpha = -34.7. This estimation was made with PlotWin. Cumulative sum of beam emittance on the X' axis (green line on the left) shows four peaks due to a bug in the acquisition interface that is still under debugging but that do not affect consistently the emittance estimation.

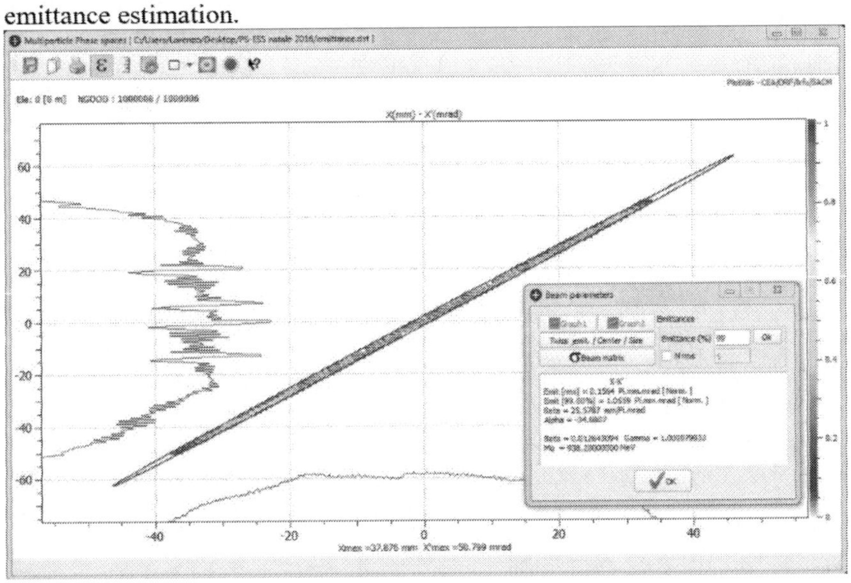

Figure 2. Measured beam emittance imported on PlotWin for Twiss parameter estimation.

2.4. Beam proton fraction verification

Because of the differential pumping obtained by the small extraction hole, the plasma chamber contains 99% of H2 gas, even if a different gas is introduced in the LEBT for beam space charge compensation. The positive ions species generated in the plasma are therefore protons, H2+ and H3+. The fraction obtained at the equilibrium is sensitive to the type of microwave to plasma coupling, which is sensitive to the magnetic configuration of the plasma chamber, and the heating microwave power used. The species fraction of the extracted beam was measured using the spectral analysis of the beam-emitted light at an angle of 25°. Due to the different velocity of the different ion species, the Doppler shift of the emitted light enable to distinguish the different species. Figure 3 shows the control sys-tem interface for the diagnostic equipment that is able to perform the background subtraction and the three Gaussian fit needed to estimate the amount of the three beam species. The proton fraction measured for a beam of 95.6 mA was 78.3%. We have observed that an increased microwave power injected in the plasma chamber increases this fraction. With this measured proton fraction amount, we can estimate an extracted proton current of 74.6 mA. The ESS request of 74 mA proton beam out of the source takes into account a loss of 5% in the LEBT. Due to the low emittance beam produced we suppose that the LEBT losses will be lower and we will reach the 68 mA of proton beam at the end of the LEBT with smaller extracted total cur-rent with respect to the case shown in Figure 1.

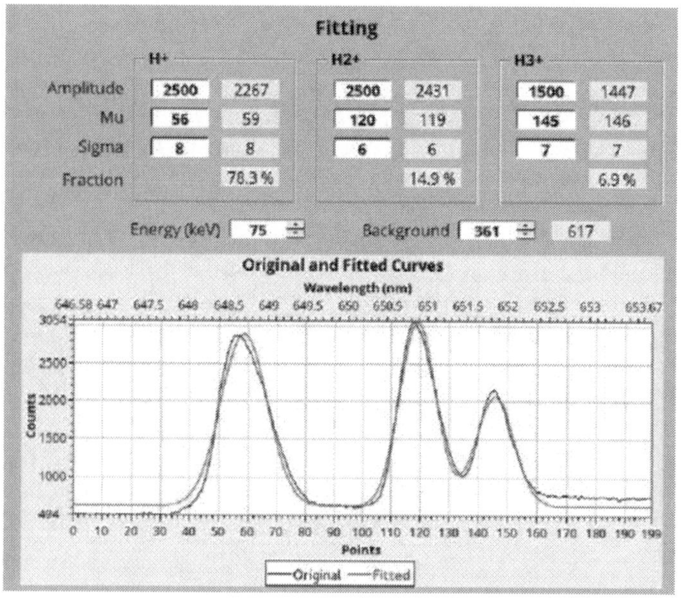

Figure 3. Species fraction estimation using Doppler shift measurement.

3. Next steps

The installation of the LEBT is ongoing without interruption of the ion source characterization. The assembly and the cabling of the full LEBT has already started from the last part of the beam line. The last step will be a fast substitution of the diagnostic cross, which was used in the first beam commissioning stage, with the first LEBT solenoid. This parallel work advances the final commissioning stage of approximately 10 weeks.

The first stage of the commissioning will define few candidates, for the nominal ion source configuration, to be used during the following stage. The LEBT com-missioning will focus on reducing the beam emittance growth by optimizing the gas type and pressure inside the LEBT. The beam chopping inside the LEBT will also require studying the beam space charge compensation during the transient. The chopper parameter will play an important role in the rise and fall time of the beam pulse formation. The Twiss parameter verification at the LEBT-RFQ interface will be characterized as function of the two solenoid fields for different apertures of the six-blades Iris that is housed after the first solenoid.

The source and LEBT configuration for the installation at ESS will be identical from the version installed at INFN-LNS except for an additional main frame with wheels. This common frame for the source and the LEBT is necessary for the assembly in the ESS tunnel, where the lifting equipment will have limited capability, and for the alignment of the whole structure with respect the theoretical beam axis. Figure 4 shows a render-view of the source and the LEBT as they will be installed in Lund, the diagnostic tank will be used for the beam commissioning before the RFQ assembly and commissioning stages.

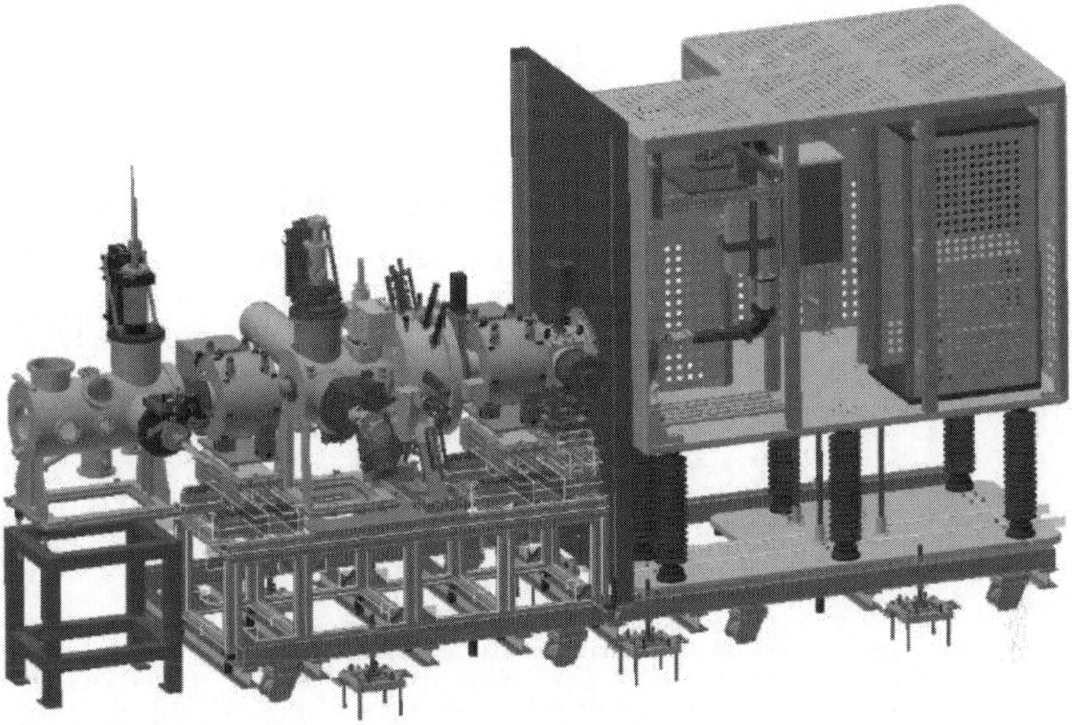

Figure 4. Render-view of the source, LEBT and diagnostic tank, for a total length of 5.8 m, as they will be installed for beam commissioning in Lund.

Acknowledgments
The authors would like to thank all the ESS accelerator division and CEA for the quality of the control system and beam diagnostics provided. The expertise provided by the SIATEL company was relevant to obtain a control system architecture that does not suffer from electromagnetic noise induced by sparks in the HV insulating column.

References

[1] Celona L *et al.* 2012 Design issues of the proton source for the ESS facility, Proc. *LINAC2012* Tel-Aviv (Israel) paper THPB076 pp 1008-1010
[2] Celona L *et al.* 2015 A high intensity proton source for the European Spallation Source facility, Proc. *IPAC2015* Richmond (USA) paper WEP-WA010
[3] Celona L *et al.* 2000 TRIPS: The high intensity proton source for the TRASCO project *Rev. Sci. Instrum.* **71** 771
[4] Gammino S *et al. 2010* Tests of the Versatile Ion Source (VIS) for high power proton beam production, Proc. *LINAC2010* Tsukuba (Japan) paper THP116
[5] Garoby R *et al. 2017* Progress on the ESS Project Construction, Proc. *IPAC2017* Copenhagen (Denmark) paper MOXBA1
[6] Neri L *et al.* 2012 The ESS Low Energy Beam Transport line design, Proc. *LINAC2012* Tel-Aviv (Israel) paper THPB028 pp 912-914
[7] Neri L *et al.* 2016 Commissioning of the High Intensity Proton Source Developed at INFN-LNS for the European Spallation Source, Proc. *LINAC2016* East Lansing (USA), paper MOPLR057

IOP Conf. Series: Journal of Physics: Conf. Series **874** (2017) 012038 doi:10.1088/1742-6596/874/1/012038

Crossbar H-mode drift-tube linac design with alternative phase focusing for muon linac

M Otani[1] K Futatsukawa[1] K Hasegawa[2] R Kitamura[3] Y Kondo[2] S Kurennoy[4]

[1]KEK, Oho, Tsukuba, 305-0801, Japan
[2]JAEA, Tokai, Naka, Ibaraki, 319-1195, Japan
[3]University of Tokyo, Hongo, 113-8654, Japan
[4]Los Alamos, NM, 87545, USA

E-mail: masashio@post.kek.jp

Abstract. We have developed a Crossbar H-mode (CH) drift-tube linac (DTL) design with an alternative phase focusing (APF) scheme for a muon linac, in order to measure the anomalous magnetic moment and electric dipole moment (EDM) of muons at the Japan Proton Accelerator Research Complex (J-PARC). The CH-DTL accelerates muons from $\beta = v/c = 0.08$ to 0.28 at an operational frequency of 324 MHz. The design and results are described in this paper.

1. Introduction

The use of a low emittance muon beam has been discussed in several scientific fields [1, 2, 3, 4]. One of those is the quest for hunting beyond the Standard Model (SM) of elementary particle physics. In the muon anomalous magnetic moment $(g-2)_\mu$, the SM prediction and the measured value with a precision of 0.54 ppm [5] differs by about three standard deviations. Since this is considered to be due to unknown interactions or particles in the SM, further investigations are desirable. The low emittance muon beam will provide more precise measurements since the dominant systematic uncertainties in the previous experiment [5] resulted from the muon beam dynamics in the muon storage ring.

We are developing a muon linac for the $(g-2)_\mu$ experiment [6] at the Japan Proton Accelerator Research Complex (J-PARC) to produce the low emittance muon beam. Details of the muon linac configuration can be found in [7]. Although conventional linacs adopt Alvarez DTLs after radio-frequency quadrupoles, an H-mode DTL is employed during the particle velocity β =0.08 to 0.28 (4.5 MeV) stage, so as to yield a higher acceleration efficiency. In order to achieve more efficient acceleration, the alternating phase focusing (APF) method is adopted.

There are two candidates for the room-temperature H-mode structure. One is an inter-digital H-mode (IH) structure that works in the TE_{11}-mode, while the other is a Crossbar H-mode (CH) operated in the TE_{21}-mode [8]. Our first effort has been devoted to the IH structure. Though our design [9] satisfies the experimental requirement, there is a substantial emittance growth in the vertical direction, generated from the dipole component of the RF electric field. The dipole field is an unavoidable issue in an IH structure and can potentially deteriorate the beam quality. On the other hand, there is no dipole field in a CH structure and better quality of the output beam is expected. However, the CH structure has never been designed and used with the APF method.

Following sections in this paper describe the design procedures and results of the APF CH-DTL for the muon linac.

2. Synchronous phase array optimization

In this step, the particle dynamics are calculated analytically using certain approximations and for a particular synchronous phase array. These calculations are performed using "LINACSapf" [10], with some modifications for the dynamics calculations and the synchronized phase array definition to accommodate the π–mode acceleration, whereas 2π–mode acceleration is assumed in the original code. Details of the approximation method can be found in [10, 9].

Table 1 shows the optimized phase array results. Gap numbers 1–2, 6–9, 15 and 16 have negative synchronous phases, during which time the beam is longitudinally focused. However, gap numbers 3–5 and 10–14 have positive phases, during which the beam is transversely focused. Because the electrostatic focusing effect is stronger in the lower-β part, the first collection of positive phase groups has a smaller number of gaps. The output energy is 4.5 MeV and the total length is 1.3 m.

Table 1. Cell Parameters for Optimized Phase Array

Cell	W [MeV]	β	ϕ [degrees]	Cell length [mm]	Total length [mm]
1	0.34	0.08	-35.9	29.5	29.5
2	0.43	0.09	-14.9	46.0	75.4
3	0.57	0.10	12.9	54.9	130
4	0.74	0.12	32.9	60.3	191
5	0.92	0.13	15.4	54.4	245
6	1.14	0.15	-13.8	56.0	301
7	1.38	0.16	-31.4	66.4	367
8	1.63	0.17	-44.3	74.1	442
9	1.86	0.19	-18.8	97.2	539
10	2.16	0.20	12.5	108	646
11	2.49	0.21	27.6	106	753
12	2.82	0.23	47.6	116	868
13	3.10	0.24	23.2	94.2	963
14	3.50	0.25	10.8	108	1070
15	3.95	0.27	-34.6	91.5	1160
16	4.30	0.28	-15.6	142	1300
exit	4.50				

Figure 1 shows the expected output beam with the analytical calculation. Here the input beam is a water-bag distribution with the expected twiss parameters. The emittance growth during the acceleration is expected to be a few percent, which satisfies our requirement.

3. CH cavity optimization

Because a CH cavity is not axially symmetric, a three-dimensional model is necessary in order to evaluate the electro-magnetic field. In addition, the electro-magnetic field and the resonant frequency depend on the structure of the cavity, and detailed information of the overall structure (including ridges, etc.) should thus be incorporated in the calculation model. Therefore, the entire CH cavity is modeled using the CST Micro Wave (MW) Studio [11] three-dimensional field solver, in order to calculate the electro-magnetic field. Figure 2 shows the three-dimensional model of the CH cavity in CST MW Studio. The CH cavity consists of a cylindrical cavity and four ridges mounted on the top, bottom, left and right of the cavity. To operate the CH cavity as

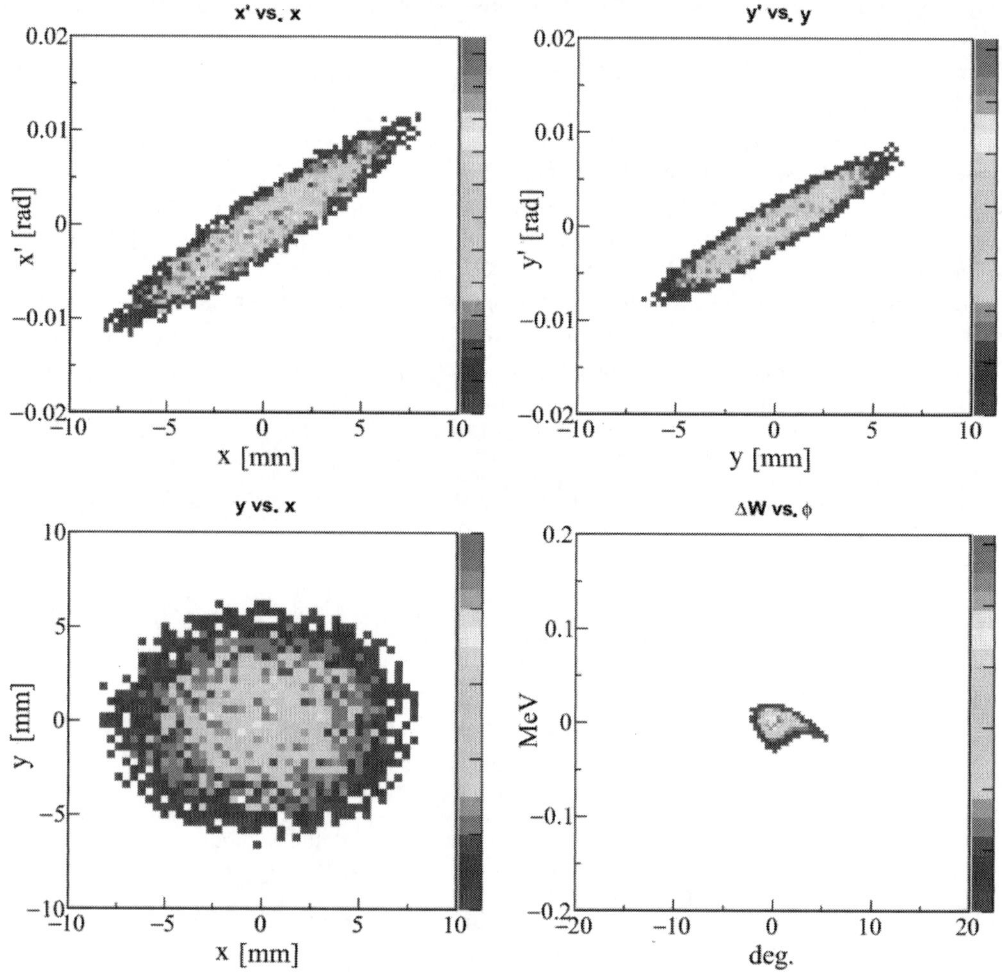

Figure 1. Expected output beam distributions based on the analytical beam dynamics calculation.

an accelerator in the TE_{21}-mode, drift tubes are mounted alternately on the top-bottom ridges and the left-right ridges via stem pairs. The stem pairs are connected to the drift tubes at the tube centers, and front and end ridge-cuts are present in all the ridges (ridge tuners). The inner radius of the cavity is tapered in the down- to upstream direction (cavity taper). The ridge tuner shape and the cavity taper are varied in order to adjust the flatness of the electric field. In the IH case, the cavity is designed in the same manner as the CH case except mounting of the drift tubes; the drift tubes are mounted alternately on the top and bottom ridges via stems and there are no left and right ridges [9].

The drift tubes and the acceleration gaps are first arranged according to the previously determined optimized parameters shown in Table 2. To optimize the acceleration field, other dimensions are optimized. Especially the length of the back ridge tuner is changed to modify the non-uniformity of the field in upstream and downstream directions. Then, small differences in the field between the gaps due to the gap length difference is adjusted by changing the drift tube outer radius.

Figure 3 shows the longitudinal (red) and vertical (green) electric field along the beam

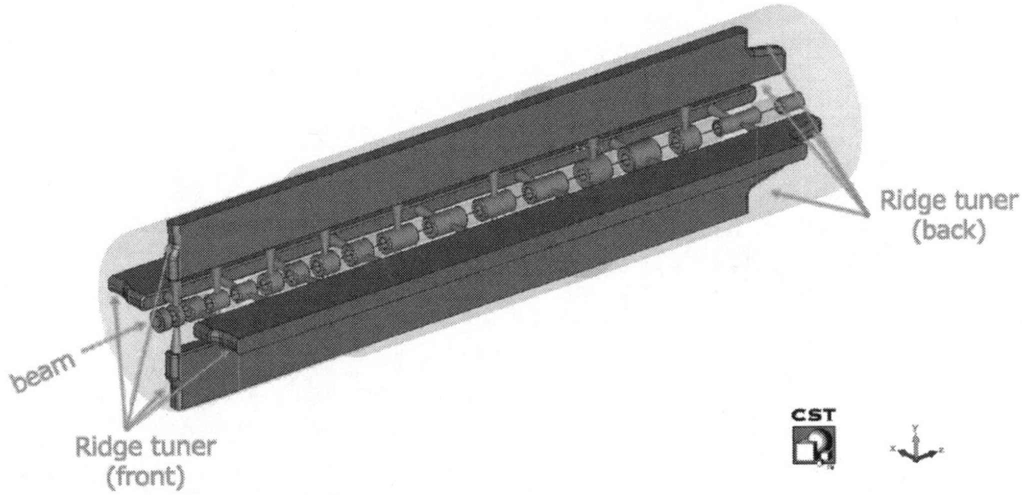

Figure 2. Three-dimensional model of the CH cavity in CST MW Studio calculation.

axis after these optimizations. The variation in the longitudinal electric field in the gaps is approximately 15%, excluding the first and last cells. Further optimization for the first and last cells will be attained by changing the gap lengths. The vertical field is less than 1% of the longitudinal field, whereas it is about 10% in the IH case.

Table 2 summarizes the basic parameters of the CH cavity. Because the acceleration field is slightly reduced in the first and last cells compared to that in the analytical design, the output beam energy is smaller then that in Table 1. The maximum surface field is calculated to be 2.1 times the Kilpatrick limit. This value is slightly higher and further reduction of the maximum field will be attained through optimization of the chamfered structure at the edge of the drift tube.

Table 2. Main Parameters of the CH Cavity

No. of gaps	16
Frequency (MHz)	323.48
Energy range (MeV)	0.3 - 4.1
Effective voltage (MV)	3.8
Q_0	14400
Power dissipation (kW)	360
Kilpatrick factor	2.1 (37.5 MV/m)

4. Particle tracking

Finally, the beam particle trajectory is simulated using the general particle tracer (GPT) [12]. The electric and magnetic fields calculated in previous steps are incorporated in the code and the particle dynamics are calculated numerically. Figure 4 shows the normalized velocity in the y-direction along the beam axis (z). Compared to the IH case, there is no meandering due to the substantial dipole field and better output beam quality is expected.

IOP Conf. Series: Journal of Physics: Conf. Series **874** (2017) 012038 doi:10.1088/1742-6596/874/1/012038

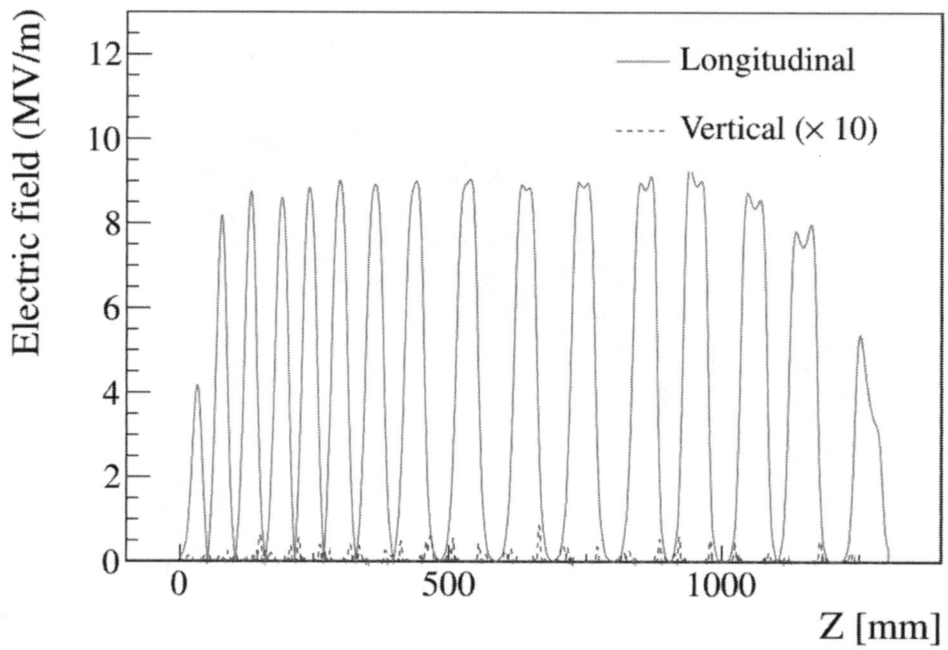

Figure 3. Longitudinal (solid red) and vertical (dotted blue) component of the electric field.

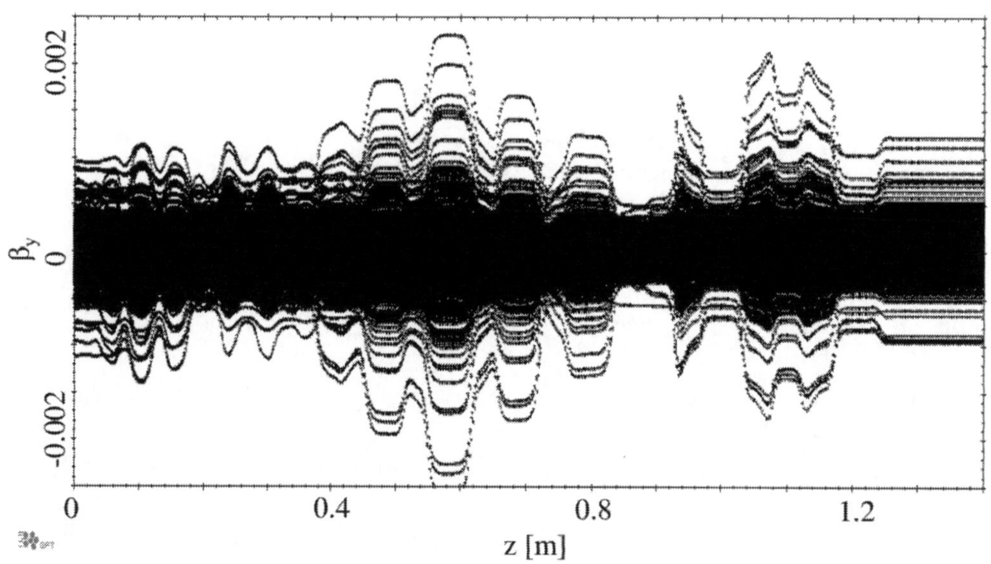

Figure 4. Normalized vertical velocity ($\beta_y = v_y/c$) along the beam axis.

Table 3 summarizes the particle simulation results. The output emittance was estimated to be 0.317π and 0.188π mm mrad in the horizontal and vertical directions, respectively. The emittance in the horizontal direction is consistent to the one obtained in the IH case within 1%. Thanks to the negligible dipole field, the emittance in the vertical direction is improved compared with the one obtained in the IH case by 4%. The transmission and survival rate are the same as those in the IH case.

Table 3. Summary of Output Beam Evaluations. IH: the output beam parameters with the IH [9]. CH: the output beam parameters with the CH.

	Input	Output	
		IH	CH
β	0.08	0.28	0.27
Energy (MeV)	0.34	4.4	4.1
ε_x [π mm mrad]	0.297	0.315	0.317
ε_y [π mm mrad]	0.168	0.195	0.188
Transmission [%]		99.9	99.9
Transient time [nsec]		25	25
Survival rate [%]		98.9	98.9
Transmission total [%]		98.7	98.7

5. Summary

In this paper, the APF CH-DTL design for the J-PARC g-2/EDM experiment has been presented. First, the synchronous phase array was optimized in order to obtain lower emittance growth, based on analytical calculations of the beam dynamics. Then, the CH cavity dimensions were optimized using finite element method calculation. Finally, the beam dynamics obtained for the calculated RF fields was evaluated via numerical calculations. Thanks to the negligible dipole field, the output beam emittance is expected to be better than the one obtained in the IH case [9].

6. acknowledgment

The authors are grateful to R. A. Jameson for useful advice on the APF linac and on modifications of the "LINACSapf" code. This work was supported by JSPS KAKENHI Grant Numbers 15H03666, 16H03987.

References

[1] Berg J S *et al.* 2011 *ICFA Beam Dynamics Newsletter* ed M. A. Palmer (Cornell U., USA) 55 p 54
[2] Miyake Y *et al.* 2013 *Hyperfine Interact* **216** 79
[3] Bogomilov M *et al.* 2012 *JINST* **7** P05009
[4] Artikova S *et al.* 2013 *Proceedings of 4th International Particle Accelerator Conference* (Shanghai) SAP045
[5] Bennett G W *et al.* 2006 *Phys. Rev.* D **73** 072003
[6] `http://g-2.kek.jp/portal/index.html`
[7] Otani M *et al.* 2016 *Proceedings of 7th International Particle Accelerator Conference* (Busan) TUPMY02
[8] Kurennoy S S *et al.* 2011 *Phys. Rev. Accel. Beams* **14** 110101
[9] Otani M *et al.* 2016 *Phys. Rev. Accel. Beams* **19** 040101
[10] Jameson R A 2014 *Preprint* arXiv:1404.5176 [physics.acc-ph]
[11] `https://www.cst.com/products/CSTMWS`
[12] `http://www.pulsar.nl/gpt/`

IOP Conf. Series: Journal of Physics: Conf. Series **874** (2017) 012039 doi:10.1088/1742-6596/874/1/012039

Prototyping high-gradient mm-wave accelerating structures

Emilio A Nanni[1], Valery A Dolgashev[1], Andrew Haase[1], Jeffrey Neilson[1], Sami Tantawi[1], Samuel C Schaub[2], Richard J Temkin[2] and Bruno Spataro[3]

[1] SLAC National Accelerator Laboratory, Menlo Park, USA
[2] Massachusetts Institute of Technology, Cambridge, USA
[3] INFN-LNF, Frascati, Rome, Italy

E-mail: nanni@slac.stanford.edu

Abstract. We present single-cell accelerating structures designed for high-gradient testing at 110 GHz. The purpose of this work is to study the basic physics of ultrahigh vacuum RF breakdown in high-gradient RF accelerators. The accelerating structures are π-mode standing-wave cavities fed with a TM_{01} circular waveguide. The structures are fabricated using precision milling out of two metal blocks, and the blocks are joined with diffusion bonding and brazing. The impact of fabrication and joining techniques on the cell geometry and RF performance will be discussed. First prototypes had a measured Q_0 of 2800, approaching the theoretical design value of 3300. The geometry of these accelerating structures are as close as practical to single-cell standing-wave X-band accelerating structures more than 40 of which were tested at SLAC. This wealth of X-band data will serve as a baseline for these 110 GHz tests. The structures will be powered with short pulses from a MW gyrotron oscillator. RF power of 1 MW may allow an accelerating gradient of 400 MeV/m to be reached.

1. Introduction

The high-gradient performance of accelerating structures is important for the selection of an accelerator's operational frequency. RF breakdown is one of the major phenomena that limits the achievable gradient in accelerating structures. Extensive studies on RF breakdown in copper accelerating structures[1] have been performed at frequencies as high as 11-12 GHz[2, 3, 4, 5, 6], 17 GHz[7] and 30 GHz[8, 9, 10].

The statistical behavior of RF breakdown became apparent during NLC/GLC work[2, 4, 11, 12]. After many pulses at the same RF power and pulse shape, for most accelerating structures the breakdown rate approached a steady state or is slowly decreasing. It is now common practice to measure and use the breakdown probability to quantify the high-gradient performance of accelerating structures.

Recently, we have expanded our experiments on the basic physics of ultrahigh vacuum RF breakdown to the mm-wave range with beam-driven accelerators at FACET[13, 14]. In this paper, we continue this study [15] using a 110 GHz MW pulsed gyrotron oscillator[16] as the RF power source. We are aiming to achieve accelerating gradients well beyond 100 MeV/m and plan to measure RF breakdown statistics.

IOP Conf. Series: Journal of Physics: Conf. Series **874** (2017) 012039 doi:10.1088/1742-6596/874/1/012039

2. Accelerating structure design

We are designing standing-wave accelerating structures with cavity geometries that allow for direct comparison with experiments at 11.424 GHz conducted at the SLAC National Accelerator Laboratory[3], where accelerating cavities with a/λ ratios of 0.105, 0.143 and 0.215 were extensively studied. Here a is the radius of the iris aperture and λ is the operational wavelength. Structures with a larger a/λ have a higher peak surface electric field for the same accelerating gradient. For our experiments we are designing cavities with the same a/λ ratios.

Table 1. RF parameters of 110 GHz single-cell accelerating structures. Fields are normalized to 1 MW of dissipated RF power. Accelerating gradient is defined as the peak surface electric field divided by K_e which was determined for the periodic solution[15].

Name	Unit	A0.286-T0.2-Cu	A0.390-T0.2-Cu	A0.586-T0.4-Cu
a/λ		0.105	0.143	0.215
Iris Aperture Radius, a	[mm]	0.286	0.390	0.586
Iris Thickness, t	[mm]	0.2	0.2	0.4
Iris Ellipticity		1	1.35	1.3
Accelerating Gradient, E_{acc}	[MeV/m]	404	361	277
Max. Surface Elec. Field, E_{max}	[MV/m]	916	827	624
Max. Surface Mag. Field, H_{max}	[MA/m]	1.13	1.13	1.16
$H_{\text{max}} Z_0 / E_{\text{acc}}$		1.04	1.13	1.47
$K_e = E_{\text{max}}/E_{\text{acc-periodic}}$		2.27	2.29	2.25
Peak Pulsed Heating, (20 ns Square Pulse)	[^{o}C]	156	156	165

We presented the detailed design of the 110 GHz structures in [15]. These structures have three cells, with the on-axis electric field in the central cell twice as high as in the adjacent cells. This was done to ensure that most of the breakdowns are in the central cell. The two end cells create fields that mimic the fields in a multi-cell standing wave structure. The on-axis electric fields for several values of a/λ are shown in Figure 1. The accelerating structure is fed through a 1.187 mm radius TM_{01} cylindrical waveguide. The beam pipe radius is 0.660 mm radius so the TM_{01} mode is below cutoff. The RF parameters of the three structures are given in Table 1. The parameters were normalized for 1 MW of dissipated RF power.

We plan to use a 110 GHz 1 MW gyrotron to power our accelerating structures. The RF power is transported to the accelerating structure in a free-space Gaussian mode. We couple the RF power into the accelerating structure using a Gaussian to TM_{01} mode converter. The mode converter is shown in Figure 2. First, the Gaussian mode is focused onto the aperture of a smooth-walled horn with ~6 mm beam waist. The horn converts the Gaussian mode into the TE_{11} mode of a circular waveguide with ~99% conversion efficiency. Following the Gaussian converter is a TE_{11} to TM_{01} mode converter, which includes a 90 degree bend. The mode converter has a 97% power conversion efficiency and a bandwidth exceeding 2 GHz. The parameters of the mode converter exceed the requirements for this experiment since its bandwidth is larger than the tunable frequency range of the gyrotron oscillator.

3. Prototype structures

For rapid prototyping we created simplified structures with geometries similar to our final accelerating structures. These structures consist of three identical cells with beam pipes at each end. The structures are made in halves, where each half is milled into a copper block as shown in Figure 3. Then the two halves are joined together with bonding and/or brazing. We call this the "split-cell" fabrication technique. The split interface is parallel to the RF currents of the operating mode, thus the ohmicQ factor of the structures is determined by the properties of the surface and not the interface.

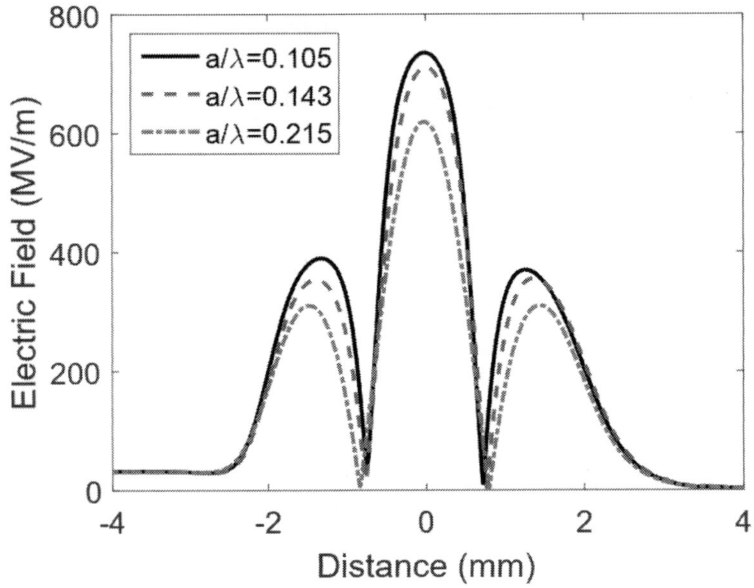

Figure 1. On-axis electric field for three cavities with different a/λ assuming 1 MW of dissipated RF power in the structure.

We measured the RF properties of the prototypes with a vector network analyzer and two coaxial probes inserted into the beam pipes. The structures were cold tested both before and after bonding.

Of particular interest for us was the Q factor value and its repeatability, the repeatability of resonant frequency between multiple structures machined in a single milling operation, and the frequency difference before and after bonding. The consistency of the resonant frequency between prototypes was considered a secondary objective, with the absolute frequency of the cavities was required to be close to our final target of 110 GHz. To date, 15 RF structures have been fabricated in 5 split-cell assemblies with three structures each.

We found that both the diffusion bonding and brazing were able to produce cavities with Q factors of 2700-2800 for the π-mode, approaching the theoretical expectation of 3300. Post bonding autopsy showed significant variation of the interface structure between brazing and diffusion bonding, see Figure 4. Due to the orientation of this interface with respect to RF currents this variation has little effect on Q value and resonant frequency. With the correct amount of braze alloy, interfaces were sealed uniformly through the iris (Figure 4(left)). However, we observed a gap in the iris with diffusion bonding (Figure 4(right)). While the gap did not result in a degradation of RF parameters, the exposed corners may result in local electric field enhancement. We speculate that this enhancement may affect the high-gradient performance of the cavity. In this particular assembly, we see a misalignment between the two diffusion bonded

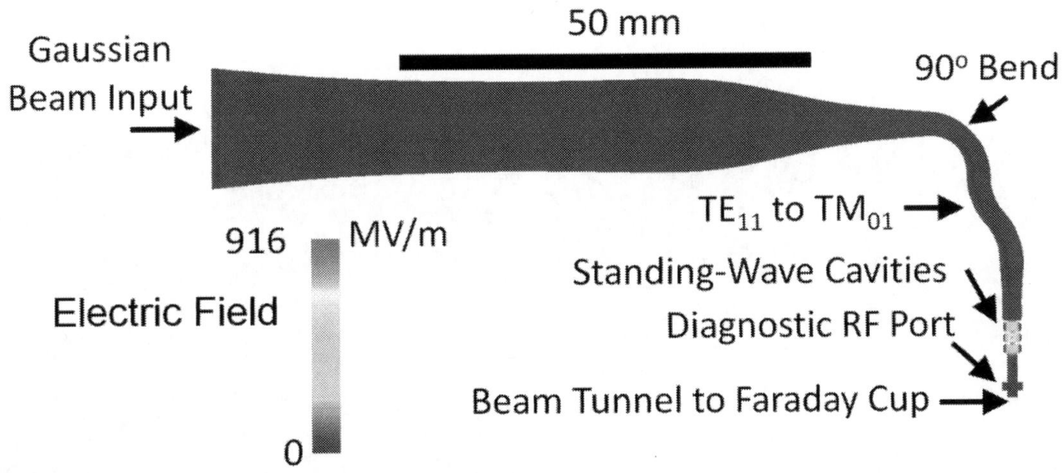

Figure 2. Model of the assembly consisting of the mode converter and the standing-wave accelerating structure A0.286-T0.2-Cu. We show the magnitude of the electric field on the symmetry plane using HFSS[17]. The field is normalized for 1 MW in the input Gaussian beam.

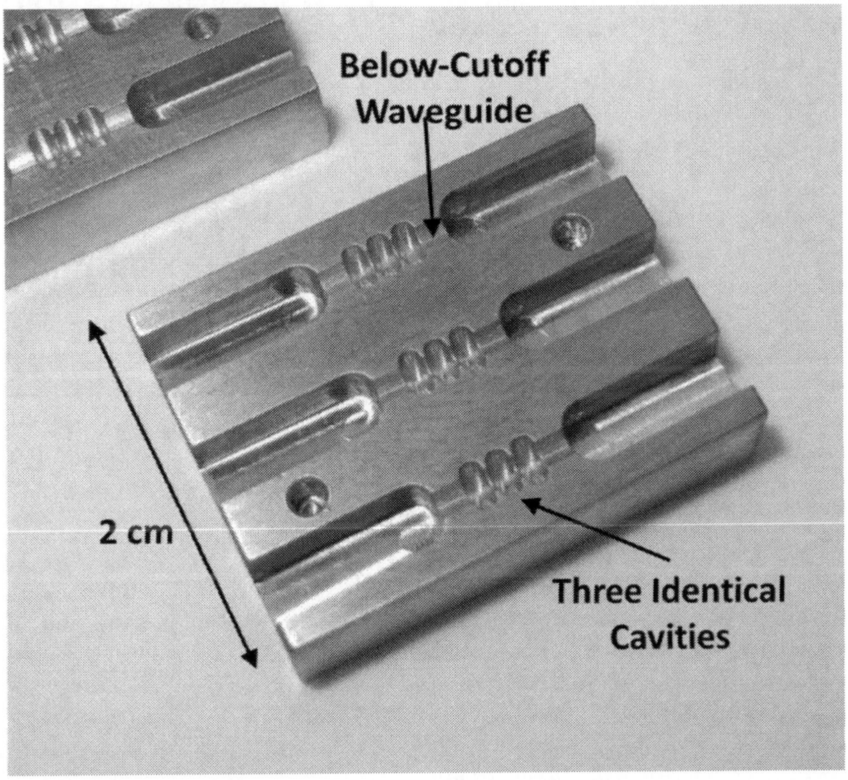

Figure 3. Photograph of a milled copper block with three halves of the accelerating structure.

blocks at the interface. Unlike prototypes, in the high power assembly we placed a stricter tolerance on the alignment between the two blocks.

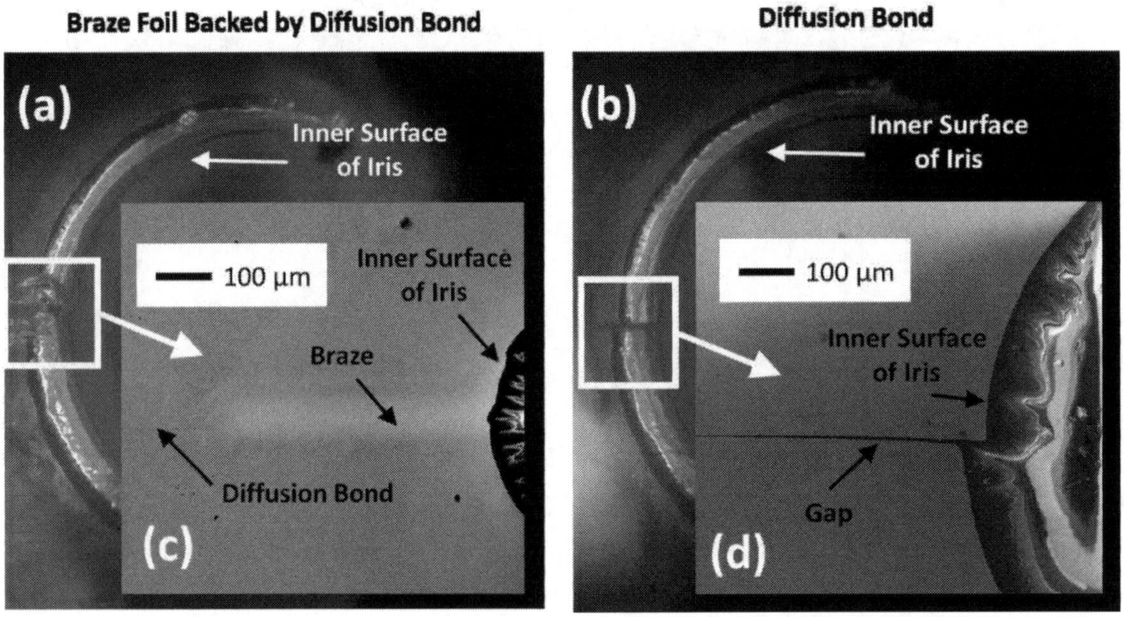

Figure 4. Photograph of (a) brazed and (b) diffusion bonded iris with a=1.02 mm. SEM image of the same (c) brazed and (d) bonded joints machined down to the middle of the iris in order to show the internal mechanical structure.

Diffusion bonded and brazed cavities had a mean π-mode resonant frequency of 107.69 GHz, with a standard deviation of 210 MHz over all 15 cavities. Between the three structures in the same split-cell assembly the deviation was much smaller, 60 MHz on average. We observed a significant frequency increase between pre-bonding and after bonding measurements, with a mean of 430 MHz and a standard deviation of 190 MHz. We did not see a significant correlation in the frequency shift with the bonding technique. We note that both the brazing and the diffusion bonding procedure include cleaning with chemical etching. We speculate that the cavity deformation from this bonding procedure is the dominant source of the frequency shift. Understanding and controlling this frequency shift is an objective of the ongoing study.

4. High power structure

Two high power RF structures from Figure 2 were fabricated in both copper (Cu) and copper-silver (CuAg) by EDM Department, Inc. In Figure 5 we show a photograph of one half of the Cu structure. These structures will be diffusion bonded. The measured π-mode frequency of the clamped structures is 109.503 GHz for Cu and 109.254 GHz for CuAg. We performed S-parameter measurements of the clamped assembly, with the S_{11} vs. frequency is shown in Figure 5. We measured a loaded Q factor of 1600 for Cu and 1200 for CuAg structures. These are in good agreement with the expected loaded Q factor of 1800. The spacing of the 0 and π modes does not match the design due to variations in cell coupling, however it is acceptable and will not impact the high power test. In future we also plan to build brazed assemblies.

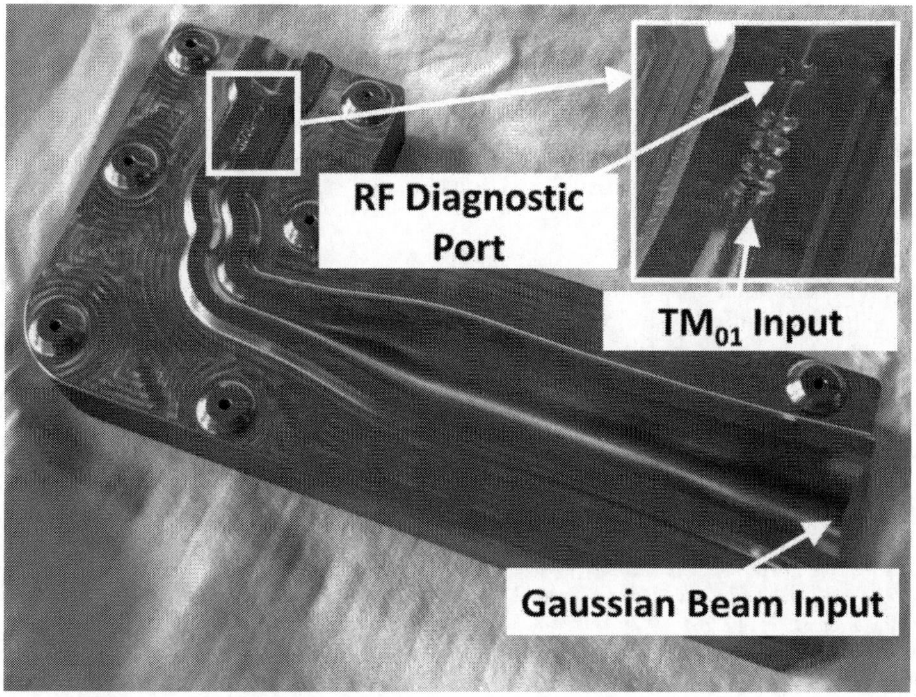

Figure 5. Photograph of one half of the high-power split-cell structure.

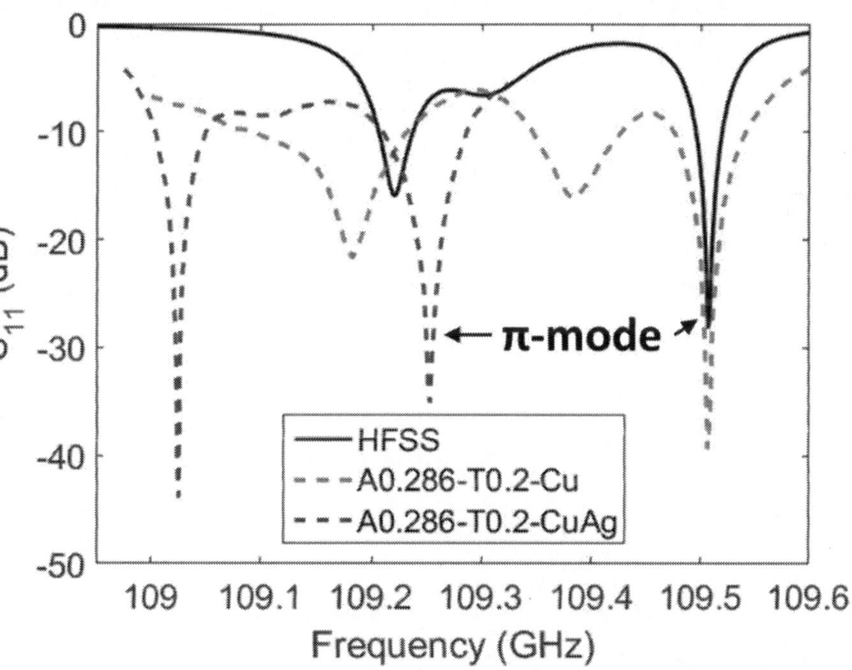

Figure 6. Simulated and measured S_{11} for the Gaussian beam mode converter and the high-power accelerating structure. The HFSS simulation is shifted down by 600 MHz from the nominal frequency to align the frequencies of the π-mode for the Cu structure. We expect that bonding will shift the π-mode frequency up by at least 400 MHz.

IOP Conf. Series: Journal of Physics: Conf. Series **874** (2017) 012039 doi:10.1088/1742-6596/874/1/012039

5. Conclusion

We have presented single-cell accelerating structures designed for high-gradient testing at 110 GHz using a gyrotron as the RF power source. We built 5 split-cell prototypes with three structures each to test manufacturing techniques. We built two high-power structures one of Cu and another one of CuAg. Cold tests indicate that the desired RF parameters could be achieved. With 1 MW these structures may reach peak accelerating gradients of 400 MeV/m.

Acknowledgments

Acknowledgment The authors thank Michael Shapiro, Michelle Gonzalez, Ann Sy and Gordon Bowden for helpful discussions and Chris Pearson for performing SEM imaging. This work was supported by Department of Energy contract DE-AC02-76SF00515.

References

[1] Chang C, *et al.* 2011 *Physics of Plasmas* **18.5** 055702
[2] Dbert S *et al.* 2005 *Proc. of IPAC 2005* (Knoxville, TN) pp. 372-374
[3] Dolgashev V et al. 2010 *Applied Physics Letters* **97.17** 171501
[4] Higo T *et al.* 2010 *Proc. of IPAC 2010* (Kyoto, Japan) THPEA013 p. 3702
[5] Marsh R A *et al.* 2011 *Physical Review Special Topics-Accelerators and Beams* **14.2** 021301
[6] Degiovanni A *et al.* 2014 *Proc. of IPAC2014* (Dresden, Germany) WEPME015
[7] Munroe B J *et al.* 2016 *Physical Review Accelerators and Beams* **19.3** 031301
[8] Achard C J. W. 2005 in *Proc. of IPAC 2005* (Knoxville, TN) pp. 1695-1697
[9] Rodriguez J *et al.* 2007 *Proc. of PAC07* (Albuquerque, NM) FROBC01
[10] Grudiev A, Calatroni S and Wuensch W 2009 *Physical Review Special Topics-Accelerators and Beams* **12.10** 102001
[11] Wang J W 2006 *High Energy Phys. Nucl. Phys.* **30** 11
[12] Adolphsen C 2003 *Proceedings of the 2003 Particle Accelerator Conference* (Portland, OR) p. 668
[13] Dal Forno M *et al. Physical Review Accelerators and Beams* **19.1** 011301
[14] Dal Forno M *et al.* 2017 *AIP Conference Proceedings* **1812** 060011
[15] Nanni, E A *et al.* 2016 *7th International Particle Accelerator Conference (IPAC'16)* (Busan, Korea) pp 3884-3887
[16] Tax D S *et al.* 2014 *Plasma Science, IEEE Transactions on* **42.5** 1128-1134
[17] ANSYS Electronics Desktop 16.2, *HFSS*

IOP Conf. Series: Journal of Physics: Conf. Series **874** (2017) 012040 doi:10.1088/1742-6596/874/1/012040

Simulation of DLA grating structures in the frequency domain

T Egenolf[1], O Boine-Frankenheim[1,2] and U Niedermayer[1]
[1] TEMF, TU Darmstadt, Schlossgartenstr. 8, 64289 Darmstadt
[2] GSI, Planckstr. 1, 64291 Darmstadt
E-mail: egenolf@temf.tu-darmstadt.de

Abstract. Dielectric laser accelerators (DLA) driven by ultrashort laser pulses can reach orders of magnitude larger gradients than contemporary RF electron accelerators. A new implemented field solver based on the finite element method in the frequency domain allows the efficient calculation of the structure constant, i.e. the ratio of energy gain to laser peak amplitude. We present the maximization of this ratio as a parameter study looking at a single grating period only. Based on this optimized shape the entire design of a beta-matched grating is completed in an iterative process. The period length of a beta-matched grating increases due to the increasing velocity of the electron when a subrelativistic beam is accelerated. The determination of the optimal length of each grating period thus requires the knowledge of the energy gain within all so far crossed periods. Furthermore, we outline to reverse the excitation in the presented solver for beam coupling impedance calculations and an estimation of the beam loading intensity limit.

1. Introduction

Several applications of particle accelerators request higher acceleration gradients than delivered by current rf-accelerators for more compact devices. Dielectric laser accelerators (DLAs) [1] are potential candidates for advanced linear electron accelerators with gradients in the range of GeV/m. In the future, such compact DLAs will provide the opportunity for compact radiotherapy devices and free electron lasers, for example.

Recently, DLA experiments have demonstrated acceleration gradients of about 376 MeV/m for subrelativistic electrons and acceleration gradients of about 690 MeV/m for relativistic beams at SLAC [2].

In these experiments, electrical near-fields of pulsed solid state lasers modulated by dielectric nanostructures are used to accelerate electrons. The structures are fabricated by lithographic techniques. Of particular interest for characterizing such periodic structures is the maximum reachable acceleration gradient expressed by the structure constant

$$SC = \frac{\max(\Delta W)}{eE_0\lambda_{gz}} = |\underline{e}_\mathrm{n}| \tag{1}$$

with the energy gain ΔW of an electron passing through period of length λ_{gz} in positive z-direction. The driving (z-polarized) laser with peak amplitude E_0 propagates in y-direction and the longitudinal electric laser field $\underline{E}_z$ in the frequency domain is expanded in a spatial Fourier series with the normalized harmonic coefficients

$$\underline{e}_n = \frac{1}{E_0 \lambda_{gz}} \int_{-\lambda_{gz}/2}^{\lambda_{gz}/2} \underline{E}_z e^{\mathrm{i}n \frac{2\pi}{\lambda_{gz}} z} dz. \tag{2}$$

In order to calculate the Fourier coefficients, we implemented a simulation tool based on the finite element method in 2D on an unstructured triangular mesh using FEniCS [3], an open-source toolbox for automated solution of PDEs. The mesh originates from Gmsh [4] and is imported via DOLFIN-CONVERT [5]. The advantage of such a mesh is a better adaptability to curved geometries than a hexahedral mesh commonly used in time domain simulations. Since most of the DLA structures are translation invariant in one direction, two-dimensional simulations suffice at the first design stage.

2. 2D FEM field solver

The fields $\underline{\mathbf{E}} : \Omega \subset \mathbb{R}^2 \to \mathbb{C}^2$ required to calculate the Fourier coefficients are solutions of the curl-curl-equation

$$\nabla \times \left[\mu_0^{-1} \nabla \times \underline{\mathbf{E}}\right] - \omega^2 \varepsilon \underline{\mathbf{E}} = -\mathrm{i}\omega \underline{\mathbf{J}} \tag{3}$$

with the permeability constant μ_0 and the permittivity $\varepsilon = \varepsilon_0 \varepsilon_\mathrm{r}(y, z)$ as a function of position. The current density $\underline{\mathbf{J}}$ models the incident laser and is imprinted as a constant on a small stripe of the mesh outside the dielectric.

Testing with functions $\underline{\mathbf{v}}_i \in \mathrm{H}_{2\mathrm{D}}^{\mathrm{curl}}$, where $\mathrm{H}_{2\mathrm{D}}^{\mathrm{curl}}$ is the Sobolev space of square integrable functions with square integrable two dimensional curl, results in the weak formulation

$$\int_\Omega \left[\mu_0^{-1} \{\nabla \times \underline{\mathbf{E}}\} \cdot \{\nabla \times \underline{\mathbf{v}}\} - \omega^2 \varepsilon \underline{\mathbf{E}} \cdot \underline{\mathbf{v}}\right] d\Omega - \int_{\partial\Omega} \left[\mu_0^{-1} \{\nabla \times \underline{\mathbf{E}}\} \times \mathbf{n}\right] \cdot \underline{\mathbf{v}} ds = -\int_\Omega \mathrm{i}\omega \underline{\mathbf{J}} \cdot \underline{\mathbf{v}} d\Omega \tag{4}$$

with the outward unit vector $\mathbf{n}$ normal to the boundary $\partial\Omega$. The Galerkin ansatz

$$\underline{\mathbf{E}} = \sum_n a_n \underline{\mathbf{v}}_n \tag{5}$$

with degrees of freedom a_n and Nédélec edge elements $\underline{\mathbf{v}}_n$ [6] provides a linear equation system which is solved numerically.

A periodic boundary condition is applied on the boundary normal to the beam path. Two different boundary conditions can be chosen on the lateral sides: a Sommerfeld boundary condition [7]

$$\mathbf{n} \times (\nabla \times \underline{\mathbf{E}}) = -\mathrm{i}\omega c^{-1} \mathbf{n} \times (\mathbf{n} \times \underline{\mathbf{E}}) \tag{6}$$

or a perfectly matched layer [8].

3. Simulation examples

The new simulation tool has various applications in the design of DLA structures. We analyze a new "slightly resonant" silicon structure ($\varepsilon_\mathrm{r} = 11.63$) which is proposed in [9]. The design is called cavity structure in the following. The structure constant depends significantly on the values of three geometry parameters depicted in the schematic drawing in Figure 1.

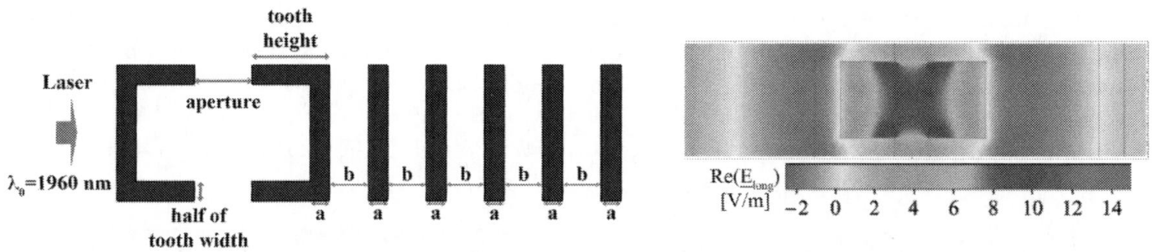

Figure 1. Schematic drawing and longitudinal electric field of the cavity structure. The thickness of the layers is given by the Bragg condition $a = \lambda_0/\left(4\sqrt{\varepsilon_r}\right)$ and $b = \lambda_0/4$ with $\varepsilon_r = 11.63$ and the beam path is in the middle of the aperture. The laser strength of the plotted field is 1 V/m.

3.1. Single period optimization

Parameter studies can be used to optimize the structure constant of a single period. Since the structure constant increases as the aperture decreases, an aperture of 200 nm is a trade-off between maximum reachable structure constant and maximum admissible transversal beam size. The absolute value of the structure constant depending on the remaining parameters, width and height of the tooth, is shown in Figure 2. Choosing a width of 200 nm and a height of 450 nm results in a maximum structure constant of about 0.74 for a period length of 620 nm (cross-checked by CST® time domain simulations, see [9]). In the choice of electron beam parameters we follow [10], i.e. $\beta = 0.3165$.

Furthermore, longer periods, corresponding to larger velocities, can reach much larger structure constants. The maximum structure constant for a period length of 1800 nm is about 2.1, for example. The geometry parameters of such a period are an aperture of 400 nm and a tooth width of 550 nm and height of 450 nm (see Figure 3).

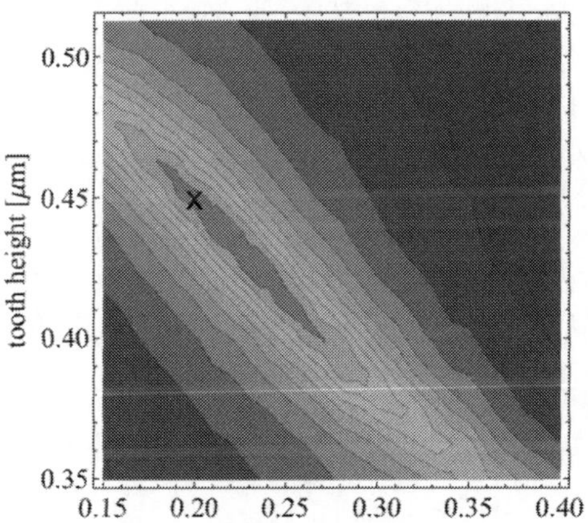

Figure 2. Parameter study of the structure constant for a period length of 620 nm. The X symbolizes the geometry parameters with the maximum structure constant.

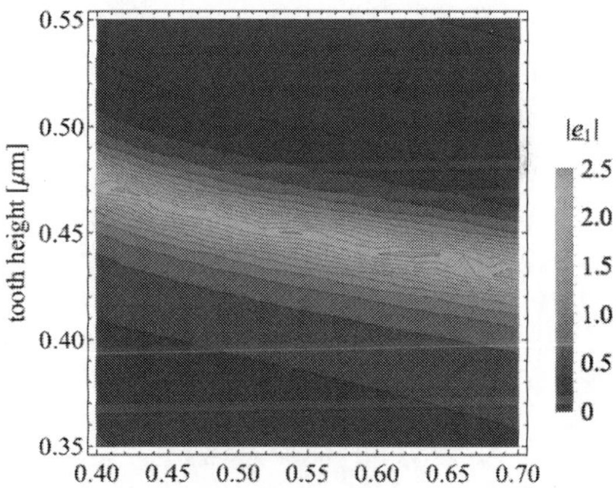

Figure 3. Parameter study of the structure constant for a period length of 1800 nm.

First prototypes of the proposed cavity structure are now being fabricated. Since 90° corners are not feasible by lithographic etching techniques, the corners in the original design have to be rounded with a minimum radius of 50 nm. Such restrictions can be integrated in simulations by modifying the mesh (see Figure 4). The structure constant is thus reduced to about 0.63 at the geometry parameters optimized before for a 620 nm period. A new maximum of 0.66 can be reached at a tooth width of 240 nm and height of 425 nm.

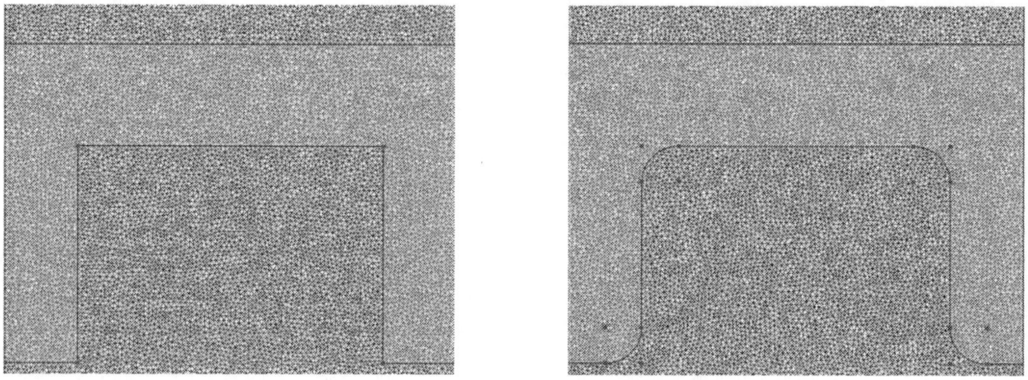

Figure 4. Comparison of the triangular mesh with 90° corners on the left and rounded corners on the right. The upper half of the cavity structure is shown.

3.2. Designing a beta-matched grating

Due to the phase synchronicity condition $\lambda_{gz} = n\beta\lambda_0$, where n is the spatial harmonic, the period length along a grating has to increase for an increasing particle velocity. The energy gain per period of the synchronous particle is

$$\Delta W = eE_0\lambda_{gz}|\underline{e}_{\mathrm{n}}|\cos(\varphi_s), \tag{7}$$

where $\varphi_s = \varphi_{\mathrm{p}} + \arg(\underline{e}_{\mathrm{n}})$ and the phase φ_{p} of the particle is referred to the laser phase. Since φ_{p} is constant along the beta-matched grating, the phase of the n-th Fourier coefficient also needs to be as constant as possible in order to accelerate in stable buckets.

3.2.1. Phase stabilization. An increase in period length along a beta-matched grating only is not sufficient since this changes the phase of the Fourier coefficient significantly. As shown in Figure 5, the change in $\arg(\underline{e}_1)$ is in the range of 30°, which may lead to particle loss. If the width of the teeth is adjusted as well, the phase drift can be limited to about 4°. Here, the tooth width is optimized as linear function of the period length

$$t(\lambda_{gz}) = t^{(0)}\left(\frac{\lambda_{gz}/\lambda_{gz}^{(0)} - 1}{\xi} + 1\right), \tag{8}$$

where $t^{(0)} = 200\,\mathrm{nm}$, $\lambda_{gz}^{(0)} = 620\,\mathrm{nm}$ and $\xi \approx 2.7$. Moreover, the decrease of the structure constant is also reduced from 33 % to 17 %.

3.2.2. Iterative approach. Finally, an iterative approach can be used to construct such a beta-matched grating. The optimized design of one period forms the basis of the approach. Since the velocity of a reference electron in a grating period depends on the energy gain in all previous periods, an iterative design process is to be favored. As an example we present a beta-matched grating based on the proposed cavity structure and further informations on the approach in [9].

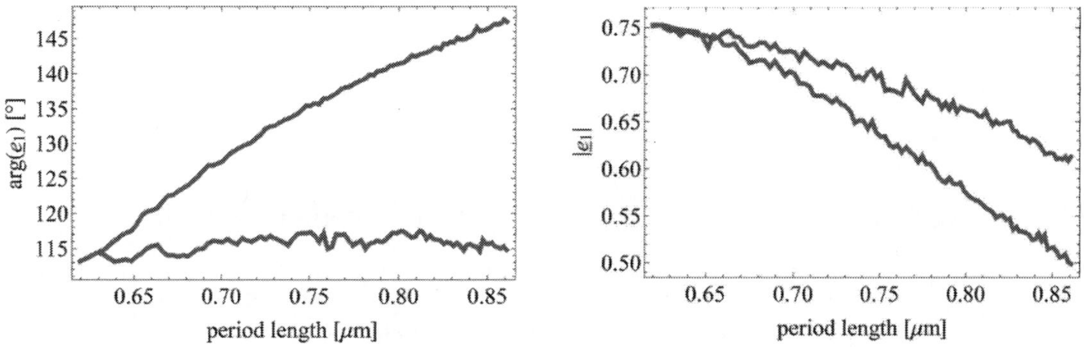

Figure 5. Phase and absolute value of the First Fourier coefficient over the period length for constant tooth width (red) and adapted tooth width according to Equation (8) (blue).

4. Conclusion

The solver described in this contribution calculates the electric field in DLA nanostructures based on the finite element method in the frequency domain and predicts the structure constant. The solver can be integrated in the design process of 2D structures, performing all necessary simulations for a beta-matched grating.

Simulations in the frequency domain require the solution of one linear matrix equation per calculation of a structure constant in contrast to simulations in the time domain, which require one matrix multiplication per time step. Furthermore, the use of a triangular mesh is favorable in the frequency domain. This makes the solver advantageous for simulations of curved geometries.

A further application of the solver is the estimation of the beam loading limit for DLA structures. Therefore, the excitation is shifted to the beam path to calculate the beam coupling impedance. The wake potential can then be obtained by a discrete Fourier transform of the impedance and a convolution with the beam distribution. This upgrade is presently under development.

Acknowledgments

This work is supported by the Gordon and Betty Moore Foundation under grant GBMF4744 and the German Federal Ministry of Education and Research under grant FKZ:05K16RDB.

References

[1] Peralta E A et al 2013 Demonstration of electron acceleration in a laser-driven dielectric microstructure *Nature* **503** 91–94
[2] Wootton K P, McNeur J and Leedle K J 2016 Dielectric laser accelerators: designs, experiments, and applications *Rev. Accl. Sci. Tech.* **9** 105–26
[3] Logg A, Mardal K-A and Wells G N 2012 *Automated Solution of Differential Equations by the Finite Element Method* (Berlin: Springer)
[4] Geuzaine C and Remacle J-F 2009 Gmsh: a 3-D finite element mesh generator with built-in pre- and post-processing facilities *Int. J. Numer. Meth. Engng.* **79** 1309–31
[5] Logg A and Wells G N 2010 DOLFIN: automated finite element computing *ACM Transactions on Mathematical Software* **37** 20
[6] Nédélec J C 1980 Mixed finite elements in $\mathbb{R}^3$ *Numerische Mathematik* **35** 315–41
[7] Jin J-M and Riley D J 2009 *Finite Element Analysis of Antennas and Arrays* (Hoboken: Wiley)
[8] Sacks Z S, Kingsland D M, Lee R and Lee J F 1995 A perfectly matched anisotropic absorber for use as an absorbing boundary condition *IEEE Trans. Antennas Propagat.* **43** 1460–64
[9] Niedermayer U, Boine-Frankenheim O and Egenolf T 2017 Designing a dielectric laser accelerator on a chip *Proc. 8th Int. Particle Accelerator Conf. (May 14–19 Copenhagen)* WEPVA003
[10] Breuer J and Hommelhoff P 2013 Laser-based acceleration of nonrelativistic electrons at a dielectric structure *Phys. Rev. Lett.* **111** 134803

Designing a Dielectric Laser Accelerator on a Chip

Uwe Niedermayer, Oliver Boine-Frankenheim, and Thilo Egenolf
TEMF, TU-Darmstadt, Schlossgartenstr. 8, 64289 Darmstadt, Germany
E-mail: niedermayer@temf.tu-darmstadt.de

Abstract. Dielectric Laser Acceleration (DLA) achieves gradients of more than 1GeV/m, which are among the highest in non-plasma accelerators. The long-term goal of the ACHIP collaboration is to provide relativistic (>1 MeV) electrons by means of a laser driven microchip accelerator. Examples of "slightly resonant" dielectric structures showing gradients in the range of 70% of the incident laser field (1 GV/m) for electrons with beta=0.32 and 200% for beta=0.91 are presented. We demonstrate the bunching and acceleration of low energy electrons in dedicated ballistic buncher and velocity matched grating structures. However, the design gradient of 500 MeV/m leads to rapid defocusing. Therefore we present a scheme to bunch the beam in stages, which does not only reduce the energy spread, but also the transverse defocusing. The designs are made with a dedicated homemade 6D particle tracking code.

1. Introduction

The Accelerator on a Chip (ACHIP) collaboration [1] aims to construct a source of relativistic electrons out of a shoe-box size Dielectric Laser Accelerator (DLA). Furthermore, the collaboration also aims at radiation generation by means of laser driven undulators. Beyond the acceleration of relativistic electrons [2], recent experiments also addressed DLA for subrelativistic electrons [3, 4]. Since only the evanescent near field which decays as $\exp(-\omega y/(\beta\gamma c))$ contributes to acceleration, both the gradients and the apertures are smaller in subrelativistic structures. In Figure 1 we present a novel DLA structure that contains both features of a side-coupled grating accelerator and a Bragg-waveguide accelerator. This structure is "slightly resonant" and thus presents a compromise between filling time and acceleration gradient. The structure constant is the ratio between the gradient and the incident laser field and can be conveniently expressed by the normalized resonant spatial Fourier coefficient $\underline{e}_m$ as

$$SC = \frac{\max_\varphi\{\Delta W\}}{eE_{z0}\lambda_g} = |\underline{e}_m|, \tag{1}$$

where E_{z0} is the incident (z-polarized) laser field and $\varphi = 2\pi s/\lambda_g$ is the phase of a particle at a distance s behind a design particle. The resonant integer m fulfills the Wideroe condition $\lambda_g = m\beta\lambda_0$, where λ_g is the structure period. Usually the fundamental $|\underline{e}_1|$ is the largest coefficient so we will restrict ourselves to $m = 1$ in the following. The thickness of the Bragg reflection layers is $d = (1/4 + n/2)\lambda_0/\sqrt{\varepsilon_r}$, where the integer n can be chosen according to practical requirements of the fabrication aspect ratio and the required height. The structure constant can be determined using time domain simulations (here: CST MWS [5]), evaluated at the center frequency of the broadband spectrum. Alternatively, we developed a dedicated finite element (FEM) code in the frequency domain [6], which was also used to optimize the $\beta = 0.91$ version of the Bragg structure.

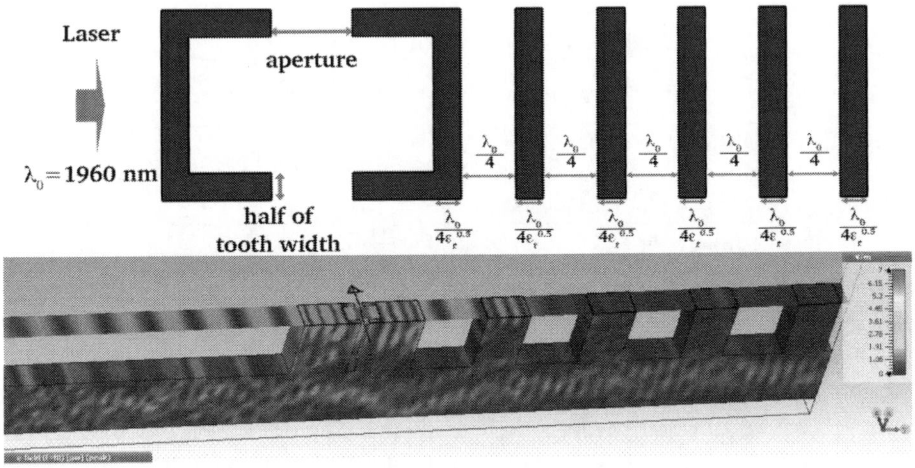

Figure 1. Bragg mirror cavity structure (Silicon $\varepsilon_r = 11.63$, length $\lambda_g = 620$nm) for n=0 (top) and 3D simulation for n=3 and height 3μm (bottom). The black arrow indicates the electron trajectory at which the structure is periodic.

2. Accelerator Design

We propose a structure that comprises ballistic bunching followed by velocity matched acceleration, see Figure 2. For simplicity we assume that the structure is driven from both lateral sides with a cw laser. In the choice of initial electron beam parameters we follow [3], i.e. $\beta = 0.3165$, $\sigma_E = 10\,$eV and the transverse emittance is disregarded at first. The parameters of the structure are summarized in Table 1 and shall be discussed in detail in the following.

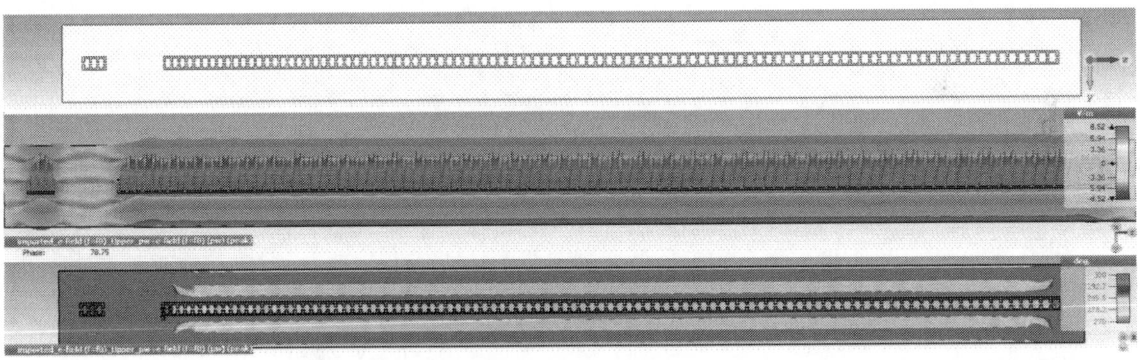

Figure 2. Chirped accelerator structure with 100 periods (top), longitudinal electric field amplitude (center), and phase (bottom) along the chirp.

2.1. Velocity bunching

Velocity bunching is well known for both ion and electron beams. The idea is to modulate a coasting beam such that it has a sinusoidal correlated energy spread pattern. A following drift section for subrelativistic beams (or a dispersive chicane for relativistic beams) will transform the energy modulation into a phase modulation of $\Delta\varphi = \pi/2$, at which the longitudinal focus is

Table 1. Accelerator parameters

Laser strength	1 GV/m
Aperture	200 nm
Buncher periods	3
Buncher period length λ_{g0}	620 nm
ΔW_{corr} (incl. fringe)	$\approx$ 1.6 keV
L_{Drift} (total)	5.06 μm
$L_{\text{Drift,int}}$	8 λ_{g0}
$L_{\text{Drift,frac}}$ (incl. ph. corr.)	0.16 λ_{g0}
Accelerator $\lvert \underline{e}_1 \rvert$	0.73 (initial)
Linear chirp Δz	3.2 nm/cell
Chirp decrement $\Delta\Delta z$	16 pm/cell
Synchronous phase φ_s	47°
Acceptance ΔE_{max}	2 keV
Synchrotron frequency f_s	4.53 THz (initial)
Gradient	500MeV/m (initial)

reached. This happens in $T = \lambda_g/(4\Delta\beta c)$ and thus the length needs to be

$$L_{\text{int}} = \beta c T = \frac{\lambda_g}{4}\frac{\beta}{\Delta\beta} = \frac{\beta^2\gamma^3}{4}\frac{m_e c^2}{\Delta W_{\text{kin}}}\lambda_g, \tag{2}$$

where the energy-velocity differential is $\mathrm{d}\beta = \mathrm{d}\gamma/(\beta\gamma^3)$. The energy modulation can be realized with more than one grating cell, since the drift between the cells is negligible, see Figure 3. The particles pile up at the zero crossing of the modulator phase and can be injected exactly at the designed synchronous phase φ_s by a fractional period drift

$$L_{\text{frac}} = \lambda_{g0}\frac{\pi/2 - \varphi_s + \Delta\varphi_B}{2\pi}, \tag{3}$$

where $\Delta\varphi_B$ is a phase correction due to the buncher fringe fields, see Figure 2 bottom. The total length of the drift is $L_{\text{Drift}} = \lambda_{g0}[L_{\text{int}}/\lambda_{g0}] + L_{\text{frac}}$, where the square brackets denote integer rounding.

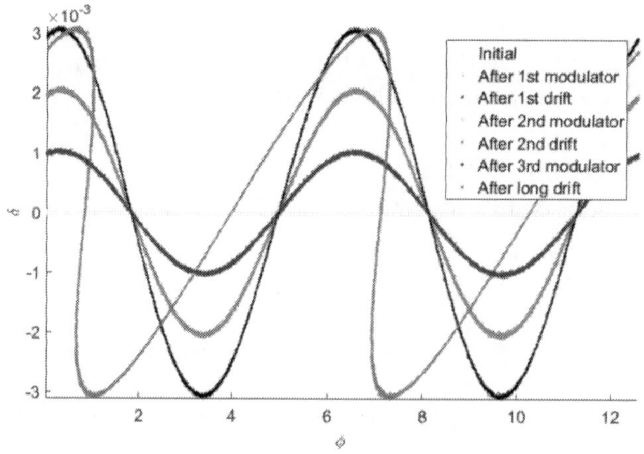

Figure 3. Buncher longitudinal phase space

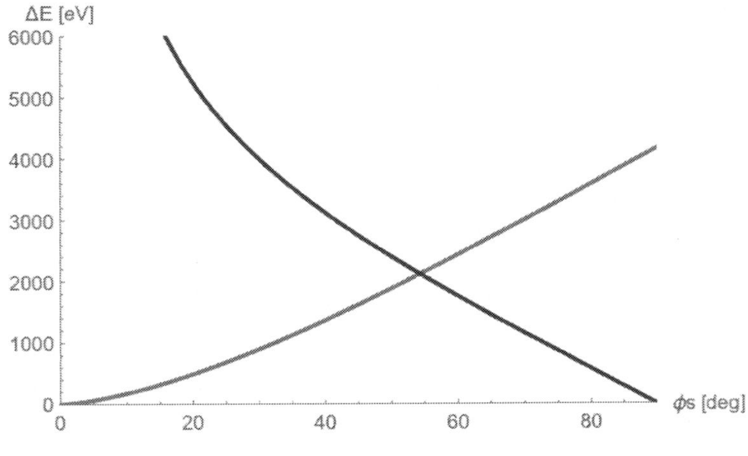

Figure 4. Energy spread acceptance (red) and energy gain per quarter synchrotron period (blue) as function of the synchronous phase φ_s

2.2. Energy spread acceptance

The energy spread acceptance and the initial synchrotron frequency are determined similarly to ordinary RF accelerators. The tracking equations can be approximated by differential equations cast in the form of Hamilton's equations, which can be integrated in the conjugate variables $\tau = \varphi/\omega_0$ and $\Delta W = \gamma m_e c^2 \delta$ to find the Hamiltonian as

$$H(\varphi, \delta) = \frac{m_e c^2}{2\beta^2\gamma}\delta^2 - eE_{z0}|\underline{e}_1|\frac{\lambda_g}{2\pi}(\sin\varphi - \varphi\cos\varphi_s). \tag{4}$$

The separatrix is found by the value of H at the saddle point $\varphi_{\text{saddle}} = -\varphi_{\text{s}}$ as

$$\delta_{\text{sep}}(\varphi) = \pm\sqrt{\frac{2\beta^2\gamma}{m_e c^2}[H(-\varphi_s, 0) - H(\varphi, 0)]}. \tag{5}$$

The bucket height gives the energy spread acceptance $\Delta E_{\max} = \gamma m_e c^2 \delta_{\text{sep}}(\varphi_s)$ as depicted in Figure 4. Moreover, from the Hamiltonian in Equation (4) the second order differential equation

$$\frac{\partial^2\varphi}{\partial t^2} = \frac{2\pi}{\gamma^3}\frac{eE_{z0}|\underline{e}_1|}{m_e\lambda_g}[\cos(\varphi) - \cos(\varphi_{\text{s}})] \tag{6}$$

can be derived. Assuming $\varphi = \varphi_s + \Delta\varphi$ the small amplitude synchrotron frequency is found as

$$f_s = \frac{\omega_s}{2\pi} = \sqrt{\frac{eE_{z0}|\underline{e}_1|}{2\pi\gamma^3 m_e\lambda_g}\sin(\varphi_s)}. \tag{7}$$

2.3. Chirped grating

In order to trap the particles with energy spread and phase spread in the bucket and accelerate, the phase of the accelerating Fourier coefficient needs to be as constant as possible along the chirped grating. This is achieved in the same manner as tuning RF cavities, namely by adjusting a geometry parameter that is still free. Here, the tooth width t is taken as

$$t = t^{(0)}\left(\frac{\lambda_g/\lambda_g^{(0)} - 1}{\xi} + 1\right), \tag{8}$$

IOP Conf. Series: Journal of Physics: Conf. Series **874** (2017) 012041 doi:10.1088/1742-6596/874/1/012041

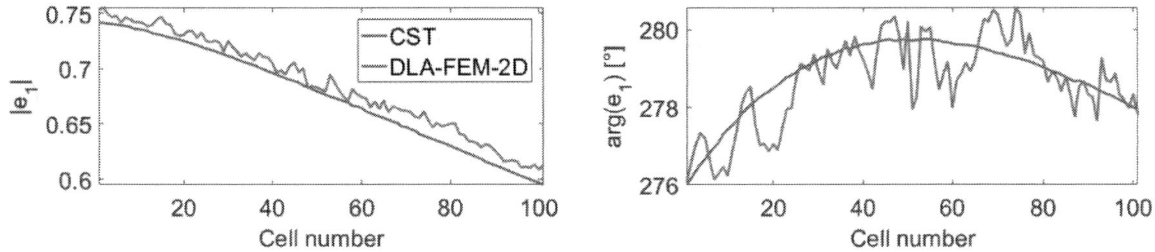

Figure 5. Establishing (almost) constant phase (4 deg jitter) by adjusting the tooth width

with $t^{(0)}$ =200 nm and optimal phase flatness for $\xi \approx 2.7$, see Figure 5. Once phase stability is established, the design of the entire structure can proceed according to the scheme in Figure 6. The linear chirp $\Delta z^{(n)} = \lambda_g^{(n+1)} - \lambda_g^{(n)}$ is given by

$$\Delta z^{(n)} = \lambda_0 \Delta\beta = \frac{\lambda_0 \Delta\gamma}{\gamma^3 \beta} = \left[\frac{\lambda_0}{\beta\gamma^2} \frac{\Delta W(\varphi_s)}{\gamma m_e c^2} \right]^{(n)}, \tag{9}$$

where $\Delta W^{(n)} = m_e c^2 \Delta\gamma^{(n)}$ is the energy gain in the n-th grating period. The decreasing amplitude of $\underline{e}_1$ is taken into account by writing $\Delta z^{(n)} = \Delta z^{(1)} - (n-1)\Delta\Delta z$. Note that Δz and $\Delta\Delta z$ are averages and are thus not requirements for the fabrication precision. The slight change in the phase of the Fourier coefficients (Figure 5) leads to an identical change of the synchronous phase, causing a small additional energy spread increase. The acceleration ramp obtained by a CST tracking simulation is shown in Figure 7, where 72 particles represent a uniform distribution. The synchrotron motion is clearly visible, its initial period agrees roughly with $\lambda_s^{\text{init}} = 21\,\mu$m calculated by Equation (7).

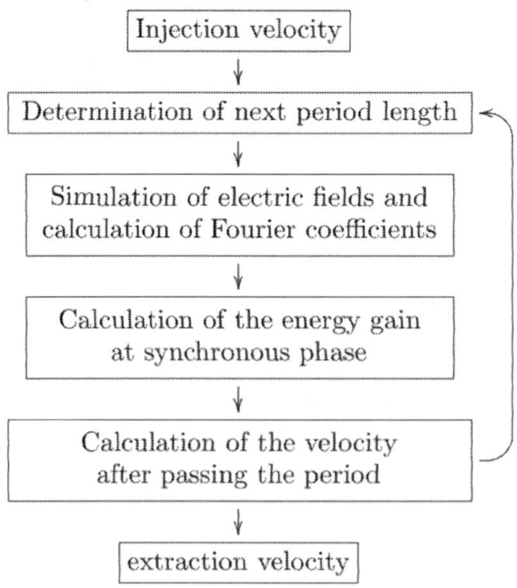

Figure 6. Iteration process for the design of a chirped grating

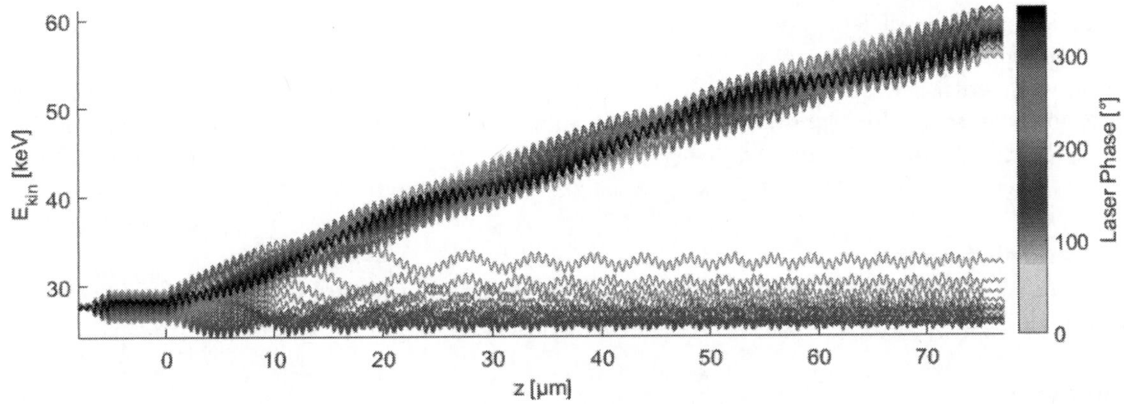

Figure 7. Energy gain along the structure for 72 particles launched at $t = 0$ sweeping over the laser phase in steps of 5 degrees. The fraction of trapped particles is 81%.

3. Optimized bunching

When taking into account finite transverse emittance, the acceleration defocusing plays a decisive role. Already in the modulator the electrons are strongly focused or defocused, depending on their arrival wrt. the laser phase. Adding a demodulator (same structure as the modulator but half a period displaced) at the end of the drift section will decrease the energy spread and the initially defocused particles will be focused and vice versa, see Figure 8. The headline of the transverse plots gives the percentage of particles that survived the aperture.

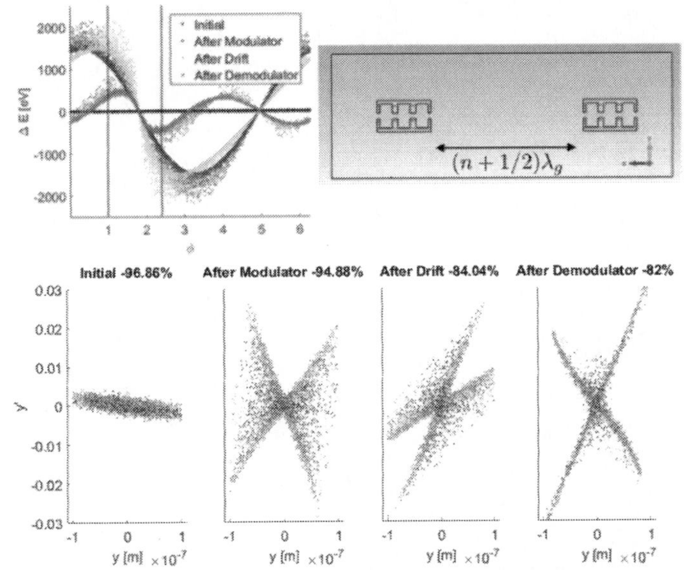

Figure 8. Longitudinal phase space and grating setup (top) and transverse phase space ($y' = \gamma\beta_y$), where the color indicates the particle phase φ (bottom).

IOP Conf. Series: Journal of Physics: Conf. Series **874** (2017) 012041 doi:10.1088/1742-6596/874/1/012041

4. Conclusion and Outlook

We introduced a novel "slightly resonant" DLA structure that combines Bragg waveguides and symmetric grating structures. We showed that the longitudinal dynamics in DLAs for low energy can be well controlled, in a similar manner as for conventional accelerators. The acceleration defocusing due to the high gradient can however not be compensated by solenoid or quadrupole magnets. Thus in future dedicated laser driven focusing schemes as outlined in the last section and approaches to the transverse dynamics have to be developed.

5. Acknowledgment

This work is funded by the Gordon and Betty Moore Foundation (Grant GBMF4744 to Stanford) and the German Federal Ministry of Education and Research (Grant FKZ:05K16RDB).

References

[1] ACHIP website: *https://achip.stanford.edu*

[2] Peralta E A *et al.* 2013 Demonstration of electron acceleration in a laser-driven dielectric microstructure *Nature* **503** 91–94

[3] Breuer J and Hommelhoff P 2013 Laser-Based Acceleration of Nonrelativistic Electrons at a Dielectric Structure *Phys. Rev. Lett.* **111** 134803

[4] Leedle K J *et al.* 2015 Laser acceleration and deflection of 96.3keV electrons with a silicon dielectric structure, *Optics Letters* **40** 18

[5] CST Microwave Studio *www.cst.com*

[6] Egenolf T, Boine-Frankenheim O and Niedermayer U 2017 Simulation of DLA grating structures in the frequency domain *Proc. 8th Int. Particle Accelerator Conf. (May 14–19 Copenhagen)* WEPVA002 (this issue)

Beam dynamics in THz dielectric-loaded waveguides for the AXSIS project*

T Vinatier[1,†], R W Assmann[1], U Dorda[1], F Lemery[2], B Marchetti[1]

[1] Deutsches Elektronen Synchrotron, Notkestrasse 85, Hamburg, Germany
[2] CFEL, Notkestrasse 85, Hamburg, Germany

† thomas.vinatier@desy.de

* Work supported by the European Research Council under the European Union's Seventh Framework Programme (FP/2007-2013)/ERC Grant Agreement n. 609920.

Abstract. In this paper, we investigate with ASTRA simulations the beam dynamics in dielectric-loaded waveguides driven by THz pulses, used as linac structure for the AXSIS project. We show that the bunch properties at the linac exit are very sensitive to the phase velocity of the THz pulse and are limited by the strong phase slippage of the bunch respective to it. We also show that the bunch properties are optimized when low frequencies (< 300 GHz) are used inside the linac, and that the longitudinal focal point can be put several tens of cm away from the linac exit thanks to ballistic bunching. However, a strong asymmetry in the bunch transverse sizes remains for which a solution is still to be found.

1. Introduction

Particle acceleration currently requires large infrastructures due to the low frequencies (a few GHz) and relatively low field amplitudes (a few tens of MV/m) used in conventional accelerating structures. One of the schemes currently studied to reduce the footprint of particle accelerators is to use structures driven by laser-generated THz fields, in which the frequencies (from 100 GHz up to 10 THz) and field amplitudes (100 MV/m to a few GV/m) are expected to be much higher than in conventional accelerating structures. This all-optical scheme would also allow avoiding the timing jitter between the photocathode laser generating the electron bunch and the accelerating field by using a unique laser as source for both. This is what the AXSIS project [1] aims for. In this project, the electron bunch will be produced by photoemission in a gun driven by single-cycle THz pulses [2, 3]. It will then be accelerated in a THz linac structure consisting in a dielectric-loaded waveguide driven by a multi-cycle THz pulse [4, 5, 6]. The goal is to obtain a 1 pC and 15 MeV electron bunch to produce attosecond X-ray pulses by inverse Compton scattering.

In this paper, we investigate with ASTRA simulations [7] the achievable bunch properties at the exit of various linac structures, using as input the bunch for which the simulated properties at the gun exit are shown in Table 1 and the longitudinal phase-space in Figure 1. In the first part, we briefly present the ASTRA model used for simulations. In the second part, we investigate the bunch properties as a function of the phase injection into the linac and of the phase velocity of the THz pulse. In the third part, the bunch properties as a function of the frequency and amplitude of the accelerating field are investigated.

Table 1. Input bunch properties used for the simulations.

Bunch property	Value at the gun exit
Charge	0.4 pC
Mean kinetic energy	1.48 MeV
Rms energy spread	82.9 keV
Rms length	43.5 μm (≡ 150.1 fs)
Rms horizontal size	28.9 μm
Rms vertical size	18.8 μm
Rms horizontal emittance	0.102 π.mm.mrad
Rms vertical emittance	0.083 π.mm.mrad

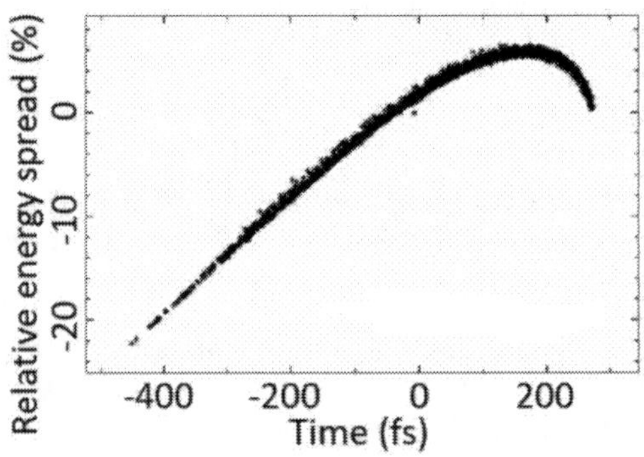

Figure 1. Bunch longitudinal phase-space at the gun exit.

2. The ASTRA model

The accelerating field in the THz linac is the TM_{01} mode [8]. It consists of a pulse travelling with group velocity v_g, having a phase velocity v_{ph} and containing a finite number N of wavelengths λ.

To simulate this field in ASTRA, we superimpose two identical standing waves phase-shifted by 90°. For a given frequency, the phase velocity of the accelerating field can then simply be adjusted by changing the wavelength of the two standing waves. In our simulations, we consider a flat-top time profile for the THz pulse. However, this can be changed by changing the profile of the two standing waves given as input to ASTRA.

The finite interaction length L of the electron bunch with the accelerating field, due to $v_g < c$ in the waveguide, is given by the following relation assuming that the bunch velocity remains close to c (speed of light in vacuum):

$$L = (N\lambda)/(1-(v_g/c))$$

This can simply be modelled in ASTRA by the length of the accelerating structure. We choose to fix it to 6.4 cm (N=27 at λ=1 mm (300 GHz) with v_g=0.62c) for the study as a function of frequency and amplitude of the field, and to 10 cm (N=42) for the study as a function of injection phase and phase velocity. Finally, the limited aperture of the dielectric loaded waveguide and the particle losses it can engender are taken into account in the ASTRA model.

The main physical effect not included in the ASTRA model is the fact that the THz pulse does not consist in a single frequency, but in a spectrum with a finite bandwidth (typically a few GHz at 300 GHz central frequency). Only the central frequency of the spectrum is considered in ASTRA, and therefore the distortion of the pulse due to group velocity dispersion is not included.

IOP Conf. Series: Journal of Physics: Conf. Series **874** (2017) 012042 doi:10.1088/1742-6596/874/1/012042

3. Influence of the injection phase and phase velocity

In order to study the bunch properties at the linac exit as a function of the injection phase into the linac and phase velocity of the THz pulse, we fix the field frequency to 300 GHz and its amplitude to 150 MV/m. Figure 2 shows the obtained bunch kinetic energy and inverse rms length, as 2D colormaps, in the phase velocity range [0.985*c*-1*c*].

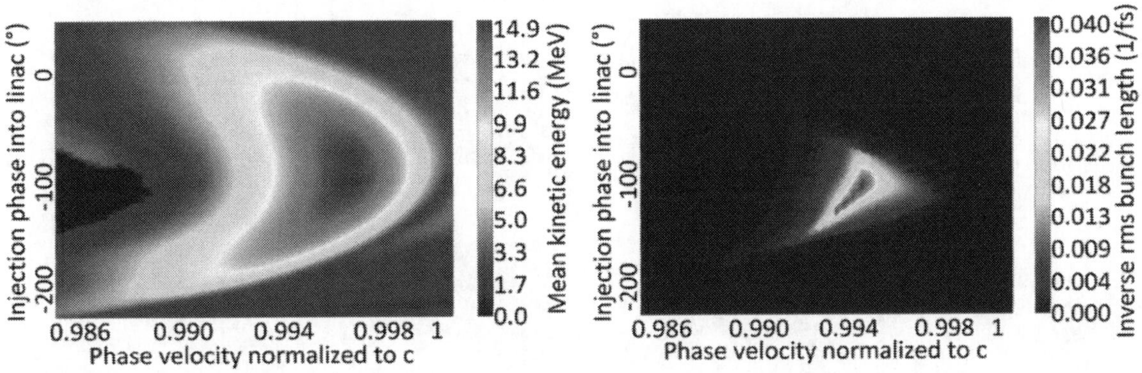

Figure 2. Bunch kinetic energy (left) and inverse rms length (right) at the linac exit as a function of the injection phase into the linac and phase velocity of the THz pulse.

Figure 2 shows that the bunch properties are highly sensitive to the phase velocity of the THz pulse, especially the bunch length, and therefore to the mechanical precision with which the dielectric-loaded waveguides can be produced. For example, a variation of the phase velocity from *c* to 0.99*c* would be generated by a variation of only 2 μm of the dielectric thickness in the linac. On the other hand, one can see that the bunch properties are less sensitive to the injection phase. Indeed one can see that, at optimal phase velocity, the energy gain remains close to the maximum (< 15% discrepancy) in a 90° window and the bunch length in a 30° window. This allows some margins for instabilities. However, the optima for these two bunch properties are obtained in different conditions. Compromises or choices have therefore to be made.

One can see in Figure 2 that the optimal phase velocity for energy gain is not *c* but lower (around 0.9965*c*). This is explained by the relatively low injection energy of the bunch into the linac (1.5 MeV), which leads to significant phase slippage respective to the accelerating field. This is visible in Figure 3, which presents the evolution of the local energy gain of the bunch reference particle along its path in the linac. This phase slippage also explains why the phase velocity for optimizing the bunch length is also below *c* (around 0.994*c*).

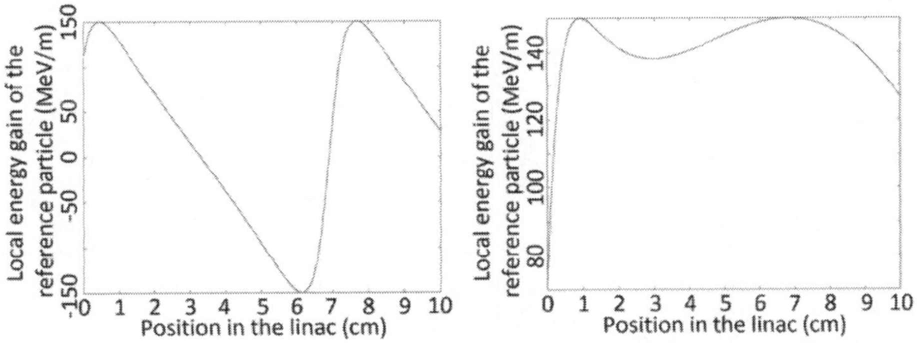

Figure 3. Local energy gain of the reference particle in the linac. Phase velocity: *c* (left) and 0.9965*c* (right). Injection phase providing the maximum energy gain.

4. Influence of the frequency and field amplitude

In order to study the bunch properties at the linac exit as a function of the frequency and amplitude of the accelerating field, we fix the phase velocity of the THz pulse to *c* and the injection phase into the linac has been for each point chosen as the one optimizing the longitudinal bunch compression at the linac exit. Figure 4 shows the obtained bunch properties, as 2D colormaps, in the frequency range [90 GHz-500 GHz] and amplitude range [50 MV/m-400 MV/m].

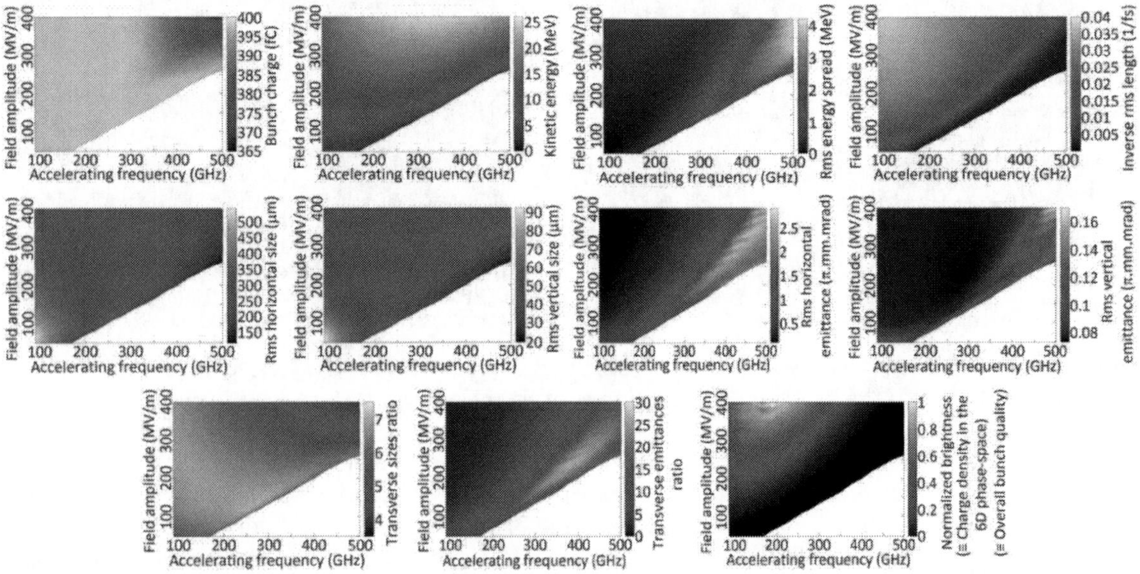

Figure 4. Bunch properties at the exit of the THz linac, simulated by ASTRA, using as input the bunch which properties are shown in Table 1. The white zone in the graphics corresponds to the case where the phase slippage of the bunch respective to the accelerating field exceeds 360° during its path inside the linac.

The results in Figure 4 show that the bunch longitudinal properties, rms length and energy spread, are optimized when low frequencies (< 300 GHz) are used. The high frequencies (> 300 GHz) provide worse longitudinal bunch properties, and high field amplitudes are required to obtain acceptable properties. There are two reasons explaining that. Firstly, the phase slippage, previously introduced (see Figure 3) is a limitation for the longitudinal bunch properties, and it is intrinsically reduced when using lower frequencies. Secondly, the initial bunch length (see Table 1) is a significant fraction of the wavelength of the accelerating field, which increases for higher frequencies. This is a major source of the induced energy spread, which can grow up to several MeV above 400 GHz, and also of the longitudinal phase-space curvature which limits the longitudinal bunch compression and therefore the final bunch length. As shown in Figure 4, the zone of high frequency and amplitude also has the drawbacks to engender particles losses due to the smaller transverse radius of the linac and the higher value of the transverse electric field which induces a transverse defocusing of the bunch.

The situation is more complicated for the transverse bunch properties, transverse sizes and emittances in the horizontal and vertical directions. Although the transverse emittances are also minimized at low frequencies, one can see that the transverse asymmetries of sizes and emittances, created in the gun (see Table 1) due to a transverse asymmetry of the field pattern, behave differently. The one for the transverse sizes is minimized for high frequencies, while the one for the transverse emittances is minimized for low frequencies. Only the combination of high frequency and field amplitude could minimize both. But as previously mentioned, this zone leads to a strong deterioration of the longitudinal bunch properties and transverse emittances, and particle losses occur.

A compromise or a choice has therefore to be made between the optimization of the longitudinal bunch properties and the minimization of particle losses and of asymmetries of transverse sizes and emittances. One solution to overcome this difficulty would be to put the gun in a strong magnetic field (several T) parallel to the bunch propagation axis. This will confine the bunch in the gun and remove the asymmetries of its transverse properties. But the non-zero magnetic field during the bunch emission from the cathode will magnetize it and strongly deteriorate its transverse emittance, making it more difficult to transport. This solution is therefore questionable. Another solution would be to use a tapered linac structure [9], which would allow dynamically matching the phase velocity of the THz pulse to the bunch velocity. The phase slippage would then be controlled, which would allow using higher frequencies and field amplitudes without deteriorating the longitudinal bunch properties.

The results presented up to now are right at the exit of the linac structure. But the goal of the AXSIS project is to achieve a short bunch length at a certain distance (typically a few tens of cm) from the linac exit, to allow a transverse focusing of the bunch after the linac in order to use it to produce short and bright X-ray pulses. The capability of the linac structure to perform ballistic bunching, therefore to adjust the position of the bunch longitudinal focal point, has also be investigated. Figure 5 displays the evolution of the position of the longitudinal focal point and of the bunch length at this point as a function of the field amplitude for several frequencies.

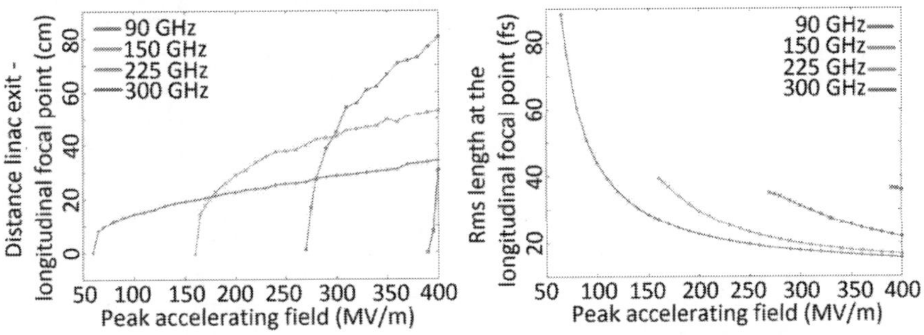

Figure 5. Longitudinal focal distance (left) and rms bunch length (right) at the longitudinal focal point as a function of the field amplitude in the linac for several frequencies.

Figure 5 shows that, for field amplitude < 400 MV/m, the ballistic bunching process can only be performed at low frequencies (< 300 GHz) and in this case longitudinal focal distances of several tens of cm can be reached. For higher frequencies, the phase slippage of the bunch respective to the accelerating field is too fast and prevents ballistic bunching to occur. This is another strong argument in favour of low operating frequencies (< 300 GHz) for the THz linac intended to be use in the AXSIS project.

5. Conclusion

Our simulations demonstrate that a proper energy gain, a control of the longitudinal focal distance and margins for instabilities of the injection phase are possible with THz-driven dielectric loaded waveguides. These results will be useful for the implementation of the AXSIS project. However, improvements have still to be made to meet the final requirements of the AXSIS project. Especially, the lowest achieved bunch length in our simulations ($\approx$ 15 fs rms) is still far from the requirements ($\leq$ 1 fs rms). This is due to the already strong curvature of the bunch longitudinal phase-space at the gun exit (see Figure 1), which limits the compression performances in the linac. This curvature is caused by the high-frequency field (300 GHz) used to accelerate the electron bunch in the gun [3]. The straightforward and most efficient way to reduce it, and therefore to improve the bunch compression in the linac, would be to use lower frequencies in the gun. For this purpose, the studies to use more

IOP Conf. Series: Journal of Physics: Conf. Series **874** (2017) 012042 doi:10.1088/1742-6596/874/1/012042

classical guns, like DC-guns or S-band RF-guns, as source to then inject the bunch into a THz linac are also currently ongoing.

References

[1] Kärtner F X *et al.* 2016 AXSIS: Exploring the frontiers in attosecond X-ray science, imaging and spectroscopy *Nucl. Instr. Meth.* A **829** pp 24–9
[2] Hoffmann M C and Fülöp J A 2011 Intense ultrashort terahertz pulses: generation and applications *J. Phys.* D **44** 083001
[3] Fallahi A, Fakhari M, Yahagi A, Arietta M and Kärtner F X 2016 Short electron bunch generation using single-cycle ultrafast electron guns *Phys. Rev. ST Accel. Beams* **19** 081302
[4] Lee Y S and Hurblut W C 2006 Generation of multicycle terahertz pulses via optical rectification in periodically inverted GaAs structures *Applied Physics Letters* **89** 181104
[5] Ravi K *et al.* 2016 Cascaded parametric amplification for highly efficient terahertz generation *Optics Letters* **41** (16) pp 3806–9
[6] Ravi K, Barre D and Kärtner F X 2016 Pulse sequences for efficient multi-cycle terahertz generation in periodically poled lithium niobate *Optics Express* **24** pp 22582–607
[7] Flöttmann K 2017 *Astra manual* DESY Hamburg Germany
[8] Zaki K A and Atia A E 1983 Modes in dielectric-loaded waveguides and resonators *IEEE Transactions on Microwave Theory and Techniques* **31** (12) pp 1039–45
[9] Lemery F, Flöttmann K, Piot P and Kärtner F X 2017 *Proc. Int. Part. Accel. Conf. (Copenhagen)* (Jacow) WEPAB 123

Laser proton accelerator with improved repeatability at Peking University

Y Shou, Y Geng, Q Liao, J Zhu, P Wang, M Wu, C Li, X Xu, R Li, H Lu, Y Zhao, W Ma, C Lin and X Yan

State Key Laboratory of Nuclear Physics and Technology, Peking University, Beijing, China

x.yan@pku.edu.cn

Abstract. The repeatability of laser proton accelerator is mainly limited by laser plasma interaction, laser target coupling and laser parameter variation. In our recent experiments performed on the Compact Laser Plasma Accelerator at Peking University, gain of proton beams with improved repeatability is demonstrated. In order to control the laser plasma interaction in pre-plasma, cross polarized-wave generation technique is employed to provide a laser pulse with an ultrahigh contrast of 10^{-10}. A semi-automatic laser and target alignment system with a sensitivity of few microns is employed. The repetition rate of the laser pro-ton accelerator is at the level of 0.1 Hz which is beneficial to decrease laser parameter variation. The shot-to-shot variation of proton energies is about 9% for a level of confidence of 0.95.

1. Introduction

Thanks to the development of laser technologies, laser plasma accelerators have made great progress since first proposed in 1979 [1]. Several main laser proton acceleration mechanisms have been studied theoretically and experimentally, such as target normal sheath acceleration (TNSA) [2-4], radiation pressure acceleration (RPA) [5-7], collisionless shock acceleration (CSA) [8, 9] and break-out afterburner acceleration (BOA) [10]. Generation of 93 MeV proton beams using femtosecond laser pulse has been reported in 2016 [7].

Laser proton accelerators are attractive for building compact particle accelerators, since they have three orders higher acceleration gradient compared to conventional accelerators. Proton beams gained from laser pro-ton accelerators have the advantages of short duration, small emittance and extreme brightness, which make them suitable for wide potential applications including cancer therapy [11], proton imaging [12], nuclear reactions [13] and injectors for conventional accelerators [14]. However, current laser-driven proton beams show wide energy spread and poor repeatability, which limit those application [15]. Great efforts have been made in improving the repeatability of laser proton accelerators [16-18]. Recently Y. Gao et al. have demonstrated experiments with 0.5 Hz repetition rate and realized shot-to-shot variation of about 27% for a level of confidence of 0.95 [19].

The shot-to-shot variations of laser-driven proton beams are mainly due to instabilities in laser plasma interaction, inaccuracies from laser target coupling and fluctuations of laser parameters. In our recent experiments performed on the Compact Laser Plasma Accelerator (CLAPA) at Peking University [20, 21], gain of proton beams with improved repeatability is demonstrated. Figure 1 shows the layout of our laser proton acceleration system. With the help of cross polarized-wave (XPW) system for pulse cleaning, the laser pulse's nanosecond contrast is at the level of 10^{-10}. A semi-automatic target alignment system with a sensitivity of 5 μm is employed to ensure the laser target coupling accuracy. Recently, the shot-to-shot variation of the cut-off proton energies has been reduced to about 9% for a level of confidence of 0.95.

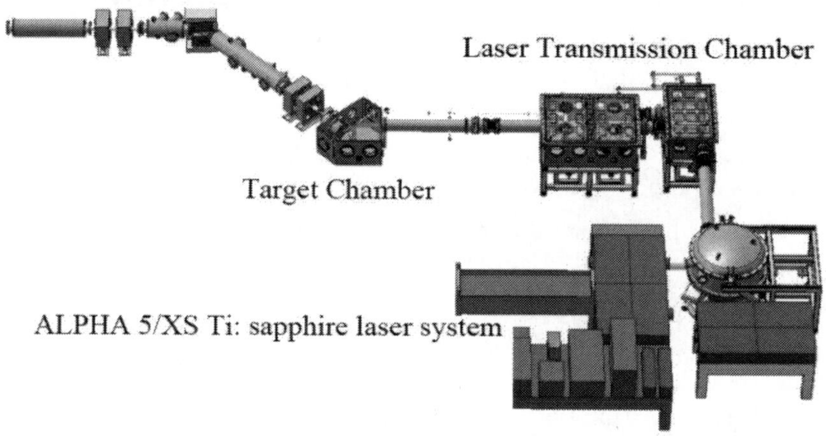

Figure 1. Layout of the CLAPA laser proton acceleration system.

2. CLAPA laser system

The CLAPA laser system is based on Ti: Sapphire with double CPA structure. The laser pulse has a central wave-length of 800 nm and a bandwidth of 45 nm. KHz pulses are generated through the first stage of CPA from an 82 MHz oscillator. In order to enhance the laser temporal contrast, XPW system for pulse cleaning is employed before the second CPA stage. A pulse with total energy of 7 J can be achieved in the output of the last amplifier, which is then sent into a vacuum compressor. A deformable mirror is mounted before the compressor to correct wavefront aberrations. Finally, a 5 J laser pulse with full width at half maximum (FWHM) pulse duration of 25 fs can be achieved. Measurements using free space InGaAs detector with a high speed oscilloscope show that the laser pulse's nanosecond contrast is at the level of 10^{-10}. Picosecond time scale contrast was measured using a third order scanning autocorrelator. The results show that the laser contrast is 10^{-10} at 40 ps before the main pulse and prepulses appear at -38 ps as well as -19 ps with contrast 10^{-9}. The shot-to-shot fluctuations of laser pulse energy are less than 2% in our 5 Hz repetition frequency tests.

3. Target alignment system

Several alignment techniques, for instance, microscope system [18], transverse optical shadowgraphy [22] and retro-focusing technique [23] have been proposed for precise laser target coupling. In our experiment, we employ a target alignment system consisting of two main components: one target part and one microscope part. The target part is a special-designed target wheel con-trolled by five-axis motorized stages. Figure 2 displays the schematic of the microscope system controlled by three-axis motorized linear stages. A 50-fold long working distance microscopic objective is employed to acquire magnifying images of the target rear surface. Two different magnifications 10-fold (CCD1) and 50-fold (CCD2) can be achieved simultaneously with the help of a beam splitter. The 10-fold camera has the advantage of larger field of view, which is beneficial for seeking the laser spot. At the same time, the 50-fold camera is employed for laser spot optimization and high precision laser target coupling.

IOP Conf. Series: Journal of Physics: Conf. Series **874** (2017) 012043 doi:10.1088/1742-6596/874/1/012043

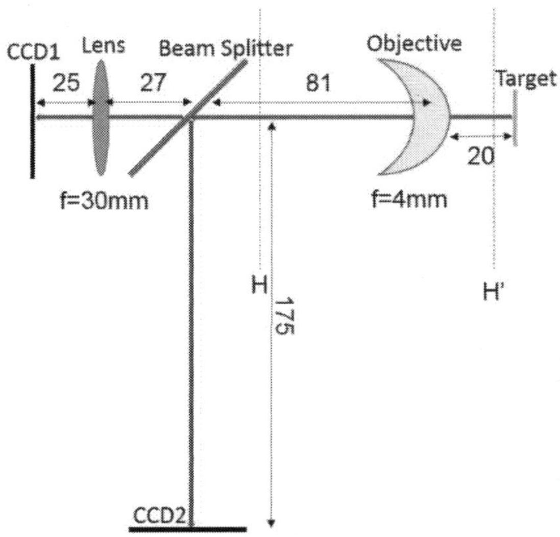

Figure 2. Schematic of the microscope system. The omitted unit is millimetre. Two different magnifications 10-fold (CCD1) and 50-fold (CCD2) can be achieved simultaneously with the help of a beam splitter.

Figure 3. Real picture of the target alignment system.

Real picture of the target alignment system is shown in Figure 3. Laser target coupling is realized by three steps. Firstly, the microscope system should be moved to a reference position based on the Thomson parabola spectrometer. Secondly, we move every target to the focal plane of the objective. Images of the target rear surface with different target thicknesses and materials are shown in Figure 4. Thirdly, the off axis parabolic (OAP) mirror is optimized to position the laser waist on the focal plane.

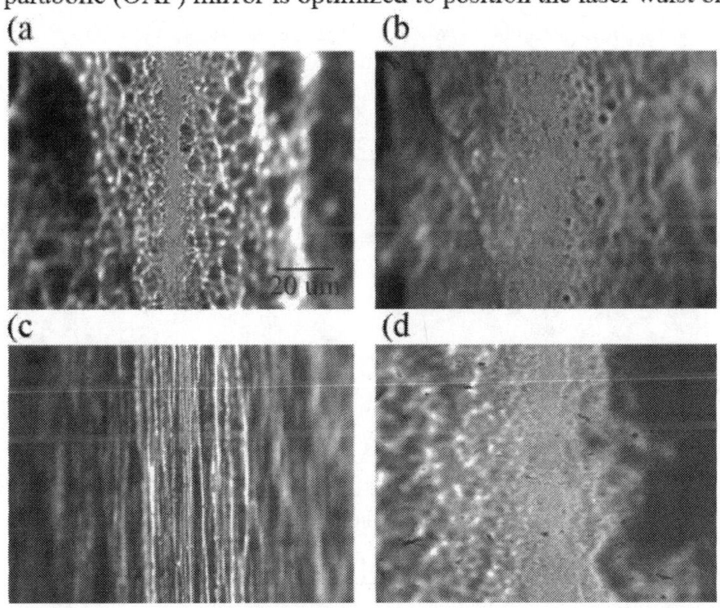

Figure 4. Images of the target rear surface. (a. 1.2 μm thick aluminium target, (b. 2.5 μm thick aluminium target, (c. 3 μm thick titanium target, (d. 50 nm thick diamond-like carbon target, respectively. The scales of the figures are the same.

The target alignment system is hidden in a safe place during shots. We have improved the repetition rate by recording the coordinates of all targets ahead of the first shot. What we should do before a shot is just moving the target wheel. However, a two-stage laser shutter system is employed for the sake of safety, which limits the further improvement of repetition rate. New shutter system is in development to resolve this problem.

The main source of uncertainty between target and laser waist is due to the objective's depth of focus and the laser's angle of incidence. The precision of the target alignment system is within 5 μm.

4. Experiment results

Preliminary experiment has been performed to study the repeatability of laser-driven proton beams. The experimental arrangement is shown in the Figure 5. A p-polarized, 25 fs and energy of nearly 0.8 J laser pulse was focused using an f/3.5 OAP on a 0.8 μm thick aluminium target at an angle of 45°. The repetition period is 30 seconds per shot.

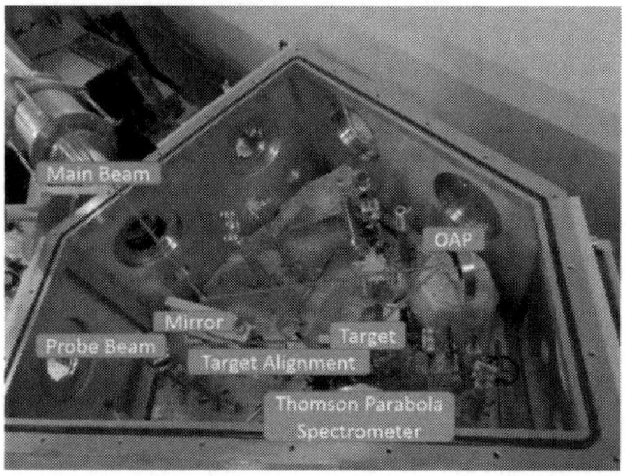

Figure 5. Target chamber for laser proton acceleration.

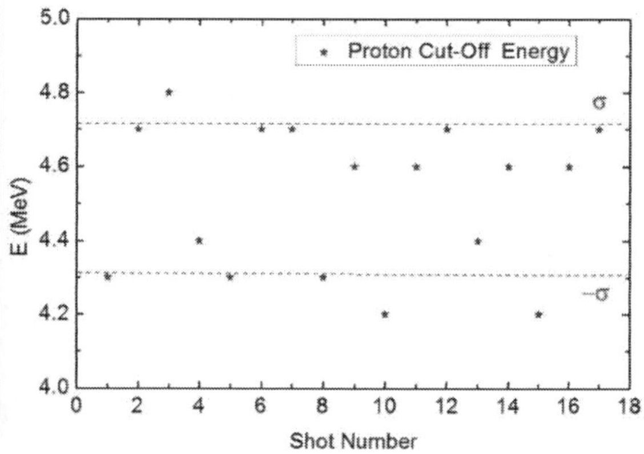

Figure 6. The shot-to-shot results of proton cut-off energy.

The results of 17 continuous shots are displayed in Figure 6. Proton beams with average cut-off energy of 4.5 MeV and shot-to-shot variation of about 9% have been demonstrated.

5. Conclusion

In conclusion, we have performed experiments with improved repeatability on the laser proton accelerator at Peking University. Thanks to the ultrahigh laser contrast, the accurate target alignment system and the high repetition rate, 9% shot-to-shot variation of proton energies has been successfully demonstrated.

Acknowledgments

The CLAPA has been designed and built with the effort of engineers and graduate students in the past few years. The authors would like to thank all the partners and contributors.

This work was supported by National Basic Research Program of China (Grant No. 2013CBA01502), National Natural Science Foundation of China (Grants No. 11575011) and National Grand Instrument Project (2012YQ030142).

References

[1] Tajima T and Dawson J M 1979 Laser Electron Accelerator *Phys. Rev. Lett.* **43** 267-70
[2] Pukhov A 2001 Three-dimensional simulations of ion acceleration from a foil irradiated by a short-pulse laser *Phys. Rev. Lett.* **86** 3562-5
[3] Wilks S *et al.* 2001 Energetic proton generation in ultra-intense laser–solid interactions *Phys. Plasmas* **8** 542-9

[4] Kim I J *et al.* 2013 Transition of proton energy scaling using an ultrathin target irradiated by linearly polarized femtosecond laser pulses *Phys. Rev. Lett.* **111** 165003

[5] Esirkepov T *et al.* 2004 Highly efficient relativistic-ion generation in the laser-piston regime *Phys. Rev. Lett.* **92** 175003

[6] Yan X *et al.* 2008 Generating high-current monoenergetic proton beams by a circularly polarized laser pulse in the phase-stable acceleration regime *Phys. Rev. Lett.* **100** 135003

[7] Kim I J *et al.* 2016 Radiation pressure acceleration of protons to 93 MeV with circularly polarized petawatt laser pulses *Phys. Plasmas* **23** 070701

[8] Silva L *et al.* 2004 Proton shock acceleration in laser-plasma interactions *Phys. Rev. Lett.* **92** 015002

[9] Habara H *et al.* 2004 Ion acceleration from the shock front induced by hole boring in ultraintense laser-plasma interactions *Phys. Rev.* E **70** 046414

[10] Yin L *et al.* 2011 Three-dimensional dynamics of breakout afterburner ion acceleration using high-contrast short-pulse laser and nanoscale targets *Phys. Rev. Lett.* **107** 045003

[11] Bolton R 2016 The integrated laser-driven ion accelerator system and the laser-driven ion beam radiotherapy challenge *Nucl. Instr. Meth.* **809** 149-155

[12] Borghesi M *et al.* 2003 Proton imaging detection of transient electromagnetic fields in laser-plasma interactions *Rev. Sci. Instrum.* **74** 1688-93

[13] Romagnani L *et al.* 2005 Dynamics of electric fields driving the laser acceleration of multi-MeV protons *Phys. Rev. Lett.* **95** 195001

[14] Busold S *et al.* 2015 Towards highest peak intensities for ultra-short MeV-range ion bunches *Sci. Rep.* **5** 12459

[15] Linz U and Alonso J 2007 What will it take for laser driven proton accelerators to be applied to tumor therapy *Phys. Rev. ST Accel. Beams* **10** 094801

[16] Tajima T, Dietrich H and Yan X 2009 Laser acceleration of ions for radiation therapy *Rev. Accl. Sci. Tech.* **2** 201-28

[17] Kraft S D *et al.* 2010 Dose-dependent biological damage of tumour cells by laser-accelerated proton beams *New J. Phys.* **12** 085003

[18] Singh K *et al.* 2016 A diagnostic for micrometer sensitive positioning of solid targets in intense laser-matter interaction *Nucl. Instr. Meth.* **829** 363-6

[19] Gao Y *et al.* 2017 An automated 1 Hz nano-target positioning system for intense laser plasma experiments submitted for publication.

[20] Liao Q *et al.* 2016 Recent progress of proton acceleration at Peking University *Proc. IPAC'16* (Busan) (Geneva: JACoW) pp 1588-91

[21] Li R *et al.* 2016 Compact laser plasma accelerator at Peking University *Proc. IPAC'16* (Busan) (Geneva: JACoW) pp 2569-71

[22] Lancaster K L *et al.* 2009 Temperature profiles derived from transverse optical shadowgraphy in ultraintense laser plasma interactions at 6×10^{20} W cm^{-2} *Phys. Plasmas* **16** 056707

[23] Booth N *et al.* 2013 Target positioning and alignment on the Astra-Gemini facility *Proc. SPIE* (San Diego) (Bellingham: SPIE) p 885001

Transport channel of secondary ion beam of experimental setup for selective laser ionization with gas cell GALS

G G Gulbekyan[1], S G Zemlyanoy[1], V V Bashevoy[1], I A Ivanenko[1], N Yu Kazarinov[1], V I Kazacha[1] and N F Osipov[1]

[1] Joint Institute for Nuclear Research, Dubna, Moscow region, 141980, Russia

nyk@jinr.ru

Abstract. GALS is the experimental setup intended for production and research of isobaric and isotopically pure heavy neutron-rich nuclei. The beam line consists of two parts. The initial part is used for transport of the primary ^{136}Xe ion beam with the energy of 4.5-9.0 MeV/amu from the FLNR cyclotron U-400M to the Pb target for production of the studying ion beams. These beams have the following design parameters: the charge Z = +1, the mass A = 180-270 and the kinetic energy W = 40 keV. The second part placed after the target consists of the SPIG (QPIG) system, the accelerating gap, the electrostatic Einzel lens, 90-degree spectrometric magnet (calculated value of the mass-resolution is equal to 1400) and the beam line for the transportation of the ions from the magnet focal plane to a particle detector. The results of simulation of the particle dynamics and the basic parameters of all elements of the beam line are presented.

1. Introduction

The experimental setup GALS is intended for research of isobaric and isotopically pure heavy neutron-rich nuclei [1].

The beam of ^{136}Xe ions, accelerated in the FLNR cyclotron U400M to the energies of 4.5-9.0 MeV/amu, is transported to the Pb target that is a source of studying secondary ion beams. The primary beam transportation line has the length of 17.9 m from the ion extraction point to the target.

The secondary ion beam has the following design parameters: the charge Z=1, the mass number A=180-270, the kinetic energy W=40 keV and the current up to 1 μA.

The transport channel of the secondary ion beam (see Figure 1) is located just after the Pb target.

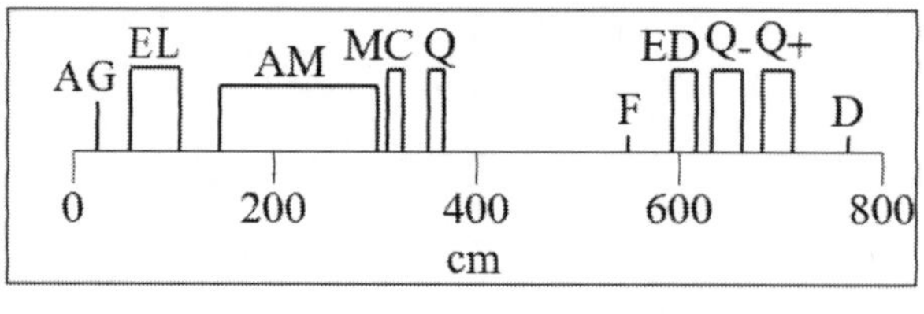

Figure 1. Scheme of the secondary ion beam line

It consists of the SextuPole Ion Guide (SPIG) system that can be added by QuadruPole Ion Guide (QPIG) system, the accelerating gap (AG), an electrostatic Einzel-lens (EL), the 90^0 spectrometric magnet (AM), the magnetic corrector (MC), the correcting quadrupole lens (Q), the electrostatic deflector (ED) and the magnetic quadrupole doublet (Q_-,Q_+).

The results of 3D simulation of the particle dynamics in the spectrometric magnet, calculations of the deflector, and the main parameters of all optical elements of the beam line transporting the secondary ions are given.

2. Primary ion beam line

Scheme of the beam line for transportation of the primary $^{136}Xe^{47+}$ ion beams is shown in Figure 2.

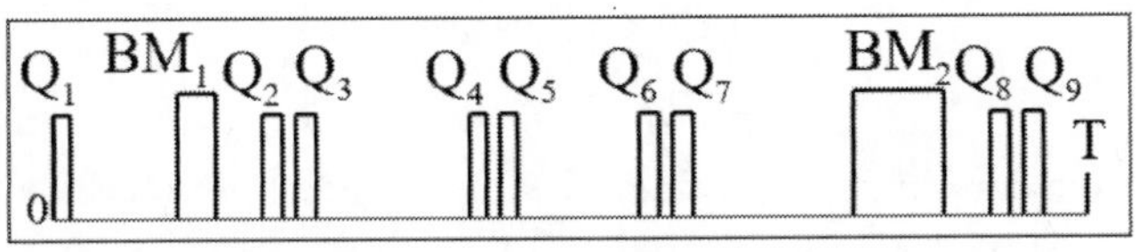

Figure 2. Scheme of the beam line for transportation of $^{136}Xe^{47+}$ primary beam. $BM_{1,2}$ – bending magnets, $Q_{1\text{-}9}$ – quadrupole lenses, T – Pb target

In calculations, the point of the beam extraction from the cyclotron (point "0" in Figure 2) coincides with the entrance of the quadrupole Q_1.

All quadrupole lenses except of lens Q_1 have the effective length l_{eff} = 35 cm and maximal gradient G_{max}=600 Gs/cm. Quadrupole Q_1 has l_{eff} = 30 cm and G_{max} = 190 Gs/cm.

BM_1 is 38^0 sector bending magnet with bending radius of ρ = 1 m. BM_2 is 90^0 bending magnet having ρ =1 m and pole face rotation angles = 21.5^0.

The horizontal and vertical RMS beam emittances are equal to 4.67 and 1.5 π mm mrad respectively. The momentum spread (4σ) of the ions is equal to 2%.

The calculated envelopes (2σ) and horizontal dispersion function of the $^{136}Xe^{47+}$ ion beam with the energy of 9 MeV/amu are given in Figure 3.

The values of envelopes do not exceed 80% of the beam line aperture. The quadrupole gradients do not exceed their designed maximum values.

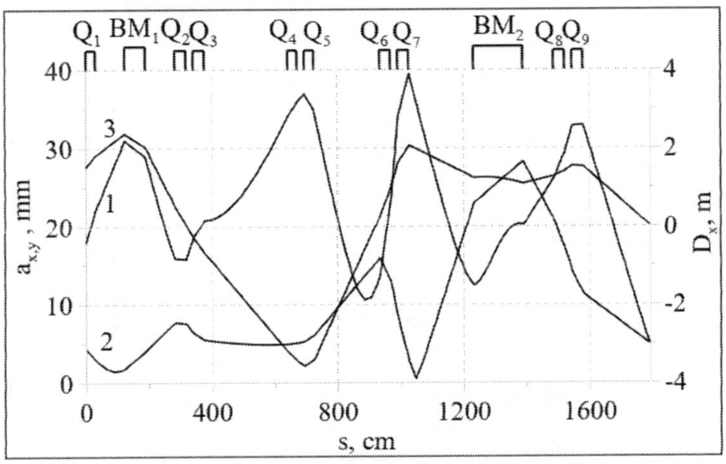

Figure 3. Horizontal a_x (curve 1), vertical a_y (curve 2) envelopes and dispersion function Dx (curve 3) of $^{136}Xe^{47+}$ beam

3. Secondary beam acceleration

Acceleration of the ion beam in the GALS spectrometer is fulfilled in two gaps. The first one with the accelerating voltage of U_0 = 1 kV is placed at the exit of the SPIG system. The second accelerating gap (AG in Figure 1) with the applied voltage of U = 40 kV is placed at the distance Δ = 40 mm from the first one.

The object point of the spectrometric magnet AM (the starting point "0" of the beam motion in the calculation) is shifted by distance L from the second gap against the beam movement:

$$L = \Delta\left(\frac{U}{U_0}\right)^{1/2} = 253 \text{ mm}$$

The initial transverse ion coordinates and velocities correspond to the SPIG system parameters.

4. Secondary beam transport

The beam diameter at the exit of the SPIG system [1] d_b is equal to 1 mm, the beam emittance for the accelerating voltage U=40 kV is evaluated as 3π mm mrad.

After acceleration, the ion beam is directed to the analyzing magnet (AM) where the ions are separated in accordance with their masses. The scheme of the beam line for the secondary ion transportation (see Figure 1) was designed taking into account the results of carried out calculations. The total beam line length is equal to 7.46 m.

4.1. Electrostatic Einzel lens EL

The total length of the electrostatic lens is equal to 50 cm, the maximum value of the voltage at the central electrode V0= 10 kV. The Einzel lens is designed to correct the position of the AM focus (the point F moves upstream the beam).

4.2. Spectrometric magnet AM

The main parameters of the spectrometric magnet were chosen as follows – rotation angle $\varphi = 90°$, ρ = 1 m, maximum magnetic rigidity $B\rho$ = 0.5 T m, the width of the horizontal working region is equal to 100 mm and the gap between the magnet poles is equal to 60 mm. The pole face rotation angles ε_{in} = +50.2^0 and ε_{out} = +6.05^0 were found with the help of the COSY INFINITY code [2].

4.3. Correcting magnet MC

The correcting magnet MC is placed at the exit of the spectrometric magnet AM. The correction is needed to eliminate the influence of the asymmetry of the AM field (see Fig. 4), resulting in a distortion of the equilibrium orbit [3]. It has the main parameters – l_{eff} =15 cm and the maximum magnetic field is equal to 400 Gs.

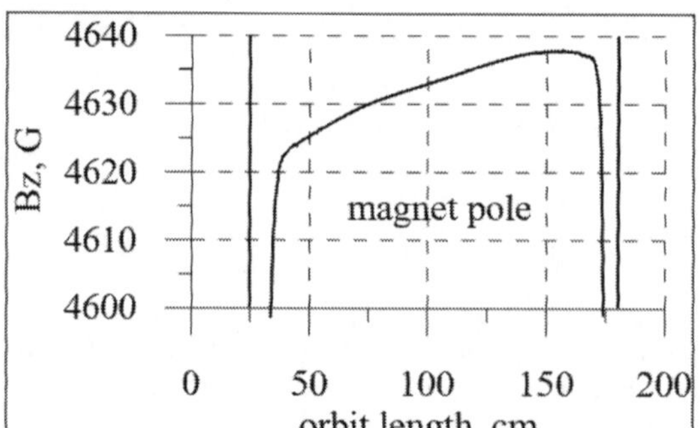

Figure 4. Detailed B_z field distribution

4.4. Magnetic quadrupole Q

The magnetic quadrupole Q , located at a distance of 50 cm from the AM exit, is designed to adjust the position of the AM focal point F and has the following main parameters: d = 80 mm; l_{eff} = 25 cm, G_{max}= 30 Gs/cm.

4.5. Electrostatic deflector ED

The cylindrical electrostatic deflector ED [4] deflects the separated ion beam with to the particle detector D. The deflector scheme is shown in Figure 5.

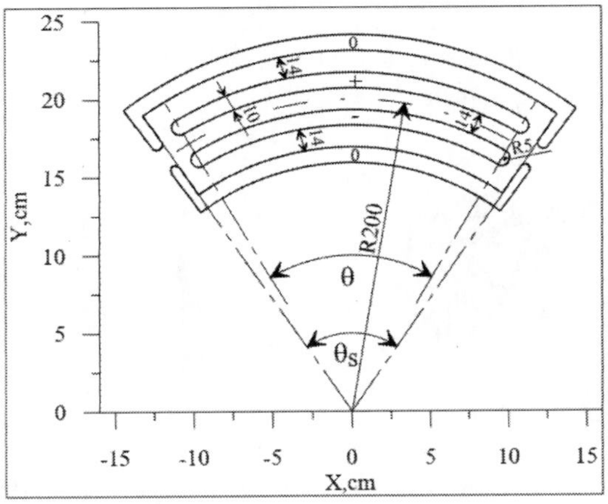

Figure 5. Scheme of the cylindrical deflector GALS

The deflector consists of two electrodes under potentials U_1 (-), U_2 (+) and two grounded screens (0). Its main parameters are – ρ =20 cm, φ =67^0 and the gap between the electrodes h=14 mm. The optimal value of the angular size of the electrodes θ is equal to 62.6^0. The optimal angular distance between the inner end face of the screens is equal to θs = θ+8^0.

The radial thickness of the electrode is equal to 10 mm; the radius of the electrode curvature is equal to 5 mm.

The voltages at the electrodes corresponding to maximum accelerator voltage of GALS setup are equal to U_1 = -2.87 kV and U_2 = 2.78 kV.

4.6. Magnetic quadrupoles Q_-,Q_+

The magnetic quadrupoles Q_-,Q_+ are intended to form an ion beam with the diameter of 5 mm at the particle detector D that is located at the distance of 216.8 cm from the focal plane F of the magnet. The parameters of quadrupoles are – d=40 mm, l_{eff} = 30 cm, the gradients G_-= 300 Gs/cm and G_+=500 Gs/cm.

5. Beam dynamics in spectrometric magnet

3D simulation of the beam dynamics in the GALS spectrometer was carried out in [3]. This simulation used 3D map of the AM magnetic field calculated using OPERA 3D code [5]. The number of particles used in this simulation was equal to 2 10^4. The computational model of the analyzing magnet is shown in Figure 6. The initial particle distribution in the configuration space is shown in Figure7. It corresponds to the steady-state exponential distribution of the ion beam with diameter of d_b in the averaged potential of the SPIG system. The distribution is truncated at the boundary value of this potential.

IOP Conf. Series: Journal of Physics: Conf. Series **874** (2017) 012044 doi:10.1088/1742-6596/874/1/012044

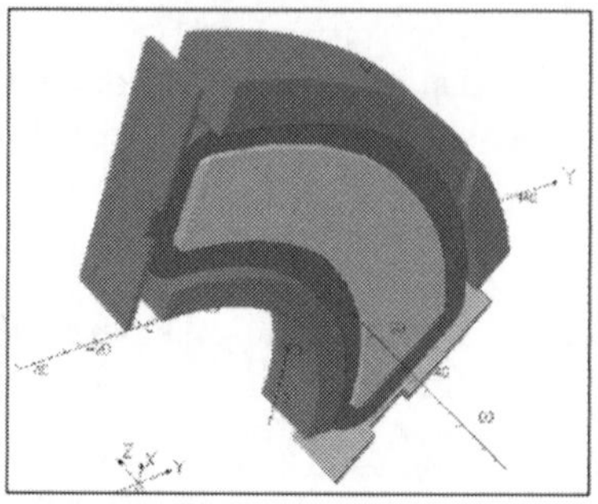

Figure 6. Computational model of the analyzing magnet

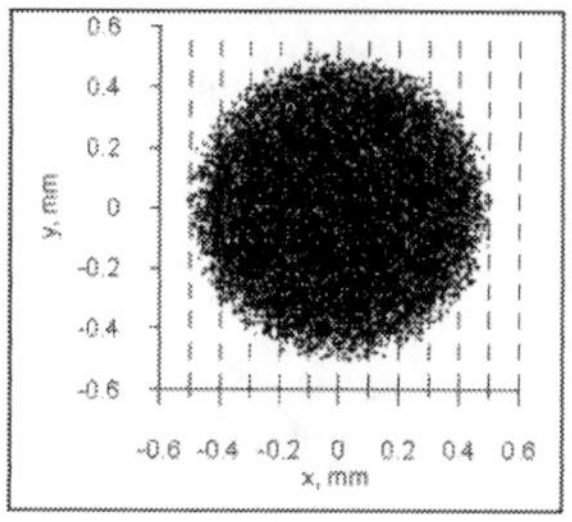

Figure 7. Initial particle distribution in plane {x,y}

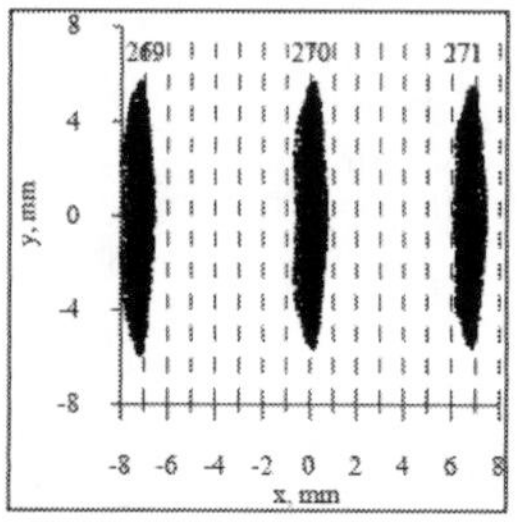

Figure 8. Ion distribution in the magnet focal plane

Figure 8 shows the calculated distribution of the particles in the AM focal plane for the ions with A=269, 270, and 271. The estimated mass resolution of the spectrometer Rm = 1400. The resolution of magnet maybe lower because of fast oscillations of the ions in the RF potential that increases the initial particle angles.

The horizontal and the vertical 100% beam envelopes of the 270^{1+} ion are shown in Figure 9. The emittance growth due to magnet field nonlinearities is acceptable and does not exceed 25%.

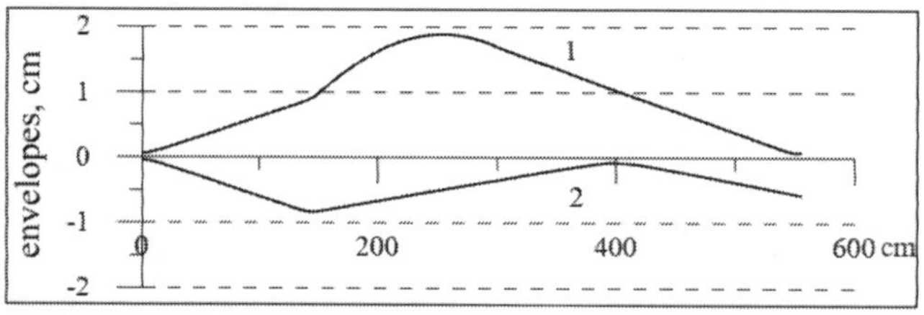

Figure 9. Horizontal (curve 1) and vertical (curve 2) 100% beam envelopes of 270^{1+} ions

6. Calculation results of secondary ion transportation

The dependences of horizontal (curve 1) and vertical (curve 2) ion beam envelopes on distance s along the beam line are shown in Figure 10.

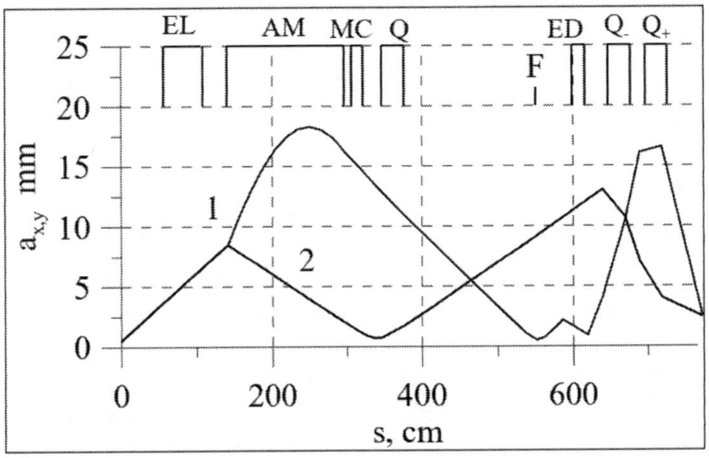

Figure 10. Horizontal (curve 1) and vertical (curve 2) beam envelopes along the beam line

References

[1] Zagrebaev V I, Zemlyanoy S G, Kozulin E M, Kudryavtsev Yu, Fedosseev V, Bark R, Janas Z and Othman H A 2014 Gas-cell-based setup for the production and study of neutron rich heavy nuclei *Hyperfine Interact.* **227** 181-9

[2] Makino K and Berz M 2005 COSY INFINITY version 9 *Nucl. Instr. and Meth.* A **558(1)** 346-50

[3] Kazarinov N Yu 2016 3D Simulation of the Ion-Beam Transport in Bending Magnets and Electrostatic Deflectors *Phys. of Part. and Nucl. Lett.* **13** N7 868-73

[4] Kazarinov N, Kalagin I and Zemlyanoy S 2016 Design and Calculation of Cylindrical Electrostatic Deflector for Transport Channel of Heavy Ions *Proc. of XXV Russian Part. Acc. Conf. RUPAC'16* (Saint-Petersburg) pp 461-3 http://www.jacow.org

[5] *Opera3D Reference Manual* 2012 (Oxford OX5 1JE, England)

IOP Conf. Series: Journal of Physics: Conf. Series **874** (2017) 012045 doi:10.1088/1742-6596/874/1/012045

Pulsed beam tests at the SANAEM RFQ beamline

G Turemen[1,2], Y Akgun[1], A Alacakir[1], I Kilic[1], B Yasatekin[1,2], E Ergenlik[3], S Ogur[3], E Sunar[3], V Yildiz[3], F Ahiska[4], E Cicek[5] and G Unel[6]

[1]Saraykoy Nuclear Research and Training Center, Turkish Atomic Energy Authority, Ankara, Turkey
[2]Department of Physics, Ankara University, Ankara, Turkey
[3]Department of Physics, Bogazici University, Istanbul, Turkey
[4]EPROM Electronic Project & Microwave Ind. and Trade Ltd. Co., Ankara, Turkey
[5]Department of Physics, Gazi University, Ankara, Turkey
[6]Department of Physics & Astronomy, University of California at Irvine, Irvine, California, USA

E-mail: gorkem.turemen@ankara.edu.tr

Abstract. A proton beamline consisting of an inductively coupled plasma (ICP) source, two solenoid magnets, two steerer magnets and a radio frequency quadrupole (RFQ) is developed at the Turkish Atomic Energy Authority's (TAEA) Saraykoy Nuclear Research and Training Center (SNRTC-SANAEM) in Ankara. In Q4 of 2016, the RFQ was installed in the beamline. The high power tests of the RF power supply and the RF transmission line were done successfully. The high power RF conditioning of the RFQ was performed recently. The 13.56 MHz ICP source was tested in two different conditions, CW and pulsed. The characterization of the proton beam was done with ACCTs, Faraday cups and a pepper-pot emittance meter. Beam transverse emittance was measured in between the two solenoids of the LEBT. The measured beam is then reconstructed at the entrance of the RFQ by using computer simulations to determine the optimum solenoid currents for acceptance matching of the beam. This paper will introduce the pulsed beam test results at the SANAEM RFQ beamline. In addition, the high power RF conditioning of the RFQ will be discussed.

1. Introduction

SANAEM is finalizing the construction of a proton beamline to be used for training and educational purposes. All of the components were designed and produced locally to train accelerator physicists and engineers on the job. The full proton beamline, which is designed for pulsed beam operation, composed of an ion source, a low energy beam transport and an RFQ operating at 352.21 MHz, together with the RF components was designed and installed. This note summarizes the ongoing commissioning work of both the proton beamline and the RF power supply unit (PSU).

2. SANAEM RFQ Beamline

The beamline of the SANAEM RFQ consists of an ion source, a LEBT (Low Energy Beam Transport) line, a 1.3 MeV RFQ and an energy spectrometer. The drawing and a view of the LEBT line and the RFQ cavity is shown in Figure 1. The ion source and the LEBT line

IOP Conf. Series: Journal of Physics: Conf. Series **874** (2017) 012045 doi:10.1088/1742-6596/874/1/012045

components were designed to match the input beam requirements of the RFQ. Since a pulsed RF operation is envisaged for the cavity, a pulsed proton beam is to be delivered to the RFQ.

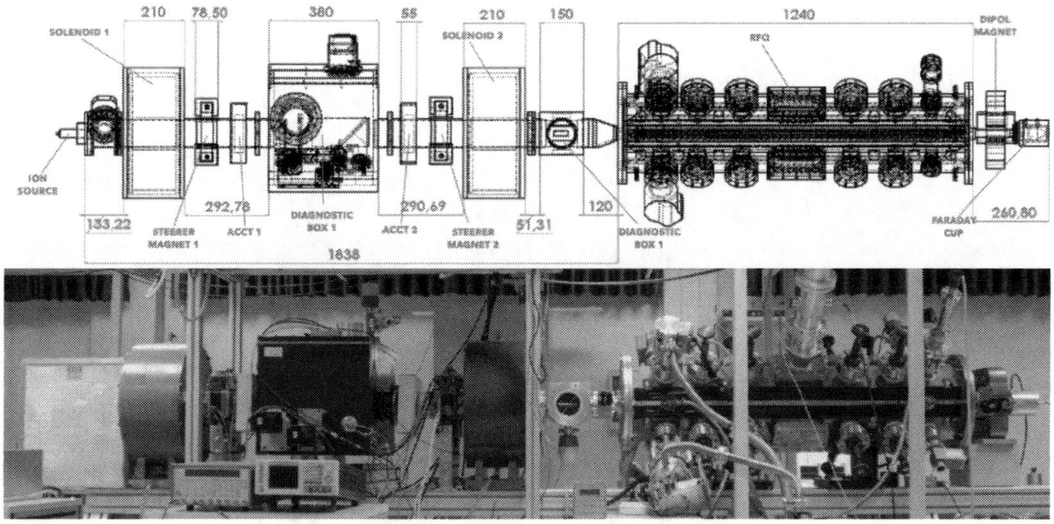

Figure 1. SANAEM RFQ beamline.

2.1. Pulsed Beam Extraction

A 1 kW, 13.56 MHz ICP ion source was developed to generate a 20 keV proton beam with a current of several hundreds of micro-amperes. The ICP source consists of a plasma chamber of 28 mm diameter, 2.5 turns RF antenna, and two electrode extraction system (Figure 2).

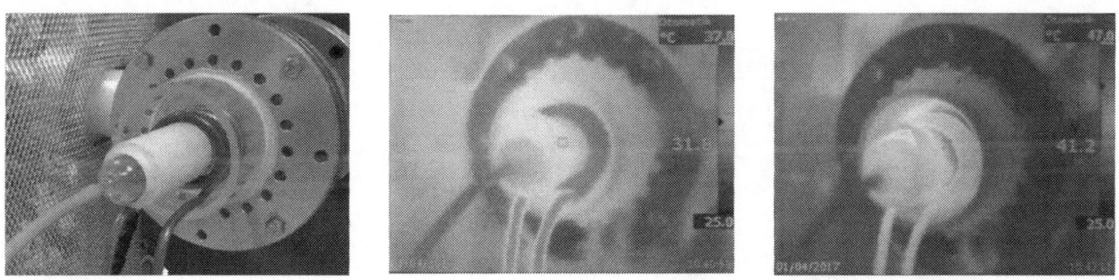

Figure 2. The measured temperatures of the plasma chamber (left) with pulsed (middle) and continuous (right) RF power.

With the current configuration, the pulsed proton beam can be produced with two different pulsing approaches. One of these, as in a previously introduced study [1], is to use a pulsed high voltage to extract the protons from the plasma. In this approach the plasma is continuous and a PFN (pulse forming network) is used to pulse the high voltage electrode of the extraction region. The other approach is pulsing the RF power, consequently the plasma, and supply a DC high voltage to the extraction electrode. In this configuration, the pulse length of the extracted beam is specified by the duty factor of the RF source. Since a low duty factor (1%) is used to excite the plasma, the temperature of the plasma chamber does not increase significantly and hence no water cooling is needed for the chamber. The temperature of the alumina plasma chamber and the copper RF antenna is measured with an infrared temperature sensor for these two different approaches (Figure 2) for 60 seconds operation time. Consequently, the second approach was found to be more convenient. On the extraction side, a plasma electrode with a 3 mm hole diameter was used to keep the transverse emittance low. The developed ICP ion source is suitable for long operation times with 1 Hz, 10 ms RF pulses.

3. Proton Beam Characterization

SANAEM RFQ's LEBT line has two diagnostic boxes and two ACCTs for beam characterization. One of the diagnostic boxes (Box-1) is located in between the LEBT solenoids (SOL-1 and SOL-2) and the other (Box-2) is located just before the RFQ cavity. The Box-2 has a Faraday cup and a phosphor screen and the Box-1 has an additional tool, a pepper-pot plate, for measuring the emittance of the beam. A CMOS camera with 85 mm-f/2.8 lens was used for the profile and emittance measurements. All the diagnostic components were developed locally, except the phosphor screen (P43) in the Box-1. Box-2 will also serve to house a beam chopper which is under production phase.

3.1. Beam Size Measurements

The beam size measurements are performed to ensure full transmission of the beam throughout the LEBT line (with a beam pipe diameter of 90 mm). The spot size of the extracted beam was measured in Box-1 and Box-2. A number of measurements were performed to characterize the beam behavior in different operating conditions, RF power and solenoid currents. The measured beam spot at Box-1 is shown in Figure 3 for 200 W RF power and 13 A SOL-1 current. Similarly, the beam spot size was measured just before the RFQ cavity (in Box-2) for 200 W RF power with a SOL-1 current of 13 A and various SOL-2 currents. As shown in Figure 3, the SOL-2 is able to over-focus the beam on the phosphor screen in Box-2 with a current of 13.7 A.

However, there is a space of 20 cm between the screen in the Box-2 and the RFQ plate, the required beam size (2.8 mm (100%) at the RFQ entrance) is provided by using the solenoids. In this case, the beam spot size always remains smaller than the beam pipe diameter thus full transmission of the low current beam is achieved.

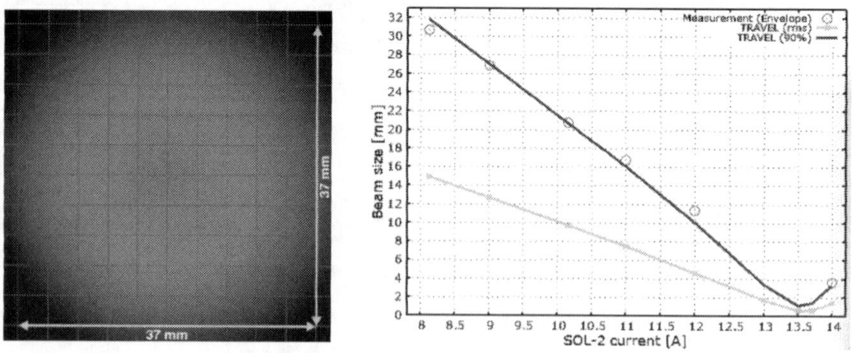

Figure 3. Measured beam spot in Box-1 (left) and simulated and measured beam spot sizes in the Box-2 (right).

3.2. Emittance Measurements

The characteristics of the transverse emittance measurement system with a pepper-pot method at Box-1 was presented in a previous study [2]. A 20 keV 10 ms proton pulse with a repetition frequency of 1 Hz was passed over the pepper-pot plate and its profile was captured with the camera on P43 screen (76 mm downstream). The captured pepper-pot image (Figure 4 on the left) was examined with a locally developed software and the transverse emittance (normalized rms) of the beam was calculated as 0.058 ∓0.001 and 0.050 ∓0.001 μm for vertical and horizontal axes, respectively. Although the whole system is cylindrical symmetric, the possible reasons for the difference between the measured emittance values are aberrations in the solenoid fields, measurement and calculation errors. As a future study, solenoid scan method will be considered to compare with the emittance values obtained with the pepper-pot.

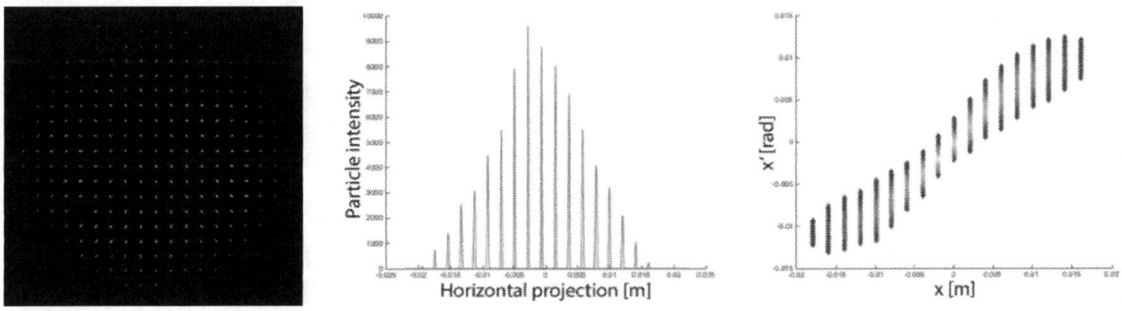

Figure 4. Captured image of the beam on the P43 screen (left), pepper-pot profile (middle) and transverse phase space (rigth) of the measured beam.

3.3. Current Measurements

Faraday cups and ACCTs are used to measure and compare the peak beam current at the LEBT line. Box-1 has a locally designed aluminum Faraday cup with an electron suppressor and an inner diameter of 80 mm. Instead of using an aperture at the entrance, the Faraday cup was designed as long as possible, 20 cm, to recapture the back-scattered protons. The Faraday cup in the Box-2 is a simple copper tube with 15 mm aperture to measure spot sizes comparable with the RFQ acceptance. The beam current was measured as 12.5 μA for 200 W RF power at Box-2 (Figure 5).

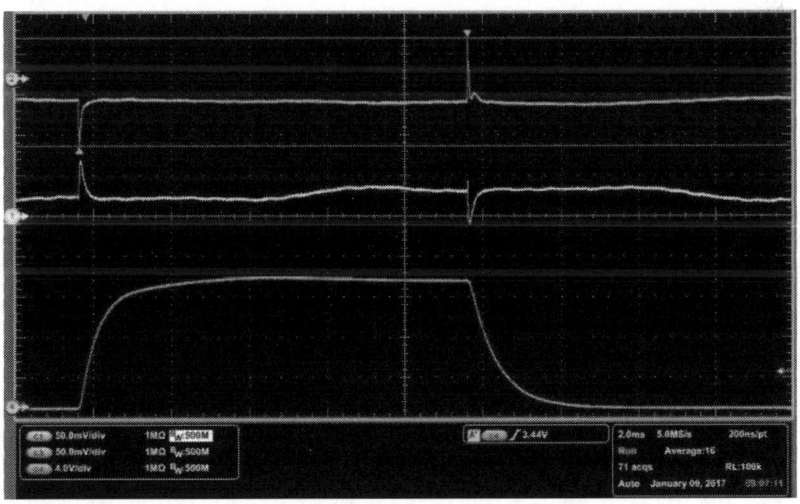

Figure 5. Beam current measurements on the LEBT line. Green: Faraday cup, yellow and blue: ACCT signals.

As an another current measurement method, two ACCTs were designed and installed in-air at input and output sides of the Box-1. The output signals of the toroids were magnified with an amplifier circuit to measurable levels. The waveforms generated by the proton beam are presented in Figure 5. The beamline calibration of the ACCTs is an ongoing process.

4. RFQ Beam Matching

The acceptance of the RFQ is calculated with LIDOS.RFQ [3] code and obtained as 0.1 μm (normalized rms). A number of simulations were performed with the TRAVEL [4] code to estimate the SOL-1 and SOL-2 currents to match the beam to the RFQ acceptance. A method based on emittance measurements was employed to reconstruct the proton beam due to the difficulty in characterizing the ICP plasma. In this method, the extracted proton beam was let to pass through the SOL-1 and its transverse emittance was measured in the Box-1. Then the beam is reconstructed at pepper-pot plane and transmitted through the SOL-2 to match the beam to the RFQ. The proton beam was reconstructed as an ellipse with TRAVEL code using the twiss parameters of the measured phase spaces. These parameters are obtained with a home-built pepper-pot code by analyzing the pepper-pot image. The field maps of the solenoids are generated by the field measurements and used in the TRAVEL code. The beam was matched to the RFQ acceptance (Figure 6) with SOL-1 and SOL-2 currents of 13.00 and 10.16 A, respectively, according to results of the TRAVEL simulations.

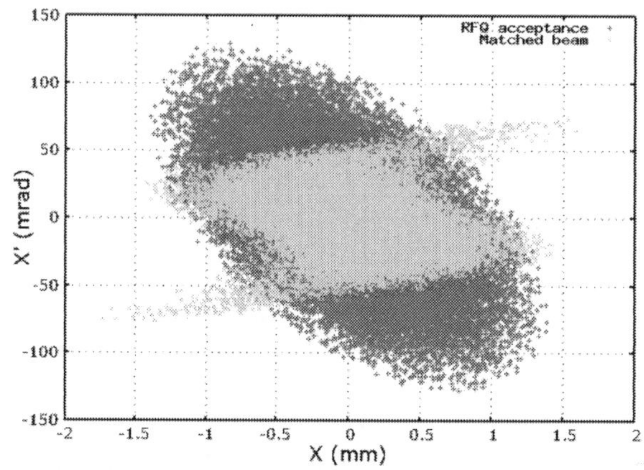

Figure 6. Matched beam on RFQ acceptance (RFQ acceptance in red and matched beam in green).

Another method for RFQ matching which is required to characterize the ICP plasma and also fully based on IBSimu [5] and TRAVEL simulations is introduced elsewhere [1]. Comparison with the results of the latter method will not be meaningful unless the ICP plasma parameters are precisely characterized.

5. High Power RF Tests

The RF power requirement of the RFQ is 94.2 kW which corresponds to a measured unloaded quality factor (Q_0) of 6426 after assembly [6]. A 60 dB RF load, with a maximum power rating of 50 kW, was used for high power RF measurements. The high power RF measurements were performed at the RF PSU, RF transmission line and RFQ cavity.

A hybrid RF power supply was developed locally using 8 solid-state amplifier (SSA) cards and a tetrode RF tube [7]. According to tests, an RF power of 5.8 kW was reached with 0.01% duty factor. An output power of 15.9 kW was measured and a gain of 13.1 dB was calculated with 780.3 W input power. Hence an input power of 4.6 kW is enough to provide the RF power required for the RFQ cavity.

A locally designed and built RF transmission line [8] was assembled for which an insertion loss of -0.34 dB was measured with a VNA. Thus, as expected, a power reduction of %7.48 was measured at 16 kW RF power level. An RF power of 47.39 kW was delivered to the end of the line which is close to the acceptable limit for the RF load in use. Beyond this power level, the power level of the RF system was measured with the RFQ cavity itself.

A coaxial loop coupler was designed, built and assembled to feed the RF power to the cavity. A pick-up loop was inserted into the RFQ cavity with -60.8 dB coupling to measure the cavity field. A HPGe detector was used to estimate the inter-vane voltage of the RFQ vanes with the x-ray spectroscopy method. Up to now 23.6 kV inter-vane voltage was reached. The power level of the cavity is gradually increased to provide 60 kV inter-vane voltage for the RFQ cavity with low spark rate.

6. Outlook

The required parameters (current, beam size and emittance) are measured to characterize the LEBT beam, except the beam energy which will be obtained by the time-of-flight method in further studies. As the results of the measurements it can be seen that the beam could match to the RFQ with determined configurations. Following the high power RF conditioning of the cavity, the low energy beam will be let to pass the RFQ. The beamline is now fully installed and is being readied for MeV level operations. The first MeV energy proton beam is expected to be available in 2017.

7. Acknowledgments

This study is funded by TAEK with a project No. A4.H4.P1. The authors would like to thank A. Lombardi for providing the TRAVEL software.

References

[1] Turemen G *et al.* 2015 Current Status of the SANAEM RFQ Accelerator Beamline *Proceedings of the IPAC'15* THPF103 p 3952

[2] Yildiz H *et al.* 2017 Compact Measurement Station for Low Energy Proton Beams *Journal of Instrumentation* vol **12** DOI:10.1088/1748-0221/12/02/T02006

[3] LIDOS.RFQ.DESIGNER Version 1.3 http://www.ghga.com/accelsoft/

[4] Perrin A *et al.* 2007 TRAVEL v4.07 *User Manual* CERN

[5] Kalvas T *et al.* 2010 IBSIMU: A Three-dimensional Simulation Software for Charged Particle Optics *Review of Scientific Instruments* vol **81** p 02B703

[6] Turemen G *et al.* 2016 Manufacturing and the LLRF Tests of the SANAEM RFQ *Proceedings of the IPAC'16* MOPOY051 p 964

[7] Ogur S *et al.* 2015 Development of a Hybrid Power Supply and RF Transmission Line for SANAEM RFQ Accelerator *Proceedings of the IPAC'15* WEPHA051 p 3228

[8] Turemen G *et al.* 2017 Design and Manufacture of the RF Power Supply and RF Transmission Line for SANAEM Project Prometheus *preprint* https://arxiv.org/abs/1504.01576

IOP Conf. Series: Journal of Physics: Conf. Series **874** (2017) 012046 doi:10.1088/1742-6596/874/1/012046

Updated cavities design for the FAIR p-linac

Ali M Almomani[1], **Marco Busch**[1], **Florian D Dziuba**[2], **Carl M Kleffner**[2], **Ulrich Ratzinger**[1] **and Rudolf Tiede**[1]

[1]Institute for Applied Physics, Goethe Universität Frankfurt, Frankfurt am Main, Germany
[2]GSI Helmholtzzentrum für Schwerionenforschung, Darmstadt, Germany

E-mail: a.almomani@iap.uni-frankfurt.de

Abstract. The research program of antiproton beams for the FAIR facility requires a dedicated 68 MeV, 70 mA proton injector. This injector will consist of an RFQ followed by six room temperature "Crossbar H-type" CH-cavities operated at 325 MHz. The beam dynamics had been revised by IAP Frankfurt in collaboration with GSI-FAIR in Darmstadt to further optimize the design. This step was followed by cavity RF design. The detailed mechanical cavity design will begin in 2017, while the quadrupole lenses are under production already. In this paper, besides an overview the RF design of the coupled cavities with integrated focusing triplets will be a main focus.

1. Introduction

The antiproton program at FAIR (Facility for Antiprotons and Ions Research) aims on producing 7×10^{10} cooled antiprotons per hour which requires a primary proton beam of about 2×10^{16} protons from the synchrotron SIS100 per hour [1-3]. The needed proton intensity from the injector linac is far beyond the capabilities of the GSI-UNILAC, therefore a high current proton linac is needed.

This new linac serves as a new injector for the SIS18 synchrotron, where the proton beams will be accelerated to the final energy of 68 MeV. Beyond this linac energy, the antiproton production and cooling chain will not profit substantially in a higher production rate [3].

It is very important to have a robust design with respect to operation at different current levels from 0 mA up to the design level of 75 mA. From beam simulations and experience with heavy ion beams from the UNILAC it was concluded that a beam current around 35 mA and within a normalized horizontal brilliance of 16.5 mA/μm should be well suited to fill the synchrotron SIS 18 up to the space charge limit [1]. The best linac current operation point will be determined during beam commissioning and injection into SIS 18.

2. FAIR proton linac

An ECR ion source will deliver protons with an energy of 95 keV [4]. After that, an RFQ structure will be used to accelerate the proton beam up to 3 MeV [5].

IOP Conf. Series: Journal of Physics: Conf. Series **874** (2017) 012046 doi:10.1088/1742-6596/874/1/012046

The main accelerator section in the proton linac consists of six room temperature "Crossbar H-type" CH – cavities operated at 325 MHz. Figure 1 shows an overview of the proton linac from the RFQ exit down to the end of the last CH cavity.

The CH–cavities consist of two sections: the first one is composed of three coupled CH – cavities (CCH1, CCH2 and CCH3) that will accelerate the beam from 3.0–33 MeV.

At 33 MeV, the space charge effects and rf defocusing are reduced. This will allow for longer lens-free sections in KONUS beam dynamics. Consequently, the high energy section consists of three lens-free CH – cavities. This helps to simplify the mechanical design and to tune the cavities besides reducing the total cost. The main parameters of the CH – cavities are summarized in Table 1.

All cavities have been designed in order to fulfill the revised beam dynamics [6]. This includes the number of gaps, voltage distribution, coupling cell lengths, additional spaces for diagnostic and vacuum elements and the total length.

Table 1. Main Parameters for the FAIR Proton Linac CH – Cavities at 75 mA Operation.

Cavity	**L (m)**	**P_{tot} (MW)**	**Z_{eff} (MΩ/m)**	**Energy (MeV)**	**Gaps No.**
CCH1	1.44	1.33	53.5	3.0 - 9.9	21
CCH2	2.56	1.87	53.2	9.9 - 21.1	27
CCH3	3.66	1.92	44.6	21.1 – 33.0	30
CH4	2.67	2.14	42.4	33.0 - 44.6	20
CH5	3.02	2.12	38.5	44.6 - 56.3	20
CH6	3.31	2.06	36.7	56.3 – 68.0	20
Magnetic Quadrupole Triplets					
Effective Length (mm): 2x				38, 60, 38	
10x				52, 95, 52	
Effective Gradients (T/m)				45 – 62	
Aperture Diameter (mm)				30	
Yoke Material				Cobalt Iron (Vacoflux50), laminated	
Beam Parameter					
Beam Current (mA)				75	
Emittance				input	Output
$\varepsilon_{n,\,95\%\,trans.}$ (mm·mrad)				1.35	2.6
$\varepsilon_{rms,\,95\%\,trans.}$ (mm·mrad)				0.266	0.455
$\varepsilon_{n,\,95\%\,long.}$ (keV·ns)				12.0	15.6
$\varepsilon_{rms,\,95\%\,long.}$ (keV·ns)				2.4	2.9

3. The coupled CH – cavity CCH2

The revised beam dynamics for the FAIR proton linac [6] was used to fix all relevant cavity parameters and to be able to start the cavity rf design and later on the construction.

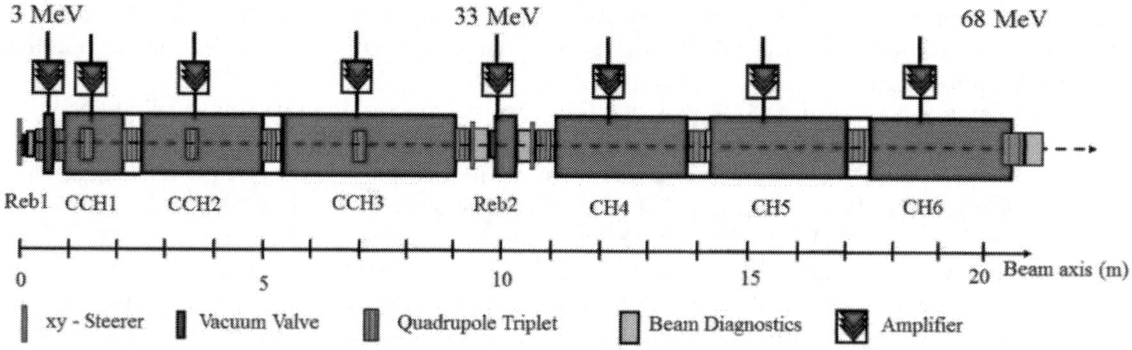

Figure 1. An overview of the FAIR proton linac from the RFQ exit down to the last CH cavity including the positions of the triplet quadrupoles, diagnostics and vacuum elements.

As an example for the cavity RF design, the final layout of the second coupled cavity (CCH2) will be presented in this section. Figure 2 shows a 3D – view of the cavity, where the two drift tube sections (low and high energy parts) are coupled by the central element that houses the quadrupole triplet. The concept of coupling two CH – sections and the physics of this kind of cavities has been discussed in detail before [2, 7].

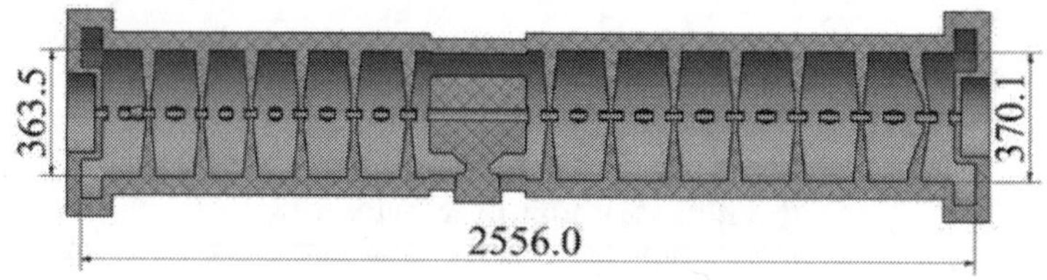

Figure 2. A cross sectional view of the second coupled CH – Cavity.

This cavity consists of 27 gaps and will accelerate the proton beam from 9.9 – 21.1 MeV within about 2.6 m. The coupling cell has a length of about $2\beta\lambda$. The rf simulations have shown that an effective coupling is achieved with this geometry as shown in Figure 3.

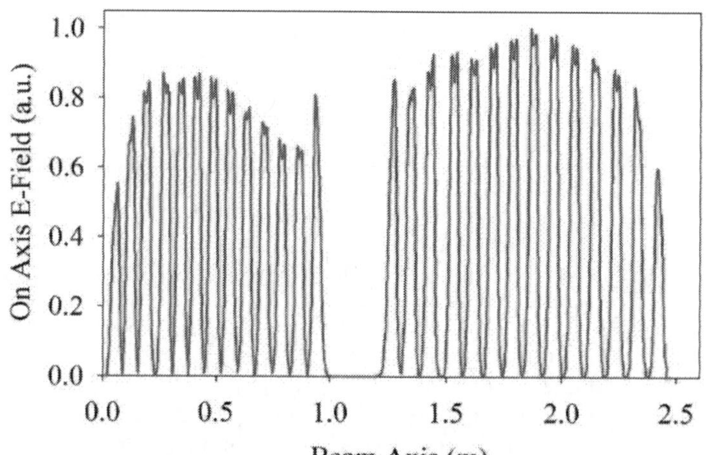

Figure 3. On axis electric field for the second coupled cavity.

IOP Conf. Series: Journal of Physics: Conf. Series **874** (2017) 012046 doi:10.1088/1742-6596/874/1/012046

The special shape at the cavity ends was needed to increase the effective voltage for the end gaps and to optimize the RF efficiency by improving the effective shunt impedance value Zeff . By that method, the matched cavity ends become short allowing for an improved mechanical integration of the inter-tank quadrupole triplets.

The effective gap voltage was calculated from the longitudinal electric field integration (Figure 3), and including the particle velocity dependence (transit time factor). These values were controlled and compared to the expected effective voltage from the beam dynamics calculation (using LORASR code). This comparison can be seen in Figure 4.

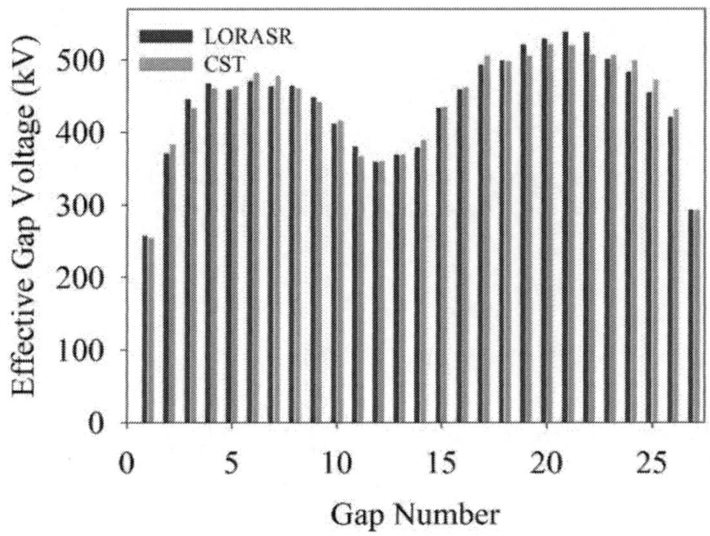

Figure 4. The calculated effective gap voltage from CST-MWS [8] simulation versus the design gap voltage from the beam dynamics with the LORASR code.

The electric field distribution along the yz - symmetry plane is shown in Figure 5. The magnetic field distribution in a xz- plane above the beam axis as indicated by Figure 5 shows that the coupling section oscillates along the central tube approximately in a 0 – mode (see Figures 6-7). The given maximum field values (as indicated in Figures 5–7) refer to the design operation level, as deduced from CST-MWS simulations.

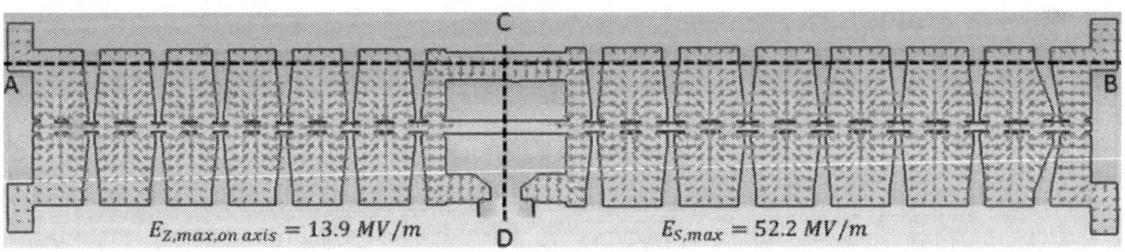

Figure 5. The electric field distribution along the xz – plane for the second coupled CH – cavity.

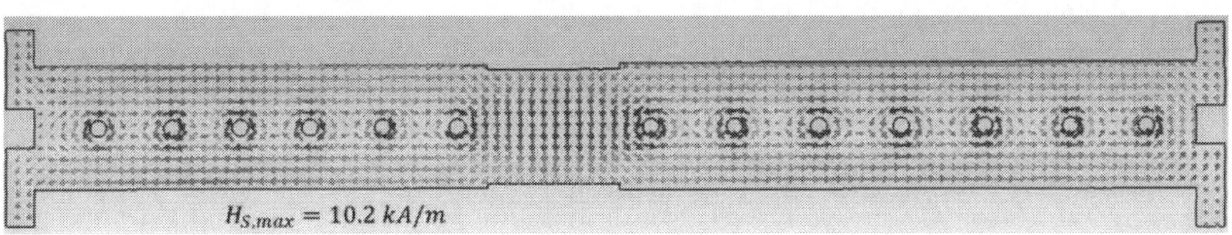

Figure 6. The magnetic field distribution along the plane A-B.

The first and third coupled CH – cavities have been designed in a similar way to the CCH2 and all the numerical results from all cavities (including the uncoupled CH-DTL's) will be used for a final beam dynamics cross check.

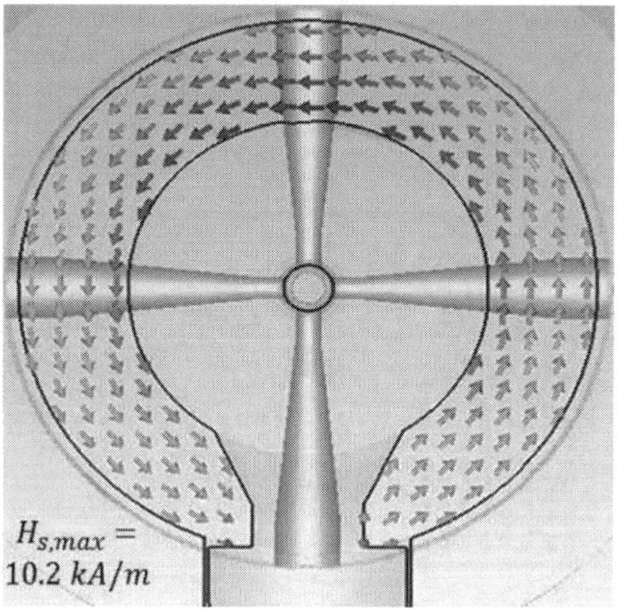

Figure 7. The magnetic field distribution in the middle plane of the central part along the plane C-D.

4. Conclusion

The proton linac for the FAIR facility will be the first one based on CH – DTL's and applying the KONUS beam dynamics.

The final layout for the cavities has almost been finished and the results will be used as the basis for the mechanical layout and cavity construction.

Besides the cavities design, the magnetic quadrupole triplets have been defined as shown by Table 1 and their integration either in the coupling cell (for the coupled cavities) or in the inter-tank section (between the cavities) are currently done in collaboration with GSI – FAIR. The quadrupole lenses are under production.

The good results from the RF power tests of the full scale CCH2 prototype cavity [9] are giving confidence to this novel linac approach and give inspiration for further improvements in the final design.

5. References

[1] Clemente G, Barth W, Groening L, Yaramyshev S, Brodhage R, Ratzinger U and Tiede R 2010 *Proceedings of HB'10 Workshop* Morschach Switzerland MOPD26 pp 115-9

[2] Clemente G, Ratzinger U, Podlech H, Groening L, Brodhage R, and Barth W 2011 *Phys. Rev. ST Accel. Beams* **14**, 110101

[3] Groening L, Barth W, Dahl L, Vinzenz W, Yaramyshev S, Clemente G, Ratzinger U, Schempp A and Tiede R 2006 *Proceedings of LINAC'06* Knoxville Tennessee USA MOP061 pp 186-8

[4] Chauvin N, Delferrière O, Gauthier Y, Girardot P, Jannin J L, Lotode A, Misiara N, Neyret J, Senée F, Simon C S, Tuske O, Berezov R, Fils J, Forck P, Hollinger R, Ivanova V, Ullmann C, Vinzenz W and Maugueret A 2014 *Proceedings of LINAC'14* Geneva Switzerland THPP015 pp 863-5

IOP Conf. Series: Journal of Physics: Conf. Series **874** (2017) 012046 doi:10.1088/1742-6596/874/1/012046

[5] Schütt M, Obermayer M A, Ratzinger U and Schnase A 2016 *Procedings of LINAC'16* East Lansing Michigan USA TUPLR053 pp 579-81

[6] Tiede R, Almomani A, Busch M, Dziuba F D and Ratzinger U 2016 *Proceedings of LINAC'16* East Lansing Michigan USA MOPRC018 pp 112-4

[7] Ratzinger U, Clemente G, Commenda C, Liebermann H, Podlech H, Tiede R, Barth W and Groening L 2006 *Proceedings of LINAC'06* Knoxville Tennessee USA TH1004 pp 526-30

[8] CST-MWS. https://www.cst.com

[9] Schreiber G, Plechov E, Salvatore J, Schlitt B, Schnase A and Vossberg M 2016 *Proceedings of LINAC'16* East Lansing Michigan USA MOPLR067 pp 288-90

The KONUS IH-DTL proposal for the GSI UNILAC poststripper linac replacement

H Hähnel, U Ratzinger and R. Tiede

Institute of Applied Physics, Goethe University, Frankfurt, Germany

E-mail: haehnel@iap.uni-frankfurt.de

Abstract. Motivated by the necessary replacement of the GSI UNILAC poststripper linac, a compact and efficient linac design based on IH-type cavities has been developed. Using KONUS beam dynamics, it was possible to design a linac consisting of only five cavities that can be operated by the existing UNILAC RF amplifier structure. The transversal focusing scheme is based on magnetic quadrupole triplet lenses. The optimized design provides full transmission and low emittance growth for the design current of 15 emA U^{28+}, accelerating the beam from 1.4 MeV/u to 11.4 MeV/u. Extensive error studies were performed to define tolerances and verify the stability of the design with respect to misalignment and injection parameters. The design provides a compact and cost effective alternative to a new Alvarez linac. With a total length of just 22.8 meters it will leave room for future energy upgrades in the UNILAC tunnel.

1. Technology and Layout

The poststripper linac of the UNILAC accelerates ions from 5.5 % to 15.5 % the speed of light. In this velocity range, the use of IH-type cavities provides high shunt impedance in combination with a period length of $L_p = \beta\lambda/2$ (see figure 1) [1, 2, 3]. The operating frequency for the UNILAC poststripper IH-DTL is 108.408 MHz. The proposed poststripper linac design consists

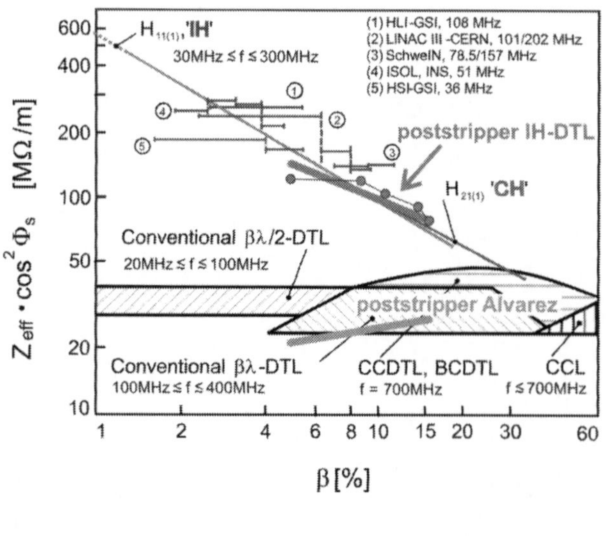

Figure 1. Effective shunt impedance for different normal conducting structures for $\beta = 1\,\% - 60\,\%$ (adapted from [4, 5])

of five IH-DTL cavities and seven quadrupole triplet lenses [6, 7], as shown in figure 2. It can be operated by the existing UNILAC RF amplifier structure well within power limits [8, 9]. To achieve this, the first cavity is designed with two internal lenses, a technology that has proven reliable in several linacs built in the last decades (e.g. HSI [10] and HLI [11] at GSI UNILAC, the HIT Heidelberg [12], NICA at JINR Dubna [13]).

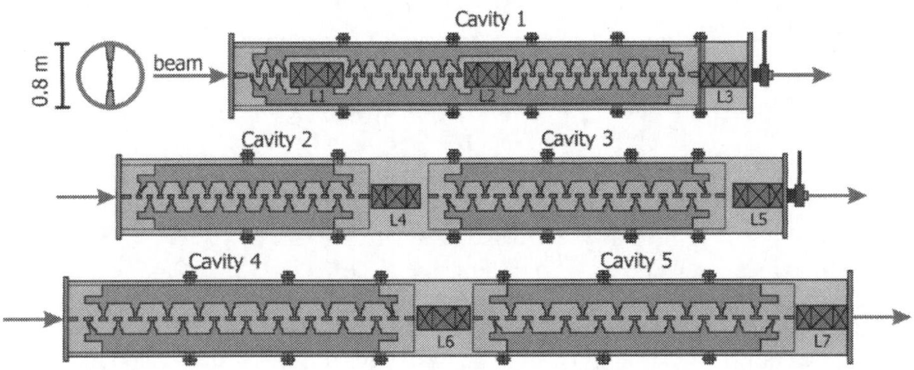

Figure 2. Layout of the proposed poststripper IH-DTL.

2. Beam Dynamics

For the design and optimization of the beam dynamics, the 3D PIC code LORASR [14, 5] was used. A waterbag particle distribution with $5 \cdot 10^5$ macroparticles was used for the final simulations. Crosscheck calculations were performed using TraceWin, confirming the beam dynamics layout performance. To achieve low longitudinal emittance growth, the KONUS lattice was optimized for an average beam current of 15 mA U^{28+}. Finally, a longitudinal emittance growth of just 11.1 % was achieved. In figure 3, the longitudinal beam envelopes show the confinement of the beam over the whole linac.

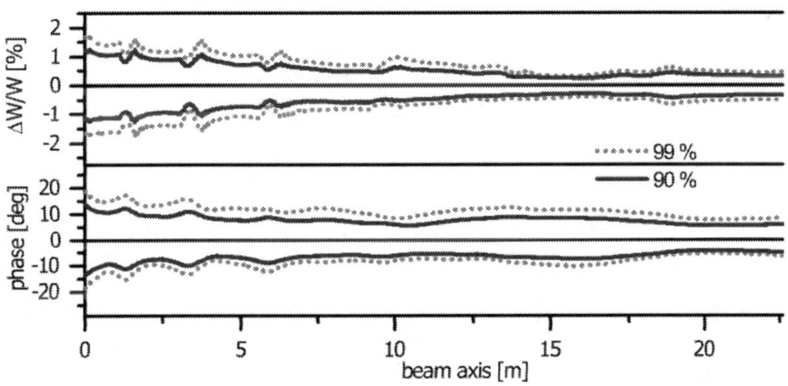

Figure 3. Longitudinal beam envelopes.

Investigations on the optimizations of KONUS lattices can be found in [7] and [15]. Transverse focusing of the beam is realized using quadrupole triplet lenses in a FDF-DFD lattice. By optimizing the phase advance along the linac, an RMS emittance growth of 27.5 % and 25.8 % in the horizontal and vertical plane, respectively, was achieved. The resulting transverse beam envelopes are shown in figure 4.

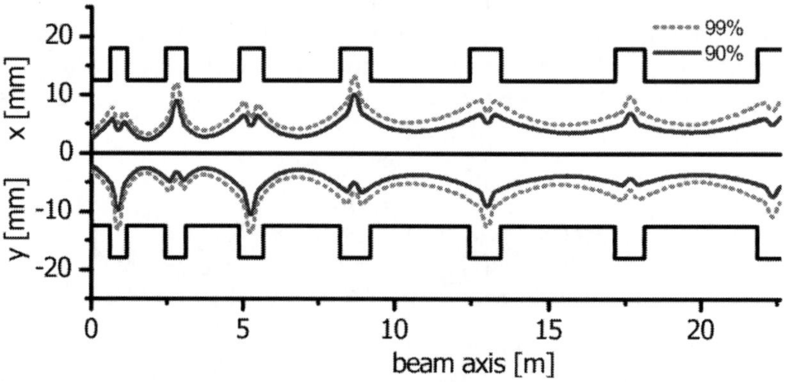

Figure 4. Transverse beam envelopes.

3. Error Studies

Error studies play an important role in the design of particle accelerators. Accounting for possible errors during fabrication, alignment and operation of the linac helps to provide a more realistic view of the linac performance. In case of the presented IH-DTL, error studies were performed using the TraceWin code. The number of simulated particles was chosen to be 10^4 to 10^5 for the sensitivity studies and combined error studies, respectively. With a total of 1000 random linacs generated using gaussian error distributions, the total number of simulated particles for each simulation run is 10^7 to 10^8. These values were chosen to achieve reproducible results.

As a first step, sensitivity studies for the individual component errors were performed to find the most critical points. Following that, error studies with combined sets of errors were performed using limits deduced from the sensitivity studies. In the final step, error compensation using corrective steering elements within the linac was investigated, showing significant improvement of the linac performance.

3.1. Error Sensitivity

A large number of individual error types was investigated to find the components that are most vulnerable to errors. The investigated errors include the displacement of the quadrupole magnets and cavities, a rotation of the magnets, field errors of the RF cavities and errors of the injected beam parameters.

For the magnetic quadrupole triplet lenses, the displacement and rotation of the individual singlets and also the triplets as a whole were investigated, as well as gradient errors. The most remarkable result is, that the displacement of the singlets within a triplet and the rotation of the triplet as a whole are the most critical effects. As shown in figure 5, significant losses already occur for singlet displacements larger than $\Delta xy = 100\,\mu\text{m}$. In contrast, quadrupole triplet displacements of up to $\Delta xy = 360\,\mu\text{m}$ are tolerated without significant emittance growth or losses. For a rotation of the elements, the situation is reversed. The rotation of singlets by up to 7 mrad leads to no significant losses and only introduces additional emittance growth. However, a triplet rotation of more than 2 mrad already leads to significant losses (see figure 6). Therefore, the rotation and displacement of quadrupole triplet lenses and their individual singlets should be treated separately in error studies.

Magnetic field gradient errors of the individual quadrupoles were investigated up to $\Delta B'/B'_0 = 2\,\%$. As a result, gradient errors of up to $0.7\,\%$ lead to no significant losses or emittance growth. A transverse displacement of the IH-type cavities of as much as $\Delta xy = 1\,\text{mm}$ only leads to $< 0.01\,\%$ losses. Therefore, cavity displacement on its own is no problem,

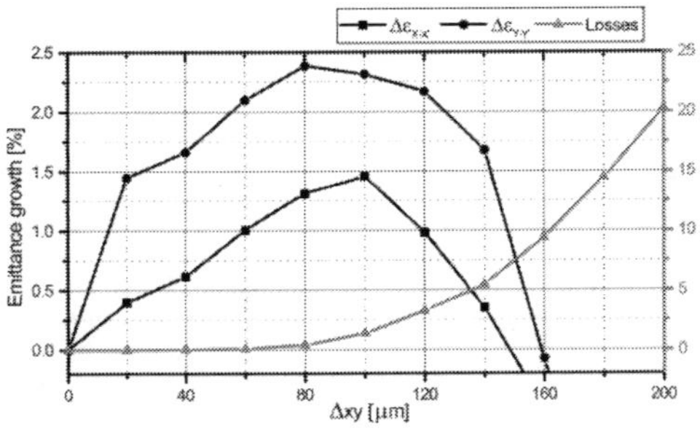

Figure 5. Losses and additional emittance growth for quadrupole singlet displacement errors.

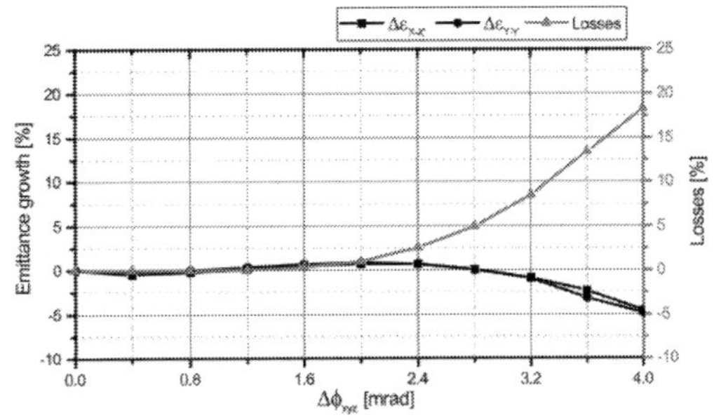

Figure 6. Losses and additional emittance growth for quadrupole triplet rotation errors.

since much better alignment accuracy can be achieved. For the electromagnetic field stability, the cavity field level, cavity RF phase and gap voltage accuracy were investigated in the range $\Delta E/E_0 = 0 - 1\%$, $\Delta\phi = 0 - 2^\circ$ and $\Delta V_{\text{gap}}/V_{\text{gap},0} = 0 - 5\%$. All three error types show an almost linear increase of the linac end energy variation with increasing error magnitude. While the cavity field level and phase only lead to longitudinal emittance growth of $< 6\%$ in the investigated range, individual gap voltage errors of $\Delta V_{\text{gap}}/V_{\text{gap},0} > 2\%$ lead to significant longitudinal emittance growth. Investigations of beam injection errors show, that beam displacements up to ± 1 mm have almost no effect. For a mismatch of the beam Twiss parameters at injection of up to 40 %, no losses are observed. However, for a Twiss mismatch of $> 20\%$, additional emittance growth arises.

3.2. Combined Errors

Following the sensitivity studies, error simulations were performed with combinations of all individual component errors. To investigate the performance of the linac, three sets of combined errors were defined. The first case resembles the error limits beyond which significant losses

Table 1. Parameters for combined error runs.

Error Type	Case A	Case B	Case C
Quadrupole Lenses			
Singlet Δxy	80 µm	80 µm	**50 µm**
Singlet $\Delta\phi_{x,y,z}$	1 mrad	1 mrad	1 mrad
$\Delta B'/B'_0$	0.7 %	0.7 %	**0.1 %**
Triplet Δxy	360 µm	**100 µm**	100 µm
Triplet $\Delta\phi_{x,y,z}$	1.6 mrad	**1 mrad**	1 mrad
Cavities			
Field $\Delta E/E_0$	0.7 %	0.7 %	**0.1 %**
Phase $\Delta\phi$	1.4°	1.4°	**0.1°**
$\Delta V_{\text{gap}}/V_{\text{gap},0}$	2 %	2 %	**1 %**
Cavity Δxy	600 µm	600 µm	**100 µm**

IOP Conf. Series: Journal of Physics: Conf. Series **874** (2017) 012047 doi:10.1088/1742-6596/874/1/012047

and/or emittance growth were observed in the sensitivity studies. This first error set is called "Case A" (see table 1). Since it was anticipated, that the relatively loose limits for quadrupole triplet displacement and rotation would lead to significant losses, when combined with the quadrupole singlet errors, a second set of errors, called "Case B", was defined. Here the magnetic triplet errors are reduced to $\Delta xy = 100\,\mu\mathrm{m}$ and $\Delta\phi_{x,y,z} = 1\,\mathrm{mrad}$. For "Case C", all errors were reduced to optimistic values that could be technically achieved, (see table 1). The resulting average transmissions for these three error sets are shown in figure 7. As expected, the "Case A" error set shows significant average losses. The strong interaction between quadrupole singlet and triplet errors is evident when comparing the average transmission of "Case A" and "Case B", where the only difference is a reduction of triplet displacement and rotation. The average transmission of "Case B" is significantly improved. Even though these combined error simulations were performed without any corrective steering in the linac, a satisfying average transmission of 99.76 % is achieved for the "Case C" scenario. The reduction of cavity RF errors in "Case C" has a big impact on the longitudinal emittance growth, which is reduced to 2.13 % (see table 2).

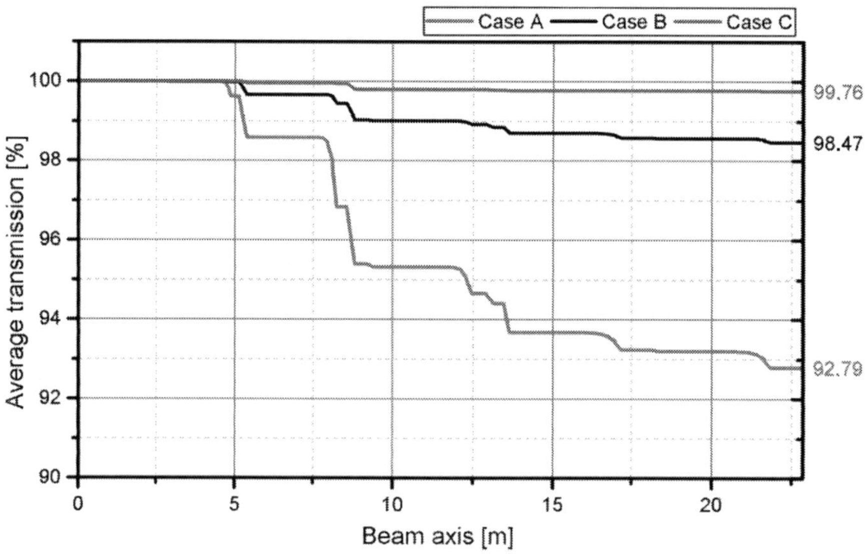

Figure 7. Average transmissions for the three cases.

3.3. Corrective Steering

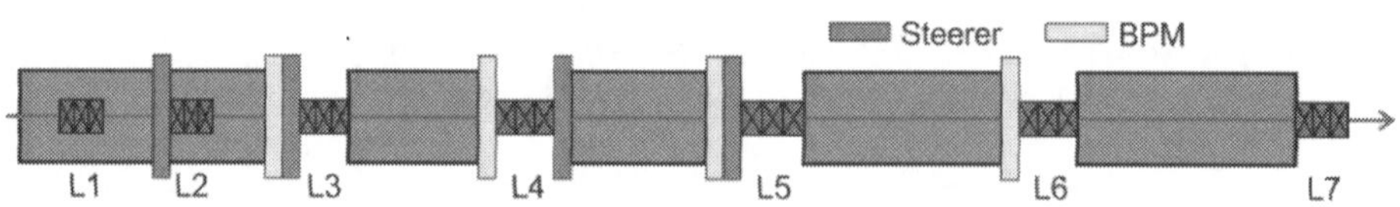

Figure 8. Layout for the final steering strategy.

To find the correct steering strategy for the proposed IH-DTL, several layouts with different numbers of steerers and different steerer positions were investigated. As a result, it was found that a total of four steerers, set up as two steerer pairs, is sufficient (see figure 8). The average losses of the linac without steerers can be reduced to 0.24 %, due to tight error tolerances in

"Case C". By using the steering strategy mentioned above, the average losses can be significantly lowered from 7.21 % to 0.03 % for "Case A" (see table 2). In combination with strict error tolerances as in "Case C", the losses can be reduced even further to $6 \cdot 10^{-8}$.

Table 2. Resulting additional emittance growth and losses of error simulations with and without steerers.

	$\Delta\epsilon_x$ [%]	$\Delta\epsilon_y$ [%]	$\Delta\epsilon_z$ [%]	Losses [%]
Case A	1.04 %	0.23 %	13.49 %	7.21 %
Case B	2.14 %	2.11 %	11.04 %	1.53 %
Case C	0.64 %	1.27 %	**2.13 %**	**0.24 %**
with steerers:				
Case A	1.43 %	2.50 %	9.98 %	**0.03 %**
Case C	0.67 %	0.82 %	**0.38 %**	**$6 \cdot 10^{-8}$**

4. Conclusion

Comprehensive error studies of the proposed poststripper IH-DTL show a high error tolerance in combination with the final steering strategy. Average losses can be reduced to below $3 \cdot 10^{-4}$ even for large errors. Additional emittance growth due to errors is in the order of a few percent. The presented design has matured and could be used as a cost effective alternative to an Alvarez DTL.

Acknowledgments

This research is funded by the German Federal Ministry of Education and Research (BMBF) within the project 05P15RFRBA.

References

[1] Nolte E, Geschonke G, Berdermann K, Oberschmid R, Zierl R, Fell M, Jahnke A, Kress M and Morinaga H 1978 The Munich heavy ion postaccelerator *Nucl. Instr. Meth.* **158** 311–24
[2] Nolte E, Geier R, Schollmeier W and Gustavson S 1982 Improved performance of the Munich heavy ion postaccelerator *Nucl. Instr. Meth.* **201** 281–5
[3] Ratzinger U, Nolte E, Geier R, Gärtner N and Morinaga H 1988 The upgraded Munich linear heavy ion postaccelerator *Nucl. Instr. Meth. Phys. Res.* A **263** 261–70
[4] Ratzinger U 1998 *Effiziente Hochfrequenz-Linearbeschleuniger für leichte und schwere Ionen* Habilitationsschrift Goethe University Frankfurt
[5] Tiede R, Mäder D, Petry N, Podlech H, Ratzinger U and Zhang C 2014 Improvements of the LORASR code and their impact on current beam dynamics designs *Proc. 27th Lin. Acc. Conf. LINAC2014 (Geneva)* (Geneva: JACoW) 569–71
[6] Hähnel H, Ratzinger U and Tiede R 2016 Efficient heavy ion acceleration with IH-type cavities for high current machines in the energy range up to 11.4 MeV/u 2016 *Proc. 28th Lin. Acc. Conf. LINAC2016 (East Lansing)* (Geneva: JACoW) 617–9
[7] Hähnel H 2017 *Development of an IH-type linac for the acceleration of high current heavy ion beams* Ph.D. thesis Goethe University Frankfurt
[8] Schlitt B, Eichler G, Hermann S, Hoerr M, Mueh M, Petit S, Schnase A, Schreiber G, Vinzenz W and Zappai J 2016 Modernisation of the 108 MHz RF systems at the GSI UNILAC *Proc. 28th Lin. Acc. Conf. LINAC2016 (East Lansing)* (Geneva: JACoW) 899–901
[9] Schlitt B, Hoerr M, Schnase A, Schreiber G and Vinzenz W 2014 RF system development for the new 108 MHz heavy ion high-energy linac at GSI *Proc. 27th Lin. Acc. Conf. LINAC2014 (Geneva)* (Geneva: JACoW) 556–8

IOP Conf. Series: Journal of Physics: Conf. Series **874** (2017) 012047 doi:10.1088/1742-6596/874/1/012047

[10] Ratzinger U 1996 The new GSI prestripper linac for high current heavy ion beams *Proc. 18th Lin. Acc. Conf. LINAC1996 (Geneva)* (Geneva: JACoW) 288–92
[11] Klabunde J 1992 The high charge state injector for GSI *Proc. 16th Lin. Acc. Conf. LINAC1992 (Ottawa)* (Geneva: JACoW) 570–4.
[12] Lu Y 2005 *Development of an IH-DTL injector for the Heidelberg cancer therapy project* Ph.D. thesis Goethe University Frankfurt
[13] Butenko A *et al.* 2017 Commissioning of the new heavy ion linac at the NICA project *Proc. 8th Int. Part. Acc. Conf. IPAC2017 (Copenhagen)* (Genveva: JACoW)
[14] Tiede R 2009 *Simulationswerkzeuge für die Berechnung hochintensiver Ionenbeschleuniger* Ph.D. thesis Goethe University Frankfurt
[15] Tiede R, Hähnel H and Ratzinger U 2017 Beam dynamics design parameters for KONUS lattices *Proc. 8th Int. Part. Acc. Conf. IPAC2017 (Copenhagen)* (Genveva: JACoW)

IOP Conf. Series: Journal of Physics: Conf. Series **874** (2017) 012048 doi:10.1088/1742-6596/874/1/012048

Status of the modulated 3 MeV 325 MHz Ladder-RFQ

Maximilian Schuett[1], Ulrich Ratzinger[1], Alexander Schnase[2], Marc Syha[1], Marcus Obermayer[1]

[1]Institute for Applied Physics, Goethe University, Frankfurt, Germany
[2]GSI, Darmstadt, Germany

E-mail: `schuett@iap.uni-frankfurt.de`

Abstract. Based on the positive results of the unmodulated 325 MHz Ladder-RFQ from 2013 to 2016, a modulated 3.3 m Ladder-RFQ is currently under construction and will be tested with beam at GSI FAIR, Darmstadt.

The unmodulated Ladder-RFQ showed a very constant voltage along the axis. It could withstand more than 3 times the operating power of which is needed in operation at a pulse length of 200 μs. That corresponds to a Kilpatrick factor of 3. The 325 MHz RFQ is designed to accelerate protons from 95 keV to 3.0 MeV according to the design parameters of the p-linac at FAIR. This particularly high frequency for a 4-Rod type RFQ creates difficulties, which are challenging in developing an adequate cavity. The results of the unmodulated prototype have shown, that the Ladder-RFQ is a suitable candidate for that frequency. Inspired by the successful RF power test, the nominal vane-vane voltage was increased from 80 kV to 96 kV.

The basic design and tendering of the RFQ has been successfully completed in 2016. Electrodynamic simulations of a modulated full structure, especially in terms of field-flatness and frequency tuning, are shown. Furthermore, the mechanical design, which includes a direct cooling of the structure for duty cycles up to about 5%, will be discussed.

1. Introduction

The idea of the Ladder-RFQ firstly came up in the late eighties [1, 2] and was realized successfully for the CERN Linac3 operating at 101 MHz [3] and for the CERN antiproton decelerator ASACUSA at 202 MHz [4]. Within the 4-Rod design the challenge is to minimize dipole components and to have geometrical dimensions which are suitable for mechanical manufacturing and assembling. At frequencies above 250 MHz the 4-Vane-type RFQ is used so far, and many versions for low and high duty factors have been realized successfully until now. Draw backs are the high costs per meter, the complexity as well as the challenging RF tuning procedure of that structure: The dipole modes tend to overlap with the quadrupole mode. Safe beam operation conditions result in ambitious mechanical vane tolerances. In the proposed Ladder-RFQ version, the ladder spokes have a significant vertical width, which increases the resonance frequency and results in an homogeneous current flow towards the mini-vanes. The mini-vanes are embedded via precisely machined carrier rings into the copper shells (cf. Figure 2). It is even possible to exchange the ring-mini-vane-structure completely by an improved electrode system as demonstrated successfully at the GSI High Current RFQ [5]. To proof the mentioned advantages and the realizability of the Ladder-RFQ, an unmodulated prototype was designed and built from

2013 until 2016. Further details of the mechanical layout can be seen in [6, 7]. The successful high power measurements are presented in this paper. The very good results motivated the design and manufacturing of a modulated prototype at its full length of approximately 3.3 m. The status of its design as well as field simulations including a modulation of the electrodes are also shown in this paper.

2. High Power Measurements of the unmodulated Prototype

Figure 1. Front view of the ring-electrode-system of the unmodulated prototype. The steel-tank and the upper half-shell are removed. Ladder dimensions: LxHxB: 650x150x365 mm.

At the GSI 325 MHz klystron test stand the unmodulated Ladder-RFQ prototype with an electrode length of 630 mm was tested with a forward power up to 487 kW at a repetition rate of 2 Hz and a pulse duration of 200 µs (Figure 2) [8] [1]. That exceeds the designed RF power at an electrode voltage of 80 kV during normal operation by a factor of 3.6. The averaged power reached 200 W, which is about 50% more than the thermal loss during normal operation. The Kilpatrick factor went up to 3, compared to 1.6 at design level. As anticipated, the stored energy increases linearly with the forward power up to 400 kW. At higher energies up to 487 kW, the stored energy begins to rise slower due to field emission. During the power tests it was clearly recognized that more time for conditioning can further reduce the dark current levels. In nine days of high power measurements the surface of the copper structures has changed its color slightly but has not been damaged. The surface remained smooth and non abrasive. The successful power tests of the prototype justify an increase of the electrode voltage from 80 kV to 96 kV for the modulated version. The low-level RF measurements such as frequency spectra, frequency tuning and field flatness for the unmodulated cavity were presented in [9]. The first HOM of the prototype is measured at 378 MHz.

[1] The authors would like to thank A. Schnase, G. Schreiber and B. Schlitt from GSI for the facilitation of the measurements.

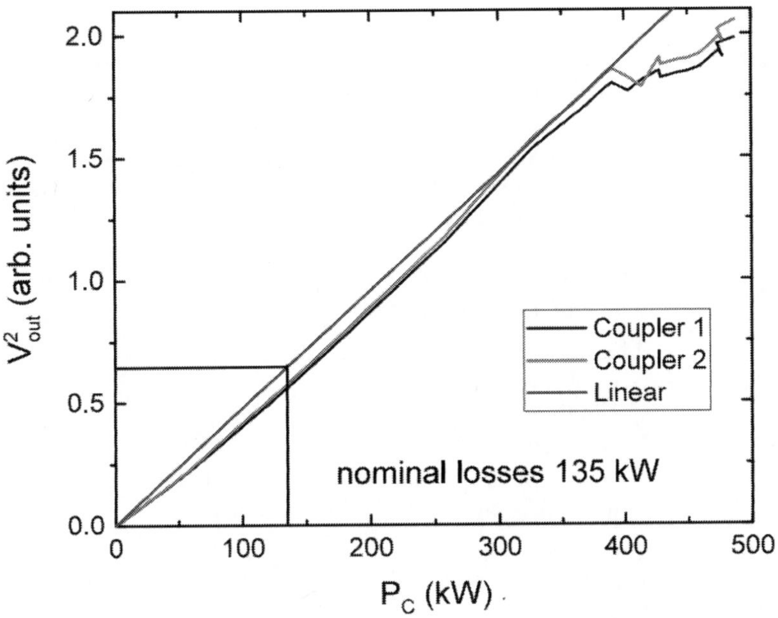

Figure 2. Results of the high power tests of the unmodulated prototype. The forward power up to 487 kW is shown in black. The RFQ accepted the power as the reflected power is stable along the flat top being lower than 3.34 kW. The RFQ is slightly over-coupled.

3. Mechanical Layout of the modulated Prototype

The mechanical design consists of an inner copper ladder structure mounted into an outer stainless steel tank. The tank is divided into three parts: A lower half-shell carrying the inner resonating structure, a middle frame and the upper half-shell (cf. Figure 3). The lower half-shell will carry and fix the position of the resonating ladder structure. The ladder structure is

Figure 3. Isometric view of the 3.3 m modulated Ladder-RFQ prototype. The copper carrier-rings guarantee the electrode positioning as well as the RF contact. The ladder structure consists of bulk copper components. Any brazing or welding processes were avoided for the assembly of the main components. In total three TMP700 at the lower side of the RFQ provide a vacuum in the 10^{-8} mbar range. The entrance and exit flanges are milled out to achieve a shorter distance for the integration of cone and steerer preventing an emittance growth.

IOP Conf. Series: Journal of Physics: Conf. Series **874** (2017) 012048 doi:10.1088/1742-6596/874/1/012048

machined from solid copper blocks. It consists of two lower and two upper half-ladder elements, which are precisely aligned via guide pins (cf. Figure 2). The RF features are mainly determined by the resonating structure, while the dimensions of the tank have no significant influence on the frequency. Based on the successful high power tests of the unmodulated prototype it was decided to develop a new beam dynamics with an increased electrode voltage of 96 kV [10]. The basic physical and mechanical parameters of the Ladder-RFQ results are shown in Table 1.

Table 1. Main RF and Geometric Parameters of the Modulated Prototype Ladder-RFQ.

No. of cells	55
Q Value (simulated)	6800
Loss (with sim. Q)	675 kW
Thermal Loss (measured Q)	1360 W
Shunt Impedance (sim.)	45 kΩm
Electrode Voltage	96 kV
Frequency	325.224 MHz
Repetition Rate	4 Hz
Pulse Duration	200 µs
Total Length	3410 mm
Cell Length	40 mm
Spoke Thickness	20 mm
Spoke Height	280 mm
Spoke Width	150 mm
Aperture	2.96 mm
Electrode Radius ρ	2.22 mm
Electrode Length	3327 mm

Besides water-cooling channels within the upper and lower tank half-shells, the RFQ is equipped with a direct cooling of the ladder side rails. Along the side of all half-ladders a groove is milled (8 channels in total). The groove will be closed by a cover bar from copper with stainless steel connection pipes at the ends. All weldings on the ladder-structure apply electron beam technology to reduce heat and deformation during manufacturing. Pursuant to temperature simulations the RFQ can be operated with the advanced cooling system up to a duty factor of 5%.,

4. Simulations

All simulations are additionally performed with including modulated electrodes, which is indispensable for the tuning process. Accordingly, the longitudinal accelerating electric field component along the beam axis could be evaluated (cf. Figure 4). The study has shown that the simulation is in accordance with the beam dynamics calculation. Furthermore, it has been confirmed that the field flatness and the electrode voltage, respectively, of the modulated model is equivalent to a model without a modulation but a varying mean aperture (cf. Figure 5). As in the case of the unmodulated RFQ it is sufficient to tune the field flatness by merely adjusting the first and last three cell heights (i.e. in cells 1-3 and cells 53-55), which is the clear distance between the side rails in a ladder cell. Nevertheless, the first and last four cells will be adjusted to reach the optimum field flatness. The first neighboring HOM is approx. 9 MHz above the operating mode.

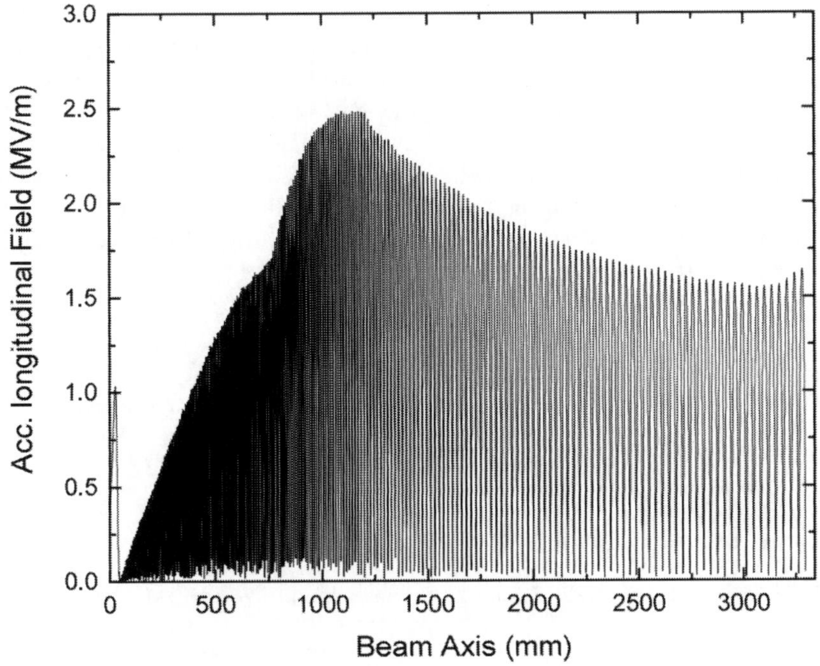

Figure 4. Longitudinal electric field distribution on the beam axis of the modulated Ladder-RFQ.

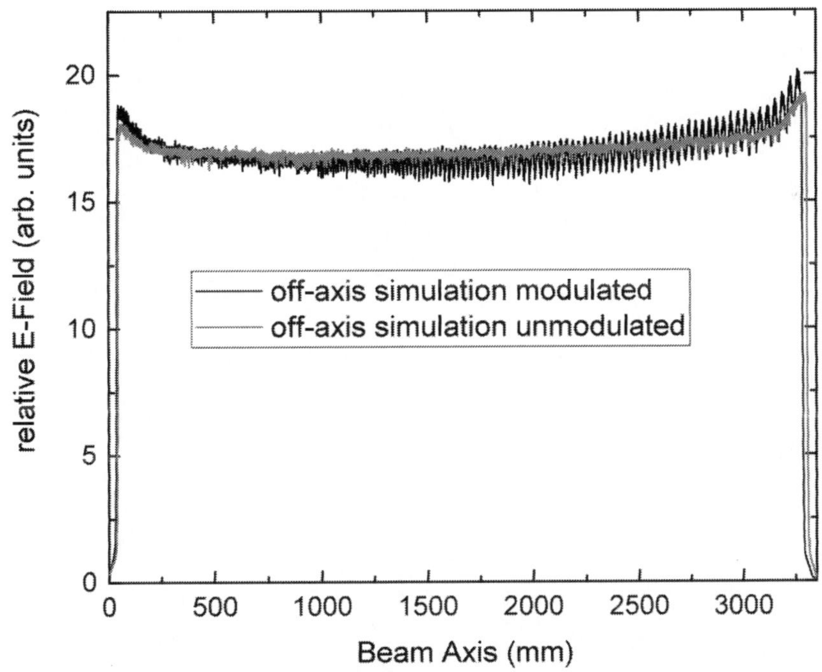

Figure 5. Field flatness along the beam axis of the modulated (black) compared to the unmodulated (red) simulation.

IOP Conf. Series: Journal of Physics: Conf. Series **874** (2017) 012048 doi:10.1088/1742-6596/874/1/012048

5. Conclusion

All measurements of the unmodulated Ladder-RFQ such as frequency spectra, tuning, field flatness are in a good agreement with the simulations. The vacuum properties and pump ability has been verified. Accordingly, the Ladder-RFQ seems to be a good candidate for the acceleration of protons at typical frequencies above 250 MHz. Based on these results, the design and construction of a full length modulated Ladder-RFQ, which will be tested with beam at GSI, is currently on-going. The current design of the modulated Ladder-RFQ remains on-schedule. Semi-finished goods are already delivered encouraging the production release in Q2/2017 once the preliminary examination has been approved. The completion of manufacturing is aimed until the end of 2017, to be ready for first RF tests in early 2018.

References

[1] Fabris A, Massarotti A and Vretenar M 1987 *New Techniques for Future Accelerators* **29** 265–269
[2] Browman M J, Spalek G, Friedrichs P B and Barts T C 1988 *Proceedings of LINAC* 119–121
[3] Bezzon G, Lombardi A, Parisi G, Pisent A and Weiss M 1994 *Proceedings of LINAC* 557–559
[4] Bylinsky Y, Lombardi A and Pirkl W 2000 *Proceedings of LINAC* 554–556
[5] Ratzinger U 2001 *Nucl. Instrum. Meth.* A **464** 636–645
[6] Schütt M, Ratzinger U and Brodhage R 2015 *Proceedings of IPAC* 3745–3747
[7] Schütt M, Ratzinger U and Zhang C 2016 *Proceedings of IPAC* 899–901
[8] Schreiber G, Plechov E, Salvatore J, Schlitt B, Schnase A and Vossberg M 2016 *Proceedings of LINAC* 287–289
[9] Schütt M, Obermayer M and Ratzinger U 2016 *Proceedings of LINAC* 578–580
[10] Syha M, Schütt M, Obermayer M and Ratzinger U 2017 *Proceedings of IPAC* TUPVA074

The phase slip factor of the electrostatic cryogenic storage ring CSR

Manfred Grieser, Robert von Hahn, Stephen Vogel, Andreas Wolf
Max-Planck Institut für Kernphysik, 69117 Heidelberg, Germany
E-mail: Manfred.Grieser@mpi-hd.mpg.de

Abstract. To determine the momentum spread of an ion beam from the measured revolution frequency distribution, the knowledge of the phase slip factor of the storage ring is necessary. The slip factor was measured for various working points of the cryogenic storage ring CSR at MPI for Nuclear Physics, Heidelberg and was compared with simulations. The predicted functional relationship of the slip factor and the horizontal tune depends on the different islands of stability, which has been experimentally verified. This behavior of the slip factor is in clear contrast to that of magnetic storage rings.

1. Introduction

The cryogenic storage ring CSR [1] shown in figure 1, is a fully electrostatic storage ring used to store atomic, molecular and cluster ion beams in the energy range of $q\cdot$(20-300) keV, where q is the absolute value of the ion charge state. The entire storage ring can be cooled down to temperatures of only a few Kelvin. This very low temperature creates an extremely low residual gas density. The observations from first cryogenic operation indicate the residual gas densities below 100 molecules/cm^3. The cooling of all ion optics and the vacuum enclosure to 10 K temperature also provides the benefit of uniquely low level of blackbody radiation where the stored molecular ion beams reach their lowest quantum states. In March 2014 the functionality of CSR was demonstrated by storing a 50 keV $^{40}\text{Ar}^+$ beam under room temperature conditions. Later, in 2015, the storage ring was cooled down to an average temperature below 10 K. At this temperature the storage time for singly charged ions achieved up to 2500 s. The details about CSR are given elsewhere [1].

2. The phase slip and momentum compaction factor

The momentum spread $(\Delta p/p)$ of an stored ion beam can be determined by using the measured revolution frequency distribution and the phase slip factor η, which is defined by:

$$\eta = \frac{\Delta f/f_0}{\Delta p/p_0}, \tag{1}$$

where f_0 is the revolution frequency and p_0 the momentum of the central ion. The momentum and frequency deviation of an ion to the central particle is given by Δp and Δf, respectively . To calculate the slip factor of the storage ring we consider an ion moving with energy E on its

Figure 1. The cryogenic storage ring CSR with 35 m circumference.

closed orbit, where the revolution time ($T = 1/f$) related to the revolution frequency f is given by:

$$T = \int_0^{C(E)} \frac{1}{\mathrm{v}(s')} ds'. \tag{2}$$

The length of the closed orbit of an ion with energy E is $C(E)$ and $\mathrm{v}(s')$ denotes the velocity ion velocity, depending on its position s'. It is favorable to describe the ion motion in the coordinate system where the central particle is moving on its central orbit. In this coordinate system the differential ds' of the variable s' is transformed to ds [2] as follows:

$$ds' = \left(1 + \frac{x(s)}{\rho(s)}\right) ds, \tag{3}$$

where s is the position on the central orbit, $x(s)$ denotes the horizontal deviation of the ion orbit to the central orbit and $\rho(s)$ is the radius of the central orbit.

In the coordinate system of the central particle the revolution time T of an ion on its closed orbit is given by:

$$T = \int_0^{C_0} \frac{1 + \frac{x(s)}{\rho(s)}}{\mathrm{v}(s)} ds. \tag{4}$$

An electrostatic storage ring can be used only for non relativistic ion velocities $\mathrm{v}(s)$ and hence $\mathrm{v}(s)$ at the potential $\phi(s)$ is given by:

$$\mathrm{v}(s) = \sqrt{\frac{2}{m}(E - Q\phi(s))}. \tag{5}$$

Since the CSR consists of electrostatic cylindrical deflectors and electrostatic quadrupoles, the potential $\phi(s)$ can be expressed by:

$$\phi(s) = V_0(s) \ln\left(\frac{\rho(s) + x(s)}{\rho(s)}\right) + K(s)(x(s)^2 - y(s)^2), \tag{6}$$

where $V_0(s) = \frac{2E_0}{Q}$ for cylindrical deflectors. E_0 denotes the energy and Q the charge of the central ion. Using the Taylor series expansion, equation (6) can be expanded in x(s) as:

$$\phi(s) = V_0(s)\frac{x(s)}{\rho(s)} - V_0(s)\frac{x(s)^2}{2\rho(s)^2} + K(s)(x(s)^2 - y(s)^2) + ... \tag{7}$$

Neglecting second order and higher terms in equation (7) results in:

$$\phi(s) = V_0(s)\frac{x(s)}{\rho(s)}. \tag{8}$$

Using equation (8) in equation (5) gives with equation (4):

$$T = \int_0^{C_0} \frac{1 + \frac{x(s)}{\rho(s)}}{\sqrt{\frac{2}{m}(E - 2E_0\frac{x(s)}{\rho(s)})}} ds. \tag{9}$$

Equation (9) can be expanded in $x(s)$. Considering only first order terms of $x(s)$ yields in:

$$T = \sqrt{\frac{m}{2E}}C_0 + \frac{E_0 + E}{E}\sqrt{\frac{m}{2E}}\int_0^{C_0} \frac{x(s)}{\rho(s)} ds. \tag{10}$$

The closed orbit deviation $x(s)$ in equation (10) is given by the dispersion ($D_p(s)$) of the storage ring and the energy deviation ($\Delta E = E - E_0$) of an ion to the central particle having an energy E_0:

$$x(s) = D_p(s)\frac{1}{2}\frac{\Delta E}{E_0}. \tag{11}$$

In equation (11) there is a factor $\frac{1}{2}$ because the dispersion function is defined via the ion momentum. The integral in equation (10) can be expressed via the momentum compaction factor (α_p) of the storage ring [3]. α_p describes the relation between circumference (C_0) of the closed orbit and the ion momentum (p) :

$$\alpha_p = \frac{\Delta C/C_0}{\Delta p/p_0} = \frac{1}{C_0}\int_0^{C_0} \frac{D_p(s)}{\rho(s)} ds. \tag{12}$$

With the introduction of the momentum compaction factor α_p, equation (10) can be rewritten as:

$$T = \sqrt{\frac{m}{2E}}C_0 + \frac{E_0 + E}{E}\sqrt{\frac{m}{2E}}C_0\frac{\alpha_p}{2}\frac{E - E_0}{E_0}. \tag{13}$$

Finally, with equation (13) we can calculate the slip factor of the storage ring, defined by equation (1):

$$\eta = -2\frac{E_0}{T_0}\frac{\partial T}{\partial E}\bigg|_{E=E0} = 1 - 2\alpha_p. \tag{14}$$

Whereas for magnetic storage rings, slip factor η_m in the non relativistic approach is given by [4] :

$$\eta_m = 1 - \alpha_p. \tag{15}$$

The difference of the slip factor of an electrostatic storage ring to a magnetic storage ring can be explained by the acceleration and deceleration processes taking place in the electrostatic deflectors of the storage ring depending on the horizontal ion position $x(s)$. The measurement of the slip factor η and the momentum compaction factor α_p for CSR is explained in the next section.

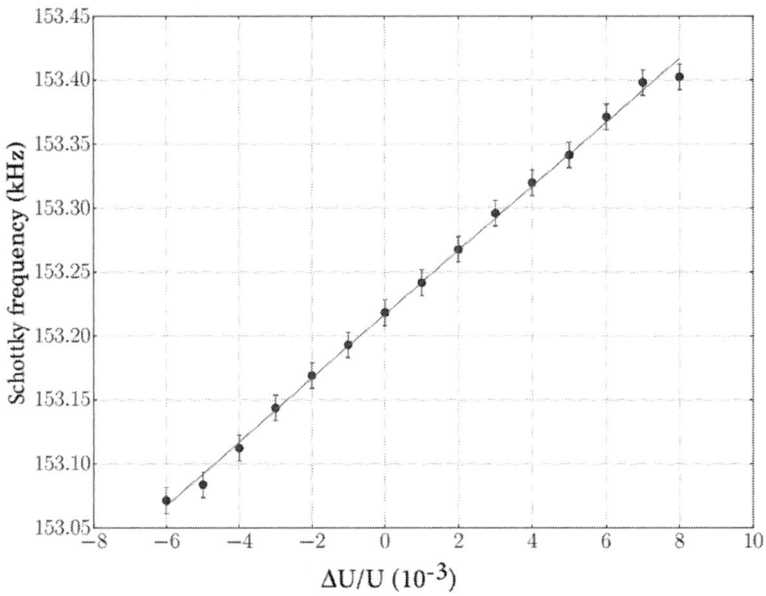

Figure 2. Schottky frequency measured at 10th harmonic of the revolution frequency as a function of the variation of the voltages of all CSR deflectors and quadrupoles expressed by $\Delta U/U_0$.

3. Measurement of α_p and η

As derived in the previous section the phase slip factor and the compaction factor α_p of an electrostatic storage ring are related to each other. The momentum compaction factor can be measured experimentally by measuring the change in the revolution frequencies (Δf) due to variation of the voltages by $\Delta U/U_0$ on all ion optical elements [6]:

$$\frac{\Delta f}{f_0} = \alpha_p \frac{\Delta U}{U_0}. \tag{16}$$

This method has been used to measure the momentum compaction factor α_p where the change of the revolution frequency was measured with Schottky noise analysis. Figure 2 shows the measured Schottky frequency f as a function of the variation of the voltages of all electrostatic elements. The frequencies shown in figure 2 correspond to the 10th harmonic of the revolution frequency of a $^{40}Ar^{+}$ beam with 60 keV energy. The frequencies were measured with the Schottky pick-up connected to a spectrum analyzer which is triggered few ms after injection. The sweep time was set to 0.5 s (corresponding to the first 7500 revolutions of the ion beam) . The errors in the determination of the Schottky frequencies are estimated to be 10 Hz. A linear fit to the data is shown as a red straight line. From its slope using equation (16) the momentum compaction factor is calculated to be α_p=0.163. Using equation (14) a slip factor of η = 0.674 can be calculated. With MAD8 [5], where each electrostatic element is described by transfer matrices, gives for the same quadrupole settings and working point a phase slip factor of η_{mad}=0.679, which is in good agreement with the experimentally determined value.

4. The slip factor η as a function of the horizontal tune Q_x

In magnetic storage rings the slip factor η is related to the horizontal tune Q_x . For example at the TSR storage ring in Heidelberg following dependency between η and the horizontal tune

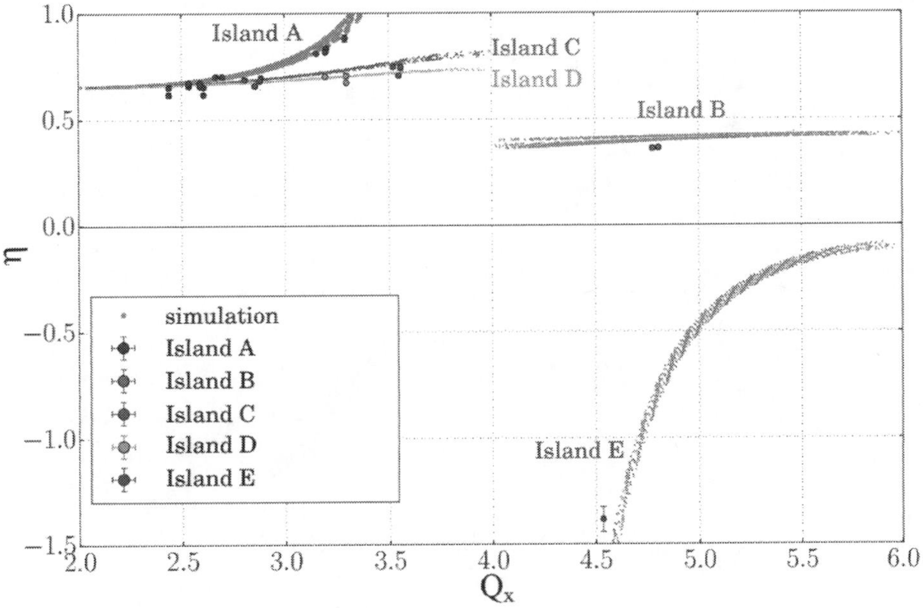

Figure 3. The measured (markers with error bars) and simulated (gray markers) phase slip factors for the cryogenic storage ring CSR as a function of the measured and simulated horizontal tune Q_x, respectively.

was found [7]:

$$\eta \approx 1/\gamma^2 - 1/Q_x^2, \tag{17}$$

where γ is the Lorentz factor. In order to study the functional relationship between η and Q_x, simulations were carried out using the MAD8 [5] program. The results of these calculations are shown in figure 3. Simulations (gray markers) predict a functional relationship $\eta = f(Q_x)$, where the functions depend on the quadrupole focusing strengths. The quadrupole focusing strengths for the stable ion orbits are used to calculate $\eta = f(Q_x)$ (figure 4). The stable working points are arranged in five islands, labeled with A-E, where island A and B are spitted into two sub islands. Each of these islands can be assigned to a relation $\eta = f(Q_x)$ shown in figure 3, which are different for the various islands. This behavior of the slip factor is different to any magnetic storage ring.

To proof the simulated relationship $\eta = f(Q_x)$, measurements were carried out for η and Q_x for the different stability islands. The experimentally measured values are shown as colored markers in figure 3. The denoted statistical errors in figure 3 are propagated from the fitting errors, whereas the errors in Q_x are too small to be visible. Although there are a few deviations between the experimental data and the simulations, the overall agreement is good and the functional relation $\eta = f(Q_x)$ predicted from the simulations is experimentally verified.

5. Conclusion

In this paper a relationship between the slip factor η and the momentum compaction α_p for an electrostatic storage ring is established. The difference of the η value from electrostatic to magnetic storage rings can be explained by acceleration and deceleration processes in the electrostatic deflectors, depending on the ion position. To measure the momentum compaction

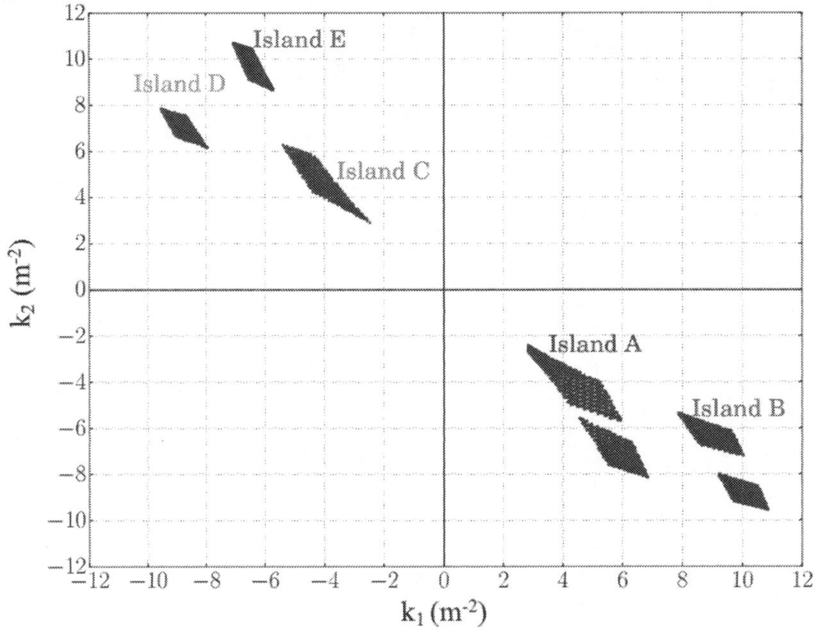

Figure 4. Stability diagramm of the CSR simulated with the MAD8 program, k_1 and k_2 are the quadrupole strengths of the two quadrupole families.

factor α_p and the phase slip factor η, a new method was introduced where the change of the revolution frequency is measured as a function of the scaled voltages of all ion optics elements. In the usual measurements of η the revolution frequencies have to be measured for different momentums of the injected ion beam. This common method is very time consuming due to the necessary energy changes of the ion source and the changes in the settings of the transfer line, consisting of many magnetic and electrostatic elements. But in this new method of measuring η, the energy of the injected ions remains constant and therefore the determination of α_p and η could be done quickly. In contrast to magnetic storage rings the phase slip factor of the electrostatic storage ring CSR does not follow the approximate dependency on the horizontal tune η (compare equation (17)). It has different functional dependencies which vary for the different islands of stability.

References

[1] von Hahn R *et al.* 2016 *Review of Scientific Instruments* **87** 063115
[2] Schmüser P 1986 *CERN accelerator school* CERN 87-10 p 33
[3] Wu Chao A and Tigner M 2002 *Handbook of Accelerator Physics and Engineering* Word Scientific second Printing p 52
[4] Le Duff J 1984 *CERN accelerator school* CERN 85-19 p 131
[5] Grote H and Iselin C 1990 *The MAD Program Version 8.13/8* CERN/SL/90-13 (AP)
[6] Vogel S 2016 Developments at an electrostatic cryogenic storage ring for electron cooled keV energy ion beams *Ph.D. thesis University of Heidelberg*
[7] Grieser M *et al.* 1993 *Workshop on beam cooling and related topics* CERN 94-03 p 257

Spin tracking for a deuteron EDM storage ring

A Skawran[1,2,3,4], **A Lehrach**[1,2,5]

[1] Institut für Kernphysik, Forschungszentrum Jülich, Jülich, Germany
[2] RWTH Aachen University, III. Physikalisches Institut B, Aachen, Germany
[5] JARA-FAME (Forces and Matter Experiments), Forschungszentrum Jülich and RWTH Aachen University, Germany

E-mail: [3] alexander.skawran@psi.ch

Abstract. The aim of the Jülich Electric Dipole moment Investigations (JEDI) collaboration is the measurement of the Electric Dipole Moment (EDM) of charged particles like protons or deuterons. There are two possible concepts under consideration for the realization of EDM measurement with deuterons; the Frozen Spin (FS) and Quasi-Frozen Spin (QFS) method. Both approaches are discussed and compared in this paper. Detailed spin- and beam dynamics simulations are performed to investigate the effect of various misalignments of ring elements and systematic effects. Furthermore, the utilization of counter rotating beams is studied and checked for its validity.

1. Introduction

Permanent EDMs of subatomic particles violate parity P and time reversal T symmetry. Assuming the CPT theorem, this leads to CP violation. The Standard Model (SM) predicts non-vanishing EDMs, their magnitudes, however, are expected to be not measurable with current experimental techniques. The discovery of a nucleon EDM larger than 10^{-31} e cm [1] would be a signal for new physics beyond the SM and could explain the matter-antimatter asymmetry observed in our Universe [2]. The measurement of an EDM is based on the observation of a spin precession in the presence of strong electric fields. So far, upper EDM limits of electrically neutral systems and charged leptons have been measured.

Charged-particle measurements have been proposed for EDM searches of protons and deuterons in storage rings [3], since the electric field also bends the particles trajectory. These experiments require a new class of high-precision storage rings to significantly improve the technology for polarized beam storage and measurement by orders of magnitude [4], as well as the knowledge of and control over the beam and spin itself.

The JEDI Collaboration is working on a series of feasibility studies at the Cooler Synchrotron COSY in Jülich and in parallel on the design study for a dedicated EDM storage ring [5,6].

2. Lattice design

The FS and QFS concepts are investigated for a deuteron EDM storage ring. In the FS concept the spins are always aligned to the momentum vector [3] which can be achieved with combined static electromagnetic beam deflectors ($E \times B$ deflectors). The QFS concept is based on the

[4] Present address: Paul Scherrer Institut, 5232 Villigen, and ETH Zürich, Switzerland

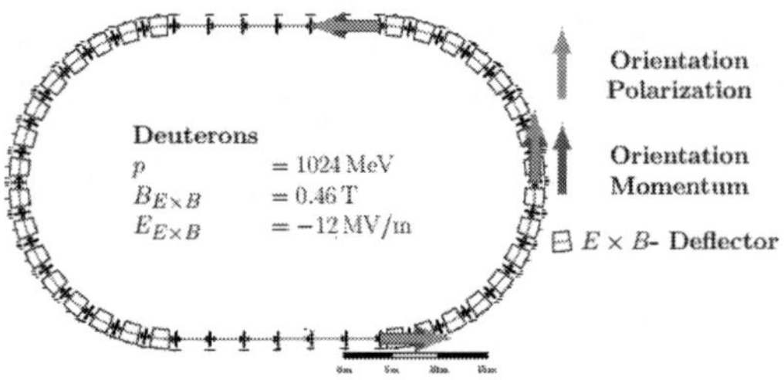

Figure 1. FS lattices with incorporated electrostatic and magnetic fields.

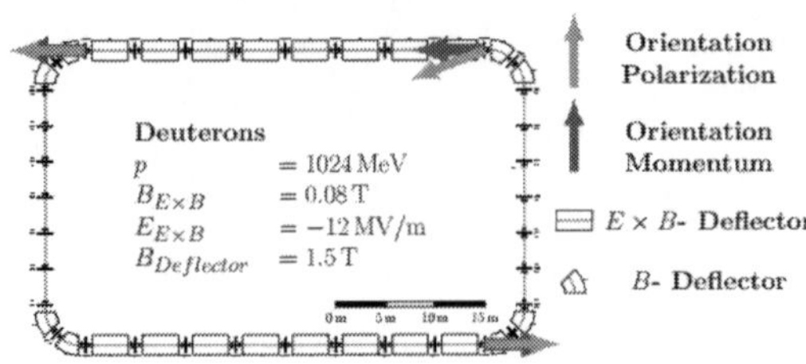

Figure 2. QFS lattice with straight electrostatic deflectors.

fact that the anomalous magnetic moment of deuterons has a small value [7]. Magnetic and electric fields can be spatially separated with magnetic arcs and either pure electric field or combined straight electromagnetic field sections connecting the two arcs. The rotation of spin relative to the momentum in two parts of ring – the arc and straight sections – can compensate each other, leading a quasi-frozen spin rotation turn by turn. This concept has the advantage that curved $E \times B$ deflectors do not have to be utilized.

Two possible lattice options are considered (see Figure 1 and 2) [8], where the elements with the combined electric and magnetic fields are used. The FS lattice has been designed based on deflectors with incorporated E and B fields with an electric field of 12 MV/m and a magnetic field of 0.46 T. The QFS lattice only uses two magnetic arcs with a magnetic field of 1.5 T and two straight sections having combined straight elements with magnetic and electric fields. The straight elements have a horizontal electric field of 12 MV/m and a vertical magnetic field of 80 mT. They provide the compensation for the spin rotation in the arc and at the same time allow having straight electric plates without higher order field components. Despite the difference in the concepts, the basic parameters, namely the size of the rings, the length of the straight sections, the number of focusing periods, and the number of deflecting and focusing elements and sextupoles, all remain approximately the same. In both options, the free dispersion straight sections are placed between the arcs. They are necessary to accommodate the polarimetry, the beam injection and extraction systems, and RF cavity [8].

3. Beam and spin tracking

3.1. Simulation program

The COSY Toolbox (COTOBO) has been utilized to perform the simulations. It is based on a C++ based interface [9] for COSY INFINITY [10]. The usability of ROOT [11] enables a fast and easy way to analyze the simulation results. $E \times B$ deflectors are the basis of the FS concept. Consequently, these elements had to be added to COTOBO. COSY INFINITY [12] contains elements to simulate $E \times B$ deflectors which are called Wien Filter. The implementation of $E \times B$ deflectors is realized such that the simulation of fringe fields and misalignments is possible. In COSY INFINITY fringe fields are approximated by

$$F(z) = \frac{1}{1 + \exp\left(\sum\limits_{n=0}^{5} A_n \cdot [z/2d]^n\right)} \quad (1)$$

where A_n are Enge coefficients, z is the coordinate on the field axis where the origin is the effective edge of the element, d is the aperture and $2d$ is the full aperture of the element. The old default Enge coefficients of electric fields in COSY INFINITY were replaced by optimized coefficients [13].

3.2. Simulation results for single beams

Detailed beam and spin tracking simulations have been performed to investigate the effect of systematic errors for the EDM build-up.

In the range of 1000 to 10000 particles have been tracked with transverse emittances of $\epsilon_x, \epsilon_y = 1$ mm mrad (rms) and a momentum spread of $\Delta p/p = 10^{-4}$. Due to the sideway component of the spin with respect to the momentum vector, the vertical polarization build-up for the QFS method will be reduced compared to FS. Simulation results indicate that this reduction is below 2% and can therefore be neglected.

As an example, the simulation results for vertical polarization build-up are shown in case of deflector rotations in Figure 3 and vertical quadrupole shifts in Figure 4 [14]. The magnitude of the EDM signal is described by the dimensionless scaling factor , with $|d_{edm}| \approx \eta \cdot 5.3 \cdot 10^{-15}$ e cm.

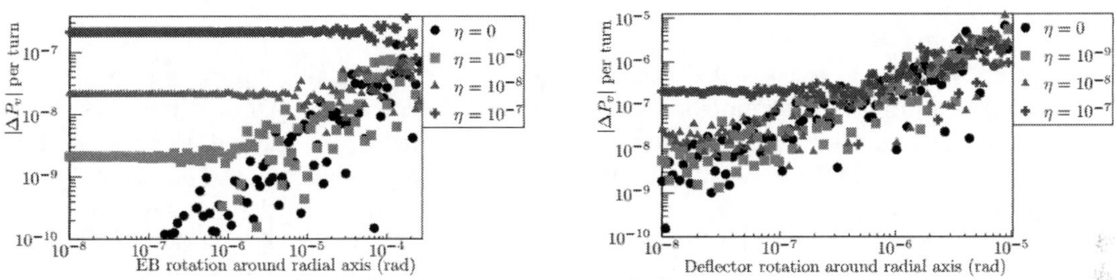

Figure 3. Vertical polarization build-up for FS (left) and QFS (right) and different magnitudes η of EDM signal and Gaussian distributed rotations of deflectors (RMS values) around the radial axis. Each simulation has different randomly generated misalignments.

These simulations indicate that the largest difference for artificial polarization build-up between FS and QFS are unwanted longitudinal fields due to misalignments of ring elements that occur if $E \times B$ deflectors are rotated around the radial axis (see Figure 3). Due to the orientation of the spin vectors parallel aligned to the beam motion, longitudinal fields do not disturb the spin motion in the FS concept in comparison to the QFS. The sideways component of the spin creates an additional artificial spin build-up, that can be derived from the Thomas-BMT equation [15] and reads

$$\frac{d\vec{s}_{MDM}}{dt} = \frac{e}{m}\left[\left(a + \frac{1}{\gamma}\right)\begin{pmatrix} -s_y B \\ 0 \\ s_x B \end{pmatrix} - \frac{a\gamma}{\gamma+1}(\beta B)\begin{pmatrix} -s_y \beta \\ 0 \\ s_x \beta \end{pmatrix}\right] \tag{2}$$

where m, e are the mass and charge and a is the anomalous magnetic moment of the particle, β, γ, the kinematic Lorentz factors, s_x, s_y the sideways spin components and B the longitudinal magnetic field. Therefore, rotations of $E \times B$ deflectors around the radial axis are more sensitive for the investigated QFS lattice compared to the presented FS lattice. Concerning unwanted radial magnetic fields the sensitivity for FS and QFS is comparable as can be seen from Figure 4 for shifts of quadrupoles in vertical direction [14].

Smaller contributions of non-frozen spin motion also show up in the FS method. Due to the motion of the particles in transverse and longitudinal phase space – even in the ideal case without misalignments or fringe field of ring elements – only the reference particle is perfectly frozen. In general, also the fringe field distribution for electric and magnetic fields have a different shape. Without any extra efforts of shapping the end regions of the elements a sideways component of the spin will also appear in the fringe field region for FS.

IOP Conf. Series: Journal of Physics: Conf. Series **874** (2017) 012050 doi:10.1088/1742-6596/874/1/012050

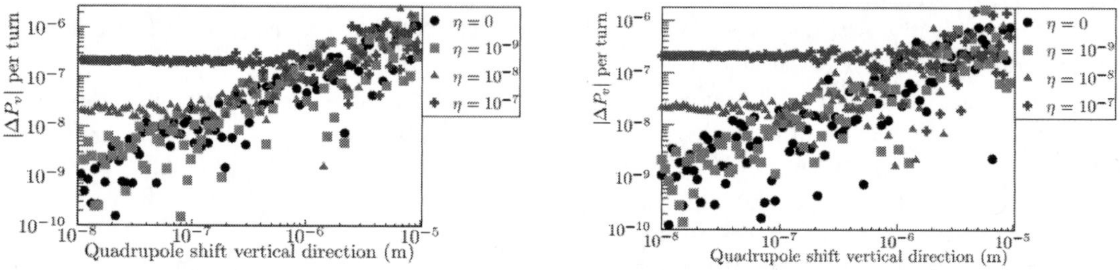

Figure 4. Vertical polarization build-up for FS (left) and QFS (right) different magnitudes of EDM and Gaussian distributed vertical shifts of quadrupoles (RMS values). Each simulation has different randomly generated misalignments.

3.3. Simulation results for CW-CCW beams

Counter-rotating beams, clockwise (CW) and counter-clockwise (CCW), enable the extraction of the magnitude of an EDM signal for the simulated misalignments of both kinds of rings. However, a small systematic contribution for rotations of deflectors around the radial axis remains utilizing the CW-CCW method. Furthermore, it was assumed that the polarity change of the magnetic field is perfect. Thus, errors regarding the field reversal must be simulated. One CW-CCW simulation is exemplary discussed [14], which includes the CW-CCW simulation results of rotation misalignments of $E \times B$ deflectors around the vertical axis of the FS lattice. If the influence of this misalignment on the spin motion is negligible the absolute value of the

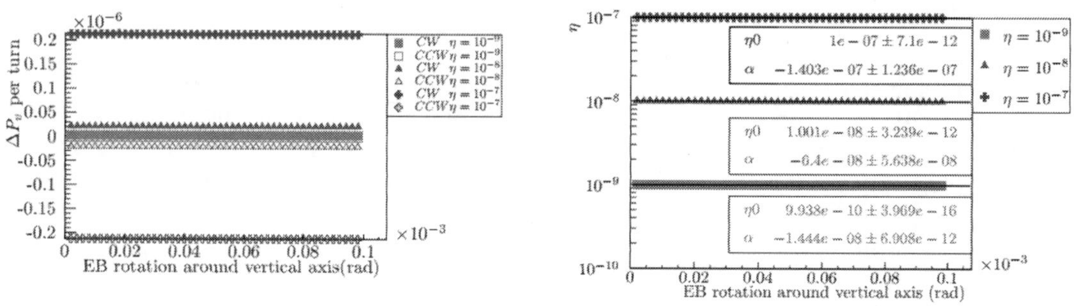

Figure 5. Frozen Spin method for CW-CCW beams for rotation misalignments of the $E \times B$ deflectors around the vertical axis. The left plot shows the vertical polarization build-up for different magnitudes η of EDM, the right plot the linear fit for the η_0 extracted from the simulation.

polarization build-up for both directions seems to be the same (see Figure 5, left plot). In Figure 5 (right plot) the scaling factor η is extracted and determined correctly within the analysis errors. The resulting values of η_0 confirm the above assumption.

4. Conclusion

For a perfect ring without misalignments the FS and QFS methods are proved and work equally well. However, unavoidable misalignments of ring elements prevent a measurement of the EDM with highest sensitivity. To counteract systematic errors CW-CCW beams are utilized. Assuming a perfect reversal of magnetic fields without influencing the electric field strength in $E \times B$ deflectors of a deuteron EDM storage ring, this procedure is a certified measure to handle errors caused by rotations of the deflectors. Furthermore, this method suppresses the misleading effect by shifts of quadrupoles.

A possibility to minimize the systematic error in QFS storage rings would be to decrease the sideways spin component. E.g. a QFS lattice should be modified in such a way, that alternating pairs of magnetic deflectors and straight $E \times B$ deflectors are used, leading to a polygon structure of the arc sections.

It is unavoidable to investigate different types of lattices (e.g. weak focusing structure) to make final conclusions for the sensitivity of EDM measurements with FS and QFS methods. In addition, 3D field maps have to be integrated in the simulation code and electric field plate and fringe field distribution of $E \times B$ deflectors optimized.

5. Acknowledgments

The authors would like to thank all members of the JEDI and srEDM collaborations for fruitful discussions. We are especially very grateful to M. Berz and K. Makino for supervised the upgrades of COSY Infinity, M. Rosenthal for developing COTOBO and Y. Senichev for providing the two ring lattices.

This work has been financially supported by an ERC Advanced-Grant (srEDM # 694390) of the European Union.

6. References

[1] Rathmann F *et al.* 2014 Search for electric dipole moments of light ions in storage rings *Physics of particles and nuclei* **45(1)** pp 229-33
[2] Sakharov A 1967 Violation of CP invariance, C asymmetry, and baryon asymmetry of the universe *JETP lett.* **5** pp 24-7
[3] Farley F J M *et al.* 2004 *Phys. Rev. Lett.* **93** 052001 hep-ex/0307006.
[4] Anastassopoulos D *et al.* (srEDM Collaboration) 2008 Search for a permanent electric dipole moment of the deuteron nucleus at the 10^{-29} e cm level proposal as submitted to the BNL PAC http://www.bnl.gov/edm/Proposal.asp
[5] JEDI Collaboration web page: http://collaborations.fz-juelich.de/ikp/jedi
[6] Lehrach A 2012 Project Overview and Computational Needs to Measure Electric Dipole Moments at Storage Rings, *Proc. of 11th International Computational Accelerator Physics Conference* (ICAP12) Rostock-Warnemünde Germany MOAAI1 p 7
[7] Senichev Y *et al.* 2015 Quasi-frozen Spin Method for EDM Deuteron Search *Proceedings of IPAC2015* Richmond VA USA MOPWA044 p 213
[8] Senichev Y *et al.* 2015 Investigation of lattice for deuteron EDM ring *Proceedings of ICAP2015* Shanghai China MODBC4 p 17
[9] Rosenthal M 2014 Development of a new control framework for lattice parameter calculations and particle tracking IKP Annual Report *Berichte des Forschungszentrums Jülich, Jül-4381 (2015)* http://www.fzjuelich.de/ikp/DE/Service/Download/Downloads/jahresbericht_2014_artikel.html
[10] Makino K 2006 *et al.* COSY Infinity version 9 *Nuclear Instruments and Methods in Physics Research Section A: Accelerators, Spectrometers, Detectors and Associated Equipment* **558(1)** pp 346-50
[11] Brun R *et al.* 1997 ROOT-an object oriented data analysis framework. *Nuclear Instruments and Methods in Physics Research Section A: Accelerators, Spectrometers, Detectors and Associated Equipment* **389(1)** pp 81-6
[12] Berz M 2015 *et al. COSY INFINTIY 9.2 Beam Physics Manual*
[13] Valetov E 2016 (MSU) private communications
[14] Skawran A 2016 Comparison of Frozen and Quasi Frozen Spin Concepts for a Deuteron Electrical Dipole Moment Storage Ring Master thesis RWTH Aachen
[15] Bargmann V *et al.*1959 Precession of the polarization of particles moving in a homogeneous electromagnetic field *Physical Review Letters* **2(10)** p 435

IOP Conf. Series: Journal of Physics: Conf. Series **874** (2017) 012051 doi:10.1088/1742-6596/874/1/012051

Analysis of closed orbit deviations for a first direct deuteron electric dipole moment measurement at the cooler synchrotron COSY

V Schmidt[1,2] **and A Lehrach**[1,2]

1 Institut für Kernphysik 4, Forschungszentrum Jülich, Leo-Brandt Str., 52425 Jülich, Germany

2 III. Physikalisches Institut B, RWTH Aachen University, Otto-Blumenthal-Str., 52056 Aachen, Germany

E-mail: v.schmidt@fz-juelich.de, a.lehrach@fz-juelich.de

Abstract. The Jülich Electric Dipole moment Investigations (JEDI) collaboration in Jülich is preparing a direct EDM measurement of protons and deuterons first at the storage ring COSY (COoler SYnchrotron) and later at a dedicated storage ring. Ensuring a precise measurement, various beam and spin manipulating effects have to be considered and investigated. A distortion of the closed orbit is one of the major sources for systematic uncertainties. Therefore misalignments of magnets and residual power supply oscillations are simulated using the MAD-X code in order to analyse their effect on the orbit. The underlying model for all simulations includes the dipoles, quadrupoles and sextupoles at COSY as well as the corrector magnets and BPMs (Beam Position Monitors). Since most sextupoles are only used during beam extraction, the sextupole strengths are set to zero resulting in a linear machine. The optics is adjusted in a way that the dispersion is zero in the straight sections. The closed orbit studies are performed for deuterons with a momentum of 970 MeV/c.

1. Introduction

The observed Matter-Antimatter asymmetry in the Universe cannot be explained by the Standard Model (SM) of Particle Physics. In order to resolve the matter dominance an additional $\mathcal{CP}$ violating phenomenon is needed. A candidate for physics beyond the SM is a non-vanishing Electric Dipole Moment (EDM) of subatomic particles. Since permanent EDMs violate parity and time reversal symmetries, they are also $\mathcal{CP}$ violating if the $\mathcal{CPT}$-theorem is assumed. Since the Standard Model (SM) predicitons for EDMs are many orders of magnitude too small to explain the dominance of matter, the discovery of larger nucleon EDMs would indicate physics beyond the SM and could give an explanation for the Matter-Antimatter asymmetry [1]. The interaction of a particles' spin with strong electric fields enables the measurement of an EDM. The underlying experiments need to be performed with high-precision storage rings and require an accurate measure and control of the spin and the beam. The JEDI collaboration therefore investigates spin and beam influencing effects in order to allow for EDM studies at COSY [2] [3].

2. Magnet misalignments

The magnet strengths and their positions mainly determine how the beam propagates through the ring. It is therefore important to investigate the effect of misaligned quadrupoles and dipoles in the model and to compare them to the actual setup of COSY [4]. Considering magnet misalignments for each dipole and quadrupole, six independent Gaussian distributed positioning errors (displacement along each axis, rotation around each axis) are generated according to

$$\Delta(x, y, s, \phi, \Theta, \Psi) = \text{Gauss}(0, \sigma_{x,y,s,\phi,\Theta,\Psi}), \tag{1}$$

with $\frac{\sigma_x}{\text{m}} = \frac{\sigma_y}{\text{m}} = \frac{\sigma_s}{\text{m}} = \frac{\sigma_\phi}{\text{rad}} = \frac{\sigma_\Theta}{\text{rad}} = \frac{\sigma_\Psi}{\text{rad}}$, where $x, (y, s)$ describes the displacement along the x- (y-, s-) axes. Rotations around the axes are indicated respectively by ϕ, Θ and Ψ.

Including these magnet misalignments in the model, the closed orbit is simulated and the closed orbit RMS[1] is calculated in both transverse directions. Te reference particle is assumed to have a vanishing momentum deviation. All transfer matrices in MAD-X [5] are based on the optical functions given in [4]. The BPMs as well as the corrector magnets are assumed to be ideal in order to investigate only the effect of magnet misalignments. The results are considered before and after applying an orbit correction[2]. For each standard deviation 1000 random samples of magnet misalignments are used. Averaging over all 1000 sample values leads to the mean closed orbit RMS in horizontal and vertical direction. The simulation results for the accumulated effect of displacements and rotations of dipoles and quadrupoles are shown in figure 1 in both transverse directions including a linear fit to the data. The error bars indicate the standard errors of the mean values. Since the model is based on a linear machine the behaviour in both transverse directions is strictly linear. The results in both directions are not equal since the optical functions in horizontal and vertical direction differ from each other and the horizontal orbit response is also influenced by dispersion. The measured uncorrected closed orbit RMS at COSY in 2016 was of the order of a couple of millimeter, which corresponds to a simulated standard deviation of about 0.5 mm. Since the simulation results after the orbit correction do not include BPM resolution constraints and corrector magnet uncertainties, they cannot be directly compared to the measured corrected closed orbit at COSY. In order to perform a high precision experiment one aims to reach a closed orbit RMS of about 100 µm in the future by improving the magnet positioning [6].

2.1. Survey at COSY

The former results show that magnet misalignments are one of the main sources of closed orbit deviations at COSY. It is therefore necessary to determine the current positions of all dipoles and quadrupoles and to correct large displacements and rotations towards the target position. A corresponding survey was conducted in April 2016. The dipoles and quadrupoles at COSY are armed with reference marks. A laser-based position measurement can be carried out according to a fixed reference point. The positioning of these marks on the magnet is sketched in figure 2 [7].

Taking the first dipole in the left arc as the reference point, the relative vertical displacement of all other dipoles is measured. The same procedure is used for quadrupoles, taking the first quadrupole after the injection point as the reference element. Given these information a best-fit-plane was estimated to which all elements afterwards should be optimally positioned. The best-fit-plane is found by taking the vertical measurement results and fitting a plane to the values which minimizes the vertical deviations. It turned out that taking only the reference marks P2 and P3 of the dipoles into account leads to the most accurate result for the fitted plane. Ignoring some outliers of the measurement points further improves the result [7]. Finally,

[1] Root Mean Square
[2] Using Singular Value Decomposition (SVD) of the response matrix

IOP Conf. Series: Journal of Physics: Conf. Series **874** (2017) 012051 doi:10.1088/1742-6596/874/1/012051

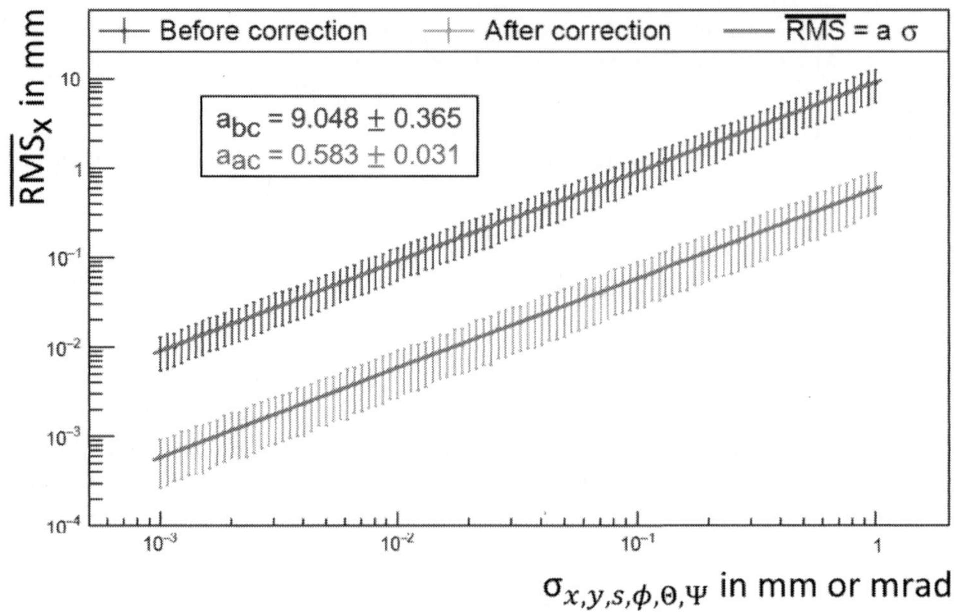

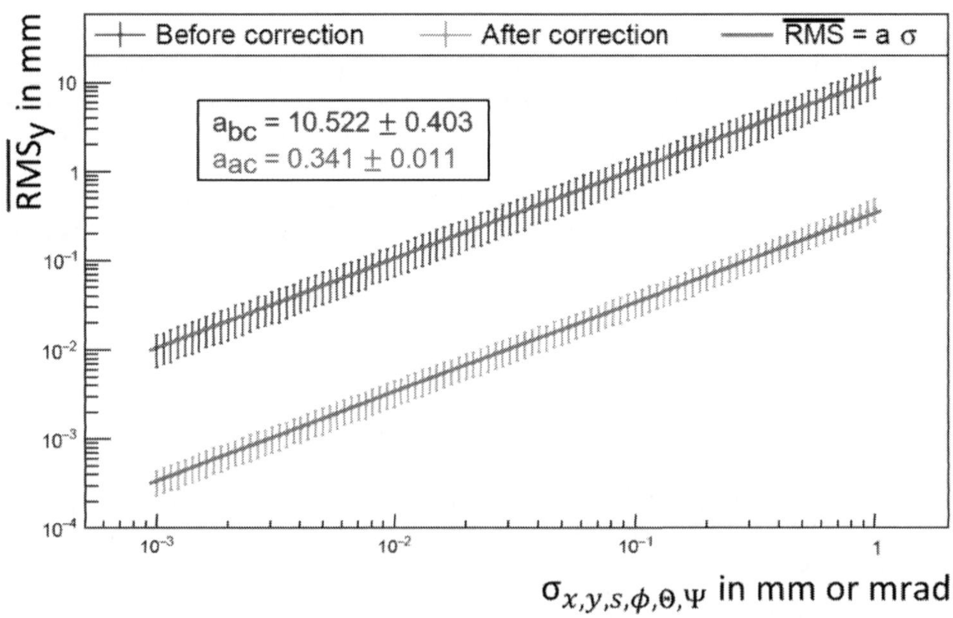

Figure 1. Mean closed orbit RMS in horizontal and transverse direction for simultaneous displacement and rotation of dipoles and quadrupoles. Blue markers: results before orbit correction. Green markers: results after the orbit correction. The values result from generating magnet misalignments using 1000 random seeds.

the deviations of all magnets from the achieved best-fit-plane were calculated in each direction using the several reference marks. Implementing the measured magnet misalignments into the

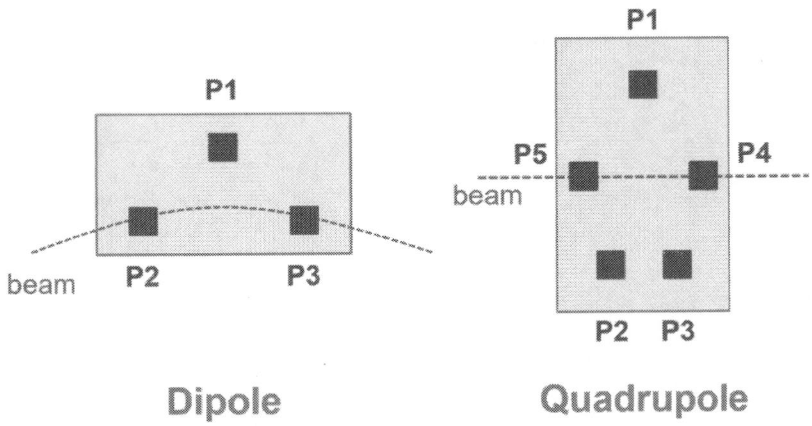

Figure 2. Topview on magnets: Reference marks on dipoles and quadrupoles. For dipoles the marks P2 and P3 are the closest to the beam path. In the case of a quadrupole, the marks P4 and P5 lie directly above the ideal beam trajectory through the magnet. [7]

COSY model leads to the uncorrected closed orbit shown in figure 3. The closed orbit RMS values are similar to the ones that are measured at COSY [8] which supports the assumption that magnet misalignments are the main source for closed orbit deviations. The comparison of the simulated corrected closed orbit and the measured one is not suitable since BPMs and correctors are assumed to be ideal in the simulation.

3. Power supply oscillations

One reason for field changes in the magnets are residual power supply oscillations which induce oscillating magnetic fields and cannot be controlled with the static orbit control system of COSY. In order to investigate the effect of residual power supply oscillations on the transverse closed orbit RMS, a sinusoidale oscillation with an amplitude of $\Delta I_{\max}/2$ is assumed, where $\Delta I_{\max}$ indicates the peak-to-peak value resulting from the relative errors and the maximum provided current of the power supplies (figure 4). In table 1 the current uncertainties as well as the maximum possible current values for each type of magnet at COSY are summarized [9]. Independently of the magnet type, its strength depends linearly on the current. Thus one can easily deduce the oscillation of the magnet strength given its variation of the current [10]. For each power supply an amplitude of the sine wave is randomly generated using a Gaussian distribution with a standard deviation of $\Delta I_{\max}/2$. The resulting value for the change in current is then given by

$$\Delta I = \text{Gauss}\left(0, \frac{\Delta I_{\max}}{2}\right). \tag{2}$$

Incorporating the generated field variations in the model and calculcating the closed orbit leads to a snapshot of the dynamic scenario. By creating various of these snapshots one can investigate the average influence of power supply oscillations on the transverse closed orbit with respect to a given reference orbit. In the simulaton 1000 snapshots, i.e. 1000 closed orbit calculations, each for a different set of current errors, are generated. For the reference orbit a fixed set of misaligned dipoles and quadrupoles was chosen. The method is sketched in figure 5.

IOP Conf. Series: Journal of Physics: Conf. Series **874** (2017) 012051 doi:10.1088/1742-6596/874/1/012051

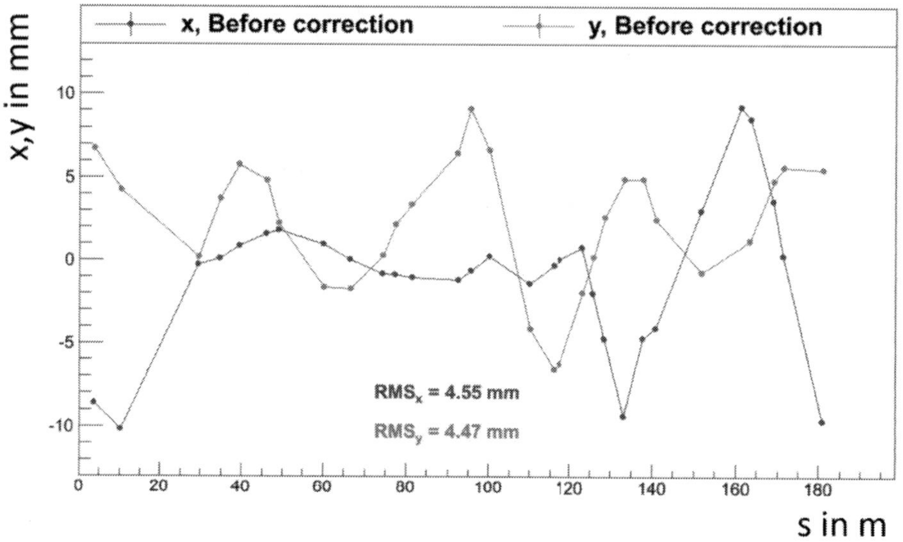

Figure 3. Simulated uncorrected closed orbit resulting from magnet misalignments taken from the survey data. The markers show the simulated BPM readings. The closed orbit RMS in the uncorrected case is similar to measured RMS values at COSY. Also the whole pathway of the uncorrected closed orbits show similarities to the measured orbits.

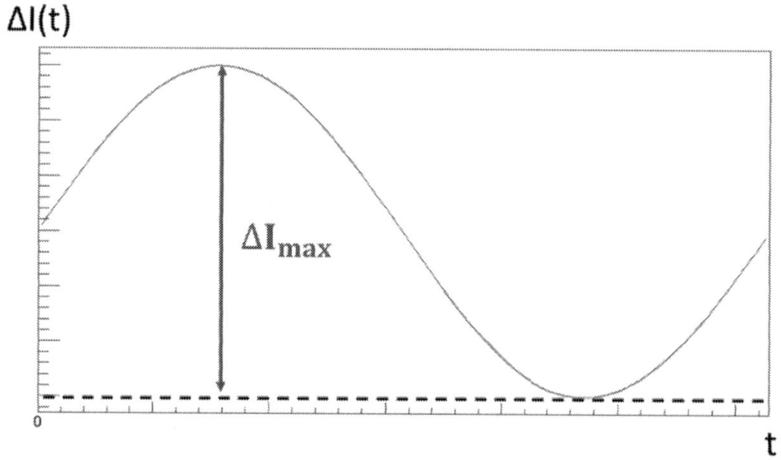

Figure 4. Sketch of sinusoidal residual power supply oscillation. The typical frequency of the power supply oscillation is approximately 600 Hz.

The deviations from the reference orbit at each BPM are collected in a histogram for all 1000 simulations and a Gaussian fit is performed to the data of each BPM. The width of the fit indicates the influence of the field oscillation at this specific BPM. The average width over all

Table 1. Relative error and maximum current of COSY magnets.

Relative error of the current of COSY magnets			
Magnet	σ [ppm]	I_{max} [A]	ΔI_{max} [A]
Dipole	20	5000	100
Quadrupole	20	550	11
Sextupole	500	275	137.5
Corrector	100	30	3

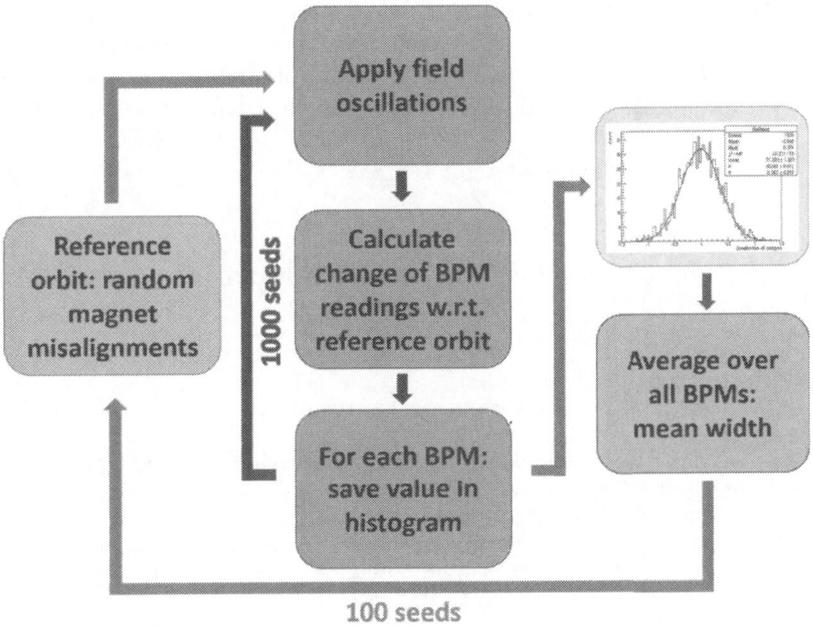

Figure 5. Simulation method of power supply oscillation influence on the transverse closed orbit of COSY.

BPMs indicates the influence on the closed orbit RMS in x- (y-) direction with respect to the given reference orbit. The procedure is repeated for 100 different reference orbits. Averaging over all reference orbits describes the global influence of field changes caused by residual power supply oscillations. The results for each magnet type are summarized in table 2. None of the RMS changes is of a larger order of magnitude than 10 µm.

4. Conclusion

The simulations show that the closed orbit at COSY is influenced much more by magnet displacements and rotations than by residual power supply oscillations. The effects of misaligend magnets dominate the effect of power supply oscillations by many orders of magnitude. Implementing the measured magnet deviations form the target positions into the simulation

IOP Conf. Series: Journal of Physics: Conf. Series **874** (2017) 012051 doi:10.1088/1742-6596/874/1/012051

Table 2. Influence of residual power supply oscillations of the COSY magnets on the transverse closed orbit RMS.

Closed orbit changes due to power supply oscillations		
Magnet	$\Delta\overline{\text{RMS}}_x$	$\Delta\overline{\text{RMS}}_y$
Dipole	(27.69 ± 0.24) µm	(9.00 ± 0.10) nm
Quadrupole	(1.11 ± 0.01) µm	(0.70 ± 0.01) µm
Sextupole	(48.70 ± 0.90) nm	(49.90 ± 0.60) nm
Corrector	(34.43 ± 0.30) µm	(28.07 ± 0.22) µm

model of COSY leads to similar closed orbit RMS values than the measured ones. In a next step also sextupole strengths are included in the model in order to provide long spin coherence time [11] and furthermore BPM and corrector magnet constraints are implemented. Regarding the future EDM experiment at COSY, the residual power supply oscillations play a minor role when trying to improve the quality of the closed orbit to reach a transverse RMS of about 100 µm [6]. Currently, the COSY magnets are realigned to their target positions with an accuracy of 0.2 mm [7] in order to reduce the closed orbit RMS.

Acknowledgments
This work has been financially supported by an ERC Advanced-Grant (srEDM # 694390) of the European Union.

References

[1] Sakharov A 1967 Violation of CP invariance, C asymmetry, and baryon asymmetry of the universe *JETP Letters* **5** 24-7
[2] Lehrach A 2011 *et al.* Precursor experiments to search for permanent electric dipole moments (EDMs) of protons and deuterons at COSY *Proc. XIV Workshop on High Energy Spin Physics (DSPIN-11)* (Dubna Russia: Joint Accelerator Conferences Website (JACoW)) pp 287-302.
[3] Lehrach A 2012 Project overview and computational needs to measure electric dipole moments at storage rings *Proc. 11th Int. Computational Accelerator Physics Conf. (ICAP12)* (Rostock-Warnemnde Germany: Joint Accelerator Conferences Website (JACoW)) p 7
[4] Weidemann C *et al.* 2015 Toward polarized antiprotons: machine development for spin-filtering experiments *Phys. Rev. ST Accel. Beams* **18** 3-5
[5] Methodical Accelerator Design, `http://mad.web.cern.ch/mad/`
[6] Rosenthal M and Lehrach A 2015 Spin tracking simulations towards electric dipole moment measurements at COSY *Proc. 6th Int. Particle Accelerator Conf. (IPAC15)* (Richmond VA USA: Joint Accelerator Conferences Website (JACoW)) pp 3764-6
[7] Vermessungsbüro Dipl.-Ing. H. J. Stollenwerk, private communication, 50126 Bergheim, Aug. 2016
[8] Hinder F *et al.* 2015 Beam Position Monitors at COSY, unpublished
[9] Retzlaff M, private communication, Forschungszentrum Jülich, Mar 2016
[10] Wille K 2000 *Linear beam optics* in *The Physics of Particle Accelerators. An Introduction* (Oxford UK :Oxford University Press) pp 50-8
[11] Guidoboni G 2015 Spin coherence time lengthening of a polarized deuteron beam using sextupole fields *Proc. 6th Int. Particle Accelerator Conf. (IPAC15)* (Richmond VA USA: Joint Accelerator Conferences Website (JACoW)) pp 4066-9

ADIGE: the radioactive ion beam injector of the SPES project

A Galatà, L Bellan, G Bisoffi, M Comunian, L Martin, M F Moisio, A Palmieri, A Pisent, G Prete, C Roncolato

INFN-Laboratori Nazionali di Legnaro, Viale dell'Università 2, 35020 Legnaro, Padova, Italy

E-mail: alessio.galata@lnl.infn.it

Abstract. The Selective Production of Exotic Species (SPES) project is presently under development at INFN-LNL: aim of this project is the production, ionization and post-acceleration of radioactive ions to perform forefront research in nuclear physics. An ECR-based charge breeder (SPES-CB) will allow post-acceleration of radioactive ions: in particular, the SPES-CB has been designed and developed by LPSC of Grenoble, based on the Phoenix booster. It will be equipped with a complete test bench totally integrated with the SPES beam line: this part of the post-accelerator, together with the newly designed RFQ, composes the so-called ADIGE injector (Acceleratore Di Ioni a Grande carica Esotici) for the superconducting linac ALPI. The injector will employ a unique Medium Resolution Mass Spectrometer (MRMS, resolving power 1/1000), mounted downstream the SPES-CB, in order to avoid the typical drawback of the ECR-based charge breeding technique, that is the beam contamination. This contribution describes the ADIGE injector, with particular attention to the analysis of possible contaminations and the performances expected for the MRMS, showing the beam dynamics calculations for a reference radioactive beam.

1. The ADIGE Injector

SPES [1] is an INFN project with the aim at developing an ISOL Radioactive Ion Beam (RIB) facility as an intermediate step toward EURISOL. SPES is under realization at INFN-LNL: its main goal is the production and post-acceleration of exotic beams to perform forefront research in nuclear physics, by studying nuclei far from stability. To boost the charge state of the radioactive beam produced, and allow post-acceleration by a newly designed RFQ [2] coupled to the ALPI [3] superconducting linac (up to 10 MeV/amu for A=130), an ECR-based charge breeding technique (ECR-CB) will be employed: in particular, it will be an updated version of the Phoenix booster developed by Laboratoire de Physique Subatomique & Cosmologie (LPSC) [4]; the SPES-CB was succesfully tested at LPSC in April 2015 and then delivered to LNL [5]. Its beam line was designed not only as a part of the final layout of the SPES project, but also as a test bench to characterize accurately this new device before operating it with radioactive ions: this part of the post-accelerator, together with the new RFQ, composes the so-called ADIGE injector, whose layout is shown in figure 1. The injector will include all the necessary elements to charaterize properly the SPES-CB and allow an optimum transport of the RIB towards the RFQ. To verify and optimize the breeding efficiency, singly charged stable beams will be produced by copies of the sources coupled to the UC_x target of SPES (S1+ in figure 1; in particular, surface ionization sources for the elements of I and II group, plasma ion sources for the others), and injected into

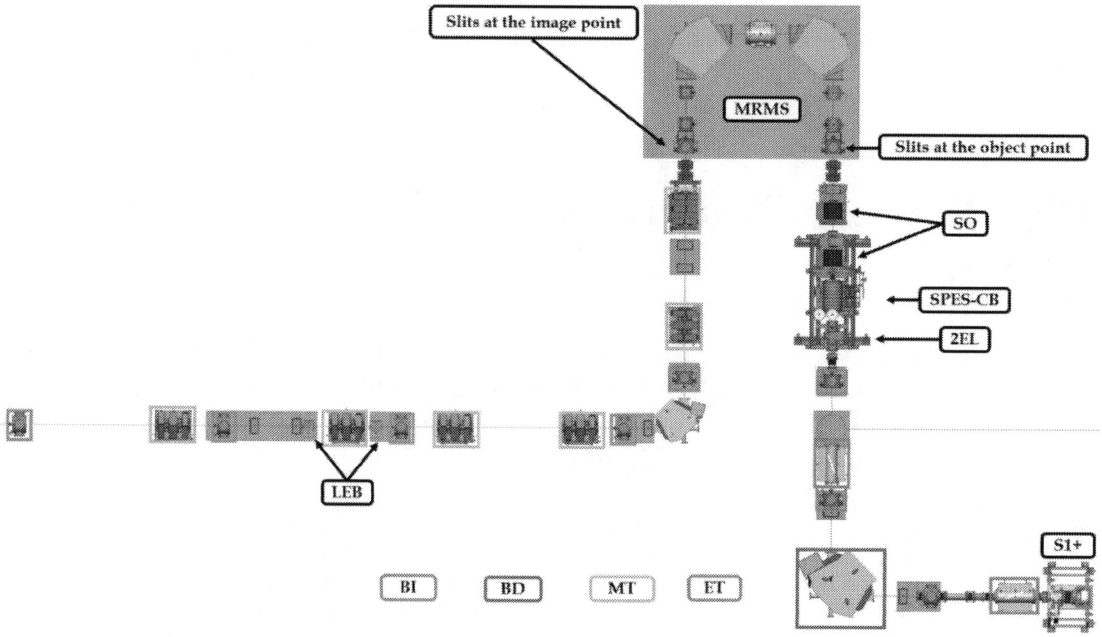

Figure 1. Layout of the ADIGE injector. S1+=1+ ion source; 2EL=double Einzel lens; SO=solenoids; LEB=low energy buncher; BD=bending dipole; BI=beam instrumentation boxes; MT=magnetic triplets; ET=electrostatic triplets.

the SPES-CB for its characterization. The 1+ beams will be magnetically selected by a 90° bending dipole (BD in figure 1; r=750 mm, $\Delta M/M\sim 1/150$), analyzed in terms of intensity and emittance through beam instrumentation boxes (BI in figure 1), and trasported through an electrostatic beam line towards the SPES-CB (ET in figure 1), where they will be focused and injected by means of an especially designed double Einzel lens incorporating electrostatic steeres (2EL in figure 1). The charge bred multiply charged RIB will be then extracted and transported in a fully magnetic beam line (SO and MT in figure 1): the desired charge state will be selected by the MRMS and finally delivered to the RFQ for post-acceleration.

One of the main concerns of an ISOL facility is the beam purity: the following sections will show an analysis of the possible contaminants induced by the breeding stage, and describe the strategies adopted to ensure pure post-accelerated radioactive beams for the future users of the facility.

2. The beam purity issue

A well known drawback of an ECR-CB is the contamination induced by the breeding stage. Impurities are usually present in the gas fed into the plasma chamber (normally oxygen), or derive from the outgassing of the surfaces exposed to vacuum: we call this source gaseous contaminants. Another source consists in the materials of the surfaces exposed to vacuum, released due to their interaction with the plasma: we call them solid contaminants. To limit the first contribution, the SPES-CB will have almost all metal sealings, except for the vacuum window (O-ring on the wave guide) and the interface between the extraction iron plug and the plasma electrode. Additianally, special attention was paid to the surface treatments, in particular of the stainless steel plasma chamber and the iron plug at extraction (ARMCO) [5]. For the former, a sequence of two thermal treatments was applied: the first are perfomed on each part comprising the plasma chamber, machined with some mm of stock material so as to

Table 1. Main radioactive species produced at SPES.

Species	Mass [amu]	Species	Mass [amu]
Al	24-26	**Cu**	69-74
Ni	78	**Kr**	88-90
Rb	86-94	**Sr**	94-100
Sn	123,125-132	**Xe**	135,137,138
Cs	131,132,134-141	**Ba**	141-143

Table 2. Possible contaminants induced by the SPES-CB.

Species	Mass [amu]	Species	Mass [amu]	Species	Mass [amu]
C	12-13	**N**	14-15	**O**	16-18
Mg	24-26	**Al**	27	**Si**	28-30
P	31	**S**	32-34,36	**Ar**	36,40
Cr	50,52-54	**Fe**	54,56-58	**Mn**	55
Ni	58,60-62,64	**Co**	59	**Cu**	63,65
Zn	64,66-68,70	**Mo**	92,94-98,100		

correct possible deformations; the second one after welding the parts together but before final machining, that was executed with isopropylic alcohol only. For what concerns the extraction plug made of ARMCO, "Normalising" and "Final Annealing" cycles were adopted, both in a vacuum atmosphere with a pressure below 10^{-5} mbar.

The contribution from solid contaminants is more problematic, mostly because consisting in heavy elements (A>40) that once ionized produce a dense spectrum of several peaks: the materials to be considered are stainless steel (AISI 316 L) for the plasma chamber and aluminum (AU4G) for the extraction electrode. As a first step, we analysed the possibile superpositions between the expected peaks of the radioactive ions produced at SPES and the contaminants (see tables 1 and 2), including the gaseous ones for the sake of completeness, and considering a charge-to-mass ratio (M/q) between 4 and 7. The analysis revealed that a resolving power $R=\Delta(M/q)/(M/q)=1/1000$ for the MRMS is a good compromise between complexity of the design on one hand, and the possibility to find a clean radioactive peak in the range of M/q of interest for most of the elements shown in table 1 on the other hand. At the same time, the analysis produced the following information:

- To have clean $^{24-26}$Al peaks, stainless steel and aluminum should be avoided, and the use of other materials must be explored.
- The isotope ^{70}Cu would show clean peaks if aluminun were not used.
- The isotope ^{78}Ni would show clean peaks if Kr were absent.
- The isotopes ^{86,93}Rb would show clean peaks if, respectively, Kr were absent and stainless steel (Mo presence) were not used.
- The isotopes $^{94-98,100}$Sr would show clean peaks if stainless steel (Mo presence) were not used.

From the results shown above it comes out that, besides the well known positive effect on the performances of an ECR source in general, presently the "cleanest" material among the ones commonly used for the plasma chamber is aluminum (except for radioactive Al and ^{70}Cu), and this will be in fact one of the first upgrades which will be carried out in the next future on the SPES-CB.

3. Cleaning effect of the MRMS

A description of the MRMS is reported in [4]. An analysis of the possible superpositions has been done considering the fact that contaminants have usually a much higher intensity, up μA in some cases, than the radioactive ones, that barely reach the nA level and only in few cases. The initial hypothesis considers two beams separated of 1/1000 in mass, with one of the two (the stable one) two orders of magnitude larger compared to the second (the radioactive or nominal one, in the present case $^{132}Sn^{19+}$), both with $M/q{\sim}7$. For both beams it has been assumed a rms normalized emittance of 0.1 π*mm*mrad (much higher compared to what has been measured during the SPES-CB accepatance tests, see [5]), and a Gaussian spatial profile up to 6σ. In the analysis, the beam is transported from the slit at the object point to the slit at the image point, both mounted on the HV platform housing the MRMS (see [4]), and the separation is carried out for a total beam energy 3.04 MeV (40 kV*q due to the charge breeder extraction plus 120 kV*q due to the HV platform; q=19+), thus limiting the effect of the energy spread induced by the SPES-CB ($\sim$ 10-15 eV) by reducing the value of $\Delta E/E{=}2\Delta p/p$. We proceed in two ways: on one hand, by doing multiparticles simulations of beam transport through the MRMS with the TraceWin code; on the other hand, by using an analytic calculation: this was done assuming two Gaussian beams with relative areas proportional to the beam intensity, and then calculating the overlap between the two. Both results showed a good agreement. Figure 2 reports the simulation results after the image slits with an aperture of $\pm$ 6 mm: the nominal beam is preserved at the 98%, while the contaminant is reduced at 0.08% the intenisty of the former. This result confirm that, with this slits aperture, it is possible to saparate the nominal beam from a contaminant two order of magnitude higher in intensity than the radioactive beam. To check

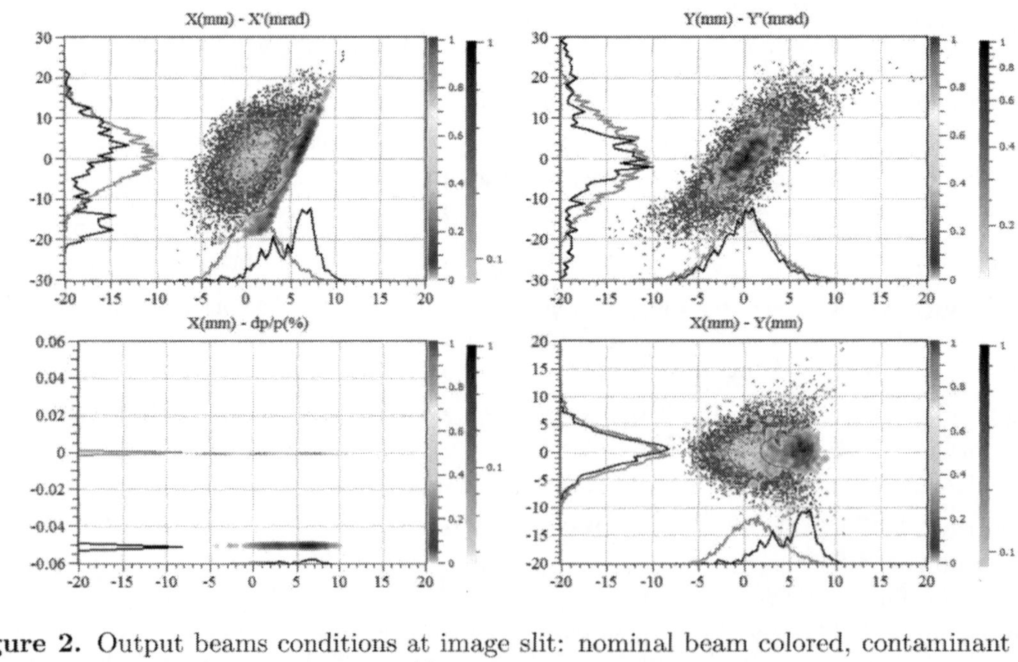

Figure 2. Output beams conditions at image slit: nominal beam colored, contaminant black; both separated of 1/1000 in mass: 98% of the peak consists in the nominal beam, 0.08% the contaminant.

the effects of a reduction of the split aperture, we considered two gaussian functions separated by the nominal dispersion of the line D=8000 that, considering a value of $\Delta p/p$=1/2000, gives a nominal separation d=4. Then, we calculated the superposition area at various contaminants amplitude assuming the effects of the separation due to the MRMS (the aforemenationed d): results are shown in figure 3. The graph shows the contamination for different slit aperture: the amount indicated on the top of the graph is the nominal beam quantity not cut by the slit. It can be seen that, by reducing the slit aperture at ±1.3 mm, even contaminants with a relative intensity 10^3 times the nominal one can be reduced below 10%, preserving the 80% of the nominal beam. In conclusion, the results showed that it is possible to significantly limit beam contamination by reducing the slits aperture, even if the contaminants are orders of magnitude higher in intensity.

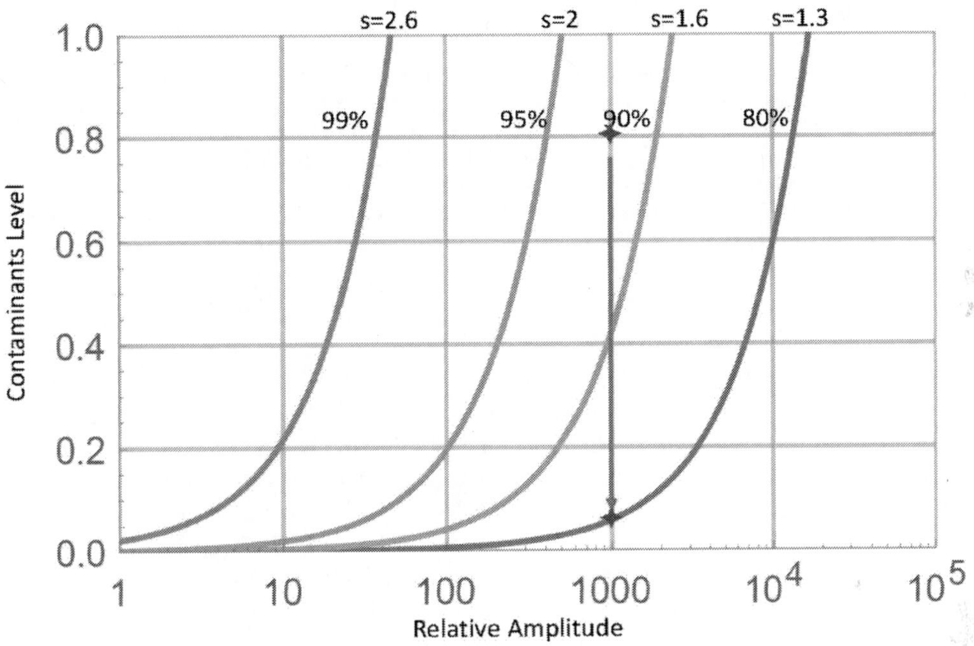

Figure 3. Level of contaminants: 1 means 100% fully contaminated beam, versus the contaminants relative amplitude, 10 means the contaminants have 10 times larger current with respect to the nominal beam. The percent figure is how much of the nominal beam survives after the slits. The "s" values indicated for each curve is the semiaperture of the slits.

References

[1] De Angelis G et al. 2016 *EPJ Web of Conferences* **107** 01001.
[2] Pisent A, Bellan L, Comunian M, Chalykh B, Russo A and Calabretta L 2014 SPES beam dynamics *Proceedings of HB2014* East-Lansing MI USA.
[3] Bisoffi G *et al.* 2016 *Nuclear Instruments and Methods in Physics Research* B **376** 402-407.
[4] Galatà A *et al.* 2016 *Nuclear Instruments and Methods in Physics Research* B **376** 329-333.
[5] Galatà A, Patti G, Roncolato C, Angot J and Lamy T 2016 *Rev. Sci. Instr.* **87** 02B503

Preliminary design of a high-intensity continuous-wave deuteron RFQ

X Liu, O Kamigaito, N Sakamoto and K Yamada
RIKEN Nishina Center for Accelerator-Based Science, Wako, Saitama, Japan
E-mail: xingguang.liu@riken.jp

Abstract. A high-intensity deuteron linear accelerator is currently being studied as a promising candidate to treat high-level radioactive waste through the nuclear transmutation process. This paper presents the study on a design of a 75.5 MHz, 400 mA, continuous-wave deuteron radio-frequency quadrupole (RFQ), which is proposed as the front-end of such a linear accelerator. The results of the beam dynamics simulation suggest that the designed RFQ can accelerate a 400-mA deuteron beam from 100 keV to 2.5 MeV with a transmission rate of 92.0 $\sim$ 93.3%, depending on the assumed input transverse emittance.

1. Introduction

To reduce nuclear waste, we are proposing a linear accelerator (LINAC) system that delivers high-intensity deuteron (D) beams to the waste target. The reaction between D and the elements in the waste has been under investigation[1, 2]. The required beam intensity has been estimated to be 1000 mA[3], which is beyond the capability of existing accelerators. Thus, a continuous-wave (CW) 400 mA D^+ LINAC is being considered. Multiple LINACs can be built to meet the intensity requirement. Superconducting cavities are also being designed [4] for the high-energy part of such a LINAC. The low-energy part is also critical, for which we propose a radio-frequency quadrupole (RFQ) structure.

High intensity RFQs have become a popular low-energy solution since the 1980s[5]. Especially for the CW beam, RFQs can be used as a buncher as well as an accelerator. The most recent high-intensity D^+ RFQ is the RFQ of the IFMIF project (IFMIF-RFQ) [6], which is scheduled for beam commissioning this year [7]. Our RFQ is similar to the IFMIF-RFQ in many ways except that in our case, we are considering a higher-intensity, 400 mA beam. In this paper, we will present our conceptual beam dynamics design of the RFQ by leaving some of the challenging issues for future work.

In the next section, the RFQ parameters and estimation of the design parameters are presented. The simulation results are discussed after that, followed by a summary.

2. Design process and optimization

2.1. RFQ Parameters

The design specifications of this deuteron RFQ (RFQ-D) are listed in Table 1. The input beam energy is 100 keV. The frequency is chosen to be identical to the superconducting cavities[4], which can accept D^+ with a $\beta(= v/c$, where v is the particle velocity and c is the speed of light) as low as 0.05. Thus, the output energy is chosen to be 2.5 MeV. For the emittance, there is

Table 1. Design specifications.

Beam	D^+
Frequency [MHz]	75.5
Input current [mA]	400
Duty	100% (CW)
RMS input emittance (normalized) [π.mm.mrad]	$0.25 \sim 1$
Input total energy [MeV]	0.1
Output total energy [MeV]	2.5

currently not enough information. In the study of the IFMIF-RFQ design, this value is chosen to be 0.25 π.mm.mrad. The commissioning results of the IFMIF ion source indicate that the actual emittance is 0.34 π.mm.mrad in the case of their 109 mA D^+ beam in CW mode [7]. Based on the simulation carried out in this study, the results with a normalized emittance of $0.25 \sim 1$ π.mm.mrad will be compared.

2.2. Design Parameters Estimation

The maximum surface electric field is determined by the well-known Kilpatrick criterion [8] which can be written as,

$$f = 1.643E^2 e^{-8.5/E} \tag{1}$$

The maximum electric field allowed for the RFQ-D is 18.67 MV/m, if given the same $K_p = 1.82$ as the IFMIF-RFQ. Both RFQs are marked on Figure 1, where RFQ-D is the RFQ in this paper.

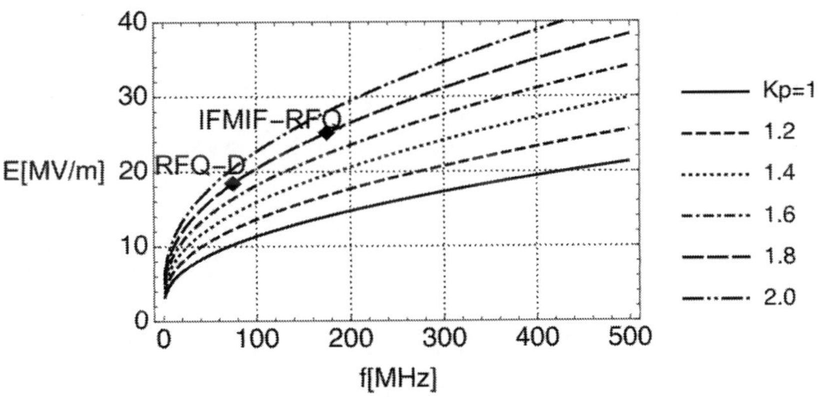

Figure 1. Maximum surface electric field in terms of f and K_p.

In the RFQ, the transverse motion is well-approximated by Mathieu's equation when the space charge is negligible,

$$\frac{d^2x}{d\eta^2} + (B\sin(2\pi\eta) + \Delta_{\rm rf})\, x = 0 \tag{2}$$

where B is the focusing factor, $\Delta_{\rm rf}$ is the RF defocussing factor, and $d\eta = dz/\pi$ is the longitudinal coordinate.

The focusing factor B is,

$$B = \frac{q\lambda^2 V_0}{m_0 c^2 r_0^2} \tag{3}$$

where q is the particle charge, λ is the wavelength, V_0 is the vane voltage, m_0 is the particle mass, and r_0 is the average aperture.

The RF defocussing factor is,

$$\Delta_{\text{rf}} = \frac{q\pi^2 V_0}{2m_0 c^2 \beta^2} A \sin\phi_s \tag{4}$$

where A is the acceleration efficiency, which can be expressed in terms of the modulation m, and k and a are the wave number and small aperture of a Bessel function, respectively.

$$A = \frac{m^2 - 1}{m^2 I_0(ka) - I_0(mka)} \tag{5}$$

The electric field on the pole tip is $E_s = V_0/r_0$ and the maximum electric field is $E_{max} = 1.36E_s$. With $K_p = 1.82$, E_s is 13.72 MV/m.

With Equation (3), the average aperture r_0 and surface field E_s can be written in terms of B and V_0,

$$r_0^2 = \frac{q\lambda^2 V_0}{m_0 c^2 B} \tag{6}$$

$$E_s^2 = \frac{m_0 c^2 B V_0}{q\lambda^2} \tag{7}$$

Figure 2 shows the relationship of B, V_0, E_s and r_0 when $f = 75.5$ MHz, given by Equation (6) and Figure 2. For such a high-intensity, we can expect a large B to provide sufficient transverse focusing. In Figure 2, when E_s is about 14 MV/m, B will be about 10, the vane voltage V_0 will be between 100 kV and 200 kV, and the average aperture r_0 will be 10 mm. A higher vane voltage is advantageous in reducing the total RFQ length. In our design, 150 kV is considered a practical maximum. By choosing B to be about 10 and with Δ_{rf} being a typical negative value, the synchronous phase will be less than $\pi/2$ to avoid envelope instability [9].

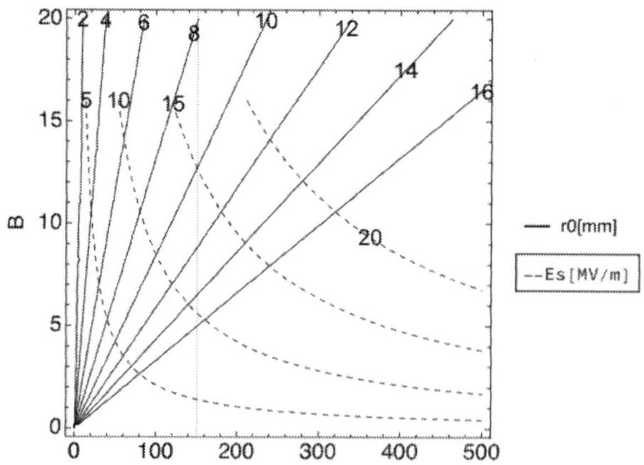

Figure 2. Relationship between B, V_0, E_s and r_0.

3. Results and Discussion

A tool-set provided by Los Alamos National Laboratory (LANL) has commonly been used for the RFQ design [10, 11]. An RFQ designed with this tool set is divided into four sections: Radial Matching Section (RMS), Shaper (SH), Gentle Buncher (GB) and Accelerator Section (ACC).

Designs with different parameters are then optimized by scanning the ranges of the parameters estimated in the previous section. The transmission rate given by PARMTEQM is used to judge the goodness of a design. Figure 3 shows the cell parameters of the designed RFQ, where W_s is the energy of the synchronous particle, from 0.1 keV to 2.5 MeV. The synchronous phase is -62° at the GB end and gradually shifts to -45° at the RFQ end. If we increase the vane voltage in the ACC region, the synchronous phase will shift to a more typical value of -30°. However, 150 kV is already a large value, so the final synchronous phase is -45° and the total RFQ length is about 11 meters. While r_0 and V_0 are kept constant along the RFQ, B and A are slightly modified by PARI along the RFQ, around 10.6 and 0.29, respectively.

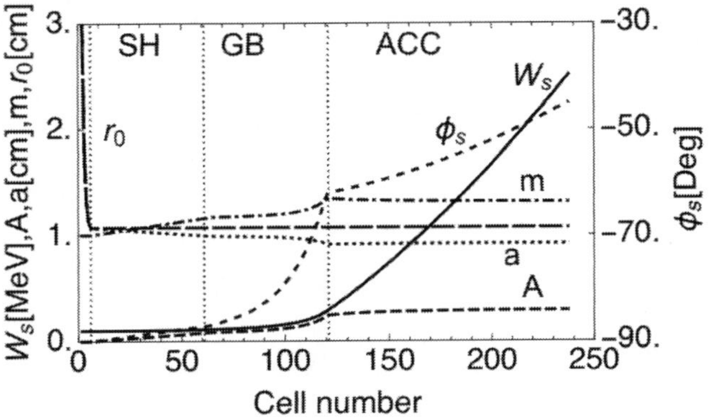

Figure 3. Cell parameters of RFQ-D.

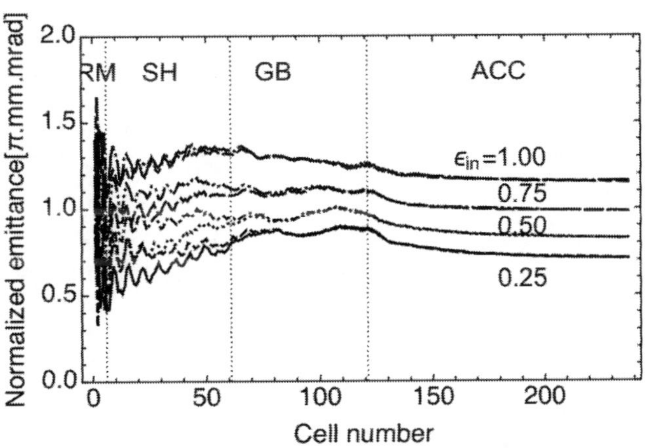

Figure 4. Transverse RMS emittance along the RFQ.

The dependence on the input emittance is also investigated. Figure 4 shows the transverse RMS emittance evolution along the RFQ. Four cases of emittance have been assumed, as shown in this figure. The horizontal and vertical emittance values are equal at the input, and both directions are shown in the figure. The results suggest that a smaller input transverse emittance leads to a smaller output emittance. However, if we look at the emittance blowup in terms of the ratio, as presented in Table 2, under the same beam intensity, a higher input emittance will have a smaller emittance blowup. There is no need to suppress the transverse emittance to an extremely small level for such an RFQ.

Table 2. Emittance and transmission rate at RFQ end.

ϵ_{in}	0.25	0.50	0.75	1.00
$\epsilon_x(/\epsilon_{x0})$	0.72(2.9)	0.82(1.6)	1.00(1.3)	0.11(1.1)
$\epsilon_y(/\epsilon_{y0})$	0.70(2.8)	0.84(1.7)	0.99(1.3)	0.11(1.1)
Trans.%	93.3	92.5	92.3	92.1

The transmission rate also depends on the input transverse emittance as shown in Figure 5. Table 2 presents the values of the transmission rate at the RFQ end. Most of the beam loss occurs in the bunching process, where the energy is still low, thus, the power loss to the RFQ chamber is low. The beam loss continues gradually in the ACC region.

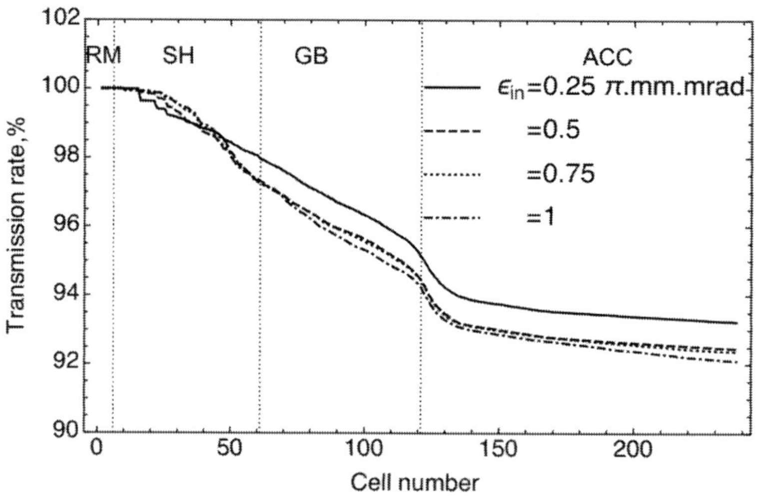

Figure 5. Transmission rate depending on the input transverse emittance.

4. Summary

A CW high-intensity D^+ RFQ is designed for the purpose of nuclear transmutation. The beam dynamics design is carried out in a traditional manner while considering the convenience of fabrication. Both the average aperture and vane voltage are kept constant along the RFQ. PARMTEQM is used to perform the particle tracking simulation. The simulation results suggest that the designed RFQ is capable of accelerating the 400 mA D^+ beam to an energy of 2.5 MeV with a 92.0 $\sim$ 93.3% transmission, depending on the input emittance of 0.25 $\sim$ 1.00 π.mm.mrad. The cavity design is scheduled to estimate the power consumption and verify the construction feasibility of such a machine.

IOP Conf. Series: Journal of Physics: Conf. Series **874** (2017) 012053 doi:10.1088/1742-6596/874/1/012053

Acknowledgments

This work has been funded by ImPACT Program of Council for Science, Technology and Innovation (Cabinet Office, Government of Japan)

References

[1] Wang H *et al* 2016 *Phys. Lett.* B **754** 104108
[2] Wang H *et al* 2017 *Prog. Theor. Exp. Phys.* **2** 021D01
[3] ImPACT: Reduction and Resource Recycling of High-level Radioactive Wastes through Nuclear Transmutation `http://www.jst.go.jp/impact/en/program/08.html`
[4] Sakamoto N, Kamigaito O, Okuno H, Ozeki K, Suda K, Watanabe Y and Yamada K 2015 *Proc. SRF2015* (Whistler) WEBA06
[5] Pisent A 2009 *Proc. 23rd Int. Particle Accelerator Conf. (PAC09)* (Vancouver) MO3GRI04
[6] Comunian M and Pisent A 2008 in *Proc. EPAC08* (Genoa) THPP075
[7] Ichimiya R 2017 *3rd IFMIF study meeting* (Rokkasho)
[8] Kilpatrick W 1957 *Rev. Sci. Instrum.* **28** 10 824-826
[9] Reiser M 2008 *Theory and Design of Charged Particle Beams 2nd Edition* (John Wiley & Sons)
[10] Crandall K R, Stokes R H and Wangler T P 1979 *RF quadrupole beam dynamics design studies* LA-UR-79-2499
[11] Crandall K R, Wangler T P, Young L M, Billen J H, Neuschaefer G H and Schrage D L 2005 *RFQ Design Codes* LA-UR-96-1836

Beam dynamics design of the muon linac high-beta section

Y Kondo[1], K Hasegawa[1], M Otani[2], T Mibe[2], M Yoshida[2] and R Kitamura[3]

[1] JAEA, Tokai, Naka, Ibaraki, 319-1195, Japan
[2] KEK, Oho, Tsukuba, Ibaraki, 305-0801, Japan
[3] Univ. of Tokyo, Hongo, Tokyo, 113-8654, Japan

E-mail: yasuhiro.kondo@j-parc.jp

Abstract. A muon linac development for a new muon g-2 experiment is now going on at J-PARC. Muons from the muon beam line (H line) at the J-PARC muon science facility are once stopped in a silica-aerogel target, and room temperature muoniums are evaporated from the aerogel. They are dissociated with lasers, then accelerated up to 212 MeV using a linear accelerator. For the accelerating structure from 40 MeV, disk-loaded traveling-wave structure is applicable because the particle beta is more than 0.7. The structure itself is similar to that for electron linacs, however, the cell length should be harmonic to the increase of the particle velocity. In this paper, the beam dynamics design of this muon linac using the disk-loaded structure (DLS) is described.

1. Introduction

The muon anomalous magnetic moment $(g-2)_\mu$ is one of the most promising probe to explore the elementary particle physics beyond the standard model (SM). Currently, the most precise $(g-2)_\mu$ experiment is E821 of Brookhaven national laboratory [1]. The precision is 0.54 ppm and the measured value indicates approximately three standard deviations from the SM prediction. The J-PARC E34 experiment aims to measure the $(g-2)_\mu$ with a precision of 0.1 ppm. In addition, the electric dipole moment also can be measured with a precision of $1 \times 10^{-21} e{\cdot}\mathrm{cm}$ [2]. The experimental method of E34 is completely different from that of the previous experiments. The previous experiments directory used decay muons from the secondary pions generated on the production target. The emittance of such muon beam is very large (typically, 1000π mm mrad); this is a major source of uncertainty of the measurement. On the other hand, E34 will use a low emittance muon beam to improve the precision. The required beam divergence $\Delta p_t/p$ is less than 10^{-5}, and assumed transverse emittance is 1.5π mm mrad. To satisfy this requirement, we are planning to use ultra-slow muons (USMs) generated by laser-dissociation of thermal muoniums (Mu: μ^+e^-) from a silica-aerogel target[3]. The room temperature USMs (25 meV) should be accelerated to 212 MeV to obtain the required $\Delta p_t/p$. A linac realizes rapid acceleration required to accelerate muons, whose lifetime is very short (2.2 μs). In Figure 1, the configuration of the muon linac [4] is shown.

The muon linac will be constructed at the H line [5] of the J-PARC muon science facility. The USMs are bunched and accelerated to 0.34 MeV by a radio frequency quadrupole linac (RFQ). Following the RFQ, an interdigital H-mode drift tube linac [6] is used to accelerate to

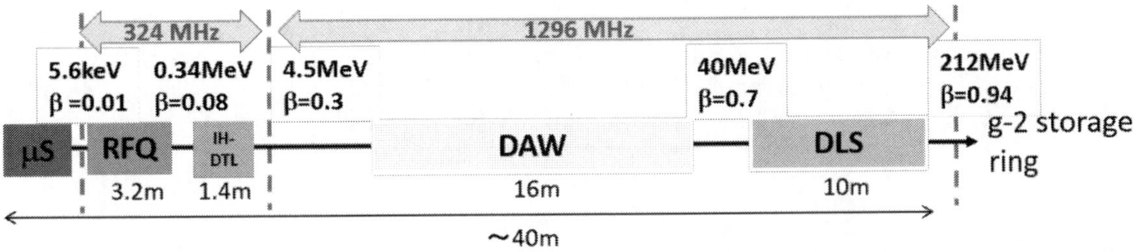

Figure 1. Configuration of the muon linac.

4.5 MeV. Then, muons are accelerated to 40 MeV through a disk and washer (DAW) coupled cavity linac (CCL) section. Because the accelerating gradient of the CCLs is typically less than 10 MV/m, the total length of the muon linac will exceed the available space of the J-PARC site if the muons are accelerated up to 212 MeV using a CCL: higher gradient is necessary. Above 40 MeV, the velocity β of the muon is more than 0.7, therefore, a disk-loaded structure (DLS) traveling-wave (TW) linac is applicable. Another choice of the high gradient structure is a superconducting linac. However, the pulse width of our muon linac is short (~10 ns), thus the power efficiency of the superconducting system will be very low. Moreover, the disk-loaded TW structure is quite mature technique widely used for electron linacs, therefore we chose this structure for the high-β section. Table 1 summarizes the main parameters of the DLS section.

Table 1. Main parameters of the DLS section.

Input energy	40 MeV
Output energy	212 MeV
Beam intensity	1×10^6 /s
Beam pulse width	10 ns
Number of bunches	3 /pulse
Repetition rate	25 Hz
Normalized transverse emittance	1.5π mm mrad
Momentum spread	0.1%

In this paper, present status of the beam dynamics design of the high-β section (DLS section) of the muon linac is described.

2. Cell Design

Table 2 shows the assumed parameters of the accelerating tubes.

Because the muons are gradually accelerated compared to the electrons, cell length D is varied cell by cell according to the velocity of the synchronous particle β_s. We adopt $2\pi/3$ mode operation, thus D is derived as

$$D = \frac{\beta_s \lambda}{3}. \quad (1)$$

IOP Conf. Series: Journal of Physics: Conf. Series **874** (2017) 012054 doi:10.1088/1742-6596/874/1/012054

Table 2. Assumed design parameters of the accelerating structure.

Structure	Disk loaded traveling wave
Frequency	1296 MHz
Accelerating mode	$2\pi/3$
Accelerating tube length	~2 m
Accelerating gradient	20 MV/m

Energy gain ΔW in one cell is determined as

$$\Delta W = E_0 D cos\phi, \tag{2}$$

where E_0 is the accelerating gradient, and ϕ is the phase relative to the crest of the wave. In the current design, ϕ is set to be -10 deg. to assure enough longitudinal acceptance.

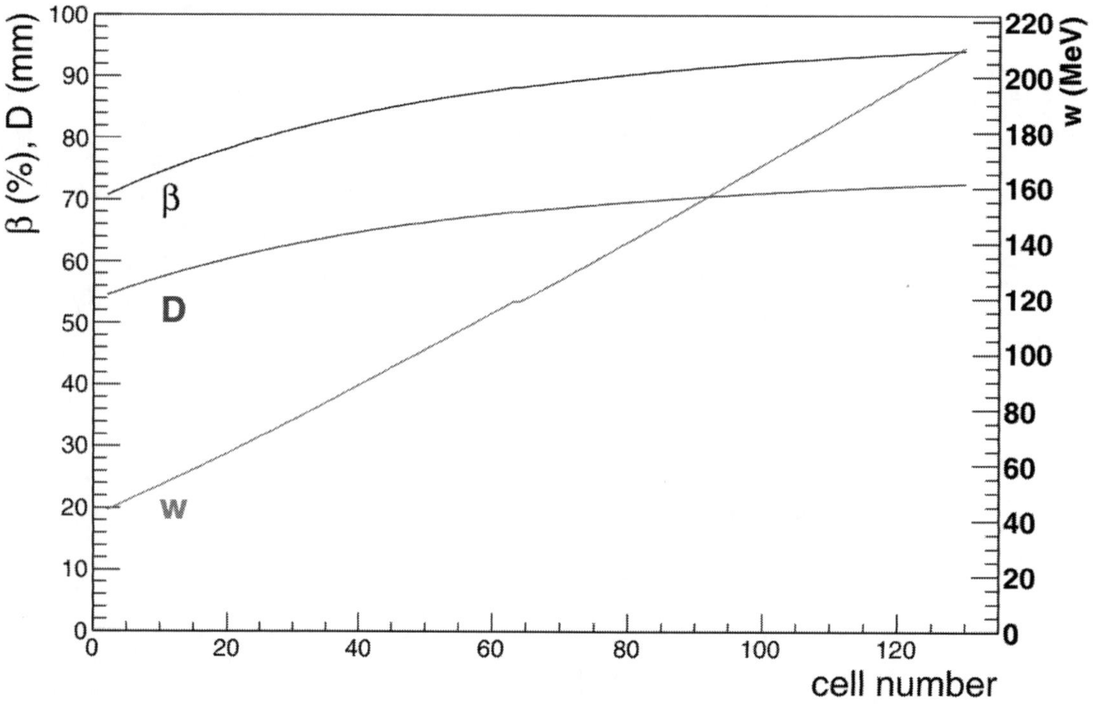

Figure 2. Calculated cell parameters of the DLS section.

Figure 2 shows the derived β, D, and the particle energy W through the DLS section as functions of the cell number. The disk-loaded stricture is designed using this D.

3. Structure Design

Design of the DLS is conducted by using a tool package for simulation of traveling-wave accelerating tubes [7]. This tool utilizes SUPERFISH [8] to determine the geometrical parameters of the structure. SUPERFISH generates the standing-wave-mode electric-field maps of each cell with open-open and short-short boundary conditions. The electric field of the traveling wave is represented by superposing these two maps with a phase difference of $\pi/2$, and used for particle simulation using General Particle Tracer (GPT) [9]. In this study, constant impedance design is adopted for simplicity, that is, the aperture radius a is fixed as $2a = 40$ mm. As described in previous section, the calculated D from β_s is used as an input parameter. The inner diameters of each cell $2b$ is automatically tuned by the code to be 179.4 to 180.3 mm to obtain the desired resonant frequency with both boundary conditions.

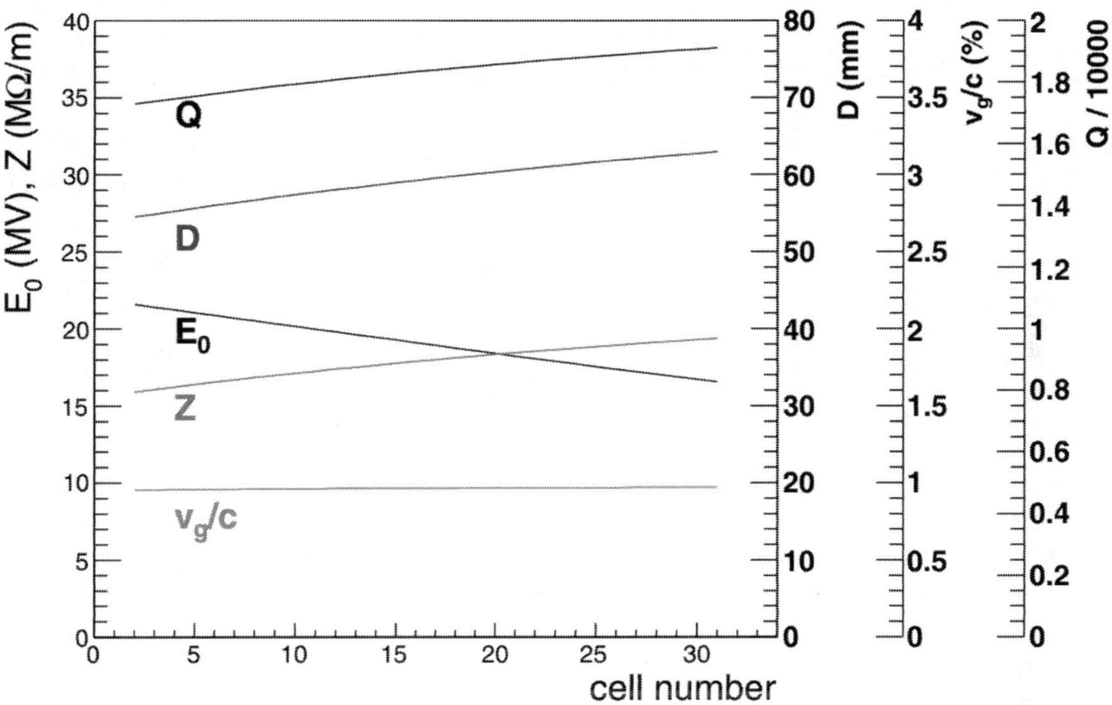

Figure 3. Structure parameters of the DLS1.

Figure 3 shows the parameters of the first disk-loaded structure (DLS1) as an example. The D is the input parameter, and E_0, shunt impedance per unit length Z, group velocity v_g, and quality factor Q are the calculated values by using SUPERFISH.

4. Particle Simulation

Particle distribution obtained by the simulation of the DAW [4] is used for the DLS simulation. In the reference [4], matching between the DAW and DLS was not considered. In this study, matching parameters of the first DLS periodic unit is derived by using TRACE3D [10], as shown in Figure 4. One periodic unit consists of an accelerating tube and a doublet of quadrupole magnets.

To realize these matching parameters, a matching section with tree quadrupole magnets is designed using TRACE3D, as shown in Figure 5. The particles are transported by using PARMILA [11], then the obtained particle distribution is inputted into the GPT simulation.

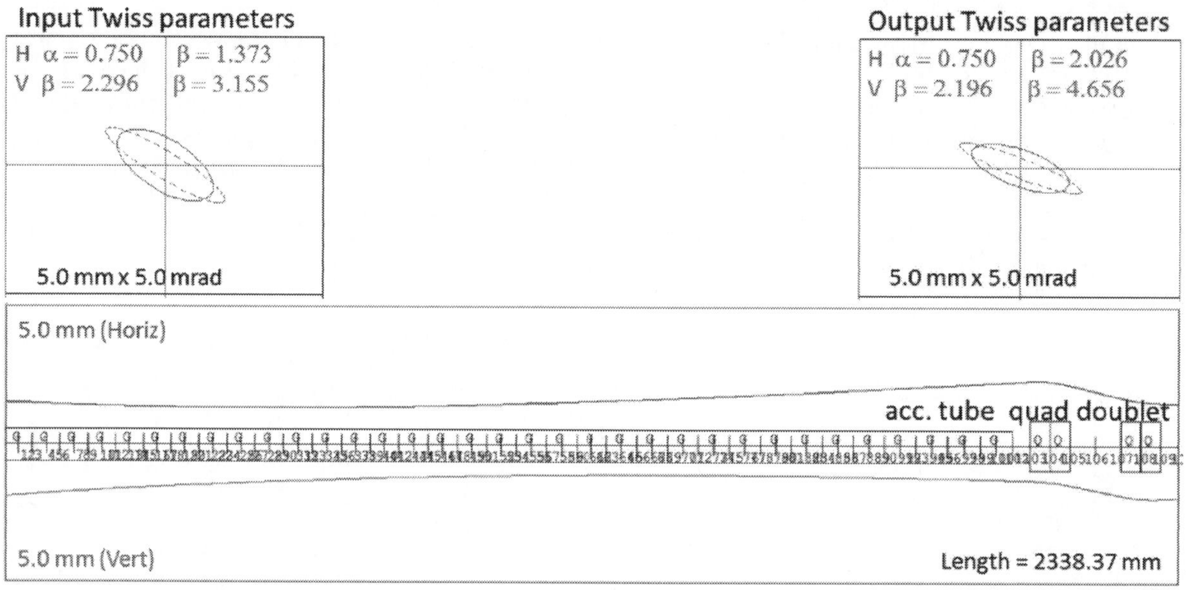

Figure 4. Injection matching of the DLS section using TRACE3D.

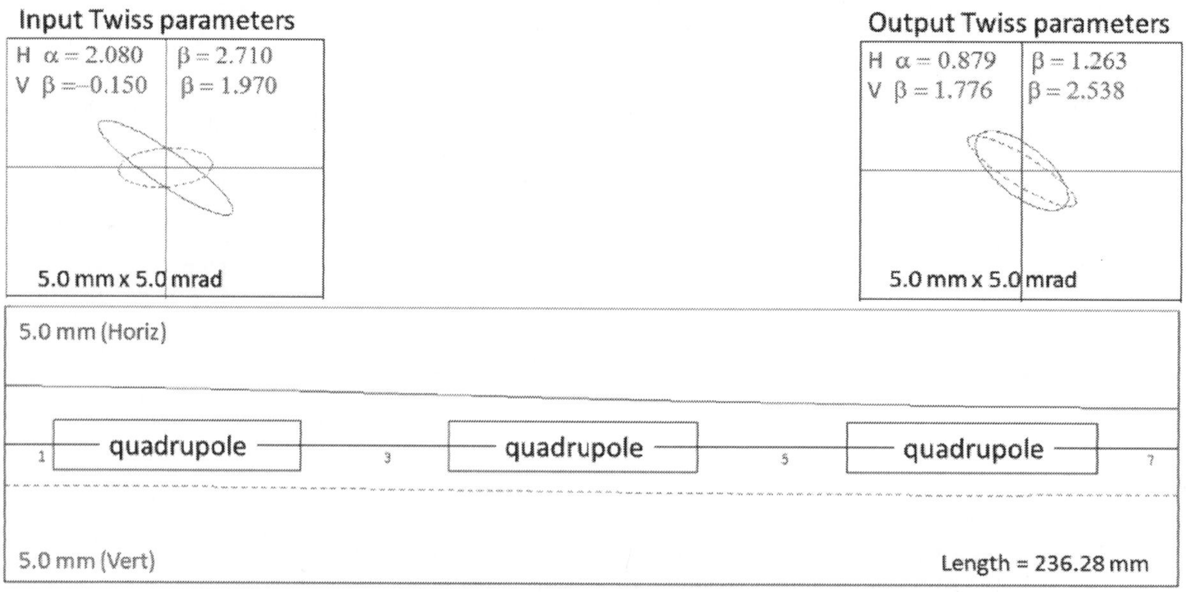

Figure 5. Matching section between the DAW and the DLS sections.

Figure 6 shows the phase-space distribution at the exit of the DLS section, and Figure 7 represents the transverse emittance evolution through the DLS section. Almost no emittance growth is observed. The transmission through the DLS section is 100%, and the loss due to the muon decay is estimated to be 1%. The horizontal and vertical normalized rms transverse emittances at the DLS exit are $\varepsilon_{x,n,rms} = 0.33\pi$ mm mrad and $\varepsilon_{y,n,rms} = 0.21\pi$ mm mrad, respectively. The rms momentum spread is 0.04%. Table 3 is a summary of the DLS simulation.

The beam quality matches the requirements. However, one issue is that the shunt impedance of the accelerating tube is low due to the short cell length and large aperture. Therefore, the

IOP Conf. Series: Journal of Physics: Conf. Series **874** (2017) 012054 doi:10.1088/1742-6596/874/1/012054

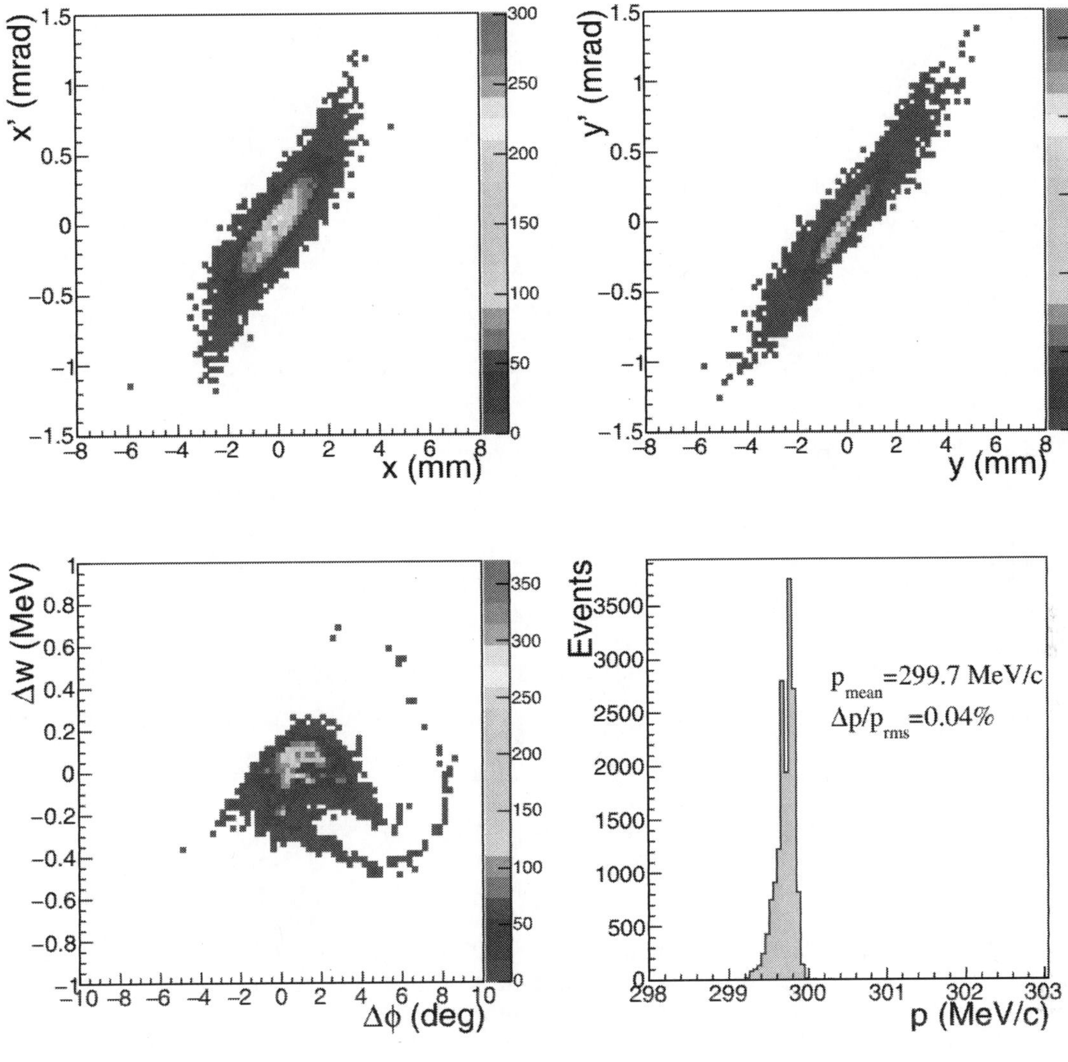

Figure 6. Simulated particle distribution at the DLS exit. Lower right figure shows the momentum histogram.

Table 3. Result of the DLS simulation.

Transmission	100%
Decay loss	1%
$\varepsilon_{x,n,rms}$	0.33π mm mrad
$\varepsilon_{y,n,rms}$	0.21π mm mrad
Momentum spread	0.04%

required power for one tube is approximately 80 MW; this is rather high for actual RF system. Figure 8 shows the beam envelopes through the DLS section, and the vertical axis represents

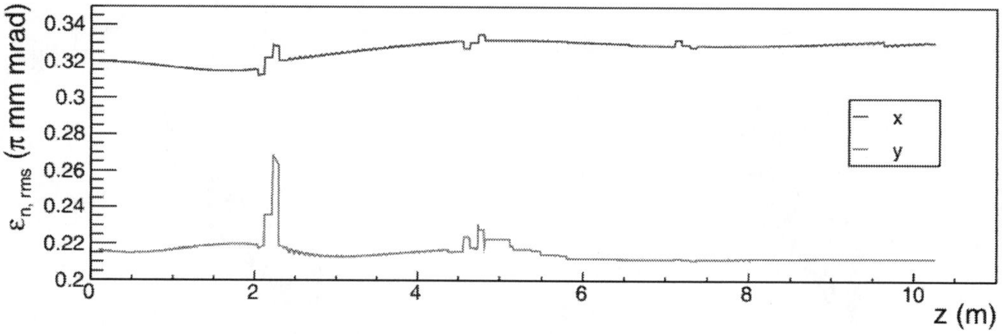

Figure 7. Emittance evolution through the DLS section.

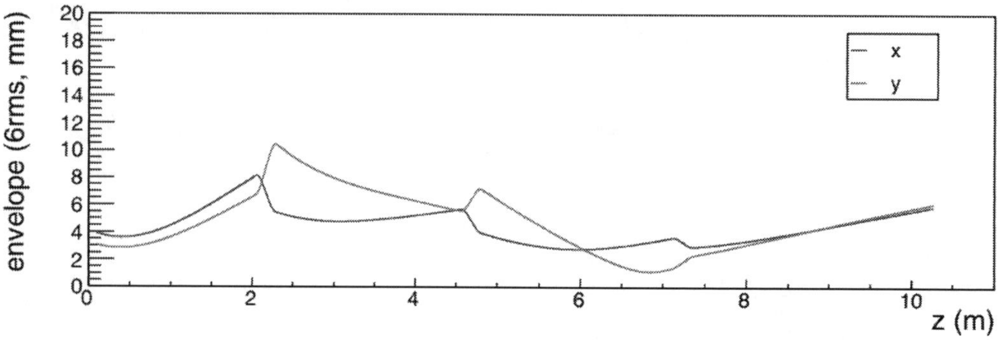

Figure 8. Beam envelope of the DLS section.

the six times the rms beam width; all particles are within this envelope in this simulation. This figure shows that the aperture of the structure ($a = 20$ mm) has enough margin to the beam envelope. Thus, we expect to improve the shunt impedance by reducing the aperture. Also, the design will be improved by adopting the constant gradient structure. Moreover, considering the S-band structure is another possibility. We continue the design work to optimize the DLS system.

5. Summary

We conducted the beam dynamics design of the muon linac high-beta section. The first reference design is obtained and the simulated transverse emittances are 0.33π mm mrad and 0.21π mm mrad in horizontal and vertical phase space, respectively. The momentum spread is 0.04%. These parameters satisfy the requirement, however, the nominal RF power of the present design is rather high. We will do further improvement of the design.

Acknowledgments

This work is supported by JSPS KAKENHI Grant Numbers JP15H0366 and JP16H03987.

References

[1] Bennett G W *et al.* 2006 *Phys. Rev.* D **73** 072003
[2] Mibe T *et al.* 2011 *J-PARC E34 Conceptual Design Report*
[3] Bakule P *et al.* 2013 *Prog. Theor. Exp. Phys.* 103C01

IOP Conf. Series: Journal of Physics: Conf. Series **874** (2017) 012054 doi:10.1088/1742-6596/874/1/012054

[4] Otani M *et al.* 2016 *Proc. LINAC2016 (East Lansing)* 1037-41
[5] Kawamura N, Toyoda A, Mibe T, Saito N, Aoki M, Shimomura K and Miyake Y 2014 *Proc. USM2013(Matsue)* 010112
[6] Otani M, Mibe T, Yoshida M, Hasegawa K, Kondo Y, Hayashizaki N, Iwashita Y, Iwata Y, Kitamura R and Saito N 2016 *Phys. Rev. Accel. Beams* **19** 040101
[7] Yamamoto M `http://www.yamamo10.jp/yamamoto/study/accelerator/GPT/TW_structure/`
[8] Billen J H and Young L M 1996 *Poisson Superfish* LA-UR-96-1834
[9] Pulser Physics *General Particle Tracer*
[10] Crandall K R and Rusthoi D P 1997 *Trace 3-D Documentation* LA-UR-97-886
[11] Takeda H 1998 *Parmila* LA-UR-98-4478

IOP Conf. Series: Journal of Physics: Conf. Series **874** (2017) 012055 doi:10.1088/1742-6596/874/1/012055

First trial of the muon acceleration for J-PARC muon g-2/EDM experiment

R Kitamura[1], M Otani[2], Y Fukao[2], N Kawamura[2], T Mibe[2], Y Miyake[2], K Shimomura[2], Y Kondo[3], K Hasegawa[3], S Bae[4], B Kim[4], G Razuvaev[5], H Iinuma[6], K Ishida[7] and N Saito[8]

[1]University of Tokyo, Hongo, Tokyo, Japan
[2]KEK, Tsukuba, Ibaraki, Japan
[3]JAEA, Tokai, Naka, Ibaraki, Japan
[4]SNU, Seoul, Korea
[5]BINP SB RAS, Novosibirsk, Russia
[6]Ibaraki University, Mito, Ibaraki, Japan
[7]RIKEN, Wako, Saitama, Japan
[8]J-PARC center, Tokai, Naka, Ibaraki, Japan

E-mail: rkita@post.kek.jp

Abstract. Muon acceleration is an important technique in exploring the new frontier of physics. A new measurement of the muon dipole moments is planned in J-PARC using the muon linear accelerator. The low-energy (LE) muon source using the thin metal foil target and beam diagnostic system were developed for the world's first muon acceleration. Negative muonium ions from the thin metal foil target as the LE muon source was successfully observed. Also the beam profile of the LE positive muon was measured by the LE-dedicated beam profile monitor. The muon acceleration test using a Radio-Frequency Quadrupole linac (RFQ) is being prepared as the first step of the muon accelerator development. In this paper, the latest status of the first muon acceleration test is described.

1. Introduction

E821 experiment of Brookhaven National Laboratory measured the muon $(g-2)_\mu$ with the precision of 0.54 ppm and reported the discrepancy of more than 3 σ for the muon $(g-2)_\mu$[1] from the theoretical prediction. More precise measurement is required in order to search the physics beyond the standard model. The J-PARC E34 experiment which aims to precisely measure the muon anomalous magnetic moment $(g-2)_\mu$ and electric dipole moment (EDM) is planned with brand new techniques[2]. The goals of the precision are 0.1 ppm for the muon $(g-2)_\mu$ and 1×10^{-21} e·cm for the muon EDM. The muon $(g-2)_\mu$ and EDM will be measured at the same time under the no electric focusing in the muon storage ring, using the ultra-cold muons (UCMs) which are produced by the new muon cooling method and the muon linear accelerator (Muon linac). The electric focusing isn't necessary for UCMs, because UCMs have a very small transverse momentum compared with the surface muon beam. The surface muons having a large emittance are stopped the muoniums (Mu: μ^+e^-) production target and become the thermal Mu. Mu are ionized by the muon ionization laser after Mu are evaporated from the surface of the production target to the vacuum. The ultra-slow muons (USMs) with the room temperature (25 meV) are produced by this procedure. The UCMs are obtained by accelerating the USMs with the muon linac.

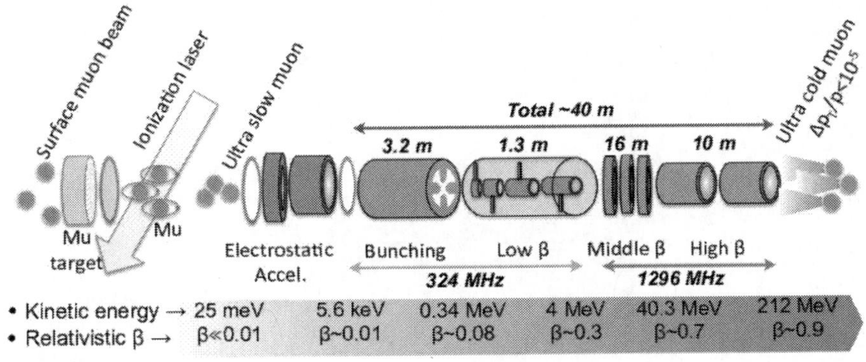

Figure 1. The conceptual diagram of the muon linac.

Figure 1. shows the conceptual diagram of the muon linac. The muons are finally accelerated to 212 MeV by the muon linac.

2. Low-energy muon source

The LE muon source with energy less than 5.6 keV is required in order to test the muon accelerator, since the design input energy of the initial RF accelerator (RFQ) is 5.6 keV. Negative muoniums (Mu^-: $\mu^+e^-e^-$) that are generated by the muons passing through a thin metal foil is a good candidate of the epi-thermal muon source for the muon acceleration, because the energy dispersion of the Mu^- is smaller than one of the decelerated positive muons from the metal foil[3]. Using the LE muon source like Mu^-, the muon accelerator can be tested even before the ionization laser is ready.

The beam parameters of the Mu^- should be measured in order to use the Mu^- as the LE muon source for the muon acceleration. Figure 2. shows the setup of the beam diagnostic beam line and detectors to develop the LE muon source. The incident surface muons with 4 MeV hit the thin aluminum foil target. Mu^- are produced by the electron capture at the downstream surface of the foil target. Produced Mu^- are reaccelerated to 20 keV and focused by the electrostatic accelerator called "SOA lens". Accelerated Mu^- are transported by the beam line which consists of the electrostatic deflector, electric quadrupoles and bending magnet. LE electrons produced around the electrodes of the SOA lens and transported by the beam line can become the serious background for the Mu^- diagnostics. However the LE electrons can be rejected by the electrostatic deflector and the bending magnet since the energy and momentum of the particles are selected by them. The Mu^- can be identified by the Time-Of-Flight (TOF) measurement. The typical TOF of the LE muons with 20 keV is about 1 μsec.

Figure 3. shows the TOF distributions with the positive and negative polarity setting of the diagnostic beam line. The Mu^- peak of TOF can be observed in the bottom plot. The peak position of the Mu^- was very close to that of the decelerated positive muons. So the first observation of Mu^- in J-PARC was successful.

The beam profile should be measured in order to measure the beam emittance. The beam emittance can be measured by the Q scan method[4]. In the Q scan method, a quadrupole is set in

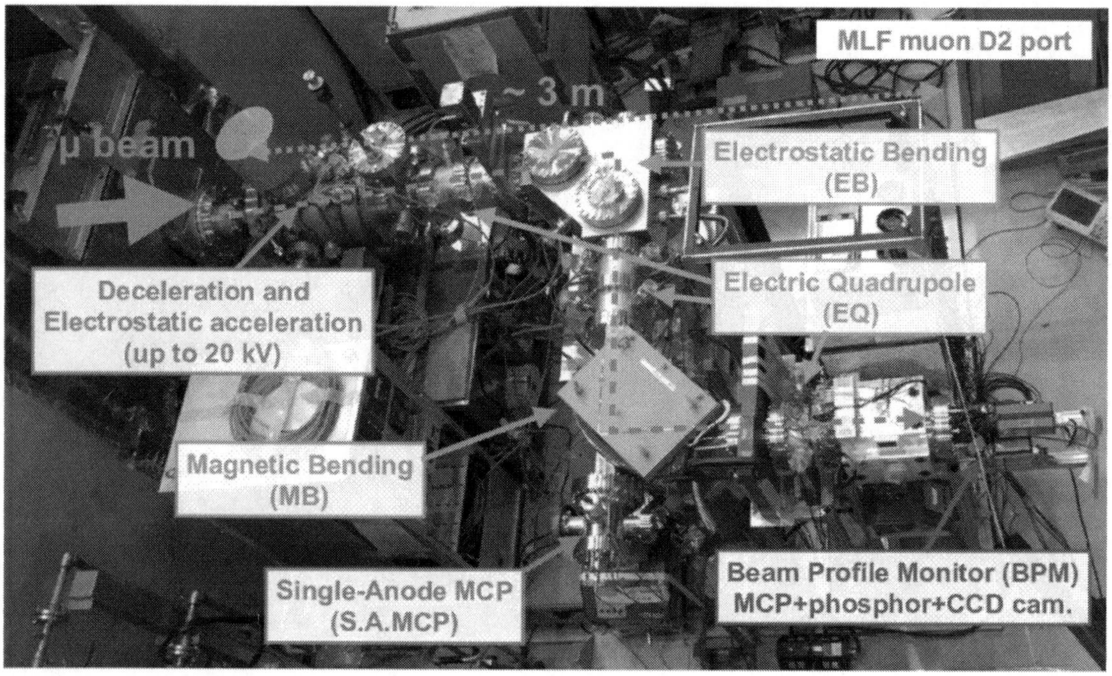

Figure 2. The diagnostic beam line for the LE muon beam using the metal foil target.

front of the beam profile monitor and the beam profiles are measured with changing the focusing power of the quadrupole. From the relation between the focusing power of the quadrupole and the width of the beam profile, the beam emittance can be estimated by parameter fitting. Because the LE muons have too low energy to be detected with the plastic scintillator commonly used, a LE muon-dedicated beam profile monitor (LE BPM) was developed. The LE BPM consists of a micro channel plate (MCP) and a CCD camera; the MCP has good sensitivity for the LE particle and CCD camera has good position resolution. Incident LE muons are amplified by the MCP and produce secondary electrons. The produced secondary electrons are converted to photons by the phosphor screen after the MCP. Finally the converted photons are detected by the CCD camera. The position resolution of the LE BPM is better than 1 mm. The gain of LE BPM was calibrated by the several sources including LE e^-, gamma ray, UV light, H^- and surface muons.

The commissioning of the beam diagnostic line was carried out by the ion beam ahead of the muon beam profile measurement. In the commissioning with the ion beam, the misalignment of the beam line was found to be a serious problem for the beam profile measurement and had to be corrected. The misalignment of a few mm at the SOA lens caused the about 10 mm of the beam position shift at the LE BPM. Eventually the beam was transported effectively and focused at the position of the LE BPM, after tuning the alignment of the components in the beam line and voltages of the beam line .

Because the beam intensity of the LE positive muon was higher than that of the Mu$^-$, the

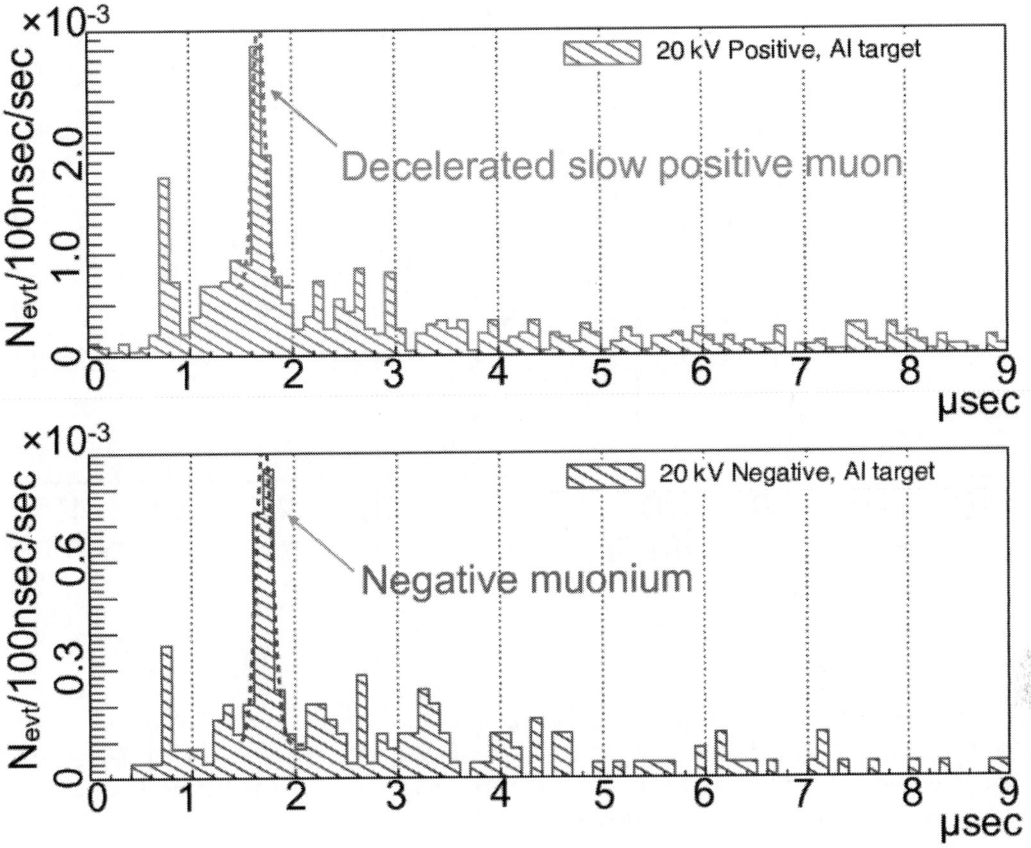

Figure 3. The top figure shows the TOF distribution of the decelerated positive muons. The bottom figure shows the TOF distribution of the negative muonium.

beam profile of the LE positive muons was measured first as shown in Figure 4.. In conclusion, the beam diagnostic system including the beam line and LE BPM was demonstrated for the muon acceleration test.

3. First muon acceleration

The first trial of the muon acceleration using the muon source and the LE beam diagnostic system is planed at the muon test beam line (D line) in the J-PARC Material and Life science Facility (MLF). The prototype RFQ (Figure 5.) of the J-PARC linac will be used for the muon acceleration test[5].

The input energy of the prototype RFQ is 5.6 keV. The structure of the prototype RFQ is similar to the first half of the RFQ which will be used for the muon linac. The surface muons are decelerated by the thin metal foil and reaccelerated to 5.6 keV by the SOA lens. Muons with 5.6 keV energy are accelerated to about 100 keV by the prototype RFQ. The diagnostic beam line after the prototype RFQ consists of the two quadrupole magnets, a bending magnet and LE BPM.

Figure 6. shows the latest drawing for the muon acceleration test at the D line. Figure 7. shows the simulation result by PARMTEQM assuming the developed LE muon source at the D line[6]. Detailed particle tracking simulations based on the actual setting are being developed. The setup will be constructed in this summer. The first trial of the muon acceleration is planned

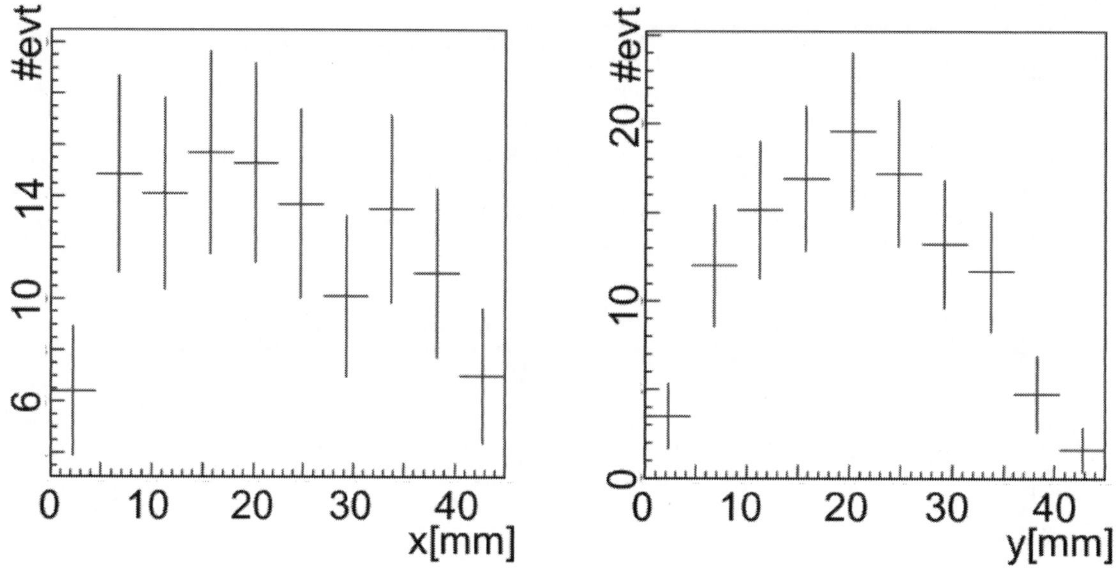

Figure 4. The projections of the beam profile of the LE positive muons decelerated by the thin metal foil. The left and right figures are for the horizontal and vertical direction, respectively. [The projections were corrected based on the geometric shape of the LE BPM. The error was only the statistical error based on Poisson distribution.]

Figure 5. J-PARC linac prototype RFQ.

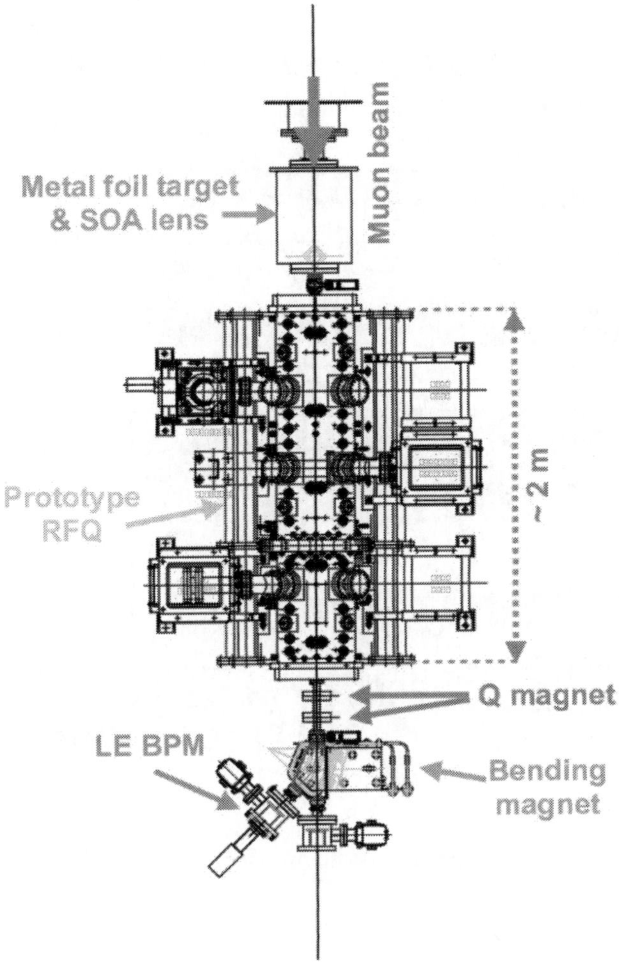

Figure 6. The latest drawing of the muon acceleration test using the prototype RFQ at the muon D line.

in this autumn.

4. Summary

The world's first muon acceleration will open the door of new frontier of accelerator physics and particle physics. The LE muon source to be used for the commissioning of the muon accelerator was developed. The diagnostic system of the LE muons was also prepared and the measurement of the LE positive muon beam profile was successfully done. The first trial of the muon acceleration with the prototype RFQ will be demonstrated in this year.

5. Acknowledgement

We would like to deeply thank to RIKEN Advanced Meson Science Laboratory and J-PARC linac group for their significant supports . This work is supported by JSPS KAKENHI Grant Numbers JP15H03666, JP16H03987, J15H05742 and JP16J07784. This muon experiment at the Materials and Life Science Experimental Facility of the J-PARC was performed under user programs (Proposal No. 2016A0067 and 2016B0214).

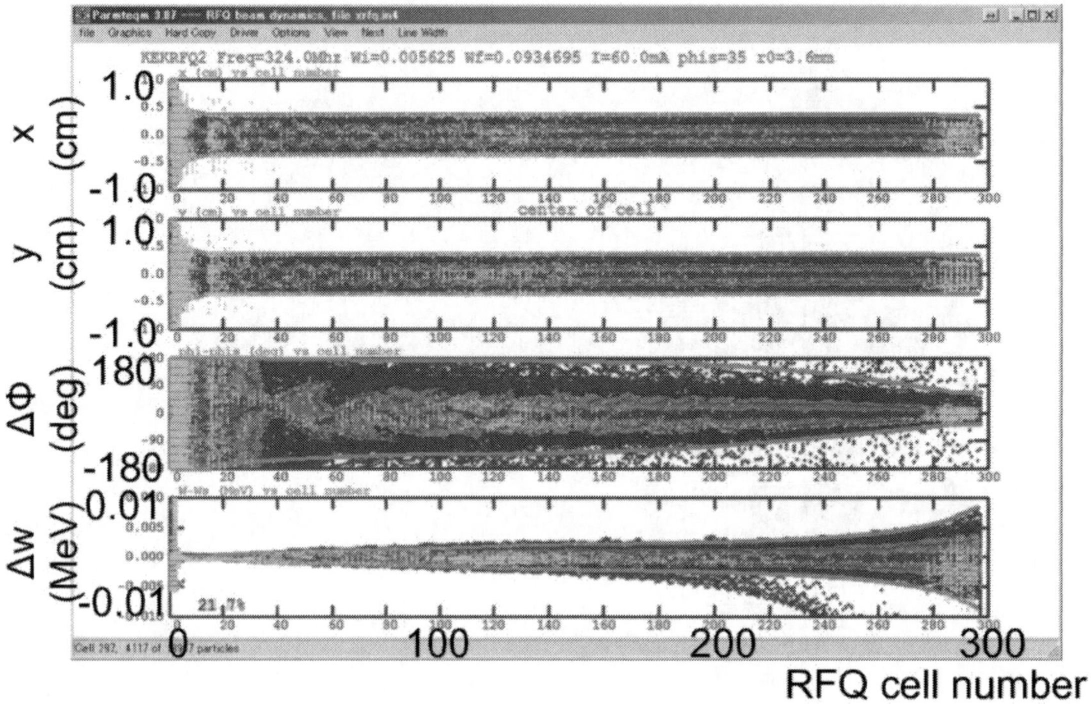

Figure 7. The particle tracking simulation of the prototype RFQ assuming the Mu^-.

References

[1] Bennett W G *et al.* 2006 *Phys. Rev.* D **73** 072003
[2] Aoki M *et al.* 2011 *J-PARC E34 conceptual design report*
[3] Kuang Y *et al.* 1989 *Phys. Rev.* A **39** 6109
[4] Wiedemann H 2007 *Particle Accelerator Physics 3rd Edition* (Berlin: Springer) chapter 5 pp 164–6
[5] Kondo Y, Hasegawa K and Ueno A 2006 *Proc. of LINAC2006* (Knoxville: JACoW) pp 749–51
[6] Crandall R K *et al.* 1996 *RFQ design codes* (Los Alamos: Los Alamos National Laboratory Report) LA-UR-96-1836

IOP Conf. Series: Journal of Physics: Conf. Series **874** (2017) 012056 doi:10.1088/1742-6596/874/1/012056

The beam lines design for the CERN neutrino platform in the CERN north area and an outlook on their expected performance

N Charitonidis[1,*], M Brugger[1], I Efthymiopoulos[1], L Gatignon[1], Y Karyotakis[2], P R Sala[1,3], E Nowak[1,4], I Ortega-Ruiz[1,5]

1 CERN, 1211 Geneva 23, Switzerland
2 LAPP, Annecy, France
3 INFN, Italy
4 AGH University of Science and Technology, Poland
5 EPFL, Lausanne, Switzerland

* ncharito@cern.ch

Abstract. In the framework of the CERN Neutrino Platform project, extensions to the existing SPS North Area secondary beam lines "H2" and "H4", able to provide low-energy charged particles in the momentum range from 0.4 to 12 GeV/c, have been designed. The parameters of these "very low energy" beam lines, the expected beam composition as seen by the experiments as well as an outlook on their expected performance are summarized in this paper. Results from Monte-Carlo simulations, important for the optimization of the future instrumentation of the beam lines (serving both the purpose of beam tuning and the experiments' needs for particle identification and momentum measurements), are also presented.

1. Introduction

The CERN's North Area Experimental Halls are multi-purpose facilities designed to deliver primary ion and attenuated proton beams, as well as a wide spectrum of secondary and tertiary particle beams of varying and flexible composition (hadrons and leptons) to fixed-target experiments. In the framework of the CERN Neutrino Platform project (CENF), the EHN1 experimental hall is extended to house two large-size detector prototypes for future neutrino experiments, designated 'ProtoDUNE-Dual Phase' (NP-02) and 'ProtoDUNE - Single Phase' (NP-04), served by extensions of the H2 and H4 beam lines, respectively. The required beam parameters for these experiments are summarized in Table 1, while the layout of the extension can be seen in Figure 1.

The existing H2 and H4 beam lines were designed to transport high-energy particles in the order of several hundreds of GeV. The magnet's power supplies are therefore out of their stability region when operating at very low currents. In addition, the overall length of the existing secondary beamlines exceeds by far the decay length of pions and kaons at the low momenta required by the experiments. Therefore, a new design of a 'third generation particle" beam line had to be devised towards the extension of the experimental hall in order to produce the required low-energy beams requested by the experiments.

Table 1. The main design beam parameters as described in the experimental proposals, for ProtoDune –SP (NP-04) [1] and ProtoDune – DP (NP-02) [2].

Parameter	**ProtoDUNE – DP**	**ProtoDUNE - SP**
Particle Type	π^{+-}, μ^{+-}, e^{+-}, K, p	
Momentum	< 12 GeV/c	< 7 GeV/c
Δp/p	< 5 %	
Beam Size Rms	~ 10 cm at the entrance of the cryostats	
Max Rate	< 100 Hz	< 100 Hz

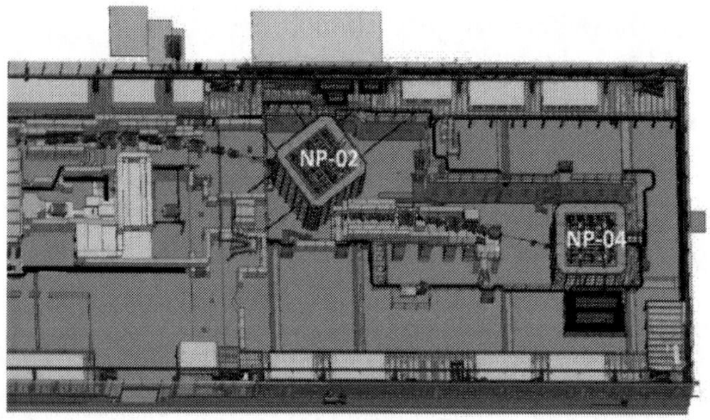

Figure 1. Layout of the extension of the EHN1 building. The two new detector prototypes (ProtoDUNE-NP02 and protoDUNE-NP-04), along with the extensions of the H2 and H4 beam lines is depicted.

2. Beam line design & performance

2.1. Layout & Optics

In order to overcome the aforementioned challenges, a two stage production, momentum selection and transport beam line has been designed and incorporated in the current infrastructure: a secondary beam produced at the primary and existing beryllium target ('T2') of North Area, with medium intensity of approximately 106 particles/spill and a chosen momentum of 80 GeV/c is being transported to a 'secondary target', where the lower momenta particles are produced via hadronic and electromagnetic interactions. After this target, a magnetic spectrometer four (two for H4-VLE) large-aperture dipoles and a momentum selection station with a field-lens quadrupole and a two-jaw collimator selects the desired momenta. The beam is focused on the experiment by a quadrupole doublet. Since the beam impact point on both cryostats had to be displaced horizontally and vertically from the trajectory of the upstream secondary beam line, the bending and quadrupole magnets, as well as the beam instrumentation had to be tilted by 34.30 (56.75 for H4) degrees with respect to the floor. The beam optics visualisation for H2-VLE, calculated with TRANSPORT [3] can be seen in Figure 2.

Both beam lines have been optimized for a maximum angular acceptance from the production target equal to 10 mrad. The beam envelope is then focused between the four (two for H4-VLE) bends and the momentum analysis stage is performed with a collimator having the jaws parallel to the beam bending plane. The field-lens positioned between the bending magnets, recombines the momentum before the final focusing at the experiment with the last quadrupole doublet. The spot size at the position of the experiment has been calculated with G4BeamLine [4] for H2-VLE [5] and is shown in Figure 3. The spot-size in H4-VLE is of similar dimensions.

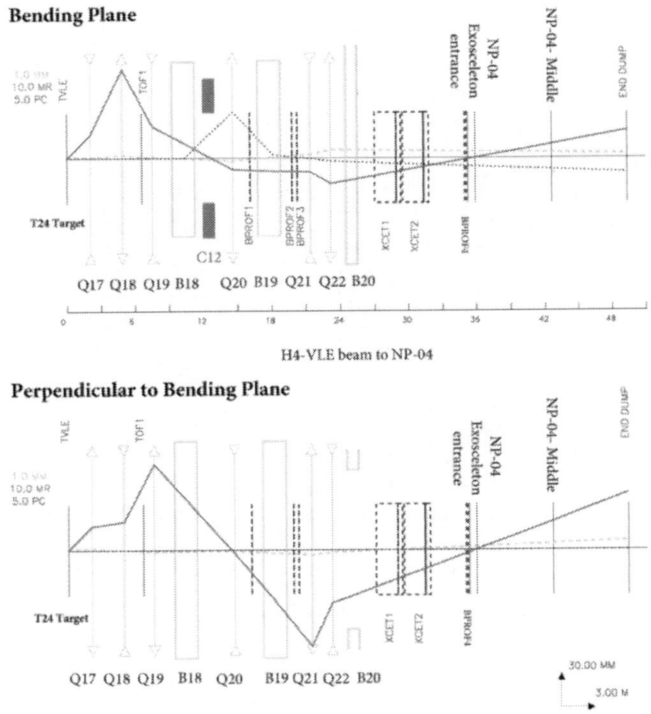

Figure 2. Optics diagram for the H2-VLE beam line. The red line corresponds to the sine-like ray, the blue dotted line represents the dispersive ray and the green line the magnification term.

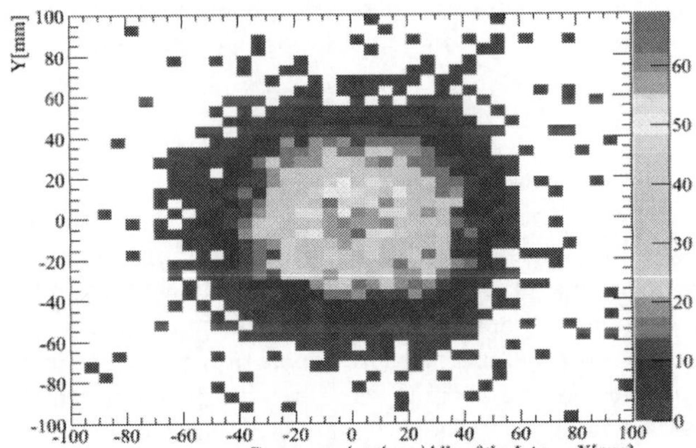

Figure 3. Spot size of the beam in the centre of NP-02 detector.

2.2. Beam composition

The beam can be operated in a "hadron" or "electron" mode. In the hadron mode, the composition of the beam calculated with G4BeamLine is shown in Table 2, for the H2 beam line. For the momenta up to 3 GeV/c a W target was used, since it absorbs more electrons and increases the relative amount of hadrons in this very low momentum regime, while for momenta > 3 GeV/c a Cu target is used. In the "electron" mode, a Pb target of a few radiation lengths will be used, and the resulting electron beam will have a purity of about 99.95 %.

Table 2. Beam composition for the hadron beams, including all the beam line material and magnetic elements. Below and equal to 3 GeV the W-target was used.

Momentum [GeV/c]	e+	K+	mu+	p	pi+
0.4	97.61%	0.00%	0.00%	0.48%	1.91%
1	74.20%	0.00%	0.00%	14.94%	10.86%
2	45.83%	0.67%	0.96%	20.04%	32.50%
3	25.16%	1.56%	1.56%	17.56%	54,17%
3	68.29%	0.64%	0.42%	7.72%	22.94%
4	53.72%	1.46%	0.65%	7.56%	36.61%
5	42.38%	2.47%	0.83%	9.18%	45.14%
6	31.42%	3.83%	0.73%	10.10%	53.92%
7	24.70%	4.08%	0.85%	9.92%	60.46%
8	19.36%	5.11%	0.97%	11.33%	63.24%
9	15.12%	5.67%	0.82%	11.10%	67.29%
10	12.36%	6.02%	0.71%	12.25%	68.66%
11	10.46%	6.95%	0.82%	13.57%	68.20%
12	8.90%	6.89%	0.66%	14.26%	69.30%

2.3. Momentum resolution
The beam line is equipped with a momentum selection and recombination station. The effect of the C12 collimator is defined by the optics and verified with G4BeamLine. It is demonstrated in [5] that for a small slit-size the momentum spread can be reduced down to ~2% at an expense of the rate.

3. Instrumentation
The VLE beam line will be equipped with several types of beam instrumentation devices with the aim at fulfilling different requirements, serving both the purpose of beam tuning and particle identification for the experiments.

3.1. Beam profile monitors
A principal requirement of the instrumentation is to provide beam profile measurements at various locations along the beam line. Towards this purpose, special beam profile monitors are being developed by CERN's Beam Instrumentation (BE/BI) group [6]. Each detector consists of one or two layers of thin scintillating fibers of square shape and 1 mm width, covering a large active area of 200 x 200 mm^2, therefore measuring the horizontal and vertical position of the beam particles, providing at the same time certain timing information. Two profile monitors will be installed in both extension beam lines, to serve as profile monitors. Three more per beam line are installed in a spectrometer configuration, as described in the next section. Thin scintillating tiles will also be used in order to provide trigger to the experiment.

3.2. Momentum measurement
To achieve a better momentum definition without reducing the count rate, the three beam profile monitors described in the previous section, placed around the fourth (second) bending magnet of the line are used as a spectrometer. The technique has been used in the past with success [7], and the here applied implementation is described in detail in [5]. The performance of the spectrometer, was validated with high statistics Monte-Carlo. The use of Euclidean geometry, allows the computation of

the particle position in the three spectrometers, which allows the Δp/p calculation. For the 2 GeV/c beam configuration, the reconstructed momentum resolution can be as low as 1.5%, as shown in Figure 4.

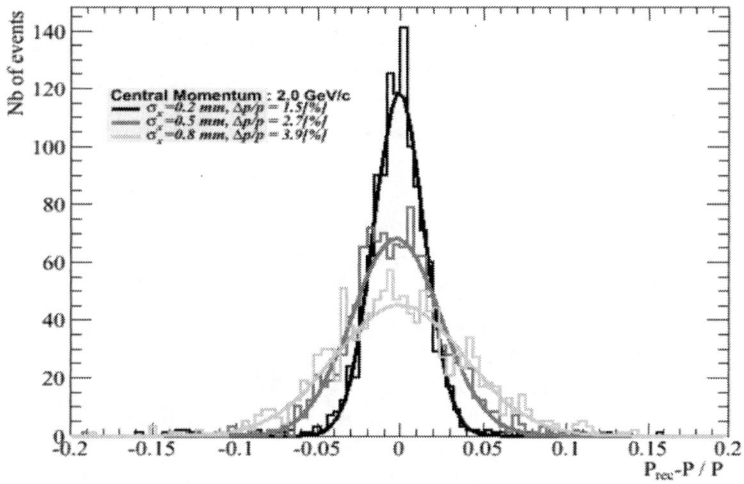

Figure 4. Reconstructed momentum spread for a beam of 2 GeV, taking into account the material on the beam line

3.3. Effect of instrumentation material

When choosing the solution for the beam instrumentation, care has been taken in limiting the material budget that the particles will transverse until they reach the detectors. The impact of beam instrumentation and detector beam windows has been cross-checked with a detailed FLUKA [9, 10] simulation. As an example, Figure 5 shows the proton spectra at the end of the beam line and at the entrance of the NP04 active volume, for a monochromatic 1GeV/c proton beam.

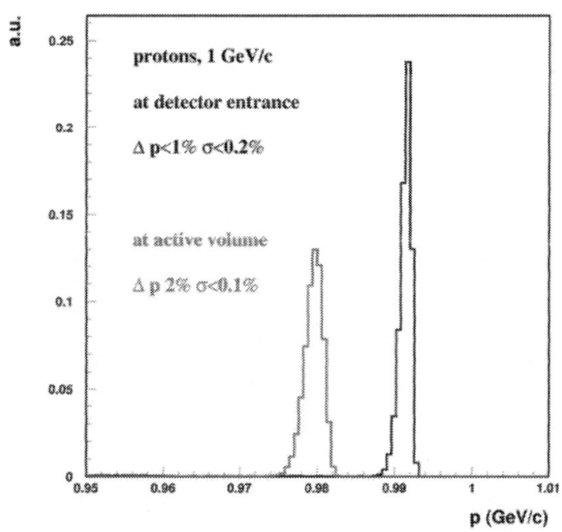

Figure 5. Momentum distribution of protons at the end of the beam line and at the entrance in the NP04 active volume, for a monochromatic 1GeV/c proton beam. Note the highly zoomed momentum range. The area under the peaks differ by 7% only.

The momentum degradation is limited, below 2%, with a negligible spread. The beam intensity with respect to a theoretical "no material" configuration is reduced by 15% only in the beam line and an additional 7% at the detector window.

4. Summary and next steps

The beam line design and an outlook on their expected performance has been reported in this paper. Additionally, a PID system for both beam lines is being currently studied, using a combination of two threshold Cherenkov counters (XCET). R134a is considered [10] as radiating gas in combination with CO_2, possibly combined with a Time-of-Flight system based on scintillating fibers.

References

[1] WA105 – TDR, CERN-SPSC-2014-013, SPSC-TDR-004
[2] ProtoDUNE – proposal, CERN-SPSC-2015-020; SPSC-P-351
[3] Brown K L *et al.* CERN 80-04
[4] Roberts T *et al.* 2008 G4BeamLine Particle Tracking in Matter Dominated Beam Lines *Proc. EPAC08* Genoa (Italy) WEPP120
[5] Charitonidis N *et al.* 2016 CERN-ACC-NOTE-2016-0052
[6] Ortega I *et al.* 2016 A scintillating fibre beam profile monitor for the experimental areas of the SPS at CERN *Journal of Physics: Conference Series* **763** No. 1 (IOP Publishing)
[7] Efthymiopoulos I and Fabich A The Very Low 1-9 GeV/c Tertiary Beam Extension of the H8 Beam Line of CERN SPS *arXiv:1206.2184*
[8] Ferrari A *et al.* 2005 *FLUKA: a multi-particle transport code* CERN 2005-10
[9] Böhlen T T *et al.* 2014 *Nuclear Data Sheets* **120** 211–4
[10] Charitonidis N *et al.* Estimation of the R134a gas refractive index for use as a Cherenkov Radiator, using a high energy charged particle beam, submitted for publication in *Nucl. Instr. Meth.* B

IOP Conf. Series: Journal of Physics: Conf. Series **874** (2017) 012057 doi:10.1088/1742-6596/874/1/012057

LINAC5 - A Quasi-Alvarez Linac for BioLEIR

J M Garland, J-B Lallement, A Lombardi
BE-ABP-HSL, CERN,
CH-1211 Geneva 23, Switzerland
E-mail: jimmy.garland@cern.ch

Abstract. LINAC5 is a new linac proposed for the acceleration of light ions with $Q/A = 1/3$ to $1/4$ for medical applications within the BioLEIR (Low Energy Ion Ring) design study at CERN. We propose a novel quasi-Alvarez drift-tube linac (DTL) accelerating structure design for LINAC5, which can reduce the length of a more conventional DTL structure, yet allows better beam focussing control and flexibility than the inter-digital H (IH) structures typically used for modern ion acceleration. We present the main sections of the linac with total length ~12 m, including a 202 MHz radio frequency quadrupole (RFQ) a matching medium energy beam transport (MEBT) and a 405 MHz quasi-Alvarez accelerating section with an output energy of 4.2 MeV/u. Permanent magnet quadrupoles are proposed for use in the quasi-Alvarez structure to improve the compactness of the design and increase the efficiency. Lattice design considerations, multi-particle beam dynamics simulations and RFQ and radio frequency (RF) cavity designs are presented.

1. Introduction

BioLEIR is a design study at CERN [1] for a multidisciplinary, biomedical research facility based on LEIR to provide light ion beams suitable for clinically oriented, fundamental research on cell cultures and for radiation instrumentation development. Within the BioLEIR framework, a new linac is required to accelerate ions with a charge to mass ratio Q/A between 1/3 and 1/4, up to 4.2 MeV/u for injection into the existing LEIR accelerator [2]. LINAC5 has been designed to fulfil this purpose. Following acceleration in LEIR to between 50 and 440 MeV/u, ions will be extracted to several beam-lines terminating in a variety of experimental stations.

A quasi-Avarez DTL accelerating structure [3, 4] is proposed for LINAC5, enabling the possibility for a shorter linac than those employing the more commonly used DTL, but with more focusing control than an inter-digital H structure (IH). We present a study of the complete linac with length ~12 m, including an electron cyclotron resonance (ECR) ion source, a LEBT (Low Energy Beam Transport), a 202 MHz RFQ, a MEBT and a 405 MHz quasi-Alvarez DTL. This report focuses on the lattice design and multi-particle beam dynamics simulations of the novel quasi-Alvarez accelerating section, the MEBT and RFQ.

2. ECR Ion Source, RFQ and Matching Sections

A schematic overview of LINAC5 is shown in Figure 1. An ECR ion source is envisaged for the production of a 2 mA ion beam [5]. The extraction energy was selected as 5.0 keV/u to maximise the extraction efficiency over the range of ions, as shown in Table 4.4 in reference [5]. A LEBT is being studied for matching the ions from the source to the RFQ. Two concepts are being investigated: the first comprises one or two solenoids and a spectrometer for charge selection; the second envisages charge selection carried out by the RFQ instead of a spectrometer. The latter is realisable due to relatively low beam current where the RFQ can be used to select the desired ions.

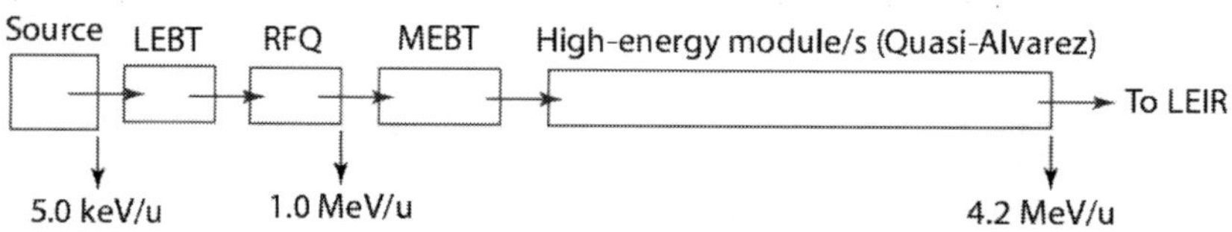

Figure 1. The basic schematic layout of LINAC5.

A 202 MHz RFQ was designed for the acceleration of ions with $Q/A = 1/4$ from an energy of 5 keV/u to 1.0 MeV/u. The main parameters of the RFQ are summarised in Table 1 and the aperture, modulation and phase are shown in Figure 2.

Table 1. The main parameters of the RFQ.

Parameter	Value
Length	4 m
RF Frequency	202.56 MHz
In/out energy	0.005 - 1.0 MeV/u
Vane voltage	60 kV
Average aperture (r_0)	3 mm
Max. surface field Es	27 MV/m
In/out. trans. $\epsilon_{rms,norm}$	0.1 - 0.1 π.mm.mrad
Out. long. $\epsilon_{rms,norm}$	0.086 π.MeV.deg

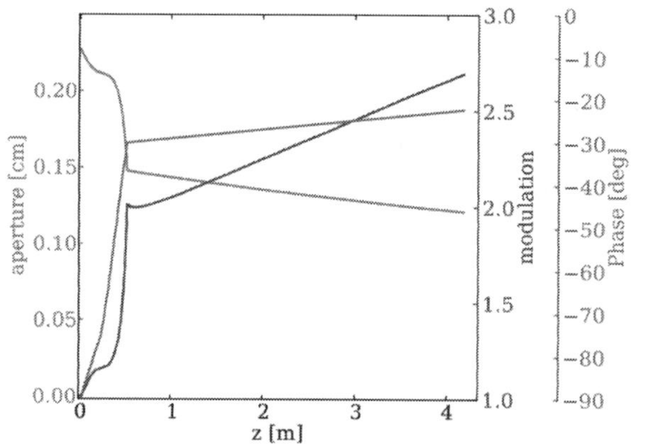

Figure 2. The aperture (red), modulation (blue) and phase (green) as a function of the distance along the RFQ.

A MEBT was studied for matching the beam from the RFQ to the quasi-Alvarez DTL. Simulations of the different ion species from the RFQ exit show that in order to control the longitudinal phase spread, two 202 MHz bunching cavities are required. Four electromagnetic quadrupole magnets are also necessary to allow flexible matching to the down-stream focusing channel.

3. Acceleration from 1.0 - 4.2 MeV/u

Three different types of RF accelerating structure meeting the LEIR injection specifications were studied: a 202 MHz two-module DTL with an FFDD focusing channel, a 202 MHz three-module IH with two inter-tank triplet focusing sections and a single-module 405 MHz quasi-Alvarez DTL. A quasi-Alvarez DTL [3, 4] is similar to a conventional 2π mode DTL structure, where a series of drift-tubes are separated longitudinally by RF gaps at a spacing of $\beta\lambda$. However in a quasi-Alvarez, quadrupole focusing magnets (quadrupoles) are not placed inside all drift tubes, but instead are placed inside the nth drift-tube where the periodicity n is a free parameter of the design. This allows the empty drift-tubes to be shortened and hence the frequency of the structure may be increased beyond that what is usually suitable for a conventional DTL. This allows a shorter over-all structure for the LINAC5 energy range. Table 2 shows a comparison of the length and output beam characteristics of each design.

Table 2. Comparison of the three types of accelerating structure studied for LINAC5. The percentage change in emittance between the end and start of the structures is shown.

Structure	Length	Trans. $\Delta\epsilon_{rms,norm,4D}$	Long. $\Delta\epsilon_{rms,norm}$
DTL	8.2 m	~0 %	1.8 %
IH	3.3 m	55 %	96 %
Q-A	5.2 m	10 %	0.2 %

Whilst all the ideal (non-error) structures studied had good transmission (>98 %), the final beam characteristics show the weakness of the IH and the superiority of the DTL for this particular case of 2 mA of current and $Q/A = 1/4$. However, the quasi-Alvarez is around 40 % shorter than the DTL, giving it an advantage without compromising greatly on the final beam emittance; the output characteristics are still comfortably within the LEIR specifications [2].

To optimise the focusing strength required for the desired ion species with the length of the linac, the geometry was chosen such that longer drift-tubes are separated by two smaller drift-tubes (n=3) and three respective gaps as shown in Figure 3. A FODO focusing channel was selected and the quadrupole dimensions and strengths were optimised to deliver focusing with a maximum beam size of 3.5 mm for the range of ions. A realistic normal-conducting pole-tip field of 0.56 - 0.77 T, a bore radius of 6 mm and a length of 50 mm were chosen. In order to reduce the drift-tube and stem sizes, and increase the shunt impedance, permanent magnet quadrupoles (PMQs) are proposed. PMQs have high reliability, do not require powering and reduce drift-tube cooling requirements. It has been shown that PMQs can be utilised effectively in light-ion accelerators with small drift-tube sizes, within low-beta accelerating structures and with similar field strengths to those proposed for LINAC5 [6, 7].

In the present design, one accelerating module is proposed with an RF field designed to accelerate a minimum $Q/A = 1/4$. The voltage in the cavities will be reduced by 25 % for $Q/A = 1/3$ to preserve synchronicity and respectively for ions with 1/3 >Q/A> 1/4. To reduce the length, we chose to double the frequency with respect to the RFQ, from 202 MHz to 405 MHz. The RF cavity structure design was optimised in SuperFish [8]. Table 3 shows the main parameters of the quasi-Alvarez DTL structure design.

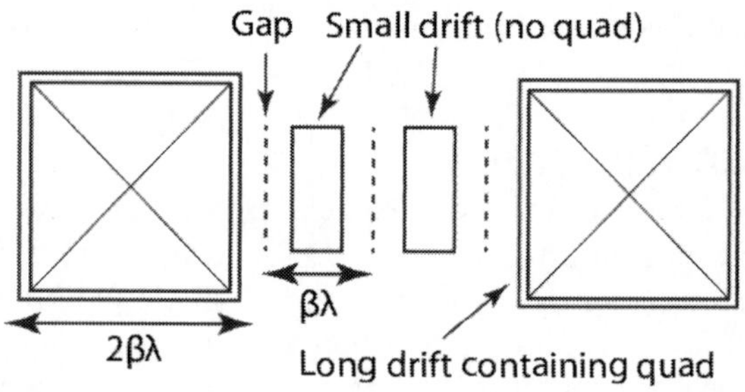

Figure 3. A schematic representation of one cell (half period) of the selected quasi-Alvarez design.

A 2 mA beam was tracked through the RFQ using the Parmteq code [9] and the output distribution was used as the input for the following tracking. The Travel code [10] was then used to track this distribution through the MEBT and quasi-Alvarez section. The transverse beam envelopes for $Q/A = 1/4$ and $1/3$ are shown in Figure 4 (a) and (b) respectively.

Table 3. The main parameters of the quasi-Alvarez accelerating module without errors. The parameter described by $A_{trans.}/\epsilon_{rms,norm,out}$ shows the transverse acceptance divided by the output transverse emittance and E_0TL/cell is the integrated RF field in a cell.

Parameter	Value
Length	5.2 m
In/out E_k	1.0 - 4.2 MeV/u
Frequency	405.12 MHz
Current	2.0 mA
Bore radius	6 mm
No. of periods	12
Quad. length	50 mm
Max/min integrated grad.	6.4 - 4.7 T
In/out. 4D $\epsilon_{rms,norm}$	0.01 - 0.011 π.mm.mrad^2
In/out. long. $\epsilon_{rms,norm}$	0.022 - 0.022 π.deg.MeV/u
$A_{trans.}/\epsilon_{rms,norm,out}$	1.5
Transmission	100 %
Tank power	1300 kW
L_{RFgap}/L_{cell} (start-end)	0.18 - 0.25
Kilpatrick (start-end)	1.5 - 1.2
E_0TL/cell (start-end)	0.09 - 0.24 MV
Sync. phase (ramped)	-30 to -20 deg.
Shunt Impedance	84 MOhm/m

4. Error Studies

The sensitivity to transverse and longitudinal errors was studied independently using the Travel code. Realistic errors were chosen based on studies of LINAC4 [11]. In the transverse case, over 7000 randomly generated lattices were created with errors in: beam jitter using Gaussian

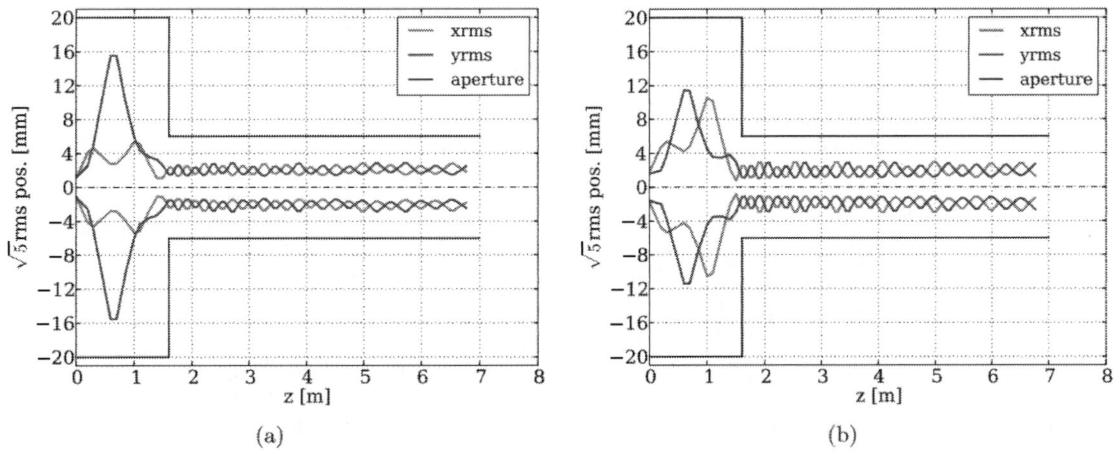

Figure 4. The horizontal (red) and vertical (blue) beam envelope in the MEBT and quasi-Alvarez sections for ions of $Q/A = 1/4$ (a) and $Q/A = 1/3$ (b). The beam-pipe aperture is shown in black.

distributions in x, y ($\sigma_{x,y} = 0.1$ mm), x' and y' ($\sigma_{x',y'} = 0.1$ mrad); quadrupole gradient errors with a Gaussian distribution of $\sigma = 0.5\,\%$; horizontal and vertical quadrupole misalignments with Gaussian distributions of $\sigma = 0.1$ mm. All had a 3σ cut-off. Of the sample cases, 98 % had > 90 % transmission and 85 % had > 99 % transmission, see Figure 5. Figure 5 also shows that > 99 % of cases had < 5 % transverse emittance growth beyond the nominal (see Table 3 for nominal values).

Static and dynamic longitudinal errors were considered separately, with 10,000 and 1,000 samples respectively. The former are defined as individual gap errors in phase and amplitude and the latter are defined as constant amplitude errors applied to all cavities equally to simulate fluctuations in the RF power supply. The random error distributions were given by: beam jitter in energy and phase with uniform distribution and cut-off at 1 % and 1 deg. respectively; static errors in gap phase and amplitude with uniform distribution and cut-off at 1 deg. and 1 % respectively; dynamic amplitude errors applied equally to each cavity from a uniform distribution with cut-off at 1 %. Simulated particles remaining in the stable RF phase-space region were counted, where 100 % had > 95 % transmission for the static errors and 100 % had > 90 % transmission for the dynamic errors.

From the results of the transverse (see Figure 5) and longitudinal error studies, it is clear that the magnitude of the errors employed do not have a significant effect on the performance of the linac. However, it should be noted that these errors are reasonably realistic, for example max. 0.3 mm magnet misalignment error is easily achievable with current technology.

5. Conclusions

A design study was carried out for LINAC5 within the BioLEIR framework. A quasi-Alvarez DTL was shown to have advantages over both a conventional DTL and IH structure. More detailed designs of the RFQ, MEBT and a quasi-Alvarez DTL were successfully completed to meet the LEIR injection parameters. Beam dynamics simulations from the RFQ entrance to the end of the quasi-Alvarez DTL, including with realistic errors, show that > 99 % of cases have a maximum transverse emittance growth of 10 %, where transmission remained > 90 % in 98 % of cases.

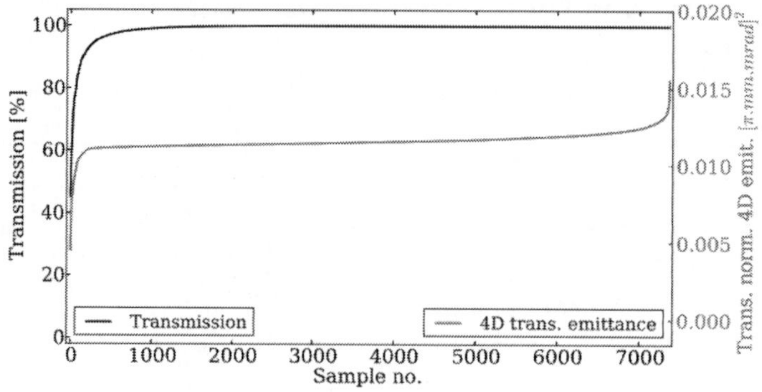

Figure 5. Transmission and 4D transverse rms emittance for different random error lattices. Sample number is ordered separately for the two curves and is not correlated.

6. Acknowledgments
The authors would like to thanks Suitbert Ramberger for his help with the RF design in SuperFish and Django Maglunki for information regarding LEIR.

References
[1] Ghithan S *et. al.* 2017 Feasibility study for BioLEIR *CERN Yellow Reports* CERN-2017-001-M
[2] Brning O S *et. al.* 2004 LHC design report vol.3: The LHC injector chain *CERN Yellow Reports* CERN-2004-003-V-3
[3] Warner D J 1988 Heavy ion acceleration using drift-tube structures with optimised focusing *Proc. Linear Acc. Conf. 1988, Williamsburg, VA, USA* pp 109-11
[4] Lapostolle P, Tanke E, Vretenar M and Warner D J 1990 Computer design and dynamics of the quasi-alvarez linac *Proc. Linear Acc. Conf. 1990, Albuquerque, NM, USA* pp 99-101
[5] Ghithan S *et. al.* 2017 Feasibility study for BioLEIR *CERN Yellow Reports* CERN-2017-001-M Chpt. 4
[6] Tommasini D, Buzio M, Thonet P A and Vorozhtsov A 2011 Design, manufacture and measurements of permanent quadrupole magnets for Linac4 *IEEE Trans. on App. Supercon.* **22(3)** p 4000704
[7] Kerennoy S S *et. al.* 2012 H-mode accelerating structures with permanent-magnet quadrupole beam focusing *Phys. Rev. ST Accel. Beams* **15(9)** 090101
[8] Holsinger R F and Halbach K 1976 SuperFish - a computer program for evaluation of rf cavities with cylindrical symmetry *Part. Acc.* **7(4)** pp 213-22
[9] Crandall K R and Wrangler T P 1988 Parmteqm - a beam-dynamics code for the RFQ linear accelerator *AIP Conf. Proc.* **177** pp 22-8
[10] Perrin A and Amand J F 2003 *Travel v 4.06 user manual* CERN
[11] Bellodi G *et. al.* 2011 Alignment and field error tolerance in Linac4 *CERN-ATS-Note-2011-021* CERN

Energy efficiency and saving potential analysis of the high intensity proton accelerator HIPA at PSI

A Kovach[†], A Parfenova, J Grillenberger and M Seidel

Paul Scherrer Institut, 5232 Villigen PSI, Switzerland

Abstract. High power proton machines consume a large amount of energy. Thus, the energy efficiency of grid to beam power conversion is particularly important for the overall power consumption of such facilities. In this study, we analyse the energy efficiency of PSI's cyclotron-based HIPA facility, which presently delivers a maximum of 1.4 MW beam power. The total power consumption of the entire facility is 12.5 MW at 2.2 mA beam current (1.3 MW). Main power consumers are: RF systems, electromagnets, water cooling and auxiliary systems including infrastructure, each consuming 5.3 MW, 3.6 MW, 1.65 MW and 1.95 MW, respectively. HIPA's grid to beam efficiency is 18.3% when considering only those parts of any subsystems (RF components, magnets, cooling, and auxiliary systems), which are minimally required to produce a full 1.3 MW beam. The dependency of individual subsystems on beam power was also studied. These findings serve as a basis for further optimizations of the HIPA facility and give a reference of the efficiency estimate for the cyclotron-based high power machines.

1. Introduction

The HIPA cyclotron-based facility commissioned in 1974 is a high intensity accelerator that presently produces a 1.3 MW proton beam. The two isochronous separated sector cyclotrons Injector-2 and Ring form the facility's core (figure 1). The 2.2 mA beam is extracted from Injector-2 at 72 MeV, and then accelerated to the final kinetic energy of 590 MeV in the Ring. After extracting from the Ring Cyclotron the proton beam is guided to experimental areas and production targets for secondary particle generation (pions, muons and neutrons) [1]. The facility's consumption was previously estimated to be roughly 10 MW [2]. It was long desired to verify this estimate by measurements and system-wide power analysis.

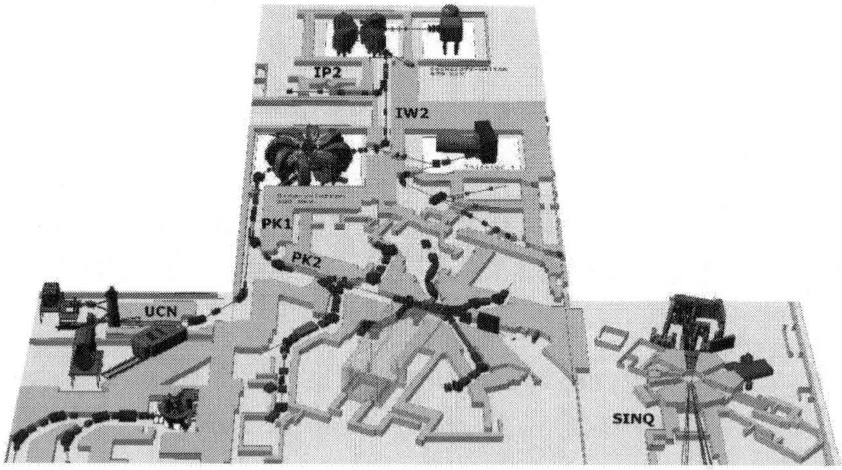

Figure 1. Schematic view of the HIPA facility.

The EnEfficient network of the EuCARD-2 integrated activity project focuses on the energy and cost efficient utilization of accelerator facilities. Following these principles, the present study extensively analyses the power consumption of high intensity accelerator facilities on the example of HIPA at PSI. Knowing exactly where and how electricity is consumed is the very first step in the optimisation cycle of a facility.

2. Distribution of Power

Figure 2 reveals the power consumption of every sub-system as well as the total 12.5 MW power consumption of the facility. At a full load HIPA accounts for more than the half of PSI's ca. 20 MW of total power consumption. It shall be emphasised that reaching the consumption of 12.5 MW requires all experimental areas (IP2, UCN, SINQ, and all secondary beamline experiments) to run concurrently.

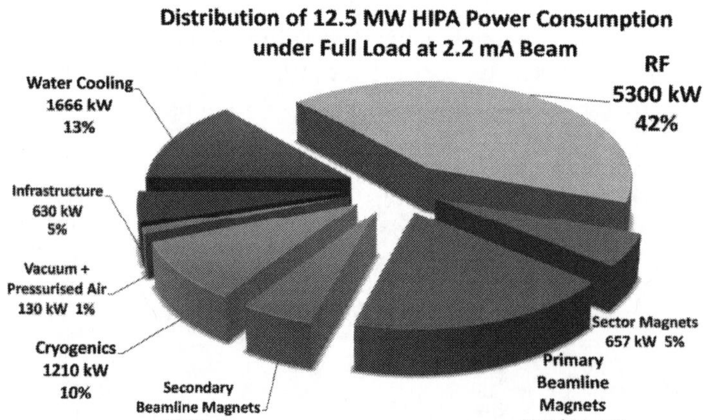

Figure 2. Distribution of 12.5 MW HIPA power consumption under full load at 2.2 mA beam current.

3. Power Consumption of Sub-systems

3.1. Efficiency of the RF System

The Injector-2 and the Ring cyclotron each have their own amplifier chains for every cavity and resonator. During the analysis process power consumption was studied on the example of the Ring machine. RF is the single largest power consumer with 42% of all HIPA power. Each of the four accelerating cavities of the Ring machine has a chain of four tetrode tube amplifiers. The first step in the energy flow is AC-DC rectification, with an efficiency of about 90%. Then DC voltage is fed to the amplifier chain, where RF power is produced. The combination of all amplifications stages has an efficiency of roughly 60%. The Ring cyclotron's RF system power consumption (and hence the forward RF power) are linear functions of the beam current in the operating region up to 2.4 mA. At 0 mA beam the system still has a minimal consumption of approx. 2.2 MW from the grid. This 'base load' is required for running the system at its operational point and to keep it ready for producing RF power. The grid to RF efficiency stays approximately 55% in this region of operation.

3.2. Magnet Power Consumption
The total 2.3 MW consumption of the beam lines was measured using a dedicated program SLEEP (Table 1). The in-house developed energy saving SLEEP program was also designed to monitor the power consumption of beam lines in real time and to put beam lines into standby during outages with no beam longer than 30 minutes.

Table 1. Power Consumption of Beam Line Magnets

Magnets	Power (kW)
Magnets outside SLEEP	217
Cyclotron sector magnets	657
Secondary beam lines	635
IW2 beam line	358
IP2 beam line	59
P-channel	746
SINQ beam line	565
UCN beam line	125

The SLEEP program was commissioned in April 2016 and successfully ran over the year. The saving potential of primary beam line magnets was forecasted to be 980 MWh/annum. The SLEEP program has saved 1171 MWh in 2016. Taking inefficiency of power supplies into account as factor 1.1 (90% efficiency) the saving in 2016 on the primary beam line magnets was 1288 MWh.

For the next operational year the SLEEP program has been upgraded with an RF module. It notifies operators that cavities and resonators can be put to a power saving mode. It shall assist operators to make another 370 MWh of additional documented savings.

3.3. Water Cooling Circuits
A considerable power consumer is the water cooling at an average electrical power of approx. 1.67 MW, which is roughly 11-14% of its cooling capacity. There are 3 different cooling methods used: water cooling from the river Aare, ground water cooling and forced air cooling. Most of HIPA is cooled with water from the Aare through heat exchangers. The cooling system has 3 levels of cooling circuits. The primary open loop cooling circuit circulates river water, whereas secondary and tertiary circuits are closed loop and are exposed to radiation.

Primary ground water, river water and forced air cooling circuits (figure 3) only consume a total of 18% (300 kW). Secondary and tertiary circuits require the remaining 1.4 MW (82%).

The power required for cooling showed very light dependency on the produced beam power: only a 5% increase in power between a 1.2 and a 2.2 mA beam. This can be justified by the large fixed load in the form of magnets.

The primary cooling circuit is highly dependent on inlet and outlet river water temperature, which varies notably with seasons 4+°C in winter 25+°C in summer. This translates to an up to 50% fluctuation in primary pump electrical power.

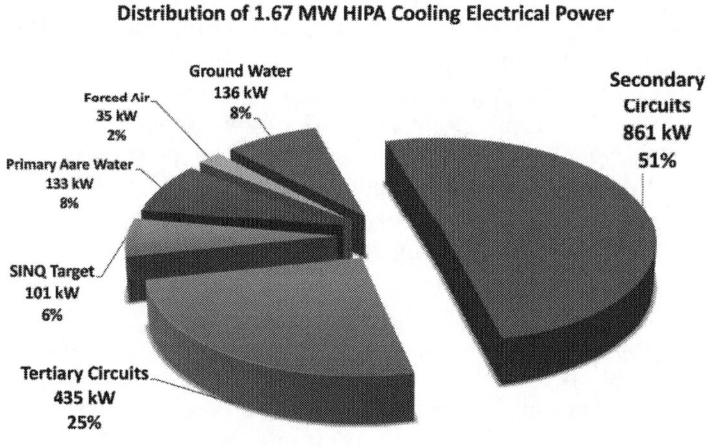

Figure 3. Distribution of cooling electrical power.

The total power consumption of water cooling could be reduced by approx. 20% if pumps were upgraded to Variable Frequency Drive (VFD) equivalents. The estimated saving potential is 1.7 GWh. Note this is a rough estimation; to find the precise saving potential, it is advised to separately study the optimisation of the cooling system.

3.4. Consumption of Auxiliary Systems

Auxiliary systems include vacuum and pressurised air (130 kW), infrastructure (630 kW; ventilation, lighting, air conditioning, etc) and cryogenic cooling.

The three cryogenic systems (1.21 MW) of HIPA are required for the cold moderators of both neutron sources, for a superconducting muon channel and for several experiment stations. Two out of the three Helium compressors will be upgraded to screw compressors technology in 2018 and hence their electrical power consumption is expected to reduce by from 780 kW to 624 kW (20%), resulting in a saving of 900 MWh per annum.

3.5. Saving Potential

The areas and amounts of saving potential were identified as shown in figure 4.

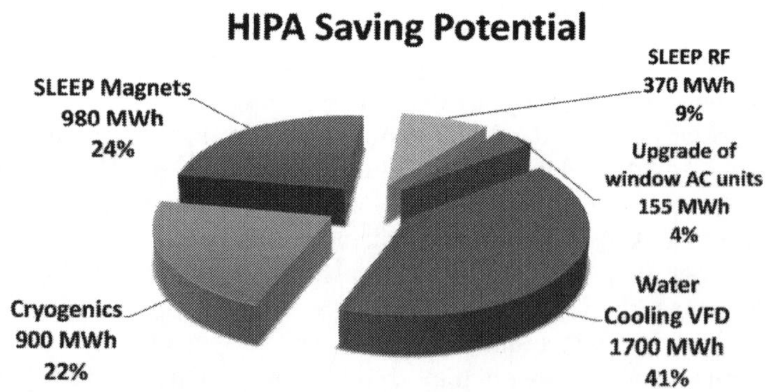

Figure 4. Distribution of 4 GWh HIPA's saving potential.

4. Efficiency of Minimum Full beam

To evaluate the performance of a facility, it is oftentimes a good practice to look at its grid to beam conversion efficiency. The grid to beam efficiency can be defined as a ratio of total power to the useful (beam) power:

$$\eta_{acc} = \frac{P_{useful}}{P_{total}} \tag{1}$$

Under full-load conditions the grid to beam efficiency of the entire facility is η=1.3/12.5*100=10.4%. Although this is a good evaluation of the facility, it does not reflect the performance of the accelerator accurately. The branch-like beam line structure and the large number of experiment sites pose a considerably varying load on top of the accelerator power. In order to assess the accelerator's efficiency, the consumption of a minimum full beam has to be determined. A minimum full beam setup only includes those consumers, which are required to produce a 2.2 mA beam extracted from the Ring cyclotron.

Accelerator related consumers, including all RF components, the Cockcroft-Walton accelerator, Injector-2, the Ring cyclotron and the beam lines between the cyclotrons are shown in figure 5. The necessary cooling circuits need 379 kW of power. As far as infrastructure is concerned, ventilation and lighting are taken into account as the most significant consumers in terms of infrastructure.

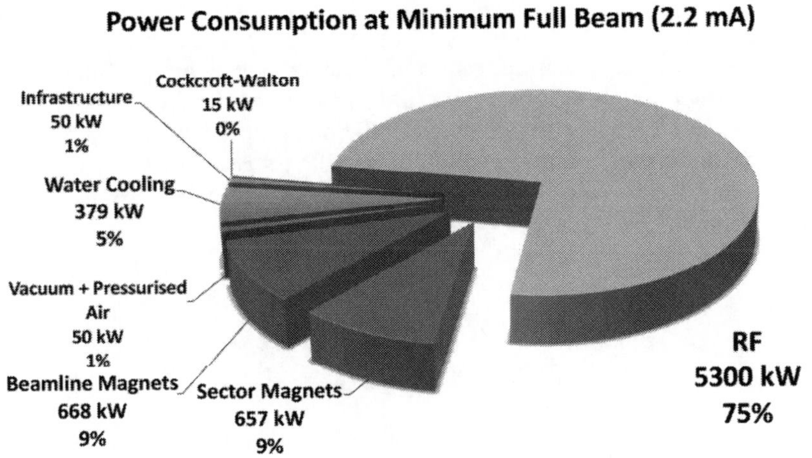

Figure 5. Power consumption at minimum full beam.

Hence, the minimum power consumption for a full beam is 7.12 MW. This in turn yields a grid to beam efficiency of $\eta = \frac{1.3}{7.12} * 100 = 18.3\ \%$. The 7.12 MW requirements and the 18.3% efficiency can be used as the most accurate measurement on HIPA's performance so far.

5. Efficiency and Beam Power

The grid to beam efficiency η_{acc} can be further ex-pressed via RF efficiency and dissipated power as:

$$\eta_{acc} = \frac{P_{useful}}{P_{total}} = \frac{P_{beam}}{P_{loss} + P_{beam}/\eta_{RF} + P_{aux}} \tag{2}$$

where η_{RF} is the overall efficiency of the amplifier chain and P_{loss} are the ohmic wall losses in the cavities P_{aux} represents the base load required for magnets, vacuum, and conventional systems etc. The beam power P_{beam} scales with the maximum extracted beam current Imax and the kinetic energy E_{kin} of the particles q:

$$P_{beam} = \frac{I_{max} \cdot E_{kin}}{q} \tag{3}$$

The wall losses are proportional to the square of the peak voltage U_{acc} of the cavities, where R is the impedance given by the cavities' geometry and material parameters:

$$P_{loss} = \frac{|U_{acc}|^2}{2\,\mathrm{R}} \tag{4}$$

According to the empirical law of Joho [3] the maximum beam current that can be extracted from the cyclotron scales with the acceleration voltage to the cube:

$$U_{acc} = \frac{\mathrm{E}_{kin}}{\mathrm{q}} \cdot \sqrt[3]{I_{max}/c} \tag{5}$$

Substituting (5) into (4); and then (4) and (3) into (2) yields:

$$\eta_{acc} = \frac{\mathrm{I}_{max} \cdot \mathrm{E}_{kin}/q}{\frac{(1/c)^{\frac{2}{3}} \cdot E_{kin}^2}{2 \cdot R \cdot q^2} + \frac{\mathrm{I} \cdot \mathrm{E}_{kin}}{q \cdot \eta_{RF}} + P_{aux}} \tag{6}$$

The relationship between the beam current and accelerator efficiency (figure 6) reveals that the efficiency could be further improved, however a limit is posed by the fact that ohmic wall losses scale with the square of the accelerating voltage, whereas the maximum current scales with the cube of the accelerating voltage [3]. At the present operation point HIPA's accelerator efficiency is 18.3%. A 3.0 mA beam will theoretically be feasible after the upgrades of Injector-2, the Ring cyclotron's flattop cavity, SINQ, and beam dump.

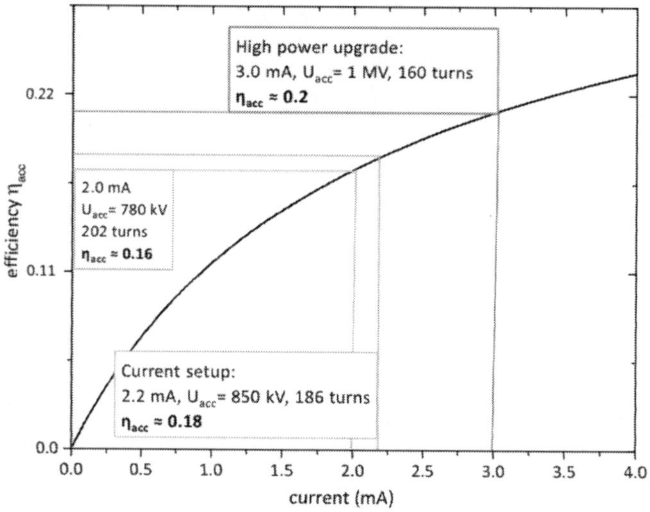

Figure 6. Accelerator efficiency vs. beam current [4].

6. Conclusion

The energy consumption of the HIPA facility was analysed. By evaluating each sub-consumer of the HIPA facility (RF, electromagnets, water cooling and auxiliary) and by summing up their power consumption, the 12.5 MW full-load power and the saving potential of the facility was identified. The power consumption of a minimum full beam setup was also found to be 7.12 MW and hence the 18.3% accelerator efficiency was calculated.

References

[1] Seidel M *et al.* 2010 Production of a 1.3 MW proton beam at PSI *Proc. 1st Int. Particle Accelerator Conf. (IPAC'10)* Japan TUYRA03 pp. 1309–13

[2] Seidel M, Gehring R, Jensen E, Parker T I, Spiller P J and Stadlmann J 2015 Improving the energy efficiency of accelerator facilities *Proc. 6th Int. Particle Accelerator Conf. (IPAC'15)* USA WEXC3 pp 2428–33

[3] Joho W 1981 *Proceedings of the 9th International Conference on Cyclotrons and their Applications* France pp 337–47

[4] Grillenberger J *et al.* 2016 The energy efficiency of cyclotrons *Cyclotrons Conference* Switzerland

IOP Conf. Series: Journal of Physics: Conf. Series **874** (2017) 012059 doi:10.1088/1742-6596/874/1/012059

Realizing a high-intensity low-emittance beam in the J-PARC 3-GeV RCS

H Hotchi#, H Harada, S Kato, K Okabe, P K Saha, Y Shobuda, F Tamura, N Tani, Y Watanabe and M Yoshimoto

J-PARC Center, Japan Atomic Energy Agency, Tokai, Naka, Ibaraki, 319-1195 Japan

#hotchi.hideaki@jaea.go.jp

Abstract. The J-PARC 3-GeV rapid cycling synchrotron is now developing beam studies to realize a high-intensity low-emittance beam with less beam halo. This paper presents the recent experimental results while discussing emittance growth and its mitigation mechanisms.

1. Introduction

The J-PARC 3-GeV rapid cycling synchrotron (RCS) is the world's highest class of high-power pulsed proton driver aiming for an output beam power of 1 MW [1-3]. A 400-MeV H^- beam from the injector linac is multi-turn charge-exchange injected into RCS through a carbon foil over a period of 0.5 ms. RCS accelerates the injected protons up to 3 GeV with a repetition rate of 25 Hz. Most of the RCS beam pulses are delivered to the materials and life science experimental facility (MLF), while only four pulses every several second are injected to the following main ring synchrotron (MR). The requirements for the beam operations to MLF and MR are different. Thus, different parameter optimizations are required for each.

Machine activations of RCS are mainly determined by the beam operation to MLF. Hence, sufficient beam loss mitigation has to be achieved in this operation mode. In addition, from MLF, a wide-emittance beam with low charge density is required to mitigate a shockwave on their neutron production target, which is essential to ensure a sufficient lifetime of the target. As already reported in IPAC'16 [4], the requirement for the 1-MW beam operation to MLF was accomplished finally by introducing wide-ranging transverse injection painting with a painting emittance of $\varepsilon_{tp}=200\pi$ mm mrad in combination with betatron resonance correction [5, 6].

On the other hand, from MR, a low-emittance beam with less beam halo is required contrary to the MLF case, which is essential to mitigate beam loss in MR. RCS is now intensively developing high-intensity beam tests (~7.0 x 10^{13} protons per pulse, corresponding to ~84% of the full intensity) to realize such a high-intensity low-emittance beam required from MR. This paper presents the recent experimental results while discussing emittance growth and its mitigation mechanisms.

2. Optimization of transverse injection painting

We first conducted the optimization of transverse injection painting [7] to minimize emittance growth during injection.

Figure 1 shows the beam widths at 1 ms right after injection, measured as a function of the painting emittance. As shown in the figure, the minimum beam width was achieved for a small painting

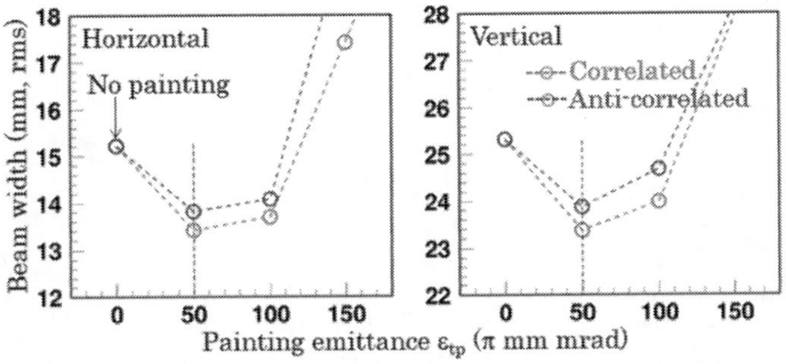

Figure 1. Beam widths at 1 ms right after injection, measured as a function of the painting emittance.

emittance of ε_{tp}=50π mm mrad. Transverse painting mitigates space-charge induced emittance growth by widely distributing beam particles on the transverse phase space. This painting parameter dependence of beam width is ascribed mainly to the balance of painting emittance and its resultant space charge mitigation. In addition, in the figure, one can find that correlated painting rather than anti-correlated painting provides narrower beam width. This empirical situation can be comprehended by taking into account the effect of the $2\nu_x-2\nu_y=0$ resonance.

The $2\nu_x-2\nu_y=0$ resonance is a 4th order systematic resonance and it is excited mainly by the octupole component in the space-charge field. As is well known, this resonance causes emittance exchange between the horizontal and the vertical planes [8]. This emittance exchange leads to the following two effects, (1) and (2), during the beam painting process.

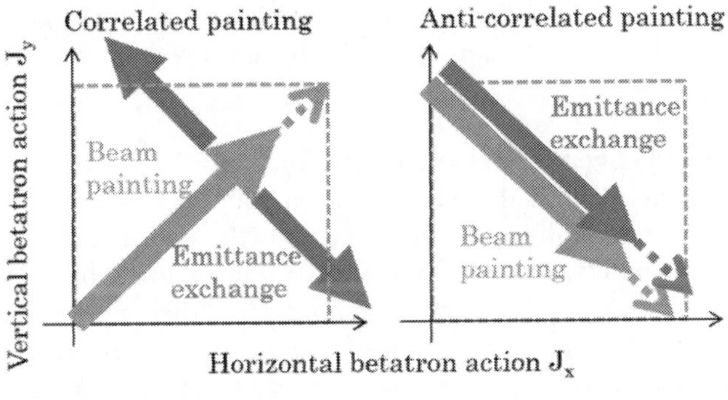

Figure 2. Schematic diagrams of injection painting and emittance exchange.

One (1) is additional emittance growth caused by the direct effect of emittance exchange, which is more enhanced in correlated painting. As shown in the left plot in figure 2, in correlated painting, the injection beam is painted along the yellow arrow, namely from the middle to the outside on both the horizontal and the vertical phase spaces. To this direction of beam painting, the emittance exchange occurs in the orthogonal direction, as shown by the red arrow. Therefore, in this case, the emittance exchange is directly connected to emittance growth over the painting area.

Another (2) is additional emittance growth caused by the secondary effect of emittance exchange, namely, by a modulation of the charge density, which is more enhanced in anti-correlated painting. As shown in the right plot in figure 2, in anti-correlated painting, the painting process on the vertical plane is reversed; the injection beam is filled from the middle to the outside on the horizontal plane, while, from the outside to the middle on the vertical plane. This direction of beam painting is the same as the direction of emittance exchange. Thus, in this case, additional emittance growth caused by the

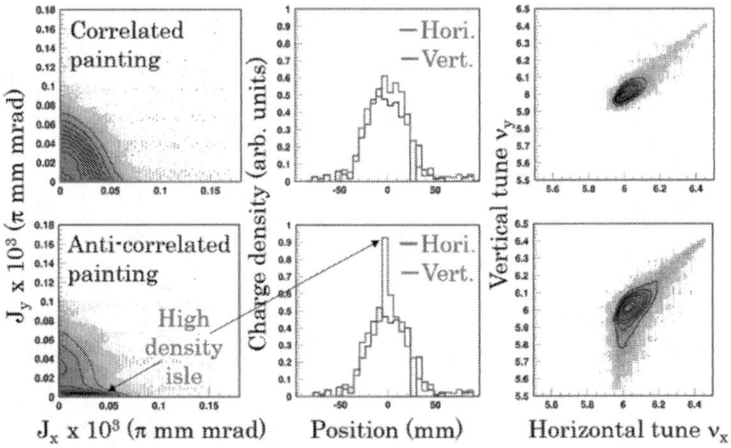

Figure 3. (Left) 2d plots of betatron actions calculated at the end of injection (at 0.5 ms). (Middle) Beam profiles measured at 0.5 ms. (Right) Tune footprints calculated at 0.5 ms.

direct effect of emittance exchange is well suppressed. But, this situation causes a significant modulation of the charge density by synchronism between beam painting and emittance exchange, as shown in the lower-left plot in figure 3. This characteristic behaviour of charge distribution was confirmed experimentally, as shown in the lower-middle plot in figure 3. This high density isle produces large space-charge detuning, as shown in the lower-right plot in figure 3, leading to significant emittance growth afterwards.

In the present operational condition, the emittance growth caused by (2) is more critical. Correlated painting avoids the effect. This is the main reason why narrower beam emittance is achieved for correlated painting.

3. Optimization of tune and chromaticity manipulations

Next, we attempted to mitigate additional emittance growth during acceleration after injection.

As is shown later, significant additional emittance growth was observed for the first 6 ms after injection. The numerical simulation conformed the emittance growth is mainly caused by three systematic resonances; $\nu_{x,y}=6$ and $2\nu_x-2\nu_y=0$ (see figure 6 shown later). In figure 4, one can see characteristic emittance growth on the vertical plane caused by the resonances; the emittance growth from (I) to (II) is caused by $\nu_y=6$, and it is further enhanced from (II) to (III) through emittance

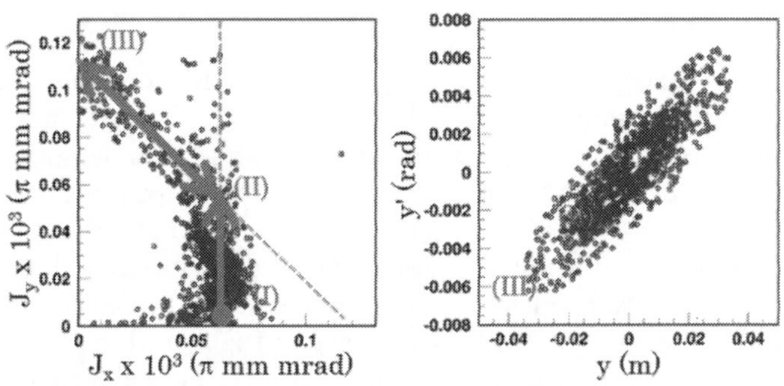

Figure 4. Single-particle behaviour of one macro-particle causing emittance growth. (Left) Turn-by-turn betatron actions. (Right) Transverse phase space coordinates.

	Tune	Chromaticity correction
(A) :	ID 1	Off
(B) :	ID 2	Off
(C) :	ID 2	On (with dc sextupoles)
(D) :	ID 3	On (with dc sextupoles)

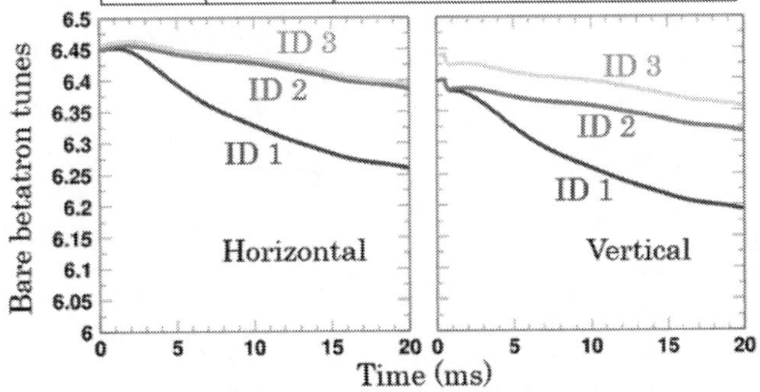

Figure 5. (Upper) Operational parameter settings tested; (A)–(D). (Lower) Tune variations from injection to extraction applied for (A)–(D); ID1–3.

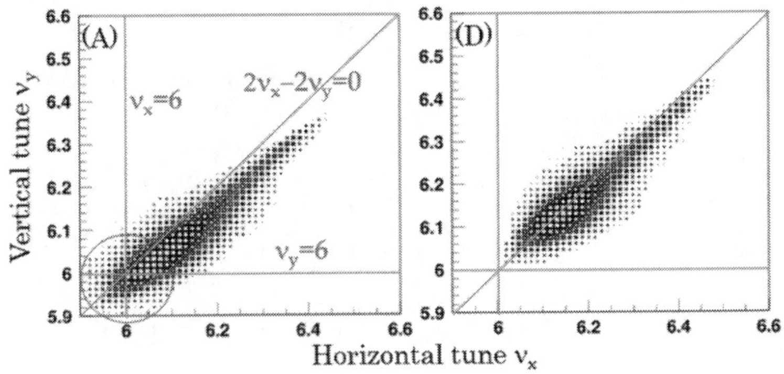

Figure 6. Tune footprints calculated at 4.7 ms.

exchange caused by $2\nu_x-2\nu_y=0$. Similar emittance growth is also generated on the horizontal plane by $\nu_x=6$ and $2\nu_x-2\nu_y=0$.

Based on the above numerical simulation result, we tried to mitigate the emittance growth originating from $\nu_{x,y}=6$. The operational parameter settings tested are listed in figure 5. The parameter (A) is the original one, in which the tune is set to ID 1 with no chromaticity correction. In this case, a portion of core particles reaches $\nu_{x,y}=6$ (see left plot in figure 6), causing large emittance growth (see blue curves in figure 7). To improve this situation, the operational parameter was modified from (A) to (D) step by step. At first, (B) the tune was modified to ID 2, and then, (C) the chromaticity correction was added with dc sextupole fields, and finally, (D) the tune was further modified to ID 3. As shown in figure 6, the separation of beam particles from $\nu_{x,y}=6$ is improved in the order from (A) to (D). The experimental results shown on the left side in figure 7 confirmed the emittance growth declines in the order from (A) to (D), as predicted by the numerical simulations displayed on the right side in figure 7.

But, at this stage, there still remained one issue to be solved for the use of the new parameters (C) and (D). In RCS, the extraction pulse kicker is the most dominant impedance source, causing horizontal beam instability depending on operational parameters [9, 10]. As shown in the left plot in

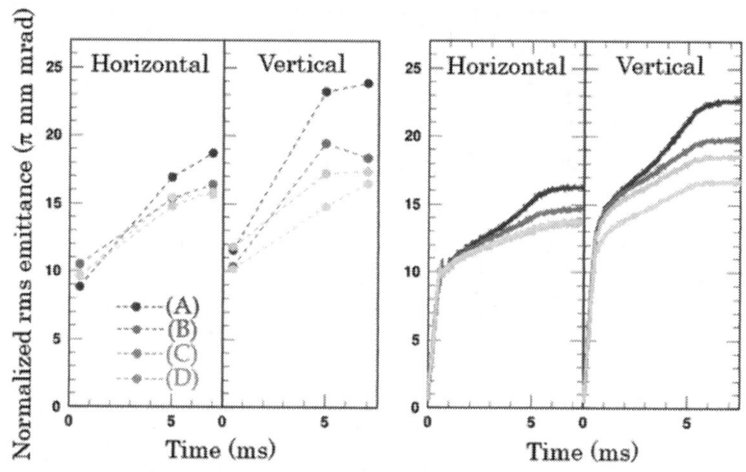

Figure 7. (Left) Normalized rms emittances measured over the first 7 ms. (Right) Corresponding numerical simulation results.

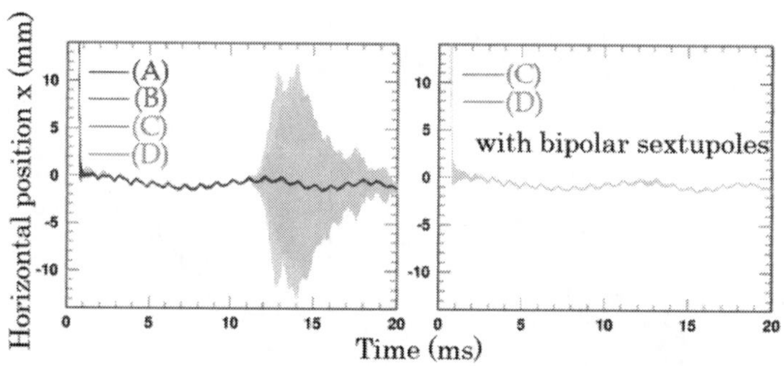

Figure 8. Horizontal beam positions measured over the whole acceleration time of 20 ms.

figure 8, the parameters (C) and (D) enhance beam instability after 10 ms, whereas they well mitigate emittance growth for the first 6 ms.

A practical way to solve the beam instability is to dynamically manipulate chromaticity during acceleration, as shown in figure 9. In this method, the chromaticity is adjusted to be small for the first 6 ms with dc sextupole fields, which is required for mitigating the emittance growth in this period. Then, after 6 ms, the sextupole fields gradually fall down to zero, and thereafter, they are excited in the opposite polarity to further increase the chromaticity. By this way, the chromaticity after 10 ms is increased by 17% over the natural value. This large negative chromaticity after 10 ms enhances Landau damping through momentum spread, which favours the suppression of beam instability.

In order to realize such a bipolar excitation of sextupole field, we improved the sextupole magnet power supply in the summer maintenance period in 2016. The following beam experiment confirmed that the beam instability is sufficiently damped as expected by this chromaticity manipulation, as shown in the right plot in figure 8.

The result of the profile measurements at the extraction energy (at 20 ms) is given in figure 10. As clearly shown in the figure, the extraction beam emittance including its tail part was successfully decreased from (A) to (D), where the bipolar sextupole field patterns were applied for (C) and (D). This improvement of the extraction beam quality reflects the emittance growth mitigation achieved for the first 6 ms.

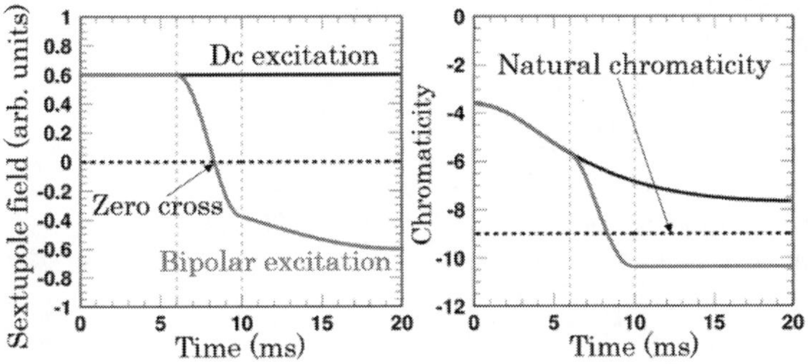

Figure 9. (Left) Sextupole field patterns over the whole acceleration time of 20 ms. (Right) Time dependences of chromaticity.

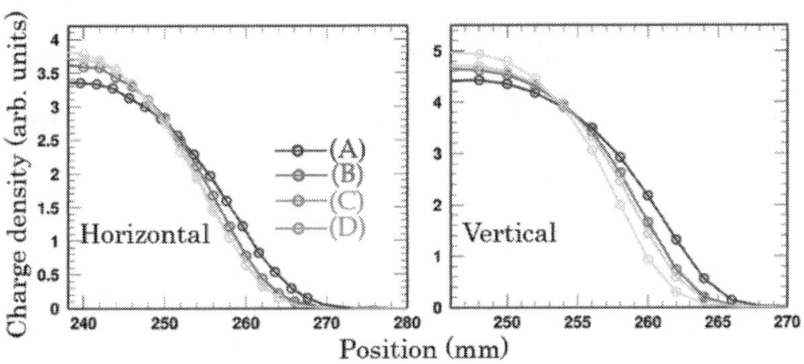

Figure 10. Beam profiles measured at the extraction energy (at 20 ms).

By this success of beam tuning, the beam loss at the beam scraping system located between RCS and MR was drastically reduced to less than half. MR beam tuning with the improved RCS beam is now ongoing.

4. Summary

For this past year, a large fraction of our effort was focused on realizing a high-intensity low-emittance beam. The extraction beam emittance including its tail part was successfully decreased by optimizing transverse injection painting, and tune and chromaticity manipulations, where bipolar sextupole field patterns were newly introduced to simultaneously achieve emittance growth mitigation at the early stage of acceleration and beam instability suppression after the middle stage of acceleration. In addition, in this work, characteristic behaviours of beam particles during injection painting, appearing coupled with emittance exchange, was revealed.

Acknowledgments

The authors acknowledge Dr. K. Yamamoto, Dr. N. Hayashi, Dr. M. Kinsho, Dr. K. Hasegawa, Prof. T. Koseki, and all the RCS group members for their continuous support and encouragement for the present work. We are also grateful to Dr. S. Meigo, Dr. M. Ooi, and Mr. S. Fukuta for their indispensable support during the beam test.

References

[1] High-intensity Proton Accelerator Project Team 2003 *JAERI Report* No. JAERI-Tech 2003-044
[2] Hotchi H *et al* 2009 *Phys. Rev. ST Accel. Beams* **12** 040402
[3] Hotchi H *et al* 2012 *Prog. Theor. Exp. Phys.* **2012** 02B003

[4] Hotchi H *et al* 2016 *Proc. 7th International Particle Accelerator Conference* (Busan: JACOW) pp 592–594
[5] Hotchi H *et al* 2015 *Nucl. Instrum. Methods Phys. Res.* Sect. A **778** 102
[6] Hotchi H *et al* 2016 *Phys. Rev. Accel. Beams* **19** 010401
[7] Hotchi H *et al* 2012 *Phys. Rev. ST Accel. Beams* **15** 040402
[8] Montague B W 1968 *CERN Report* No. 68–38
[9] Shobuda Y *et al* 2017 *Prog. Theor. Exp. Phys.* **2017** 013G01
[10] Saha P K *et al* 2016 *Proc. 7th International Particle Accelerator Conference* (Busan: JACOW) pp 589–591

IOP Conf. Series: Journal of Physics: Conf. Series **874** (2017) 012060 doi:10.1088/1742-6596/874/1/012060

Preliminary study of injection transients in the TPS storage ring

C H Chen[1,a], Y C Liu[1], J Y Chen[1], M S Chiu[1], F H Tseng[1], S Fann[1], C C Liang[1], C S Huang[1], T Y Lee[1], B Y Chen[1], H J Tsai[1], G H Luo[1] and C C Kuo[1]

[1] National Synchrotron Radiation Research Center, Hsinchu, 30076, Taiwan
[a] Corresponding's author: chen.chiahsiang@nsrrc.org.tw

Abstract. An optimized injection efficiency is related to a perfect match between the pulsed magnetic fields in the storage ring and transfer line extraction in the TPS. However, misalignment errors, hardware output errors and leakage fields are unavoidable. We study the influence of injection transients on the stored TPS beam and discuss solutions to compensate these. Related simulations and measurements will be presented.

1. Introduction

The Taiwan Photon source (TPS) is one of the most advanced synchrotron light source around the world and has operated since completion of the phase-I commissioning with insertion devices, superconducting cavities and a beam current up to 400 mA [1]. The TPS is a 3 GeV synchrotron facility with design parameters compiled in Table 1. Its booster (BR) and storage rings (SR) share the same tunnel. The septum magnets in the transfer line between the BR and SR have to deflect the beam by about 120 mrad within very limited space. The septa parameters are listed in Table 2. The pulsed magnetic fields from two septum magnets create significant leakage fields and eddy currents which can influence the stored beam during beam injection. The basic requirements of a top-up injection are perfect injection efficiency to reduce the excess radiation and supress orbit fluctuation during injection. In order to reach these two requirements as much as possible, we will present the efforts in reduction stray fields from the septa and minimization the perturbation of stored beam by optimizing the operation condition of injection kicker.

Synchrotron light sources utilizing pulsed magnets in their injection schemes face beam perturbations during injection. At the NewSUBARU, they found eddy currents in injection beam pipes, in support arms and circulating beam duct. Y. Shoji et al. [2] broke the current loops to reduce these stray currents, which are interfering with the horizontal deflection of the injection beam. P. Lebasque et al. [3] applied a full-sine shape current pulse to simultaneously compensate the delayed eddy currents at SOLEIL in addition to shielding screens on the stored beam path to reduce magnetic stray fields. They further installed mu-metal foils surrounding the stored beam chamber and ground connected it with braided copper conductors in order to eliminate induction currents from septum magnets [4].

Table 1. Design Parameters of TPS Storage Ring

energy	3 GeV
circumference	518.4 m
RF frequency	499.655 MHz
natural emittance	1.6 nm-rad
beam current	500 mA
lifetime time	> 10 hours
tune ν_x / ν_y	26.16/14.24

Table 2. Operational Parameters of SR Septum Magnets

	septum 1 (upstream)	septum 2 (downstream)
septum length (m)	1	
pulse duration (μsec)	370	
pulse shape	half-sine	
peak voltage (V)	397	392
peak current (A)	8733	8706
peak operation field (T)	0.7274	0.6243
peak design field (T)	0.6283	0.5550
septum thickness (mm)	3	
inner chamber size (mm^2)	20×13	
chamber thickness (mm)	0.4	

2. Septum magnet and eddy current

The septum magnets of the TPS are excited by a half-sine pulse to deflect the electron beam into the storage ring. There is an up- and down-stream septum magnet in the transfer line bending the beam by 3.6° and 3.18°, respectively. The operational septum field is larger than the design value due to the stray field counteracting the main field by inducing eddy currents in the septum chamber. The specifications and performance of the storage ring injection (SRI) septa are listed in Table 2.

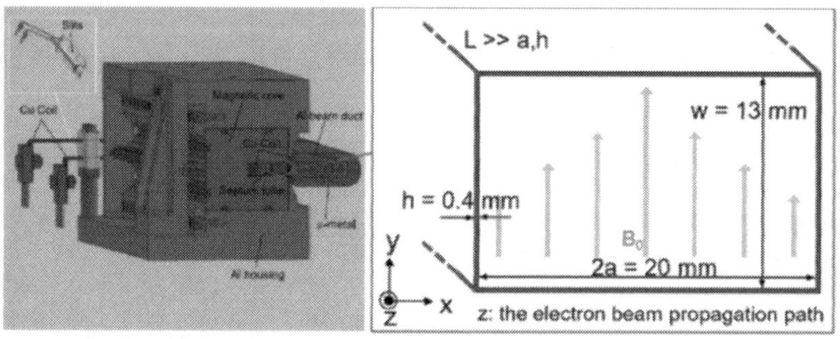

Figure 1. Schematic view of the septum magnet in the transfer line of the TPS [5], the septum vacuum chamber is shown enlarged on the right-hand side.

The septum field follows a half-sine wave and the magnetic field B_y is a function of time expressed by Equation (1)

$$B_y(t) = B_0 \sin\left(\frac{\pi t}{T}\right), \quad 0 < x < 2a, 0 < t < T, \tag{1}$$

where B_0 is the peak magnetic field in the septum, T is the width of the half-sine, $2a$ is the long side dimension of the vacuum chamber, w is the height of the vacuum chamber and h is the thickness of the vacuum chamber walls.

The eddy current is induced by the pulsed septum field forming a loop in the vacuum chamber top or bottom wall and is from Faraday's law of electromagnetic induction Equation (2)-(7) [6] given by

$$\oint \vec{E} \cdot d\vec{l} = -\frac{\partial}{\partial t}\int \vec{B} \cdot d\vec{A} \tag{2}$$

$$\oint \vec{E} \cdot d\vec{l} = 2\int_0^{L_B} \vec{E} \cdot d\vec{l} + 2\int_{-a}^{a} \vec{E} \cdot d\vec{l} = 2EL_B + 4Ea \approx 2EL_B \quad (L.H.S.) \tag{3}$$

$$-\frac{\partial}{\partial t}\int \vec{B}_y(t) \cdot d\vec{A} = 2\dot{B}_y aL \ (R.H.S.), \tag{4}$$

where L_B is the vacuum chamber length, L is the septum magnet length, and $L >> a, w, h$ is the vacuum chamber length. Because of the similarity between Equation (3) and (4), the induction electric field and the current can be expressed by Equation (5) and (6), respectively. Opposite eddy currents, induced both on top and bottom, are counteracted symmetrically. The total current is equal to the product of the field with the conductivity and chamber wall cross section.

$$E = \dot{B}a\frac{L}{L_B}, \tag{5}$$

$$I(t) = \sigma\dot{B}_y a(2wh + 4h^2)\frac{L}{L_B} = \sigma a(2wh + 4h^2)B_0\frac{\pi}{T}\frac{L}{L_B}\cos\left(\frac{\pi t}{T}\right) \tag{6}$$

The septum vacuum chamber is made of stainless steel type 304, with an electric resistivity of $7.2\times10^{-7}\ \Omega\cdot\mathrm{m}$ and a conductivity σ of 1.39×10^{6} A/V$\cdot$ m. The current loop is simplified as a loop in the upstream septum magnet in Figure 2(b), where L_B is about 1 m and the peak to peak eddy current can be calculated by Equation (7).

$$I_p = \sigma a(2wh + 4h^2)B_0\frac{\pi}{T}\frac{L}{L_B}, \qquad I_{p-p} = 2I_p \sim 1520\ A \tag{7}$$

IOP Conf. Series: Journal of Physics: Conf. Series **874** (2017) 012060 doi:10.1088/1742-6596/874/1/012060

3. Stray field measurements and simulation

Top-up mode operation of the TPS will face injection transient fields from four-bump kicker and two septum magnets. The kicker magnets excite the stored beam during ~ 6 μsec at the peak of the septa half wave of ~ 370 μsec. The betatron motion of the circulating beam will be excited at the 105th turn along the septum pulse at the rising edge of the kicker magnet pulse, if the leakage field of septums can be neglected. Figure 2(a) illustrates the durations and triggers relationship between the kicker and septum magnets. The black and blue lines show the pulsed signals from kicker and septum magnets, respectively, while Figure 2(b) shows the layout of the injection section and the eddy current loop in the red frame. Figure 3(a) shows a top-view photo of the TPS injection section, where the right-hand side is the SR and the left-hand side is the transfer line. We utilized mu-metal foils to wrap up the SR beam chamber andminimize the leakage fields to the SR chamber. Figure 3(b) is a photo of eddy current measurements and 3(c) shows the measured eddy currents using two current transformers. We detected a bipolar transmitted current up to 758 amperes by using a current transformer around the vacuum chamber near the upstream septum magnet (SP1) shown as the blue curve in Figure 3(c). The blue curve shows partial ground currents of 160 A via a short copper cable. Nevertheless, the induction currents are induced by the septum magnet acting like a power source. More parallel current loops could extract larger total currents from the septum magnet power supply. The current loop near the septum magnet is clearly shown on the eddy current measurement.

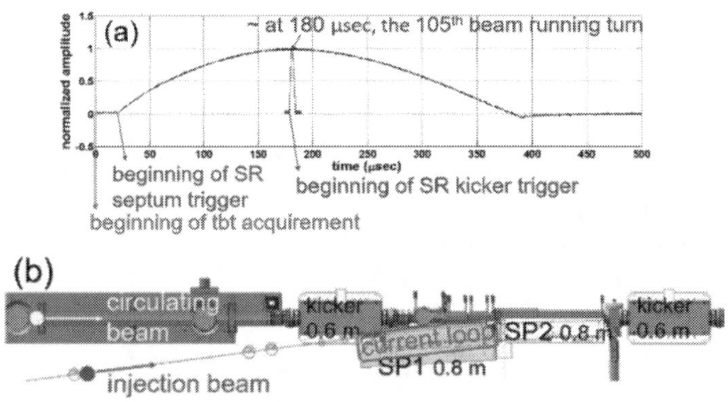

Figure 2. (a) time-relation scheme of septum (black) and kicker magnets (blue), (b) layout of the injection section.

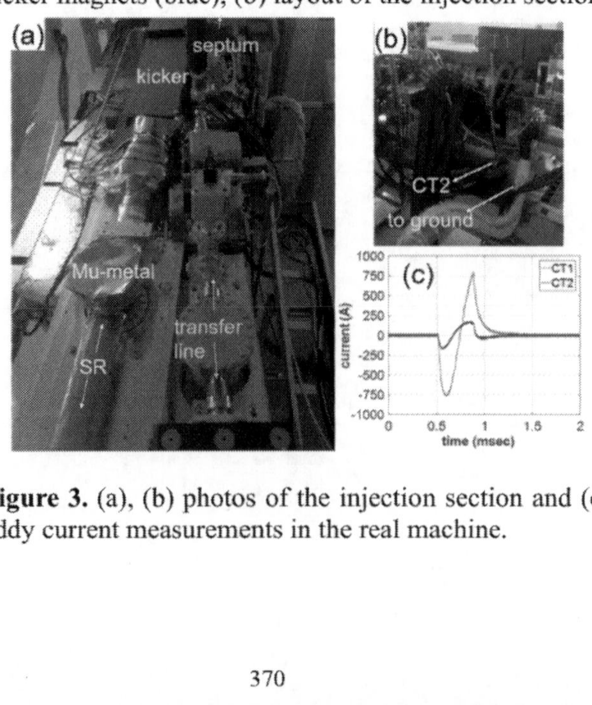

Figure 3. (a), (b) photos of the injection section and (c) eddy current measurements in the real machine.

We mimic the arrangement of a closed circuit using the same septum magnet in the laboratory. A copper wire is connected to the septum and SR vacuum chamber. The pulsed excitation current induces an eddy current on the chamber forming a current loop. The total excited current increases with the numbers of possible current loops due to the lower resistance of parallel resistors. Figure 4 shows that three equal length current loops connected in parallel (blue)increasing the eddy current by 50 % in this experiment.

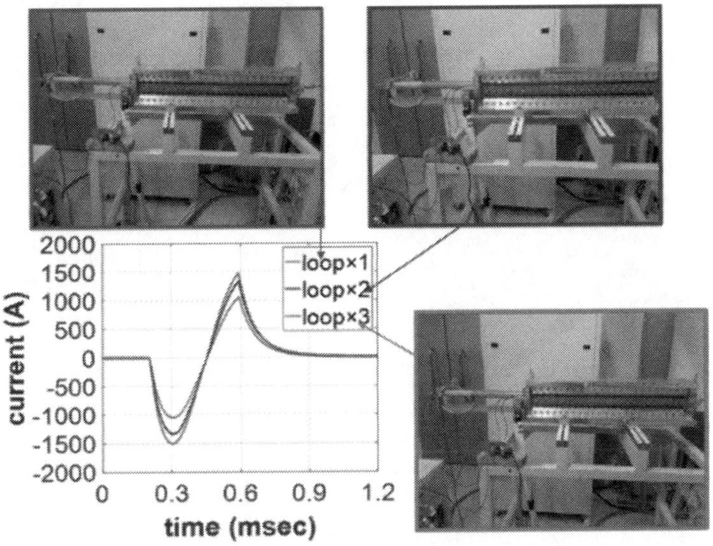

Figure 4. eddy current simulation for different numbers of parallel current loops.

4. Beam COD and field measurement after improvement

The circulating beam orbit will be periodically perturbed during top-up injection. Taking turnby-turn BPM signals is a typical way to further diagnose the betatron motion and stray magnetic fields. We recorded the beam displacements with best matched kickers' fields during the stray field improvement process. In Figure 5, the measurement employed one of the 172 SR BPMs near the injection section. We recorded the betatron variation before improvement, after mu-metal shielding with cable grounding and current loop isolation. The data were taken during 500 turns within the septum magnet half cycle. During the initial TPS commissioning, there was little shielding outside of the SR vacuum chamber and little grounding to block stray fields causing serious betatron motions as shown in the green line of Figure 5. We first wrapped mu-metal foils on the SR chamber near the kicker and septum magnets which substantially blocked the leakage field into the SR chamber. The girder is a large mechanical support structure with a low resistivity between beam devices and ground and we therefore use short copper cables to connect the vacuum chambers of the beam transfer line to the girders near the septum magnet. From Equation (7), we concluded that a long current path in the vacuum chamber should be established to achieve low induced eddy currents, which could be reduced three-fold. Furthermore, we eliminated the possibility of current loops and isolated the current path in the septum chamber. This scheme strongly reduced the eddy current by a factor of 250 in the septum chamber. The insert in Figure 5 shows that the eddy current is now less than 3-A in the septum chamber. After improving the stray field, the red line in Figure 5 shows the horizontal and vertical beam displacements to be within peak to peak $\pm$ 400 μm and $\pm$100 μm, respectively. Comparison of beam orbit perturbations before and after improvements show that the betatron horizontal and vertical displacements decreased by a factor of ten and four, respectively.

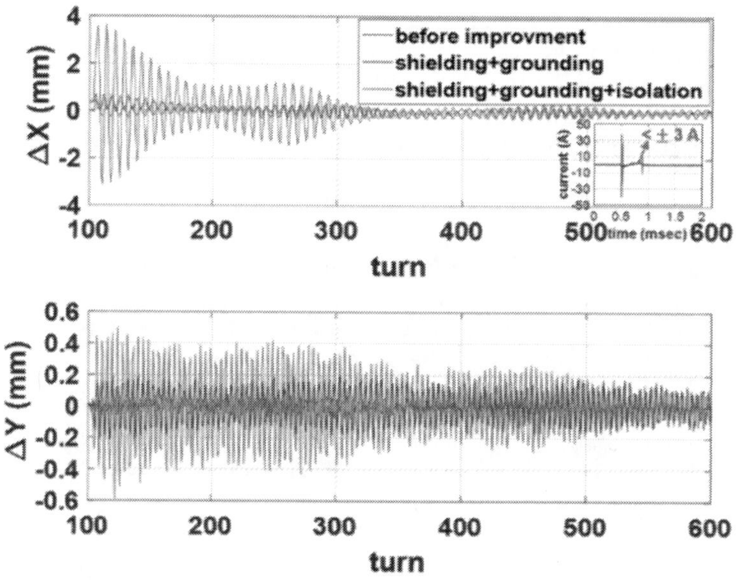

Figure 5. Turn-by-turn BPM data with best matched kicker show betatron oscillations before improvement (green), after mu-metal shielding and grounding with copper cables (blue) and after isolation of current loops (red). The figure insert shows the eddy current after improvements.

5. Conclusion

From a simple model for eddy currents based on Ampere's law, we estimated a 1520 A peak to peak eddy current in the septum chamber. This estimate is approximately equal to the measured eddy current of 1516 A in the real machine.

In order to retain the quality of low photon intensity fluctuations, preserving a good beam orbit stability is necessary. We presented methods to reduce stray fields from septum magnets: by shielding materials surrounding the SR chamber, by adding ground cables from the septum chamber and by breaking multiple current loops, thus improving beam distortion during injection. Preliminary experimental results show an improvement of the beam displacement down to peak to peak $\pm$ 400 μm in the horizontal and peak to peak $\pm$ 100 μm in the vertical plane so far. We will slightly adjust the rotation of the kicker magnets to compensate for vertical betatron displacements. Optimization of kicker-matching and shimming kicker magnets are future plans towards better beam stability during injection.

Acknowledgments

Authors highly appreciate the technical supports from Instrumentation & Control, Vacuum, Magnet and Precision Mechanical Engineering Groups at NSRRC. We also thank Prof. Humlt Wiedemann to proofread this paper.

References

[1] Chiu M S, Tseng F H, Chung T Y, Huang J C, Hwang C S, Chen Jenny, Chiu P C, Wu C Y, Cheng Y S, Hu K H, Chen J Y, Chen C H, Lee T Y, Huang C H, Chang C H, Fann Sam, Liang C C, , Tsai H J, Kuo C C, Chou P J and Liu Y C 2016 The commissioning of phase1 insertion devices in TPS *in Proc. IPAC'16* paper THPMB050

[2] Shoji Y and Kumagai K 2004 Stray field of a pulse septum induced by eddy currents *IEEE Transactions on Applied Superconductivity* **14** 2

[3] Lebasque P, Gros P, Lavieville J P, Da Silva Castro J, Mary A and Muller D 2006 Eddy current

septum magnets for booster injection and extraction, and storage ring injection at synchrotron SOLEIL *in Proc. EPAC06* paper THPLS101

[4] Lebasque P, Loulergue A, Ben El Fekih R, Lavieville J P, Bol M, Muller D and Marlats J L 2008 Improvement on pulsed magnetic systems at SOLEIL *in Proc. EPAC08* paper WEPC081

[5] Chan C K, Chang C C, Chen C L, Yang C S, Chen C S, Lin F Y and Chen J R 2014 Design and fabrication of the vacuum systems for TPS pulsed septum magnets *Nuclear Instruments and Methods in Physics Research* A **763** 388–393

[6] Edwards D A and Syphers M J 1993 *An introduction to the physics of high energy accelerators* John Wiley & Sons, Inc. chapter 4 pp 111-113

IOP Conf. Series: Journal of Physics: Conf. Series **874** (2017) 012061 doi:10.1088/1742-6596/874/1/012061

Limiting effects in double EEX beamline

G Ha[1], J G Power[2], M Conde[2], D S Doran[2], and W Gai[2]

[1]Pohang Accelerator Laboratory, Pohang, Gyeongbuk, 790-784, KOREA
[2]Argonne National Laboratory, Argonne, IL 60439, USA

E-mail: homehe@postech.ac.kr

Abstract. The double emittance exchange (EEX) beamline is suggested to overcome the large horizontal emittance and transverse jitter issues associated with the single EEX beamline while preserving its powerful phase-space manipulation capability. However, the double EEX beamline also has potential limitations due to coherent synchrotron radiation (CSR) and transverse jitter. The former limitation arises because double EEX uses twice as many bending magnets as single EEX which means stronger CSR effects degrading the beam quality. The latter limitation arises because a longitudinal jitter in front of the first EEX beamline is converted into a transverse jitter in the middle section (between the EEX beamlines) which can cause beam loss or beam degradation. In this paper, we numerically explore the effects of these two limitations on the emittance and beam transport.

1. Introduction

A program is underway at the Argonne Wakefield Accelerator (AWA) facility to develop an exchange-based method of longitudinal bunch shaping [1]. The current single emittance exchange (EEX) beamline is being upgraded to a double EEX beamline. After installation and commissioning is complete (in early 2018) it will first be used to study the capabilities and limitations of double EEX and, second, for applications requiring longitudinal bunch shaping. In this paper, we study the emittance growth from coherent synchrotron radiation (CSR) and jitter issues since they are expected to be the two biggest limitations of the double EEX beamline.

The first limitation we study arises from CSR which is an unavoidable side effect of all beamlines using dispersive elements (e.g. a dipole magnet). Since the EEX beamline uses multiple dipole magnets the beam has multiple chances to interact with CSR leading to emittance dilution. We note that the exchange process leads to a growth of the slice emittance in contrast to other dispersive beamlines (e.g. chicane) [2].

The second limitation we study is due to longitudinal jitter at the entrance to an EEX beamline that gets converted into the transverse jitter (i.e. horizontal offset) at the exit. This transverse jitter is critical because it can generate serious effects on the transverse optics and the emittance. Fortunately, since the double EEX beamline restores the phase space relationship, the converted transverse jitter after the first EEX becomes longitudinal jitters after the second EEX.

In this paper, we study these limiting issues in the AWA double EEX beamline using the particle tracking code GPT [3] with a CSR module [4]. The double EEX beamlines consists of two identical single EEX beamline and transverse manipulation section in the middle (See Figure 1). A double EEX beamline is simulated to study the jitter issue, but a single EEX beamline is used for the CSR study for the simplicity.

Figure 1. The schematic of a double EEX beamline. The beam is moving from the right to the left in this figure.

2. Emittance growth in the EEX beamline

The current AWA single EEX beamline [1] is used in the simulation study of emittance growth due to CSR. A 100 pC beam is numerically generated at the entrance to the double EEX beamline for the study. A horizontal emittance ($\varepsilon_{x,i}$) of this beam is 1 μm. Bunch length (σ_z) and the chirp ($S_z \equiv \langle z\delta\rangle/\langle z^2\rangle$) are varied around 2 mm and -2 m^{-1} respectively. The horizontal beam size (σ_x) and slope ($S_x \equiv \langle xx'\rangle\langle x^2\rangle$) are varied in 0.5–3 mm for the size and -1 m^{-1} to 1 m^{-1} for the slope.

The emittance growth without CSR is shown in Figure 2a as a function of the incoming beam parameters. The y-axis of this figure shows the ratio of the final longitudinal emittance ($\varepsilon_{z,f}$) to the initial horizontal emittance ($\varepsilon_{x,i}$) for the single EEX beamline. In this case, the 2nd-order or higher order terms are the only sources of the emittance growth. A smaller beam size normally generates a smaller emittance growth and an appropriate slope minimizes the high-order growth. Figure 2a follows the expected tendency. The horizontal slope near -0.2 m^{-1} suppresses this growth and a smaller transverse beam size induces a smaller emittance growth. The minimum emittance growth ($\varepsilon_{z,f}/\varepsilon_{x,i}$) is 1.05 with the initial beam size ($\sigma_{x,i}$) of 0.5 mm and the slope ($S_{x,i}$) of -0.2 m^{-1}.

The emittance growth with CSR is shown in Figure 2b as a function of the same incoming beam parameters as above. The overall patterns are similar to Figure 2a, since the higher order effects are dominant for this low charge, but the minimum emittance growths for each initial beam size increases due to CSR. This can be explained by Equation (1) which shows the emittance growth with the integrated momentum kick concept introduced in Ref. [5].

$$\Delta\varepsilon = \langle x_i^2\rangle \left[\{\kappa L\Delta_z - (\eta + \kappa\xi L)\Delta_\delta\}S_x + \kappa(\Delta_z - \xi\Delta_\delta)\right]^2 + \frac{\varepsilon_{x,i}^2}{\langle x_i^2\rangle}\{\kappa L\Delta_z - (\eta + \kappa\xi L)\Delta_\delta\}^2. \quad (1)$$

Figure 2c and d show the emittance growth plot for the bunch lengths of 1 mm and 0.6 mm respectively. A short incoming bunch length shortens the bunch length along the beamline. Due to the short bunch length, CSR along the beamline is stronger than the long bunch case (b). Since the higher-order effect competes with the CSR induced emittance growth (Equation (1)), the pattern and slope that minimizes the emittance growth change (Figure 2c and d). The optimal slope is 0 m^{-1} which yields a growth of 1.24 for $\sigma_z = 1$ mm case, and 0.2 m^{-1} provides 1.41 for $\sigma_z = 0.6$ mm case.

Figure 3 shows the emittance growth plot for different chirps. The top rows are the simulation results without CSR and the bottom is the results with CSR. Left column of the figure used the chirp of -2 m^{-1} and the right column used 0 m^{-1} while the bunch length is fixed to 2 mm for all cases. The emittance growth sources related to the longitudinal chirp are thick-lens effect [6] and CSR. The chirp of $-1/\xi$ minimizes the thick-lens effect but it also minimizes the bunch length in the middle of the EEX beamline (i.e. maximum CSR). Since the charge is low, the thick-lens effect looks stronger than CSR, so the chirp closer to $-1/\xi$ ($-1/\xi \approx -2$ m^{-1}) reduces the emittance growth. Note, this trend depends on the charge level or the initial bunch length.

IOP Conf. Series: Journal of Physics: Conf. Series **874** (2017) 012061 doi:10.1088/1742-6596/874/1/012061

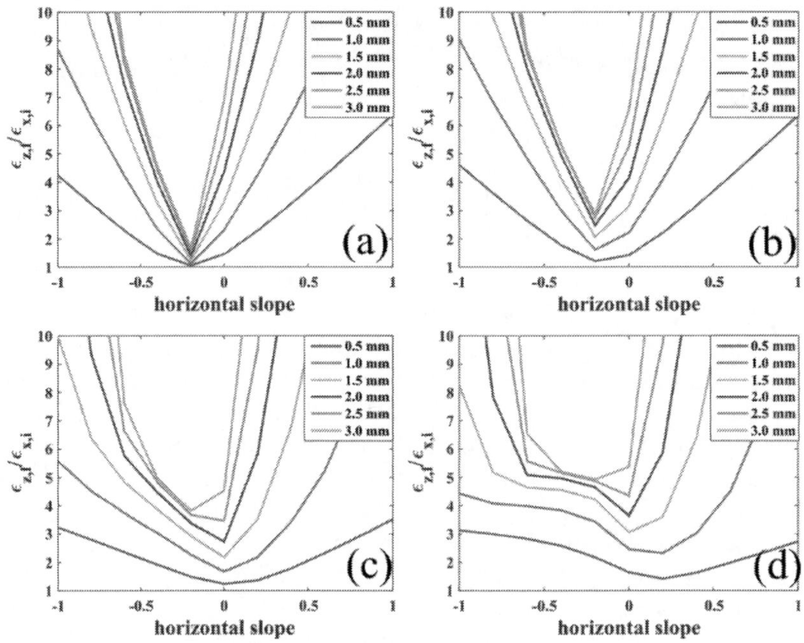

Figure 2. Emittance growth for various incoming beam parameters. The parameters $\varepsilon_{x,i} = 1\ \mu$m and $S_z = -2\ m^{-1}$ are fixed and the colors correspond to different values of σ_x. (a) $\sigma_z = 2$ mm (without CSR), (b) $\sigma_z = 2$ mm (with CSR), (c) $\sigma_z = 1$ mm (with CSR), and (d) $\sigma_z = 0.6$ mm (with CSR).

3. CSR suppression

Since CSR is one of the major issues in modern accelerators, CSR suppression for the bunch compressor has long been studied. However, the study on CSR for EEX is started very recently. Here we tried to simulate two well-known ideas as a start: CSR shielding and low bending angle.

Figure 4 shows the simulation comparison with and without a shielding effect. Figure 4a is the result without the shielding effect, and Figure 4b includes the shielding effect. The plate gap size of 2 cm is applied to the dipole magnet area while the 5 cm gap is applied to the remainder of the beamline. For the $\sigma_x = 0.5$ mm case, the minimum emittance growths are 1.22 without the shielding and 1.12 with the shielding. Although the shielding gap limits the transverse optics, it is very effective way to suppress the emittance growth from CSR.

We also tried to minimize the emittance growth even further using both a smaller gap for the shielding and a low bending angle for the EEX beamline. The bending angle is changed from 20° to 5°, and the drift length of the dogleg is changed from 2 m to 5 m to keep the dispersion in a reasonable level ($\eta_{\alpha=20°} = 0.89$ m and $\eta_{\alpha=5°} = 0.46$ m). The shielding uses 2 cm gaps for the whole beamline.

Figure 5 shows the simulation results with a low bend angle EEX beamline. Figure 5a and b use different initial bunch lengths (0.3 mm and 2 mm, respectively). The higher order effect generates a similar pattern as in Figure 2c and d. However, the short bunch now provides a smaller emittance growth than the long bunch. Since CSR is weak in this low bend angle case, the thick-lens effect dominates and the long bunch has the larger emittance growth. The minimum emittance growth in Figure 5a is only 1.06. Note, the thick-lens effect can be eliminated by introducing a fundamental mode cavity [6].

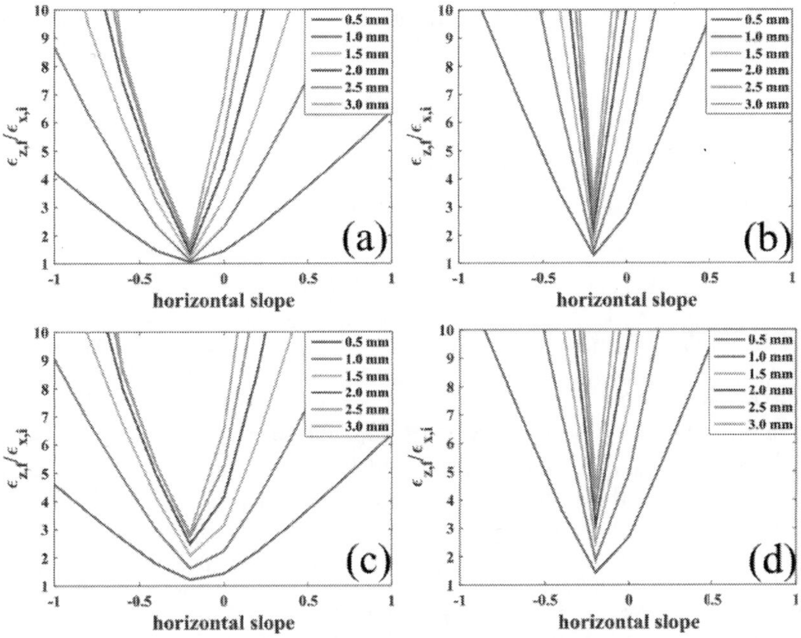

Figure 3. Emittance growth for various incoming beam parameters. Colors correspond to the initial beam size, and the initial emittance is 1 μm. The top row are the simulation result without CSR and the bottom row are the result with CSR. The bunch length is fixed to 2 mm while the chirp is -2 m^{-1} for the left column and 0 m^{-1} for the right column.

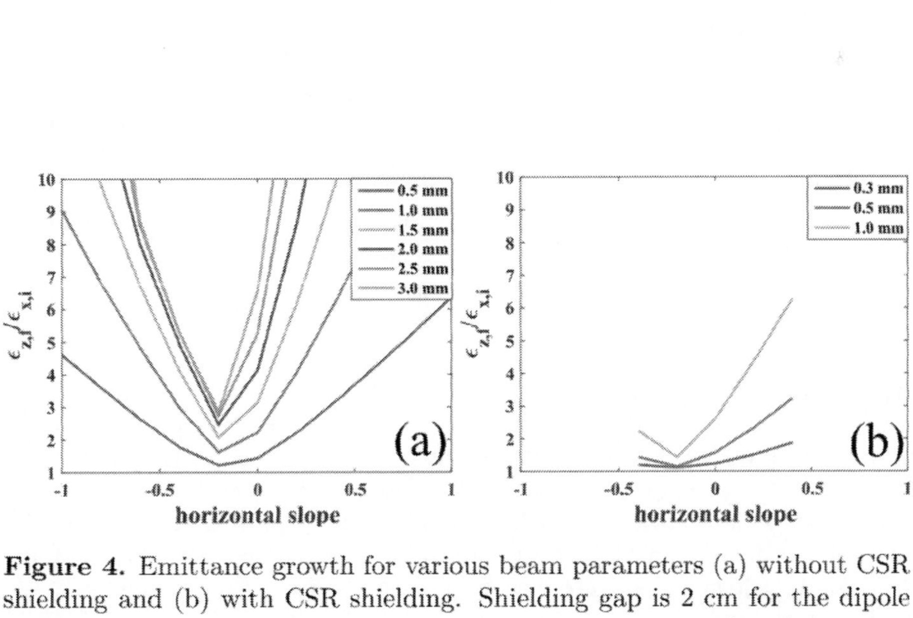

Figure 4. Emittance growth for various beam parameters (a) without CSR shielding and (b) with CSR shielding. Shielding gap is 2 cm for the dipole magnets and 5 cm for remainder. Initial emittance is 1 μm, bunch length is 2 mm, and the chirp is -2 m^{-1}.

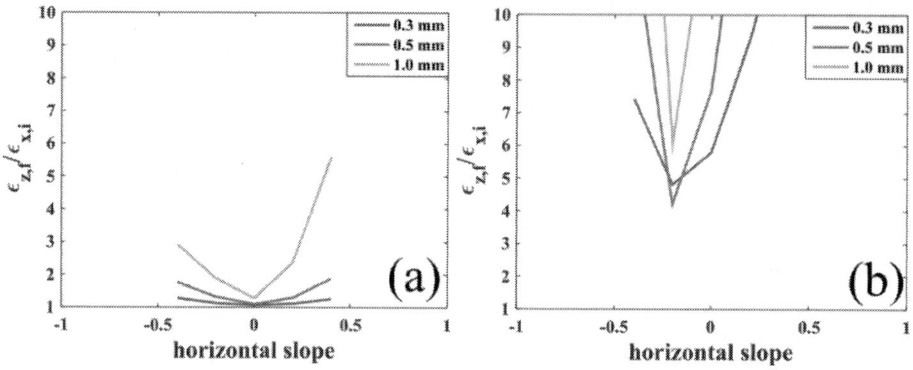

Figure 5. Emittance growth for various beam parameters with low angle EEX beamline. The dogleg parameters are changed to 5° bending and 5 m separation. Initial emittance of 1 μm and the chirp of -2 m^{-1} are fixed. The bunch length is (a) 0.3 mm and (b) 2 mm.

4. Longitudinal-to-transverse jitter conversion

As described earlier, any longitudinal jitter before the EEX beamline is converted to the horizontal jitter after a single EEX beamline due to the phase space exchange. In the double EEX beamline, however, the second EEX beamline brings all properties back to its original phase space, including jitter.

The jitter conversion along the double EEX beamline is shown in Figure 6a and b. For the artificially generated 100 pC beam, initial timing offsets are applied as a longitudinal jitter. When these beams arrive at the middle section (~8 m), the timing jitter is fully eliminated (See Figure 6b) and the beam has horizontal offsets (See Figure 6a). Once the beam passes the second EEX beamline, all horizontal offsets in the middle section disappear and the timing jitter appears again. Since the AWA-EEX beamline, with weak focusing, provides a little bunch compression, the final timing jitter is slightly reduced.

Figure 6c shows the emittance difference along the double EEX beamline due to timing jitter. At the end of the beamline, the largest emittance difference between the reference and the one with timing jitter is only 0.15 μm for -5 ps offset (initial emittance is 1 μm). This emittance growth may come from the mismatch of the EEX condition in the second EEX beamline. (As a proof, even the transverse offset is not fully eliminated.) Another possibility is the emittance growth in the middle section. A large transverse offset in the middle section makes the beam pass the off-axis of the quadrupole magnets which introduce extra nonlinear fields. Even in this case, the emittance growth is reasonably small compared to the large timing offset, so it is not a major issue.

5. Summary and future work

We performed simulations to study the emittance growth due to CSR and the jitter conversion issue in the EEX beamline. We observed the trend of emittance growth originated from the higher order effect, thick-lens effect and CSR. We showed that the shielding method and the low bending angle suppress the emittance growth very effectively. The longitudinal jitter before EEX is converted to the horizontal offset in the middle section, and it is re-converted to longitudinal jitter by the second EEX beamline. During this process, the emittance is almost preserved and the jitter is slightly reduced.

Since this work was a simple scan-type simulation to study emittance growth and jitter issues, a theoretical study is required for further insight. We are also exploring the emittance growth

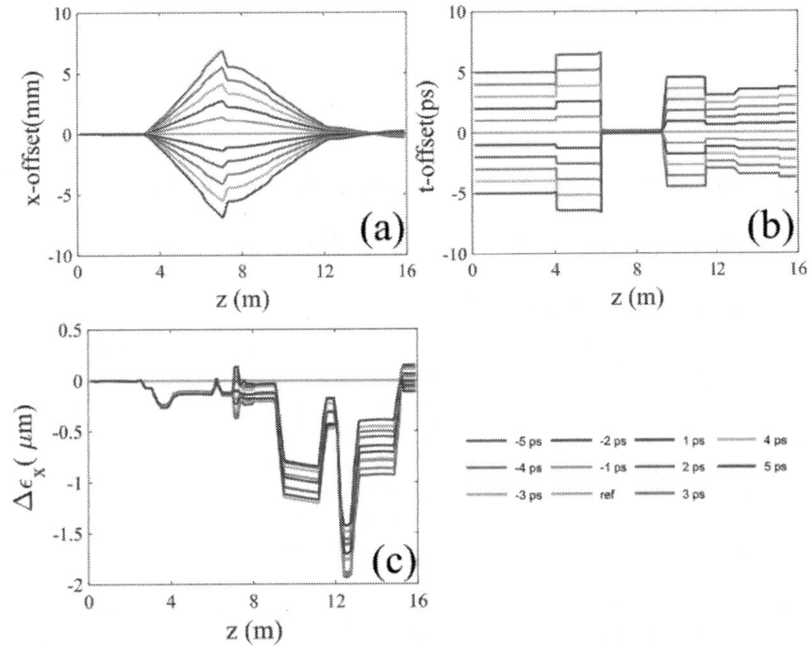

Figure 6. (a) Horizontal position, (b) timing, and (c) transverse emittance differences between the reference beam and the beam having initial timing offset. Simulation is done with 100 pC beam along the double EEX beamlie. ~8 m is the middle section.

for high charge beam and trying to find other effective methods to suppress CSR.

Acknowledgement

This work is supported by Department of Energy, Office of Science, under Contract No. DE-AC02-06CH11357.

References

[1] Ha G *et al.* 2017 *Phys. Rev. Lett.* **118** 104801
[2] Guetg M W, Beutner B and Reiche S 2015 *Phys. Rev. Accel. Beams* **18** 030701
[3] Www.pulsar.nl/gpt
[4] Bazarov I V and Miyajima T 2008 *Proc. European Partical Accelerator Conference 2008* (Mulhouse: EPS-AG) p 118
[5] Ha G, Cho M H, Gai W, Kim K J, Namkung W and Power J G 2016 *Phys. Rev. Accel. Beams* **19** 121301
[6] Zholents A and Zolotorev M S 2011 A new type of bunch compressor and seeding of a short wave length coherent radiation (*Preprint* `ANL-APS-LS-327`)

IOP Conf. Series: Journal of Physics: Conf. Series **874** (2017) 012062 doi:10.1088/1742-6596/874/1/012062

Photoinjector optimization using a derivative-free, model-based trust-region algorithm for the Argonne Wakefield Accelerator

N Neveu,[1,2] J Larson,[1] J G Power[1] and L Spentzouris[2]
[1]Argonne National Laboratory, Lemont, IL
[2]Illinois Institute of Technology, Chicago, IL
E-mail: nneveu@hawk.iit.edu, jmlarson@anl.gov

Abstract. Model-based, derivative-free, trust-region algorithms are increasingly popular for optimizing computationally expensive numerical simulations. A strength of such methods is their efficient use of function evaluations. In this paper, we use one such algorithm to optimize the beam dynamics in two cases of interest at the Argonne Wakefield Accelerator (AWA) facility. First, we minimize the emittance of a 1 nC electron bunch produced by the AWA rf photocathode gun by adjusting three parameters: rf gun phase, solenoid strength, and laser radius. The algorithm converges to a set of parameters that yield an emittance of 1.08 µm. Second, we expand the number of optimization parameters to model the complete AWA rf photoinjector (the gun and six accelerating cavities) at 40 nC. The optimization algorithm is used in a Pareto study that compares the trade-off between emittance and bunch length for the AWA 70 MeV photoinjector.

1. The AWA Facility

The 70 MeV rf photoinjector at the Argonne Wakefield Accelerator (AWA) facility [1] consists of an rf gun followed by six rf accelerating cavities, hereafter referred to as the linac. See Figure 1 for the beam line layout. The 1.5 cell rf gun operates at 1.3 GHz with three solenoids and a Cs_2Te photocathode excited by a 248 nm UV laser. Solenoid 1 (S_1) is used to buck the field at the cathode, while the other two solenoids (S_2 and S_3) are used for emittance compensation. The accelerating cavities, also operated at 1.3 GHz, are 7 cell standing-wave cavities [2] each with independently controllable phase. The cavities are labeled L_1-L_6 in Figure 1.

2. Optimization Algorithm

Model-based, derivative-free algorithms are frequently used to optimize computationally expensive simulations due to their judicious use of function evaluations. In cases specific to accelerator physics, beam properties at different operational parameters are observed; these methods then build models of the unknown function and minimize these models to identify candidate parameters to evaluate. BOBYQA [3] is one such method that is available via the NLopt [4] package and was used in this study. Given a candidate set of optimal parameters v^k, BOBYQA constructs a quadratic model using function values of points near v^k. This model is minimized in a neighborhood of v^k in order to produce a point $\hat{v}$. If $\hat{v}$ has a smaller objective function value than v^k, the estimate of the optimum is updated to $\hat{v}$, and a new model is

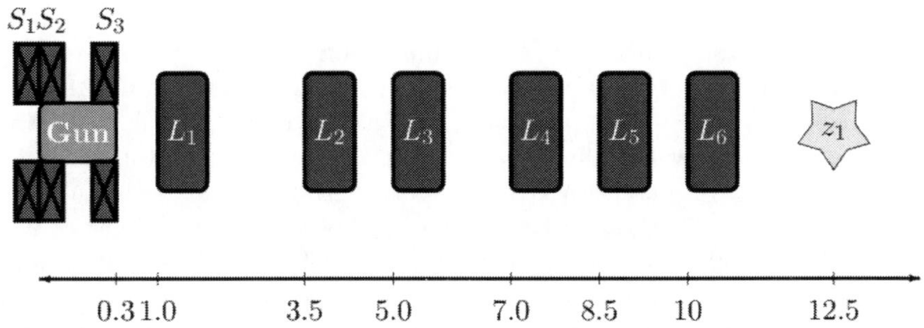

Figure 1. Layout of the AWA linac. The gun is enlarged to show solenoid detail. The physical length is 0.3 m. The cathode is located at $z = 0$ m. Linac cavities are 0.85 m long. Tick marks are located at the exit of the gun, entrance of each accelerating cavity, and location of optimization.

constructed. If $\hat{v}$ is not a sufficient improvement over v^k, the model around v^k is improved. For more information about derivative-free optimization, see [5].

The parameters v^k are generated and supplied to the open source particle-in-cell code OPAL-t [6]. This code was chosen in part because it models the 3D space charge necessary to accurately simulate the linac. The optimization package NLopt and OPAL-t were used in combination with Python code written at Argonne National Laboratory to perform simulation evaluations and optimization. All the files needed to replicate the results in this paper are available at

www.mcs.anl.gov/~jlarson/AWA

Interested parties are welcome to adapt the code to their needs and suggest improvements.

3. Optimization Parameters

When optimizing the gun, three parameters were varied: solenoid strength (S_3), gun phase (ϕ_g), and laser radius (R) of a uniform pulse. The minimized objective was emittance (ϵ_x). (The phase is defined as 0° at maximum energy gain.) When optimizing the entire linac, seven additional parameters were varied: the longitudinal laser full width at half maximum (T) and accelerating cavity phases (ϕ_L). The optimization parameters and bounds are given in table 1; we denote the set of ten optimization parameters as $v = [S_3, \phi_g, R, T, \phi_L]$, where $\phi_L = [\phi_{L1}, \ldots, \phi_{L6}]$ represents the phase of each linac cavity L_1-L_6.

Table 1. Parameter bounds for gun and linac optimization.

Variable	Range	Unit
Solenoid Strength	$0 \leq S_3 \leq 440$	amps
Phase of Gun	$-60 \leq \phi_g \leq 60$	degrees
Laser Radius	$0.1 \leq R \leq 30$	mm
Laser FWHM [1]	$2 \leq T \leq 10$	ps
Cavity Phase [1,2]	$-20 \leq \phi_L \leq 20$	degrees

[1] not varied during gun optimization
[2] $\phi_L = [\phi_{L_1}, \ldots, \phi_{L_6}]$

4. Gun Optimization

Much work has been done to optimize 1.5 cell rf guns at 1 nC [7]. This known solution was used as a baseline test of BOBYQA when applied to an accelerator application. An optimization of the single objective emittance (ϵ_x) was performed over a length of 5 m. All linacs were turned off and only gun parameters were varied. Non-varying parameters are listed in table 2; their values are based on work done at PITZ and AWA [7, 8]. Local optimization runs were started from five points with various distances from the optimum value. The optimization runs converged (in less than 100 function evaluations) to a parameter set ($S_3 = 269$ A, $\phi_g = -3.0\,°$, and $R = 0.6$ mm) with an emittance of 1.08 µm. An exhaustive search of the parameter space was not done, and there may be other local minima that were not found. However, the results match expectations based on the literature.

Table 2. Non-varying parameters for gun optimization.

Parameter	**Value**
Charge	1 nC
Gradient	60 MV/m
Laser FWHM	20 ps
Laser Rise and Fall Time	6 ps
Kinetic Energy at Cathode	0.55 eV
S_1 and S_2	550 A

5. Linac Optimization

Next we performed a multiobjective optimization of the linac (Figure 1), by adjusting the ten parameters in table 1. The charge was set to 40 nC and was chosen for upcoming two-beam acceleration experiments [9]. Two objectives were considered: emittance, and bunch length, σ_z. The location of interest is $z_1 = 12.51$ m, as this is the entrance of the first quadrupole magnet after the linac. We optimize ϵ_x instead of ϵ_{xy} because no asymmetric focusing elements were used in the linac. The non-varying parameters for all linac simulation runs are shown in table 3. The model used simulated emission from a Cs_2Te cathode using a laser with initial kinetic energy of 4 eV. These are typical operating conditions at AWA.

Table 3. Non-varying parameters for linac optimization.

Parameter	**Value**
Charge	40 nC
Laser Rise and Fall Time	1.0 ps
Gun Gradient	70 MV/m
S_1 and S_2	550 A
Cavity Gradient L_1–L_4	25 MV/m
Cavity Gradient L_5–L_6	27 MV/m

A 1,000 point sample of linac parameters were drawn from the domain in table 1 and simulated. Of these, 132 simulations completed without error, and the emittance and bunch length at $z_1 = 12.51\,\text{m}$ was recorded for each of these points. From the sample, the minimum and maximum values of emittance and bunch length were found (i.e: ϵ_{min} and ϵ_{max}). The raw $\epsilon_x(v, z_1)$ and $\sigma_z(v, z_1)$ sample values were then shifted and scaled to produce $\bar{\epsilon}_x(v, z_1)$, and $\bar{\sigma}_z(v, z_1)$, which have a minimum value of 0 and a maximum value of 1 over the 132-point sample set. That is,

$$\bar{\epsilon}_x(v, z_1) = \frac{\epsilon_x(v, z_1) - \epsilon_{\text{min}}}{\epsilon_{\text{max}} - \epsilon_{\text{min}}} \tag{1}$$

and $\bar{\sigma}_z(v, z_1)$ is defined similarly. This scaling is done in order to remove the difference in the units between emittance and bunch length when optimizing.

With the scaled values of $\bar{\epsilon}_x$ and $\bar{\sigma}_z$, a sequence of eleven optimization problems were solved by minimizing

$$f(v, w) = w\,\bar{\epsilon}_x(v, z_1) + (1 - w)\,\bar{\sigma}_z(v, z_1) \tag{2}$$

for $w \in \{0, 0.1, \ldots, 1\}$. For each weight w, BOBYQA was started from the sample point with the smallest value of $f(v, w)$. From the initial random sample, six unique starting points were chosen. (There were fewer starting points than weights because some samples had the smallest objective value for multiple weights. For example, the smallest values of $f(v, 0.4)$, $f(v, 0.5)$, and $f(v, 0.6)$ occurred when at the eighteenth sample point.) Some $f(v, w)$ values also had one or more linac phases near the initial $\pm 20°$ boundary. The ϕ_L boundary was expanded to $\pm 40°$ for those BOBYQA runs.

6. Pareto Front for AWA Linac

Since multiple objectives are under consideration in this case, a trade-off analysis is necessary. This can be aided by examining a Pareto front: the set of parameters for which no other point

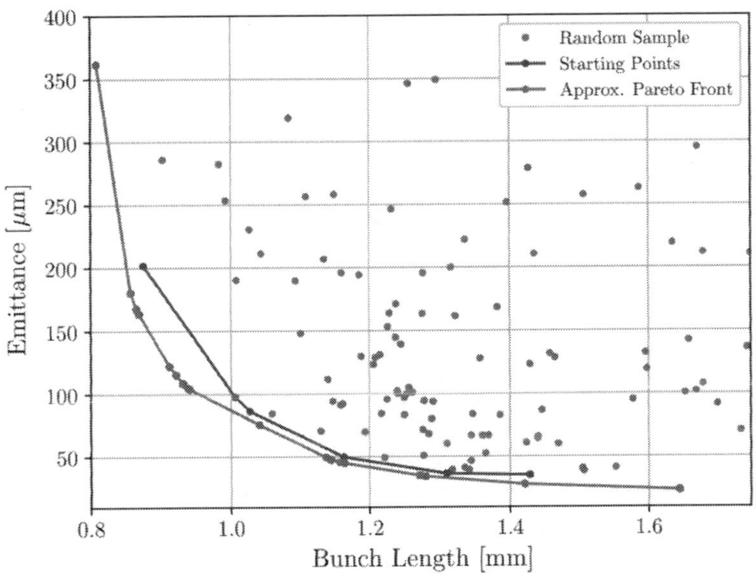

Figure 2. Random sample results, starting sample points, and resulting approximate Pareto front for the linac at 40 nC. The Pareto front is the result of all BOBYQA evaluations.

exists that is better with respect to both objectives [10]. In Figure 2, blue dots show the emittance and bunch length for the evaluated random sample. The sample points for which no other point has better emittance and bunch length are connected with a blue line. BOBYQA was started from these points, as described above, producing the green approximate Pareto front.

The number of simulation evaluations needed to obtain convergence in the BOBYQA runs varied from a minimum of 107 evaluations to a maximum of 208 evaluations. In order to generate the Pareto front in Figure 2, a total of 2,492 simulation evaluations were completed. Most simulation evaluations took approximately 7 minutes, using 16 cores and 100,000 particles. Runs with phases near the domain boundary (table 1) took approximately 10 minutes. These numbers are driven by the amount of time OPAL-t needs to simulate the beam line.

The best-found objective value through each BOBYQA run is shown in Figure 3. Weight 0.0 and 1.0 are negative due to the scaling in Equation (1). While the weights, w, are always nonnegative, $\bar{\epsilon}_x(v, z_1)$ or $\bar{\sigma}_z(v, z_1)$ will be negative if BOBYQA finds a point with a value of $\epsilon_x(v, z_1)$ or $\sigma_z(v, z_1)$ that is less than $\epsilon_{\min}$ or $\sigma_{\min}$ from the initial sample.

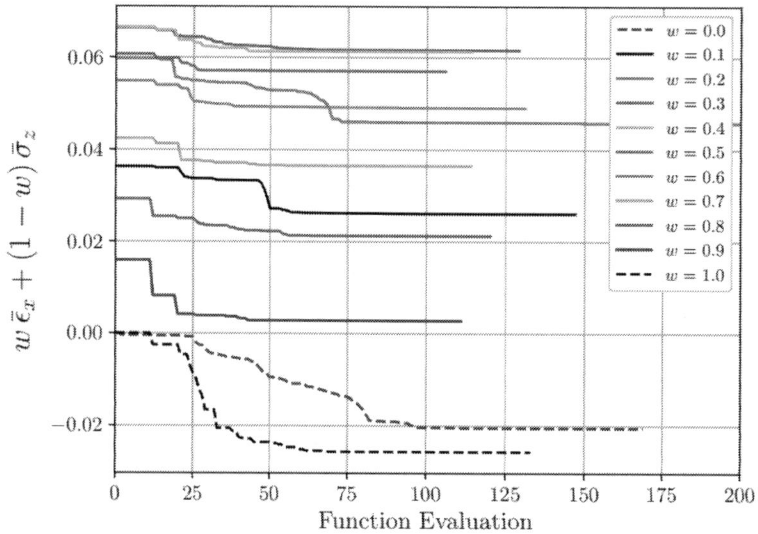

Figure 3. Minimum observed objective function values during eleven BOBYQA runs at 40 nC.

We note that seven of the BOBYQA runs converged to emittance values between 20 μm and 50 μm, which is shown in Figure 4 along with gun phase and bunch length for each of the 11 optimized points. We annotate Figure 4 with T because that parameter shows strong correlation with the gun phase. Other optimized parameters such as the laser radius were found to stay within a narrow range (10 mm–16 mm).

The optimized linac phases maximize energy gain while minimizing the energy spread. The energy spread of the beam exiting the gun depends strongly on ϕ_g. There are three distinct regions where the optimized points had similar gun phases (less than $\pm 10°$) which resulted in nearly identical linac phases. For example: all six phases, ϕ_L, varied by less than $10°$ for $w \in \{0, 0.1\}$, less than $5°$ for $w \in \{0.4, 0.5, 0.6\}$, and less than $10°$ for $w \in \{0.7, 0.8, 0.9\}$. This indicates optimized linac phases may benefit a range of gun settings during operation. AWA will conduct experimental measurements in order to verify these results.

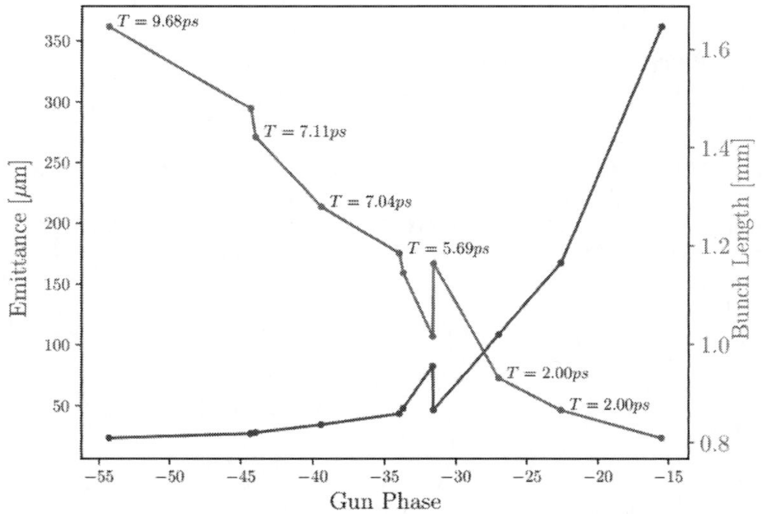

Figure 4. Bunch length and emittance vs. gun phase at each optimized point in the Pareto front for the linac at 40 nC. The phase of the maximum energy gain is 0°.

7. Conclusion

Using an AWA beam line as the simulation model, we used the BOBYQA algorithm to optimize the emittance produced by the gun at 1 nC. Using the same algorithm, we performed a multiobjective analysis for the linac at 40 nC. A Pareto front comparing the trade-off between bunch length and emittance was generated for the linac. This analysis will be used to decide future operating parameters at the AWA during high-charge experiments. In total, only 2,492 simulation evaluations were needed to produce the approximate Pareto front. Future work will include a refinement of results using 3D field maps for all cavities, experimental measurements to verify the Pareto front, and a comparison with evolutionary algorithms.

Acknowledgments

We gratefully acknowledge the computing resources provided on Blues, a high-performance computing cluster operated by the LCRC at Argonne National Laboratory. This material is based upon work supported by the U.S. Department of Energy, Office of Science, under contract number DE-AC02-06CH11357 and grant number DE-SC0015479. Travel to IPAC'17 supported by the Division of Physics of the United States National Science Foundation Accelerator Science Program and the Division of Beam Physics of the American Physical Society.

References

[1] Conde M *et al.* 2017 Research program and recent results at the Argonne Wakefield Accelerator facility (AWA) *Proc. of IPAC'17* (Copenhagen, Denmark) p WEPAB132
[2] Power J, Conde M, Gai W, Li Z and Mihalcea D 2010 Upgrade of the drive LINAC for the AWA facility dielectric two-beam accelerator *Proc. of IPAC'10* (Kyoto, Japan) p THPD016
[3] Powell M 2009 *The BOBYQA algorithm for bound constrained optimization without derivatives* University of Cambridge report NA2009/06 University of Cambridge
[4] Johnson S G 2017 *The NLopt nonlinear-optimization package* `http://ab-initio.mit.edu/nlopt`
[5] Conn A R, Scheinberg K and Vicente L N 2009 *Introduction to Derivative-Free Optimization* MPS/SIAM Series on Optimization (SIAM)

[6] Adelmann A *et al.* 2008-2017 *The OPAL (Object Oriented Parallel Accelerator Library) Framework* Tech. Rep. PSI-PR-08-02 Paul Scherrer Institut URL `https://amas.psi.ch/OPAL`
[7] Stephan F *et al.* 2010 *Phys. Rev. ST Accel. Beams* **13** 020704
[8] Neveu N, Spentzouris L, Power J G, Piot P, Metzger-Kraus C, Russell S J, Adelmann A and Ha G 2016 Benchmark of rf photoinjector and dipole using ASTRA, GPT, and OPAL *Proc. of NAPAC'16* (Chicago, IL)
[9] Jing C *et al.* 2015 The two beam acceleration staging experiment at Argonne Wakefield Accelerator facility *Proc. of IPAC'15* (Richmond, VA, USA) p WEPJE020 URL `http://accelconf.web.cern.ch/AccelConf/IPAC2015/papers/wepje020.pdf`
[10] Ehrgott M 2005 *Multicriteria Optimization* (Springer Science & Business Media)

IOP Conf. Series: Journal of Physics: Conf. Series **874** (2017) 012063 doi:10.1088/1742-6596/874/1/012063

Numerical investigation of beam halo from beam gas scattering in KEK-ATF

R Yang[1], P Bambade[1], K Kubo[2], T Okugi[2], N Terunuma[2] and D Zhou[2]

1. Laboratoire de l'Accélérateur Linéaire (LAL), Univ. Paris-Sud, CNRS/IN2P3, Universit Paris-Saclay, Orsay, France
2. High Energy Accelerator Research Organization (KEK), Tsukuba, Japan

E-mail: ryang@lal.in2p3.fr

Abstract. To demonstrate the final focus schemes of the Future Linear Collider (FLC), the Accelerator Test Facility 2 (ATF2) at KEK is devoted to focus the beam to a RMS size of a few tens of nanometers (nm) vertically and to provide stability at the nm level at the virtual Interaction Point (IP). However, the loss of halo particles upstream will introduce background to the diagnostic instrument measuring the ultra-small beam, using a laser interferometer monitor. To help the realization of the above goals and beam operation, understanding and mitigation of beam halo are crucial. In this paper, we present the systematical simulation of beam halo formation from beam gas Coulomb scattering (BGS) in the ATF damping ring. The behavior of beam halo with various machine parameters is also discussed.

1. Introduction

As a test beam line for the Final Focus Systems based on the local chromaticity correction scheme, ATF2 aims to achieve a 37 nm vertical beam size at the virtual Interaction Point (IP), with a stability at the nm level [1, 2]. To visualize the vertical spot size at the IP, a laser interferometer monitor, *Shintake Monitor*, was installed [3]. Beam size is measured from the modulation of the Compton scattering signal. Accuracy reconstruction of this modulation can be affected by background caused by loss of halo particles upstream.

Beam gas Coulomb scattering (BGS), Intra-Beam Scattering (IBS) and the residual alignment errors are considered the main sources of beam halo in the storage ring. Besides, optical mismatch and the wakefield effects in the ATF2 beam line can also contribute. Considerable efforts have been devoted to revealing the primary halo sources, the main one being BGS, theoretically and experimentally [4, 7, 8, 5, 6, 9] (for a recent experimental study, see also Reference [10]). An analytical method has been established to express the distribution function in the storage ring in the presence of BGS, considered as an incoherent stochastic process [4]. This method has not yet been compared in detail to simulation and measurement.

Accurate and systematic simulation of halo generation is essential to understand the interplay between radiation damping, quantum excitation, residual dispersion, xy coupling and BGS. In this paper, the simulation (based on SAD [11]) of beam halo from BGS in ATF is described, which verifies and complements the analytic estimation.

2. Theoretical approximation

Particles are transversely deflected when scattering in the Coulomb field of the nucleus of a residual gas atom, which leads to an increased betatron oscillation amplitude. Considering the optics of the ring in the smooth approximation, Reference [4] provides an analytical method to approximate the deviation of the beam distribution from a Gaussian shape due to such scatterings. The distribution function in equilibrium is approximated with the scaled single scattering distribution

$$\rho(X) = \int \frac{d\omega}{(2\pi)^2} e^{-i\omega X} \tilde{\psi}_0(\omega)\tilde{\psi}_f(\omega) \tag{1}$$

where X is the normalized coordinate, $X = x/\sigma_x$, $\tilde{\psi}_0(\omega)$ the characteristic function due to synchrotron radiation, $\tilde{\psi}_f(\omega)$ the characteristic function for BGS. $\tilde{\psi}_f(\omega)$ is a function of the scattering rate and of the time over which the evolution of the distribution is considered. Beam profiles can be approximated as Gaussian core plus exponential halo, see Figure 1.

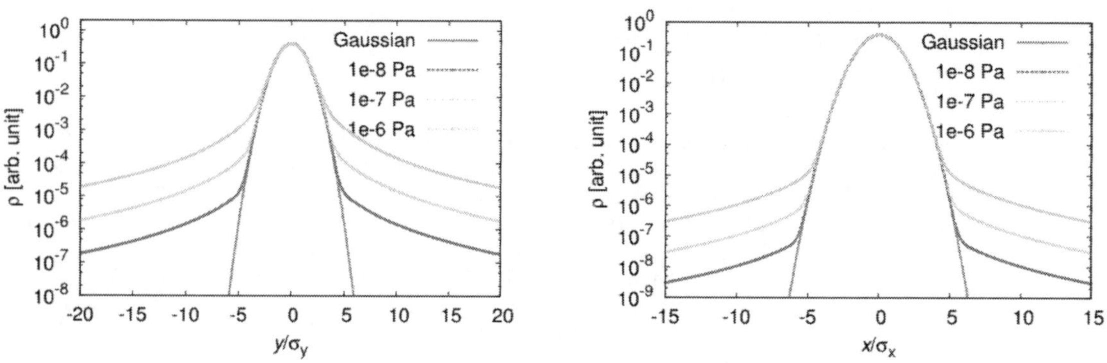

Figure 1. Analytic solution to the beam profile in equilibrium, vertically (left) and horizontally (right).

Here and after, only elastic BGS is taken into account for halo estimation since the probability of inelastic BGS scattering (Bremsstrahlung and inelastic scattering off the e^- of an atom) is 10^{-3} less than elastic scattering in ATF [6].

3. Simulation study in ATF

3.1. Description of simulation method

Tracking of parent and daughter particles is performed turn-by-turn, separately, taking into account synchrotron radiation and quantum excitation. The algorithm adopted to simulate the BGS effect can be summarized as follows: Firstly, in each turn, the number of BGS events is calculated according to the total cross section σ_s and density of gas atoms. The generated events are distributed randomly along the ring. Secondly, the perturbations to the transverse momentum ($\delta x'$, $\delta y'$) of the daughter particles are generated by a Monte-Carlo method according to the cross section. The initial coordinates of the daughter particles are determined by the local Twiss parameters, closed orbit distortion (COD) and the beam parameters (taking into account the radiation damping). Thirdly, the scattered particles in the present turn are transported to the observation point, to be combined with the other scattered particles accumulated from the previous turns. The above process is then repeated until extraction.

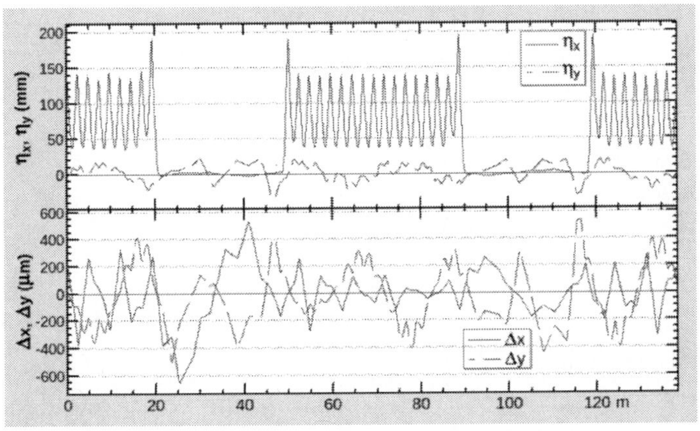

Figure 2. Dispersion functions and COD of ATF ring.

3.2. Equilibrium parameters

The equilibrium vertical emittance ϵ_y is mainly determined by the residual vertical dispersion η_y and betatron coupling, both of which strongly depend on the residual alignment errors [12]. The vertical emittance which is expected in the presence of realistic residual errors after applying standard corrections is achieved by introducing efficient random vertical displacements to quadrupoles and sextuplets (20μm and 70μm, RMS) and rotations to quadrupoles (2 mrad, RMS). The residual COD following this procedure is less than 600 μm, while η_y is under 30 mm, as shown in Figure 2. The equilibrium emittance ϵ_y, obtained for various seeds, ranges from 5 pm to 30 pm, as shown in Table 1.

Table 1. ATF beam parameters

Beam energy [GeV]	E_0	1.3
Intensity [/pulse]	N	1-10$\times 10^9$
Vertical emittance [pm]	ϵ_y	5-30
Horizontal emittance [nm]	ϵ_x	1.17
Energy spread [%]	σ_δ	0.056
Bunch length [mm]	σ_z	5.3
Damping time [ms]	$\tau_x/\tau_y/\tau_z$	27.0/19.8/20.6

3.3. Benchmarking of simulation

Estimation of the vacuum lifetime τ_v, which depends on the BGS directly, supplies a benchmarking of the simulation. Considering the elastic and inelastic scatterings, the theoretical prediction of τ_v can be expressed as

$$\tau_v^{-1} = \rho_v(\sigma_e + \sigma_i)\beta\gamma \tag{2}$$

where ρ_v is the volume density of the residual gas atoms, σ_e and σ_i the cross sections of particles loss due to elastic and inelastic BGS, respectively, β the electron velocity and γ the Lorentz factor. σ_e is limited by the physical aperture, and σ_i depends on the momentum acceptance (~1%) and the maximum acceptance of the instantaneous orbit. Assuming the residual gas mainly consists of CO and the averaged gas pressure is 1×10^{-6} Pa [13], vacuum lifetime evaluated analytically is 83 min. Meanwhile, the simulated value is 87 min using equilibrium beam parameters and realistic physical apertures [14]. Beam loss due to elastic scattering is more severe than inelastic scattering, by a factor of 2.85, see Figure 3. The former mainly happens in the large β regions, while the latter is concentrated at where η_x is maximal.

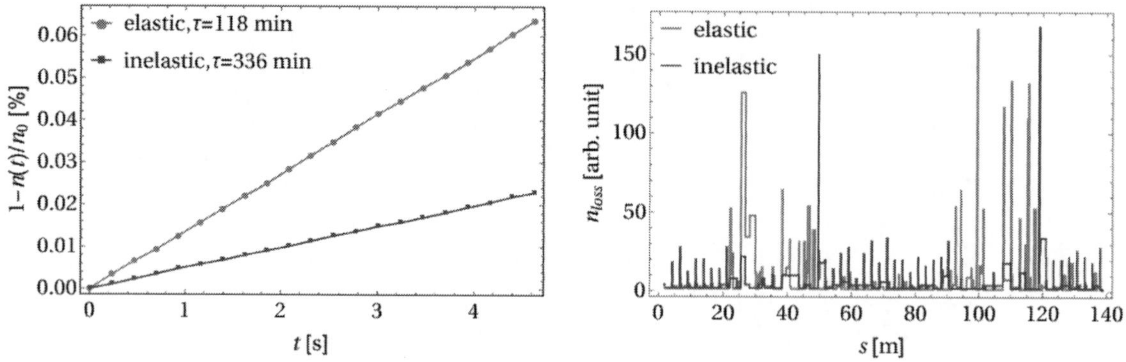

Figure 3. Loss rate (left) and loss map (right) of the scattered particles.

3.4. Comparison with analytic calculation

Due to the radiation damping, the beam distortion from BGS is mainly determined by the collisions in the time interval just before the extraction. In the theoretical estimation, the distribution is characterized taking into account only scatterings in the last damping time. In the simulation, it takes more than two damping times to obtain the equilibrium distribution, even if differences among profiles for successive damping times are not very large, as shown in Figure 4.

For the typical vacuum level of 5×10^{-7} Pa, satisfactory agreement between the analytical approximation and simulation is observed, see Figure 4, where the distribution is normalized to the RMS beam size. Horizontal tail/halo is lower than vertical halo, by around two orders of magnitude, after such a normalization, due to the large aspect ratio $\epsilon_x/\epsilon_y \approx 100$.

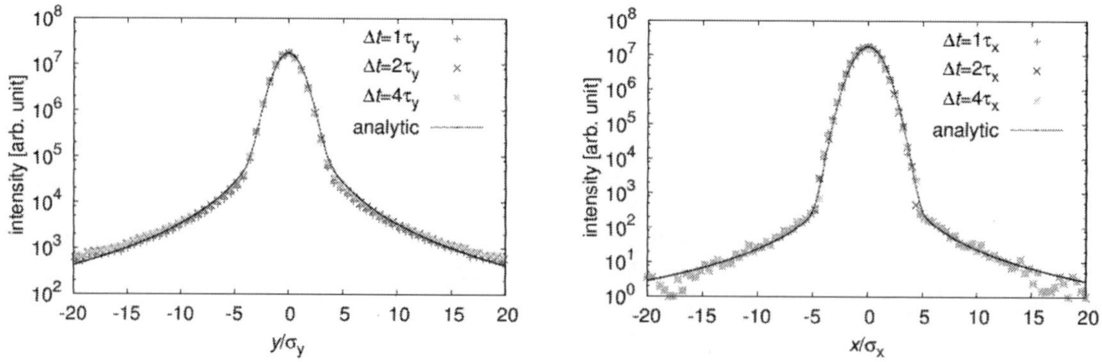

Figure 4. Comparison of vertical (left) and horizontal (right) beam distortion between analytic calculation and simulation.

In the analytic calculation, radiation damping isn't treated dynamically during the considered time interval. Instead, $\tilde{\psi}_0(\omega)$ and $\tilde{\psi}_f(\omega)$ are approximated as being in the final state. Since the distribution function has no assumption whether or not equilibrium has been reached, Equation

(1) can in principle also be used for calculations of the profiles during the damping process. To verify this, beam profiles predicted analytically and by simulation are compared at different times, showing a good agreement, see Figure 5

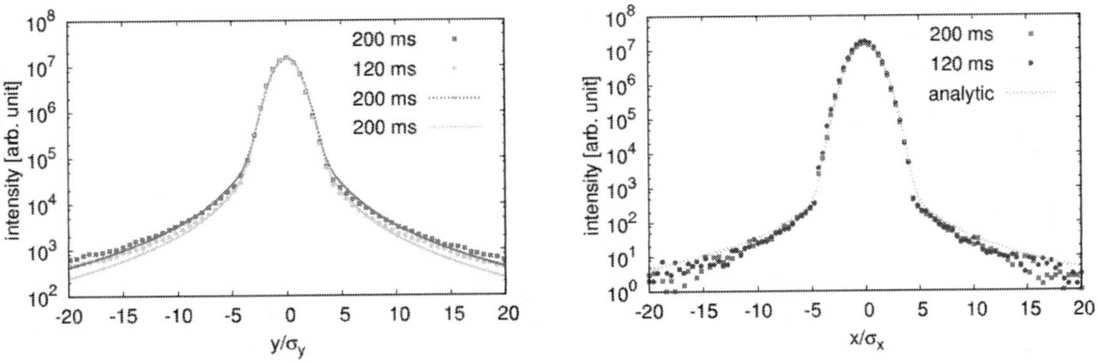

Figure 5. Beam profiles visualized at different moments, vertically (left, symbol:simulation and line:analytic calculation) and horizontally (right, with ϵ_x in equilibrium).

3.5. Vacuum dependence

The probability of BGS is in proportion to the density of residual molecules, and therefore, beam halo varies according to average vacuum in the ring. Presently, the averaged gas pressure is 2×10^{-7} Pa, which can be adjusted by turning on/off part of the ions pumps in the arc section (up to 2×10^{-6} Pa). Simulation is performed for the typical vacuum levels achieved in operation. Significant increases of the beam tail/halo are observed for the worsened vacuum conditions, as shown in Figure 6.

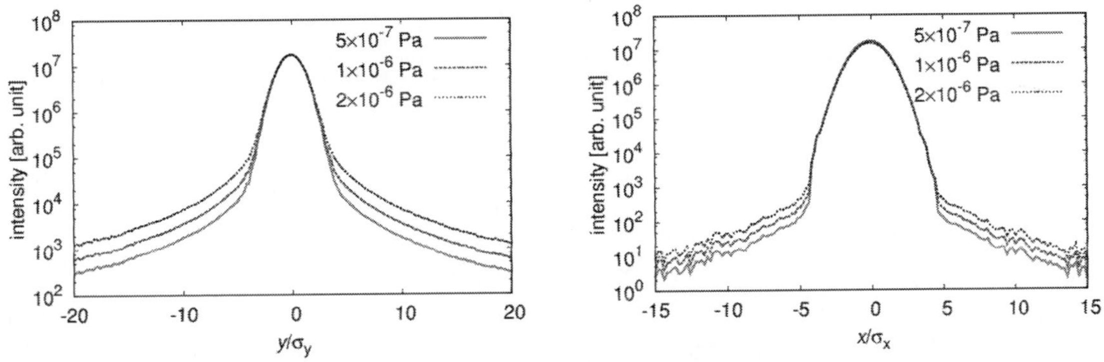

Figure 6. Vacuum dependence of beam halo, vertically (left) and horizontally (right).

3.6. Emittance dependence and emittance dilution

Notice that halo from BGS mainly depends on the accumulated stochastic transverse kicks, whereas equilibrium emittance determines beam size (beam core). Therefore, different behaviors of core and halo parts of beam profile due to BGS with various equilibrium emittances are expected. In the physical coordinate system, the beam core is broadened for increasing

IOP Conf. Series: Journal of Physics: Conf. Series **874** (2017) 012063 doi:10.1088/1742-6596/874/1/012063

emittances, while the halo distribution is not affected. An enhancement of beam tail/halo as a function of emittance is therefore observed in the normalized coordinate, see Figure 7. Here, only the evolution with varying ϵ_y is considered since ϵ_y is sensitive to tuning, and halo from BGS is more significant in the vertical plane. The averaged gas pressure is assumed to be 1×10^{-6} Pa. If BGS dominates vertical halo, the relations between emittance and beam tail/halo could be used as an alternative method for vertical ultra-small emittance measurement, without a knowledge of the β value or scanning the betatron phase.

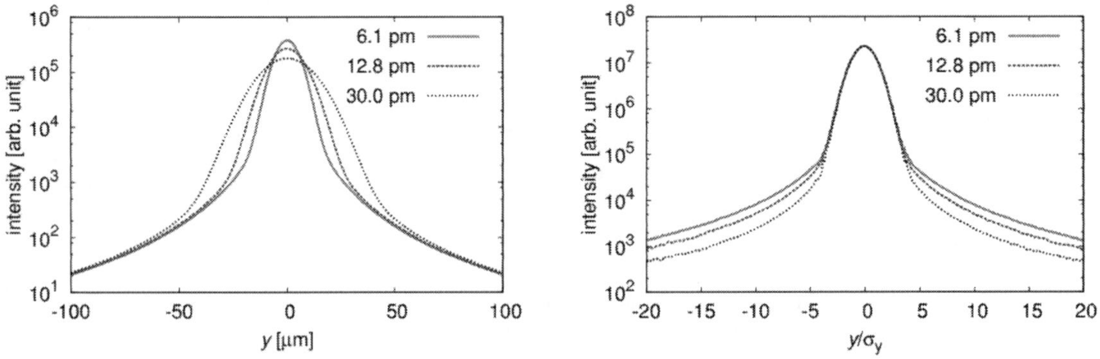

Figure 7. Vertical beam profile with different emittances, in physical frame (left) and the normalized coordinate (right).

If the BGS rate is large enough, not only halo tails will be produced in the transverse beam distribution, but also some emittance growth can result. With an initial vertical emittance ϵ_{y0} =12.8 pm, an increase of 30% for an averaged gas pressure of $5{\times}10^{-6}$ Pa is found in the simulation. This exceeds the acceptable value of 10 % [14], but is less than the analytical prediction of 22.8 pm, as shown in Figure 8. We think that this difference arises because radiation damping is not treated as a parallel process in the analytic approximation.

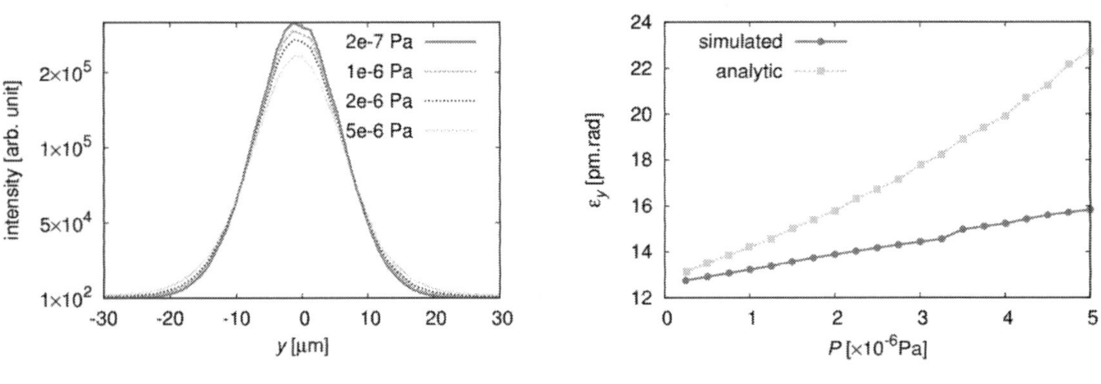

Figure 8. (Left) Vertical distribution for various vacuum levels; (Right) Emittance growth predicted by simulation of scatterings in 4 τ_y and analytic calculation.

4. Conclusion and prospects

To understand the generation of beam halo and its dynamics at ATF, a detailed simulation of halo due to BGS was performed. The equilibrium beam parameters are obtained by introducing artificial alignment errors, and the simulation is benchmarked using the vacuum lifetime. Satisfactory agreement between analytical estimations and simulation results of beam halo distributions for different storage time is achieved, which implies the feasibility to extend the convenient analytical approximation along the damping process. Evolution of halo corresponding to various averaged gas pressure is observed, which provides an approach to check if BGS is dominating. Higher tail/halo levels are observed for reduced vertical emittance, for the typical vacuum levels, which indicates a potential model independent to estimate the vertical emittance based on the profile distortion from BGS. The accumulated scatterings can results in some vertical emittance growth for cases of badly deteriorated vacuum levels, however the prediction form simulation and analytical calculation are different, with some improvements apparently needed in the analytical approach. Additional studies are underway, also involving the IBS process, which may play a role in the horizontal halo distribution.

5. Acknowledgement

We thank S. Bai, X. Li, T. Naito and K. Oide for helpful discussions, comments and assistance.

References

[1] Grishanov *et al.* 2005 *ATF2 proposal*
[2] White G *et al.* 2014 Experimental validation of a novel compact focusing scheme for future energy-frontier linear lepton colliders *Phys. Rev. Lett.* **112** 034802
[3] Suehara T *et al.* 2010 A nanometer beam size monitor for ATF2 *Nucl. Inst. Meth.* A **616** D(1)
[4] Kohji H and Yokoya K 1992 *Non-Gaussian distribution of electron beams due to incoherent stochastic processes* KEK
[5] Raubenheimer T 1992 *Emittance growth due to beam-gas scattering* KEK
[6] Raubenheimer T 1991 *The generation and acceleration of low emittance flat beams for future linear colliders* Stanford University
[7] Naito T and Mitsuhashi T 2015 Beam halo measurement utilizing YAG:Ce screen *Proc. IBIC'15* pp 373-376
[8] Yang R *et al.* 2016 Modeling and experimental studies of beam halo at ATF2 *Proc. IPAC'16* p 88.
[9] Wang D *et al.* 2013 Analytical estimation of ATF beam halo distribution *Chinese Physics* C **38**
[10] Yang R *et al.* 2017 Experimental study of halo formation at ATF2 *Proc. IPAC'17* MOPAB029.
[11] SAD is a computer program for accelerator design. http:/acc-physics.kek.jp/SAD/sad.html
[12] Raubenheimer T *et al.* 1993 The vertical emittance in the ATF damping ring *Nucl. Inst. Meth.* A **335**
[13] Okugi T *et al.* 2000 Evaluation of vertical emittance in KEK-ATF by utilizing lifetime measurement *Nucl. Inst. Meth.* A **455**(1)
[14] Hinode F *et al.* 1995 *ATF Design and Study report* KEK Internal 95.4

Pushing the space charge limit: electron lenses in high-intensity synchrotrons?

William D Stem and Oliver Boine-Frankenheim

TEMF, Technische Universität Darmstadt, Schlossgartenstr. 8, 64289 Darmstadt, Germany

E-mail: stem@temf.tu-darmstadt.de

Abstract. Several accelerator projects require an increase in the number of particles per bunch, which is constrained by the space charge limit. Above this limit the transverse space charge force in combination with the lattice structure causes beam quality degradation and beam loss. Proposed devices to mitigate this beam loss in ion beams are electron lenses. An electron lens imparts a nonlinear, localized focusing kick to counteract the (global) space-charge forces in the primary beam. This effort is met with many challenges, including a reduced dynamic aperture (DA), resonance crossing, and beam-beam alignment. This contribution provides a detailed study of idealized electron lens use in high-intensity particle accelerators, including a comparison between analytical calculations and pyORBIT particle-in-cell (PIC) simulations.

1. Introduction

Space charge represents a major intensity limitation in booster synchrotrons operating at low or medium energies. The limitation arises due to stopbands caused by envelope instabilities driven by the lattice structure [1, 2] or due to the space charge induced tune spread and its overlap with incoherent, nonlinear resonances [3]. Depending on the chosen working point and the intensity, coherent or incoherent effects can dominate or occur together. The space charge limit in booster synchrotrons is still an active field of research, both numerically as well as experimentally. After the success of electron lenses for beam-beam compensation in RHIC [4] the potential of such lenses to also compensate, at least partially, for space charge would allow to push the intensity in booster synchrotrons. However, one should keep in mind that space charge acts as a distributed defocusing error, whereas the beam-beam tune shift has a local source. Therefore localized lenses cannot be expected to be as efficient for space charge as they are for beam-beam compensation. One has to install several lenses around the ring instead of placing one lens close to the beam-beam interaction section. Furthermore, in order to be effective for space charge the electron current profile has to match the bunch profiles of the ions. For very short, relativistic bunches the optimum bunch overlap was studied in [5]. For booster synchrotrons with typically long bunches compared to the length of an electron lens, the matching of the electron current profile with the ion bunch profile might be easier to achieve experimentally. Different concepts for space charge compensation were studied in [6]. Electron lenses for space charge compensation were also discussed in [7]. The possibility of space charge compensation in booster synchrotrons was studied in [8] using analytical models for the envelope instability and the coherent dipole tune. The limitation of the compensation degree by the upward shift of the coherent dipole tune was pointed out. For the FNAL booster synchrotron a minimum number of lenses was estimated.

For the SIS synchrotron, used as a reference machine in this study, the incoherent resonances induced by the space charge field of the electron beam in the cooling section were analyzed in [9]. For electron cooling the tune shift induced by the electron beam is usually well below 0.1 and the ion beam currents are low. Therefore incoherent resonances dominate and were identified up to order 6 in the tracking studies together with a severe emittance growth. The present study relies on a simulation model for intense beams including the linear synchrotron lattice, the 2D self-consistent space charge force and simple kicks for the lenses. We compare the results of extended simulation scans to analytical models.

2. Space charge compensation

2.1. 2D Model

The compensation of space charge in bunches is a 3D problem. The transverse profile of the electron beams and its current profile ideally have to both match the transverse profile and the longitudinal bunch profile of the circulating ion bunches. In this study we assume ion bunches that are long compared to the interaction length with the electron beam. The current profile of the electron beam in the interaction section is assumed to follow exactly the ion bunch profile. If a certain space charge compensation degree is chosen for the bunch center, it will therefore hold also at the bunch ends. If we further assume that the relevant effects are faster than the synchrotron oscillation period, a 2D model for the bunch center slice is sufficient.

The space charge tune shift in the bunch center is (vertical plane):

$$\Delta Q_y \approx -\frac{NZ^2 r_p}{2\pi\varepsilon_y\beta_0^2\gamma_0^3 A B_f}, \tag{1}$$

where β_0 and γ_0 are the relativistic parameters for the ion beam, N is the total particle number in the ring, Z the charge state of the ions, A the mass number, B_f is the bunching factor, ε_y is the unnormalized emittance of the equivalent KV beam.

The beam-beam tune shift (co-propagating: $-$, counter-propagating: $+$) induced by one electron lens is

$$\Delta Q_y^e = \frac{1 \mp \beta_e\beta_0}{\beta_e}\frac{Z I_e L r_p}{2\pi A e c \varepsilon_y \beta_0^2\gamma_0}. \tag{2}$$

Hereby we assume that the transverse profiles of both beams overlap ideally. I_e is the current of the electron beam, L is the length of the interaction section, β_e is the velocity of the electrons divided by the speed of light.

The degree of space charge compensation we define as

$$\alpha = \frac{\Delta Q^e}{\Delta Q}, \tag{3}$$

where ΔQ is the space charge tune shift and ΔQ^e the total beam-beam tune shift generated by N_e electron lenses. The GSI SIS heavy-ion synchrotron [11] is used as a reference case for the study. The SIS has a circumference of $C = 216$ m and $S = 12$ periodic sectors. The injection energy is 11.4 MeV/u. As a simple example, we assume each sector is an ideal FODO cell, and the electron lenses are distributed symmetrically around the ring and treated as thin lenses.

2.2. Stability considerations: Orbits and Envelopes

The transverse space charge force acts as a distributed (de)focusing error around the ring. In periodic focusing lattices, space charge alone only leads to incoherent structural or coherent parametric resonances. On the contrary, the electron lenses act as localized focusing errors. If they are spaced symmetrically around the synchrotron they define new periodic structure cells (for $N_e < S$). In addition the lenses result in additional nonlinear error resonances.

As a first-order effect, the limitation of the compensation degree due to the shift of the coherent tune has been pointed out in [8]. The focusing electron lenses cause a positive shift of the coherent tune (the frequency of the beam centroid oscillations)

$$Q_c = Q_0 + \alpha \Delta Q, \tag{4}$$

where Q_c is the coherent tune and Q_0 the bare tune. This has been experimentally shown at the SIS for both coasting and bunched ion beams in [12]. An important limitation for the total strength of the lenses is that Q_c should stay well below integers n in order for the closed orbit to remain stable. This limitation only depends on the total beam-beam tune shift and not on the number of lenses.

The number of lenses N_e is important for the perturbation of the optical functions. The maximum beating of the β-function caused by an electron lens j, treated here as a localized focusing error is (see e.g.[10])

$$\left(\frac{\Delta\beta}{\beta}\right)_{\max} \approx 2\pi \Delta Q^e / N_e, \tag{5}$$

where ΔQ^e is the total beam-beam tune shift. If we require that the maximum beta beating remains well below 1, only very small space charge tune shifts can be compensated with one lens (well below 0.1 in the SIS). For the typical space charge tune shifts in high current synchrotrons, of the order of $0.4 - 0.5$, several (at least $N_e \gtrsim 3$ for the SIS) lenses are required to at least partially compensate the space charge tune shift, without causing a large beating amplitude of the optical functions.

It is important to note that for dc electron beams the beta beating could be reduced by corrector quadrupoles. However, for the desired pulsed electron beam operation a correction with fixed gradient quadrupoles is not possible.

The required number of lenses can also be estimated from the stability criteria for betatron oscillations in periodic focusing lattices:

$$|\cos(2\pi Q/N_e) - 2\pi \frac{\Delta Q^e}{N_e} \sin(2\pi Q/N_e)| < 1, \tag{6}$$

where the tune Q is the particle tune and ΔQ^e is the total beam-beam tune shift induced by the N_e electron lenses. The area in tune space with $|\ldots| \geq 1$ are stopbands in which the amplitude of linear betatron oscillations grows exponentially. Including the space charge tune shift in the particle tune $Q = Q_0 + \Delta Q$ causes a shift of the stopbands, but leaves the stopband width unchanged. The stopband width is determined by the beam-beam tune shift and the density of bands by N_e. The distance between stopbands is $N_e/2$, which is also why we call those stopbands "180°". The stopband width increases as $\delta Q \approx 2\Delta Q^e$.

The 2D envelope equations (see for example [13]) can be solved numerically including the electron lenses, treated as thin, linear focusing elements. Thereby one can obtain space charge structural instabilities and their modification due to the lenses. The condition for a parametric resonance or envelope instability including a partial incoherent space charge compensation is

$$2(Q_0 - \alpha\Delta Q) - \Delta Q_2 = \frac{n}{2} S, \tag{7}$$

where Q_0 is the bare tune, ΔQ_2 is the coherent tune shift for the envelope modes, n is the harmonic number and $S = N_e$ is the number of structure cells, which in this case equals the number of electron lenses. It is important to note that the stopbands arising from (coherent) envelope instabilities, also called 90° stopbands, are different from the 180° stopbands due to localized gradient errors.

Both stopbands can overlap, as is the case in the SIS close to $Q_y = 3$ (see Figure 1). In this case the 180° stopband dominates, as it defines the linear stability for the single particle and envelope betatron oscillations.

IOP Conf. Series: Journal of Physics: Conf. Series **874** (2017) 012064 doi:10.1088/1742-6596/874/1/012064

3. Particle-In-Cell (PIC) tracking

The 2D envelope model, discussed in the previous section, allows for very fast tune scans and helps to separate coherent and incoherent effects, which occur together in Particle-In-Cell (PIC) tracking simulations. In our simplified model, incoherent resonances are excited by the (static) nonlinear beam-beam forces and by the $n = 0$ harmonic of the nonlinear space charge force. We use an initial waterbag distribution for the PIC ion macro-particles, which might be closer to the actual distribution of the injected beam in a booster synchrotron than a Gaussian. A 2D tune scan performed with the PIC simulation code pyORBIT [14] is shown in Figure 1. The elliptical transverse profile of the ion beam at the location of the lens is met with the round, Gaussian beam of the electron lens. The emittance growth observed after 500 cells is shown for $N_e = 3$ lenses. The reference high-intensity working point of the SIS is indicated as a white diamond. The overlapping 90° and (lens-induced) 180° stopbands as described in the previous section are shown clearly near $Q_y = 3$. Sixth order resonance lines appear near $Q_{x,y} = 3.75$, arising from localized, nonlinear focusing errors in the electron lenses. It should be noted that the order and location of these nonlinear resonance lines depend greatly on the number of lenses and beam profile matching. An example ion beam profile displaying sixth order resonance after 100 cells is shown in the bottom right-hand corner of the figure. Without the lenses, the simulation results only in the 90° stopbands, shifted upwards according to the tune shift in Equation (1). The widths of the 90° stopbands are unaffected by the lenses.

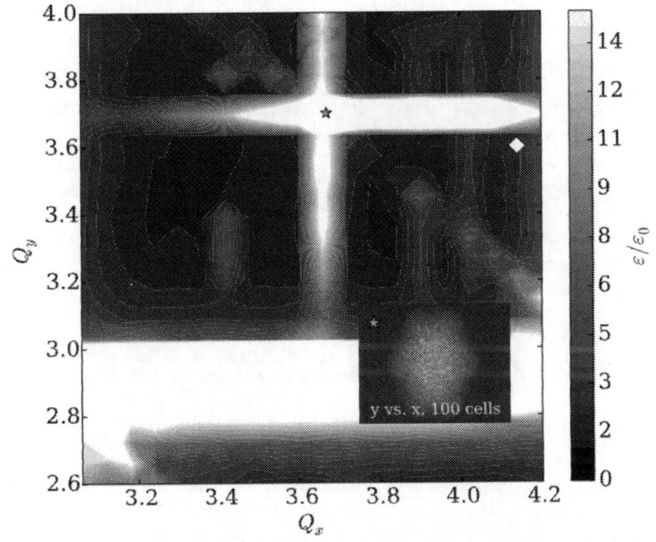

Figure 1. PIC simulation scan for $N_e = 3$, $\Delta Q_{x,y} = -0.3, 0.8$ and $\alpha = 1$. The SIS high-intensity working point is represented by a white diamond. 6th-order nonlinear resonance is shown after 100 cells (star).

4. Conclusions and Outlook

Within a simplified 2D beam dynamics model we studied the (partial) compensation of large space charge tune shifts, typical for booster synchrotrons, by localized electron lenses. The model includes self-consistent 2D space charge and fixed (nonlinear) kicks from the electron lenses. We use the GSI SIS18 heavy-ion synchrotron without bends as the reference case, and only the transverse space charge and lenses account for the 'error' sources in our simulation model. From the envelope model we obtain the beta beating amplitudes, the 180° stopbands (caused by the localized gradient errors) as well as the envelope instability or 90° stopbands (caused by space charge structural instabilities). The PIC simulation provides additional information on nonlinear resonances excited by the lenses, which cannot be resolved in the envelope model.

For any $N_e < S$ the unwanted effects caused by the lenses seem to dominate in our example study. One has to keep in mind that in our model many additional sources of errors are not included, for example systematic and random errors in the alignment of the lenses and in the 3D overlap between the ion and electron beams. The alignment errors of the lenses will add to the closed orbit instability at integer coherent tunes. The errors in the electron current for different lenses will lead to additional gradient errors stopbands. Also we did not include any lattice errors, which are usually the reason for the space charge limit without the lenses. Therefore we expect our model to be very optimistic.

Still there is room for further studies and possible improvements. For example, instead of trying to ideally overlap both beams one could use a transverse McMillan lens profile [15, 16]. This could reduce the effect of nonlinear resonances induced by the lenses, but would not affect the error stopbands or the beta beating. The compensation might work better for lower space charge tune shifts (0.1 or lower) and a space charge limit dominated by nonlinear lattice errors. This would require studies with a 3D simulation model including synchrotron motion over much longer time scales than in this study. We plan to conduct such long-term simulation studies for the FAIR SIS100 synchrotron, first within a frozen 3D space charge model.

5. Acknowledgements

This work is supported by the Bundesministerium für Bildung und Forschung (BMBF) under contract FKZ:05P15RDRBA. We would like to thank Sabrina Appel for her support of the pyORBIT code.

6. References

[1] Baartman R 1998 Betatron resonances with space charge *AIP Conference Proceedings* **448** 56
[2] Hofmann I and Boine-Frankenheim O 2017 Parametric instabilities in 3D periodically focused beams with space charge *Phys. Rev. Accel. Beams* **20** 014202
[3] Franchetti G *et al.* 2010 Experiment on space charge driven nonlinear resonance crossing in an ion synchrotron *Phys. Rev. ST Accel. Beams* **13** 114203
[4] Gu X *et al.* 2017 Electron lenses for head-on beam-beam compensation in RHIC *Phys. Rev. Accel. Beams* **20** 023501
[5] Litvinenko V N and Wang G 2014 Compensating tune spread induced by space charge in bunched beams *Phys. Rev. ST Accel. Beams* **17** 114401
[6] Aiba M *et al.* 2007 Space-charge compensation options for the LHC injector complex *Proc. of Particle Accelerator Conf. (New Mexico)* THPAN074
[7] Shiltsev V D 2016 *Electron Lenses for Super-Colliders* (New York: Springer)
[8] Burov A V, Foster G W, and Shiltsev V D 2000 Space-charge compensation in high-intensity proton rings *FERMILAB-TM-2125*
[9] Sorge S, Boine-Frankenheim O, and Franchetti G 2007 Analysis of resonances induced by the SIS-18 electron cooler *Proc. of COOL (September 9-14 Bad Kreuznach, Germany)*, 121–124
[10] Lee S Y 2004 *Accelerator Physics, Second Edition*, (New Jersey: World Scientific)
[11] Appel S, Boine-Frankenheim O, and Petrov F 2017 Injection optimization in a heavy-ion synchrotron using genetic algorithms *Nucl. Instr. Meth.* A **852** 73
[12] Stem W D and Boine-Frankenheim O 2016 Using an electron cooler for space charge compensation in the GSI synchrotron SIS18 *Proc. of HB2016 (July 3–8 Mälmo, Sweden)* THPM5X01
[13] Yuan Y S, Boine-Frankenheim O, Franchetti G, and Hofmann I 2017 Dispersion-induced beam instability in circular accelerators *Physical Review Letters* **118** 154801
[14] Shishlo A, Cousineau S, Holmes J, Gorlov T 2015 The particle accelerator simulation code PyORBIT *Proc. Comput. Sci.* **51** 1272–1281 doi:10.1016/j.procs.2015.05.312, `https://github.com/PyORBIT-Collaboration`

[15] Stancari G *et al.* 2015 Electron lenses for experiments on nonlinear dynamics with wide stable tune spreads in the fermilab integrable optics test accelerator *Proc. 6th Int. Particle Accelerator Conf. (May 3–8 Richmond)* MOBC3
[16] McMillan E M 1967 Some thoughts on stability in nonlinear periodic focusing systems *UCRL–17795*

IOP Conf. Series: Journal of Physics: Conf. Series **874** (2017) 012065 doi:10.1088/1742-6596/874/1/012065

Electron cloud simulations for the main ring of J-PARC

Bruce Yee-Rendon, Ryotaro Muto, Kazuhito Ohmi, Kenichirou Satou, Masahito Tomizawa and Takeshi Toyama

KEK, High Energy Accelerator Research Organization, Tsukuba, Ibaraki 305-0801, Japan

E-mail: byee@post.j-parc.jp

Abstract. The simulation of beam instabilities is a helpful tool to evaluate potential threats against the machine protection of the high intensity beams. At Main Ring (MR) of J-PARC, signals related to the electron cloud have been observed during the slow beam extraction mode. Hence, several studies were conducted to investigate the mechanism that produces it, the results confirmed a strong dependence on the beam intensity and the bunch structure in the formation of the electron cloud, however, the precise explanation of its trigger conditions remains incomplete. To shed light on the problem, electron cloud simulations were done using an updated version of the computational model developed from previous works at KEK. The code employed the signals of the measurements to reproduce the events seen during the surveys.

1. Introduction

In the past, several electron cloud simulations for the J-PARC MR were done to estimate the conditions of its formation and its effect in the beam stability [1, 2, 3]. In those cases, the longitudinal distributions used were bunched or coasting beams and the proton loss was the main source of primary electrons (seeds).

The MR accelerates protons up to energy of 30 GeV, it operates in two modes: Fast extraction (FX) and Slow extraction (SX) [4]. Electron cloud occurred during both modes. At FX, the phenomenon was mitigated by beam operation (scrubbing effect) [5]. On the contrary, the electron cloud remains at SX, therefore several surveys were done to investigate the conditions for its build-up [6, 7]. It was observed that the electron cloud appeared during the debunching process, reaching the peak about 75 ms after P3 (the time when debunching starts). Indeed, the presence of a strong microbunch structure for the cases with electron cloud was also reported, thus, it is believed that this particular beam shape reinforce the electron multiplication. Finally, the low level of beam loss detected, before the electron cloud began to form, which indicates that residual gas ionization as the principal source of primary electrons.

Figure 1 presents a map of the electron cloud as a function of the beam intensity and phase difference at injection [7]. The work is focus to reproduce the electron cloud formation seen at SX, with emphasis in the cases of zero phase difference shown in Figure 1. This study used the longitudinal beam distribution obtained by using a fast current transformer monitor and the pressure measured with vacuum gauges installed in the ring.

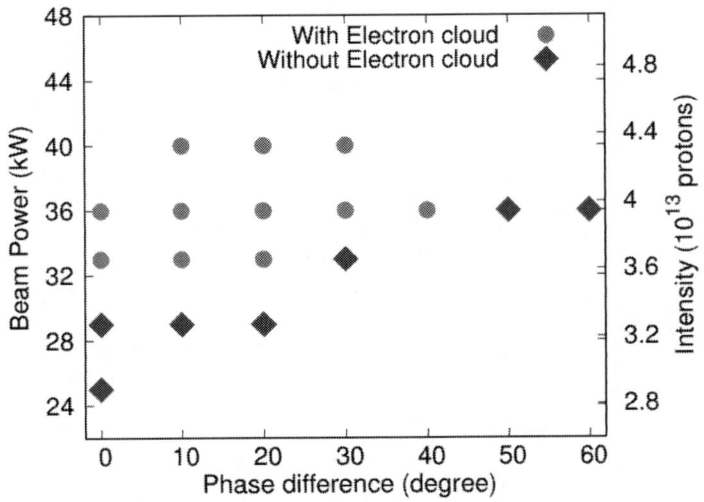

Figure 1. The presence (red circles) or absence (blue diamond) of the electron cloud as a function of the intensity and injection settings (phase difference between the beam and acceleration RF).

Additionally, the frequency spectrum of the electron cloud detector and the fast current transformer monitor in the electron cloud cases presented a large amplitude for the components between 20 to 40 MHz [6], which corresponds to the bouncing frequency of the electrons in the MR vacuum chamber [8]. Therefore, a sinusoidal distributions is employed to corroborate that the frequencies in the same range can increase the electron production in the simulations.

2. Simulation setup

The update model presented here includes new features such as the debunching process and the use of the data of the beam measurements as inputs. The simulation code is described elsewhere [1]. The relevant features of the model are:

- The longitudinal distribution and electrons are simulated by macro-particles. The beam profile is represented by 5231 macro-particles to describe the entire proton distribution around the ring (synchrotron period at the MR is 5.231 μs at 30 GeV).
- The effect of the microbunch structures is taking into account in the simulations. The beams distribution corresponded to a two different times in the debunching: 5 ms (the bunch has smooth shape) and 75 ms (the microwave structure in the beam is severe). Figure 2 shows the difference in the bunch shape for the same beam at these two times.
- CO and H_2 were used as the residual gas. Their ionization cross-section and the corresponding pressure, for the different intensities and times, are shown in Table 1.
- $\delta = 1.6$ and $E_{max} = 200$ eV were adopted for the Secondary Emission Yield (SEY) model. These parameters were based on the MR J-PARC beam pipes measurements [9].

To verify if the simulations present a similar bouncing frequency as in the measurements, the model had the next changes:

- The longitudinal beam profile was a sinusoidal distribution, $A(I) \mid Sin(\pi f_0 t) \mid$, where $A(I)$ was a normalized factor such as the integration over all the circumference produced the corresponding intensity I for each case, and f_0 took values in a range from 10 to 100 MHz in an step of 10 MHz.

- The SEY model used $\delta = 1.1$ and $E_{max} = 200$ eV.

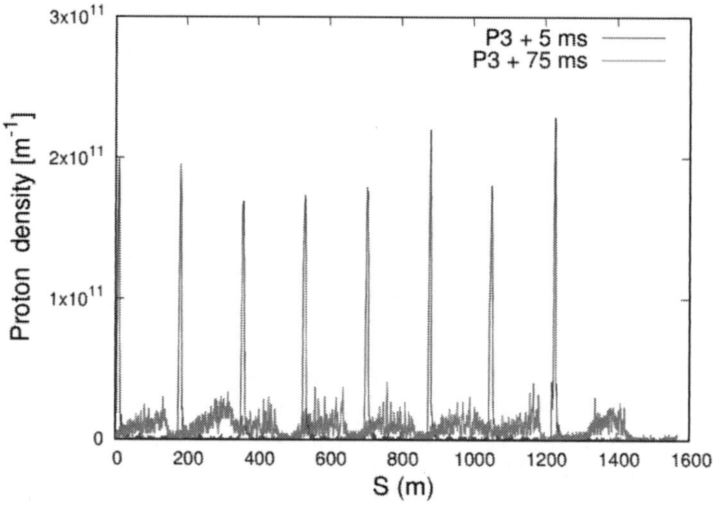

Figure 2. Longitudinal beam profile at two different times during the debunching process. A remarkable microbunch structure is presented at P3+75 ms.

Table 1. Simulations parameters for the JPARC MR.

Parameters	Units	Value
Energy	GeV	30
Bunch population[a]	10^{13} ppp	4.2, 3.8, 3.4, 2.8
Circumference	m	1567.5
Beam pipe radius	cm	6.5
rms bunch size	cm	0.5
Ionization cross section	Mbarn	2
Vacuum pressure at 5 ms[b]	μPa	0.2
Vacuum pressure at 75 ms[a,b]	μPa	0.9, 0.7, 0.5, 0.4
Time step	ns	1

[a] Each intensity has a corresponding vacuum pressure at 75 ms.
[b] The time after debunching starts (P3).

3. Results

Figure 3 shows an example of the longitudinal beam profile employed by our model (top) and its electron density generated (bottom) at 5 ms after P3. The largest amount of the secondary electron (electrons gained) after the bunch peak suggests the trailing edge multipactor is the main mechanism of secondary electron production. Each simulation case was repeated five times and the average value was reported with their corresponding standard errors (Figures 4 and 6).

IOP Conf. Series: Journal of Physics: Conf. Series **874** (2017) 012065 doi:10.1088/1742-6596/874/1/012065

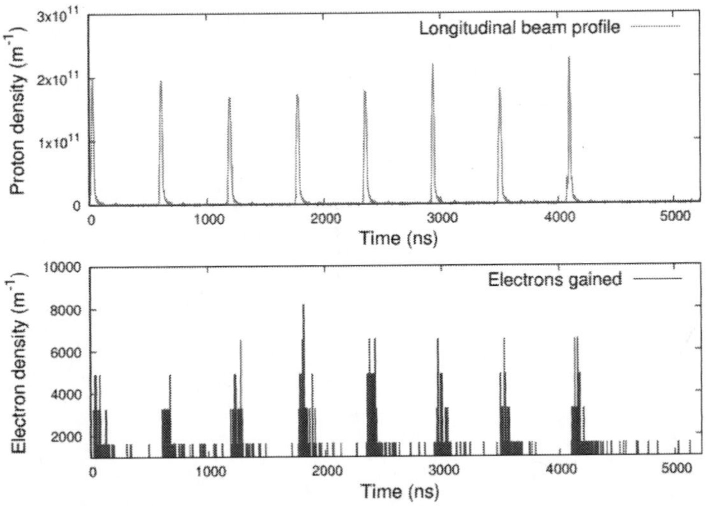

Figure 3. The longitudinal beam distribution (top) and the corresponding electron density (bottom). The data length corresponds to one turn of the J-PARC MR (5.231 μs).

The summary for the simulations cases appears in Figure 4. The numbers of electrons generated were similar for the cases using longitudinal beam distributions at 5 ms after debunching starts (Figure 4 blue bars). In contrast, the results for the time of 75 ms presented a clear difference between the higher and lower intensities (Figure 4 red bars). Indeed, the lowest intensity (2.8 x 10^{13} ppp) produced almost the same amount of electrons for both times.

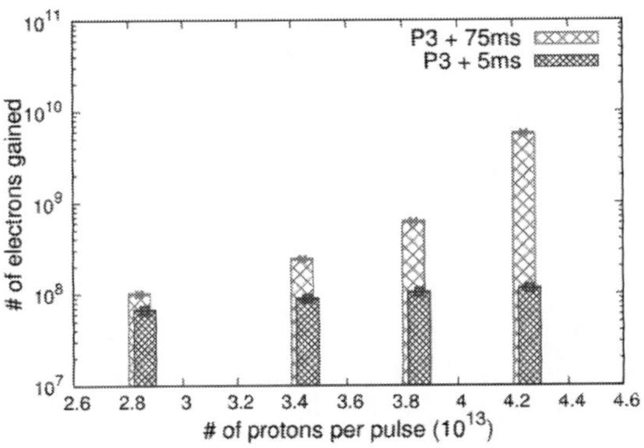

Figure 4. The comparison of the total electrons generated by the four intensities simulated at the two different times (5 ms and 75 ms after P3). The model used a $\delta = 1.6$ and $E_{max} = 200$ eV.

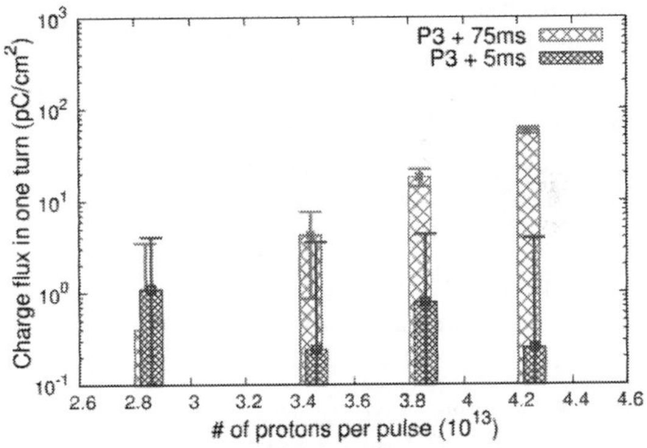

Figure 5. The total flux charge of the signal measured by the electron cloud detector in one turn. The signal was average over 100 turns, the beam fluctuations and detector noises are included in the error bars.

Additionally, Figure 5 presents the total flux charge measured by the electron cloud detector for the same cases as in the simulations. The charge flux per turn for the time of 5 ms was in the same range for all the cases (Figure 5 blue bars). At 75 ms, an increase of the signal detected as a function of the intensity appeared. In addition, similar amounts of charge flux for the two time cases were observed in the lower intensities [7].

Finally, Figure 6 shows the electrons gained as a function of the frequency of the sinusoidal distribution. A remarkable production of electrons were obtained at the high intensities for frequencies between 30 to 50 MHz. Due to the largest number of electron generated at certain frequencies δ was 1.1 for this study.

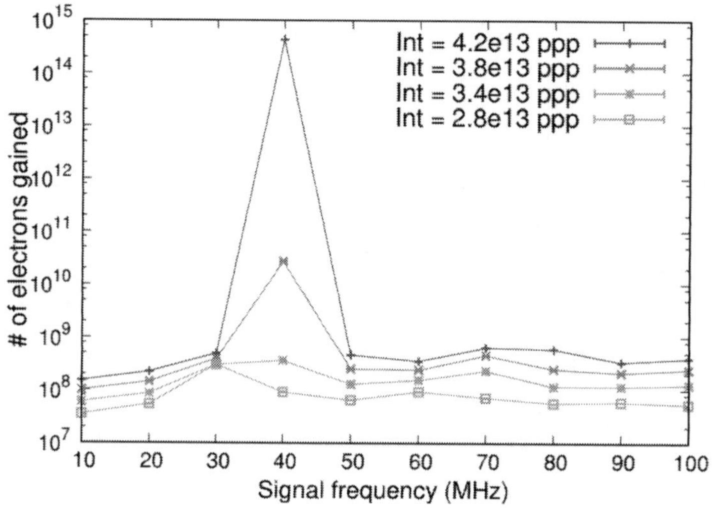

Figure 6. Electrons produced in the simulations vs. the frequency of the sine signal at different intensities, using $\delta = 1.1$ and $E_{max} = 200\ eV$.

4. Conclusions

The map of Figure 1 shows that for the first column (phase difference equal to zero degrees), the electron cloud did not appear in the lower intensities, on the contrary, it developed in the higher cases. The same conclusion can be obtained from the values presented in Figures 4 and 5 using the next interpretation: at the beginning of debunching the electron cloud does not appear (smooth bunch shape), nevertheless, exists some number of electrons. After the beam becomes coasting, a microbunch structure appears (which became severe for high intensity beams and eventually disappear when the beam is totally unbunched), then, if the multipactor condition is reached, the number of electron increases and the electron cloud occurs. In Figures 4 and 5 only the higher intensities presented a significant increase of the electron signal with respect to the early stage, therefore, in these cases the electron cloud occurred. Thus, the simulations results are consistent with the measurements.

Finally, Figure 6 presents a remarkable enhancement of the electron generation for the sinusoidal frequency in the range of 20 to 50 MHz, this result is in agreement with the frequency bounce observed by the the electron cloud detector and the fast current transformer monitor in the cases with electron cloud [6].

5. Acknowledgment

The authors thank the members of J-PARC Main Ring, in specially to M. Okada, M. Uota and C. Yee-Rendon for the fruitful discussions. This work was partially supported by JSPS KAKENHI Grant Number JP16H06288, Grant-in-Aid for Specially promoted Research titled Measurement of CP symmetry of neutrino by upgrading T2K experiment.

IOP Conf. Series: Journal of Physics: Conf. Series **874** (2017) 012065 doi:10.1088/1742-6596/874/1/012065

References

[1] Ohmi K *et al.* *Phys. Rev. ST Accel. Beams* **5** 114402
[2] Toyama T *et al.* 2004 *Proc. of the ECLOUD (Geneva)* vol 4
[3] Ohmi K *et al.* 2003 *Proc. of the Part. Acc. Conf. (Oregon)* p 3083
[4] Tomizawa M *et al.* 2006 *Proc. of the European Part. Acc. Conf. (Edinburgh)* p 1987
[5] Uota M *et al.* 2012 *Proc. of the Part. Acc. Society of Japan (Osaka)* p 462
[6] Yee-Rendon B *et al.* 2016 *Proc. of the Int. Part. Acc. Conf. (Busan)* p 4175
[7] Yee-Rendon B *et al.* 2016 *Proc. of the Part. Acc. Society of Japan (Chiba)* p 149
[8] Toyama T private communication
[9] Okada M *et al.* 2010 *Proc. of the Part. Acc. Society of Japan (Nagoya)* p 1176

IOP Conf. Series: Journal of Physics: Conf. Series **874** (2017) 012066 doi:10.1088/1742-6596/874/1/012066

Controlled longitudinal emittance blow-up using band-limited phase noise in CERN PSB

D Quartullo[1,2], E Shaposhnikova[1] and H Timko[1]
[1] CERN, Geneva, Switzerland
[2] Università di Roma La Sapienza, Rome, Italy
E-mail: danilo.quartullo@cern.ch

Abstract. Controlled longitudinal emittance blow-up (from 1 eVs to 1.4 eVs) for LHC beams in the CERN PS Booster is currently achievied using sinusoidal phase modulation of a dedicated high-harmonic RF system. In 2021, after the LHC injectors upgrade, 3 eVs should be extracted to the PS. Even if the current method may satisfy the new requirements, it relies on low-power level RF improvements. In this paper another method of blow-up was considered, that is the injection of band-limited phase noise in the main RF system (h=1), never tried in PSB but already used in CERN SPS and LHC, under different conditions (longer cycles). This technique, which lowers the peak line density and therefore the impact of intensity effects in the PSB and the PS, can also be complementary to the present method. The longitudinal space charge, dominant in the PSB, causes significant synchrotron frequency shifts with intensity, and its effect should be taken into account. Another complication arises from the interaction of the phase loop with the injected noise, since both act on the RF phase. All these elements were studied in simulations of the PSB cycle with the BLonD code, and the required blow-up was achieved.

1. Introduction

In 2021, after Long Shutdown 2 (LS2), all the LHC injectors will be upgraded to fulfil the specifications of the LIU (LHC Injectors Upgrade) project [1].

The CERN PS Booster is the first synchrotron in the LHC proton injection chain. Currently it accelerates protons from 50 MeV to 1.4 GeV kinetic energy E_k in about 500 ms, while the longitudinal emittance is increased from 1 eVs to 1.4 eVs in a controlled way to provide stability of the beam during the ramp and to reduce space charge effects at PS injection. After LS2, the new injection and extraction energies will be 160 MeV and 2 GeV respectively, but the cycle duration will not change. The beam intensity will be a factor of two higher than now and the emittance needs to be increased from 1 eVs to 3 eVs.

Three RF systems are used in the PSB, one for acceleration (C02, h=1, V_1 = 8 kV), one for bunch shaping (C04, h=2, V_2 = 8 kV), and one for controlled longitudinal emittance blow-up (C16, h>6, V_3 = 6 kV). All these narrow-band and tunable ferrite RF systems will be replaced by a broad-band Finemet system [2], to allow for higher peak voltages and flexibility. The Finemet gaps will be modular, each of them will be set to the needed harmonic and peak voltage, for a total of 24 kV available.

Simulations studies have shown that the required emittance blow-up can be achieved using an optimised phase modulation of the C16 RF system [3]. This paper will deal with a possible alternative to this technique, namely the injection of band-limited RF noise in the main RF harmonic cavity. Bunch shaping and blow-up using band-limited noise have already been studied [4]. Longitudinal emittance blow-up using phase noise, never tried in the PSB, is successfully implemented in CERN SPS and LHC [5,6], where there are no dedicated RF systems for blow-up and cycles are longer.

There are three main reasons to propose an alternative to the current method of blow-up. The RF phases between the different PSB RF systems are not accurately known in operation and there can be issues of reproducibility. In addition, the present method for blow-up implies the use of a dedicated RF system that can be removed with noise injection in main RF system. Finally, the proposed technique, in addition to blow-up, is able to lower the peak line density of the bunch decreasing the impact of intensity effects.

The phase noise can be injected through phase loop at a limited sampling rate. Simulations using the CERN BLonD code [7] with a corresponding longitudinal impedance model have been used to verify the feasibility of this method in upgraded PSB (after LS2).

2. The band-limited phase noise

The effect of band-limited phase noise lies in-between the diffusion generated by white phase noise acting on all frequencies and the resonant excitation created by sinusoidal phase modulation. Indeed only the particles with a synchrotron frequency inside a certain band are affected (see Figure 1). The band upper limit f_{up} is above the bunch central synchrotron frequency f_s (with intensity effects) to affect fully the bunch core and its lower limit f_{down} is related to the target bunch length (or emittance) to be reached. Figure 1 shows also that space charge, defocusing in the PSB (below transition), lowers the zero-intensity synchrotron frequency f_{s0} and the noise band should follow this shift.

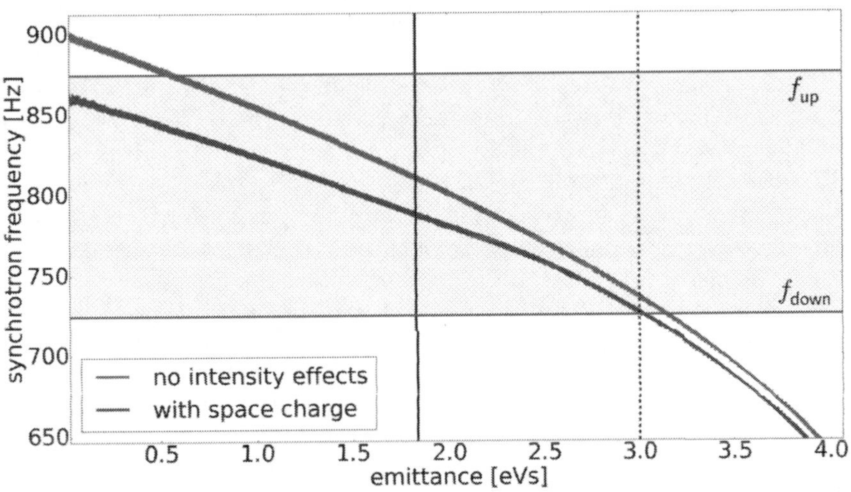

Figure 1. Synchrotron frequency distribution in single RF calculated with (blue line) and without (green line) space charge effect (N = 3.6×10^{12} ppb) in PSB at $E_k = 1$ GeV. The bunch emittance increases from 1.8 eVs to 3 eVs applying phase noise in the band [725 Hz, 875 Hz] (magenta lines).

The RF phase noise $\phi(t)$ was constructed similarly to the LHC implementation. It was obtained colouring white phase noise in frequency domain with the desired probability density and Fourier-transforming the result to time domain [8]. The double-sided noise power spectral density $S(f)$ [rad^2/Hz] (see Figure 2) determines the rms phase noise as $\phi_{rms} = \sqrt{f_0 < S(f) >}$ (f_0 is the revolution frequency).

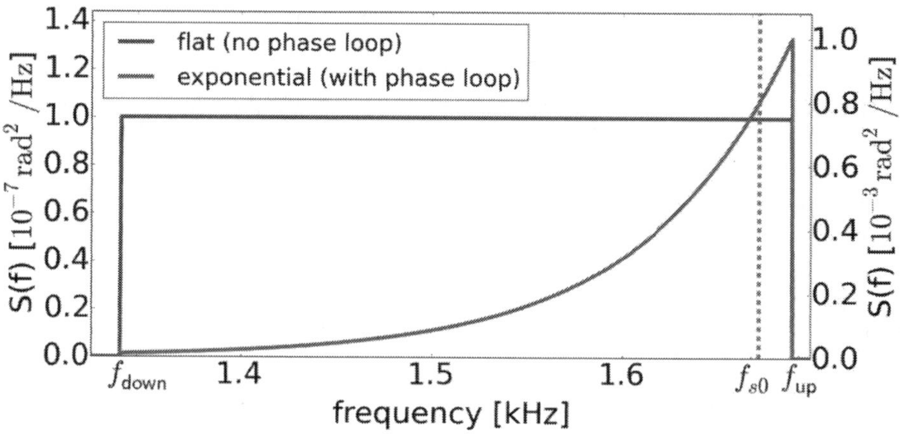

Figure 2. Flat (left axis) and exponential (right axis) spectral densities used in simulations without and with phase loop respectively at $E_k = 425$ MeV (start of blow-up).

3. Simulations in single RF

The effect of band-limited phase noise on emittance blow-up during acceleration ramp was simulated using the BLonD code [7]. The longitudinal space charge was carefully estimated at injection energy and then rescaled through the cycle with $\beta\gamma^2$, where β and γ are the relativistic parameters. The full PSB impedance model, consisting of 36 Finemet gaps, extraction kickers and cables, KSW kicker magnets, resistive wall and beam pipe step transitions, was included in simulations [2,9]. The Finemet resistive impedance is reduced by RF feedback. The effect of the feedback is included in simulations (as opposed to results presented in [10]) using two different models. The first consists in subtracting resonator impedances from the Finemet impedance at the revolution harmonics [11] to reproduce the measured notches (36 dB reduction), the second reduces the whole Finemet impedance at all frequencies by 11 dB (this value gives similar results for bunch stability during ramp). In the first case the values of f_s during blow-up are slightly higher (the resonators increase the total capacitive impedance), but the same parameters for blow-up were suitable in both cases.

The simulations were done for constant 8 kV and 16 kV voltages at h=1 with a realistic momentum program [12]. The 8 kV case shows no losses, but there is little time for a smooth blow-up. With 16 kV it was possible to increase the emittance up to the requested value of 3 eVs (Figure 3).

The noise was generated at regular intervals to follow the changes of the synchrotron and revolution frequencies during the blow-up interval (450-600) ms (f_{s0} decreases from 1.7 kHz to 0.9 kHz while f_0 increases from 1.4 MHz to 1.65 MHz). Indeed, since f_0 and f_{s0} vary significantly along the PSB cycle, ideally the noise should be often regenerated to follow these changes. On the other hand if the regeneration is done too often, then the resolution in frequency domain is too low and $S(f)$ will cover a band different than the chosen one. A good compromise was found regenerating the noise sample every 10000 turns.

The spectral density $S(f)$ was chosen constant in the band $[0.8\ f_{s0},\ f_{s0}]$. Indeed, for the examined intensity (LHC beams, N = 3.6×10^{12} ppb), choosing $f_{up} = f_{s0}$ allows to affect fully the bunch core since $f_s < f_{s0}$ during blow-up. The value $f_{down} = 0.8 f_{s0}$ was chosen examining

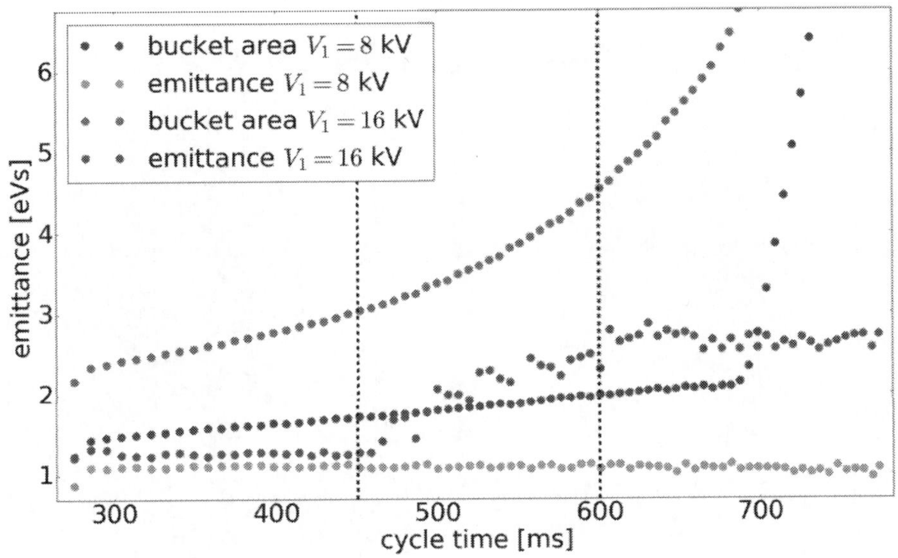

Figure 3. Emittance blow-up using $V_1 = 16$ kV. Small time margin with $V_1 = 8$ kV. The PSB injection and extraction times are at 275 ms and 775 ms. The bunch length (emittance) is measured at 5% of the line density.

the synchrotron frequency distribution with intensity effects during the blow-up time interval for the emittance increase from 2 eVs to 3 eVs. Then the noise amplitude at 450 ms, rescaled with f_{s0} during the blow-up to have the same noise strength ϕ_{rms}, was gradually risen to a value leading to the required emittance at 600 ms (S = 10^{-7}rad^2/Hz, Figure 2).

Figure 4 shows that intensity effects help for particle filamentation in phase space during and after controlled emittance blow-up with phase noise which produces islands in phase space modulating the bunch profile at PSB extraction.

4. Effect of phase and radial loops

In the simulations shown above, it was assumed that the noise is injected turn after turn directly in the main RF cavity. However, the operational scenario is that the noise will be introduced in the beam phase loop. This causes two complications. Firstly, the noise will be counteracted by phase loop. Additionally, the noise signal will only be sampled every 10 μs, the interval between two consecutive phase loop triggers [13]. This should not cause any issue since the sampling frequency of 100 kHz is well above the range of synchrotron frequencies (0.9-1.7) kHz during the blow-up.

The main goal of the phase loop is to damp the rigid-bunch dipole oscillations. This is usually done measuring at turn n the difference between the beam and designed synchronous phases $\Delta\phi$ and changing accordingly the RF frequency ω_{RF} by $\Delta\omega_{PL}$ at turn n+1. This passage from $\Delta\phi$ to $\Delta\omega$ is done since phase measurements could contain unwanted high frequency components that are filtered in this integration.

This frequency shift leads also to a change of the bunch orbit radius R from the design machine radius R_d. The aim of the radial loop is to maintain the orbit at the design one in the long run, reducing $|\Delta R| = |R - R_d|$. This is done giving a second contribution $\Delta\omega_{RL}$ to ω_{RF}. Thus, for each turn, $\Delta\omega_{RF} = \Delta\omega_{PL} + \Delta\omega_{RL}$, where the two contributions have usually opposite signs and $|\omega_{PL}| > |\omega_{RL}|$.

If the phase noise is introduced in the phase loop (as in the PSB), then its contribution will

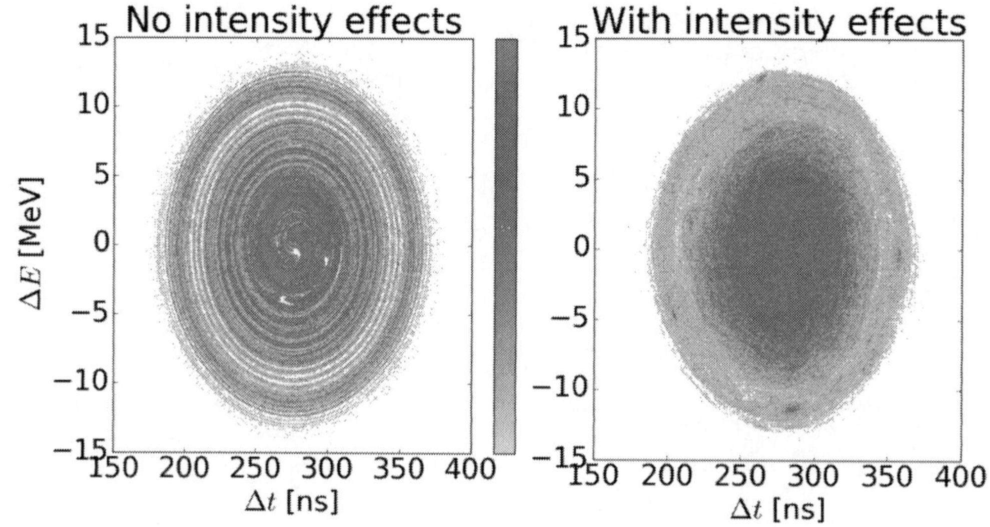

Figure 4. Bunch in longitudinal phase space after emittance blow-up at PSB extraction without intensity effects and with the full impedance model (Finemet impedance reduction with notches). The colour bar indicates the particles density.

be summed to $\Delta\phi$. The usually flat band-limited spectrum will then have a notch close to f_{s0} that can slow down the diffusion of the core. A compensation for this effect can be adopted changing the shape of the spectrum, as for PSB, where an exponentially growing $S(f)$ in the band $[f_{\text{down}}, f_{\text{up}}]$ was chosen (Figure 2).

The phase loop in the PSB is characterized by a Cascaded integrator-comb (CIC) filter [13] which yields

$$\Delta\omega_{PL}^{n+1} = a_1 \Delta\omega_{PL}^n + g_{PL}(b_0 \Delta\phi^n - b_1 \Delta\phi^{n-1}), \tag{1}$$

where $g_{PL} > 0$ is the gain and b_0, b_1 and a_1 are constants $\lesssim 1$. The value of g_{PL} has to be lower than f_0 to have a stable phase loop.

The relative error $\Delta R/R_d$ in the radial loop is filtered with a PI (Proportional-Integrator) corrector [13], and

$$\Delta\omega_{RL}^{n+1} = \Delta\omega_{RL}^n + (g_P + g_I)\left(\frac{\Delta R}{R_d}\right)^n - g_P\left(\frac{\Delta R}{R_d}\right)^{n-1}, \tag{2}$$

where g_P and g_I are positive gains. In reality ΔR is measured through transverse pick-ups in PSB. In simulations, ΔR was estimated assuming a constant magnetic field for all particles at a given turn

$$\frac{\Delta R}{R_d} = \frac{\Delta\omega_{RF}}{\omega_{d,RF}} \frac{\gamma^2}{\gamma_T^2 - \gamma^2}, \tag{3}$$

$\omega_{d,RF}$ and γ_T being the design RF frequency and the relativistic factor at transition respectively.

Starting from a 1 eVs bunch it was possible to blow up the emittance to 3 eVs increasing by four orders of magnitude the amplitude of the noise in f_{s0} compared to the case without loops, see Figure 2. The emittance evolution was similar to Figure 3. The phase and radial loop gains ($g_{PL} = 330$ 1/s, $g_P = 10^7$ rad/s, $g_I = 10^6$ rad/s) were chosen in such a way that $\Delta\omega_{PL}$ and $\Delta\omega_{RL}$ had a stable evolution during and after the blow-up interval (Figure 5).

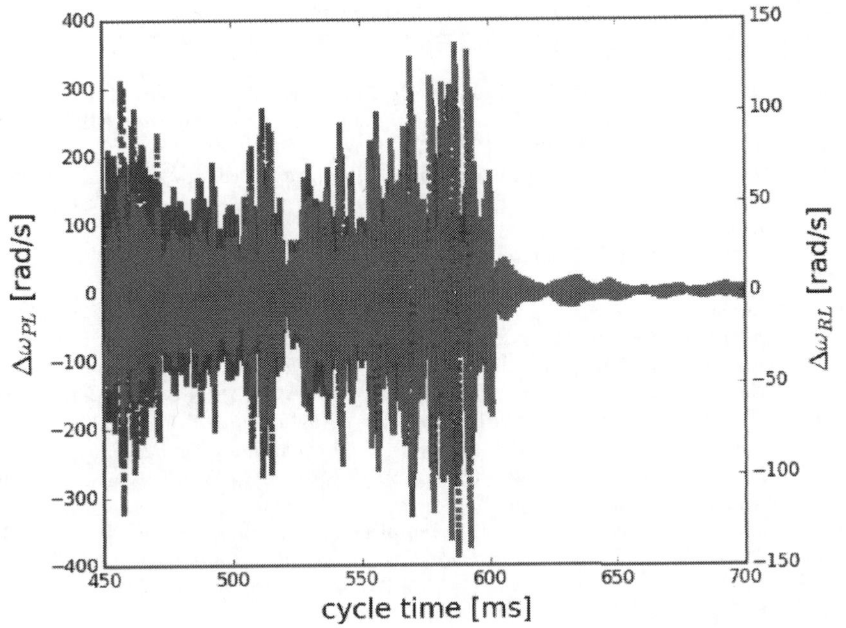

Figure 5. $\Delta\omega_{PL}$ (blue, left axis) and $\Delta\omega_{RL}$ (green, right axis) evolution during cycle time (450-700) ms. There is no injected noise after 600 ms, the loops are always on.

This simulation showed that the phase noise could be applied together with the phase loop and that the 10 μs sampling rate should not be an issue.

5. Conclusion

These studies have shown that bunch excitation with band-limited phase noise in the main RF system can provide significant controlled longitudinal emittance blow-up in a fast-cycling machine with strong space charge like the CERN PSB. For an optimised set of parameters it was possible to increase by factor 3 the emittance applying the noise only for 150 ms during the ramp. The requested blow-up was also achieved injecting the noise through the phase loop with limited sampling rate. This method, which could replace the currently used one based on phase modulation in a dedicated high harmonic RF system, will be tested in PSB in 2017 after minor low-level RF modifications.

Acknowledgments

Special thanks go to our CERN colleagues S. Albright, M. E. Angoletta, P. Baudrenghien, A. Butterworth, M. Jaussi, J. Molendijk and M. Paoluzzi for their precious support.

References

[1] Damerau H, Funken A, Garoby R, Gilardoni S, Goddard B, Hanke K, Lombardi A, Manglunki D, Meddahi M, Mikulec B, Rumolo G, Shaposhnikova E, Vretenar M and Coupard J 2014 *LHC Injectors Upgrade, Technical Design Report, Vol. I: Protons* Tech. Rep. CERN-ACC-2014-0337 URL `https://cds.cern.ch/record/1976692`

[2] Paoluzzi M *et al.* Design of the New Wideband RF System for the CERN PS Booster 2016 *Proc. of International Particle Accelerator Conference (IPAC'16), Busan, Korea, May 8-13, 2016* (*International Particle Accelerator Conference* no 7) (Geneva, Switzerland: JACoW) ISBN 978-3-95450-147-2 doi:10.18429/JACoW-IPAC2016-MOPMW024 URL `http://jacow.org/ipac2016/papers/mopmw024.pdf`

[3] Albright S 2017 private communication (CERN, Geneva, Switzerland)
[4] Toyama T Uniform bunch formation by RF voltage modulation with a band limited white signal 2000 *Nucl. Instrum. Meth.* A **447** 317–27
[5] Papotti G, Bohl T, Linnecar T, Shaposhnikova E and Tückmantel J 2008 *Study of Controlled Longitudinal Emittance Blow-Up for High Intensity LHC Beams in the CERN SPS* Tech. Rep. LHC-PROJECT-Report-1117 URL `https://cds.cern.ch/record/1123692`
[6] Baudrenghien P, Butterworth A, Jaussi M, Mastoridis T, Papotti G, Shaposhnikova E and Tückmantel J 2011 *Longitudinal emittance blow-up in the LHC* Tech. Rep. CERN-ATS-2011-034 URL `https://cds.cern.ch/record/1378464`
[7] CERN BLonD code URL `https://blond.web.cern.ch/`
[8] Tückmantel J Digital generation of noise-signals with arbitrary constant or time-varying spectra 2008 *Particle accelerator. Proceedings, 11th European Conference, EPAC 2008, Genoa, Italy, June 23-27, 2008* vol C0806233 URL `https://cds.cern.ch/record/1182456`
[9] Zannini C 2014 private communication (CERN, Geneva, Switzerland)
[10] Quartullo D, Albright S, Shaposhnikova E and Timko H CERN PS Booster Longitudinal Dynamics Simulations for the Post-LS2 Scenario 2016 *Proc. of ICFA Advanced Beam Dynamics Workshop on High-Intensity and High-Brightness Hadron Beams (HB'16), Malmö, Sweden, July 3-8, 2016* (Geneva, Switzerland: JACoW) ISBN 978-3-95450-185-4 doi:10.18429/JACoW-HB2016-MOPR028 URL `http://jacow.org/hb2016/papers/mopr028.pdf`
[11] Quartullo D, Albright S and Shaposhnikova E Studies of Longitudinal Beam Stability in CERN PS Booster After Upgrade 2017 *Proc. of International Particle Accelerator Conference (IPAC'17), Copenhagen, Denmark, May 14-19, 2017* (*International Particle Accelerator Conference* no 8) (Geneva, Switzerland: JACoW) ISBN 978-3-95450-182-3
[12] Albright S *PSB Magnetic Cycles to 1.4 and 2 GeV After LS2* (CERN EDMS, to be published)
[13] Angoletta M E, Blas A, Bento J, De Long J H, Findlay A, Matuszkiewicz P, Pedersen F and Salom-Sarasqueta A 2005 *PS Booster Beam Tests of the New Digital Beam Control System For LEIR* Tech. Rep. AB-Note-2005-017. CERN-AB-Note-2005-017 CERN Geneva URL `http://cds.cern.ch/record/832882`

IOP Conf. Series: Journal of Physics: Conf. Series **874** (2017) 012067 doi:10.1088/1742-6596/874/1/012067

Commissioning and first results of the Intense Beam EXperiment (IBEX) linear Paul trap

S L Sheehy[1,2], E J Carr[1], L K Martin[1], K Budzik[1,3], D J Kelliher[2], S Machida[2] and C R Prior[2]

[1] University of Oxford, Denys Wilkinson Building, Keble Rd, Oxford OX1 3RH, UK
[2] STFC, ISIS Neutron and Muon Source, Harwell, Didcot, OX11 0QX, UK

E-mail: suzie.sheehy@physics.ox.ac.uk

Abstract. The Intense Beam Experiment (IBEX) is a linear Paul trap designed to replicate the dynamics of intense particle beams in accelerators. Similar to the S-POD apparatus at Hiroshima University, IBEX is a small scale experiment which has been constructed and recently commissioned at the STFC Rutherford Appleton Laboratory in the UK. The aim of the experiment is to support theoretical studies of next-generation high intensity proton and ion accelerators, complementing existing computer simulation approaches. Here we report on the status of commissioning and first results obtained.

1. Introduction

The Intense Beam Experiment is a linear Paul trap similar to the Simulator of Particle Orbit Dynamics (S-POD) system at Hiroshima University. Using an electric rf quadrupole which focuses and defocuses a bunch of ions in time, the transverse dynamics of this system are equivalent to the strong focusing systems in quadrupole focusing channels of accelerators. The equivalence between this system and that of a linear focusing channel has previously been well established [1, 2, 3, 4]. This method of rf electric focusing produces equations of motion which take the form of Matthieu equations, analogous to those in a strong focusing accelerator channel.

Previous studies with linear Paul traps have addressed topics in high intensity beam dynamics such as halo formation and long term stability [5], parametric resonances [6] and resonance crossing [7, 8]. IBEX will address questions pertinent to next generation high intensity hadron accelerators, particularly those of fundamental interest in accelerator physics that are challenging to study in existing facilities due to the requirement for a wide parameter range or the combination of high intensity with intentional losses. For example, studies of integer and half integer resonance mechanisms at low and high intensity, lattice stability and collective effects and in the future implementations, tests of novel accelerator concepts such as non-linear integrable optics (NIO).

1.1. Experiment Overview

Ions are created *in situ* inside the vessel by means of electron ionisation of low pressure Argon gas.

[3] Currently at University of Warsaw, Krakowskie Przedmieście 26/28, 00-927 Warsaw, Poland.

1.1.1. Ion Number. The total number of ions N_i created from the electron gun per unit time is a function of the effective interaction length L, the electron current i.e. electron number per unit time dN_e/dt and the number density of the Argon gas. Assuming a pressure of 2.7×10^{-7} mbar, so $n_a = 6.678 \times 10^{15}\text{m}^{-3}$, an electron current of $100\,\mu$A, an effective interaction length of 5 mm which is the inscribed radius between the four rods and a cross section of $\sigma = 2.33 \times 10^{-16}\text{cm}^2$, Equation (1) gives 4.86×10^8 ions created per second.

$$\frac{dN_i}{dt} = \frac{dN_e}{dt} n_a L \sigma \tag{1}$$

Not all of the ions produced will be trapped, but this is more than sufficient given an expected ion number range of $10^5 - 10^7$ required to cover the range from low to high intensity for accelerator physics studies.

1.1.2. Ion Trapping. The produced Ar^+ ions are confined longitudinally along the trap length with a DC potential applied to two sets of shorter rods, referred to as the end caps. Ions are confined transversely using a 1 MHz sinusoidal focusing waveform. This is supplied using two synchronised channels from a LeCroy WS2052 arbitrary waveform generator, 180 degrees out of phase. The waveforms are amplified using Falco WMA300 amplifiers and applied to the two pairs of rods in the central region in quadrupole formation. In full operation mode, this waveform will also be applied to the end cap rods.

1.1.3. Ion Distribution. After trapping, the ions are left for a period of up to 50 ms to stabilise to an equilibrium distribution. It is useful to estimate the measurable signal expected on extraction from the trap. We estimate the current expected at the Faraday cup by considering the case where 10^6 ions leave the trap at the thermal velocity. From the experience of S-POD, a temperature of roughly 0.5 eV can be expected giving a thermal velocity v_{th} in the longitudinal direction of 1095 m s^{-1}. The time taken to extract is then $2L_{\text{trap}}/v_{\text{th}}$ where L_{trap} is the length of the Paul trap, since ions with initial velocity away from the Faraday cup will need to make a round trip before they escape the central region. The extraction time is then 136 μs. The total charge divided by this time results in a current of 1.2 nA. This is a very low current to measure using the Faraday cup, so to improve the signal quality of the nA level ion signal, a transimpedance amplifier was implemented with a factor of 1×10^6 to allow conversion the small current to a reasonable voltage that can be observed by an oscilloscope.

2. Commissioning

2.1. Vacuum and Gas Pressure

The vacuum vessel routinely achieves a UHV base pressure lower than 2.0×10^{-10} mbar after a bakeout period of 24 to 36 hours at 150°C. The vessel has ample space and many ports to allow future upgrades to the experiment. Gas flow into the vessel is achieved using a remote controlled VAT Series 59 variable leak valve. This works in combination with a regulated air-conditioned room temperature to ±2°C to control temperature and pressure fluctuations in the system which would otherwise affect ion number.

2.2. Electron Gun and Ionisation

The electron gun is a low cost thermionic gun from Tuoma (Japan). It is expected to produce a current of around 10 to $100\,\mu A$ depending on the filament current and grid voltages. It was modelled in CST studio, as shown in Figure 2, to ascertain appropriate grid voltage settings and then tested experimentally in a separate vacuum vessel with a phosphor screen to determine the output beam size. This established experimental control over the beam diameter from around $3 - 5$ mm.

Figure 1. Close up view of the IBEX Paul trap inside vacuum vessel. The electron gun points downward between the main quadrupole rods (75 mm in length, 11.5 mm diameter), beneath which sits the electron faraday cup. To the RHS is the faraday cup and the MCP detector will be mounted on the LHS.

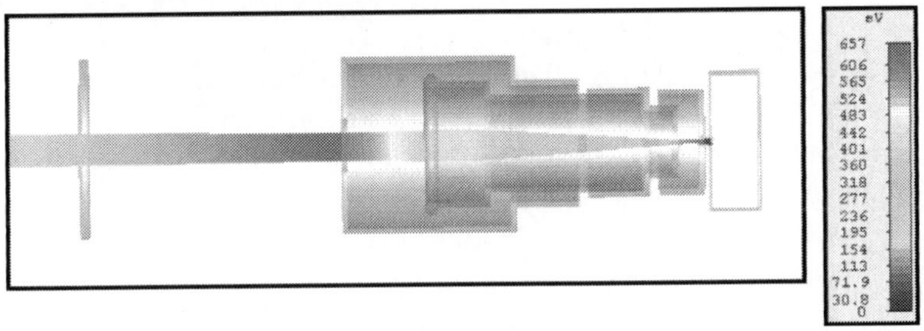

Figure 2. CST model of the IBEX electron gun in operating mode.

Once mounted in the IBEX vacuum vessel, the electron beam current can be measured by a small Faraday cup located beneath the rods. Note that even when the rods are unpowered, only a fraction of the beam current hits the Faraday cup, so the resulting measurement is lower than the full output current of the electron gun. Some electrons also hit the rods, so these can also be used like a Faraday cup to read electron current to provide additional insight. The electron gun faraday cup provides a relative measurement of the current for different grid voltage settings and is used to ensure the electron gun is operating consistently for each experimental run. Experimentally, it was found that the electron gun output parameters are dependent on the argon gas pressure in the vessel. An example measurement to optimise the gun output current as a function of filament current, for varying gas pressures, is shown in Figure 3. Note that the output current is reliably reproduced as long as consistent operating conditions are employed. In the operating region where the electron beam current changes rapidly with filament current, some systematic time-based drift in current can occur otherwise.

2.3. Confinement Waveforms and DC Potentials

The main 1 MHz confinement waveform is provided by a LeCroy Wavestation 2052. After confinement, a fast switching circuit drops the DC potential on one end cap to zero to extract ions. A base level DC voltage on the main rods provides an additional push longitudinally toward the faraday cup.

During initial commissioning, it was observed that an electrical connection exists between two of the end-cap rods and the vacuum vessel, which should be electrically isolated from ground.

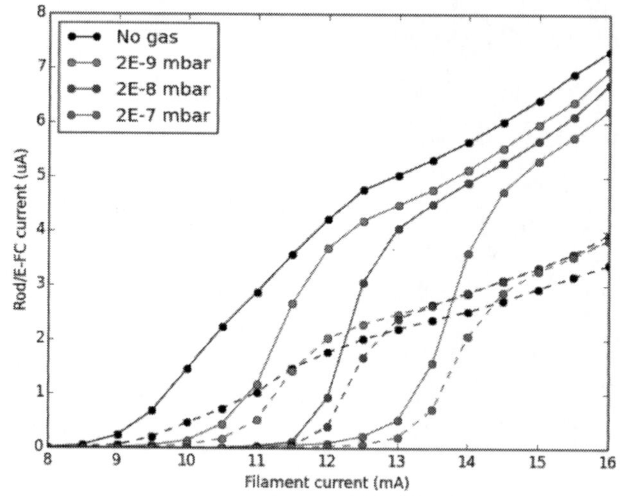

Figure 3. Variation of electron gun current as a function of filament current, measured on the electron faraday cup (dashed) and from the main rods (solid line). The output current varies with argon gas pressure.

While not preventing operation, this is expected to introduce some unwanted effects into the dynamics which reduces ion trapping efficiency and will be rectified in the near future.

2.4. Measurement of Ions on Faraday Cup

The two main diagnostics in the present system include the main Faraday cup for ion counting and at the opposing end an MCP detector and phosphor screen setup for imaging the ions directly. During the current commissioning phase only the Faraday cup was installed. First ions were trapped, extracted and observed in January 2017. Due to fast switching gate noise, the signal is determined from a difference between FC signals with and without argon gas. An example trace of the ion signal is shown in Figure 4.

The ion signal will vary depending on the operating point at which trapping occurs. The operating tune can be controlled using either the waveform amplitude (voltage) V_0 or frequency f, according to Equation (2).

$$\nu_0 = \frac{\sqrt{2}eV_0g}{\pi^3 m}\left(\frac{1}{fr_0}\right)^2 \tag{2}$$

where $r_0 = 5\,\text{mm}$ is the inscribed radius of the trapping region and g is a geometrical factor roughly equal to unity [9]. At each operating point a different number of ions will be confined stably, as shown in Figure 5. At some values one expects almost no ions (i.e. tune of 0.25) as the value is near to a resonance driven by any trap imperfections.

3. Discussion

The experimental commissioning has proceeded well. However, a few adjustments will be made to improve operation going forward. At present, there is noise detected on the FC from the fast switching of the end caps. In addition to the signal amplifier, adjustment of the position of the

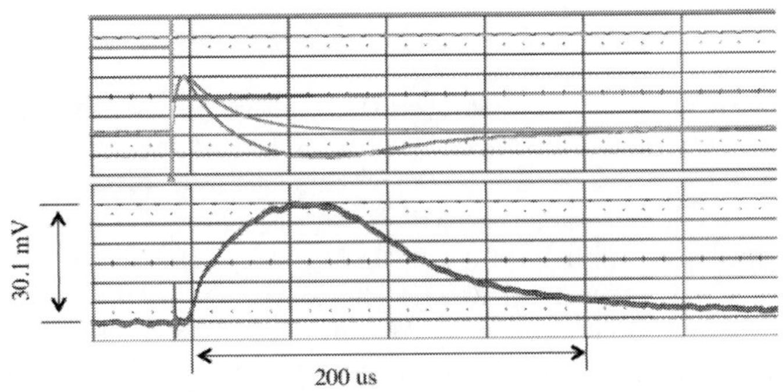

4in

Figure 4. Oscilloscope trace of ion signal on the Faraday cup after extraction from the IBEX Paul trap. The upper trace shows timing signal (blue) and signals in yellow without (upper) and with argon gas (lower), the difference between these gives the ion signal (pink) shown on the lower trace. The noise arising from the fast gate switch can also be observed. Note that 1 mV corresponds to 0.2 nA on the vertical axis.

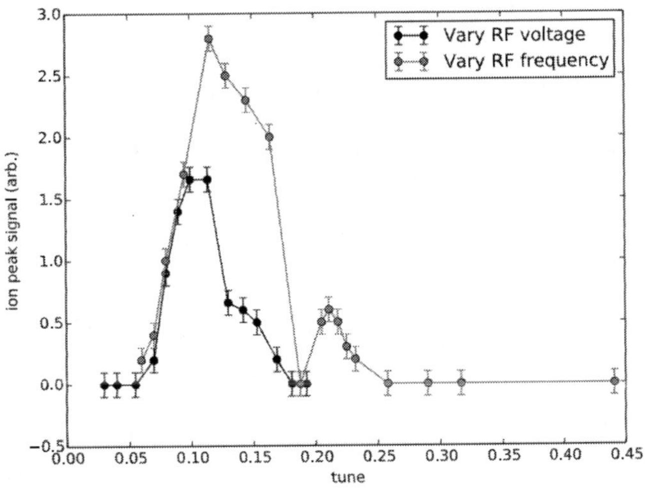

Figure 5. Variation of ion number with varying tune using either voltage or frequency to shift the operating point. The tune is the number of betatron oscillations per rf period (at 1MHz). Note that for the frequency (red) data, an additional DC bias was applied to all four rods, which increases the proportion of ions that reach the Faraday cup by providing a longitudinal velocity. Uncertainty is an estimate of oscilloscope measurement error.

faraday cup, additional shielding of cables and adjusting the switching speed should all improve this scenario to improve signal to noise ratio.

To provide an additional diagnostic and information of the transverse ion distribution, an MCP detector with phosphor screen is being installed. As the MCP gain is known to saturate for high ion number, an analysis method for correcting this and extracting ion number from the resulting images has been developed [10]. The MCP will be calibrated against ion signal from the faraday cup to determine ion number.

To stabilise the number of ions trapped during the ionisation period, the usual operating scenario will include trapping at one tune value, before shifting the experiment to a chosen working point for an experimental period prior to extraction. In this way the starting ion number can be made predictable between each run of an automated set of experiments, allowing a true comparison of ion number data. This scenario will be implemented in the near future. The LabView controls will be further developed to ensure the experiment is automated, such that large parameter variations can be reached without intervention from the experimentalist.

Acknowledgments
We would like to express our gratitude to our collaborators Prof. Okamoto and his group at Hiroshima University, to Prof. Iwashita from Kyoto University and Prof. Reichold from Oxford University for valuable and insightful discussions and to STFC Daresbury Laboratory Technology Division, ISIS Neutron and Muon Source and Oxford Physics Central Electronics for their contributions to this experiment.

References
[1] Okamoto H 2002 *Nucl. Instr. Meth.* A **485** 244-54
[2] Takai R *et al.* 2004 *Nucl. Instr. Meth.* A **532** 508-12
[3] Fukushima K, Ito K, Okamoto H *et al.* 2014 *Nucl. Instr. Meth.* A **733** 18-24
[4] Davidson R C, Qin H and Shvets G 2000 *Phys. Plasmas* **7** no. 3 1020
[5] Gilson E P *et al.* 2006 *Nucl. Instr. Meth.* A **544** 171-78
[6] Okamoto H and Yokoya K 2002 *Nucl. Instr. Meth.* A **482** 51-64
[7] Takeuchi H, Fukushima K, Ito K, Moriya K, Okamoto H, and Sugimoto H 2012 *Phys. Rev. ST Accel. Beams* **15** no. 7 074201
[8] Moriya K *et al.* 2015 *Phys. Rev. ST Accel. Beams* **18** 034001
[9] Kelliher D J *et al.* 2014 *in Proc. HB2014 workshop (East Lansing, USA)*, MOPAB26
[10] Ito K, Nakayama K, Ohtsubo S, Higaki H and Okamoto H 2008 *Jpn. J. Appl. Phys.* **47** 8017-25

Energy acceptance and on momentum aperture optimization for the Sirius project

P S Dester, F H Sá and L Liu

Brazilian Synchrotron Light Laboratory (LNLS), Campinas, Brazil

E-mail: liu@lnls.br

Abstract. A fast objective function to calculate Touschek lifetime and on momentum aperture is essential to explore the vast search space of strength of quadrupole and sextupole families in Sirius. Touschek lifetime is estimated by using the energy aperture (dynamic and physical), RF system parameters and driving terms. Non-linear induced betatron oscillations are considered to determine the energy aperture. On momentum aperture is estimated by using a chaos indicator and resonance crossing considerations. Touschek lifetime and on momentum aperture constitute the objective function, which was used in a multi-objective genetic algorithm to perform an optimization for Sirius.

1. Introduction

The adopted approach to optimize the non linear optics of Sirius storage ring is to use a Multi-Objective Genetic Algorithm, where the objective function is an estimator of on momentum aperture area and Touschek lifetime. Details about Sirius storage ring lattice can be found in [1] and [2]. On momentum aperture calculation is performed in key regions using chaos indicators, and energy acceptance calculation is performed using the concept of energy aperture (dynamic and physical). The Touschek lifetime is calculated through the energy acceptance. The idea is to have a fast algorithm to estimate the on momentum aperture area and Touschek lifetime, instead of performing a long and precise calculation, which in the case of the detailed lattice of Sirius takes around 7 hours. The goal is to take between 2 and 10 minutes per ring model. Therefore, we do not want to simulate lattice errors and correct them for each storage ring model tested, since it takes a considerable amount of time. Furthermore, for a lattice without errors, it is possible to take advantage of the five-fold symmetry in the Sirius optics, by simply using one superperiod for *tracking*. Regarding this strategy, care should be taken for 6D *tracking* to not change the synchrotron frequency nor the RF bucket. Our approach was to adjust the RF cavity frequency f_{RF}, so that $f_{\mathrm{RF}} = kf_0$, where k is the harmonic number and f_0 is the revolution frequency in a superperiod.

To rigorously test a ring model, a set of twenty lattices with different configuration errors is generated. For each lattice, orbit, tune, coupling and symmetry are corrected. Then, on and off momentum apertures are carefully determined for each one of them. Throughout the paper, we always refer to this test when introducing lattice errors; the average of the results represents the estimation.

See Table 1 for definition of notations and symbols used in the paper. 4D map implies that cavity and radiation are not considered. When not explicitly stated, the derivative is implied to

IOP Conf. Series: Journal of Physics: Conf. Series **874** (2017) 012068 doi:10.1088/1742-6596/874/1/012068

be with respect to the longitudinal position s. The majority of the calculations was performed with the MatLab Accelerator Toolbox.

Table 1. Notations and Symbols Used in the Paper

Symbol	Definitions/explanation
L	circumference (518.4 m)
s	$\in [0, L)$, longitudinal position
δ	energy deviation
τ	delay in respect to the synchrotron particle
β	energy dependant betatron function
$\boldsymbol{\nu}$	horizontal and vertical tunes
$\boldsymbol{x}$	$= (x, x', y, y', \delta, \tau)$, space coordinates
$\boldsymbol{x}_{co}^{\text{XD}}$	4D or 6D closed orbit coordinates
$\Phi_N^{\text{XD}}(\boldsymbol{x})$	N turns 4D or 6D map of a particle,
	$\boldsymbol{x}$ represents the initial coordinates
$\|\|\cdot\|\|_2$	is the euclidean norm

2. Energy Acceptance Calculation

Let us define physical aperture A_{phys} as the smallest invariant betatron amplitude for which a particle with energy deviation δ collides with the vacuum chamber. Then,

$$A_{\text{phys}}(\delta) = \min_{s\in[0,L)} \left\{ \frac{(X_{VC}(s) - x_{co}^{4\text{D}}(s,\delta))^2}{\beta_x(s,\delta)} \right\}, \tag{1}$$

where X_{VC} is the vacuum chamber half-width.

Dynamic aperture represents the smallest invariant betatron amplitude for which the particle will eventually be lost. For the case of Sirius, where we do not have an x-plane symmetry, the calculation of dynamic aperture A_{dyn} is performed as follows

$$A_{\text{dyn}}(\delta) = \beta_x(s_0,\delta) \max_{x'\in\Omega(\delta)} \left\{ (x' - x_{co}'^{4\text{D}}(s_0,\delta))^2 \right\}, \tag{2}$$

where $\Omega(\delta)$ is the greatest interval containing $x_{co}'^{4\text{D}}(s_0,\delta)$ and satisfying

$$\Omega(\delta) \subset \left\{ x' \in \mathbb{R} \mid ||\Phi_\infty^{6\text{D}}(x_{co}^{4\text{D}}, x', y_{co}^{4\text{D}}, y_{co}'^{4\text{D}}, \delta, \tau_{co}^{6\text{D}})||_2 < \infty \right\}$$

for a fixed longitudinal position s_0, which in our case it was chosen in a straight section. All the above 4D closed orbit coordinates depend on s_0 and δ. $||\Phi_\infty^{6\text{D}}(\boldsymbol{x})||_2 < \infty$ means that the particle trajectory does not diverge. In practice, we used $N = 131$ turns for a rough estimation and $N = 900$ for a more precise determination of A_{dyn}. The next step is to calculate the invariant betatron amplitude induced by Touschek scattering,

$$a_{\text{ind}}(s^*,\delta) = \frac{(\Delta x(s^*,\delta))^2}{\beta_x(s^*,\delta)} + \beta_x(s^*,\delta)\left(\Delta x'(s^*,\delta) - \frac{\beta_x'(s^*,\delta)}{2\beta_x(s^*,\delta)}\Delta x(s^*,\delta)\right)^2,$$

$\forall s^* \in [0, L)$, where $\Delta\boldsymbol{x}(s^*,\delta) = \boldsymbol{x}_{co}^{4\text{D}}(s^*,\delta) - \boldsymbol{x}_{co}^{6\text{D}}(s^*)$. A more detailed explanation of the equation used to calculate a_{ind} is discussed in [3]. For each s^*, we solve for δ,

$$a_{\text{ind}}(s^*,\delta) = \min\left\{ A_{\text{dyn}}(\delta), \min_{\delta'\in[-\delta,\delta]} \{A_{\text{phys}}(\delta')\} \right\}. \tag{3}$$

Equation (3) has two solutions $\delta_t^+(s^*)$ and $\delta_t^-(s^*)$, which are the positive and negative local transverse acceptances, respectively. The overall energy acceptance is given by

$$\delta_{\mathrm{acc}}^{\pm}(s) = \pm\min\{\delta_{\mathrm{RF}}, |\delta_{\mathrm{w}}^{\pm}|, |\delta_t^{\pm}(s)|\}, \quad s \in [0, L), \tag{4}$$

where δ_{RF} is the acceptance of the RF system and $\delta_{\mathrm{w}}^{\pm}$ is the energy deviation (positive and negative) for which the tune cross a resonance. An analogous approach to calculate energy acceptance is also used in [4]. Finally, Touschek lifetime is calculated by the method described in [5].

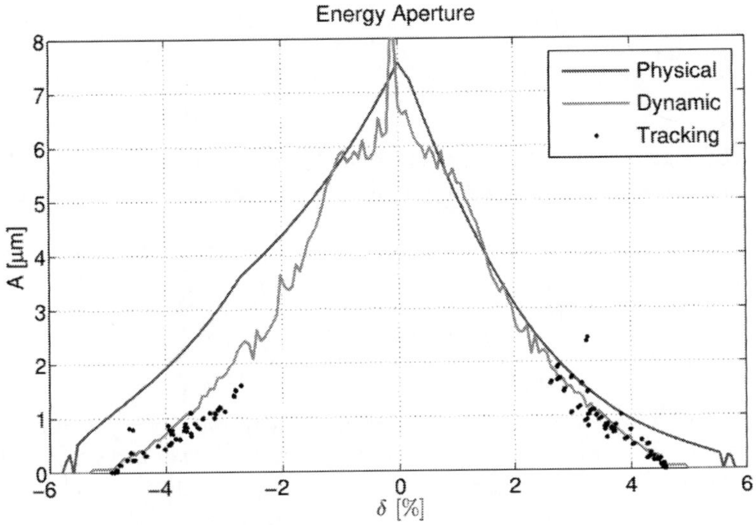

Figure 1. Dynamic and physical apertures for Sirius. The black dots correspond to the aperture calculated by tracking when lattice errors are introduced.

Figure 1 shows that the limiting aperture for Sirius is the dynamic aperture. Black dots represent the corresponding invariant betatron amplitude of a particle lost during *tracking* in the rigorous test with errors. In this case, the acceptance estimation is a good approximation of the Touschek acceptance. However, the method presented to calculate Touschek lifetime only guarantees an upper bound, which mostly is a satisfactory approximation.

3. On Momentum Aperture Calculation

For Sirius project, the horizontal plane aperture at the negative side is important, since injection will occur at $x \approx$ -8 mm [6]. We use chaos indicators to estimate this aperture. When introducing errors in the lattice, particles in regions where there is a significant trace of chaos are usually lost. In this section, we compare the performance of two chaos indicators, the well known Diffusion and the proposed one ASDR (Average Square Distance Ratio).

3.1. Diffusion

Diffusion is calculated using NAFF (Numerical Analysis of Fundamental Frequencies) [7], a fast and precise algorithm to calculate fundamental frequencies of a motion. As it is explained in [8], let us denote NAFF by the operator

$$\begin{aligned}\mathcal{F}_N :\quad & \mathbb{R}^6 \longrightarrow \mathbb{R}^2 \\ & \boldsymbol{x}_0 \longmapsto \boldsymbol{\nu} - \lfloor \boldsymbol{\nu} \rfloor\end{aligned}$$

This operator calculates the fractional part of the horizontal and vertical tune, based on 4D *tracking* of N turns in the ring, starting from coordinate $\boldsymbol{x}_0$. The diffusion vector is defined as

$$\boldsymbol{D} = \mathcal{F}_N(\boldsymbol{x}_0) - \mathcal{F}_N\left(\Phi_N^{4D}(\boldsymbol{x}_0)\right). \tag{5}$$

A chaos indicator is, then, given by $||\boldsymbol{D}||_2$. We observed that $||\boldsymbol{D}||_2 > 10^{-4}$ represents a significant probability of losing the particle, when lattice errors are introduced.

3.2. ASDR

Let $\{\boldsymbol{x}_{i,0}\}_{1\le i\le M}$ be a set of initial conditions and $\boldsymbol{x}_{i,n} = \Phi_n^{4D}(\boldsymbol{x}_{i,0})$. Then, we define

$$\left\langle \Delta\zeta_{i,n}^2 \right\rangle_{0\le n\le N} = \frac{1}{N+1}\sum_{n=0}^{N}\left(\zeta_{i+1,n} - \zeta_{i,n}\right)^2,$$

for $1 \le i < M$, where ζ represents any space coordinate in $\mathcal{R} = \{x, x', y, y'\}$. For example, $\zeta = x$, gives $x_{i,n}$, which is the horizontal position coordinate of the vector $\boldsymbol{x}_{i,n}$. Finally, the ASDR indicator is given by

$$\text{ASDR}_i^2 = \frac{1}{4\,\#(\mathcal{R})}\sum_{\zeta\in\mathcal{R}}\frac{\left\langle \Delta\zeta_{i,n}^2\right\rangle_{0\le n\le 2N}}{\left\langle \Delta\zeta_{i,n}^2\right\rangle_{0\le n\le N}}, \tag{6}$$

where $\#(\mathcal{R}) = 4$ is the cardinality of $\mathcal{R}$. To simplify the analysis, let us consider only the horizontal dynamic, i. e., $\mathcal{R} = \{x, x'\}$. If (x, x') is of the form $(A_i\cos(\omega_i n+\phi_i), -A_i'\sin(\omega_i n+\phi_i))$, then one can show that

$$\text{ASDR}_i^2 = 1 - \frac{1}{4N} + o\left(N^2\Delta\omega_i^2, \frac{1}{N}, \frac{\Delta A_i}{A_i}, \frac{\Delta A_i'}{A_i'}, \Delta\phi_i\right),$$

where $\Delta A_i = A_{i+1} - A_i$, $\Delta A_i' = A_{i+1}' - A_i'$, $\Delta\omega_i = \omega_{i+1} - \omega_i$ and $\Delta\phi_i = \phi_{i+1} - \phi_i$. Notice that we must have the successive initial conditions close to each other, in particular $\Delta\omega_i << 1/N << 1$; and β_x' must be small, in order to obtain a $\pi/2$ phase shift between x and x', which allow us to cancel oscillating terms. Furthermore, $\omega_i/(2\pi)$ must be far from any integer, in order to the oscillating terms remain small. It is important to note that, if we are in a region where the frequency shift $\Delta\omega_i << (\Delta A_i/A_i)/N$, then $\text{ASDR}_i^2 \approx 1/4$.

Analogous reasoning can be done when added the vertical dynamics. For a well-behaved motion, it is expected that $1/4 < \text{ASDR}_i^2 < 1$. Therefore, for sufficiently close initial conditions, $\text{ASDR}_i^2 > 1$ indicates a motion different from the one proposed, which suggests a chaotic behavior.

3.3. Comparison

We have compared the performance of chaos indicators over 62 ring models, on which we had already performed the rigorous test including errors. Figure 2 shows the Relative Mean Square Error (RMSE) of the indicator as a function of the chosen *threshold*. As expected, if the chaos indicator *threshold* is too small or too large, then the prediction fail.

For both indicators the number of turns per initial condition was $N = 130$ and the step between initial conditions for x^- was 20 m and for δ^- was $8.5 \cdot 10^{-5}$.

For this comparison, indicator ASDR with a relative error of approximately 10% at *threshold* around 1.05 performs better than Diffusion with a relative error of approximately 15% at *threshold* around 10^{-4}. In fact, we observed that for these *thresholds*, ASDR generally predicts a smaller aperture, which is more precise than Diffusion.

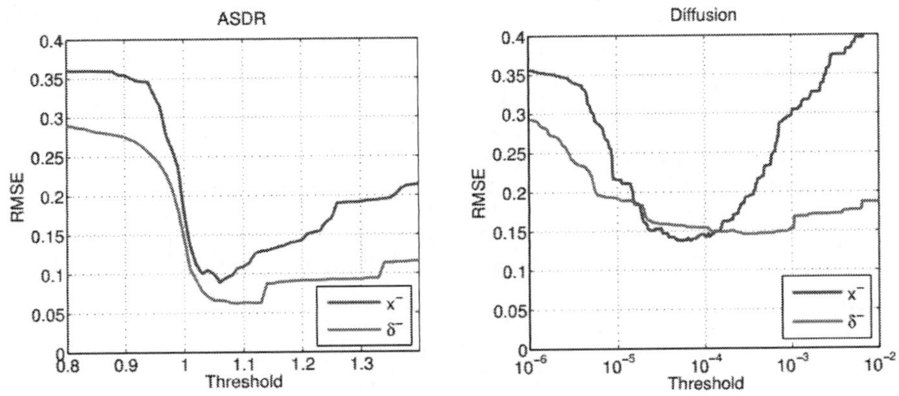

Figure 2. Blue and green curves correspond to RMSE of horizontal and energy negative apertures, respectively.

4. Optimization Results

The described methods to calculate energy acceptance and on momentum aperture are sufficient to form an objective function, which calculates Touschek lifetime τ_{TC} and on momentum aperture area $A_{x\times y}$, as a function of the ring model. This function takes between 2 to 10 minutes per model, while the standard technique takes around 7 hours. Let us represent the objective function as

$$f: \quad \mathbb{R}^{N_S+2} \longrightarrow \mathbb{R}^2$$
$$(\boldsymbol{S}, \boldsymbol{\nu}) \longmapsto (\tau_{TC}, A_{x\times y}),$$

where $\boldsymbol{S}$ is the vector of strength of sextupole families, $N_S = \dim(\boldsymbol{S})$ (In case of Sirius, $N_S = 21$) and the quadrupole strengths are calculated in order to reach the specified tunes $\boldsymbol{\nu}$ with minimal modification. Singular Value Decomposition (SVD) was used to perform this task. This objective function is used in a Multi-Objective Genetic Algorithm (MOGA), which in our case was NSGA-II. A detailed explanation of this algorithm is given in [9].

To enhance search time, some restrictions are imposed, which are much faster to calculate than the objective-function. The feasible region was restricted to model rings with small positive chromaticity and a minimum upper bound for Touschek lifetime, which is calculated using only the physical aperture, see Eq. (3). The gain is to provide a fast way to avoid rings which does not have the minimum lifetime required.

At certain point in the optimization of Sirius, NSGA-II was exploring solutions for which some *driving terms* (see [10]) were being increased to enhance on momentum aperture (h_{21000}, h_{30000}) or energy acceptance (h_{10110}), however when lattice errors were added, the particles did not survive. Therefore, these *driving terms* were added as constraints to remain smaller than a certain quantity. Thus, avoiding this kind of problems. Another strategic constraint we have added was a specific tune resonance of fourth order for the x-plane, which aided to predict the on momentum aperture, along with the chaos indicators.

The best results of the optimization, which were submitted to the rigorous test with errors are listed in Table 2, where we used a multi-bunch uniform filling operation mode, with 1% coupling and total current of 100 mA. Despite model R11G70M024 having a gain of 2.3 hours in Touschek lifetime, the loss in on momentum aperture may be risky for injection. The most safe result is model R11G16M019, where we have a significant gain in every aspect and a small standard deviation.

Table 2. Performance of Models

Model	$A_{x\times y}$ [mm^2]	$x_{\min}$ [mm]	τ_{TC} [h]
original	40.0 ± 2.6	-10.0 ± 0.4	12.2 ± 0.4
R11G16M019	43.6 ± 1.6	-10.2 ± 0.4	13.4 ± 0.4
R11G23M017	42.0 ± 2.2	-10.4 ± 0.3	13.3 ± 0.4
R11G70M024	37.3 ± 2.3	-9.5 ± 0.8	14.5 ± 0.6
R13G38M018	38.9 ± 2.0	-11.0 ± 0.7	11.5 ± 0.7

5. Conclusion

Using the energy acceptance and on momentum aperture methods, we only had an upper bound estimation of the parameters. However, adding constraints related to *driving terms* and resonances, enhanced the methods to be reasonable approximations.

The best ranked machines from the optimization were selected to perform a rigorous test with lattice errors. Then, we could detect the best configuration models among them. These results are shown in Table 2.

The optimization method still needs improvement. The next step is to study the possibility of using another optimization algorithm with faster convergence than NSGA-II, for example the algorithm presented in [11].

Acknowledgments

We thank Ximenes R. Resende for the suggestions and fruitful discussions; the reviewers for improving the readability of the paper and the members of the LNLS Mechanical Design Group for conceding the spare time of their computers to our calculations.

References

[1] Liu L, Sá F and Resende X 2016 *7th International Particle Accelerator Conference (IPAC'16)* (Busan, Korea) pp 3413–3416
[2] Sá F, Liu L and Resende X 2016 *7th International Particle Accelerator Conference (IPAC'16)* (Busan, Korea) pp 3409–3412
[3] Nadji A *et al.* 1997 *Particle Accelerator Conference* vol 2 (IEEE) pp 1517–1519
[4] Tsai H *et al.* 2011 *IPAC'10* pp 2621–2623
[5] Piwinski A 1999 *arXiv preprint physics/9903034*
[6] Liu L, Resende X, Rodrigues A and Sá F 2016 *7th International Particle Accelerator Conference (IPAC'16)* (Busan, Korea) pp 3406–3408
[7] Laskar J 1999 *Hamiltonian systems with three or more degrees of freedom* (Springer) pp 134–150
[8] Papaphilippou Y 2014 *Chaos: An Interdisciplinary Journal of Nonlinear Science* **24** 024412
[9] Deb K, Pratap A, Agarwal S and Meyarivan T 2002 *IEEE transactions on evolutionary computation* **6** 182–197
[10] Dragt A J 1997 *Lie methods for nonlinear dynamics with applications to accelerator physics* (University of Maryland, Center for Theoretical Physics, Department of Physics)
[11] Huang X and Safranek J 2015 *Physical Review Special Topics-Accelerators and Beams* **18** 084001

Resistive-wall impedance effects for the new KEK Light Source

N Nakamura[1,2]

[1] High Energy Accelerator Research Organization (KEK), Tsukuba, Japan
[2] The Graduate University for Advanced Studies (SOKENDAI), Tsukuba, Japan

E-mail : norio.nakamura@kek.jp

Abstract. Effects of resistive-wall (RW) impedance on a 3-GeV storage ring of the KEK Light Source (KEK-LS) are presented. Cu sheets used for in-vacuum undulators (IVUs) are regarded as the main source of the RW impedance. Although the calculated heating power per unit length due to the longitudinal impedance is more than 20 W, it is not serious for the water-cooled IVUs. The maximum growth rate of the coupled-bunch instability caused by the transverse impedance is calculated and as a result, a transverse feedback system with the damping rate of more than 10^4 s^{-1} is required for the instability suppression. NEG coating of about 1 μm can be used for the KEK-LS vacuum pipe because increase of the heating power due to the NEG coating impedance is small and the effect on the transverse coupled-bunch instability is negligible.

1. Introduction

The KEK-LS is a 3-GeV storage ring of 20-cell HMBA (Hybrid Multi-Bend Achromat) lattice [1], which is planned to be constructed as a successor of the two existing Photon Factory storage rings (PF ring and PF-AR) in the KEK Tsukuba Campus. In this ring, a lot of IVUs with a small magnetic gap (4 mm at minimum) will be installed and Cu vacuum pipes of a small aperture (12.5 mm in inner radius) will be normally used. In addition, NEG coating, having a low electric conductivity, will be utilized for the Cu vacuum pipe to ensure a sufficient beam lifetime early in the machine commissioning. In this paper, the heating power due to the longitudinal RW impedance and the maximum growth rate of coupled-bunch instability caused by the transverse RW impedance are calculated and the effects of the RW impedance on KEK-LS are discussed.

2. Heating due to longitudinal impedance

The longitudinal RW impedance of a flat or circular metal pipe with the pipe wall thicker than the skin depth δ is approximately given by

$$Z_\ell(\omega) = \frac{\omega Z_0 \delta(\omega) L}{4\pi bc}\{\mathrm{sgn}(\omega) - i\} \tag{1}$$

$$\delta(\omega) = \left(\frac{2}{\sigma_c \mu_0 |\omega|}\right)^{1/2}. \tag{2}$$

Here ω, c, b, L, σ_c, μ_0, and Z_0 are angular frequency, the velocity of light, the inner radius for a circular pipe or half gap for a flat pipe, length, electric conductivity and permeability of the pipe and the characteristic impedance of vacuum. The permittivity and permeability of the metal are assumed to be equal to those of vacuum. A flat pipe corresponds to the IVU Cu sheet and a circular pipe to the Cu vacuum pipe. The heating power of the resistive-wall pipe is given by

$$P_{RW} = k_{loss} Q_b^2 f_b = \frac{k_{loss} I_{total}^2}{f_b} , \tag{3}$$

where k_{loss}, I_{total}, Q_b, and f_b are the loss factor, total stored current and bunch charge and frequency. For a Gaussian bunch with the bunch length σ_t, the loss factor is expressed by

$$k_{loss} = \frac{1}{\pi}\int_0^{\infty} \mathrm{Re}\{Z_\ell^{eff}(\omega)\}d\omega \tag{4}$$

$$Z_\ell^{eff}(\omega) \equiv Z_\ell(\omega)\cdot \exp\left\{-(\sigma_t\omega)^2\right\} . \tag{5}$$

Here Z_ℓ^{eff} is the longitudinal effective impedance, which includes the effect of the longitudinal bunch profile. The loss factor for the impedance of equation (1) is obtained by use of equations (4) and (5) and given by the following expression:

$$k_{loss} = \frac{Z_0}{8\pi^2 bc}\Gamma\left(\frac{3}{4}\right)\left(\frac{2}{\sigma_c\mu_0}\right)^{1/2}\sigma_t^{-3/2} . \tag{6}$$

The heating power per unit length is calculated for the IVU Cu sheet with a minimum gap of g_u=2b=4 mm and the circular Cu pipe with a inner radius of b=12.5 mm.

$$\begin{aligned} P_{RW}/L &= 25.3[\mathrm{W/m}]\,(\text{IVU Cu sheet}) \\ &= 4.05[\mathrm{W/m}]\,(\text{Cu pipe}) \end{aligned} \tag{7}$$

In equation (7), the electric conductivity of copper is 5.9x10^7 Ω^{-1}m^{-1} and the bunch length is assumed to be 10 ps for the KEK-LS [1]. The total stored current and the bunch frequency are 500 mA and 500 MHz in the multi-bunch operation where all the bunches are identical and the bunch number k_b is equal to the harmonic number of 952. The relatively high heating of the IVU Cu sheet requires a water cooling with the same power as in the PF ring.

3. Coupled-bunch instability due to transverse impedance

The transverse RW impedance of a flat metal pipe with the pipe wall thicker than the skin depth δ is approximately given by

$$Z_t(\omega) = \frac{\pi Z_0\delta(\omega)L}{16b^3}\{\mathrm{sgn}(\omega) - i\} . \tag{8}$$

In the KEK-LS ring, 5m-class and 1m-class IVUs are installed in the long and short straight sections respectively.

The growth rate of the coupled-bunch instability in the lowest mode (the rigid dipole mode) is given by [2]

$$g_\mu^t = -\frac{eI_{total}\omega_0\beta_t}{4\pi E}\cdot\sum_{p=-\infty}^{\infty}\mathrm{Re}\left[Z_t^{eff}(\omega_p^t)\right]_\mu \tag{9}$$

$$Z_t^{eff}(\omega) \equiv Z_t(\omega)\cdot\exp\left\{-(\sigma_t\omega)^2\right\} \tag{10}$$

$$\omega_p^t = \left(pk_b + \mu + \nu_t\right)\omega_0,\ \omega_0 = 2\pi f_0 \tag{11}$$

Here E, f_0, ν_t and β_t are the beam energy, revolution frequency, betatron tune and betatron function at the impedance source. For k_b bunches, there are k_b coupled-bunch oscillation modes characterized by a number μ=0, 1, …, (k_b-1) specifying the phase shift between bunches. The chromaticity is assumed to be zero here.

Figure 1 shows the maximum growth rates of a 5-m IVU in the long straight section with the averaged betatron function of 3.25 m and a 1-m IVU in the short straight section with the averaged betatron function of 1.39 m as a function of vertical betatron tune. The transverse RW impedance dependence on frequency results in the maximum growth rate to be inversely proportional to the square root of 1-$\Delta\nu_y$, with $\Delta\nu_y$ the fractional part of the vertical betatron tune. The sum of the two maximum growth rates is 1050 s^{-1} for 1 cell and 21000 s^{-1} for 20 cells at the present vertical betatron tune of 17.62. Therefore a transverse feedback system with a damping rate of more than 10000 s^{-1} is needed to suppress the transverse coupled-bunch instability. Such a fast transverse feedback system is operated in the PF ring for suppression of the beam instability due to ion trapping [3]. Increasing the chromaticity reduces the coupled bunch instability growth rate, but can increase higher modes instabilities grow rate, which cannot be damped by the feedback system.

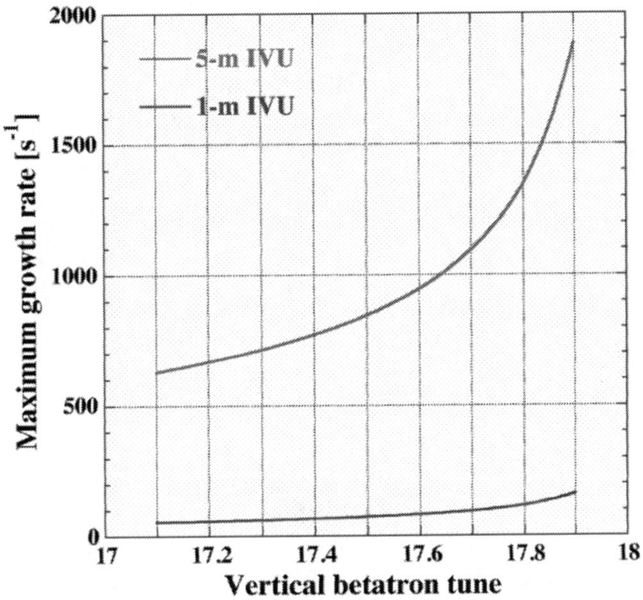

Figure 1. Maximum growth rate of the transverse instability due to a 5-m IVU in the long straight section and a 1-m IVU in the short straight section for the magnetic gap of 4 mm and the total current of 500 mA as a function of vertical betatron tune.

4. Effects of NEG coating

Longitudinal resistive-wall impedance of a two-layer circular metal pipe should be used to evaluate the effect of NEG coating on the resistive-wall heating of the KEK-LS vacuum pipe and can be expressed analytically as equation (12) [4].

$$Z_\ell(\omega) = \frac{-iL}{2\pi\varepsilon_0 bc\left\{\left(\frac{\omega}{\lambda_1 c} + \frac{\lambda_1 c}{\omega}\right)\alpha_{l2} - \frac{b\omega}{2c}\right\}}. \tag{12}$$

$$\alpha_{12} = \frac{J_1(\lambda_1 b) + \kappa N_1(\lambda_1 b)}{J_0(\lambda_1 b) + \kappa N_0(\lambda_1 b)}$$

$$\lambda_{1,2} = \frac{i + \mathrm{sgn}(\omega)}{\delta_{1,2}}, \quad \delta_{1,2} = \left(\frac{2}{\sigma_{1,2}\mu_0|\omega|}\right)^{1/2}$$

$$\kappa = \frac{\left(\frac{\omega}{\lambda_1 c} + \frac{\lambda_1 c}{\omega}\right) H_0^{(1)}(\lambda_2(b+d)) J_1(\lambda_1(b+d)) - \left(\frac{\omega}{\lambda_2 c} + \frac{\lambda_2 c}{\omega}\right) H_1^{(1)}(\lambda_2(b+d)) J_0(\lambda_1(b+d))}{\left(\frac{\omega}{\lambda_2 c} + \frac{\lambda_2 c}{\omega}\right) H_1^{(1)}(\lambda_2(b+d)) N_0(\lambda_1(b+d)) - \left(\frac{\omega}{\lambda_1 c} + \frac{\lambda_1 c}{\omega}\right) H_0^{(1)}(\lambda_2(b+d)) N_1(\lambda_1(b+d))}$$

Here $b, L, d, \sigma_{1,2}$ and $\delta_{1,2}$ are the inner radius and length of the pipe, the thickness of the inner layer, the electric conductivities and the skin depths of the inner and outer layers. The thickness of the outer layer is assumed to be infinite or thicker than the skin depth. J_0, J_1, N_0, N_1, $H_0^{(1)}$ and $H_1^{(1)}$ are the 0th-order and 1st-order Bessel functions of the 1st and 2nd kinds and the 0th-order and 1st-order Hankel functions of the 1st kind. The inner and outer layers correspond to the NEG coating and the Cu pipe.

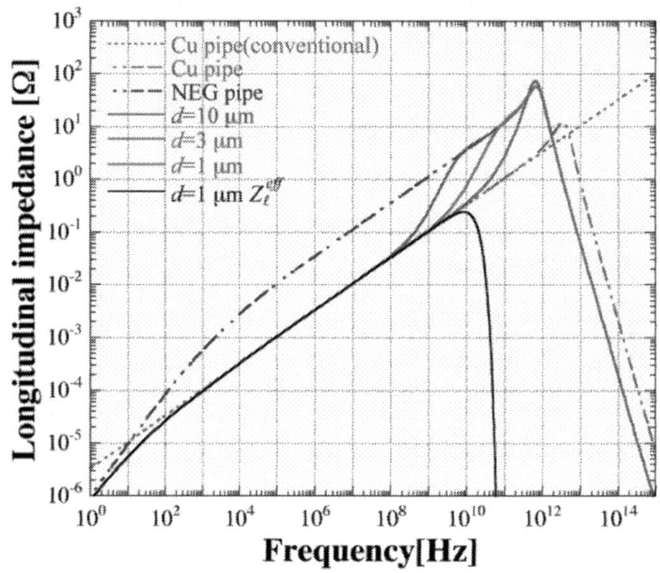

Figure 2. Longitudinal impedance per unit length of NEG-coated Cu pipes (d=1, 3 and 10 μm) with b=12.5 mm and effective impedance of the NEG-coated Cu pipe (d=1 μm) shown by solid lines. The broken lines show longitudinal impedances for pure Cu and NEG pipes with b=12.5 mm and the conventional expression of equation (1) for the pure Cu pipe.

Figure 2 shows the real part of the resistive-wall impedances of NEG-coated Cu pipes calculated from equation (12). The DC electric conductivity of NEG is assumed to be $5\times10^5\ \Omega^{-1}\mathrm{m}^{-1}$ [5]. The effective longitudinal impedance in equation (5) for the NEG-coating thickness of 1 μm and the bunch length of 10 ps is also shown in the same figure. The broken lines show longitudinal impedances for pure Cu and NEG pipes with the same inner radius and the conventional expression of equation (1) for the pure Cu pipe.

These results show that the longitudinal impedances of the NEG-coated Cu pipes deviate at high frequencies from that of the pure Cu pipe(d=0 μm) and approach to that of the pure NEG pipe ($d\rightarrow\infty$) around the frequency f_d where the skin depth of NEG is equal to the NEG coating thickness. However the effective impedance of the NEG-coated Cu pipe with d=1 μm is not so changed from that of the pure Cu pipe because the bunch length of 10 ps is longer than $1/2\pi f_d$ and consequently the loss factor and heating power are expected to only slightly change.

Figure 3 shows the calculated heating power per unit length as a function of NEG-coating thickness. The heating power significantly increases above the NEG thickness of ~ 2 μm and saturates around 10 μm. On the other hand, as expected from the effective impedance in figure 2, the heating power for NEG coating thickness of 1 μm is almost unchanged and is higher by only less than 10 percent than that of the uncoated Cu pipe (d=0 μm).

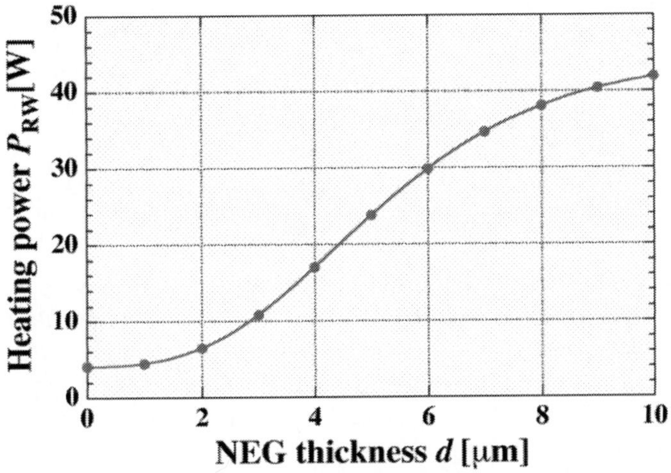

Figure 3. Heating power per unit length of the NEG-coated Cu pipe with b=12.5 mm as a function of NEG thickness d.

The transverse impedance of a two-layer circular pipe can also be obtained with an analytical expression [6], not shown here for lack of space. Figure 4 shows the real part of the calculated transverse impedances of NEG-coated Cu pipes (d=1, 3 and 10 μm) and the effective impedance for d=1 μm and σ_t=10 ps. As for the longitudinal impedance, the transverse impedances of the NEG-coated pipes change from the one of the pure Cu pipe to the one of the pure NEG pipe at frequencies around f_d. The effective impedance of the NEG-coated Cu pipe with the coating thickness of 1 μm remains almost unchanged from that of the uncoated Cu pipe because of the relatively long 10 ps bunch length. Moreover the dominant frequency for the growth rate of the transverse RW coupled-bunch instability is $(1-\Delta\nu_y)f_0$ and much lower than f_d. Therefore the effect of NEG coating of 1 μm at least on the transverse coupled-bunch resistive-wall instability is neglected.

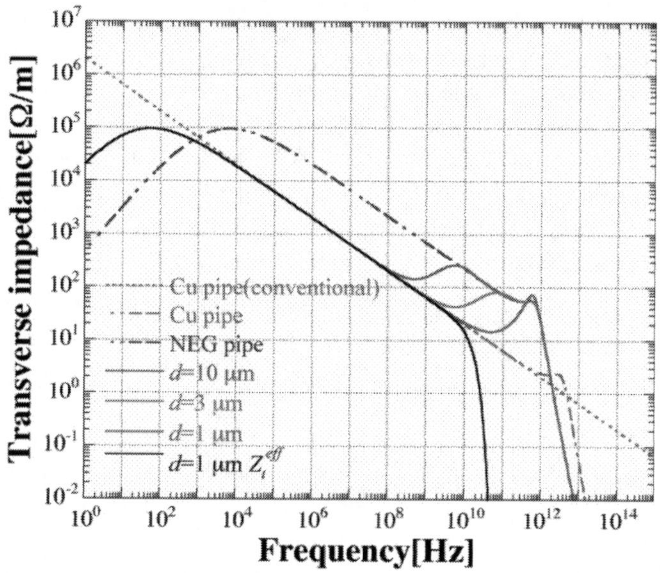

Figure 4. Transverse impedance per unit length of NEG-coated Cu pipes (*d*=1, 3 and 10 μm) with *b*=12.5 mm and effective impedance of the NEG-coated Cu pipe (*d*=1 μm) shown by solid lines. The broken lines show transverse impedances for pure Cu and NEG pipes with *b*=12.5 mm and the conventional expression for the pure Cu pipe.

5. Conclusions

The heating power per unit length reaches about 25 W at the IVUs of KEK-LS. However it is tolerable in the normal multi-bunch operation for the water-cooled IVU system. From the calculated maximum growth rate of the transverse coupled-bunch instability, a transverse feedback system with the maximum damping rate of higher than 10^4 s^{-1} is needed for suppressing the instability. NEG coating of about 1 μm or less can be applied to the vacuum pipe of the KEK-LS ring to achieve ultra high vacuum against the small aperture of the vacuum pipe, because the consequent increase of the heating power and of the instability growth rate can be considered negligible.

References

[1] Harada K *et al.* 2016 *Proc. IPAC2016* (Busan Korea) pp 3251-3253
[2] Zisman M S, Chattopadhyay S and Bisognano J J 1986 *LBL-21270/UC28*
[3] Takai R *et al.* 2009 *Proc. DIPAC2009* (Basel Switzerland) pp 59-61
[4] Nakamura N 2009 *Proc. ERL2009* (Ithaca USA) pp 85-89
[5] Koukovini-Platia E *et al.* 2015 *Proc. IPAC2015* (Richmond USA) pp 3097-3099
[6] Nakamura N unpublished

Approximating nonlinear forces with phase-space decoupling

B Folsom[1,2], **E Laface**[1,2]
[1]Lund University, Particle Physics Division, Lund, Sweden
[2]European Spallation Source ERIC, Lund, Sweden

E-mail: ben.folsom@esss.se

Abstract. Beam tracking software for accelerators typically falls into two categories: fast envelope simulations limited to linear beam optics, and slower multiparticle simulations that can model nonlinear effects. To find a middle ground between these approaches, we introduce virtual coordinates in position and momentum which have a cross-dependency (i.e. $p^* = f(x_0)$ where x_0 is an initial position and p^* is a virtual projection of momentum onto the position axis). This technique approximates multiparticle simulations with a significant reduction in calculation cost.

1. Introduction

The software for predicting and correcting beam dynamics in real time is largely dependent on approximations that treat the phase-space density as a single envelope [1, 2], while multiparticle tracking codes that can account for nonlinear forces are more CPU-intensive and only suitable for machine design or offline diagnostics [3, 4]. In developing the ESS Linac Simulator (ELS), our group intends to incorporate nonlinear tracking without sacrificing real-time diagnostics capability.

Expanding on prior works [5, 6], we introduce a multiparticle optimization method by using "virtual" phase-space coordinates that allow for the independent calculation of position and momentum densities. It is important to classify these coordinates as non-physical: they are derived assuming a cross-dependency exists (e.g. $p^* = f(x_0)$, where p^* and x_0 are virtual momentum and real position, respectively).

When used in conjunction with standard techniques for multiparticle tracking, these virtual coordinates allow for the use of monovariate polynomials, which present a significant reduction in the number of required calculations when compared with the bivariate polynomials normally needed.

Although the results to follow consider only multiparticle tests, these techniques may be integrated into beam-envelope simulations by building contour maps from low-particle-count samples.

2. Theory

For simplicity, we will begin by considering only transverse motion along a single axis with an initial Gaussian distribution (though we will test the resulting approximation in 2D on various initial distributions).

The starting point for this method, outlined in [6], involves taking particle count N as an invariant as the position and momentum envelopes ρ_x and ρ_p evolve:

$$\int \rho_x^L dx_L = \int \rho_x^0 dx_0 = \int \rho_p^L dp_L = \int \rho_p^0 dp_0 = N \ . \tag{1}$$

where ρ_x^L and ρ_x^0 are the respective final and initial position densities, and likewise for the momentum densities.

To exploit this identity, ρ_x^L must be independent of p_0. This can be accomplished using the approximation [7, 8]

$$p_0 \approx -\frac{\alpha}{\beta} x_0 \ , \tag{2}$$

where α and β are the well-known Twiss parameters. Unfortunately, this approximation is only valid if the previous history of the beam is linear (thus maintaining elliptical phase space densities). Since we want an algorithm that remains accurate for iterated nonlinear kicks (which develop irregular density profiles), a new approximation is needed.

Proceeding under the constraint that the initial distribution is Gaussian in both x and p, we can solve equation (1) for x_0 and p_0.

$$\begin{aligned} \int \rho_x^0 dx_0 &= \int \rho_p^0 dp_0 \\ \frac{1}{2} \operatorname{erf}\left(\frac{\sqrt{2} x_0}{2\,\sigma_{x_0}}\right) &= \frac{1}{2} \operatorname{erf}\left(\frac{\sqrt{2} p_0}{2\,\sigma_{p_0}}\right) \ , \end{aligned} \tag{3}$$

which yields

$$p_0 = \sqrt{2}\sigma_{p_0} \operatorname{erfinv}\left(\operatorname{erf}\left(\frac{\sqrt{2} x_0}{2\sigma_{x_0}}\right)\right) = x_0 \frac{\sigma_{p_0}}{\sigma_{x_0}} \ . \tag{4}$$

This solution does produce a bigaussian phase-space ellipse, but is unsuitable for approximating p_0 with distributions of an irregular shape.

We continue by guessing that a solution exists for $p_0 = f(x_0)$ for irregularly shaped distributions. Denoting these solutions as p^* and ρ_p^* for momentum and momentum density, respectively, we have

$$\int \rho_p^*(p_0)\ dp^*(x_0) = \int \rho_p^0 dp_0 \ , \tag{5}$$

where it is critical to note that p^* is solely dependent on x_0, while ρ_p^* remains a function of p_0 (and likewise for x^* and ρ_x^*). With the left-hand-side integrand and integration variables decoupled, it follows (for both x^* and p^*):

$$\begin{aligned} N = x^* \rho_x^* &= \int \rho_x^0 dx_0 \\ N = p^* \rho_p^* &= \int \rho_p^0 dp_0 \ . \end{aligned} \tag{6}$$

Exploiting particle-count invariance again, and squaring N, we can assert that

$$N^2 = \mathcal{C} = \rho_x^* \rho_p^* x^* p^* \ , \tag{7}$$

and thus

$$\frac{d}{dx_0}\mathcal{C} + \frac{d}{dp_0}\mathcal{C} = 0\ . \tag{8}$$

We can then simplify, treating all p_0-dependent terms as $f(p_0)$:

$$\begin{aligned} 0 &= \frac{d}{dx_0}\mathcal{C} + \frac{d}{dp_0}\mathcal{C} \\ &= \frac{\partial \rho_x^*}{\partial x_0}\left(\rho_p^* x^* p^*\right) + \frac{\partial p^*}{\partial x_0}\left(\rho_x^* \rho_p^* x^*\right) + f(p_0) \\ &= \frac{\partial \rho_x^*}{\partial x_0}\left(\int \rho_p^0 dp_0\ x^*\right) \\ &+ \frac{\partial p^*}{\partial x_0}\left(\int \rho_x^0 dx_0\ \frac{\int \rho_p^0 dp_0}{p^*}\right) + f(p_0)\ . \end{aligned} \tag{9}$$

Then, dividing by $\int \rho_p^0 dp_0$, the expanded $f(p_0)$ terms become zero and we have:

$$\begin{aligned} 0 &= \frac{\partial \rho_x^{\mathcal{I}}}{\partial x_0}\ x^* + \frac{\partial p^*}{\partial x_0}\ \frac{1}{p^*}\int \rho_x^0 dx_0 \\ &= \frac{\partial \rho_x^*}{\partial x_0}\ x^* + \frac{\partial p^*}{\partial x_0}\ \frac{1}{p^*} x^* \rho_x^* \\ &= \left(\frac{\partial \rho_x^*}{\partial x_0} + \frac{\partial p^*}{\partial x_0}\ \frac{1}{p^*}\rho_x^*\right)\ . \end{aligned} \tag{10}$$

By reusing equation (6), all p_0 dependence can be eliminated, leaving

$$\frac{\partial p^*}{\partial x_0} = -p^* \frac{\rho_x^0}{\int \rho_x^0 dx_0} = -\frac{p^* \rho_x^0}{\Upsilon_x}, -\frac{2p^* \rho_x^0}{\Upsilon_x} \tag{11}$$

where the second solution can be obtained integrating by parts, and, in the case of a Gaussian initial distribution, the placeholder in the denominator is defined as

$$\Upsilon_x \equiv \frac{1}{2}\,\mathrm{erf}\left(\frac{\sqrt{2}x_0}{2\sigma_x}\right)\ . \tag{12}$$

Thus, in contrast with equation (4), we have an expression where $\frac{\partial p}{\partial x}$ is no longer constant.

We now check the following approximation:

$$p^* \ \approx\ -2\sinh\left(\frac{\rho_x^0}{\Upsilon_x}x_0\right)\mathcal{D} \tag{13}$$

Where we normalize $\mathcal{D}$ using equation (4); setting to $p^* \approx -p_0$ near $|x_0| = 0$, leaving

$$p^* \ \approx\ 2\sinh\left(\frac{\rho_x^0}{\Upsilon_x}x_0\right)x_0\frac{\sigma_p}{\sigma_x} \tag{14}$$

which can be shown numerically to agree with equation (11) for $|x_0| \lesssim 6\,\sigma_x$.

At this point, the updated particle postition x_L can be calculated using an exponential Lie-operator method [9]:

$$x_L(x_0, p^*) = \{\exp[\ -t : \mathcal{H}(x_0, p_0) :\ ]\,x_0\}|_{p_0 \to p^*}\ . \tag{15}$$

IOP Conf. Series: Journal of Physics: Conf. Series **874** (2017) 012070 doi:10.1088/1742-6596/874/1/012070

where t is elapsed time in the lab frame and the Hamiltonian for a normal multipole magnet in the transverse plane is

$$\mathcal{H} = \frac{e}{p}\frac{k \cdot \mathfrak{Re}(x_0 + iy_0)^n}{a_0^{n-1}\, n!} + \frac{(p_0)^2}{2m} \,. \tag{16}$$

Here, e, p, m, and a_0 are the fundamental charge, reference longitudinal momentum, particle mass, and magnet-pole raidus, respectively; $n = 3, 4, 5...$ for sextupoles, octupoles, decapoles, etc; and k has units of $[\mathrm{T}\cdot\mathrm{m}^{-1}]$. In the following sections, longitudinal momentum is normalized to 1 GeV/c and a_0 is set to 20 mm unless otherwise noted.

In implementing equation (15), $\mathcal{H}$ must be calculated symbolically first for each element. Then, $p^*(x_0)$ and x_0 are substituted in at each step, reducing the bivariate $x_L(x_0, p_0)$ to a monovariate $x_L(x_0, \sigma_x, \sigma_p)$, where σ_x and σ_p remain constant for a given timestep.

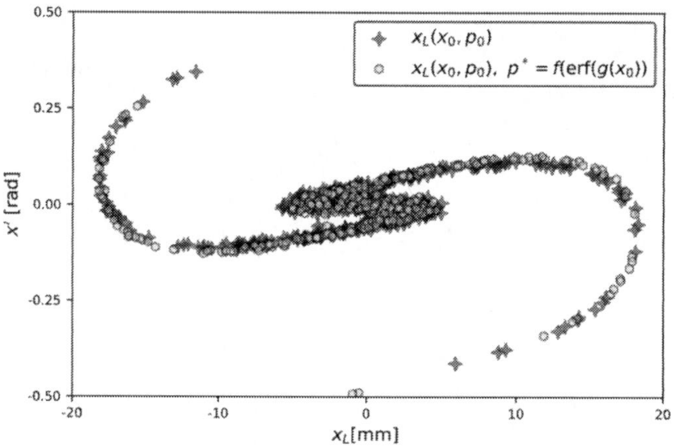

(a) Error-function Approximation

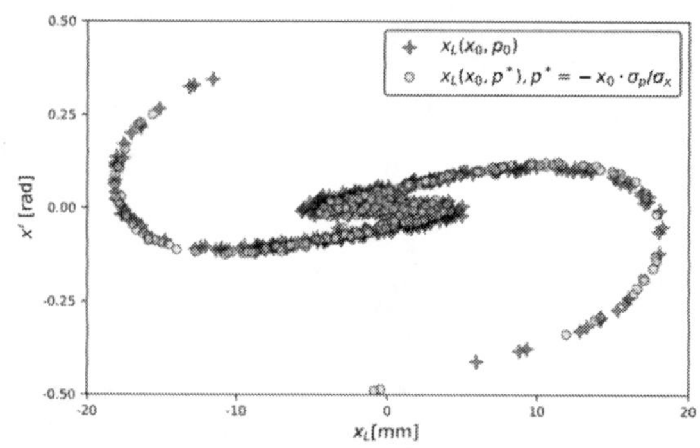

(b) Naive Approximation

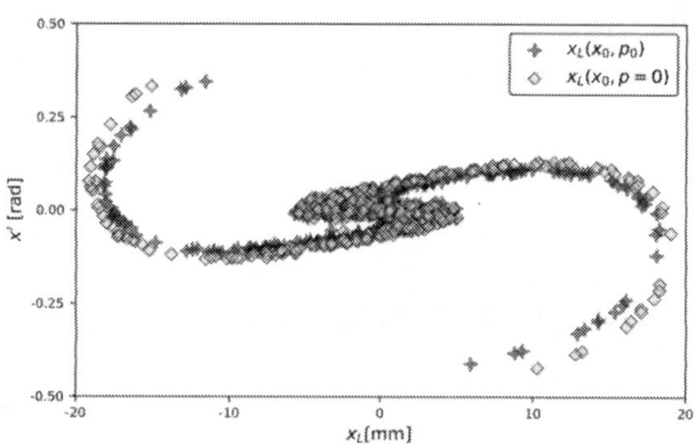

(c) Null-momentum Approximation

Figure 1. Iterated use of equation (14) versus standard Lie transport results in 1D for 100,000 protons with initial Gaussian distributions of σ_x = 10 mm, σ_p = 0.01 rad (a). Also shown are two alternate p^* approximations: $p^* = x_0\frac{\sigma_p}{\sigma_x}$ in (b) and $p^* = 0$ in (c). The transport map consists of 200 octupole–drift sections: $B_0 = 10$ [T], L_{oct} = 0.1 [mm], L_{drift} = 1.0 [mm]. Lie transforms are truncated to fifth order.

Although an analogous $x^*(p_0)$ can be derived, it is not useful in practice. Specifically, in calculating equation (16) in 2D for position and momentum – $x_L(x_0, y_0, p_x^*, p_y^*)$ and

$p_{xL}(x^*, y^*, p_{x0}, p_{y0})$ – the resulting x_L expression is dependent on σ_x and σ_{px}, while p_L is dependent on σ_x, σ_y, σ_{px}, and σ_{py}, rendering it computationally inefficient. Other schema involving alternate forms such as $p_L(x_0, p^*)$ have been checked, but the following is found to be most stable, with notable performance gains:

$$\begin{aligned} x_1, y_1, p_{x1}, p_{y1} &\rightarrow x_L(x_0, y_0, p_x^*, p_y^*), y_L(x_0, y_0, p_x^*, p_y^*) \\ & \quad\; p_{xL}(x_0, y_0, p_{x0}, p_{x0}), p_{yL}(x_0, y_0, p_{x0}, p_{y0}) \\ &\downarrow \\ x_2, y_2, p_{x2}, p_{y2} &\rightarrow x_D(x_1, y_1, p_{x1}, p_{y1}), y_D(x_1, y_1, p_{x1}, p_{y1}) \\ & \quad\; p_{xD}(x_1, y_1, p_{x1}, p_{y1}), p_{yD}(x_1, y_1, p_{x1}, p_{y1}) \end{aligned} \tag{17}$$

Where the D subscript denotes a drift space of at least five times the kick length. This effectively limits the technique to a thin-lens approximation. Such drift spaces can be reserved for incorporating space-charge effects, leading to a comparable number of calculation steps using equation (17) versus a standard nonlinear beam-physics code.

3. Multiparticle Simulation

Figure 1 compares the accuracy of multiparticle transformations following equation (17) with and without using p^*. Also shown are tests for $p^* = 0$ and a "naive" approximation, where $p^* \approx p_0$ from equation (4) is used in the low $|x_0|$ limit of equation (14):

$$p^* \approx -x_0 \frac{\sigma_p}{\sigma_x} \; . \tag{18}$$

To emphasize visible discrepancies, the results shown have their σ values updated after each timestep by taking a new standard deviation. However, if mean absolute deviations are taken instead, an improved matching with the baseline can be observed.

Both the naive and null-momentum approximations fail at $\sigma_p \gtrsim \sigma_x$ (i.e. at energies exceeding 1 GeV). Figure 2 illustrates such a case for 2D Gaussian proton distributions with a kinetic energy of 8 GeV passing through an octupole magnet. In the 2D case, beam parameters were derived relativistically from Twiss parameters and B_0 by normalizing the kinetic term in equation (16) to the beam's average kinetic energy then verified against Tracewin [10].

4. Conclusion

For the non-null p^* approximations, performance improves with increasing particle count, with increasing magnetic pole count, and particularly with increased order of Lie-transform series truncation (figure 3). Since trajectory variations are negligible beyond a 6th-order truncation in most cases, the average reduction in CPU overhead using equation (14) is roughly 15%.

Similar results were obtained for sextupoles, decapoles, and high-order magnets, as well as with waterbag distributions, despite the assumption of a Gaussian shape in deriving equation (14). At low energies (or specifically, any low $\frac{\sigma_p}{\sigma_x}$ ratio), all three approximations tested have essentially identical results, with escalating performance in the following order: $p^* = (f[\text{erf}])$; $p^* = -x_0 \cdot \sigma_p/\sigma_x$; $p^* = 0$.

For all the approximations tested, trajectories only became unstable in cases where the momentum of the baseline exceeded $\sim 100\,\sigma_p$. Thus, the major limitation to this technique is its large drift–kick ratio requirement.

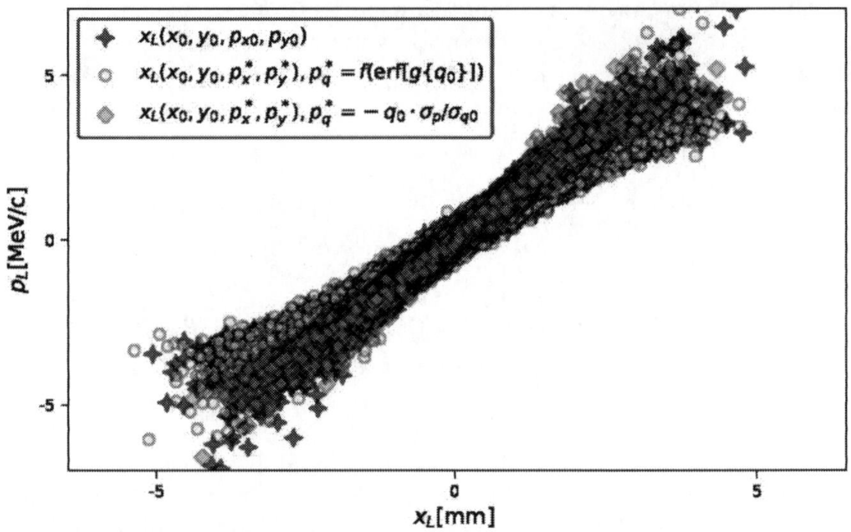

Figure 2. Iterated octupole transforms for 2D Gaussian bunches of 100,000 protons at 8 GeV: $a_0 = 15$ [mm], $\epsilon_\perp = 0.25$ [$\pi \cdot$ mm $\cdot$ mrad], $\beta_\perp = 1$ [mm/($\pi \cdot$ mrad)]. The map consists of 400 kick–drift sections: $B_0 = 12$ T, $L_{oct} = 0.2$ [mm], $\mathrm{L_{drift}} = 2$ [mm], for an integrated field strength of 0.211 [T/m^2].

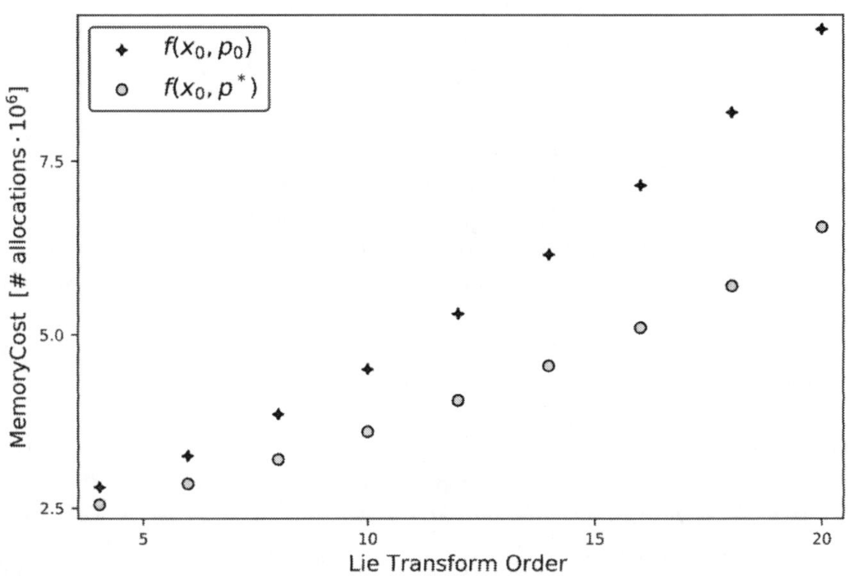

Figure 3. Raw memory-allocation cost for multiparticle simulations with k and L parameters matching those of figure 1. Number of particles: 10,000. Number of simulated segments: 5.

References

[1] Grote H and Schmidt F 2003 *Proc. PAC 2003, Portland, OR, USA*. **5** 3497–99
[2] Pelaia T *et al.* 2015 *Proc. IPAC 2015, Richmond, VA, USA* 1270–72
[3] Forest E, Schmidt F and McIntosh E 2002 *Introduction to the polymorphic tracking code. Fibre bundles, polymorphic taylor types and "Exact Tracking"* Tech. Rep. (No. KEK–2002-3) High Energy Accelerator

Research Organization
[4] Ryne R *et al.* 2006 *Recent Progress on the Marylie/Impact Beam Dynamics Code* Tech. rep. Ernest Orlando Lawrence Berkeley NationalLaboratory, Berkeley, CA, USA
[5] Laface E 2015 *Proc. IPAC 2015, Richmond VA, USA*
[6] Folsom B and Laface E 2016 *Proc. IPAC 2016, Busan, Korea, 2016* 3080–82
[7] Yuri Y, Miyawaki N, Kamiya T, Yokota W, Arakawa K and Fukuda M 2007 *Physical Review Special Topics-Accelerators and Beams* **10** 104001
[8] Meot F and Aniel T 1996 *Nuclear Instruments and Methods in Physics Research Section A: Accelerators, Spectrometers, Detectors and Associated Equipment* **379** 196–205
[9] Dragt A J 2015 *Lie methods for nonlinear dynamics with applications to accelerator physics* (University of Maryland, Center for Theoretical Physics, Department of Physics)
[10] Uriot D and Pichoff N 2011 *CEA internap report CEA/DSM/DAPNIA/SEA/2000/45*

Beam shaping with 4n-order multipole magnets

B Folsom[1,2], **E Laface**[1,2]
[1]Lund University, Particle Physics Division, Lund, Sweden
[2]European Spallation Source ERIC, Lund, Sweden

E-mail: ben.folsom@esss.se

Abstract. A uniformly irradiating beam is beneficial in spallation for preventing irregular wear on the target. For octupoles (n = 4) and higher-order (n = 4N) magnets, passing charged-particle bunches undergo symmetric shaping effects along the x and y axes. Using a Lie-mapping formalism, we illustrate how well Gaussian distributions can be flattened symmetrically in 2D with single, dual-pulse, and RF magnets of 4N order. Incidental shaping effects are also discussed.

1. Introduction

For charged particle or ion bunches, an octupole displaces the outermost particles, reshaping a Gaussian transverse profile to a flattened or concave one [1]. Inherent to this flattening property is a four-fold rotational symmetry octupoles impart on transverse beam distributions (specifically, symmetry along each x–y diagonal).

Because of this, it is standard practice to introduce octupoles in pairs with respective quadrupoles so that their effects are only appreciable along one axis. This negates the defocusing of off-center particles near the diagonals. Such an approach is common when correcting high-order errors in beam-focusing or beam-steering optics [1–4].

With this single-axis flattening, however, a number of transverse and longitudinal shaping phenomena are disregarded. This study considers such effects, beginning with a discussion of the octupole and octupole-like Hamiltonians in a symplectic Lie-algebra formalism. It is shown that x–y decoupled, alternating-gradient, and dual-pulse multipoles can act on transversely symmetric 2D bunches to induce a variety of shaping effects such as low-loss flattening, isotropic focusing, transverse trapping, and longitudinal momentum dilation.

2. 4N-Pole Symmetry

In previous works, we reviewed the Lie-algebra formalism which implicitly preserves symplecticity in calculating the trajectories of bunched particles under nonlinear potentials [5,6]. In this approach, a position or momentum coordinate can be tracked as

$$c = e^{-t:H:}c_0 \ , \tag{1}$$

where t is traversal time through the element, H is the machine-element's Hamiltonian, and the colons denote Lie brackets: $\left(\frac{\partial f(x,p)}{\partial x}\frac{\partial g(x,p)}{\partial p} - \frac{\partial g(x,p)}{\partial x}\frac{\partial f(x,p)}{\partial p}\right)$. These absorb any variable falling to the right as the $g(x,p)$ term. Here, the exponential can be expanded in series and truncated at a desired precision (resulting in nested Lie brackets) or simplified with Yoshida's symplectic integration approach [7].

The component of the Hamiltonian corresponding to the external potential from a multipole magnet can be represented as a complex binomial:

$$K(x + iy)^n \ , \tag{2}$$

where the real and imaginary terms correspond to normal and skew orientations, respectively. To derive the field gradient K, we begin with [8,9]:

$$\kappa_n = \frac{e \cdot B_0}{\vec{p}_{\|}(n-1)!\ a_0^{(n-1)}} [\mathrm{m}^{-n}] \ , \tag{3}$$

where e is particle charge, $\vec{p}_{\|}$ is longitudinal momentum, and B_0 and a_0 are the pole-tip magnetic flux and radius, respectively. Noting that any binomial expansion results in terms of cumulative order n, the potential term for a normal multipole

$$\kappa_n \cdot \mathfrak{Re}(x + iy)^n \ , \tag{4}$$

is thus unitless. We can then normalize to

$$K_n = \kappa_n \cdot \mathrm{K_E} \ , \tag{5}$$

where $\mathrm{K_E}$ is kinetic energy. This results in a Hamiltonian with units of energy:

$$H_n = K_n \cdot \mathfrak{Re}(x + iy)^n + \frac{\vec{p}_{\perp}^{\,2}}{2m} \ . \tag{6}$$

Lie-algebra tracking with this normalization has been cross-checked with the beam-physics simulation package Tracewin [10] for the results to follow, unless otherwise noted.

For a normal octupole, the potential term is

$$K_4(x^4 - 6x^2y^2 + y^4) \ , \tag{7}$$

What is noteworthy here is that the leading-order terms are both positive (since $\mathrm{i}^{4n} = 1$), resulting in the observed four-fold rotational symmetry. This carries for higher-order 4n-pole magnets, where the sign parity of like-order terms ensures identical distributions along the x and y axes. For example, the potential term for the hexadecapole is

$$K_8(x^8 - 28x^6y^2 + 70x^4y^4 - 28x^2y^6 + y^8) \ . \tag{8}$$

To generalize, any n = 4N pole magnet could be considered "octupole like", with all other n = 2N pole magnets as "quadrupole like". In the latter case, x and y profiles may have symmetry about their own axes, but their profile shapes are always dissimilar. For both cases, as the number of poles is increased, the shaping effect converges toward a circular 2D profile, with particles along the circumference forming narrowing cusps.

3. Decoupling

Recent works have demonstrated that octupoles can be virtually decoupled in a periodic lattice by using quadrupole and dipole inserts on either side of an octupole [11,12]. That is,

$$H_4 = K_4(x^4 - \cancel{6x^2y^2} + y^4) \ . \tag{9}$$

Figure 1 illustrates that such a magnet may be used for improved beam flattening. Since decoupled multipoles are not implemented into Tracewin, these simulations were carried out solely with an ab-initio Lie-algebra code, using rudimentary space-charge kicks on either side of the magnet:

$$p = p_0 + \frac{C}{2}\,\mathrm{erf}\left(\frac{x}{\sigma_x}\right) , \tag{10}$$

with

$$C = \frac{It}{6\pi} , \tag{11}$$

where I is beam current and the factor of 1/2 in equation (10) reflects a halving of the magnet length and "erf" is the error function. This equation was verified to match Tracewin for Gaussian distributions in drift spaces with wide enough apertures to avoid image effects.

A key trait of the decoupled octupole is its virtually lossless flattening: a coupled magnet of matched strength can only produce a similarly flattened profile by ejecting a significant percentage of its outermost particles. For example, the coupled octupole in figure 1a loses 1% of its particles to beyond $6\,\sigma_x$ whereas the decoupled octupole's losses are 0.1%.

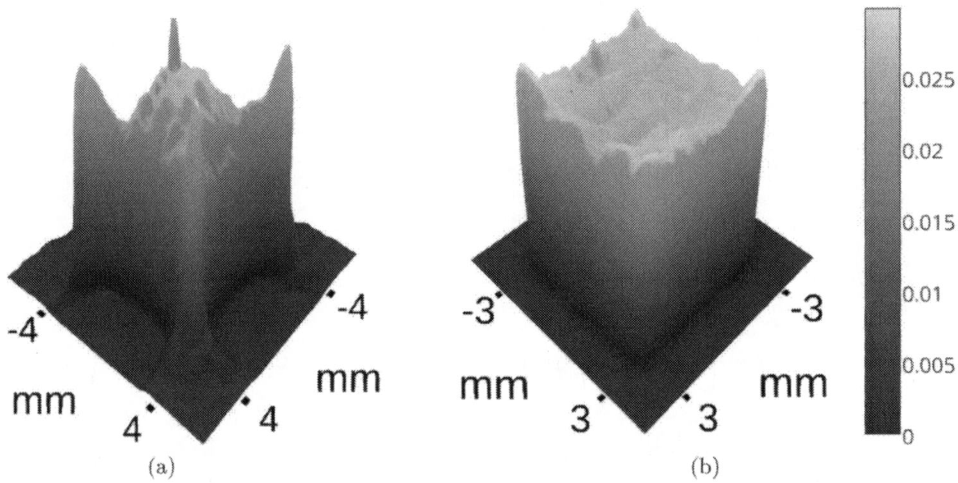

Figure 1. Surface plot of Gaussian distributions (100k protons) passing through (a) coupled and (b) uncoupled octupoles, normalized to maximum bin occupancy. Beam and magnet parameters are otherwise identical: $a_0 = 25$ [mm], $\epsilon_\perp = 0.5$ [$\pi \cdot$ mm $\cdot$ mrad], $\beta_\perp = 1$ [mm/($\pi \cdot$ mrad)], $B_0 = 3$ T, $K_E = 4$ MeV, I = 30 mA, and magnet length $L = 500$ mm.

As field strength is increased, focusing can also be observed in decoupled magnets with a reduced loss rate (i.e. a reduction of σ_x by 50% with ~7% ejecta for decoupled magnets and $\gtrsim$30% for coupled magnets).

However, the use of equation (9) in obtaining these results must be considered an idealization, since in practice, the inserts required for virtual decoupling are calibrated via thin-lens approximation.

4. RF Octupoles

Although better flattening can be achieved with coupled octupoles than that of figure 1a, it typically requires a significant dilation of the transverse dimensions. This can be improved if we consider alternating-current $n = 4N$ pole magnets operating between 1 MHz and 1 THz. For the present work, this effect was simulated with stepped waves (i.e. successive kicks of $-9, -6, -3, 0, 3, 6, 9$ T as one half-cycle).

For low field strength–emittance ratios, this effectively creates a virtual aperture: Initially, a shell of the outermost particles is lost in the first few cycles. After a few hundred cycles (~ 100 mm), the beam distribution tends toward a conical shape and with a reduced loss rate. This effect holds for Gaussian distributions, but is more easily observed with waterbag or KV distributions. When accounting for space charge, the shaping effects are similar, with losses occurring at a constant rate proportional to beam current. At higher field strengths, losses can be eliminated altogether, with transverse positions effectively trapped after an initial dilation.

Figure 2 shows the longitudinal effects of trapping a 50 MeV electron beam. With an initial $\sigma_{z'}$ of 0.0005 mrad, the longitudinal momentum quickly dilates symmetrically (figure 2a). Particles with high transverse momenta then begin to recirculate in the positive z' direction (figure 2b). Here, an extremely small pole-tip aperture is used to exaggerate the effect — with more realistic parameters, a dilation in momentum of a factor of 10 can be observed, with a similar concentration of the electron population in the positive z' direction.

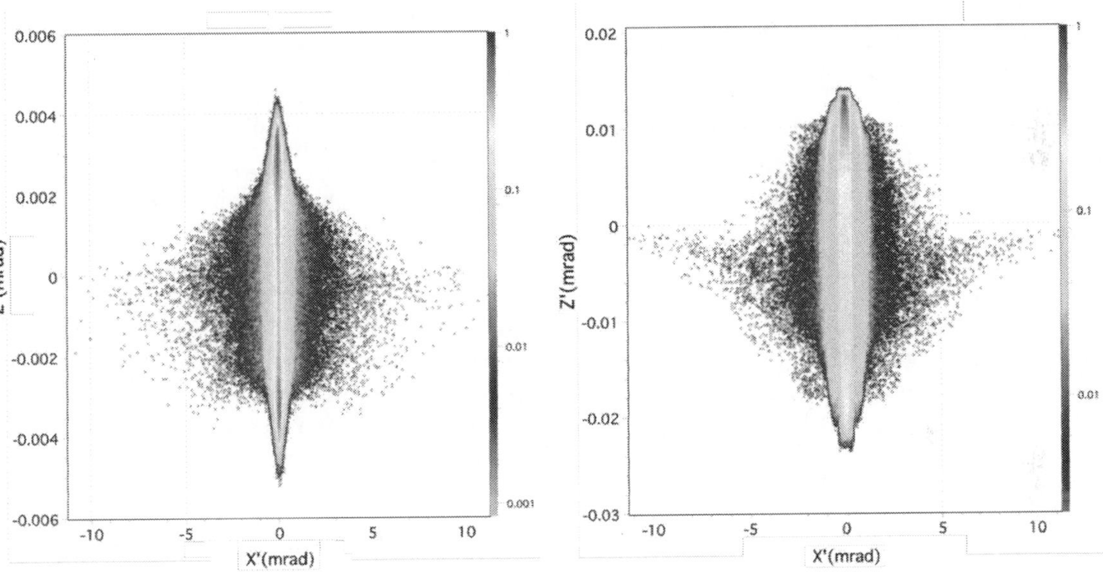

(a) (b)

Figure 2. Transverse–Longitudinal density mapping for 500k electrons in a 30 GHz RF octupole at (a) 40 mm and (b) 400 mm; $a_0 = 1$ [mm], $\epsilon_\perp = 0.25$ [$\pi \cdot$ mm $\cdot$ mrad], $\epsilon_\parallel = 0.35$ [$\pi \cdot$ mm $\cdot$ mrad], $\beta_\perp = 4$ [mm/($\pi \cdot$ mrad)], $\beta_\parallel = 5$ [mm/($\pi \cdot$ mrad)], $B_0 = 10$ T, $K_E = 25$ MeV, I = 50 mA.

The cost of such a shift is incurred as emittance gains (proportional to beam current). Here, the longitudinal emittance increases steadily, but transverse emittance levels off quickly as the particles become trapped.

At higher frequencies, hollowing can be observed (figure 3a), with the bunch collapsing centrally as the magnet length is increased (figure 3b). Similar effects were reported in [13–15].

In this case, hollowing can be seen as an extension of beam flattening—a beam becomes approximately flat before its central population is forced toward the halo.

Moreover, at these frequencies but increased field strengths, focusing can be achieved with zero ejecta (which forgoes hollowing and rapidly condenses the central region).

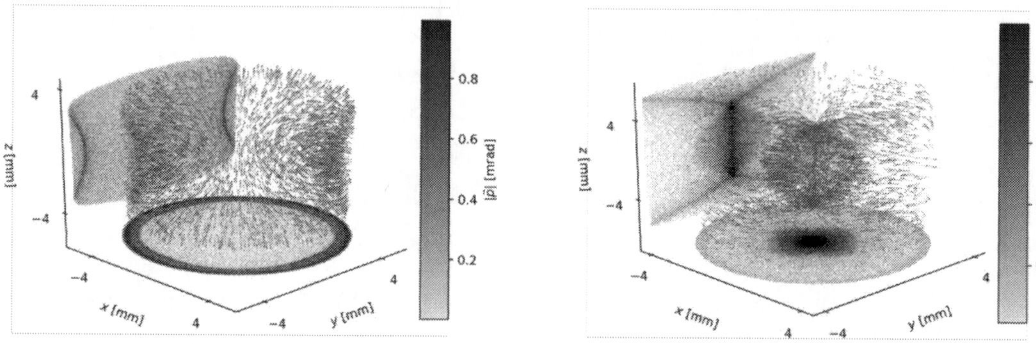

(a) (b)

Figure 3. Beam hollowing and subsequent collapse of electrons (100k, Gaussian) in a 1 THz, $B_{max} = 9$ T, RF octupole: $a_0 = 20$ [mm], $\epsilon_\perp = 0.25$ [$\pi \cdot$ mm $\cdot$ mrad], $\beta_\perp = 4$ [mm/($\pi \cdot$ mrad)], $K_E = 0.75$ MeV, I = 62 mA. Magnets lengths are 0.375 and 0.778 m and momenta are scaled to maxima of 20.9 and 24.5 mrad for (a) and (b) respectively.

5. Double-Pulse Shaping

The aforementioned traits of alternating-gradient multipoles may be worth investigating further, but the fabrication of such devices is likely to require prohibitive stress tolerances or materials costs.

For a more practical design, an initial octupole can provide a brief high-strength pulse so that a passing bunch arrives at the second octupole with a compacted momentum distribution but before its position distribution has been significantly altered. The second magnet then gives a longer, weaker pulse which effectively traps the perturbed central trajectories. This results in a sharper beam flattening than can be achieved with individual magnets, with an improvement comparable to that shown in figure 1. This scheme induces negligible losses when shaping to a bunch perimeter of $\sim 3\sigma_x$. By increasing the initial pulse strength, a width reduction to $\sim 2\sigma_x$ can be achieved with ~1% losses.

6. Conclusions

Octupole and octupole-like magnets are shown to act with four-fold rotational symmetry in the transverse plane. While the effect of such magnets is a flattening of the transverse momentum and position distributions in turn, any rapid, pronounced shaping carries inherent particle losses. To address this, decoupling, high-frequency alternating gradients, and dual-pulse kicks may be beneficial. The incidental phenomena of transverse-beam trapping, longitudinal momentum dilation, and beam hollowing may also merit further study.

References

[1] Tsoupas N, Zucker M S, Ward T E and C L Snead J 1997 *Nuclear Science and Engineering* **126** 71–79 URL `http://epubs.ans.org/?a=24458`
[2] Meigo S i, Ooi M, Ikezaki K, Akutsu A and Fujimori H 2014 *Proc. IPAC2014, Dresden, Germany* 15–20
[3] Ho D D M, Haber I, Crandall K R and Brandon S T 1991 *Part. Accel.* **36** 141–160 URL `http://cds.cern.ch/record/1054399`
[4] Brinkmann R and Raimondi P 2001 *Proc. PAC 2001. Chicago, IL, USA* **5** 3828–3830
[5] Laface E 2015 *Proc. IPAC2015, Richmond, VA, USA* 345–347

[6] Folsom B and Laface E 2016 *Proc. IPAC2016, Busan, Korea* 3080–3082
[7] Yoshida H 1990 *Physics Letters A* **150** 262–268 ISSN 0375-9601 URL http://www.sciencedirect.com/science/article/pii/0375960190900923
[8] Wiedemann H 2015 *Particle accelerator physics* (Springer)
[9] Wille K 2000 *The physics of particle accelerators: an introduction* (Clarendon Press)
[10] Uriot D and Pichoff N 2011 *CEA internap report CEA/DSM/DAPNIA/SEA/2000/45*
[11] Antipov S A, Nagaitsev S and Valishev A 2016 *arXiv preprint arXiv:1604.08565*
[12] Nagaitsev S, Valishev A and Danilov V 2012 *arXiv preprint arXiv:1207.5529*
[13] Neuner U *et al.* 2000 *Physical Review Letters* **85** 4518–4521 URL https://link.aps.org/doi/10.1103/PhysRevLett.85.4518
[14] Stancari G, Valishev A, Annala G, Kuznetsov G, Shiltsev V, Still D A and Vorobiev L G 2011 *Physical Review Letters* **107** 084802 URL https://link.aps.org/doi/10.1103/PhysRevLett.107.084802
[15] Blaugrund A E and Cooperstein G 1975 *Physical Review Letters* **34** 461–464 URL https://link.aps.org/doi/10.1103/PhysRevLett.34.461

IOP Conf. Series: Journal of Physics: Conf. Series **874** (2017) 012072 doi:10.1088/1742-6596/874/1/012072

Analysis of performance fluctuations for the CERN Proton Synchrotron multi-turn extraction

A Huschauer, M Giovannozzi, O Michels, A Nicoletti, G Sterbini
CERN, 1211 Geneva 23, Switzerland
E-mail: alexander.huschauer@cern.ch

Abstract. After the successful beam commissioning and tests in 2015, the Multi-Turn Extraction (MTE) has been put in operation in 2016. In this paper, the remaining issues related with fluctuation of the MTE performance are evaluated and correlation studies are presented in view of estimating the impact of planned improvements.

1. Introduction

To provide high-intensity beams for fixed-target physics at the Super Proton Synchrotron (SPS), the beam delivered by the PS has to be fully de-bunched and uniform in intensity. Therefore, the Continuous Transfer (CT) process was proposed in 1973 [1]. This extraction technique, which occurs over five turns at 14 GeV/c, allows to optimize the duty cycle as only two subsequent extractions from the PS are necessary to fill the SPS. On the downside, the CT extraction comes with the major drawback of significant beam loss occurring at multiple locations around the ring [2], leading to high dose to personnel during accelerator maintenance and repair, as well as to long cool down times.

The Multi-Turn Extraction (MTE) technique was proposed to replace the CT process in 2001 [3] in view of mitigating the shortcomings of CT. MTE is a resonant extraction mechanism, which exploits advanced concepts of non-linear beam dynamics and applies a fourth-order stable resonance to perform beam splitting in the horizontal phase space. The resulting beamlets - four islands and one core - are then extracted over five subsequent turns (see [4] for the implementation and [5] for some theory of adiabatic trapping).

Due to the complexity of the MTE scheme, its operational implementation, which came to a successful close in September 2015 (see [6, 7] for more detail), has had to overcome many challenges and, particularly, significant fluctuations in the efficiency of the transverse splitting, defined as

$$\eta_{\text{MTE}} = \frac{\langle I_{\text{Island}} \rangle}{I_{\text{Total}}}, \tag{1}$$

where $\langle I_{\text{Island}} \rangle$ and I_{Total} stand for the average intensity in each island and the total beam intensity, respectively. The nominal efficiency is 0.20, corresponding to an equal beam sharing between islands and core. This figure of merit is derived from the signal of the beam intensity in the transfer line, just downstream of the PS extraction point.

Figure 1 shows the distribution of η_{MTE} during the 2016 physics run. The shape is quasi-Gaussian, skewed towards low values of η_{MTE}. Understanding the fine detail of this distribution and finding means to improve it, i.e., reducing its spread, are the goal of the studies discussed in this paper.

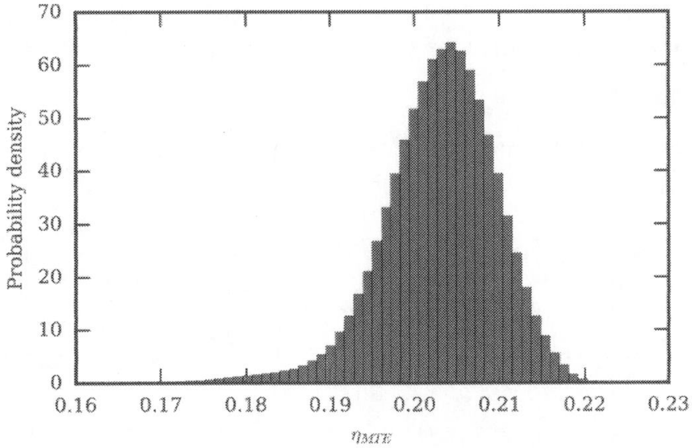

Figure 1. Distribution of η_{MTE} for the 2016 physics run.

2. Analysis of fluctuations

Figure 2 (left) shows the time-evolution of η_{MTE} over five days. In addition to the raw data, the time series sampled over non-overlapping time intervals of 30 min is shown as well. The comparison of these signals reveals low- and high-frequency structures.

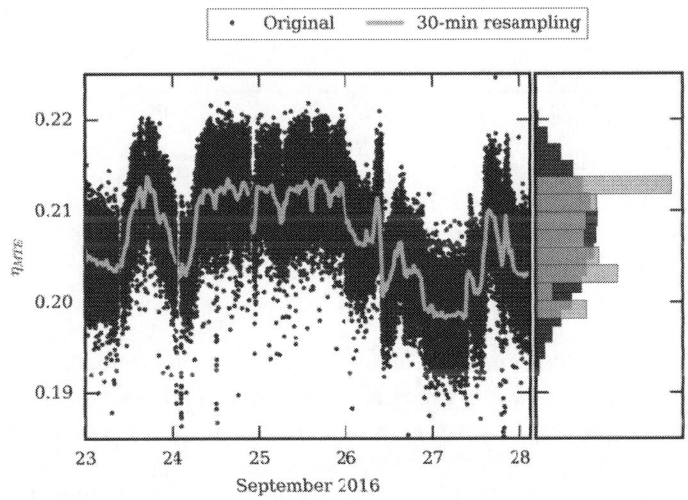

Figure 2. Time evolution of η_{MTE} and of its re-sampled version over non-overlapping intervals of 30 min (left) and their distributions (right).

The corresponding distributions are shown in Figure 2 (right). While the distribution of the raw data is quasi-Gaussian, this is by far not the case for the low-frequency contribution. These features suggest a different physical source and, therefore, different mitigation strategies to the fluctuations of η_{MTE}. In fact, the low-frequency variations can be cured by a slow feedback or by a human intervention, while the high-frequency components cannot. Therefore, most of the efforts have been devoted to finding the sources of the high-frequency variation in view of correcting them at the source.

An example of the time-variations that can be corrected by a human intervention is given in Figure 3 (upper), where a slow drift of the PS transverse tunes is shown together with the change in η_{MTE} (lower).

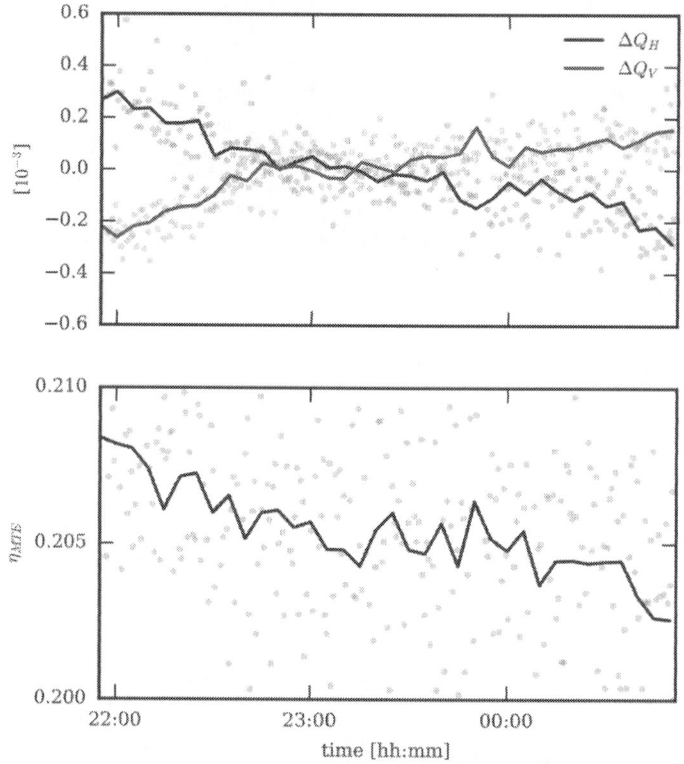

Figure 3. Upper: Time evolution of ΔQ_H and ΔQ_V. Lower: Time evolution of η_{MTE}. The continuous lines represent the re-sampled version over non-overlapping intervals of 5 min.

Even though the origin of the tune drift is not known, the time scale is such that the variation can be compensated by means of the PS ring tuning quadrupoles. Another example is given by the impact of the change of the PS magnetic configuration on η_{MTE}, shown in Figure 4. A step variation is clearly visible, which is likely to be generated by a change in the hysteresis of the combined-function main magnets. Also in this case the change in η_{MTE} can be easily compensated by acting on the horizontal tune.

3. Correlation analysis

The main source of the high-frequency fluctuations was identified as the 5 kHz ripple component of the converters powering the additional coils installed in the main magnets [6]. The clocks of these six power converters are not synchronised, thus generating a time-dependent ripple component. The excellent correlation between the amplitude variation of this ripple component and η_{MTE} is shown in Figure 5 (upper). The measurement device installed in a reference magnet provides the value of dB/dt.

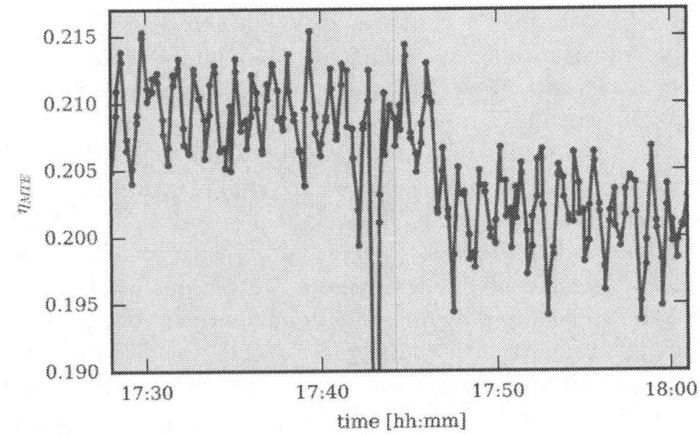

Figure 4. Time evolution of η_{MTE} during a change of PS magnetic configuration (indicated by the change of colour of the background).

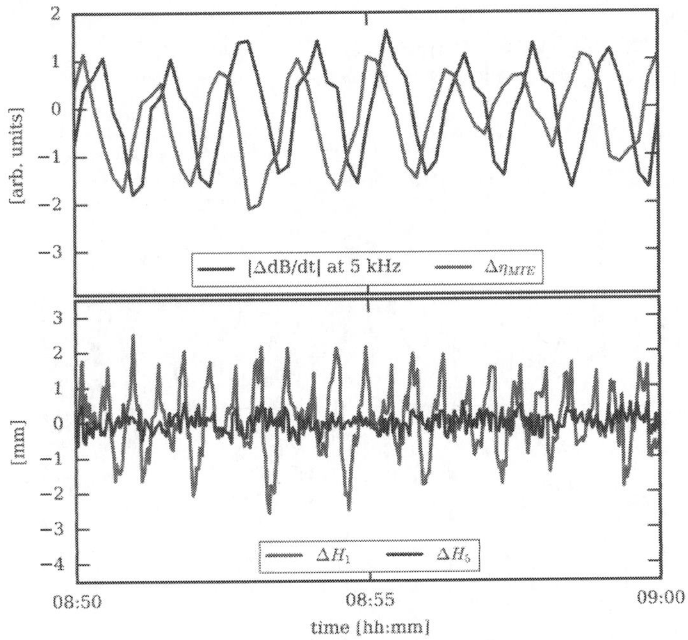

Figure 5. Upper: Time evolution of η_{MTE} and of the amplitude of the 5 kHz current ripple component on the power supplies. Lower: Time evolution of the variation of extraction conditions of first island (ΔH_1) and of core (ΔH_5).

Hence, the observed $\pi/2$ phase difference between η_{MTE} and dB/dt turns into a π phase difference with the B-field, which then is in counter phase with η_{MTE}. All this is a sign of the good correlation between the two quantities, not to mention that the two quantities feature a very similar time variation. It is also interesting to inspect the behaviour of the horizontal extraction position as a function of time. To analyse the trajectory stability of the extracted beam, a pick-up in the transfer line just downstream of the PS extraction septum is used to determine the position of each extracted turn. The average position over several extractions is subtracted from the measured value and the result is indicated by $\Delta H_i, 1 \leq i \leq 5$. The analysis showed that $\Delta H_i, 1 \leq i \leq 4$ are all well correlated between them, as expected, so that it is sufficient to consider in the analysis only ΔH_1 and ΔH_5, i.e., the horizontal variation of the extraction condition of the first island and of the core, respectively. Their time evolution is shown in Figure 5 (lower): while ΔH_5 is almost constant, with a variation close to the pick-up resolution, ΔH_1 changes considerably. Furthermore, the frequency content appears different with respect to the 5 kHz ripple component. This behaviour, however, does not imply that the fluctuations of ΔH_1 are not related with the ripple. In fact, the ripple affects η_{MTE} during the resonance crossing stage extending over several tens of ms, so that its phase is irrelevant. On the other hand, the extraction conditions are indeed sensitive not only to the amplitude, but also to the phase of the 5 kHz component with respect to the extraction time.

This analysis can be further pursued by checking the autocorrelation of η_{MTE} and ΔH_1, as shown in Figure 6. The former quantity features a rather regular pattern, showing that a strong correlation (positive or negative) occurs for time intervals slightly over one minute, while ΔH_1 features a slightly weaker autocorrelation, much richer in frequency content. The striking point is the close resemblance of the autocorrelation of η_{MTE} and ΔH_1 with that of the amplitude and phase of the 5 kHz component, respectively. All this is clearly visible in Figure 6, thus confirming that the ripple is the source of both η_{MTE} fluctuations and ΔH_1 variations.

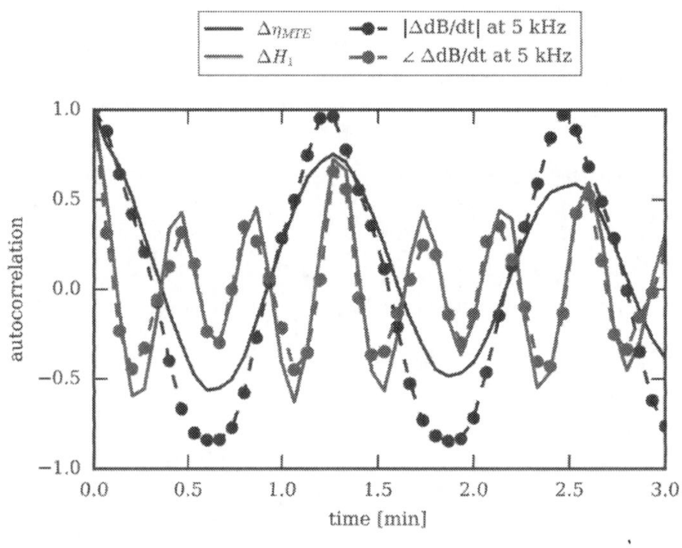

Figure 6. Autocorrelation of η_{MTE} (blue) and ΔH_1 (red).

From an operational point of view, given that two extractions from the PS are needed to fill the SPS ring, the performance of consecutive extractions should be very similar. This turns out to be the case as can be seen in Figure 7, where the correlation between the two extractions for η_{MTE} and ΔH_1 is clearly visible.

The last aspect analysed is the correlation between fluctuations and extraction losses. To this aim, the signal from a beam loss monitors (BLM) installed close to the extraction septum (SS16) has been used (see also [6, 8]). The BLM detects two loss spikes generated by the continuous beam during the rise time of the extraction kickers, when the islands and then the core are

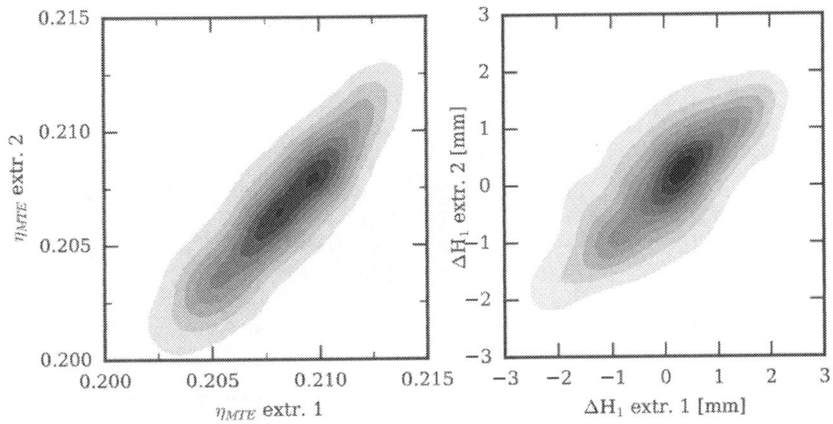

Figure 7. Correlation plot of two consecutive extractions for η_{MTE} (left) and for ΔH_1 (right). The correlation coefficients for η_{MTE} and ΔH_1 are 0.84 and 0.74, respectively.

extracted. These two spikes are individually integrated and correlated with η_{MTE} in Figure 8. The best correlation between η_{MTE} and beam losses is observed for the core.

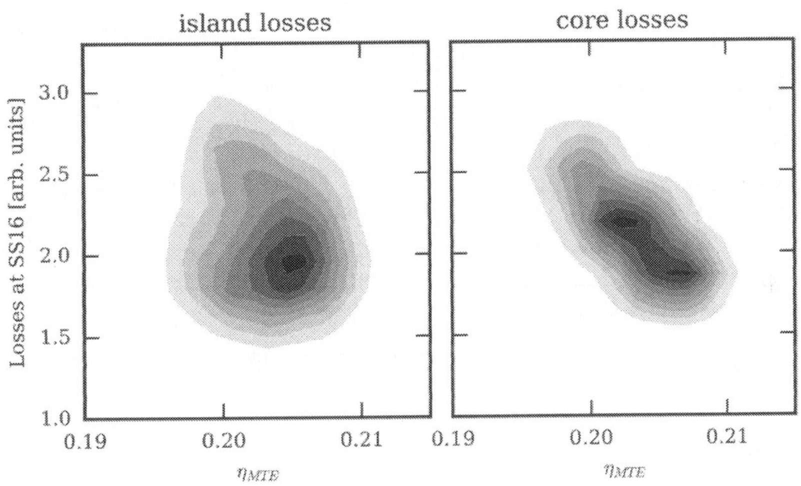

Figure 8. Correlation plot of losses for islands (left) and core (right) and η_{MTE}. The correlation coefficients for islands and core losses are −0.19 and −0.75, respectively.

The observation that the core-induced beam losses are closely connected with the fluctuations of η_{MTE} is confirmed by the result that the core width is well correlated with η_{MTE} as shown in Figure 9. The core width can be estimated by means of a diamond BLM also installed in SS16. In fact, the time-response of a diamond BLM is fast enough to ensure that the FWHM of the loss spike is proportional to the beam width.

The results shown in Figure 9 indicate that a smaller core size implies a larger η_{MTE} and hence lower losses at extraction.

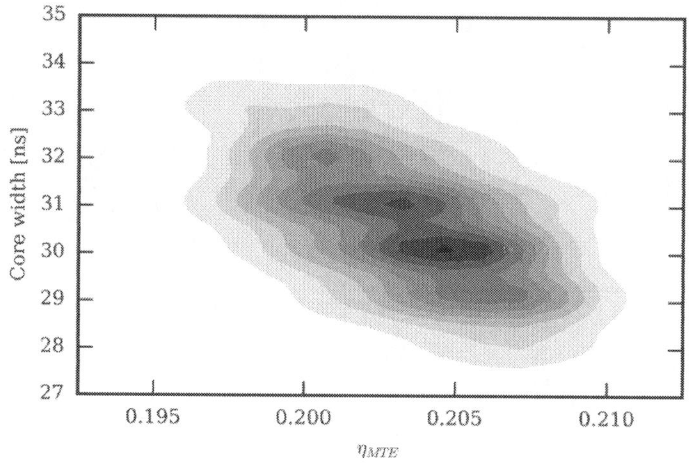

Figure 9. Correlation between core size and η_{MTE}. The correlation coefficient is -0.54.

4. Conclusions and outlook

The efforts devoted to the understanding of the fluctuations of η_{MTE} are paying off. The mitigation measures put in place in 2015 to reduce the amplitude of the 5 kHz ripple of the power converters of the auxiliary coils in the PS main magnets made MTE a suitable operational replacement of CT. The measurements and analysis presented in this paper confirm that η_{MTE} correlates well with the power converter ripple. For this reason the controls of the power converters have been upgraded during the 2016-17 winter shut down so to allow synchronising the clocks and to double the ripple frequency, thus shifting it outside of the beam spectrum [8]. New current transformers have been installed to provide a direct current measurement for the auxiliary circuits and should provide additional information to asses which field component affects η_{MTE} and ΔH_1. The new observations presented here indicate that curing the ripple should have a beneficial impact on the reproducibility of the extraction trajectories and should lower extraction losses, whose fluctuations are linked with the beam core size. Finally, after the successful reduction of the high-frequency η_{MTE} fluctuations, only the low-frequency components will remain.

Acknowledgments

Our warm thanks go to the PS Operations crew, and to G. Arduini, J.-P. Burnet, K. Kahle, M. Martino, R. Steerenberg for continuous support, encouragement and help.

References

[1] Bovet C, Fiander C, Henny L, Krusche A and Plass G 1973 *IEEE Trans. Nucl. Sci.* **20** 438
[2] Barranco García J and Gilardoni S 2011 *Phys. Rev. ST Accel. Beams* **14**, 030101
[3] Cappi R and Giovannozzi M 2002 *Phys. Rev. Lett.* **88**, 104801
[4] Barnes M J, Berrig O E, Beuret A, Borburgh J, Bourquin P, Brown R, Burnet J-P, Caspers F, Cravero J-M, Dobers T, Fowler T, Gilardoni S, Giovannozzi M, Hourican M, Kalbreier W, Kroyer T, Di Maio F, Martini M, Mertens V, Métral E, Metzmacher K-D, Rossi C, Royer J-P, Sermeus L, Steerenberg R, Villiger G, Zickler T 2006 CERN-2006-011
[5] Bazzani A, Frye C, Giovannozzi M, and Hernalsteens C 2014 *Phys. Rev.* E **89**, 042915
[6] Borburgh J, Damjanovic S, Gilardoni S, Giovannozzi M, Hernalsteens C, Hourican M, Huschauer A, Kahle K, Le Godec G, Michels O, and Sterbini G 2016 *EPL* **113**, 34001
[7] Abernethy S, Akroh A, Bartosik H, Blas A, Bohl T, Cettour-Cave S, Cornelis K, Damerau H, Gilardoni S,

Giovannozzi M, Hernalsteens C, Huschauer A, Kain V, Manglunki D, Métral G, Mikulec B, Salvant B, Sanchez Alvarez J-L, Steerenberg R, Sterbini G, and Wu Y 2017 *Phys. Rev. Accel. Beams* **20**, 014001

[8] Huschauer A, Blas A, Borburgh J, Damjanovic S, Gilardoni S, Giovannozzi M, Hourican M, Kahle K, Le Godec G, Michels O, Sterbini G, and Hernalsteens C, *Transverse beam splitting made operational: key features of the multi-turn extraction at the CERN Proton Synchrotron*, accepted for publication in Phys. Rev. Accel. Beams.

Characterization of resonant impedances of CERN-SPS gate valves

Thomas Kaltenbacher[1], **Joel Repond**[1], **Christine Vollinger**[1]
[1]CERN, Geneva, Switzerland
E-mail: Thomas.Kaltenbacher@cern.ch

Abstract. For the CERN High Luminosity LHC project, a doubling of bunch intensity is foreseen. However, this intensity increase is currently limited by the LHC injector chain, in part due to longitudinal multi-bunch instabilities in the SPS. Therefore, the implementation of an accurate SPS impedance model was started some time ago in order to obtain a better understanding of instability sources and develop mitigation measures. In this paper, we present the electromagnetic characterization of commonly used all-metal gate valves with respect to their contribution to the SPS longitudinal impedance. The valve impedance was evaluated with commercially available EM-field simulation programs and verified with RF-bench measurements. Using this input, it was possible to obtain in particle simulations the dependence of the multi-bunch stability threshold on the number of these valves. The threshold can be increased by using commercially available impedance shielded valves. Consequently, we present the associated reduction in beam coupling impedance and the expected gain in beam stability if all existing unshielded valves are replaced by shielded ones.

1. Introduction

The upgrade of the CERN Super Proton Synchrotron (SPS) for the high luminosity LHC project (HL-LHC) requires a significant improvement in longitudinal beam coupling impedance to allow operation at the desired bunch intensities. One important part of the required impedance reduction is expected from the implementation of impedance shields in the vacuum flanges [1]. Vacuum flanges in the SPS were found to be mainly responsible for certain microwave instabilities [2]. Hence, the LIU (LHC Injector Upgrade) project baseline [1] foresees an impedance reduction campaign based on the shielding of several flanges interfacing vacuum chambers and bellows. Besides this, it was recently seen from beam dynamics simulations that the existing unshielded sector valves also give a significant contribution to the total beam coupling impedance and the respective beam instabilities [3]. In this paper, we concentrate on the evaluation of the most frequently used types of valves, labelled as *VVSA* and *VVSB*.

Table 1 gives the number of valves that are currently installed in the SPS, and their bore sizes. They are all-metal valves with a stainless steel body, differing only in bore size. Their main purpose is to allow a vacuum sectorization of the entire ring. Images of the installed VVSA (left) and VVSB (right) valves are shown in Figure 1.

Table 1. Number of Sector Valves Installed and Planned for Installation in the SPS. Note that DN100 and DN150 represents the inner bore diameter in mm.

Sector Valves	Bore Size	Number of Valves Installed
VVSA	DN100	19 (+11 planned)
VVSB	DN150	35

Figure 1. Images of installed sector valves; left: VVSA type and right: VVSB type.

2. Sector Valves Characterization

2.1. EM-Simulation and Measurement

For the EM-evaluation of these sector valves, the CAD model was imported into the EM-simulation programs for calculation of eigenmodes (HFSS [4] and CST [5]), S-parameter transmission (CST) in frequency domain to benchmark the RF-measurements and wakefields (CST) for coupling impedance calculation in time domain.

Figure 2.1 shows CAD models of a VVSA valve, with and without a shielding in the bore opening. The purpose of the impedance shield is to screen the valve body and its mechanism by providing continuity of the beam pipe via the RF-fingers (yellow in the figure). Note the larger body of the shielded valve version that is required for housing the impedance shield if the valve closes. The CAD model was provided by the valve supplier [6] and simplified for the EM-calculations. S-parameter transmission measurements with RF-probes were carried out to benchmark the simulation model for geometrical correctness. Figure 2.1 shows the schematics of the RF-measurement set-up where on-axis RF-probes were used to excite modes in the structure. In general, transmission measurements have superior sensitivity of the measured traces, and their evaluation for quality factor Q has a smaller systematic error than measurement of S_{11}. Note that this RF-probe (tip shown in Figure 2.1 inset) will not excite all modes supported by the structure, thus some resonances cannot be detected with this method. Figure 2.1 shows

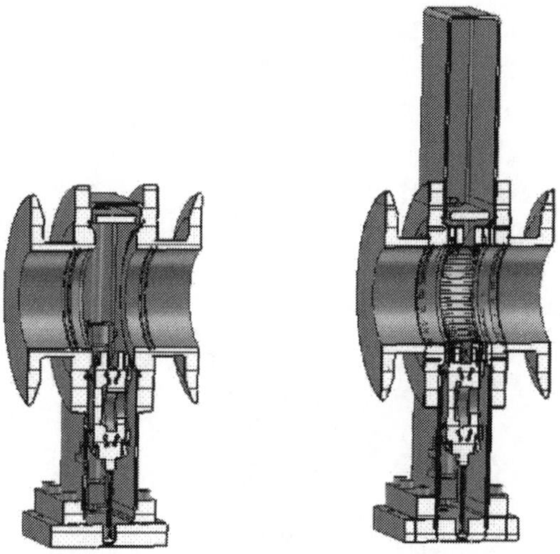

Figure 2. CAD model of a VVSA sector valve (left) and the impedance shielded version (right). The body of the shielded valve version is significantly larger than the unshielded one. The extended size is required for housing the impedance shield if the valve closes.

a comparison of S_{21}measurement with calculated longitudinal coupling impedance which is in very good agreement for the modes that can be excited with these RF-probes. For instance, the mode at ~650 MHz couples very weakly to the probes and the resonance at ~1320 MHz is not excited at all. Moreover, due to the coupling of the probes to the EM-fields of the mode at ~1270 MHz, the resonant peak is slightly shifted in frequency.

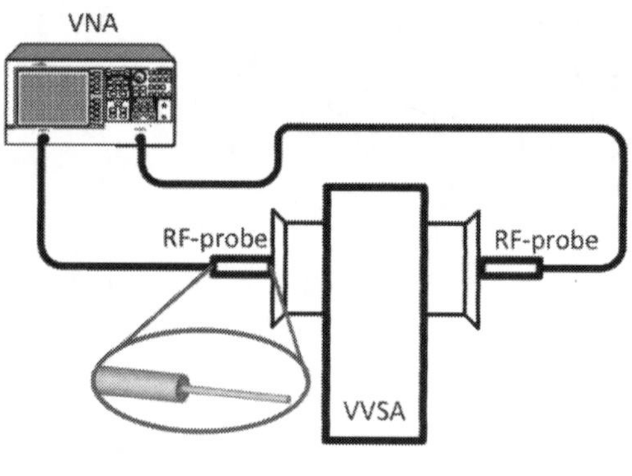

Figure 3. Schematic of the measurement set-up for benchmarking the EM-simulation model with S-parameter measurements via a vector network analyser (VNA). The inset shows the tip of the RF-probe.

2.2. Impedance Calculation

Figure 2.2 shows the simulated longitudinal beam impedances with and without impedance shield for the two valve types VVSA and VVSB.

In both cases, the traces indicate a strong reduction in resonance peaks starting from low frequencies to about 2.2 GHz. For the VVSA valve, resonances above 2.2 GHz are also slightly reduced, except the resonance peak at about 2.35 GHz which is enhanced if the impedance shield is implemented (blue arrow in Figure 2.2, top). Analysis shows that this resonance is caused by the cross-sectional change taking place when the DN100 bore of the valve reduces to the diameter of the beam pipe of 83 mm. The VVSB valve has no such a step in the cross-section, hence an impedance reduction over the entire frequency range can be achieved by implementing the impedance shield. These results were used as input to the beam dynamics calculations carried out with the Beam Longitudinal Dynamics code (BLonD) [7, 8] in order to evaluate the impact of the valve impedance on beam stability.

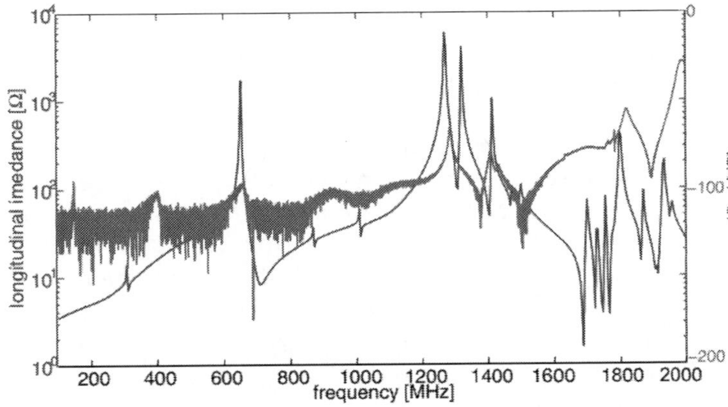

Figure 4. Comparison of S-parameter transmission measurement with calculated impedance spectrum of the VVSA.

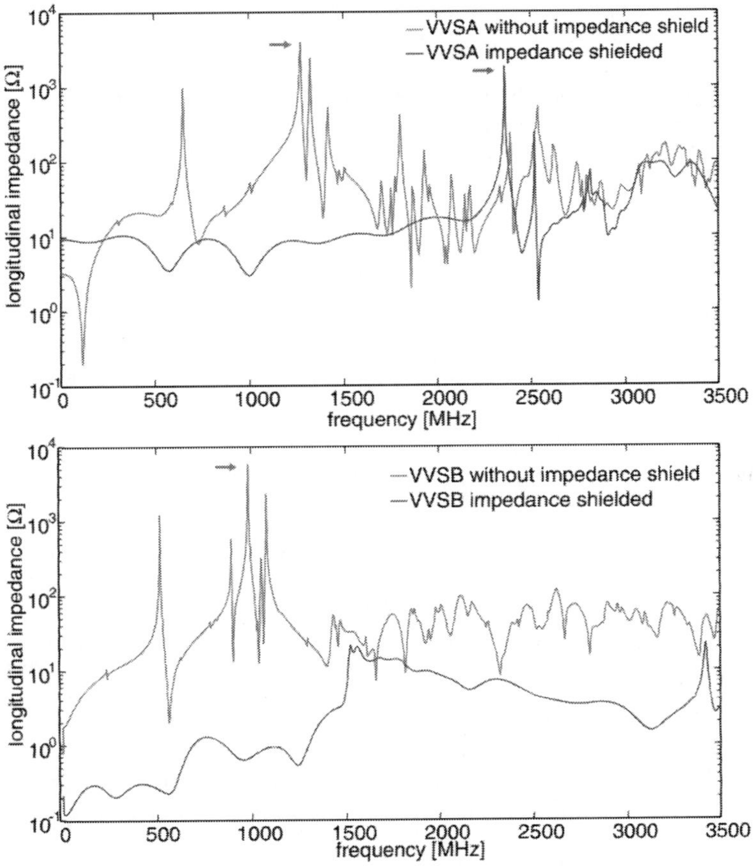

Figure 5. Calculated longitudinal beam-coupling impedance of VVSA valve (top) and VVSB valve (bottom) vs. frequency for cases without (red) and with (blue) impedance shields.

3. Beam Stability

Since the sector valves are significant contributors to the current longitudinal beam coupling impedance model of the SPS, the impact on longitudinal beam stability with various options for shielding have been studied. The simulations were carried out with BLonD at SPS flat top for 72 proton bunches. The SPS impedance model is obtained from simulations of the main machine elements that were benchmarked by RF-measurements; then the impedance model was cross-checked via measurements with beam [9]. For the situation in the accelerator after the Long Shutdown 2 (LS2 in 2019-2020), a reduction of flanges impedance by a factor 20 in the 1.4 GHz range is assumed [10]. As discussed above, this reduction can be achieved by the installation of impedance shields in the SPS vacuum flanges, a measure that is to be implemented during LS2. In addition, a 26 dB reduction on main harmonic of the 200 MHz Travelling Wave Accelerating system (TWC) of the SPS is to be achieved via the low-level RF correction loops [11]. With these assumptions, a scan in intensity and bunch length is performed using one million macroparticles per bunch over 100'000 turns (~2.3 sec) that also allows the observation of slowly growing instabilities. After its foreseen upgrade, the double RF system of the SPS is assumed to deliver after upgrade a voltage of 10 MV in the 200 MHz TWC and 1 MV in the 800 MHz cavities. A bunch length of 1.65 ns (4σ), with parabolic profile and a spread of up to $\pm$ 10%, and 2.4 $\times 10^{11}$ protons per bunch are considered nominal parameters for HL-LHC [11].

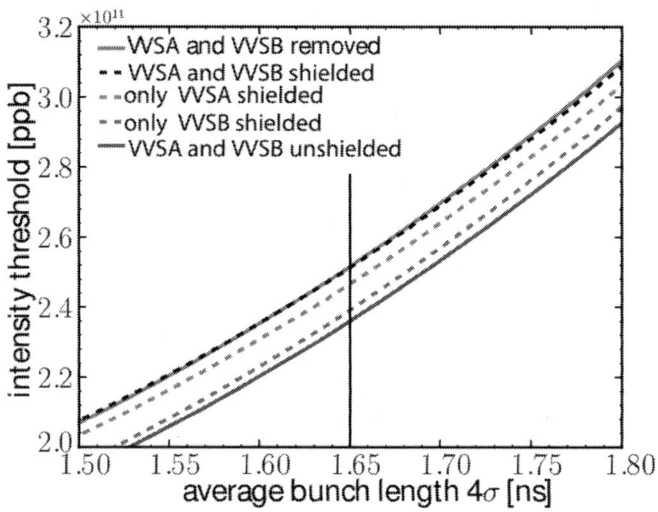

Figure 6. Results of macroparticle simulations (BlonD) for 72 bunches at SPS flat top in a double RF system (10 MV at 200 MHz and 1 MV at 800 MHz in bunch shortening mode) for the situation as expected after LS2. The plot shows intensity thresholds vs. average bunch length, for the situation with shielded and unshielded sector valves in the SPS ring. Three different scenarios for shielded valves are shown together with the case of no sector valves at all as well as the current status quo.

Figure 3 shows the intensity threshold versus average bunch length for the different configurations of sector valves for the SPS machine situation as expected after LS2:

- Current situation with existing unshielded valves, including the 11 additional valves that will be installed for the SPS sectorisation, (red continuous line);
- Simulated situation in the SPS without impedance contribution of the sector valves VVSA and VVSB, (blue continuous line);
- Configurations with alternately VVSA sector valves and VVSB sector valves equipped with impedance shields, (dashed lines).

Assuming the nominal beam parameters, shielding only one type of valve allows increasing the stability threshold by 4% or 1% for valves VVSA and VVSB, respectively (see Figure 3). If both type of valves are shielded, the the stability thresholds are very close to the case of no valves at all. A dedicated study has shown that the resonances affecting more the intensity threshold are at 1.27 GHz in case of the VVSA and at 0.992 GHz for the VVSB (see red arrows in Figure 2.2). Note that for small average bunch lengths, intensity thresholds calculated for the situation without valves and shielded valves overlap. This leads to the conclusion that shielded valves are electromagnetically transparent to the beam with the given bunch length and suggests that any new all-metal sector valve installed in the SPS should be shielded.

4. Conclusion

After accounting for the general SPS impedance reduction foreseen in LS2, the SPS all-metal sector valves were identified to reduce the multi-bunch instability threshold significantly. This translates to an intensity limitation lower then the required 2.4×10^{11} p/b at a 4σ bunch length of 1.65 ns for the High Luminosity LHC project. Even if new installations requiring VVSA or VVSB valves will be of the shielded version, it is not yet decided if the already installed sector valves will be replaced. However, based on the existing data, it seems that all newly approved installations which require VVSA or VVSB valves, shall be of the shielded type.

5. Outlook

The functionality of the VVSA and VVSB valves requires the connection of a pumping port in certain locations. Since both valve types are available from suppliers with equipped pumping ports, a further reduction of impedance could be achieved by removing existing pumping ports next to the valves and using the the valve pumping port instead, as this allows to reduce the overall number of pumping ports. Such an approach would need to be examined in detail as part of an impedance consolidation of the long straight sections of the SPS.

6. Acknowledgement

We would like to thank E. Shaposhnikova for advice and fruitful discussions, and P. Kramer for contributing to the RF-measurements as well as A. Harrison and J. Kortesmaa for kindly providing a VVSA for RF-measurements.

References

[1] Coupard J *et al.* 2014 *LHC Injectors Upgrade Technical Design Report, Vol. I: Protons* European Organization for Nuclear Research Report
[2] Argyropoulos T, Bohl T, Damerau H, Esteban Müller J F, Shaposhnikova E and Timko H 2013 Identification of the SPS impedance at 1.4 GHz *Proc. 4th Int. Particle Accelerator Conf. (Shanghai)* p 1793
[3] Shaposhnikova E, Argyropoulos T, Bohl T, Caspers F, Esteban Müller J F, Lasheen A, Salvant B, Timko H, Varela J E 2014 Identification of high-frequency resonant impedance in the CERN SPS *Proc. 5th Int. Particle Accelerator Conf. (Dresden)* p 1416
[4] ANSYS, Canonsburg, USA, `http://www.ansys.com/products/electronics/ansys-hfss/`
[5] CST AG, Darmstadt, Germany, `http://www.cst.com`

[6] VAT Group AG, Switzerland, `http://www.vatvalve.com`
[7] CERN BLonD code, Beam Longitudianl Dynamics Code, `https://blond.web.cern.ch`
[8] Timko H, Esteban Müller J E, Lasheen A and Quartullo D 2016 Benchmarking the Beam Longitudinal Dynamics code BLonD *Proc. 7th Int. Particle Accelerator Conf. (Busan)* p 3094
[9] Varela J E, Argyropoulos T, Bohl T, Caspers F, Esteban Müller J E, Ghini J, Lasheen A, Quartullo D, Salvant B, Shaposhnikova E and Zannini C 2015 An extended SPS longitudinal impedance model *Proc. 6th Int. Particle Accelerator Conf. (Richmond)* p 360
[10] Shaposhnikova E, Argyropoulos T, Bohl T, Cruikshank P, Goddard B, Kaltenbacher T, Lasheen A, Perez Espinos J, Repond J, Salvant B and C Vollinger 2016 Removing known SPS intensity limitations for high luminosity LHC goals *Proc. 7th Int. Particle Accelerator Conf. (Busan)* p 989
[11] Lasheen A, Argyropoulos T, Repond J and Shaposhnikova E 2016 Effect of the various impedances on longitudinal beam stability in the CERN SPS *Proc. 7th Int. Particle Accelerator Conf. (Busan)* p 1666

IOP Conf. Series: Journal of Physics: Conf. Series **874** (2017) 012074 doi:10.1088/1742-6596/874/1/012074

Review of stripline beam impedance: application to the extraction kicker for the CLIC damping rings

C Belver-Aguilar and M J Barnes

CERN, Geneva, Switzerland

E-mail: carolina.belver.aguilar@cern.ch

Abstract. The beam coupling impedance of the stripline kicker for beam extraction from the CLIC Damping Rings (DRs) has been studied analytically, numerically with Computer Simulation Technology (CST) Particle Studio (PS) and measured in the laboratory, although not all the results were previously understood. In order to have a better knowledge about the beam coupling impedance of a stripline kicker, a simple model has now been studied, with flat electrodes and a cylindrical beam pipe. From this preliminary study, a new approach for the dipolar component of the horizontal impedance has been derived, when considering both odd and even operating modes of the striplines. This new approach has been used to understand the differences found between the predicted transverse impedance and the two wires measurements carried out in the laboratory for the prototype CLIC DR striplines. Future tests of beam coupling impedance with beam in the ALBA Synchrotron Light Source will complete this study.

1. Preliminary study: a simple approximation

Several approaches for calculating the beam coupling impedance of stripline pickups and kickers can be found in the literature [1, 2, 3, 4, 5], most of them based on the image current analysis of Lambertson and Goldberg, which is only valid at frequencies below c/L [3], where L is the striplines length and c is the speed of light in vacuum. These approaches are similar, but two main differences can be noted: some of them use an analytical approach for the geometrical factors [1, 2], with an accuracy greater than 95% for electrodes with $h > 2a$, where h is the electrodes height and a is the aperture, i.e. the distance between the electrodes. Other authors use a more precise geometric factor calculated numerically with an electrostatic solver [3, 4, 5], which can be applied to striplines of any shape and dimensions. In addition, in [1, 2, 3] the two different operation modes of striplines (differential and common, also known as odd and even, respectively) are not considered, whereas in [4, 5] the difference between the odd and even characteristic impedances has been taken into account. However, the question of which operation mode is related to the longitudinal and transverse beam coupling impedances, for a stripline kicker, is not well documented and is studied in more detail in this paper.

In [5], analytical equations for the longitudinal and transverse beam coupling impedances were compared with numerical simulations and laboratory measurements: a good agreement was found for the longitudinal beam coupling impedance. However there was a disagreement in the dipolar component of the transverse beam coupling impedance, which was not understood. In [1, 2, 3], the transverse beam coupling impedance is related to the even mode characteristic impedance, whereas in [4, 5] it is related to the odd mode. Hence, the starting point of this preliminary study was to analyse the horizontal beam coupling impedance in the two operation modes, and compare these results with numerical simulations and measurements, to better understand which operation mode was excited.

The simple model used (Figure 1), consists of flat electrodes of 500 mm length and 25 mm height, with an aperture of 20 mm and a cylindrical beam pipe radius of 25 mm. For these striplines, the predicted odd and even mode characteristic impedances are 44.6 Ω and 60.6 Ω, respectively.

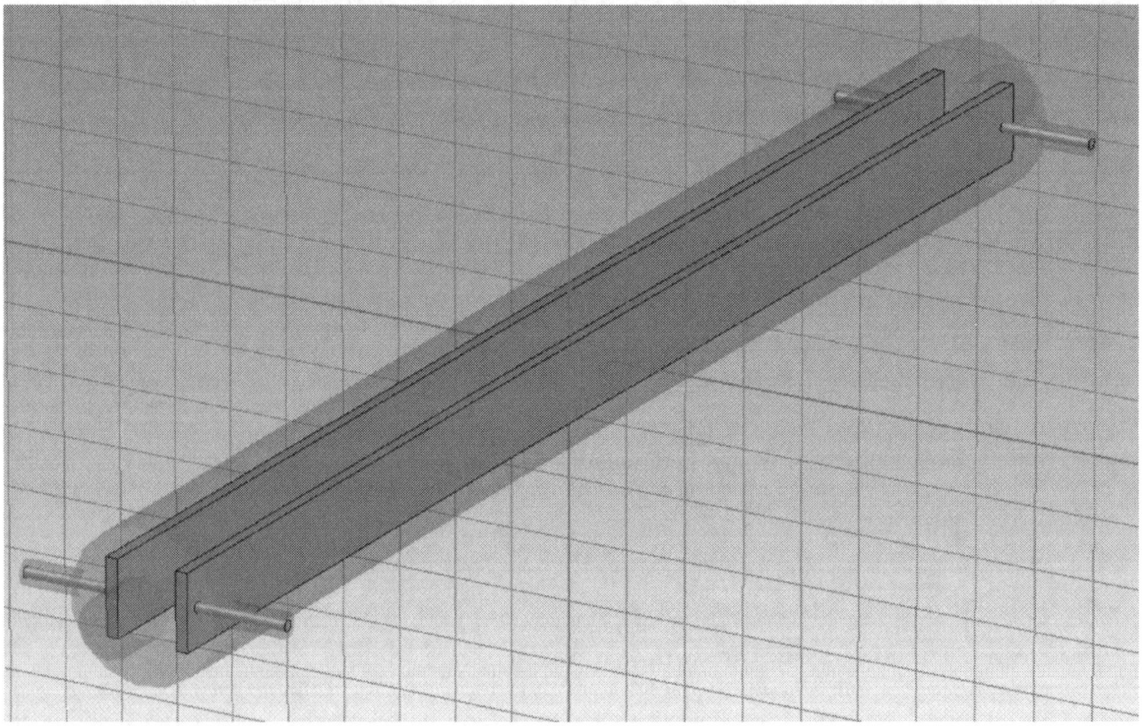

Figure 1. Simple model used for studying the beam coupling impedance and its relationship to the two operation modes of a stripline kicker.

1.1. Longitudinal beam coupling impedance

The longitudinal beam coupling impedance can be analytically calculated from the image currents in the electrodes, created by a beam passing through the aperture of the striplines, i.e. in the common or even mode configuration, with the following equation [4, 5]:

$$Z_{||} = Z_{comm} g_{||}^2 \Big(sin^2\Big(\frac{\omega L}{c}\Big) + j sin\Big(\frac{\omega L}{c}\Big) cos\Big(\frac{\omega L}{c}\Big)\Big) \tag{1}$$

where Z_{comm} ($= Z_{even}/2$) is the common mode characteristic impedance, $g_{||}$ ($= 0.89$) is the longitudinal geometric factor, and ω is the angular frequency. This equation is valid under the assumption that the beam is passing through the centre of the stripline aperture, or with an offset $x_0/b \ll 1$ [2], where x_0 is the beam offset and b is the half-aperture of the striplines. The longitudinal geometrical factor has been calculated with the electrostatic solver of Opera 2D as $g_{||} = V(0,0)$, i.e. it is equal to the voltage at the centre of the stripline aperture, when considering that the electrodes are driven by a voltage of 1 V each. The result has been compared with the longitudinal beam coupling impedance predicted with CST PS, when modelling a bunch of 20 mm length passing through the centre of the stripline aperture, and found to be in good agreement, as will be shown below.

In addition, the longitudinal beam coupling impedance can be measured in the laboratory, by measuring the transmission through a wire placed in the centre of the stripline aperture. The longitudinal beam coupling impedance can be calculated using the formula [8]:

$$Z_{||} = -2Z_{line}ln(S_{21}) \tag{2}$$

where Z_{line} is the characteristic impedance of the transmission line formed by the stripline and the wire and $S_{21} = S_{21,DUT}/S_{21,REF}$ is the corrected transmission parameter. A more detailed explanation of this measurement can be found in [5]. Although the results for the longitudinal beam coupling impedance agreed well, for this preliminary study, we have now included predictions from a new model: the wire setup has been simulated in CST Microwave Studio (MWS), and the longitudinal beam coupling impedance has been calculated from the transmission parameter through the wire, as would be done in the laboratory. For this simulation, a Z_{line} of 220.7 Ω has been numerically calculated with CST MWS. The results of this model have been compared with Equation (1) and the CST PS simulation, as shown in Figure 2.

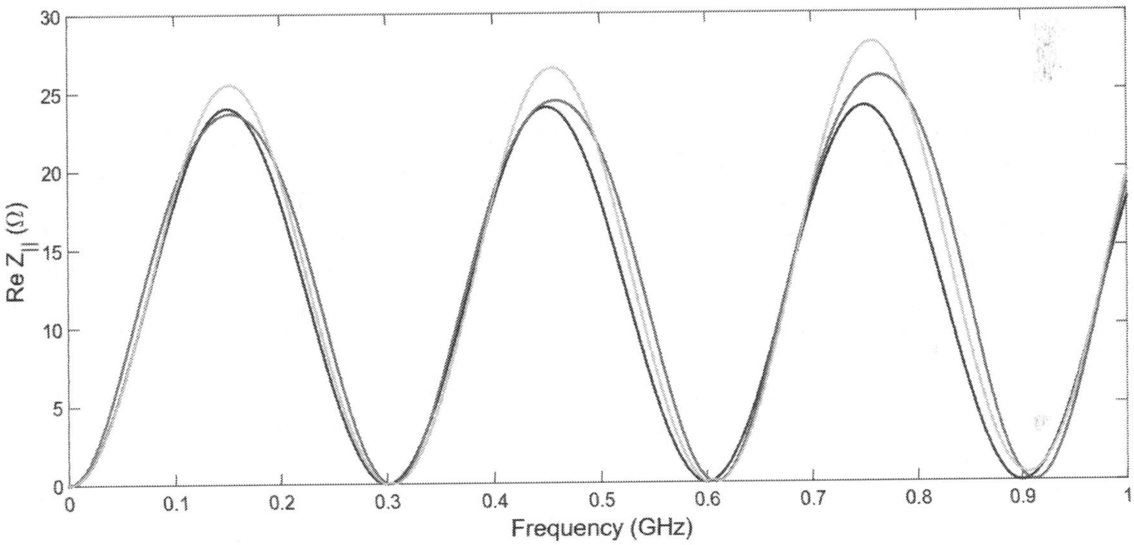

Figure 2. Real component of longitudinal beam coupling impedance results for Equation (1) (blue line), the CST PS simulation without a wire (red line), and the CST MWS simulation with a wire (green line).

Figure 2 shows the real component of the longitudinal beam coupling impedance. Only the real components of the beam coupling impedance are shown to keep the same notation through the paper –with the two wire method only the real component of the transverse beam coupling impedance was derived and studied. Equation (1) and the CST PS simulation match well in the frequency range studied, although some differences are observed already below ∼ 200 MHz. The peaks resulting from the wire method simulation (CST MWS) are slightly larger, but can still be considered a good approximation.

1.2. Horizontal beam coupling impedance

The transverse (horizontal or vertical) impedance of striplines can be divided into two different components: the dipolar component and the quadrupolar component [8]. The dipolar component of the horizontal beam coupling impedance can be directly calculated from the longitudinal beam coupling impedance (Equation (1)), with the Panofsky-Wenzel theorem [3]:

$$Z_{dip,x} = \frac{cZ_{||}}{\omega b^2}\left(\frac{g_{\perp}}{g_{||}}\right)^2 \tag{3}$$

In Equation (3) b (= $10mm$) is half of the stripline aperture, and $g_{\perp}$ (= 0.98) is the transverse geometrical factor. The transverse geometrical factor has been calculated with the electrostatic solver of Opera 2D, as $g_{\perp} = V(x,0)b/x$, where x is the considered beam displacement, when driving one electrode with 1 V and the other one with -1 V. The analytical results have been compared with the CST PS predictions, when offsetting the beam horizontally by 1 mm from the centre of the stripline aperture. The results of the comparison are shown below.

The two wire measurement set-up has been modelled with CST MWS, where a dipolar field is created by driving the two wires with equal but opposite polarity voltages. The longitudinal beam coupling impedance is calculated using Equation (2), where now Z_{line} (= 369.2 Ω) is the differential characteristic impedance of the transmission line formed by the stripline and the two wires, calculated with CST MWS. The dipolar component of the horizontal beam coupling impedance is [8]:

$$Z_{dip,x} = \frac{cZ_{||}}{\omega \Delta^2} \tag{4}$$

In this equation Δ corresponds to the half-aperture of the striplines, instead of being the distance between the two wires, as it was supposed in [5]: for striplines with small aperture, the dipolar field extends along the whole striplines aperture. The authors suggest that results from Equation (4) cannot be directly compared with Equation (3) and CST PS simulation results, because the two wire measurement excite the odd (or differential) mode, whereas both Equation (3) and the CST PS simulation results refer to the horizontal beam coupling impedance in the even (or common) mode. Nevertheless, a modified version of Equation (3) is proposed by the authors to consider the dipolar component of the horizontal beam coupling impedance in the odd mode. For this mode, the Panofsky-Wenzel theorem cannot be directly applied to Equation (1), since it is related to the even mode. To consider the odd mode, the additional factor Z_{diff}/Z_{comm} should be added:

$$Z_{dip,x} = \frac{cZ_{||}}{\omega b^2}\left(\frac{g_{\perp}}{g_{||}}\right)^2\left(\frac{Z_{diff}}{Z_{comm}}\right) \tag{5}$$

where Z_{diff} (= $2Z_{odd}$) is the differential characteristic impedance. All the results are shown and compared in Figure 3.

CST PS calculates the transverse beam coupling impedance from the transverse wakefields created by the image currents of the beam, i.e. in the even mode configuration of the striplines, whereas the two wire simulation has been made by considering currents with opposite direction in each wire, i.e. the odd mode configuration of the striplines. Figure 3 shows an excellent agreement between Equation (3) for the even mode and the CST PS simulation, and between Equation (5) for the odd mode and the 2 wire simulation.

2. Application to the compact linear collider damping ring striplines

2.1. Longitudinal beam coupling impedance

The longitudinal beam coupling impedance of the extraction kicker for the CLIC DRs was calculated and also measured in the laboratory, with a good agreement. The simulation of the

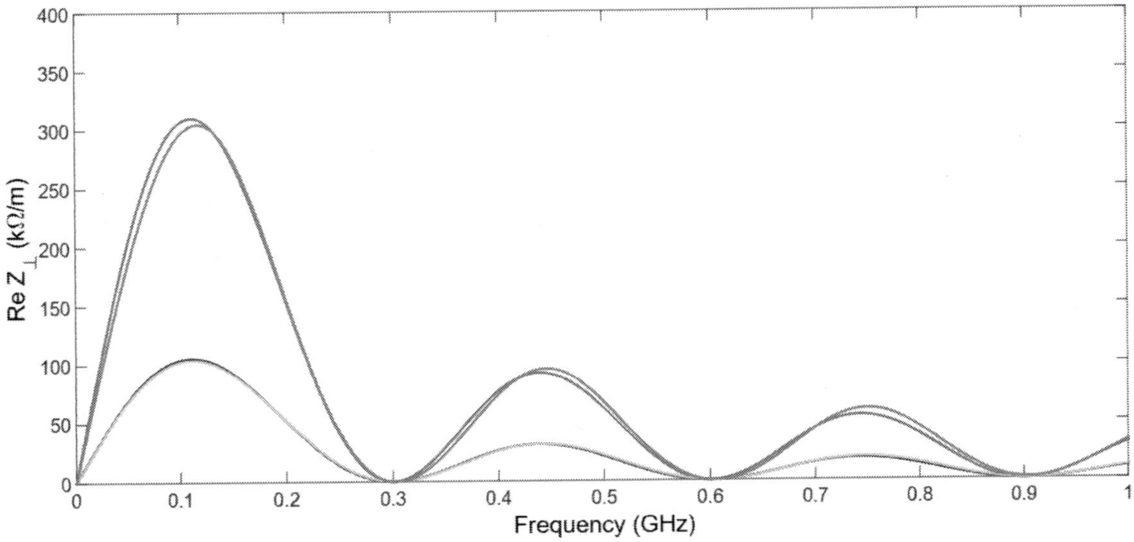

Figure 3. Dipolar component of the horizontal beam coupling impedance from Equation (3) (blue line) and Equation (5) (red line), the CST PS simulation (green line) and the CST MWS simulation of the two wire measurement set-up (magenta line).

single wire set-up with CST MWS has recently been included: results are shown in Figure 4. For good visualization of the beam impedance behaviour, the frequency range is plotted up to 500 MHz.

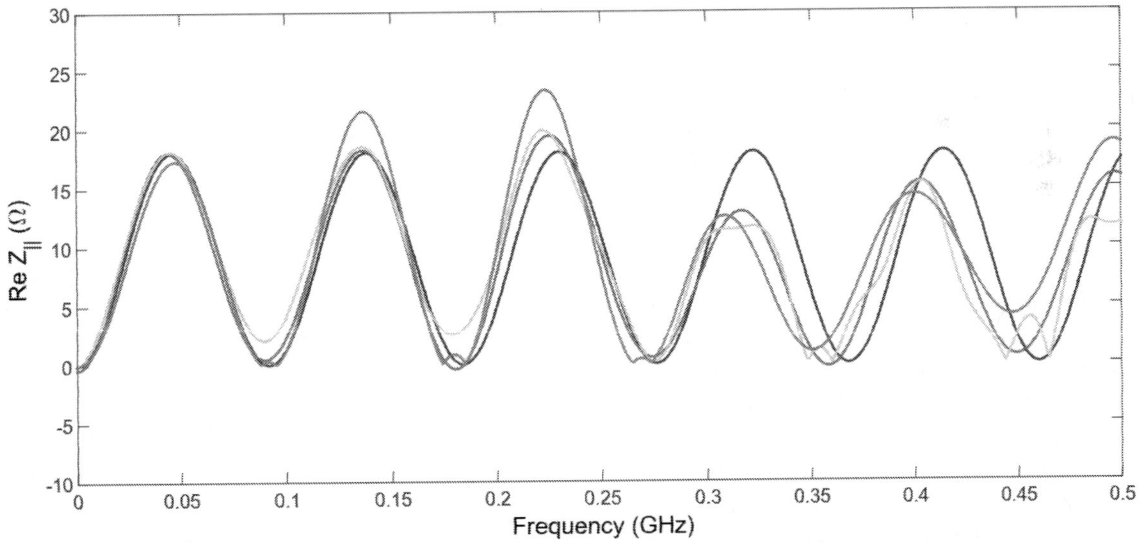

Figure 4. Longitudinal beam coupling impedance from Equation (1) (blue line), the CST PS simulation (red line), the single wire measurement (green line) and the CST MWS single wire simulation (magenta line).

The longitudinal beam coupling impedance measured (green line) agrees well with the CST

PS (red line) up to around 450 MHz, whereas the single wire simulation (magenta line) is slightly different but still quite good. In addition, Equation (1) (blue line) is a good approximation up to 300 MHz.

2.2. Horizontal beam coupling impedance
The dipolar component of the horizontal beam coupling impedance of the extraction kicker for the CLIC DRs was calculated and also measured in the laboratory, although a disagreement was previously found [5]. We have made new calculations considering the approaches defined in the previous section, and also included a simulation of the single wire set-up with CST MWS. The results are shown in Figure 5.

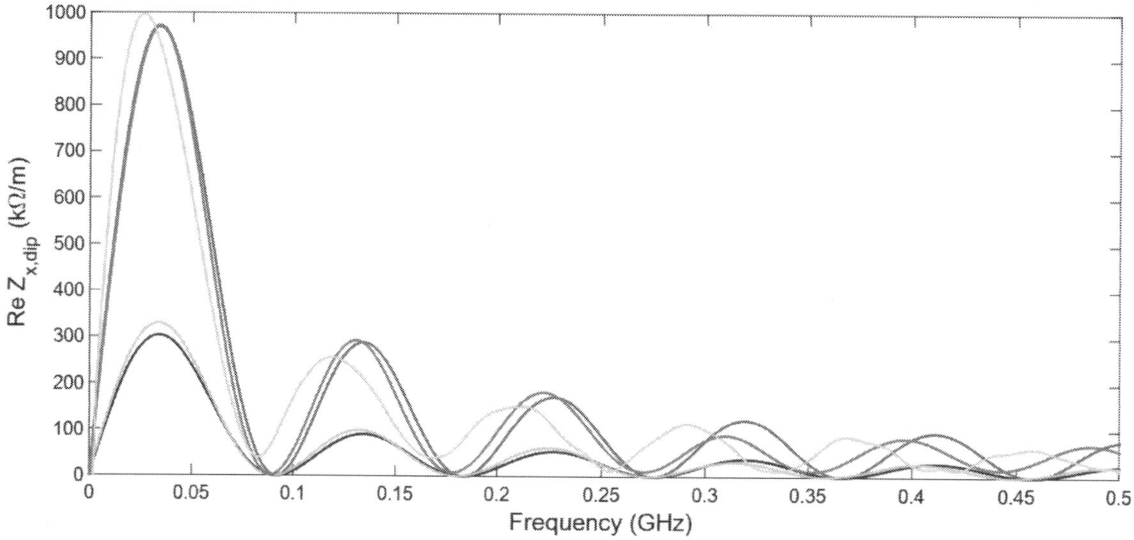

Figure 5. Dipolar component of the horizontal beam coupling impedance from Equation (3) (blue line) and Equation (5) (red line), the CST PS simulation (green line), the simulation of the two wire measurement set-up (magenta line) and the two wire measurement result (cyan line).

Figure 5 shows that Equation (3) (blue line) agrees very well with the CST PS simulation (green line): both are for the even mode. Furthermore, Equation (5) agrees with the 2 wire simulation (cyan line) and the 2 wire measurement (magenta line): all three are for the odd mode. There is a frequency shift in the measurements, probably due to the reduced velocity of the current through the wires, which were considered to be lossless in the simulations.

3. Conclusions

Analytical equations for estimating the beam coupling impedance of stripline pickups and kickers have been widely used in the past. However, some confusion arises from the different notation and approximations which depend upon the author. In this paper we have tried to solve one of the main issues, i.e. to identify which stripline mode is excited when the beam passes through the centre of the aperture, when the beam is displaced from the centre, and when one or two wires are inserted into the stripline aperture. The conclusion of the preliminary study is that the beam excites the even mode of the stripline, whereas the two wire measurement excites the odd mode. Therefore, these two results cannot be compared directly, and a modified equation

IOP Conf. Series: Journal of Physics: Conf. Series **874** (2017) 012074 doi:10.1088/1742-6596/874/1/012074

(Equation (5)) has been presented to consider this difference. Further studies of the quadrupolar contribution to the transverse impedance will be done in the future, as well as measurements with beam in ALBA.

References

[1] Shafer R E 1985 *IEEE Trans. on Nucl. Sci.* **NS-32, 5** pp 1933–37.
[2] Ng K Y 1987 *Proc. of American Inst. of Phys. Conf. (Batavia)* **184** pp 472-524.
[3] Goldberg D A and Lambertson G R 1992 *Proc. of American Inst. of Phy. Conf. (Upton)* **249** pp 537–600.
[4] Blednykh A, Cheng W and Krinsky S 2013 *Proc. of Part. Acc. Conf. (Pasadena)* 895–7.
[5] Belver-Aguilar C, Faus-Golfe A, Toral F, Barnes M J and Day H 2015 *Proc. of Int. Part. Acc. Conf. (Richmond)* pp 3058–61.
[6] *Opera2D User Guide, Version 17R1.* Cobham Technical Services, Vector Fields Software 2014 (Oxfordshire).
[7] CST - Computer Simulation Technology, http://www.cst.com
[8] Kroyer T, Caspers F and Gaxiola E 2007 *CERN AB-Note-2007-028 (Geneva).*

IOP Conf. Series: Journal of Physics: Conf. Series **874** (2017) 012075 doi:10.1088/1742-6596/874/1/012075

Studies and observations of beam dynamics near a sum resonance

G Franchetti*, S Gilardoni†, A Huschauer†, F Schmidt†, R Wasef†

* GSI Helmholtzzentrum für Schwerionenforschung GmbH, 64291, Darmstadt, Germany

† CERN, 1211 Geneva 23, Switzerland

Abstract. The effect of space charge on bunches stored for long term in a synchrotron can be severe for beam survival. This may be the case in projects as SIS100 at GSI or LIU at CERN. In the past decade systematic simulation studies and experiments performed at CERN and GSI have highlighted the space charge induced periodic crossing of one dimensional resonances as an underlying mechanism of long term beam loss or emittance growth. However only in 2012, for the first time, the effect of space charge on a normal third order coupled resonance was investigated at the CERN-PS. The experimental results have highlighted an unprecedented asymmetric beam response where in the vertical plane the beam exhibits a thick halo, whereas the horizontal profile has only core growth. The quest for explaining these results requires a journey through the 4 dimensional dynamics of the coupled resonance investigating the fixed-lines, and requires a detailed code-experiment benchmarking also including beam profile benchmarking. This study shows that the experimental results of the 2012 PS measurements can be explained by the dynamics the of the fixed-lines with space charge also including the effect of the chromaticity.

1. The accelerator case

Space charge effects on beam circulating in a synchrotron becomes stronger for an increased beam intensity, and the problems arising with the increased intensity define an intensity limit of performance. The space charge limit is a figure of merit used to characterize the maximum intensity tolerated by a machine. It is usually defined with respect to the beam loss, although other criteria can be used to define the limit performance: for example, emittance growth, halo formation.

For projects the space charge limit plays a significant role, because it defines the ultimate accelerator performance in terms of stored particles. Usually the space charge limit is defined by a constraint on the maximum allowed incoherent space charge tune-shift. This is dictated by the working point the machine for avoiding the overlapping of the space charge tune-spread with machine resonances. For a well placed machine working point, the maximum allowed incoherent tune-shift is usually $\Delta Q_{sc} \sim -0.25$ to avoid second, third, and fourth order resonances. Clearly this argument mainly depends on which machine resonances are significant for beam loss. If the 3rd and 4th order resonances are very weak one can use more optimistically relax the constraint to $\Delta Q_{sc} \sim -0.5$. The assessment of the real effect of the space charge on machine resonances requires a detailed study of the dynamics, and it is strongly related to the storage time.

The investigation of the effect of space charge for long term storage of a bunched beam has been motivated from the planned operations with U^{28+} ions in the SIS100 synchrotron of the FAIR project [1]. In this scenario SIS18 will operate with harmonics 2, and 4 cycles will be injected into SIS100 for a total of 8 bunches transferred in 10 buckets. An amount of 6×10^{11} ions will be accumulated during 1 second at the injection energy of 200 MeV/u. This means that the first 2 bunches from the SIS18 cycle injected into SIS100 will be stored for 1 second, namely for 1.5×10^5 turns. During this time the beam with transverse rms emittances of $\epsilon_{x,rms} = 8.75$ mm-mrad, $\epsilon_{y,rms} = 3$ mm-mrad, will be bunched with bunching factor $B_f = 0.33$, generating an incoherent space charge tune-shift of $\Delta Q_x \simeq -0.2, \Delta Q_y \simeq -0.33$. In the SIS100 RF buckets U^{28+} ions follow a longitudinal dynamics with a synchrotron tune of $Q_s \sim 4.27 \times 10^{-3}$, and over 1 second storage, a particle will perform 643 or less synchrotron oscillations. This accelerator regime is very unique with no established operational experience in the accelerator community.

The synchrotron motion pushes periodically bunched particles from regions of small transverse space charge, to the region of large space charge tune-shift (at $z = 0$). Therefore, according to the initial condition, a particle may cross a machine resonance each quarter of a synchrotron oscillation.

A fast single crossing of a weak resonance typically produces a small emittance growth: in SIS100 for U^{28+} ions a quarter of synchrotron oscillation happens in $\sim$ 58 turns. However, over one second storage the number of resonance crossings becomes as high as $\sim$ 2700. The integrated effect arising from this large number of repeated resonance crossing may create a large impact generating a beam diffusion to large transverse amplitudes, hence may lead to a steady beam loss throughout the storage time. Note that for this effect to be significant requires many resonance crossings, hence it disappears, or becomes insignificant for short term storage.

The assessment of the beam dynamics in this regime is challenging. The simulations are difficult: standard PIC codes are subjected to artificial noise, which affects the long term result of the simulations [2, 3, 4, 5]. In order to mitigate this artifact one should use a very large number of macro-particles [5], which makes unfeasible systematic studies to understand the properties of the dynamics. Especially it is difficult to develop tools which may be used to predict the effects, or to develop a theory that may allow to assess how the different beam and machine features influence the emittance growth.

2. Space charge and one dimensional resonances

In the past decade, the interplay of space charge and machine nonlinearities has been studied for 1D resonances, i.e. type $n_x Q_x = N$. Experimental and numerical studies on beam survival and emittance growth in this regime have been carried out in the CERN-PS in the year 2003 for the resonance $4Q_x = 25$ (see Reference [6, 7]), and later in the SIS18 at GSI for the resonance $3Q_x = 13$ (see Reference [8]).

The mechanism leading to emittance growth and beam loss is explained in terms of instantaneous stable islands in the two-dimensional phase space and their crossing the particle orbits because of the combined effect of space charge and synchrotron motion [9]. The one dimensional resonances allow to discuss the dynamics in one plane.

The key features understood from past studies are summarized in the following qualitative points:

a) The space charge detuning determines the position of the fixed-points generated by the 1D resonance, which is approximately given by the condition

$$n[Q_{x0} + \Delta Q_x(X)] = N, \tag{1}$$

where here X is the amplitude of the resonant particle, and $\Delta Q_x(X)$ is the incoherent amplitude dependent space charge detuning. For $X = 0$ the quantity $\Delta Q_x(0)$ corresponds to the usual incoherent tune-shift.

b) The strength of the resonance determines the tune of particles around the fixed-points, and consequently how large is the island. The island size has folded-in also the detuning created by space charge: stronger gradients in the amplitude dependent detuning leads to smaller islands.

c) The synchrotron tune determines how fast is the resonance crossing. A figure of merit on how fast the resonance crossing is, is given by the parameter T obtained as the ratio between the speed of migration of the fixed-points to the maximum speed of rotation of the particle in the island. If this ratio is small ($T \ll 1$) it means that the motion is adiabatic, and the particles remain locked to the island (trapping); If instead $T > 1$ then a single resonance crossing produces a "kick" to the particle invariant (scattering).

The standard accelerators settings bring the dynamics mainly to scattering processes, which are highly dependent from the strength of the resonance, space charge tune-shift, distance from the resonance, and synchrotron tune.

3. The difficulty of the two dimensional resonances

For a two dimensional resonance as $n_x Q_x + n_y Q_y = N$, the dynamics are coupled. That means that a two dimensional analysis of the Poincaré section of the coordinates x, x' does not provides the proper understanding. The major difficulty is in understanding how the space charge modify the tori embedded in the four dimensional phase space. Usually the tune-spread is not the same in horizontal and vertical planes, hence it is particularly difficult to conclude on the process of halo formation, and on its extension in the $x - y$ plane. The meaning of trapping/scattering is also not straightforward to define.

4. Measurements on the 3rd order coupled resonance

Recently the LIU project at CERN [10] has engaged in studies to explore space charge effects [11], and in investigating the long term storage for high intensity bunches as well. In the framework of the CERN-GSI collaboration, studies on the effect of space charge on the resonance $Q_x + 2Q_y = 19$, have been carried out in the CERN-PS in 2012. We present here a short summary of the full experiment analysis. A more comprehensive presentation of the experimental results and discussion is part of a future publication in Reference [12].

In the experiment, in a resonance-free region of the tune diagram the third order resonance $Q_x + 2Q_y = 19$ was excited with two nearby sextupoles of strength $K_2 \simeq 0.015$ m^{-2}. The experimental campaign used a beam with 55×10^{10} protons per bunch, which produced an incoherent space charge tune-shift of $\Delta Q_x \simeq -0.05, \Delta Q_y \simeq -0.07$. The beam was stored at the energy of 2 GeV for 1.1 seconds corresponding to $\sim 0.5 \times 10^6$ turns. During this time, the synchrotron tune $Q_s \sim 8.57 \times 10^{-4}$ induces ~ 1700 possible resonance crossings, and their effect on the beam was assessed by measuring the beam profiles at the beginning and at the end of the storage time.

The measurements were repeated for several machine tunes so to explore the dependence of the effects on the distance from the resonance. The comparison of initial and final beam profiles revealed an unexpected beam response to the distance from the resonance. When the space charge tune-spread overlaps the third order resonance and the machine tune is close to the resonance without beam loss, the beam profiles evolves differently in the transverse planes: in the horizontal plane the beam exhibits core growth, whereas in the other plane a large halo is formed. This is shown in Figure 1a,b for the tunes $Q_{x0} = 6.104, Q_{y0} = 6.476$. The x/y asymmetry of the beam profile is quite evident and shows that a new and more complex dynamics is driving the beam halo formation. In the picture are also shown the profiles from computer simulation as obtained from MADX, and MICROMAP (see Reference [13] for a general code benchmarking discussion). Figure 1c shows the overall beam response in the experiment. The profiles in Figure 1a,b correspond to the largest emittance growth.

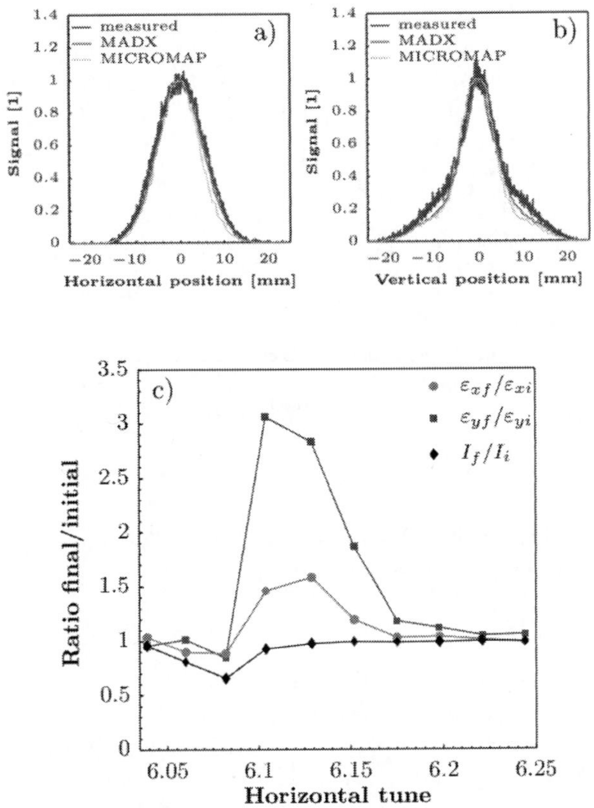

Figure 1. Part a, and b: horizontal and vertical beam profiles after 1.1 second storage of the beam in the CERN-PS. The asymmetry of the beam response is evident. In part c is shown the full beam response to all working points investigated.

5. Analysis with detuning

The explanation of the halo vs. core finding can be approached from a detuning analysis. The theory of the resonances (see Reference [14, 15]) shows that the distance from the third order resonance is $\Delta_{r0} = Q_{x0} + 2Q_{y0} - 19$ so that when $\Delta_{r0} = 0$ the machine tunes sit on the resonance. This definition valid for the machine tunes can be extended to any arbitrary particle as $\Delta_r = Q_x + 2Q_y - 19$, where now the tunes Q_x, Q_y are the effective tunes experienced by a test particle, which is affected by space charge, chromaticity, and any other effects.

For the case of pure space charge, these tunes are depending on the particle amplitudes X, Y. The amplitude dependent detuning $\Delta Q_{sc,x}(X,Y), \Delta Q_{sc,y}(X,Y)$ is folded in the single particle tune as $Q_x = Q_{x0} + \Delta Q_{sc,x}(X,Y)$,and $Q_y = Q_{y0} + \Delta Q_{sc,y}(X,Y)$, so that

$$\Delta_r = \Delta_{r0} + \Delta Q_{sc,x}(X,Y) + 2\Delta Q_{sc,y}(X,Y). \tag{2}$$

This relation allows for an approximate finding of the amplitudes X, Y of resonant particles. Given the machine tunes Q_{x0}, Q_{y0} we find Δ_{r0}, and the resonant transverse amplitudes X, Y which satisfy the equation $\Delta_r = 0$. More in general Equation (2) shows that Δ_r is dependent on

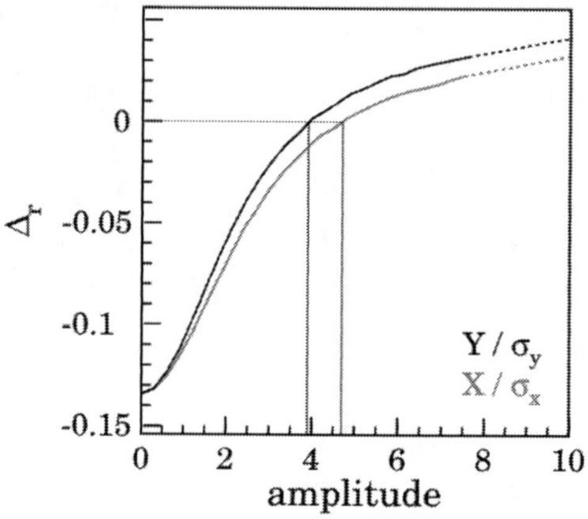

Figure 2. Resonance detuning Δ_r as function of the particle amplitude.

X, Y, and can be regarded as a "resonance detuning" that incorporates the coupled character of the resonance $Q_x + 2Q_y = 19$ and space charge.

In Figure 2 the two curves show the dependence of Δ_r for two types of particle amplitudes. The red curve is obtained for amplitudes type $(X, 0)$, while the black curve is obtained for amplitudes $(0, Y)$. The horizontal line of height $\Delta_r = 0$ intercepts the two curves at the resonant amplitudes, which in our case are $X \sim 5\sigma_x$ for the red curve, and $Y \sim 4\sigma_y$ for the black curve.

From the periodic resonance crossing induced by space charge we expect that particle diffusion does not exceed the outer position of the "resonant particles", which is $X \sim 5\sigma_x$, and $Y \sim 4\sigma_y$. However, a comparison with a multi-particle simulation in absence of chromaticity, shows that no halo with amplitude $X \sim 5\sigma_x$ is found.

By including into Equation (2) the effect of the chromaticity we investigate the role of the chromaticity on the resonant amplitudes. For example, for particles with maximum $\delta p/p = 2.4 \times 10^{-3}$ we can construct an equivalent graphic of Figure 2, and search for the resonant particles at largest amplitudes (condition $\Delta_r = 0$). We find that the halo predicted by this analysis is $X > 9\sigma_x$, and $Y \sim 9\sigma_y$. Once more, this result contradicts the experimental findings, in which the halo is found at $Y \simeq 5.5\sigma_y \simeq 20$ mm.

6. Interpretation with the fixed-lines

The previous analysis based on the detuning ignores the resonant process itself, which clearly is included in the beam profiles observed and retrieved from simulations in Figure 1. In the absence of space charge, a particle satisfying the resonance condition $\Delta_{r0} = 0$ has dynamics locked to the resonance. Contrary to the 1D resonances, the 2D resonance makes the particle orbit at the Poincaré section to be bounded to a one dimensional closed curve extending in the full 4D phase space. This curve, in analogy to the fixed-points for one dimensional resonances, is a fix-line [16, 17, 15]. The analytic form of these closed lines is parameterized as

$$x = \sqrt{\beta_x a_x} \cos(-2t - \alpha + \pi M), y = \sqrt{\beta_y a_y} \cos(t). \tag{3}$$

The coordinates x', y' are readily derived by differentiation. The fixed-line emittances a_x, a_y are determined by the distance to the resonance Δ_{r0}. β_x, β_y are the beta functions at the Poincaré section location. The variable t parameterizes this one dimensional curve. The coefficient M is 0, or 1 according to the condition of existence of the fixed-line. In Figure 3 we show an example of the projections of one fixed-line plotted in normalized coordinates. The $x - y$ projection has a distinct "8" shape. The $x' - y'$ projection has instead a "C" shape. The dependence of α in Equation (3) makes the shape "C" and the shape "8" exchangeable according to the location of the Poincaré section: the parameter α is the phase of the driving term of the third order resonance, and it depends on the difference of phase advance between the location of the sextupoles used to excite the resonance with respect to the observation point along the machine (the flying wire position).

The interplay of the resonant dynamics with the space charge is the key issue, and is here presented with a numerical study. The role of the fixed-lines in the bunched beam is investigated

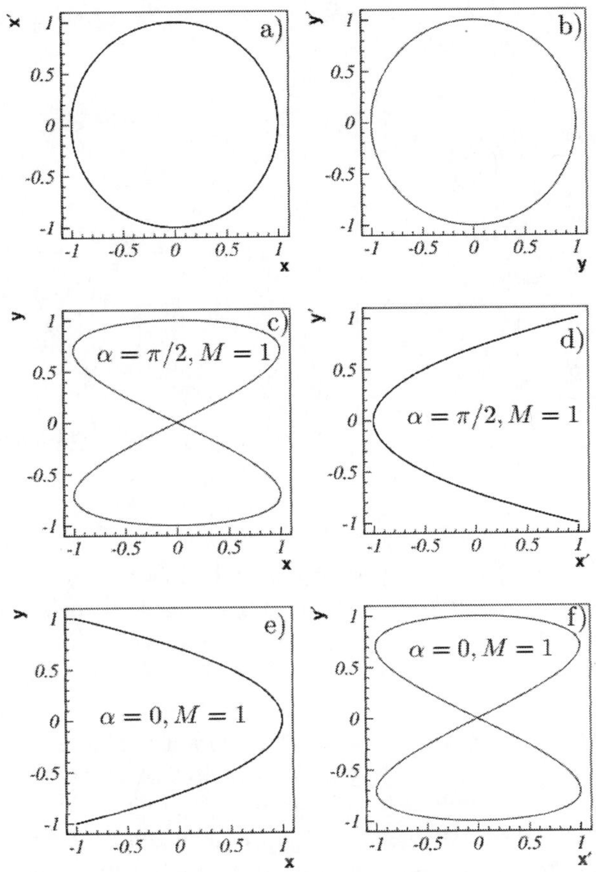

Figure 3. Fixed-line projections in normalized coordinates. The parameters in Equation (3) are indicated in each picture. Note that the topology of the $x - y$ projection of a fixed-line depends substantially on the angle α.

by freezing artificially the longitudinal motion in the simulation modeling the experimental beam profiles of Figure 1, and by constructing a tune footprint of a set of test particles. The resonant particles are easily identified using an FFT method. In fact, the main peak of the FFT spectrum yields the nonlinear tunes Q_x, Q_y of the particles satisfying $Q_x + 2Q_y - 19 = 0$. We then plot the $x - y$ projection of the resonant orbits and compare it with the $x - y$ projections in Figure 3 searching for indications of a fixed-line dynamics. In part a of Figure 4 we show two resonant orbits for test particles located at $z/\sigma_z = 0$. The topology of the orbit projection leaves no doubt that we see two particles locked to two distinct fixed-lines having the "8" shape in $x - y$ projection. The asymmetric form of the orbits is remarkable although the resonant particle is not sitting exactly on the fixed-line (we see more the projection of a resonant tori). This result simply shows that the presence of the beam space charge detuning does not destroy the dynamics that creates the fixed-lines. By repeating the same analysis for particles at $z/\sigma_z = 1.5$ we can visualize the resonant orbits at a location of the bunch with smaller transverse space charge. These resonant orbits are shown in part b of Figure 4. Even in this case we clearly distinguish in the resonant orbits the pattern of fixed-lines. The aspect ratio is now different, and the extension of the orbits is smaller.

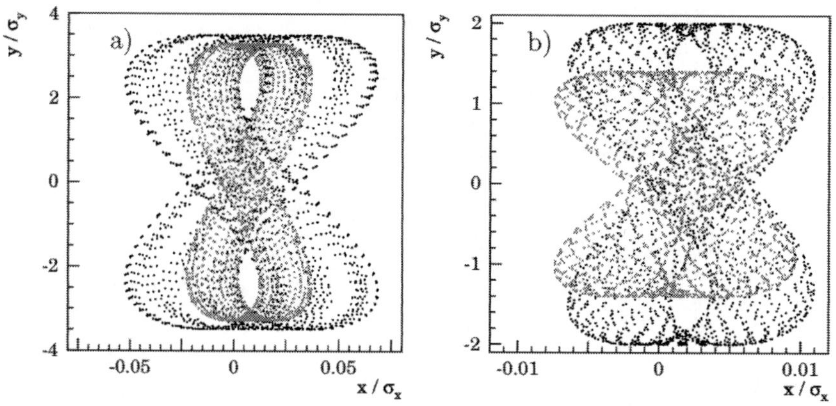

Figure 4. Part a). $x - y$ projection of the two largest resonant orbits at $z/\sigma_z = 0$; Part b). The two largest resonant orbits now at $z/\sigma_z = 1.5$.

The extension of the "frozen" fixed-lines along the bunch is consistent with the studies on one dimensional resonances. In Reference [9] it is shown that the frozen transverse islands have the maximum size and their fixed-points have maximum amplitudes at longitudinal positions $z/\sigma_z = 0$. For other longitudinal positions within the bunch the fixed-points are found at smaller amplitudes. Eventually at locations far enough from the bunch center, the fixed-points merge to the transverse origin and disappear. This pattern is at the base of the periodic resonance crossing mechanism. Figure 4a,b convey the same information as they show that instantaneous fixed-lines have amplitude function of the longitudinal particle coordinate z. This pattern which exhibits the largest fixed-lines at $z/\sigma_z = 0$, with amplitude decreasing for increasing the longitudinal amplitude, will create phenomena of periodic crossing of the fixed-lines. This can be seen in Figure 5 where the evolution the emittance of one test particle is shown during one synchrotron oscillation. The test particle is one of those used in the simulation for obtaining the profiles a), b) of Figure 1, where the effect of chromaticity is included. The scattering of the single particle emittance is clearly visible with 4 kicks per synchrotron oscillation.

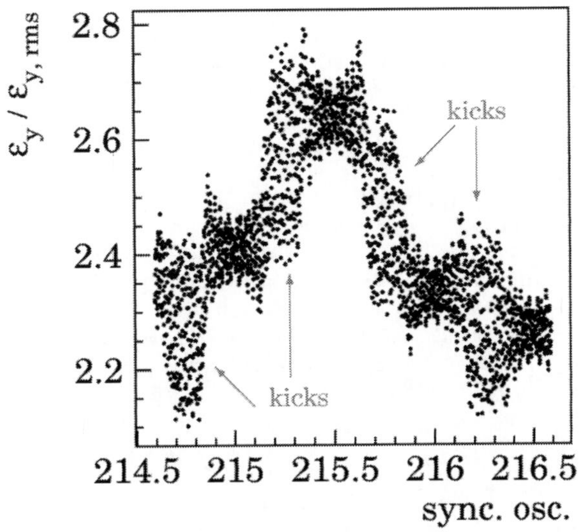

Figure 5. Emittance of one test particle during storage. The picture reveals the 4 kicks exerted by the fixed-lines during one synchrotron oscillation.

7. Conclusion and Outlook

We have here shortly summarized the main findings of the PS experiment performed in 2012. We find in simulations the evidence that the dynamics creating the halo is controlled by the fixed-lines. We find that the fixed-lines amplitude is a function of the instantaneous space charge tune-spread, which depends on the particle longitudinal position within the bunch. This induces a phenomena of periodic crossing of the fixed-lines with the particle orbit. Scattering phenomena are found in the simulation as a clear trace of this fundamental mechanism. We conclude that the x/y asymmetry of the measured beam response (Figure 1 a,b) is the result of the asymmetric shape of the instantaneous fixed-lines (Figure 4) provided the effect of chromaticity is included.

It remains open and is not discussed in this proceeding the influence of the many fixed-lines on the particle motion, and the explanation of why the halo in the experiment extends only up to $5.5\sigma_y$ although simulations clearly show the existence of many fixed-lines which extends on a larger surface, reaching up to $9\sigma_y$ when the effect of the chromaticity is included. The understanding of this mechanism is also of relevance on the issue of the resonance compensation [18]. The discussion of all these aspects as well as of the open question here presented is left to future studies.

8. Acknowledgment

The research leading to these results has received funding from the European Commission under the FP7 Research Infrastructures project EuCARD-2, grant agreement no. 312453.

References

[1] Kester O, Spiller P, and Stoecker H 2016 in *Challenges and Goals for Accelerators in the XXI Century* edited by Brüning O and Myers S (World Scientific Singapore) p 611
[2] Struckmeier J 2000 *Phys. Rev. ST Accel. Beams* **3** 034202
[3] Boine-Frankenheim O, Hofmann I, Struckmeier J and Appel S 2015 *Nucl. Instrum. Methods Phys. Res.* Sect. A **770** 164

[4] Hofmann I and Boine-Frankenheim O 2014 *Phys. Rev. ST Accel. Beams* **17** 124201
[5] Kesting F and Franchetti G 2015 *Phys. Rev. ST Accel. Beams* **18** 1142201
[6] Franchetti G, Hofmann I, Giovannozzi M, Martini M and Métral E 2003 *Phys. Rev. ST Accel. Beams* **6** 124201
[7] Métral E, Franchetti G, Giovannozzi M, Hofmann I, Martini M and Steerenberg R 2006 *Nucl. Instr. and Meth.* A 561 257-265
[8] Franchetti G, Chorniy O, Hofmann I, Bayer W, Becker F, Forck P, Giacomini T, Kirk M, Mohite T, Omet C, Parfenova A and Schuett P 2010 *Phys. Rev. ST Accel. Beams* **13** 114203
[9] Franchetti G and Hofmann I 2006 *Nucl. Instr. and Meth.* A **561** 195-202
[10] Coupard J *et al.* 2014 *LHC Injectors Upgrade Technical Design Report Vol I: Protons* LIU Technical Design Report (TDR) CERN-ACC-2014-0337
[11] Forte V, Benedetto E, and McAteer M 2016 *Phys. Rev. Accel. Beams* **19** 124202
[12] Franchetti G, Gilardoni S, Huschauer A, Schmidt F and Wasef R submitted to PRAB.
[13] Bartosik H *et al.* 2016 in *Proc. HB2016* Malmö Sweden paper WEAM1X01 p 357
[14] Schoch A 1958 *Theory of Linear and Non-Linear Perturbations of Betatron Oscillations in Alternating Gradient Synchrotrons* CERN 57-21 (Proton Synchrotron Division) Section 14
[15] Franchetti G and Schmidt F 2015 *Phys. Rev. Lett.* **114** 234801
[16] Franchetti G and Schmidt F `http://arxiv.org/abs/1504.04389`
[17] Schmidt F 1988 *Untersuchungen zur dynamischen Akzeptanz von Protonenbeschleunigern und ihre Begrenzung durch chaotische Bewegung* PhD thesis DESY HERA 88-02
[18] Franchetti G, Aumon S, Kesting F, Liebermann H, Omet C, Ondreka D, Singh R, Gilardoni S, Huschauer A, Schmidt F and Wasef R 2014 in *Proc. HB2014* East Lansing MI USA paper THO1LR03 p 330

Proposal to observe half-bare electrons on a 45-MeV linac

S.V. Trofymenko[1,2], N.F. Shul'ga[1,2], N. Delerue[3], S. Jenzer[3], V. Khodnevych[3], A. Migayron[3]

[1]NSC "Kharkov Institute of Physics and Technology", Akademicheskaya st. 1, 61108 Kharkov, Ukraine

[2]Karazin Kharkov National University, Svoboda sq. 4, 61022 Kharkov, Ukraine

[3]LAL, Université Paris-Sud, CNRS/IN2P3, Université Paris-Saclay, Orsay, France

E-mail: strofymenko@gmail.com

Abstract. The experimental investigation of the transition radiation (TR) generated by a "half-bare" electron having the proper field different from the Coulomb one is proposed. The electrons in half-bare state are intended to be obtained in the result of their crossing of a conducting screen. We propose to investigate the influence of the half-bare state of electron in this process upon TR generated by such electron on a downstream OTR screen situated on some distance along the direction of the electron beam from the upstream screen which "undresses" the particle. Calculations are presented for the case of a 45 MeV linac and the distance between the screens in the region between 100 mm and 300 mm. The proposed experiment is expected to reveal new features of TR signal in such process comparing to previous measurements.

1. Introduction

In normal conditions electrons are surrounded by a Coulomb field. When a relativistic electron interacts with matter or external fields it can lose part of its electric field and become "half-bare". In such state the characteristics of its electromagnetic field are significantly modified and this can be observed when the particle interacts with matter again: the properties of the emitted radiation will be different. For example, in the case of relativistic electron multiple scattering in amorphous medium such modification results in the Landau-Pomeranchuk-Migdal and Ternovsky-Shul'ga-Fomin effects (see [1–3]) of the electron bremsstrahlung suppression.

Presently we propose to study the influence of the half-bare state of electron upon its transition radiation (TR). For the case when the electron becomes half-bare in the result of scattering or deflection in magnetic field the existence of such influence was theoretically predicted in [4]. In [5] it was shown that the electron also appears in the analogous state after crossing a conducting screen such as an optical transition radiation (OTR) screen. Such way of obtaining relativistic electrons in half-bare state provides certain advantages, from the experimental point of view, comparing to the case of electron deflection (e.g. unnecessity of the use of bending magnets). Therefore it is such approach that we adopt in the present paper and discuss the TR produced by electron impinging upon a downstream screen after having traversed an upstream one.

Let us note that previously the experimental investigation of TR produced by electron beam crossing a system of two screens was reported in [6]. This experiment was made at three times higher energy (150 MeV) than the one provided at CLIO [7–9]. Here the increase of the TR signal at certain wavelengths with the increase of the distance between the screens was observed. In the present paper we show that the conditions available at the CLIO facility provide the possibility of investigating new features of half-bare electron TR, comparing to the ones studied in [6]. The proposed investigation also aims at answering certain questions which the study [6] remained obscure. Particularly, it concerns the measurement of the ratio of TR signal from half-bare particle to the one from the particle having Coulomb field.

Let us also note that the example of direct experimental observation of gradual recovery of the Coulomb field of the electron after its traversal of the conducting screen (based on the measurement of the electric field spatial distribution) is presented in [10].

2. Theory

Let us consider the traversal by an ultrarelativistic electron of a system of two conducting screens (figure 1). Presently we will consider the case of ideally conducting screens. Such approximation works well in the waveband applied in the proposed study. As follows from [5] (see also the discussion in [11]), the scalar potential of the electromagnetic field of the electron moving along the z axis after its traversal at the moment of time $t = 0$ of the upstream (left) target can be approximately presented in the form:

$$\varphi(\mathbf{r}, t) = \frac{e}{\sqrt{\rho^2\gamma^{-2} + (z - vt)^2}}\theta(ct - r), \tag{1}$$

where $\theta(x)$ is the θ-function, which equals unit for $x > 0$, and zero for $x < 0$. Here also $\mathbf{r} = (\rho, z)$ is the radius-vector of the observation point, e, v and γ are the electron charge, velocity and Lorenz-factor, c is the speed of light.

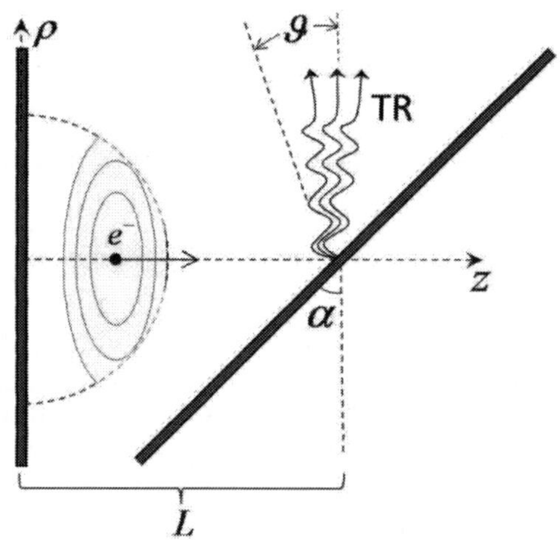

Figure 1. Incidence of the half-bare electron upon the downstream screen.

The expression (1) shows that after the electron traversal of an ideally conducting target its electromagnetic field is different from zero only inside the hemisphere of radius $r = ct$ with the center in the point of the electron emission from the target. Here it coincides with the Coulomb field of a relativistic charged particle. The part of the field outside the hemisphere is absent, which makes the electron half-bare. The radius of the hemisphere increases with the speed of light and the electron gradually restores its Coulomb field.

The strength of the electric field around the electron, which in ultrarelativistic case can be approximately considered as having just the component perpendicular to z-axis, can be defined as $E \approx -\partial\varphi/\partial\rho$. In the experiment it is more convenient to study the considered electromagnetic process in a narrow frequency range. A single-frequency component of the electric field strength around the half-bare electron can be derived from (1) in the following form:

$$E_\omega(\mathbf{r}) = \frac{2e}{v} e^{i\frac{\omega z}{v}} \int\limits_0^\infty \frac{dx x^2 J_1(x\rho)}{x^2 + \omega^2/(v^2\gamma^2)} \left[1 - \exp\left\{i\omega z[\gamma^{-2} + x^2 v^2/\omega^2]/(2v)\right\}\right], \tag{2}$$

where the first term in the parentheses is associated with the electron proper Coulomb field while the second one – with TR (which comes from taking $\partial/\partial\rho$ in (1) from the θ-function). Here $J_1(x)$ is the Bessel function. Expression (2) shows that the Fourier component with the frequency ω is suppressed in the field surrounding the electron at distances from the upstream screen $z \ll z_0 = 2\gamma^2 c/\omega \sim \gamma^2\lambda$, where λ is the wavelength.

The simplest case of the considered problem corresponds to infinite screens and small value of the inclination angle α of the downstream one (see figure 1). In this case the spectral-angular density of TR produced by electron with the field (2) impinging on the downstream screen, registered in the far field (wave) zone, is defined by the following expression:

$$\frac{d^2W}{d\omega do} = \frac{e^2}{\pi^2 c} \frac{\vartheta^2}{(\vartheta^2 + \gamma^{-2})^2} 4\sin^2\left(\frac{\pi L(\vartheta^2 + \gamma^{-2})}{2\lambda}\right), \tag{3}$$

where ϑ is the observation angle (see figure 1). The interference term consisting of the sine squared, together with the coefficient 4 preceding it, differs this quantity from the corresponding distribution of TR generated by the electron with Coulomb field.

3. The CLIO 45-MeV and the proposed experiment

To study the half-bare state we propose to use the linac of the CLIO Free Electron Laser at Orsay [7–9]. This linac can deliver pulses of up to 45 MeV and with a charge per beam of more than 0.5 nC. An experiment using a transition radiation screen has already been installed at the exit of the accelerating cavity [12] and is being upgraded.

We propose to use this experimental setup to insert two transition radiation screens at the same time as shown on figure 2. One of these screens will have a 45° angle with respect to the beam axis and will be imaged through a fused silica viewport. It will be used to measure the transition radiation. The second screen will be located slightly upstream and will have a shallow angle with respect to the beam. By moving this screen up or down the position at which it will be intersected by the beam will change and therefore the distance travelled by the beam between the two screens will also change.

By completely removing the upstream screen it will be possible to measure the normal transition radiation whereas by inserting this screen it will be possible to measure the TR by electrons in half-bare state. By moving this screen up and down the distance travelled by these electrons will vary between 100 mm and 300 mm allowing to observe the partial recovery of their Coulomb field.

4. Signal prediction at CLIO

In order to estimate the signal in the situation of the experiment proposed to be performed at CLIO several complicating factors (neglected by the expression (3)) should be taken into account. They are associated, particularly, with the fact that the measurement is proposed to be made with the use of a collecting parabolic mirror of finite size situated in the near-field

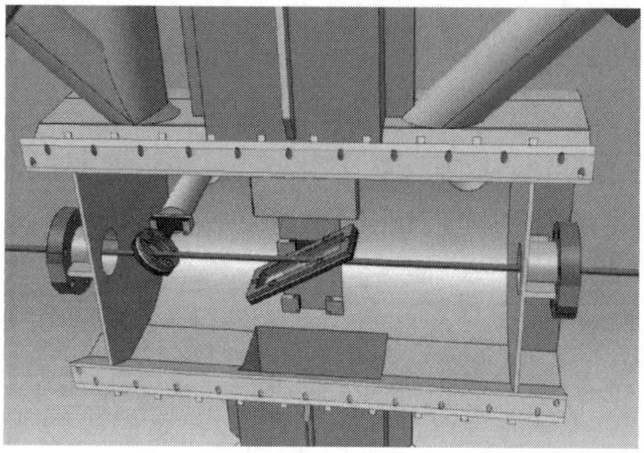

Figure 2. Conceptual view of the screens layout in the chamber for the half-bare electrons experiment. The beam comes from the right. The right screen is used to strip the electrons from their Coulomb field and the left one is used to measure the TR. The left screen will move back and forth and the right screen up and down (actuators are not shown).

(prewave) zone of the TR process. Further we present the numerical estimations of the expected results of TR characteristics measurements taking this into account.

Figure 3 represents the expected integrated yield of TR for a single electron in the wavelength range 1 mm< λ <5 mm. It shows significant increase of the signal with the increase of the distance L between the screens in the available range 10 cm< L <25 cm. The TR signal is suppressed comparing to the corresponding signal from the electron with Coulomb field (dashed line).

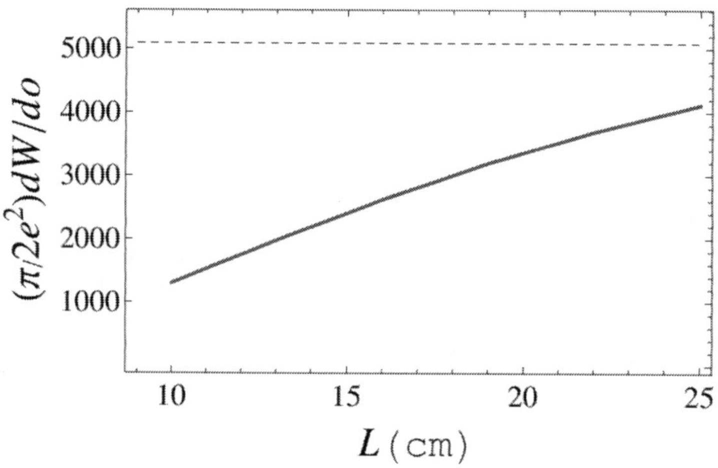

Figure 3. Half-bare electron TR yield (solid line) integrated over the wavelength region 1 mm< λ <5 mm. Dashed line – corresponding signal from the electron with Coulomb field.

Figure 4 demonstrates the calculated TR spectral-angular distribution for the wavelength $\lambda = 0.5 \pm 0.1$ mm (averaged over such interval) and 3° inclination β of the parabolic mirror axis with respect to the direction orthogonal to the beam. The figure shows the possibility of

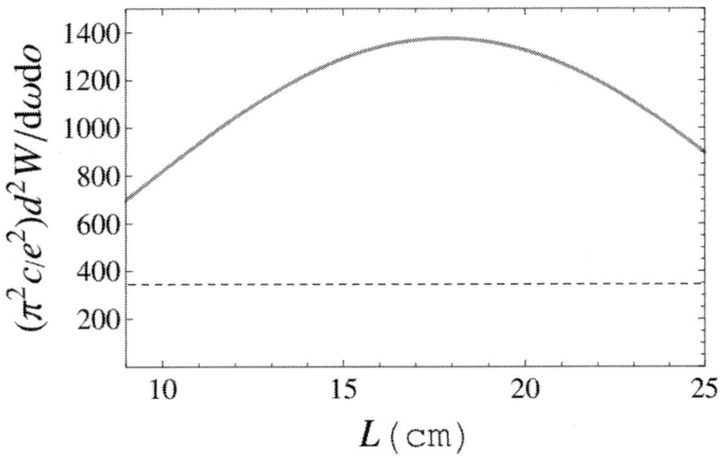

Figure 4. Half-bare electron TR yield for $\lambda = 0.5 \pm 0.1$ mm and parabolic mirror inclination $\beta = 3°$. Dashed line – corresponding signal from the electron with Coulomb field.

observation of something like an antieffect in half-bare electron TR. It is the enhancement of the TR signal comparing to the one from the electron with Coulomb field and its further decrease with the increase of separation between the screens. Let us note that the presented calculations do not take into account the finiteness of the screens size and certain (45°) inclination of the downstream screen. The first factor may reduce radiation intensity in the long wavelength region. The second one should not lead to significant modification of the results at considered electron energies. Additionally the TR coherence may considerably increase the bunch signal (up to 10^9 times).

5. Discussion

Summarizing the discussion presented above, let us outline the main new features of the present experimental proposal and the half-bare electron TR properties which it deals with. Firstly, we propose to make the comparison of TR signals from half-bare electron and the one from the particle with Coulomb field. It could allow to judge of the magnitude of the influence of the half-bare state of electron upon its TR and was not made in [6]. Secondly, calculations show that at parameters available at CLIO facility it is possible to observe the effect opposite to the one, reported in [6], which is the decrease of the TR signal with the increase of the distance between the screens. Thirdly, we propose to study the evolution of the integrated signal (with respect to wide waveband) with the change of L, which provides more complete information (comparing to narrow waveband study) about the process of the electron field regeneration after crossing the screen. Finally, under the discussed conditions the study of the prewave zone effect in half-bare electron TR could be performed as well by varying the detector aperture and its distance from the screen.

The proposed investigation is expected to reveal valuable for TR-based diagnostics information about TR properties in the case when interference effects in radiation are significant. It is also relevant for the TR-based diagnostics of extracted beams (with the use of bent crystal or magnetic field) which under certain conditions present the example of half-bare particle beams.

Acknowledgements

Work supported by CNRS/IN2P3, the P2I department of Paris-Saclay University and the IDEATE International Associated Laboratory (LIA) between France and Ukraine. The work of

S. Trofymenko and N. Shul'ga is partially supported by project no. 0117U004866 of the Ministry of Education and Science of Ukraine

References

[1] Akhiezer A I, Shul'ga N F and Fomin S P 2005 *Landau-Pomeranchuk-Migdal Effect* (*Physics Reviews* vol 22) (Cambridge Sci. Publ.)
[2] Klein S 1999 *Rev. Mod. Phys.* **71** 1501
[3] Thomsen H D, Esberg J, Andersen K K *et al.* 2010 *Phys. Rev. D* **81** 052003
[4] Shul'ga N F, Trofymenko S V and Syshchenko V V 2011 *JETP Letters* **93** 3–7
[5] Shul'ga N F, Trofymenko S V and Syshchenko V V 2011 *Il Nuovo Cimento C* **34** 327
[6] Shibata Y, Ishi K, Tokahashi T *et al.* 1994 *Phys. Rev. E* **49** 785
[7] Ageron M *et al.* 1989 *LAL-RT-89-04*
[8] Bourdon J, Belbeoch R, Bernard M C, Brunet P, Leblond B, Omeich M, Plouviez E, Rodier J, Chaput R, Bergher M and Ortega J 1990 *Free Electron Laser International Conference 12* vol 304 (Paris, France: Elsevier) pp 322–328. URL `http://hal.in2p3.fr/in2p3-00020744`
[9] Glotin F *et al.* 1992 *Conf. Proc.* (C920324) pp 620–622
[10] Naumenko G, Popov Y and Shevelev M 2012 *J. of Phys.: Conf. Series* **357** 012005
[11] Bolotovskii B M and Serov A V 2009 *Sov. Phys. Usp.* **52** 487–493
[12] Delerue N, Jenzer S, Khodnevych V, Berthet J P, Glotin F, Ortega J M and Prazeres R 2016 *IPAC2016* (Proceedings of IPAC2016. Busan, South Korea) p MOPMB005 URL `http://hal.in2p3.fr/in2p3-01322166`

IOP Conf. Series: Journal of Physics: Conf. Series **874** (2017) 012077 doi:10.1088/1742-6596/874/1/012077

Reconstruction of the 3D charge distribution of an electron bunch using a novel variable-polarization transverse deflecting structure (TDS)

D Marx[1,2], R Assmann[1], P Craievich[3], U Dorda[1], A Grudiev[4] and B Marchetti[1]

[1] DESY, Notkestraße 85, 22607 Hamburg, Germany
[2] Department of Physics, Universität Hamburg, Jungiusstraße 9-11, 20355 Hamburg, Germany
[3] Paul Scherrer Institut, 5232 Villigen PSI, Switzerland
[4] CERN, 1211 Geneva 23, Switzerland

E-mail: daniel.marx@desy.de

Abstract. A TDS is a well-known device for the characterization of the longitudinal properties of an electron bunch in a linear accelerator. So far, the correlation of the slice properties in the horizontal/vertical planes of the electron bunch distribution has been characterized by using a TDS system deflecting in the vertical/horizontal directions respectively and analysing the image on a subsequent screen. Recently, an innovative design for a TDS structure has been proposed, which includes the possibility of continuously varying the angle of the transverse streaking field inside a TDS structure. This allows the beam distribution to be characterized in all transverse directions. By collecting measurements of bunches streaked at different angles and combining them using tomographic techniques, it is possible to retrieve 3D distributions of the charge density. In this paper, a method is proposed and simulation results are presented to show the feasibility of such an approach at the upcoming accelerator R&D facility, SINBAD, at DESY.

1. Introduction

A transverse deflecting structure (TDS) is a commonly-used diagnostic device for investigating the longitudinal dependence of beam properties in electron linacs [1]. The induced EM mode in the structure exerts a time-dependent transverse kick on the beam, which converts the longitudinal plane of the bunch into a transverse axis on a screen placed downstream. From the screen image, direct measurements can be made of the bunch length and the charge distribution in the transverse direction perpendicular to the kick. A TDS can also be combined with other components, such as a dipole for slice-by-slice energy spread measurements or a quadrupole for slice-by-slice emittance measurements.

Using recently-designed RF components [2], a new X-band TDS with variable polarization is being developed for use at several DESY facilities and at SwissFEL, allowing the beam to be streaked at any angle in the transverse plane [3]. This opens up the possibility of reconstructing the full 3D charge distribution in a multi-shot measurement by combining profiles of the beam streaked at several different angles using tomographic reconstruction techniques.

At the upcoming SINBAD facility at DESY, short bunches with length ranging from subfemtosecond to a few femtoseconds and charge in the picocoulomb range will be produced for potential use as a witness beam in laser-driven plasma wake-field acceleration (LWFA) experiments and for testing dielectric accelerating structures [4]. The planned X-band TDS will allow characterization of the 3D charge density profile of the input beam to femtosecond longitudinal resolution. For tests involving dielectric accelerating structures, it is planned to inject asymmetric beams [5]. In this case, characterizing the full profile of the beam can be important for matching and aperture considerations. For LWFA experiments, it may be important to know the transverse beam profile over the length of the beam as the fields in the plasma, and therefore the matching conditions, have longitudinal dependence in certain regimes [6]. The accelerating fields are also influenced by beam offsets, therefore having a complete charge profile reconstruction can help, for instance, in the case of beam loading.

In the following sections, the reconstruction procedure is described and simulation results are presented. Finally, there is a discussion of the limitations of the process and further investigations that could be performed. Other measurements involving the TDS and considerations for a lattice design are presented in a separate contribution [7].

2. Reconstruction Methodology

The aim of this procedure is to combine 2D screen profiles of a bunch streaked at several angles into a 3D charge density profile of the complete bunch. In what follows, a rotating coordinate system is considered with the y-axis defined parallel to the streaking axis and Δy defined relative to the bunch centre at the screen. Each screen profile can therefore be viewed as a projection of the complete charge distribution on the corresponding x-axis.

The first step is to transform the Δy-coordinate into a Δt-coordinate, where Δt is the arrival time normalized to the centre of the bunch. This can be achieved directly using the equation [8]

$$\Delta y(t) \approx Sc\Delta t, \tag{1}$$

where the shear parameter, S, is defined as

$$S = M^y_{1,2} \frac{2\pi f e V_0}{c^2 |p|}. \tag{2}$$

In this equation, M^y is the vertical transfer matrix from the centre of the TDS to the screen, f and V_0 are the frequency and total kicking voltage respectively and p is the particle momentum. The shear parameter for each direction of streaking can be determined by scanning over the bunch arrival time at the TDS and measuring the centroid offset at the screen.

The xt profiles for different streaking angles are then combined to reconstruct the complete charge density profile of the bunch using tomographic reconstruction techniques. They are first split into slices in Δt, and the corresponding time slice from each profile is then used as an input to the reconstruction algorithm. In this paper, the Simultaneous Algebraic Reconstruction Technique (SART) [9], using the Kaczmarz iterative solver method [10], has been applied with two iterations. Due to the effects discussed below, the reconstruction does not provide a completely faithful reconstruction of the bunch, however it is observed that the key features are reproduced.

It is clear that the charge density distribution will evolve, both within the TDS and within the following drift section before the screen. The transverse information in the reconstruction pertains to the bunch at the screen position, whereas the longitudinal information pertains to the bunch at the TDS where the bunch is sheared. The bunch length at these two locations may be different in regimes in which velocity bunching or space charge are significant. Although in the ideal case calculations could be performed to transfer the transverse charge distribution back to the TDS, this would require knowledge of the transverse phase space distribution. For this, slice emittance measurements for each streaking angle would be needed, making the measurement much more involved.

IOP Conf. Series: Journal of Physics: Conf. Series **874** (2017) 012077 doi:10.1088/1742-6596/874/1/012077

3. Simulation results

Simulations were carried out using *elegant* [11] for particle tracking and a script written in *Python*, using the *scikit-image* package [12], for the tomographic reconstruction. The bunches used were generated in simulations of the SINBAD-ARES linac using the magnetic compression chicane [13]. They are not symmetrical and contain interesting features that make them useful examples to study with this technique. Table 1 shows the properties of the bunch presented in this paper. Space charge effects, wakefields, timing jitter and misalignments have not been taken into account in these simulations (although space charge was included in the generation of the input bunch).

Table 1. Bunch statistical properties at TDS entrance (here, x and y correspond to axes with 0 rotation).

Energy [MeV]	84.2
Charge [pC]	3
σ_t [fs]	5.15
$\sigma_{x/y}$ [µm]	87.9 / 96.8
$\epsilon_{x/y}^{\text{norm}}$ [mm mrad]	0.224 / 0.190
$\beta_{x/y}$ [m]	5.69 / 8.14
$\alpha_{x/y}$	-0.377 / 0.258

The setup used in the simulations comprises two 0.8-metre TDS cavities, each providing a maximum voltage kick of 20 MV[1], a 5-metre drift space and finally a screen. The resolution of the screen is assumed to be 20 µm, which should be achievable with screen stations based on a similar design to those at the European XFEL [14]. The temporal resolution depends on the direction of streaking, as it is a function of the beam size at the screen with the TDS off (σ_y^{off}) [7]:

$$R_{\text{t}} = \frac{\sigma_y^{\text{off}} |p| c}{2\pi f e V_0 L}. \tag{3}$$

Here, L is the length from the centre of the TDS to the screen. In this paper, slice lengths of 0.85 fs have been used – this is just over the largest calculated resolution. The reconstruction has been carried out using profiles from 16 evenly-spaced streaking angles. As the TDS is operated at zero voltage-crossing, the bunches are streaked in both the positive and negative directions.

Figure 1 shows sample screen images of the bunch streaked in different planes and the way in which they have been divided up into time slices. Figure 2 shows the 3D charge density profile at the screen obtained by tracking with the TDS off, and the reconstructed profile below.It is clear that the key features, such as the offset in the x distribution, have been successfully reproduced, although the reconstruction is limited by the perturbation of the bunch in the TDS and the inherent inaccuracy of a tomographic reconstruction with a finite number of projections.

There is significant variation in bunch length between the TDS and the screen, as shown in Fig. 3. The reason for the bunch length decrease when the TDS is off is that the initial chirp results in velocity compression in the drift, as the electrons are not ultrarelativistic. The increase in bunch length when the beam is streaked arises from the induced momentum spread

[1] For the TDS cavities, the *RFDF* element in *elegant* has been used. The frequency of the structure is 11.9916 GHz. The voltage kick per cavity has been halved for the phase scan to obtain the shear parameters due to the induced energy spread. In order to account for the bunch length change within the TDS cavities due to energy spread (which is not currently accounted for automatically in *elegant*), 100 zero-length *RFDF* elements separated by drift spaces have been used to simulate each cavity.

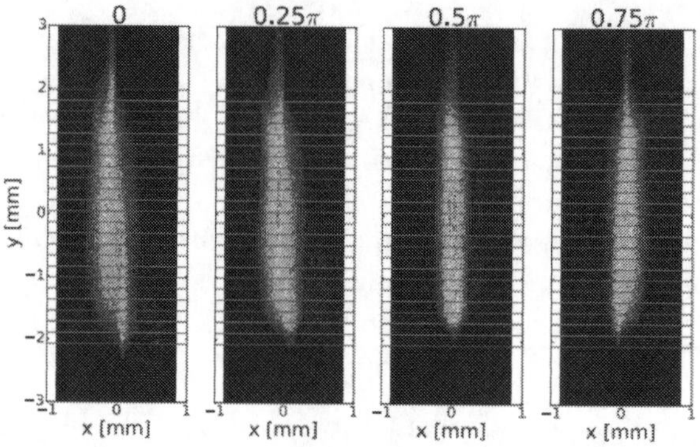

Figure 1. Sample screen images of the streaked bunch with red lines showing the temporal slices. The axes rotate so that y is always in the direction of streaking.

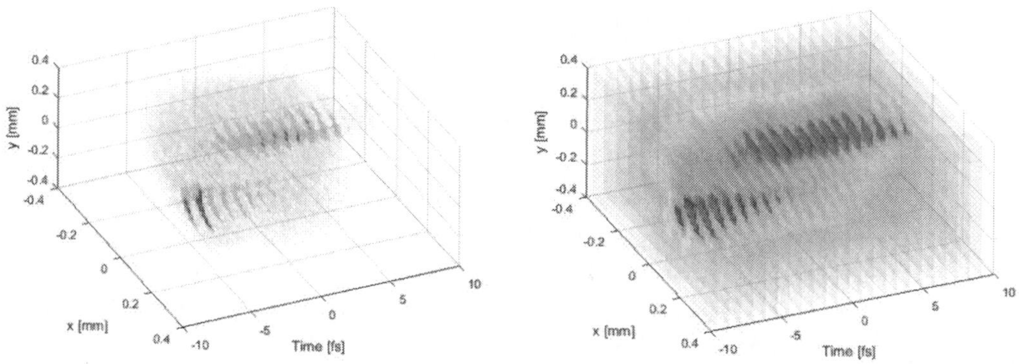

Figure 2. Actual (left) and reconstructed (right) view of the beam divided into temporal slices according to the calculated resolution. The colour map illustrates the charge density profile and has been normalized to the maximum intensity in each plot.

in the TDS, which is illustrated in Fig. 4. This induced momentum spread is a consequence of the Panofsky-Wenzel theorem [15], which states that a transverse deflection is only possible if there is a transverse gradient in the longitudinal electric field. Due to the bunch length decrease between the TDS and screen with the TDS off, the simulated bunch slices at the screen location and the reconstructed slices do not match up.

4. Conclusion & Outlook

A new method for reconstructing the 3D charge density profile using a novel X-band TDS with variable polarization has been described and simulations have been carried out using simulated beams from the SINBAD-ARES linac. This technique reconstructs the transverse beam profile at the screen, although the time slices correspond to the bunch at the TDS. The simulations show that this reconstruction technique can reproduce key features in the slice-by-slice transverse distributions, such as offset from the beam axes and the general shape of the distribution. The induced momentum spread in the TDS perturbs the beam and causes an increase in bunch length in the drift section preceding the screen.

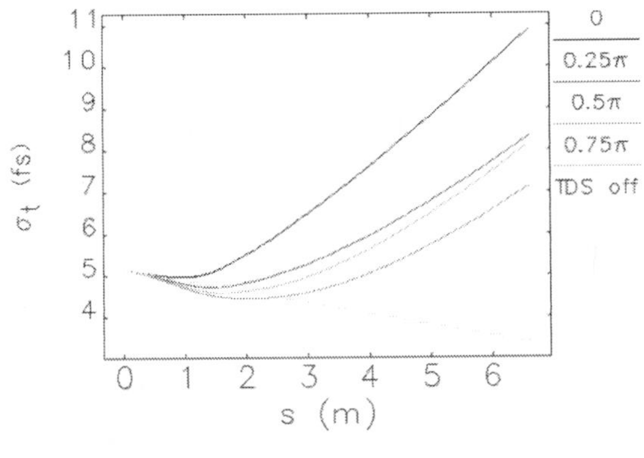

Figure 3. Variation in bunch length from the TDS to the screen for a selection of streaking angles. The exit of the TDS is at $s = 1.6\,\mathrm{m}$ and the screen is at $s = 6.6\,\mathrm{m}$.

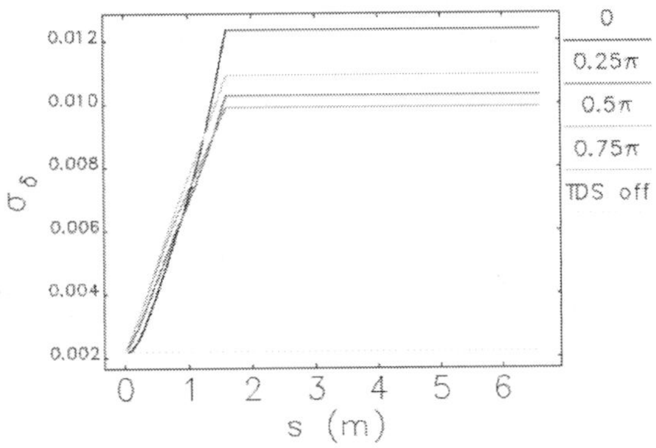

Figure 4. Change in fractional momentum spread from the TDS to the screen for a selection of streaking angles.

Although the principle of this reconstruction method has been presented in this paper, further studies are required to investigate space charge effects, wakefields, timing jitter and misalignments, as well as the resolution limit for specific applications.

Possible applications for characterizing the input beam to LWFA and dielectric experiments at the SINBAD facility have been discussed in the introduction. As this reconstruction technique relies on shot-to-shot stability, it would be more difficult to use it to characterize a beam output from an experiment. Nevertheless, by taking many measurements for each streaking direction, it could potentially be possible in some cases to statistically reconstruct a typical bunch. Further tests would be needed to investigate this.

Acknowledgments

The authors would like to thank J Zhu for providing the initial beam distributions and for many helpful discussions. In addition, A Ferran Pousa, K Flöttmann, W Hillert, W Kuropka, F Mayet and M Weikum provided useful background information and/or feedback.

References

[1] Emma P, Frisch J and Krejcik P 2000 *A Transverse RF Deflecting Structure for Bunch Length and Phase Space Diagnostics* (SLAC National Accelerator Laboratory, California) LCLS-TN-00-12
[2] Grudiev A 2016 *Design of Compact High Power RF Components at X-band* (CERN, Geneva) CLIC-Note-1067
[3] Marchetti B *et al.* 2017 *Proc. 8th Int. Particle Accelerator Conf.* (Copenhagen) MOPAB044
[4] Dorda U *et al.* 2017 *Proc. 8th Int. Particle Accelerator Conf.* (Copenhagen) MOPVA012
[5] Mayet F, Assmann A, Dorda U, Kuropka W, Marchetti B and Zhu J 2017 *Proc. 8th Int. Particle Accelerator Conf.* (Copenhagen) WEPVA007
[6] Weikum M, Assmann A, Svystun E, Ferran Pousa A, Li F Y, Dorda U, Marchetti B and Sheng Z M 2017 *Proc. 8th Int. Particle Accelerator Conf.* (Copenhagen) TUPIK013
[7] Marx D, Assmann R, Dorda U, Marchetti B and Mayet F 2017 Lattice considerations for the use of an X-Band transverse deflecting structure (TDS) at SINBAD *J. Phys.: Conf. Series* this volume
[8] Röhrs M, Gerth C, Schlarb H, Schmidt B and Schmüser P 2009 Time-resolved electron beam phase space tomography at a soft x-ray free-electron laser *Phys. Rev. ST Accel. Beams* **12** 050704
[9] Andersen A H and Kak A C 1984 Simultaneous algebraic reconstruction technique (SART): a superior implementation of the ART algorithm *Ultrasonic Imaging* **6** 81-94
[10] Kaczmarz S 1937 Angenäherte Auflösung von Systemen linearer Gleichungen *Bulletin International de l'Académie Polonaise des Sciences et des Lettres* **35** 355-357
[11] Borland M 2000 *elegant: A Flexible SDDS-Compliant Code for Accelerator Simulation* Advanced Photon Source LS-287

[12] van der Walt S, Schönberger J L, Nunez-Iglesias J, Boulogne F, Warner J D, Yager N, Gouillart E, Yu T and the scikit-image contributors 2014 *scikit-image: image processing in Python* PeerJ **2** e453
[13] B. Marchetti *et al.* 2017 *Proc. 8th Int. Particle Accelerator Conf.* (Copenhagen) TUPAB041
[14] Wiebers Ch, Holz M, Kube G, Nölle D, Priebe G and Schröder H-Ch 2013 *Proc. 2nd Int. Beam Instrumentation Conf.* (Oxford) p 807
[15] Panofsky W K H and Wenzel W A 1956 *Rev. Sci. Instrum.* **27** 967

IOP Conf. Series: Journal of Physics: Conf. Series **874** (2017) 012078 doi:10.1088/1742-6596/874/1/012078

Lattice considerations for the use of an X-band transverse deflecting structure (TDS) at SINBAD, DESY

D Marx[1,2], R Assmann[1], U Dorda[1], B Marchetti[1] and F Mayet[1,2]

[1] DESY, Notkestraße 85, 22607 Hamburg, Germany

[2] Department of Physics, Universität Hamburg, Jungiusstraße 9-11, 20355 Hamburg, Germany

E-mail: daniel.marx@desy.de

Abstract. An X-band TDS is a well-known device for the characterization of the longitudinal properties of an electron bunch in a linear accelerator. It is planned that a novel X-band TDS with variable polarization will be installed within the next few years at SINBAD, an upcoming accelerator R&D facility at DESY. There are several measurements that can be performed with the TDS, each with specific optics requirements to reach the highest possible resolution and keep induced energy spread to a tolerable level. Quadrupoles will be installed between the TDS and the screen to help satisfy these conditions. In this paper, the requirements for the bunch length measurements, a novel 3D charge density reconstruction technique and slice energy measurements are discussed and some simulation results for the slice energy measurement using example lattices are presented.

1. Introduction

A transverse deflecting structure (TDS) is a commonly-used diagnostic device for investigating the longitudinal dependence of beam properties in electron linacs [1]. The induced field in the cavity exerts a time-dependent transverse kick on the beam, which translates the time coordinate into a transverse coordinate on a screen placed downstream. Using this setup, the beam properties in the transverse direction perpendicular to the kick can be probed. From the screen image, direct measurements can be made of the bunch length and beam distribution perpendicular to the kick. A TDS can also be combined with other components, such as a dipole for slice energy spread measurements or a quadrupole for slice emittance measurements.

Based on novel X-band technology proposed recently [2], a TDS with variable polarization is currently being designed for installation at the FLASH and SINBAD facilities at DESY and the SwissFEL facility at PSI in the next few years [3]. This TDS will allow beams to be streaked at any angle in the transverse plane, thus opening up the possibility of a 3D reconstruction of the charge density distribution, as detailed in a separate contribution [4].

At the upcoming SINBAD facility at DESY, short bunches with length ranging from subfemtosecond to a few femtoseconds and charge in the picocoulomb range will be produced in the ARES linac [5, 6]. They will be compressed, either by velocity bunching using the two travelling wave structures or by magnetic compression using a four-dipole chicane further downstream. A likely location for the two 0.8-metre TDS cavities is just after the chicane, and a screen station will be placed a few metres further downstream. There is currently flexibility

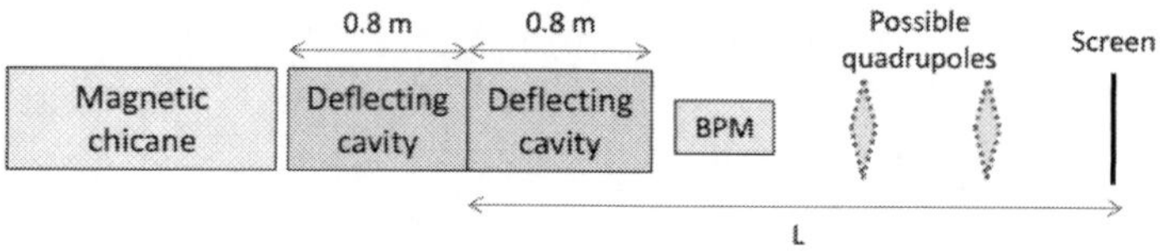

Figure 1. Schematic of beamline setup (not to scale).

over the design of the lattice between the TDS and the screen with ongoing studies into the optimal setup. A schematic of the setup being considered is shown in Figure 1.

Measuring the bunch properties of such short bunches is challenging and proper consideration must be given to the longitudinal resolution, as well as to the resolution of the other quantity being measured, e.g. energy. A key difficulty is that the TDS induces a considerable energy spread, which increases as the longitudinal resolution improves. In the following sections, the optimization of the resolution is discussed for bunch length measurements, a novel 3D charge density reconstruction technique and slice energy measurements.

2. Theory

2.1. Longitudinal Resolution

A TDS imparts a time-dependent transverse kick on an electron bunch. Assuming the TDS operates at zero voltage crossing, this kick contributes to an increase in the particle spread in the direction of the kick (defined here as the y direction). The RMS spot size on a screen located at s_1 is therefore a combination of the spot size at the screen when the TDS is switched off, σ_y^{off}, and a contribution due to the TDS streaking:

$$\sigma_y(s_1) = \sqrt{(\sigma_y^{\text{off}})^2 + (S\sigma_z)^2}. \tag{1}$$

Here, σ_z is the bunch length at the TDS and S is the shear parameter, defined as

$$S = M_{1,2}^y \frac{2\pi f e V_0}{c^2|p|}, \tag{2}$$

where M^y is the vertical transfer matrix from the centre of the TDS to the screen, f and V_0 are the cavity frequency and peak voltage respectively, and p is the mean momentum [7]. The shear parameter relates the shift in position of a particle on the screen, Δy, to its position along the bunch, ζ [7]:

$$\Delta y \approx S\zeta. \tag{3}$$

The longitudinal resolution is defined here as the bunch length when the two terms on the right-hand side of Equation (1) are equal, i.e.

$$|S|\sigma^{\text{long}} = \sigma_y^{\text{off}}. \tag{4}$$

In general,

$$M_{1,2}^y = \sqrt{\beta_y(s_0)\beta_y(s_1)} \sin \Delta\phi_y, \tag{5}$$

where $\beta_y(s_0)$ and $\beta_y(s_1)$ are the β-functions at the centre of the TDS and the screen respectively, and $\Delta\phi_y$ is the phase advance between these two locations. The temporal resolution can therefore be expressed as

$$R_{\text{t}} = \frac{\sqrt{\epsilon_y}}{\sqrt{\beta_y(s_0)}|\sin(\Delta\phi_y)|} \frac{|p|c}{2\pi f e V_0}. \tag{6}$$

In the case of a drift, the $M^y_{1,2}$ term is simply equal to the length of the drift, L, and so the resolution is

$$R_t = \frac{\sigma_y^{\rm off}|p|c}{2\pi feV_0L}. \tag{7}$$

It should be noted that this formula assumes the thin-lens approximation and the length L is therefore the length from the centre of the TDS to the screen.

It is useful to consider the meaning of the resolution defined here. If the time slices are much larger than the resolution, then the contribution to the spot size at the screen from the streaking is much more significant than the contribution from the transverse particle spread before the TDS. It is therefore only possible to resolve differences in features between time slices that are larger than the resolution.

Another important quantity is the minimum measurable bunch length. This can be calculated by considering the bunch length required so that the difference between the streaked and unstreaked bunch is equal to the screen resolution, $R_{\rm screen}$. The minimum measurable bunch length in seconds can be expressed as

$$\sigma_t^{\rm min} = \frac{\sqrt{R_{\rm screen}\left(R_{\rm screen} + 2\sigma_y^{\rm off}\right)}}{Sc}. \tag{8}$$

2.2. Induced Energy Spread

A consequence of the Panofsky-Wenzel theorem [8] is that there must be a transverse gradient in the longitudinal electric field present in any TDS. This has the undesired effect of inducing energy spread in the sheared bunch, which makes measuring the original energy spread challenging, but can also perturb other measurements due to chromatic effects, for instance. It has been shown that there is an inherent limitation in minimizing both the temporal resolution (R_t) and the induced energy spread (σ_δ) [9]:

$$R_t\sigma_\delta > \frac{\epsilon_y}{c\sin(\Delta\phi_y)}. \tag{9}$$

3. Bunch Length Measurement

The bunch length measurement is the simplest measurement involving the TDS. The bunch is streaked onto a screen and the temporal profile can be seen in the direction of streaking. By varying the arrival time of the electrons at the cavity and measuring their mean position on the screen, the shear parameter can be determined from the gradient of a plot of Δy against arrival time [7]. The shear parameter is then used to convert the Δy coordinate on the screen to a longitudinal component, and the RMS spread calculated.

In this case, the aim is to achieve the highest possible longitudinal resolution and so there should be a phase advance of $\pi/2$ (or $3\pi/2$ etc.) in the direction of streaking, as can be seen from Equation (6). This can be achieved via a drift space of an appropriate length or using one (or multiple) quadrupole(s). In addition, the highest stable voltage should be used, which is taken to be 20 MV per cavity, or 40 MV in total.

Although a drift space can satisfy the phase advance requirement of a given beam in principle, the desired drift length would be different for different beams. In addition, the spot size on the screen cannot be optimized. For the case of a sample working point bunch of length 0.461 fs, charge 2.73 pC and energy 100 MeV, a phase advance of $\pi/2$ can be achieved with a drift space of 3.99 m. Using the definition in the preceding section, the minimum measurable bunch length is calculated to be 0.37 fs for a screen resolution of 20 µm.

An important consideration is that the lattice after the TDS does not cause the beam to evolve in a way that would change the bunch length measurement to a significant extent. Specifically, the mapping from the TDS exit to the screen should be approximately linear in the direction of streaking, otherwise the longitudinal coordinate cannot be reproduced via Equation (3).

Although a strong focusing quadrupole will reduce the length of drift space required, it will also lead to undesired chromatic and space charge perturbations. The former is especially important given the significant induced energy spread in the TDS.

4. 3D Charge Density Reconstruction

The details of this novel technique to reconstruct the charge density of a bunch are presented in a separate contribution [4]. The principle is that the bunch is streaked at multiple angles and the screen intensity profiles are divided into time slices and combined using tomographic techniques. In order to achieve the highest possible resolution of the reconstruction, the longitudinal resolution should be simultaneously minimized in all directions of streaking. It is also important that the lattice does not distort the bunch in a way that depends on the direction of streaking as the information from different streaking directions needs to be combined. Finally, it is important to limit chromaticity due to the considerable energy spread induced in the TDS.

With these considerations in mind, a drift space between the TDS and the screen has been used in simulations of this measurement. The disadvantage of this approach is that the phase advance cannot be adjusted and so the longitudinal resolution cannot be optimized.

For the case simulated using a 5-metre drift space, the phase advance in the direction of streaking is approximately 0.5 rad to 0.6 rad between the centre of the TDS and the screen, which is sufficient here. However, this technique will be limited by the longitudinal resolution when probing shorter bunches.

5. Slice Energy Measurement

In this measurement, the bunch is first streaked in the y direction using the TDS and then deflected in the x direction using a dipole. The setup consists of the TDS cavities, two quadrupoles, a dipole and a screen placed perpendicular to the beam leaving the dipole. The x coordinate on the screen can be converted to energy whereas the y coordinate can be converted to a longitudinal time coordinate, as in the previous cases.

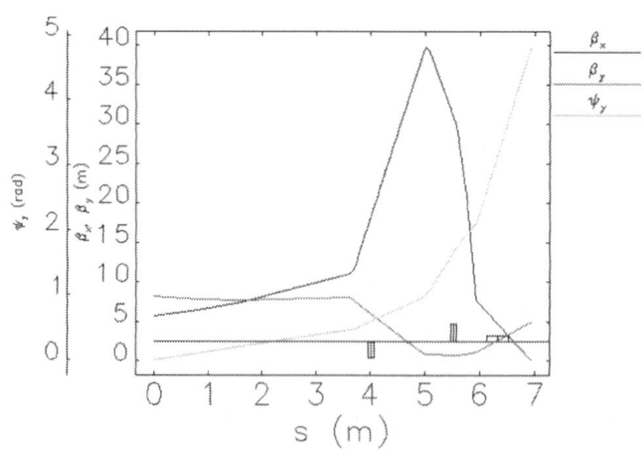

Figure 2. Evolution of β-functions from the TDS entrance to the screen. The positions of the two quadrupoles and dipole are indicated.

IOP Conf. Series: Journal of Physics: Conf. Series **874** (2017) 012078 doi:10.1088/1742-6596/874/1/012078

The theoretical resolution achievable for a dipole spectrometer depends on the emittance, β-function and dispersion at the screen, according to the relation [10]

$$\delta_{\text{min}} \geq 2\frac{\sqrt{\epsilon\beta}}{D}. \tag{10}$$

The temporal information pertains to the bunch at the TDS with the resolution given by Equation (6). A major difficulty in this measurement is that achieving a high temporal resolution and a low energy spread is not possible due to Equation (9).

In order to limit the induced energy spread to reasonable levels, a kick of just 2 MV per cavity (4 MV in total) has been used. To maximize the energy resolution, β_x should reach a maximum and then decrease to the screen as the beam is focused. For optimal longitudinal resolution, $|\sin(\Delta\phi_y)|$ should equal 1, and in this case the most appropriate phase advance is $3\pi/2$. The optics design for the lattice used in simulations is shown in Figure 2. The dipole is rectangular with a magnetic length of 0.302 m and the beam enters perpendicular to the edge. The bending angle is set to 0.93 rad.

Simulations using an input bunch from the SINBAD-ARES linac (parameters given in Table 1) were performed in *elegant* [11] and do not include space charge, wakefields, jitter or misalignment errors. Figure 3 shows the energy profile of the incoming 84.2 MeV beam and the reconstruction using this lattice. The induced energy spread is clearly visible, even at these low voltages, however the original correlated energy spread can still be discerned.

Table 1. Bunch statistical properties at TDS entrance for slice energy measurement simulations (here, x and y correspond to axes with 0 rotation).

Energy [MeV]	84.2
Charge [pC]	3
σ_t [fs]	5.15
$\sigma_{x/y}$ [µm]	87.9 / 96.8
$\epsilon_{x/y}^{\text{norm}}$ [mm mrad]	0.224 / 0.190
$\beta_{x/y}$ [m]	5.69 / 8.14
$\alpha_{x/y}$	-0.377 / 0.258

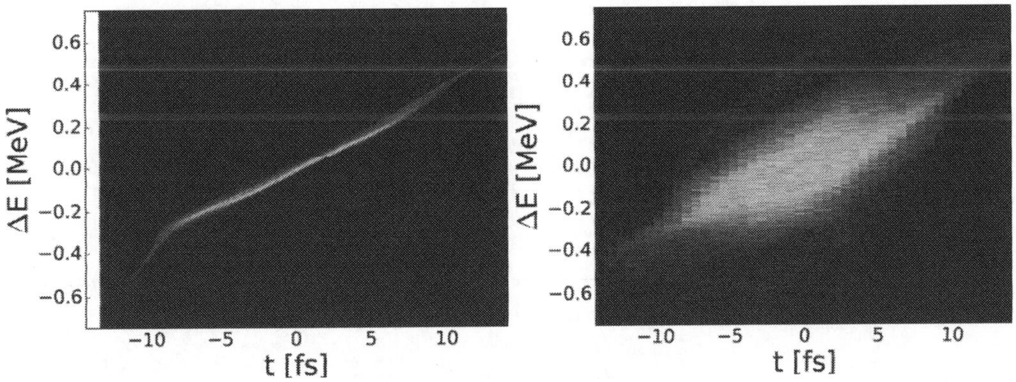

Figure 3. Original (left) and reconstructed (right) slice energy spread. There is significant induced energy spread from the TDS visible in the reconstruction.

IOP Conf. Series: Journal of Physics: Conf. Series **874** (2017) 012078 doi:10.1088/1742-6596/874/1/012078

6. Conclusion

In this paper, some considerations relating to the lattice design for various measurements involving the TDS have been discussed and some preliminary studies presented. Collective effects, timing jitter and misalignments have not been included, although they should be studied in future simulations. Further studies are required, in particular for planned slice emittance measurements and novel 6D beam characterization techniques, which will combine slice emittance measurements on different streaking axes with slice energy measurements. Finally, the best way to combine the requirements of all the different measurements to fix the quadrupole and screen positions must also be considered.

Acknowledgments

The authors would like to thank J Zhu for providing the input beam distributions and for many helpful and informative discussions.

References

[1] Emma P, Frisch J and Krejcik P 2000 *A Transverse RF Deflecting Structure for Bunch Length and Phase Space Diagnostics* (SLAC National Accelerator Laboratory, California) LCLS-TN-00-12

[2] Grudiev A 2016 *Design of Compact High Power RF Components at X-band* (CERN, Geneva) CLIC-Note-1067

[3] Marchetti B *et al.* 2017 *Proc. 8th Int. Particle Accelerator Conf.* (Copenhagen) MOPAB044

[4] Marx D, Assmann R, Craievich P, Dorda U, Grudiev A and Marchetti B 2017 Reconstruction of the 3D charge distribution of an electron bunch using a novel variable-polarization transverse deflecting structure (TDS) *J. Phys.: Conf. Series* this volume

[5] Dorda U *et al.* 2017 *Proc. 8th Int. Particle Accelerator Conf.* (Copenhagen) MOPVA012

[6] B. Marchetti *et al.* 2017 *Proc. 8th Int. Particle Accelerator Conf.* (Copenhagen) TUPAB041

[7] Röhrs M, Gerth C, Schlarb H, Schmidt B and Schmüser P 2009 Time-resolved electron beam phase space tomography at a soft x-ray free-electron laser *Phys. Rev. ST Accel. Beams* **12** 050704

[8] Panofsky W K H and Wenzel W A 1956 *Rev. Sci. Instrum.* **27** 967

[9] Behrens C and Gerth Ch 2009 On the limitations of longitudinal phase space measurements using a transverse deflecting structure, in *Proc. 9th European Workshop on Beam Diagnostics and Instrumentation for Particle Accelerators* (Basel) p 269

[10] Wiedemann H 2007 *Particle Accelerator Physics* (Berlin: Springer) p 184

[11] Borland M 2000 *elegant: A Flexible SDDS-Compliant Code for Accelerator Simulation* Advanced Photon Source LS-287

IOP Conf. Series: Journal of Physics: Conf. Series **874** (2017) 012079 doi:10.1088/1742-6596/874/1/012079

Reconstruction of sub-femtosecond longitudinal bunch profile measurement data

M K Weikum[1,2], G Andonian[3,4], R W Assmann[1], U Dorda[1] and Z M Sheng[2,5]

[1] Deutsches Elektronensynchrotron (DESY), 22607 Hamburg, Germany
[2] SUPA, Department of Physics, University of Strathclyde, Glasgow, G4 0NG, UK
[3] RadiaBeam Technologies, LLC, Santa Monica, CA 90404, USA
[4] Department of Physics and Astronomy, UCLA, Los Angeles, CA 90095, USA
[5] Laboratory of Laser Plasmas and Department of Physics and Astronomy, Shanghai Jiao Tong University, Shanghai 200240, China

E-mail: maria.weikum@strath.ac.uk

Abstract. With a current trend towards shorter electron beams with lengths on the order of few femtoseconds (fs) to sub-femtoseconds both in conventional and novel accelerator communities, the need for diagnostics with equivalent attosecond resolution is increasing. The proposed design for a sub-femtosecond diagnostic by Andonian et al. (2011, Phys. Rev. ST Accel. Beams 14 072802) is one such example that combines a laser deflector with an RF deflecting cavity to streak the electron beam in the horizontal and vertical direction. In this paper, we present a tool for the reconstruction of the longitudinal beam profile from this diagnostic data, which can be used both for the analysis of planned experiments and testing of different beam scenarios with respect to their specific setup requirements. Applying this method, the usefulness of the device for measurements in a number of example scenarios, including plasma-accelerated and ultrashort RF-accelerated electron beams, is discussed.

1. Introduction

Both the need for short, high current electron beams for application in advanced radiation generation schemes, like Free-Electron Lasers (FELs), and the natural bunch length reduction to femtoseconds in novel accelerators, such as plasma- and dielectric-based devices [1, 2], have led to an increased interest in ultrashort electron beams and their measurement in the accelerator community. The current standard for bunch length diagnostics comprises a wide range of setups including electro-optical detectors, devices based on coherent and incoherent transition radiation and transverse deflecting cavities. Most of them can resolve electron beams on the order of few to tens of femtoseconds [3, 4, 5] and are thus not suitable as diagnostics in the attosecond regime.

In this paper, we discuss a device proposed by Andonian et al. [6] which promises both sub-femtosecond resolution and a large dynamic range for the measurement of longitudinal beam profiles. The setup of this diagnostic is a combination of a laser modulator, i.e. a high power laser pulse in the TEM_{10}-mode co-propagating with the electron beam in a few-period undulator, and a transverse deflecting RF cavity (RF-TDS), followed by a drift space and a screen for measuring the transverse beam profile. While the laser pulse provides a transverse kick to the electron beam in the undulator, the strength of which has a sinusoidal dependence on the longitudinal position within the beam, the deflecting cavity behind additionally streaks the beam in the orthogonal direction. In the following drift space, the transverse kicks are converted into offsets in the horizontal and vertical directions, respectively, and the longitudinal beam profile is imprinted in the trace of the signal on the screen.

As derived in [6], the angular modulation of the beam by the laser modulator in x and the deflector in y based on the longitudinal beam coordinate s_0 is given by

$$\begin{aligned} x' &= x'_0 + S_{LM} \sin(ks_0) \\ y' &= y'_0 + S_{rf} k_{rf} s_0 \end{aligned} \quad (1)$$

where $k = 2\pi/\lambda$, $k_{rf} = 2\pi/\lambda_{rf}$ are the laser and RF wavenumbers, respectively, and $S_{LM} \propto B\lambda_u\sqrt{P_L}/\gamma^2$ and $S_{rf} \propto eV_{rf}/\gamma$ are the streaking strengths of the laser modulator and deflecting cavity. B and λ_u are the undulator peak magnetic field and period, P_L is the laser power, V_{rf} the deflecting voltage and γ is the Lorentz factor of the electron beam. The temporal resolution of the measurement is determined by the laser modulator and can be estimated based on the strength of this streaking:

$$\Delta t = \frac{\epsilon}{\sigma_x c S_{LM} k} \quad (2)$$

with ϵ the beam geometric emittance and σ_x its rms size in x. In comparison to the RF-TDS alone, the resolution of the full diagnostic hence benefits from streaking at optical instead of radio frequencies, while the second deflector acts to avoid signal overlaps due to the shorter streaking wavelength.

2. Bunch profile reconstruction

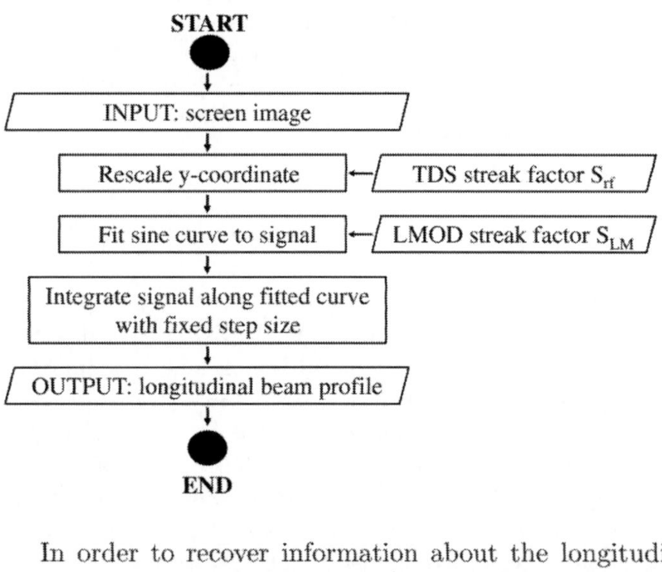

Figure 1. Flow chart of the algorithm used to reconstruct the longitudinal electron beam profile from a screen image of the sub-fs diagnostic.

In order to recover information about the longitudinal bunch length and profile from the screen image measured behind the sub-fs diagnostic setup, a reconstruction tool was developed, the algorithm for which is described in Figure 1. After re-calibrating the image vertically with the streaking strength of the deflecting cavity, a sinusoidal curve is fit to the screen signal and the longitudinal beam profile is calculated based on integration of the data along this fit.

The resolution calculated in Equation (2) is, due to the sinusoidal dependence of the streaking effect, only valid for the linear sections of the signal, whereas around the turning points it is limited to that of the RF-TDS and hence difficult to reconstruct correctly at the sub-femtosecond

IOP Conf. Series: Journal of Physics: Conf. Series **874** (2017) 012079 doi:10.1088/1742-6596/874/1/012079

level. The recovery mechanism thus also includes a feature to combine multiple screen images; if taken at a slightly different phase in the laser modulator, as would naturally occur in an experiment due to jitter, the position of the beam along the sinusoidal screen pattern changes and so the high resolution reconstructions of different parts of the beam can be combined to an overall higher quality beam profile.

Figure 2 provides an example of the reconstruction options of the developed tool: while subfigure a) depicts a measurement example of the screen image, Figure 2 b) presents the application of the algorithm with the original beam shape as the black dotted line and the reconstructed profile based on multiple (single) shots at random phase as the blue solid (red dotted) line. Not only is the error in length and distance of the microbunches around 3 % and less, one can also clearly see the improvement in the reconstructed profile, particularly the shape of the microbunches, using multiple shots.

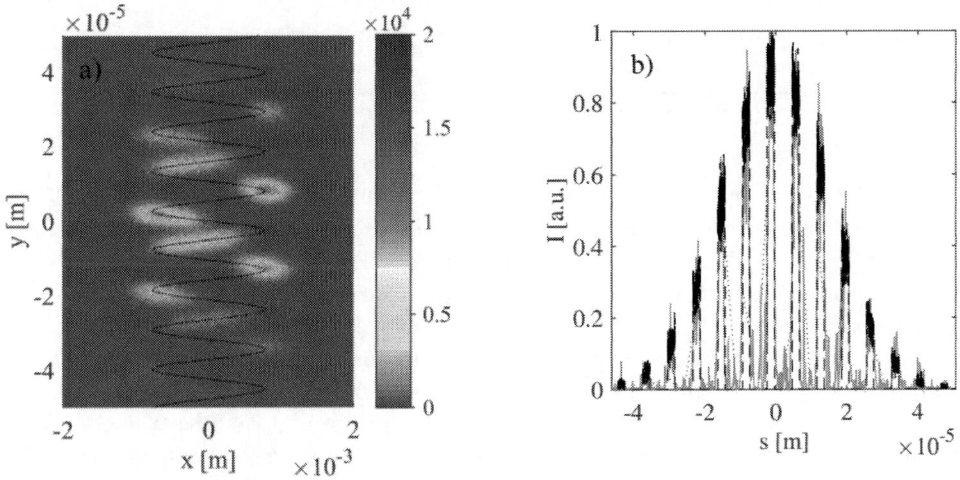

Figure 2. Reconstruction of an electron beam of 15 µm rms length consisting of 2 µm long microbunches at 5 µm distance. The reconstructed values are 15.55 µm for the full rms beam length, 1.94 µm for the microbunch length and 5.00 µm for the average microbunch distance. a): transverse beam distribution at the imaging screen with a sinusoidal fit overlaid; b): original beam distribution (dashed black) and recovered profiles based on a single sub-fs diagnostic measurement (dotted red) and five measurements with slightly varying laser phase (solid blue).

3. Application to ultrashort electron beams

Whereas previous studies have focused on applying the sub-fs diagnostic scheme to electron beams with longer overall pulse duration, but very fine substructures [6, 7], such as produced for example in FELs, in the following an application of this diagnostic to ultrashort beams in the sub-femtosecond regime is discussed. To this effect, simulations with the particle-tracking code ELEGANT [8] were performed to create sample diagnostic images that could then be analysed with the newly developed beam profile reconstruction routine.

The test case that has been simulated here is that of an RF-accelerated electron bunch from the planned SINBAD facility [9] with around 200 attoseconds rms duration, the exact properties of which are shown in Table 3. The latter also describes the design parameters for the diagnostic setup which have been chosen in order to fulfill the laser modulator resonance condition, while minimising the theoretical device resolution based on Equation (2). Note that due to the electron bunch length being shorter than half a laser wavelength in this setup, the beam is resolved over a single turn on the screen and the streaking in y is thus in principle not necessarily required. Moreover, the profile can be reconstructed in a single shot without loss in resolution.

Table 1. Design parameters for a test analysis of an ultrashort sample beam from the SINBAD facility at DESY with the sub-fs diagnostic.

Beam energy	(150.70 ± 0.38) MeV
Beam rms duration	6.32×10^{-8} m
Beam geometric emittance in x,y	0.72 nm, 0.66 nm
Laser power	350 GW
Laser wavelength	10.3 µm
Undulator peak field	1.354 T
Undulator period, no. of periods	6 cm, 3
Deflector voltage, wavelength	12 MV, 2.6 cm

Despite an expected resolution on the order of 1.48×10^{-8} m, a large error above 50 % was found in the reconstructed beam duration. This is likely due to the large initial beam size and divergence on the order of 100 µm and 10 µrad, respectively. As the final transverse beam distribution on the screen is composed of contributions from the initial transverse beam profile as well as the streaked longitudinal profile, a smearing effect of the pattern on the screen and hence an increased reconstruction error are observed if the initial beam distribution is too large. By using a collimator (radius 50 µm) in front of the laser modulator, though, the transverse beam properties can be improved. The results of the bunch profile simulation with the collimator are shown in Figure 3. The rms beam length in this case can be recovered with an error of 4.6 % improving upon the results gained from reconstruction with the deflecting cavity alone by a factor of around 20. One drawback of using a collimator is the significant reduction of the propagated beam charge to about 9 % of the initial value of 2.8 pC. Considering this charge loss as well as the relatively small size of the beam at the final measurement point, a high resolution imaging setup is necessary to resolve the beam.

A more challenging application for the diagnostic setup could be for plasma-accelerated electron beams which typically have bunch lengths on the order of one to few femtoseconds, yet a large correlated energy spread and geometric emittance of a few nanometers. While in theory the required resolution should be achievable with a medium strong streaking effect and smearing of the screen signal could be controllable through beam collimation, an important limitation in this case is likely the energy spread. Detailed simulations with a test beam show that an increased energy spread up to a few percent leads to a deterioration of the reconstructed beam quality. For a spread above around 5 %, the screen signal further becomes strongly distorted and unrecoverable. The cause for this effect is two-fold: on the one hand, the dependence of the streaking strength in x and y on electron energy, as shown in Equation (1), also results in smearing of the screen profile features with increasing spread. On the other hand, the momentum compaction effect of the undulator, quantified as $R_{56} = 2N_u\lambda$ (with N_u the number of undulator periods and λ the laser wavelength), causes a change in the length of the bunch during propagation in the laser modulator which can affect the final recovered beam profile. For ultrashort beams the latter is particularly dominant and can occur already at low values of energy spread. In order to avoid significant compression or stretching of a plasma-accelerated beam in the undulator, possible solutions could be the use of an extremely short, single period undulator in order to minimise the undulator R_{56} effect or a tapering of the undulator in order to

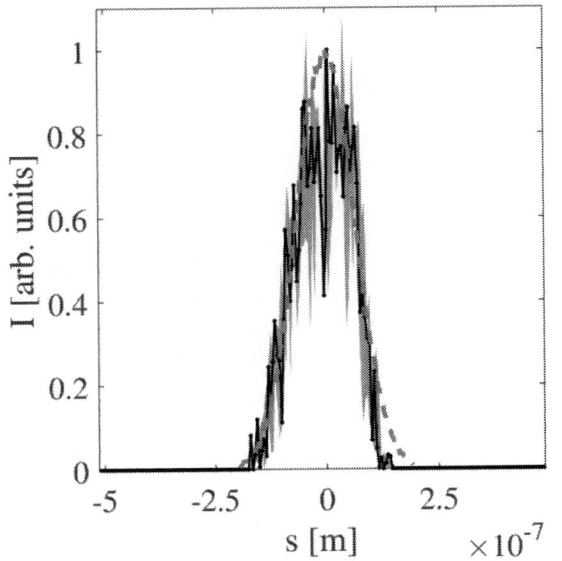

Figure 3. Reconstructed SINBAD beam with an rms beam length of 6.03×10^{-8} m (solid black) overlaid by the normalised initial beam profile (dashed red). The blue, shaded region shows the profile error, estimated based on the uncertainty in the fitting coefficients of the reconstruction algorithm.

compensate the energy chirp. Both options, however, require further study and will be reported on in the future.

4. Summary

A beam profile reconstruction tool for a sub-femtosecond longitudinal beam diagnostic based on streaking with a laser modulator and deflecting cavity was presented and its capability demonstrated, among others with improved performance due to the combination of data from multiple shots. It was further applied to the study of ultrashort beams in the attosecond regime with the sub-fs diagnostic setup. While the required design parameters are at the edge of currently available technology, possible example cases of high resolution bunch profile measurements with RF- and plasma-accelerated beams were presented. Initial beam transverse properties and energy spread were defined as two main limiting factors in reconstruction quality in the ultrashort regime. Future studies will extend the application of this diagnostic device to more challenging ultrashort electron beam accelerator setups, in particular also for the AXSIS (Frontiers in Attosecond X-Ray Science: Imaging and Spectroscopy) project at SINBAD [2].

Acknowledgments

This work is funded by the European Research Council under the European Union's Seventh Framework Programme (FP/2007-2013) / ERC Grant Agreement n. 609920.

References

[1] Lundh O *et al.* 2011 Few femtosecond, few kiloampere electron bunch produced by a laser-plasma accelerator *Nat. Physics* **7** 219-22

[2] Kaertner F X *et al.* 2016 AXSIS: Exploring the frontiers in attosecond X-ray science, imaging and spectroscopy *Nucl. Instr. Meth.* A **829** 24-9

[3] Berden G *et al.* 2007 Benchmarking of electro-optic monitors for femtosecond electron bunches *Phys. Rev. Lett.* **99** 164801

[4] Behrens C *et al.* 2014 Few-femtosecond time-resolved measurements of X-ray free-electron lasers *Nat. Commun.* **5** 3762

IOP Conf. Series: Journal of Physics: Conf. Series **874** (2017) 012079 doi:10.1088/1742-6596/874/1/012079

[5] Nozawa I, Gohdo M, Kan K, Kondoh T, Ogata A, Yang J and Yoshida Y 2015 Bunch length measurement of femtosecond electron beam by monitoring coherent transition radiation (MOPTY002) *Proc. 6th Int. Particle Accelerator Conf. (IPAC15)* (May 2015, Richmond, VA, USA) 940-3

[6] Andonian G, Hemsing E, Xiang D, Mumuseci P, Murokh A, Tochitsky S and Rosenzweig J B 2011 Longitudinal profile diagnostic scheme with subfemtosecond resolution for high-brightness electron beams *Phys. Rev. ST Accel. Beams* **14** 072802

[7] Andonian G, Hemsing E, Mumuseci P, O'Shea F, Rosenzweig J, Murokh A, Xiang D, Pogorelski I and Yakimenko V 2011 Diagnostic concept for high-resolution temporal profile measurements (WEOBB02) *Proc. 2nd Int. Particle Accelerator Conf. (IPAC11)* (Sep 2011, San Sebastián, Spain) 1967-9

[8] Borland M 2000 Elegant: a flexible SDDS-compliant code for accelerator simulation Advanced Photon Source (Argonne, USA) *APS Technical Report* LS-287

[9] Dorda U *et al.* 2016 SINBAD - The accelerator R&D facility under construction at DESY *Nucl. Instr. Meth.* A **829** 233-6

IOP Conf. Series: Journal of Physics: Conf. Series **874** (2017) 012080 doi:10.1088/1742-6596/874/1/012080

Bunch shape monitor development in J-PARC linac

A Miura[1], J Tamura[1], Y Liu[2] and T Miyao[2]

[1] J-PARC Center, Japan Atomic Energy Agency, 2-4, Shirakata-Shirane, Tokai, Ibaraki, 319-1195, JAPAN
[2] J-PARC Center, High Energy Accelerator Research Organization, 203-1, Shirakata, Tokai, Ibaraki, 319-1106, JAPAN

E-mail: akihiko.miura@j-parc.jp

Abstract. In the linac at the Japan accelerator research complex (J-PARC), we decided to use bunch shape monitors (BSMs) as phase-width monitors. Both centroid-phase set point at the frequency jump from SDTL (324 MHz) to ACS (972 MHz) and phase-width control are key issues for suppressing excess beam loss. BSM was designed and developed at the Institute for Nuclear Research, Russia. Because the BSM was first used between acceleration cavities, we need to improve it to protect it from the leakage-magnetic field of the quadrupole magnets and from outgassing impacts on the cavities. In this paper, we introduce these improvements to the BSM for the adoption of the location nearby the acceleration cavities.

1. Introduction

The linac at Japan accelerator research complex (J-PARC), which serves as the injector for the downstream 3-GeV synchrotron (RCS: Rapid Cycling Synchrotron), accelerates a negative-hydrogen-ion beam (H^-). The linac comprises an ion source, radio-frequency quadrupole linac (RFQ), drift tube linac (DTL) and separated-type DTL (SDTL). For the energy upgrade to achieve 1 MW beam power from RCS, we upgraded the beam energy and increased the output energy to 400 MeV by adding annular-ring coupled structure linac (ACS) cavities downstream of the SDTL. We use an accelerating frequency of 324 MHz for the accelerator cavities of RFQ, DTL, and SDTL and 972 MHz for ACS. Because there is a three-fold frequency jump at the SDTL-ACS transition (from 324 to 972 MHz), proper matching at that point is crucial to suppressing excess beam loss. We have decided to use three bunch shape monitors (BSMs) installed periodically at the entrance of the ACS section [1]. The amplitudes of two bunchers located between SDTL and ACS are adjusted so that the measured rms phase-widths with the BSM array become the same [2].

Since the BSM developed by A. Feschenko (INR: Institute for Nuclear Research, Russia) has been used by many accelerator facilities [3], we have developed a BSM in collaboration with INR. The development was completed in 2012. The BSMs were commissioned along with the beam, and their operability was demonstrated [4].

We have encountered some problems with the BSM measurements. The first problem involves the influence of the quadrupole magnetic fields on the ability to focus the secondary electrons. This is because the BSMs are located between and close to the quadrupole doublets. The second problem involves degradation of vacuum during measurement [5]. After we overcame above mentioned problems, we remounted one BSM in front of the upstream of the ACS cavities. We have used the BSM and the proposed tuning method at 40 mA for peak beam current operation [6]. Because the

BSM was first used between acceleration cavities, we need to protect the leakage magnetic field from the quadrupole magnets and outgassing effects on the cavities. This paper focuses on the two problems and the associated countermeasures for the BSM installed near the acceleration cavities.

2. Bunch Shape Monitor Development in J-PARC Linac

2.1. Principle of bunch shape measurement

A device for longitudinal beam-width (phase-width) measurement was designed on the basis of the observation of secondary electrons from a single wire intersecting the beam, as described in [3, 4]. A series of bunches of the beam under measurement was made to intersect a target wire of 0.1 mm diameter, and low-energy secondary electrons were emitted owing to the beam-wire interaction. The wire was held at a typical negative potential of −10 kV. The secondary electrons moved almost radially and entered the RF deflector through collimators. We applied an RF-field with the same frequency as the accelerating RF (324 MHz). The deflection of the electrons at the exit of the RF deflector depends on their phase with respect to the deflecting field. By adjusting the phase of deflecting field with respect to the accelera-tor RF reference, we could obtain the longitudinal (phase) width of bunches.

2.2. System configuration of bunch shape monitor

The BSM comprises a body, an RF deflector, a steering magnet, an actuator, and electron detector, as shown in Figure 1. The BSM body was designed for installation between the quadrupole doublets, which are located between the acceleration cavities. The RF deflector and the actuator that holds the target wire were installed vertically against the beam axis. The secondary electrons pass through the collimators of the RF deflector and the duct connected to the electron detector. The collimators and the structural parts degrade the pumping conductance. During the BSM measurement, we applied a high bias voltage to the static lens and the target wire, and controlled the RF power input as the main parameter. They are the main factors influencing the vacuum condition.

The RF deflector is composed of a pair of electrodes. We applied a constant DC voltage typically −8.45 kV to focus the trajectories of the secondary electron. The secondary electrons accelerated away from the target wire with a negative potential because −10.0 kV was applied to the collimator in front of the bending magnet.

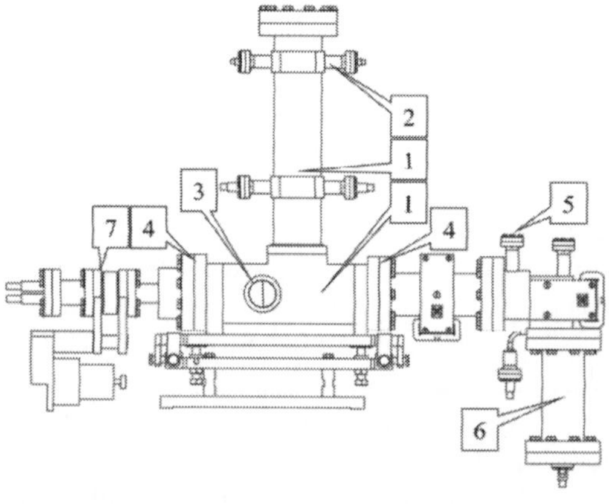

Figure 1. BSM Outline (1-Body and RF deflector, 2-SHV connector, 3-Beam duct, 4-Flange to connect body and pipe, 5-View port, 6-Detector, 7-Actuator).

3. Defocusing by Quad Fringe Field

Visual observations of the focusing of thermal electrons from the view ports with the turned-on doublet quadrupoles immediately revealed a strong influence of the quad-fields. Increasing the quadrupole current resulted in an increase in electron beam size at the output collimator installed in bending magnet. We considered the relationship between the quadrupole current and focusing voltage applied to the collimator. The typical focusing voltage was −8.45 kV, and a rapid decrease in the focusing voltage can be seen in Figure 2 (no shield). With a current of 30 A, the focusing voltage had to be decreased to zero. Therefore, we concluded that the compensation effect facilitated the focusing/defocusing effect of the quad-fields. It became clear that BSMs cannot work with the recent nominal doublet current to say nothing of higher currents needed.

The limits of BSM operability depend on the thickness of steel shields and are estimated to be as follows:

- two 0.4-mm-thick shields for up to 100 A,
- 0.4-mm-thick shield and 1.0-mm-thick shield for up to 150 A, and
- 0.4-mm-thick shield and 1.4-mm-thick shield for up to 180 A.

Based on the magnetic simulation, which treats the secondary electron trajectories affected by the excess quad-field [7], we decided to decrease the quads-fields by mounting steel shields measuring 0.4 mm in thickness on the BSM body and 1.0-mm-thick steel plates vertically near the coils (Figure 3).

After installation of the shields, focusing of the thermal electrons was reproduced (Figure 2). To evaluate the influence of the shield on the quad-field around the beam path, we measured the beam profile downstream of the beam transport by using wire scanners (WSM). The rms beam sizes in the transverse profile with magnetic shields are about 1.0% larger than those without the shields. According to an INR estimate, the magnetic shield reduced the GL product (product of the gradient of magnetic field and the length of magnet poles) by about 1.5%, which is in good agreement with the estimation.

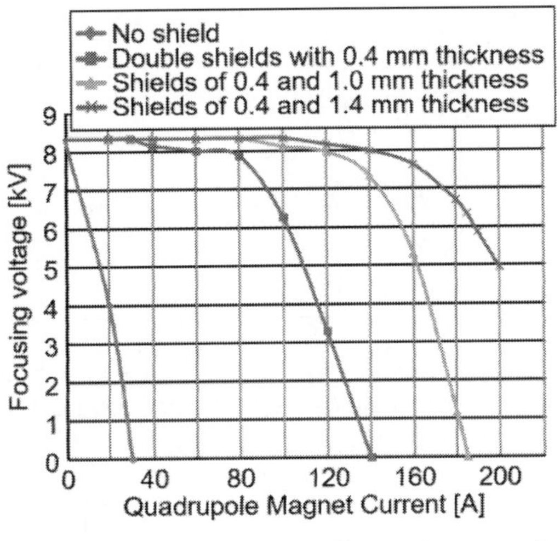

Figure 2. Focusing Voltage Dependence on the Electrical Current of Quadrupole Magnet.

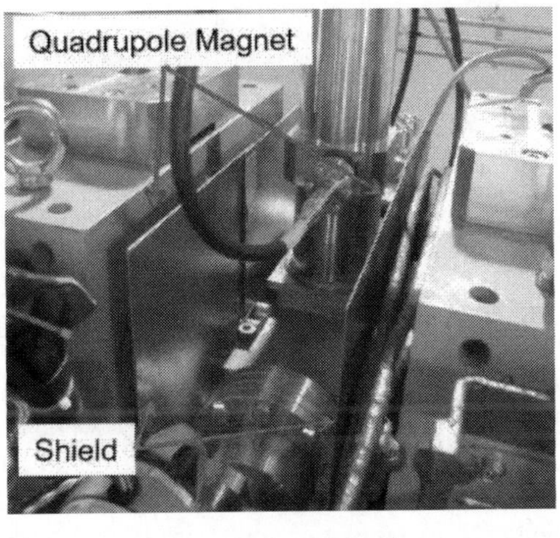

Figure 3. Installation of Shields in Quadrupole Doublets between Acceleration Cavities.

4. Vacuum Degradation

During the BSM measurement, a problem related to the influence on the vacuum was encountered. Before BSM installation, vacuum pressure in the upstream part of the ACS section was slightly under 1.0e-6 Pa. However, a vacuum of over 1.0e-4 Pa, which is the interlock level required to stop beam operation, was encountered immediately after supplying RF power.

We estimated the causes of this degradation, and the relevant countermeasures are given below [5].

a) Dark electron current from a target accelerated by bias potential was a candidate because the effect was observed immediately after supplying high voltage potential. We conditioned a target wire applying a bias voltage to it for several days continuously.

b) We examined the vacuum test with and without the BSM parameters at the beam line to estimate the contributions of BSM parameters and to estimate the pumping speed based on the vacuum trends. We decided to remove the BSMs from the beam line for off-line tests and for vacuum conditioning.

Without any BSM conditioning, vacuum speed was evaluated using an off-line setup. Based on a test that ran for a week (about 100 h), the time to reach an operational level of 5.0e-7 Pa (in Figure 4) was estimated to be 2,000 h. This means that more than a week is required to exchange a broken target wire after pumping.

We conducted baking with an additional turbo-molecular pump installed on top of the RF deflector. Notably, the RF arrangement, which is an extension of an RF deflector, affects the RF characteristics. After cooling, we measured vacuum speed and RF parameters to adjust the resonance frequency by changing the length of the deflector electrode. The required time was reduced to almost half the original time after conditioning. Therefore, effective conditioning during pumping results in quicker recovery. This means that online conditioning after installation is important to minimize the recovery time. In addition, the time for the vacuum impact to the acceleration cavities can be avoided.

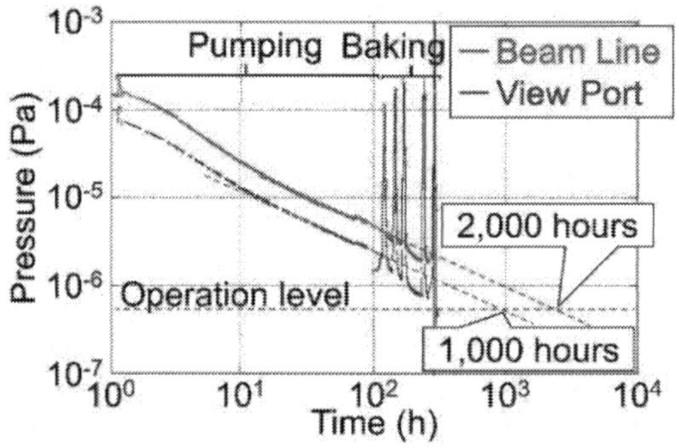

Figure 4. Initial pumping Speed of BSM and Acceleration Time for Achieving Target Vacuum Level.

5. Phase Width Tuning by Longitudinal-Transverse Coupling

5.1. Installation layout

Because of vacuum degradation, we installed one BSM downstream of the last SDTL (SDTL16) and ACS-type buncher cavities (Bunchers 1 and 2). Accelerating frequency jumps between SDTL16 and Buncher 1. The distances from SDTL16 and Buncher 1 to the BSM are 8.4 m and 16.9 m, respectively (Figure 5). Clear dependence of the amplitude and phase-width of Buncher 2 cannot be obtained because the distances are too short to obtain the phase-width within the phase resolution of BSM.

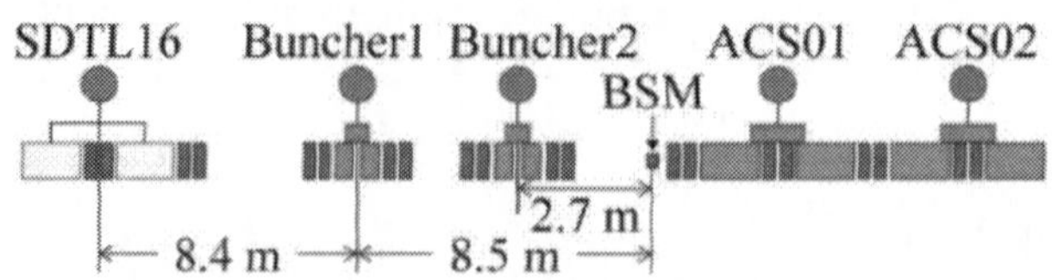

Figure 5. BSM Installation Layout after SDTL16 with Their Distances Shown.

5.2. Phase width tuning

We remounted one BSM in front of the area upstream of the ACS cavities (in Figure 5) to propose a tuning method. Buncher 1 is used as the knob for amplitude scan at the location of BSM. The amplitudes of Buncher 1 were scanned to obtain the curves shown in Figure 6. Based on the Q-scan method for matching transverse beam profiles, we proposed a tuning method that employs amplitude scan curves and the Twiss parameters obtained from transverse profiles [6, 8]. In the method, we use the following formula;

$$\sigma^2_{BSM} = \varepsilon_z[(1 + Lk)^2\beta_B - 2L(1 + Lk)\alpha_B + L^2\gamma_B] \quad (1)$$

where ε_z is phase-width emittance; L is drift length; α_B, β_B, and γ_B are Twiss parameters at the BSM position; and k is the longitudinal focusing force. We substituted σ2BSM and k obtained from Figure 6. These free parameters of ε_z, α_B, and β_B, were obtained by iterated calculation, and γ_B was obtained by 3D-PIC simulation. These parameters were adapted to the buncher and quadrupole settings, as well as the total beam loss measurements, and proper settings for minimum beam loss were determined.

The relationship between the measured horizontal beam position and the current of the bending magnets was obtained. The BPM can be used for a beam position range of approximately 20 mm, and the WSM offset values were also obtained as well.

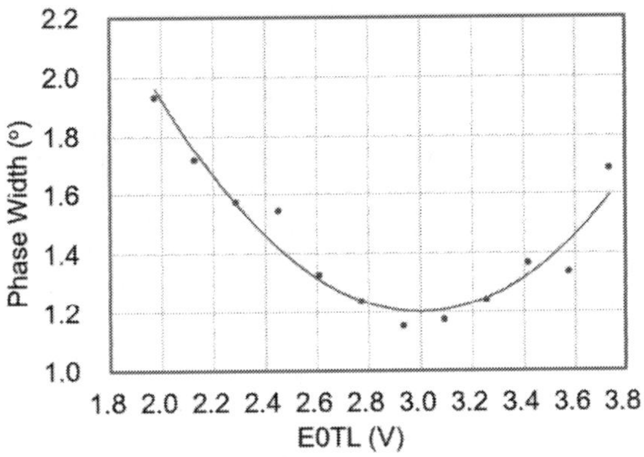

Figure 6. Amplitude Scan by Buncher 1.

6. Conclusion

BSM was successfully developed in collaboration with INR and the improvements to overcome the problems of a defocusing of static lens and a vacuum degradation. BSMs were installed between the quadrupole doublets. Therefore, we considered the influence of the quad-magnetic fields on the operational settings. We overcame the problems of the defocusing of static lens and vacuum degradation to maintain the required super-high vacuum level during BSM operation. A vacuum system with an online baking scheme was proposed to operate BSMs without significant vacuum degradation.

Presently, we use a BSM to make an amplitude scan curve of buncher 1 to find an optimum set point of amplitude. In addition, a BSM is used for resonance studies [9, 10] to determine the equipartitioning (EP) condition. Beam dynamical experiments were conducted by measuring the phase-width with BSM to make Hofmann's stability charts which indicated a region of space-charge driven transverse-longitudinal coupling resonance. BSM measurement has the advantages of being able to measure longitudinal (phase-width) emittance, which is useful for determining the EP-set point. In addition, we have commenced modification of the BSMs to be installed additionally upstream of the ACS section with deep consideration of the effect of vacuum, because we consider back to the policy to take matching of buncher cavities using three BSMs.

7. Acknowledgement

The authors would like to acknowledge the specialists to establish the BSM development, especially for Dr. A. Feschenko (INR, Russia) and the late A. Mirzojan to perform the vacuum test and new vacuum system arrangement. The authors also appreciate the J-PARC writing support group, which continuously encouraged us to write this article.

IOP Conf. Series: Journal of Physics: Conf. Series **874** (2017) 012080 doi:10.1088/1742-6596/874/1/012080

References

[1] Miura A, Maruta T, Liu Y, Miyao T, Kawane Y, Ouchi N, Oguri H, Ikegami M and Hasegawa K 2015 *JPS Conf. Proc.* vol. 8 011002 pp 011002-1–011002-6
[2] Ikegami M, Maruta T, Liu Y, Futatsukawa K, Miyao T, Fang Z, Miura A and Tamura J 2013 *Proc. IPAC2013* (Shanghai, China) THPWO028
[3] Feschenko A V 2001 *Proc. PAC2001* (Chicago, USA) pp 517–21
[4] Miura A, Feschenko A V, Mirzojan A N, Miyao T, Ouchi N, Maruta T, Liu Y, Oguri H, Ikegami M and Hasegawa K 2015 *JPS Conf. Proc.* vol. 8 011003 pp 011003-1–011003-6
[5] Miura A, Kawane Y, Ouchi N and Miyao T 2014 *Proc. IBIC2014* (Monterey, CA, USA) TUPD09
[6] Miura A, Hayashi N, Maruta T, Liu Y, Miyao T and Fukuoka S 2015 *Proc. IBIC2015* (Melbourne, Australia) TUPB027
[7] Tamura J, Ao H, Miura A, Ouchi N, Ikegami M, Maruta T, Miyao T and Takata K 2013 *Proc. IPAC2013* (Shanghai, China) THPWO035
[8] Maruta T, Liu Y, Futatsukawa K, Miyao T, Miura A and Ikegami M 2015 *Proc. 12th Annual Meeting of Part. Accel. Soc. Japan* (Tsuruga, Japan) WEP014, in Japanese
[9] Liu Y, Maruta T, Futatsukawa K, Miyao T, Ikegami M and Miura A 2015 *Proc. IPAC15* (Richmond, Virginia, USA) THPF039
[10] Plostinar C 2015 *in private communication*

A beam position monitor for the diagnostic line in MEBT2 of J-PARC linac

A Miura, J Tamura and Y Kawane

J-PARC Center, Japan Atomic Energy Agency, 2-4, Shirakata-Shirane, Tokai, Ibaraki, 319-1195, JAPAN

E-mail: akihiko.miura@j-parc.jp

Abstract. In the linac of the Japan Proton Accelerator Research Complex (J-PARC), the neutral hydrogen (H^0) beam from the negative hydrogen ion (H^-) beam is one of key issues in mitigating beam losses. To diagnose H^0 particles, we installed a set of beam-bump magnets to generate a chicane orbit of the H^- beam. The beam position monitors (BPMs) in the beam line are used for orbit correction to maintain the beam displacement within 2.0 mm from the duct center. To measure the beam displacement under different drive currents of the beam-bump magnets, a new wide-range BPM was designed and manufactured to evaluate the horizontal beam position by using a correction function to compensate for non-linearity. We also employed the beam profile monitor (WSM: wire scanner monitor) to measure the H^- beam profile, which helped us to compare the beam position measurements. In this paper, the design and the performance of the wide-range BPM are described. In addition, we present a comparison of the beam position measured by the BPM and the WSM.

1. Introduction

In the Japan Proton Accelerator Research Complex (J-PARC) linac, a diagnostics beam line incorporating a beam bump was installed in the medium energy beam transport (MEBT2) part, between the section of a separation-type drift tube linac (SDTL) and an annular-ring coupled structure linac (ACS). The diagnostics to measure beam displacement and to detect H^0 particles were prepared for the diagnostics line in the summer of 2015 [1]. The diagnostics line comprises four beam-bump magnets which are structurally identical to produce a chicane orbit in the horizontal plane, as depicted in Figure 1 and 2. The transverse profile monitor (WSM: wire scanner monitor), located in the beam line is used to measure the intensity distribution of both the H^+ and H^0 particles by using a carbon plate mounted on a sensor head of the WSM. In addition, the performances of the beam-bump magnets and their power supplies were evaluated by measuring beam displacements by using a beam position monitor (BPM). We use a rectangular beam duct with dimensions of 100 mm horizontally and 40 mm vertically for a maximum beam displacement of 30 mm horizontal to avoid a scraping of the beam core.

More than 40 BPMs have already been fabricated and installed in the MEBT2 and ACS beam line [2], and they are used for standard orbit-correction procedures. Following the J-PARC tuning policy, the beam orbit must be located within 2.0 mm of the duct center, the measurement accuracy of the beam-line BPMs are required to be within 5.0 mm from the duct center. We designed and fabricated a wide-range BPM beam pickup considering the aperture and the beam orbit variations.

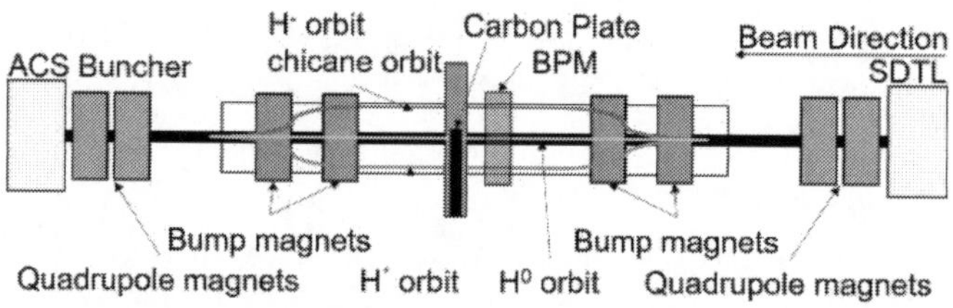

Figure 1. Device Location in MEBT2 Diagnostics Line.

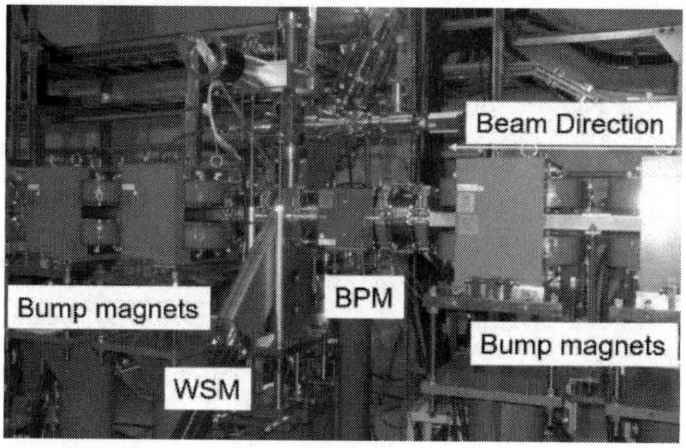

Figure 2. Installation of Bump Magnet and Diagnostics in MEBT2 Diagnostic Line.

After installation of the newly fabricated BPM in diagnostics line, the beam displacements were evaluated by comparing the magnet supply currents with the BPMs. The measured beam positions were compared with the beam center positions found in the beam profile obtained using the WSM. This paper describes the design the BPM, its calibration and evaluation results.

2. Design of the Beam Position Monitor

2.1. Design of the BPM pickup

An alternating current signal is induced on the electrode, coupled at radio frequencies (RF) and is four pick-up electrodes are installed crosswise in the beam duct and the difference in the signals of the counter-facing electrodes in the center-of-mass of the beam.

For low-β beams, the three dimensional (3D) Laplace equation must be used when the beam current modulation wavelength is comparable to the aperture. The position sensitivity of pairs of BPM pickup electrodes was estimated using Equations (1) and (2) and the Lorentz factor. Image charges in the form of logarithmic ratio in dB are given by the following functions [3]:

$$\left(\frac{I_{WR}}{I_{WL}}\right) = \frac{160}{ln(10)}(1+G)\frac{sin(\phi/2)}{\phi}\frac{x}{b} + O(x^2), \tag{1}$$

where G is given as:

$$G = 0.139\left(\frac{\omega b}{\beta\gamma c}\right)^2 - 0.0145\left(\frac{\omega b}{\beta\gamma c}\right)^3. \tag{2}$$

Here, ϕ is the electrode opening angle, b is the radius of the beam duct, x is the beam position, ω is the angular frequency, $O(x^2)$ denotes the higher-order terms in x, c is the velocity of light, and γ is the Lorentz factor.

In our application a stripline type electrode was selected for this diagnostics BPM, as it could be better matched operating at the acceleration frequency compared to the button-style BPM electrodes owing their stray capacitances [4].

Figure 3 shows a drawing of the BPM. A large aperture with an inner diameter of 120 mm was selected to ensure the measurement of displacement [1]. The signal induced on a stripline by the beam charge is proportional to the distance between the electrode and the beam, as well as to the beam intensity, and it further depends on the dimensions, for example, length and width, of the stripline. If the characteristic impedance at both the upstream and the downstream ports is equal to the characteristic impedance of the stripline (here: $Z_0 = 50\ \Omega$), the voltage induced at the upstream port of a stripline electrode of length (l) is given as follows [5] :

$$V(\omega)=\frac{\phi Z_0}{2\pi} Ib(\omega) \sin\left[\frac{\omega l}{2c}\left(\frac{1}{\beta_s}-\frac{1}{\beta_b}\right)\right], \tag{3}$$

where ϕ represents the opening angle of the stripline electrode (Figure 3); $I_b(\omega)$ the beam spectrum, here detected at the RF acceleration frequency of 324 MHz, β_s the relative velocity of the signal on the stripline electrode; and β_b the relative velocity of the beam. By using Equation (3), we optimized the electrode length to achieve the maximum output signal voltage at our operation frequency, which is 155 mm for our beam energy of 191 MeV ($\beta_b = 0.56$).

The width of the stripline was determined based on its characteristic impedance, and it was found to be 50 Ω to match the impedance of the signal cable. To determine the width of the stripline, we calculated the characteristic impedance of the stripline by using the cross-sectional distribution of the electrical field generated by the potential difference between the electrode and the beam duct [6]. The characteristic impedance was then calculated based on a numerical, two-dimensional Poisson equation calculation code called "Poisson Super Fish" [7]. An example of thd calculated electrical field is shown in Figure 4. In the narrow space, which is the gap between the stripline and the body, there is ceramic insulation, and high-electrical-field regions can be observed. After varying the width of the stripline as a parameter in those two-dimensional calculations, we fabricated prototype pickups, and finally set the width to 34.40 mm.

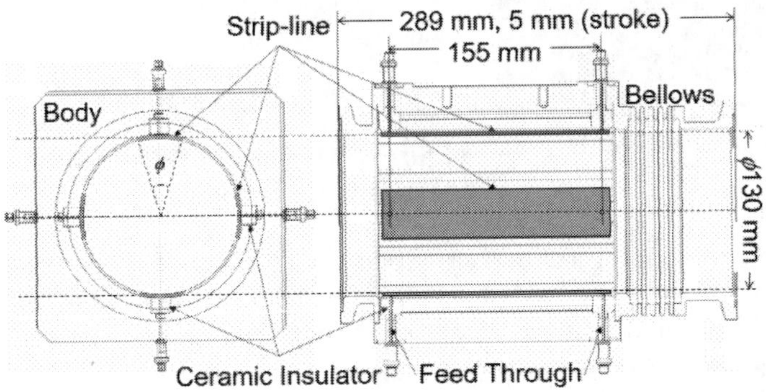

Figure 3. Drawing of BPM for diagnostic line in MEBT2 of J-PARC linac.

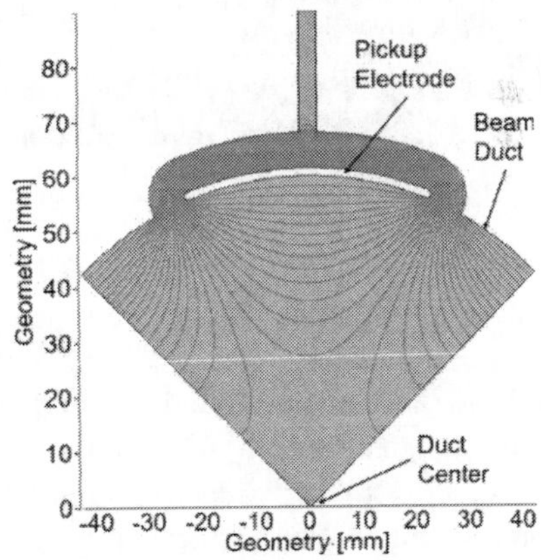

Figure 4. 2D Simulation results of electrical field in a cross section around one electrode pickup.

2.2. *Signal processing*

The logarithmic ratio method of signal processing was selected because of its robustness and wide dynamic range. The output signal from the logarithmic amplifier was low-pass-filtered for noise reduction. The beam position obtained by the logarithmic processing is as follows [3]:

$$x = \frac{1}{S} \cdot (logV_R - logV_L) = \frac{1}{S_x} \cdot log\frac{V_R}{V_L} \propto \frac{1}{S_x} \cdot V_{out} \quad (4)$$

Position sensitivity given by S in [dB/mm], and V_R and V_L are induced signal voltages [V] of the right and thw left electrodes, respectively. Sensitivity is described below in terms of stripline width W,

$$S = \frac{160}{ln\,10} \times \frac{sin(\phi/2)}{\phi} \times \frac{1}{r} = \frac{160}{ln\,10} \times \frac{W}{2\phi} \times \frac{1}{r}. \quad (5)$$

For calculation of the beam position, for example, the vertical position X [mm] is given below using Equation (4),

$$X = \frac{1}{S} \cdot 20log_{10}\left(\frac{V_R}{V_L}\right). \quad (6)$$

3. Calibration

We measured the electrical center of the BPM by using a simulated beam signal induced on a stretched wire calibration bench, as in Figure 5. To compare the position of the physical wire and its reconstructed position obtained from the electrical BPM readout, the following scheme was used in the calibration bench.

· The wire was scanned through the transverse cross section of the BPM, horizontally and vertically, by applying a 324 MHz RF stimulus signal.

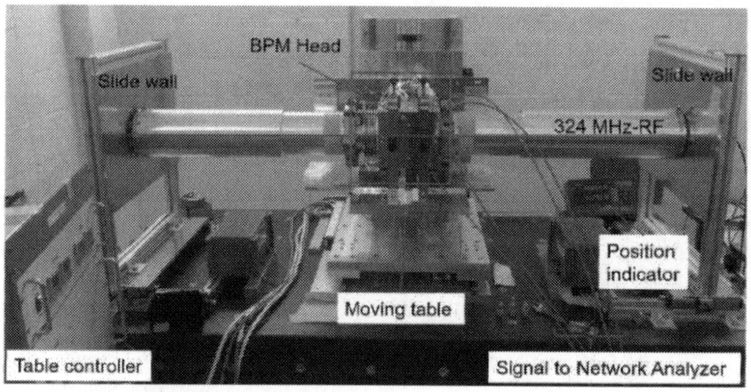

Figure 5. Overview of calibration bench for BPM.

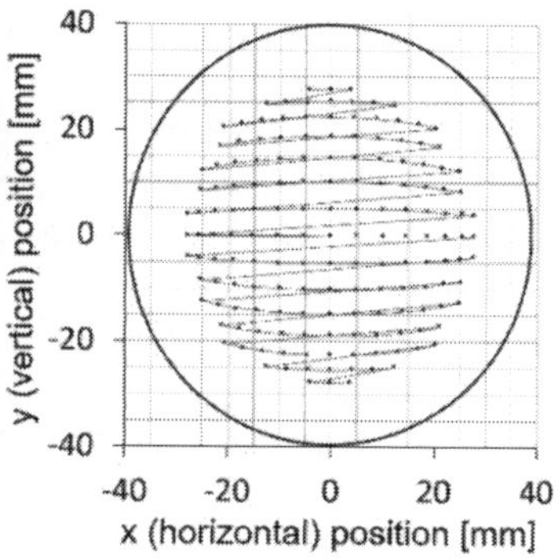

Figure 6. Mapping of BPM position characteristics moving wire.

· The induced electrode signals were processed with logarithmic amplifiers on a 50 Ω load. The ratios of the demodulated output signals between the right and the left, and the top and the bottom electrodes give the horizontal and vertical measured beam positions, respectively.

· We interpolated (curve fitting) the acquired position values. As shown in Figure 6, for large displacements of the wire (especially under 45° of the x-y coordinates), the non-linear effects become stronger. For correction, a fitting function must be derived. Moreover, fitting functions are required for displacement than 20 mm. In addition, this calibration method excited with either pulses or high-frequency sinusoidal currents, the EM wave represents the principal (TEM) mode in a coaxial transmission line. Thus, the stretched-wire method is unsuitable for the simulation of slow particle beams in high frequency diagnostic devices. Assuming position displacement by using the applicable fitting function comprising a fourth-order polynomials, we extrapolated the calibration results to the obtained BPM mapping (Figure 6).

Figure 6 shows the relationship between the mechanical and the electrical position. The red points represent the position of the wire, scanned throughout the BPM aperture. Red dots indicate the wire position on a 1.0-mm square grid, and the black ones are the obtained wire positions. The horizontal (X) and vertical position values (Y) were measured directory by the logarithmic amplifier module. For relativistic beams, we considered the fitting function to be adaptable in a region of radius 25.0 mm.

As the fitting function, we derived a fourth-order polynomial for the x and y:

$$x = -3.594e\text{-}8m^4 - 1.608e\text{-}8m^3n - 6.455e\text{-}8m^2n^2 + 5.623e\text{-}7mn^3 + 7.265e\text{-}9n^4 - 5.518e\text{-}5m^3 + 1.819e\text{-}6m^2n + 1.970e\text{-}4mn^2 + 5.375e\text{-}6n^3 + 3.305e\text{-}5m^2 - 6.008e\text{-}5mn + 1.970e\text{-}4n^2 + 9.561e\text{-}1m + 3.305e\text{-}5n + 3.498e\text{-}3 \quad (7)$$

$$y = -1.566e\text{-}8e\text{-}8m^4 - 7.372e\text{-}8m^3n + 1.785e\text{-}7m^2n^2 - 1.056e\text{-}7mn^3 + 2.165e\text{-}8n^4 - 2.635e\text{-}7m^3 + 1.845e\text{-}4m^2n + 5.204e\text{-}6mn^2 - 5.057e\text{-}5n^3 - 1.405e\text{-}6m^2 + 6.896e\text{-}5mn - 7.083e\text{-}5n^2 + 2.343e\text{-}3m + 9.572e\text{-}1n + 0.209 \quad (8)$$

where m and n are the position values reported by the logarithmic amplifier signals. m is the horizontal position and n is the vertical position. We implemented these fourth-order fitting functions, corrected the BPM pickup non-linearity, and reported the resulting beam position.

4. Orbit-Shift Evaluation by Transverse Profile

We used a WSM to evaluate beam displacement from the transverse beam profile. A tungsten wire of diameter 0.03 mm was used to measure the beam profile. The WSM is a suitable device to evaluate the beam displacement, because the scan range of WSM is strictly maintained. The results of the beam profile measurement with several current sets of the beam-bump magnet are summarized in Figure 7. The beam profile peaks corresponding to 10-A intervals of the magnet can be observed 7.5-mm apart.

5. Measurement Results

The beam positions acquired by the WSM measurements are plotted in Figure 8 along with the acquired BPM data for comparison. The WSM measurement shows a high linearity, even for large beam displacements. The measured beam position of WSM and BPM differs slightly, particular at large beam displacements. One reason is the linear fitting function only considers distance less than 20 mm, so the errors depend upon the distance from the center. The gradient of the fitting function of BSM is about 7-8 % smaller than that of WSM. It is considered to be an error of the value of sensitivity described by Equation (6). The sensitivity is defined by the geometrical parameters, if a negligible error exists, the measurement errors extend linearly. Usually, the fitting function is applied for a range of 5.0 mm from the beam center. However, this function was extended to consider larger

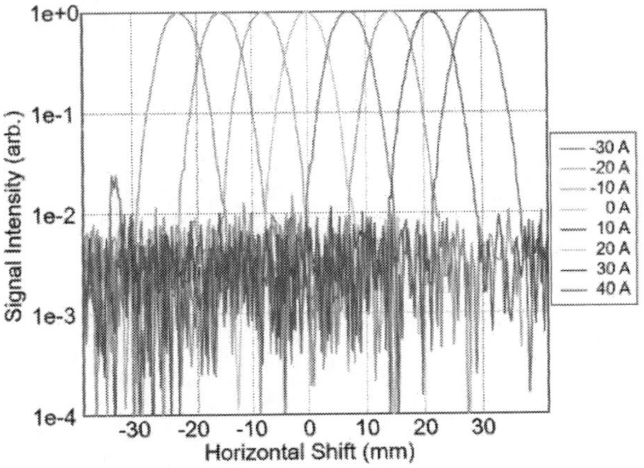

Figure 7. Beam profile measurement results with several current-sets.

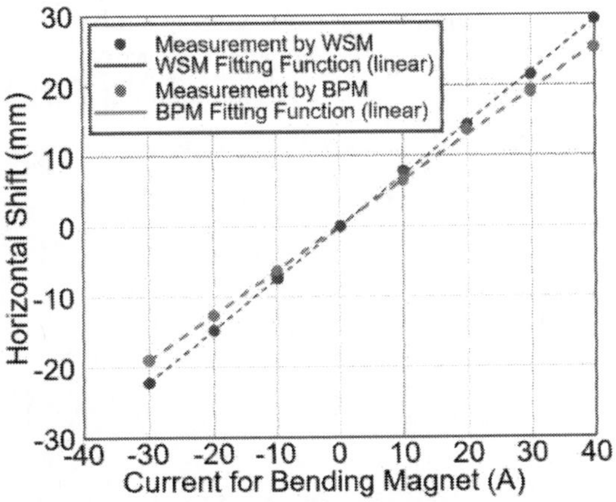

Figure 8. Beam position obtained by BPM and WSM.

beam displacements. In addition, we assumed a beam energy of 191 MeV, which corresponds to β of 0.56. The wire method is unsuitable for low-β beam fields, and errors can be found in the calibration. In the case of low-β fields, we must use the Equations (1) and (2) to fit these measurement data by considering the higher-order terms of the higher-harmonic number of the frequency (ω).

The relationship between the measured horizontal beam position and the current of the bending magnets was obtained. The BPM can be used for a beam position range of approximately 20 mm, and the WSM offset values were also obtained as well.

6. Conclusion

A diagnostic beam line was installed to measure the orbit displacement by using pairs of beam-bump magnets. Beam diagnostics, such as BPM and WSM, were designed and installed to observe horizontal beam displacement and its profile. We use both diagnostics to evaluate the beam displacement when separating H^-, H^0 and H^+ particles. In this paper, we have described the design of a wide-range BPM to extend the calibration function to a range of about 20.0 mm. A comparison of beam position measurements by BPM and WSM, revealed acceptable agreement between the two sets within a range of 20 mm, so both monitors can be utilized for large beam displacements. The newly developed BPM can perform online monitoring of the beam displacement without interrupting any operation. By constant, WSM cannot be used for online beam monitoring, but it can measure the beam orbit by performing horizontal scanning because of its high linearity over a wide aperture.

7. Acknowledgement

The authors would like to acknowledge the specialists who installed the diagnostics system arrangement, and performed the chicane test. The authors also appreciate the J-PARC writing support group, which continuously encouraged us to write this article.

References

[1] Tamura J, Miura A, Morishita T, Ao H, Maruta T, Miyao T and Nemoto Y 2016 *Proc. LINAC16* (East-Lansing, MI, USA) MOPLR063

[2] Miura A, Ouchi N, Oguri H, Hasegawa K, Miyao T and Ikegami M, 2015 *J. Korean Phys. Soc*, **66** 3, pp 364-372

[3] Shafer R E 1994 *AIP Conf. Proc.* **319** pp. 303

[4] Forck P 2012 *Lecture Notes on Beam Instrumentation and Diagnostics*, (Text. of Joint Univ. Accelerator School) pp 9-39

[5] Shafer R E1985 *IEEE Trans. Nucl. Sci.* **32** pp 1933-37

[6] Sato S, Igarashi Z, Lee S, Tomisawa T, Hiroki F, Kishiro J, Ikegami M, Kondo Y, Hasegawa K, Ueno A, Toyama T, Kamikubota N, Nigorikawa K, Tanaka M, Yoshikawa H 2004 *Proc. LINAC2004* (Lubeck, Germany) TUP70

[7] Poisson Superfish Announcement by LANL Home Page: http://laacg1.lanl.gov/laacg/

IOP Conf. Series: Journal of Physics: Conf. Series **874** (2017) 012082 doi:10.1088/1742-6596/874/1/012082

Techniques for transparent lattice measurement and correction

Weixing Cheng#, Yongjun Li, Kiman Ha

NSLS-II, Brookhaven National Laboratory, Upton, NY 11973, USA

chengwx@bnl.gov

Abstract. A novel method has been successfully demonstrated at NSLS-II to characterize the lattice parameters with gated BPM turn-by-turn (TbT) capability. This method can be used at high current operation. Conventional lattice characterization and tuning are carried out at low current in dedicated machine studies which include beam-based measurement/correction of orbit, tune, dispersion, beta-beat, phase advance, coupling etc. At the NSLS-II storage ring, we observed lattice drifting during beam accumulation in user operation. Coupling and lifetime change while insertion device (ID) gaps are moved. With the new method, dynamical lattice correction is possible to achieve reliable and productive operations. A bunch-by-bunch feedback system excites a small fraction (~1%) of bunches and gated BPMs are aligned to see those bunch motions. The gated TbT position data are used to characterize the lattice hence correction can be applied. As there are ~1% of total charges disturbed for a short period of time (several ms), this method is transparent to general user operation. We demonstrated the effectiveness of these tools during high current user operation.

1. Introduction

Ultra-low emittance is essential for high brightness operation of light sources like NSLS-II. The same is true to achieve high luminosity in lepton collider machines. During NSLS-II operation, betatron tune, coupling and lifetime can be changed while user ID gaps are moved. Orbit disturbance due to the gap movement can be minimized by the feedforward table. However, the tune and coupling variation indicate the lattice is affected by the ID gaps. Besides the ID gaps, there are apparent tune drifts observed with different storage current. This is believed due to the image current effect of flat vacuum chambers and magnet yokes.

Typical lattice characterization and correction are carried out at a low beam current and fixed ID gaps. The corrected lattice isn't reproducible as the ID gaps move and current accumulates. Therefore dynamic lattice measurement and correction are necessary to maintain good machine performance (large dynamic range, long lifetime, good injection efficiency etc.).

There are two types of methods developed to characterize the linear optics. The most common method is Linear Optics from Closed Orbit (LOCO) [1]. Some third generation light sources have used this method to achieve 1pm.rad vertical emittance [2, 3]. At NSLS-II, the more often used method is using BPM turn-by-turn (TbT) data, which is developed based on the original work at LEP [4]. Similar examples can be seen in references [5-7]. Compared to LOCO, the TbT method is much faster. This is especially true for big rings like NSLS-II, equipped with more than 200 BPMs. One iteration of lattice measurement and correction take about 2 minutes [8]. Beta-beat measurement accuracy is ~0.4% from multiple data snapshots. The TbT data has position resolution better than 10μm (7-8 μm for 20-

bunches 1.5mA fill). Two pingers kick the beam within one turn to excite betatron motions in X/Y planes. BPM TbT data acquisition is synchronized with the pinger kicks. Beta-function, phase advance, coupling and other lattice parameters can be measured from one set of BPM TbT data. Corrections can be calculated and applied as needed.

Pinger pulse shape at NSLS-II is about 2μs half sine. To minimize the de-coherence due to bunch to bunch oscillation amplitude difference, only a short bunch train (20-50 bunches) is typically filled. This limits the reliable lattice measurement from TbT data during high current operation when a train of 1000 bunches is filled and bunch to bunch kick amplitude varies.

A new method to characterize the lattice during high current operation is proposed by exciting a small portion of bunches filled in the ion gap and use gated BPM TbT data which measures the betatron motion of those bunches. We refer to those bunches as test bunches. To have a reliable BPM readings, ~1% of total charge is typically filled in the test bunches. See Figure 1 for a typical fill pattern suitable for the new method. With a total beam current of 276mA filled in the ring, the majority of the charge is distributed in the 1000-bunch long train followed by 320 empty buckets to allow ions to escape. Ten bunches were filled in the gap start at bucket #1200, these test bunches had similar per bunch charge as the long bunch train and small bunch to bunch charge variation. Current of the test bunches were ~3.5mA which is ~1.3% of the total storage current of 276mA.

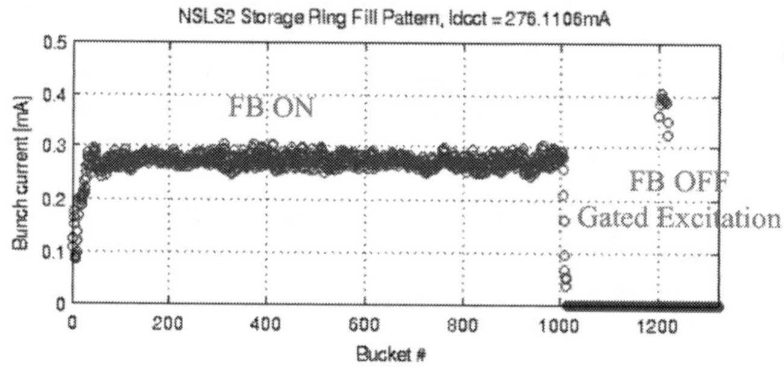

Figure 1. Fill pattern for transparent lattice characterization for 275mA user operation beam.

The 1000-bunch train has transverse bunch by bunch feedback ON to suppress instabilities while the test bunches had feedback OFF. The test bunches were verified to be stable even without feedback. Betatron tune can be measured using any of the test bunches by the integrated function of transverse feedback system, which sweeps across the betatron frequency to see the resonant peak. Once the betatron is known, the test bunches can be resonant excited at this frequency. The excitation is short and is triggered externally. The same trigger is synchronized with BPM gated TbT data which process the test bunches signals.

Digital BPM signal processing expands the capability to select signals from part of the bunches filled in the ring. A similar method has been used during commissioning stage to process the ADC raw data for first turn trajectory and suggested to improve the BPM resolution as discussed in [9]. FPGA firmware has been updated recently to have two different gates that can process the TbT/FA/SA data for different bunches filled in the ring. Timing alignments of the gates are critical to achieve proper processed data.

2. Gated BPM

The NSLS-II in-house developed BPMs incorporate the latest technology available. The same hardware design is used for the LINAC, booster, transfer lines and storage ring. BPM has met the specification and operates reliably for the past years.

The BPM electronics includes mainly two boards: analog front-end (AFE) and digital front-end (DFE). The AFE has band pass filter with center frequency of 500MHz and bandwidth +/-10MHz, this determines the impulse response of ~300ns (~30 ADC samples at 117MHz sampling rate). The DFE board processes the sampled raw ADC data to deliver TbT, 10 kHz and 10 Hz data; the data is then communicated to the control network. Further information of the NSLS-II BPM electronics developments can be found at [10-14].

Because of the +/-10MHz band pass filter and 117MHz sampling rate, NSLS-II BPM electronics are not able to resolve the bunch to bunch positions. However, with bunches separated by more than 300ns, the digitizer is capable to separate signals from those bunches. A typical fill during NSLS-II operation has ~1000 bunches train with bunch separate of 2ns, followed by 320 empty buckets to clear the ions. A signal bunch is filled in the gap for continuous tune measurement and possible single bunch users in the future. Bunch by bunch feedback system is configured to suppress any coupled bunch instabilities in the long bunch train, while keep the test single bunch without feedback.

We propose using the single bunch (or several bunches to improve the signal strength) for lattice measurement. If the BPM signal processing can select the signal only from the bunch(es) in the gap, i.e. to process ADC sampled data with programmable gate, we will be able to measure beam positions from single bunch (or several bunches nearby). The gate function was recently implemented inside the FPGA. Two separate signal processing channels with separate gates are available. Gate delay and gate width can be adjusted so that signal from different bunches can be selected and processed simultaneously. Fine timing alignment (8ns steps) ensures all the BPMs around the ring process the signal from same bunch(es).

Practically the gated bunch(es) shall be ~1% of the total charge so that clean signals can be detected and processed. With the fill pattern in Figure 1, BPM ADC raw data sampled at 117MHz (310 samples per turn) is shown in Figure 2 top image. BPM electronics RF attenuator is set to ~20dB for this current. Four button signals are overlapped as the cables are well matched. With gate, the signal processing will only include the ADC sampled data from bunches in the gap (bottom image).

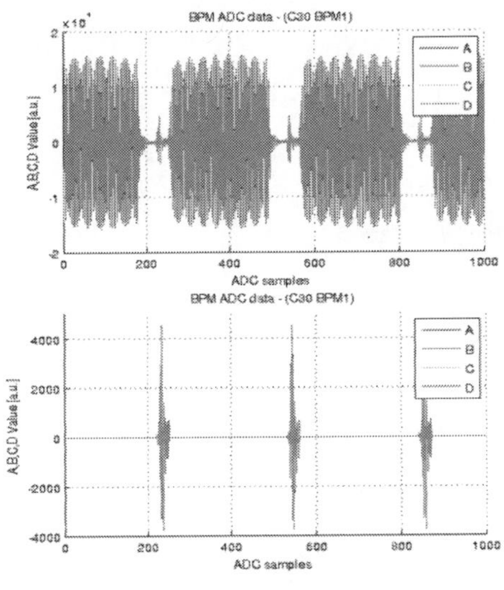

Figure 2. ADC sampled data without gate (top) and with gate (bottom).

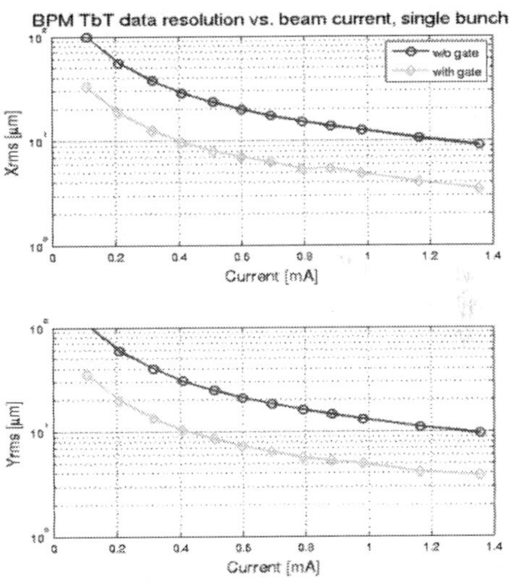

Figure 3. Measured BPM TbT data resolution, with and without gate.

As can be seen from Figure 3, gated BPM TbT data shows ~3 times improvement, compared to the same TbT data without gate. The gate width is set to 30 ADC samples, which improves the signal-noise ratio (SNR) by ~10, the 3 times resolution improvement is proportional to square root of SNR. From Figure 2, the gate width can be further decreased to have better resolution.

IOP Conf. Series: Journal of Physics: Conf. Series **874** (2017) 012082 doi:10.1088/1742-6596/874/1/012082

3. Bunch excitation

NSLS-II storage ring has transverse bunch-by-bunch (BxB) feedback system installed and commissioned in 2014 [15]. Besides its major function as a feedback controller, the digitizer [16] has extra integrated functions to be used as a powerful diagnostic tool. One such kind of function is the single bunch excitation, which is used at NSLS-II for continuous tune measurement from a test single bunch. Similarly to sweeping tune excitation, the digitizer can drive the beam a single frequency within some pre-defined durations. The transient resonant excitation can be synchronized with external trigger signal so that BPM TbT data will be updated on the same trigger. With proper timing alignment, the feedback digitizer can drive any bunches with frequency near the betatron tune. For bunches to be driven, BxB feedback is turned OFF while the rest of bunches have BxB feedback ON all the time. Bunch to bunch motions are checked with feedback digitizer, during the transient excitation, only test bunches are excited.

BxB feedback station is the only location when bunch to bunch positions can be measured. This is not sufficient to fully measure the machine optics. With bunch(es) filled in the gap with sufficient distance away from other bunches, the normal BPMs are able to separate the signals from the gap bunches, as shown in Figure 2.

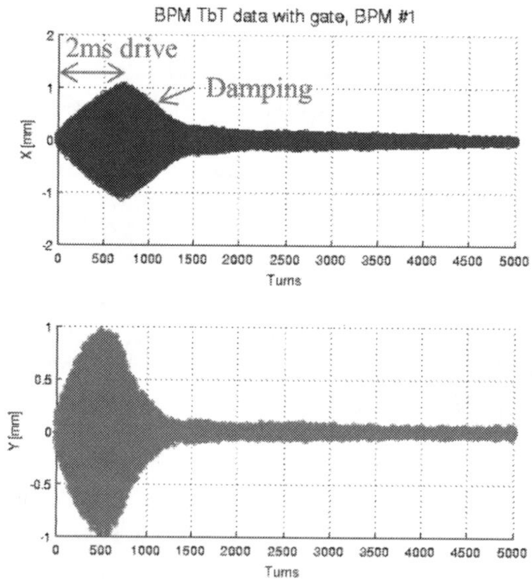

Figure 4. Gated BPM TbT data with bunches in the gap excited to ~1mm peak. The duration of excitation is 2ms.

As described earlier, with proper gates, NSLS-II BPMs around the ring are able to see the bunch motions in the gap. This allows us to measure TbT position of excited bunch(es). The gated TbT data during the free betatron damping can be used to characterize the machine lattice. Figure 4 is gated TbT data with test bunches driven to ~1mm on both planes. The driven signal is switched OFF after 2ms. The test bunches has ~1% of the total charge, it is confirmed from BPM TbT data without gate that there is ~10μm peak oscillation amplitude. The perturbation is much weaker than top-off injection disturbance (~1mm peak amplitude horizontally). As only a very small fraction of bunches oscillate for a short period of time (several ms), we consider this method is transparent to the operational beam.

4. Lattice measurement and correction

Storage ring lattice can be measured from the gated BPM TbT data, even with high current user beam. Figure 5 shows the beta-function measured along the ring. With a well-corrected lattice at low current, beta-beat was increased to 3.9% and 3.4% (X/Y) plane while filling the machine to a higher current and changing ID gaps. It was able to correct to the similar level of ~2% with the gated BPM and CW excitation techniques.

At low current and short bunch train fills, the measured lattice parameters have been compared with the pulse kick and resonant excitation with good agreement.

Phase advance, and linear coupling can also be measured with the same gated TbT data. We observed lattice variation at various beam currents and ID gaps. Applying this technique to study the lattice variation due to collective effects, even to localize impedances, is still under way.

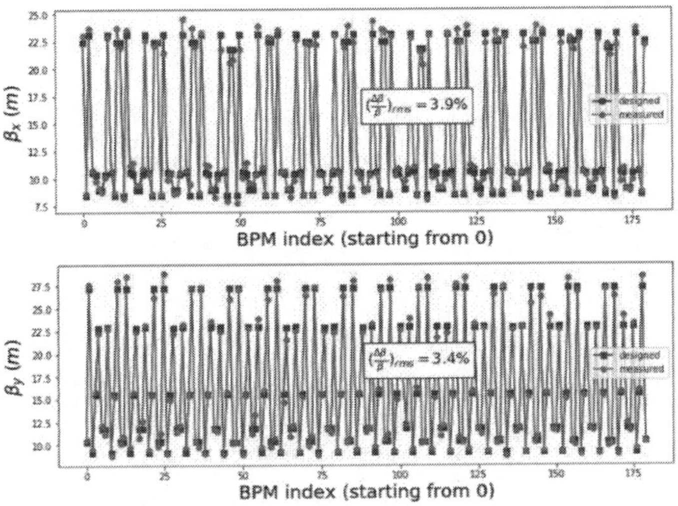

Figure 5. Beta-function measured from gated BPM TbT data. X/Y beta-beat was ~3.9%/3.4%.

5. Summary

New techniques have been proposed and tested to measure the lattice function using gated turn-by-turn data processed from several bunches containing only 1% of the total stored charge. Combine the gated BPM function recently developed at NSLS-II and single bunch excitation function integrated in the BxB feedback system, storage ring linear optics can be measured with little disturbance during high current user operation beam. The method has been successfully tested at NSLS-II storage ring to characterize the lattice and make corrections accordingly. The techniques can be used to study the collective effect and localize the impedances.

Authors thank R. Smith for proof-reading the manuscript.

References

[1] Safranek J 1997 *Nucl. Instrum. Methods Phys. Res.* A **338** 27
[2] Aiba M 2012 *Nucl. Instrum. Methods Phys. Res.* A **694** 133
[3] Dowd R 2011 *Phys. Rev. ST Accel. Beams* **14** 012804
[4] Castro P *et al.* 1993 Betatron function measurement at LEP using the BOM 1000 turns facility *Proc. of PAC'93* p 2103
[5] Aiba M *et al.* 2013 *Phys. Rev. ST Accel. Beams* **16** 012802
[6] Franchi A 2016 Error Analysis of linear optics measurements via turn-by-turn beam position data in circular accelerators https://arxiv.org/abs/1603.00281
[7] Alexahin Y, Gianfelice-Wendt E 2011 *Journal of Instrumentation* Vol **6**
[8] Li Y *et al.* 2015 Using big data in NSLS-II storage ring commissioning *Proc. of ICAP2015 (Shanghai China)* p 6
[9] Podobedov B 2016 Single micron single-bunch turn-by-turn BPM resolution achieved at NSLS-

II, *Proc.of IPAC'16 (Busan, Korea)* p 2095

[10] Vetter K 2011 NSLS-II RF beam position monitor *Proc. of PAC'11 (New York, USA)* p 495

[11] Vetter K 2012 NSLS-II RF beam position monitor update *Proc. of BIW'12 (Newport News, VA, USA)* p 238

[12] Singh O 2013 NSLS-II BPM and fast orbit feedback system *Proc. of IBIC'13 (Oxford, UK)* p 316

[13] Cheng W 2014 NSLS2 diagnostic systems commissioning and measurements *Proc. of IBIC'14 (Monterey, CA, USA)* p 16

[14] Mead J 2014 NSLS-II RF beam position monitor commissioning update *Proc. of IBIC'14 (Monterey, CA, USA)* p 500

[15] Cheng W 2014 Commissioning of bunch-by-bunch feedback system for NSLS2 storage ring *Proc. of IBIC'14 (Monterey, CA, USA)* p 707

[16] http://www.dimtel.com/

Electrostatic pickup in the CNAO Injection Line

A Parravicini, G M A Calvi, E Rojatti and C Viviani

Fondazione CNAO, Pavia, Italy

E-mail: anna.parravicini@cnao.it

Abstract. The paper is about the electrostatic pickup installed in the Medium Energy Beam Transfer (MEBT) line of the CNAO (Centro Nazionale di Adroterapia Oncologica), the Italian facility for oncological hadrontherapy. The MEBT Pickup (PUB) has been designed with the purpose of having a continuous and non-interceptive measurement of the horizontal and vertical beam position, close upstream the injection point in the synchrotron. Detector commissioning, data-analysis algorithm and first year measurements are discussed in the paper.

1. Detector Overview

Figure 1 shows the PUB assembly: the external vacuum pipe (AISI 316L), the pickup body (AISI 316L) and the electrodes (OFHC Copper) inside. The PUB is composed of two consecutive pickups: first the beam passes through the horizontal electrodes and then through the vertical ones. Electrodes are cylindrically-shaped, conveniently cut to pick left and right, or top and bottom signals. The PUB is 700-millimeters flange-to-flange long, with an inner diameter of 72 mm.

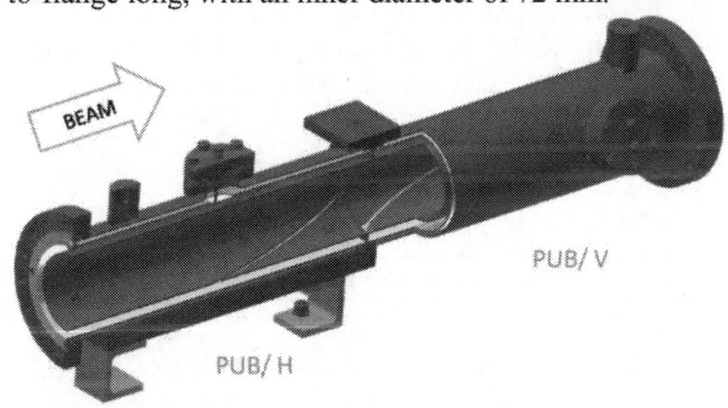

Figure 1. PUB 3D model with side section view. The beam, coming from left, enters the horizontal (H) electrodes first and the vertical (V) ones later.

The electronics has been designed with great attention to match the RC component of the 4 electrodes, in order to equalize their response to the beam passage. Each electrode signal is input in a commercial amplifier (HVA10M-60-F by FEMTO) with 1 Hz-10 MHz bandwidth and 40 dB gain; the four resulting signals are digitized by a NI6132 DAQ board (maximum sample rate 2.5 MHz; typical working rate 1 MHz) [1]. Before installation, the PUB overall bandwidth, from mechanic to amplifier output, has been measured being more than 2 MHz, using a dedicated test bench.

Figure 2 shows the layout of the beam line around the PUB. Following the beam way, it is composed of: one corrector (CORR), one profile grid (G1), two quadrupole magnets, one diagnostic tank with degrader filters and emittance plates, one corrector, the AC-current transformer, the PUB passing through an additional corrector and, finally, a second profile grid (PG2).

The beam crossing the PUB is made of 7 MeV/u protons or, alternatively, C^{6+} ions with nominal intensities of 1.4 mA and 0.2 mA, respectively. The upstream LINAC shapes the beam to form about 50-microsecond long batches, with about 2 us rise and 2 us fall time, at 0.5 Hz frequency. Table 1 reports typical beam position (X0, Y0) and dimension (σX, σY) measured at PG1 and PG2, for horizontal (X) and vertical (Y) plane, for protons (P) and C^{6+} ions (C).

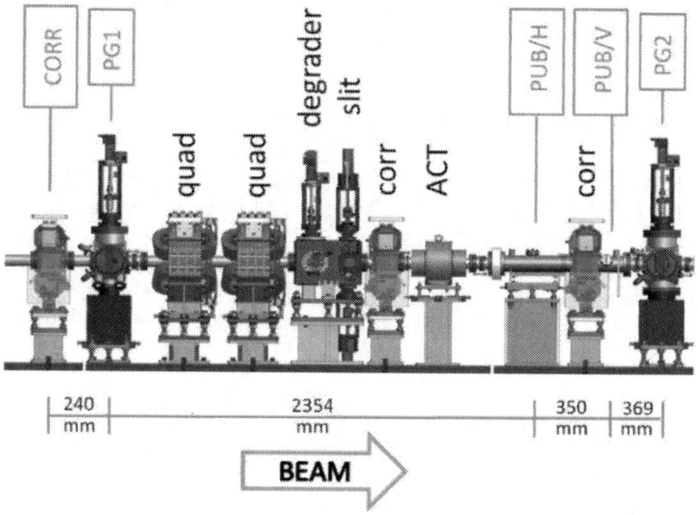

Figure 2. Line layout around the PUB, with correctors (CORR), profile grids (PG), quadrupoles (quad), the AC-current transformer (ACT), the emittance plates (slit) and the degrader filters (degrader). The outlined objects are used for the PUB calibration.

Table 1. Beam parameters measured at PG1 and PG2.

Particle	X0(mm)	σX(mm)	Y0(mm)	σY(mm)
P @PG1	2.5	10.4	0.9	4.9
P @PG2	1.4	7.5	10.2	4.6
C @PG1	4.4	6.8	0.8	3.6
C @PG2	3.9	5.4	9.6	2.7

2. Beam Commissioning

2.1. First Measurements

The first measurements pointed out that the shape of the electric signals picked up by the four (right, left, top, bottom) electrodes was often different from the expectations, indicating alternatively positive (primary ions) or negative (secondary emitted electrons) particles impinging on the same electrodes (figure 3).

This effect becomes dramatic if one element (e.g., profile grids, degrader filters or emittance plates) is moved into the beam path in the PUB proximities. If this is the case, beam trajectory is deviated, its dimension is enlarged and secondary electrons are produced by the primary ions hitting the obstacle. Consequently, the probability that positive or negative particles strike the electrodes becomes very high. If no objects are on the beam path, primary and secondary interactions are drastically reduced and electrode signals are much closer to the expectations, although some distortion effects remain evident.

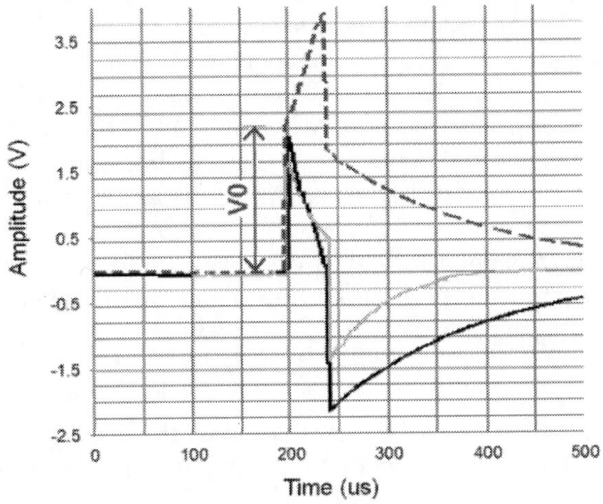

Figure 3. Expected electrode signal (light gray); signal if electrons hit the electrode (dot gray); signal if positive ions hit the electrode (black).

This observation led to restrict the use of the PUB to the situations where none of the immediately close objects is on the beam path. This is a low-impact restriction for the PUB operation, being the most common setup.

It remains the need to fix how to evaluate signal amplitude in the allowed situations to get reliable measurements of the beam transverse position.

A common solution to avoid particles on the electrodes consists in mounting a shielding / repeller ring at the pickup entrance. This solution is not easily practicable in this case, because its installation would require to dismantle a big sector of the MEBT line and consequently stop the treatments for an unacceptable long period. Waiting for a planned long shutdown period when it could be installed, it has been tuned a data analysis and post processing strategy strong against signal distortion.

2.2. The Algorithm

As usual for pickup detectors, the beam transverse position is calculated from the Delta over Sigma ratio, Delta being the difference and Sigma the sum of the signal induced on the two opposite electrodes (e.g., right and left electrodes for the horizontal beam position) by the passing through beam. The crux consists in how to define signal amplitude, the signal being far from a stepwise function.

In the ideal case, the signal, after the rise induced by the beam (e.g.: at t~200 us in figure 3), is expected to follow an exponential-decay with time constant (of about 150 us in the PUB case) defined by PUB resistance and capacitance versus ground. In case positive particles hit the electrode during the bunch transit time, the negative current, induced on the electrodes by the positive beam current, falls down faster; vice versa, if negative particles hit the electrodes, their charge sums up to the beam induced current and the electrode signal decreases less steeply or, even, increases after the beam induced rise.

Different algorithms have been evaluated in order to find a robust solution to evaluate the signal amplitude, despite these effects. The most immediate method, consisting in determining a "good" region of the signal after the beam induced rise and calculating signal amplitude as average on this samples subset, cannot work in this scenario, being significantly influenced by the signal shape after the rise.

A different approach has been preferred, based on the fact that particles lost on the electrodes change the signal shape, but don't contribute significantly to the rise amplitude, essentially determined by the current variation (from 0 to beam current, and vice versa) due to the bunch transit. Thus, signal amplitude is evaluated as the signal rise (V0 for dot-line signal in figure 3), induced by the bunch arrival. The PUB acquisition is performed at 1 MHz rate and is hardware triggered by the LINAC system: the beam bunch occurs at fixed delay from the acquisition start, consequently. Time has been spent tuning the script in order to determine the rise amplitude accurately, independently on the signal after-rise shape.

Although this solution has the important fragility of using single points rather than averages, many months of measurements have proved its success in attributing an efficient value to each electrode and a dependable beam position, consequently.

2.3. The Calibration

Once a reliable way to calculate Delta and Sigma was found, the detector has been calibrated in order to estimate A and B parameters, needed to obtain the beam position in millimetres:

$$x(mm) = \frac{1}{A}\left(\frac{Delta}{Sigma} + B\right) \tag{1}$$

Some measurements were performed with this purpose before installation, but an experimental calibration has been preferred, using PG1 and PG2. A PG is made up of two 64 wire harps, one horizontal and one vertical, with an effective resolution of 0.5 mm.

The PUB calibration is made independently for protons and carbon ions, horizontal and vertical plane. For each configuration, the procedure is the same: CORR power current is changed from a low to a high value (about ±20 A for carbon and ±10 A for proton), in 5 steps. For each step, the beam position is measured at PG1 and at PG2 and about 50 measurements are performed by the PUB, in such a way to calculate a reliable Delta/Sigma value for each corrector setting.

The A and B parameters are deduced by regression line of Delta/Sigma versus the expected position at PUB (figure 4), the last being calculated assuming a linear beam trajectory from PG1 to PG2. This assumption is rigorous only if all the magnets from PG1 to PG2 are off; in fact, all the correctors - except CORR for the scanning plane- are switched off, but the two quadrupoles in between are kept on, to avoid beam blowing up and the consequent losses. Beam optics simulations performed with MadX show that the deviation from the linear path due to quadrupoles on can be neglected except for the most peripheral positions. The most external scan positions suffer of other troubles too: when the beam is very deviated by CORR, it goes out of the PG2 sensitive region and thus beam position at PG2 cannot be calculated accurately; in addition, if the beam is very kicked, the risk of hitting PUB electrodes becomes high and the Delta/Sigma values are more likely fake. On the other hand, the beam most peripheral positions are not reached in the usual operations, thus the linear beam path from PG1 to PG2 can be retained acceptable in the interesting region of work. The R-squared regression line coefficient is very close to 1 for all the particle / plane configurations.

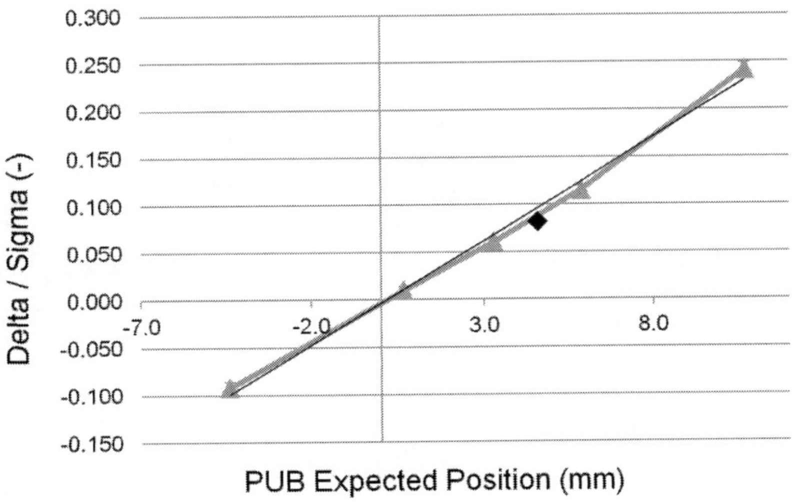

Figure 4. Delta/Sigma versus PUB expected position. Triangular markers refer to the calibration scan (carbon beam, horizontal plane); the square marker refers to the nominal magnets configuration; the black line is the regression line (y=0.0218x-0.0034 with R^2=0.9949)

3. In Operation

After a long commissioning period, the PUB started regular operation in January 2016. It is presently working in watch-dog mode: the beam horizontal and vertical barycenters, measured by the PUB, are displayed in the main control room and automatically saved in a database every machine cycle. In case they result out of a preset range, an indicator warns the operator about the probable need of reviewing the line or LINAC settings.

Despite the successful strategy to evaluate the electrode signal amplitude, the PUB provides a fake measurement if the upstream machine (sources, LINAC, magnets) produces a slightly wider or offset beam, causing swarms of ions or electrons to hit the electrodes. It regards about the 3% of the acquisitions during the regular machine activity.

So far, no time has been invested in implementing an automatic data rejection for various reasons: fake data are rare and, as such, they do not affect significantly the hour-averaged values used for long-term monitoring, that is the main use of the detector. On the other hand, the same fake data rate can be used as indicator of the machine status and its eventual increase acts as a warning.

In the rare case the PUB is used for short-term (i.e., a few minutes) analysis, the saved data are processed offline calculating distribution center and standard deviation. It has been proved that the barycenter mean value, after the rejection of eventual 2 sigma-out data, is 0.2 mm within the expected value, for a distribution of more than 20 samples and magnets at the nominal configuration. Barycenter distribution standard deviation results better than 0.6 mm for all the particle / plane configurations.

Studies are in progress about correlation of PUB with LINAC parameters and with the injected and accelerated synchrotron beam currents. The correlation results evident in particular with the IH high voltage signal, correlation coefficient absolute value being greater than 0.5, in the most of the analyzed periods. Figure 5 shows the overlapping of proton vertical position trend over the IH high voltage trend, after proper rescaling, for the month of February 2016, where correlation factor is 0.65.

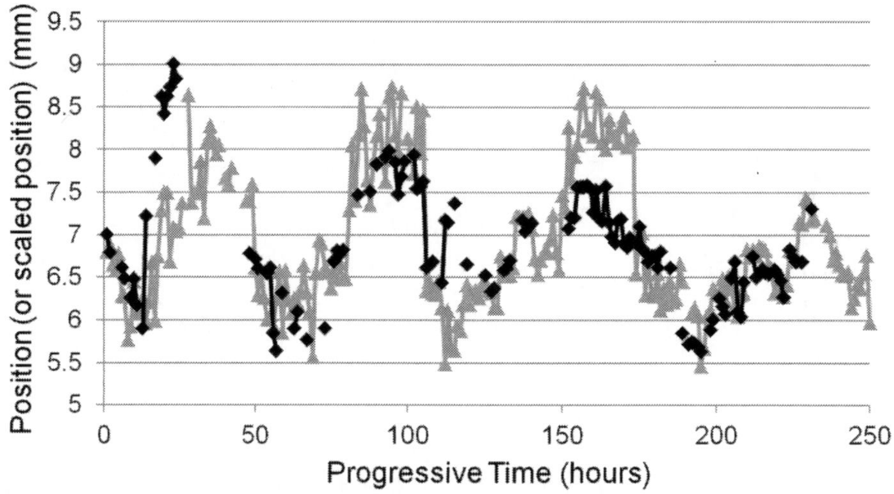

Figure 5. Vertical beam position measured by PUB (gray) with proton beam from February 2nd to 29th, 2016, compared with the IH high voltage (black) measured at the same time and properly scaled. One point is one hour averaged signal; data are not available for all the hours.

4. Conclusions

The PUB is installed at the end of the MEBT line, close to the synchrotron injection point. After a complex commissioning, it started to work as watch-dog monitor in January 2016. It permits a continuous measurement of the beam barycenter position, in the horizontal and vertical plane, just upstream the injection point, providing a global view over unavoidable instabilities of complex systems such as sources and LINAC. This information is immediate, xPUB and yPUB being available in the main control room every cycle, and with no side effects, the PUB being a non-interceptive detector.

Despite a signal sometime perturbed by ions and electrons striking the electrodes, an algorithm able to provide reliable measurements was tuned. Detector repeatability and accuracy results adequate for the purpose. Correlation between PUB measurements and LINAC parameters is evident.

Acknowledgments

Many people worked on the PUB design and commissioning in the last 10 years contributing to its present work. In particular, it's worth to mention the contribution of: G.Balbinot, J.Bosser, M.Caldara and M.A.Garella.

References

[1] Balbinot G and Caldara M, MEBT PU Measurement System, *CNAO Internal Note*

IOP Conf. Series: Journal of Physics: Conf. Series **874** (2017) 012084 doi:10.1088/1742-6596/874/1/012084

Orbit and dispersion tool at European XFEL injector

N G Ghazaryan[1], E Castro[2] and W Decking[2]

[1] CANDLE Synchrotron Research Institute, Yerevan, Armenia
[2] Deutsches Elektronen-Synchrotron (DESY), Hamburg, Germany

E-mail: ghazaryan@asls.candle.am

Abstract. Trajectory and electron beam size play an essential role in Free Electron Laser (FEL) obtainment. Since transverse dispersion changes off-energy particle trajectories and increases the effective beam size, dispersion and orbit must constantly be controlled and corrected along the whole lattice. In this paper the principles underlying the orbit and dispersion correction tool, developed at DESY, are described. The results of its testing on European XFEL injector are presented.

1. Introduction

In a magnetic lattice the dispersion is generated by any type of dipole field, caused for instance by quadrupole misalignments, dipole error fields, field errors of quadrupole magnets, etc. Inside a bunch of electrons, particles with different energies have different trajectories due to dispersion. Consequently, the transverse space, occupied by the beam, increases. For optimal FEL performance it is mandatory to ensure a good beam quality in undulators [1], so the dispersion and orbit must frequently be measured and corrected along the whole lattice.

The schematic layout of the 3-kmetre long European XFEL (X-ray Free- Electron Laser) is shown in figure. 1. The elements include 1.3 GHz RF sections (yellow), 3.9 GHz acceleration structure (red), undulators (green/red), and main dipole magnets (blue).

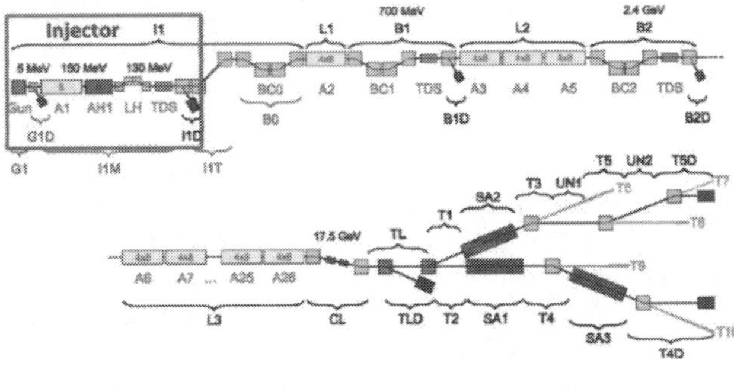

Figure 1. The schematic layout of XFEL

Currently, the XFEL is under commissioning. For a simultaneous correction of the dispersion and beam orbit a software tool has been developed. It uses a set of corrector magnets; including quadrupoles (adjusted by movers) and the laser heater chicane dipole magnets to minimize the excursions with respect to ideal orbit/dispersion (Dispersion Free Steering), and a set of beam position monitors to measure the orbit changes. It operates based on the Singular Value Decomposition algorithm to calculate corrector angles (strengths).

This paper presents the performance details of the software tool written in MATLAB code, as well as, the results of its testing.

2. Dispersion free steering principle and formalism

The dispersion free steering correction algorithm [2] guarantees that the beam orbit stays within acceptable values, minimizing the dispersion at the same time. The resulting orbit and dispersion change is achieved through the variation of the corrector's strength, which is given by:

$$\begin{pmatrix}(1-\alpha)\cdot\Delta\vec{u}\\ \alpha\cdot\Delta\vec{d}\end{pmatrix}=\begin{pmatrix}(1-\alpha)\cdot OR\\ \alpha\cdot DR\end{pmatrix}\cdot\Delta\vec{\theta} \tag{1}$$

where $\Delta\vec{u}$ and $\Delta\vec{d}$ are the vectors formed by the orbit and dispersion changes at all N monitors, respectively, $\Delta\vec{\theta}$ is the vector formed by all the M correctors' strengths, OR is the N×M orbit response matrix, DR is the N×M dispersion response matrix and α is a weight factor. In general, the optimal closed orbit and dispersion are not of the same magnitude and the weight factor α must correspondingly be adjusted. Depending on its value we can get either pure orbit (α = 0) or pure dispersion (α = 1) corrections. Appling the least square algorithm to equation (1), we minimize the following expression:

$$(1-\alpha)^2\left\|\vec{u}_{measured}-OR\cdot\Delta\vec{\theta}\right\|^2+\alpha^2\left\|\vec{d}_{measured}-DR\cdot\Delta\vec{\theta}\right\|^2 \tag{2}$$

However, the solutions given by this approach might exceed the limits of acceptable corrector strengths. Since these settings are not feasible, the standard solution would be to disable a subset of correctors and remove the corresponding lines from the linear system of equation (1). Regularization can also be obtained by extending equation (1) to constrain sizes of corrector kicks. Moreover, when more correctors are available than BPMs, null vectors can be used to reduce corrector strengths. This leads to the following system:

$$\begin{pmatrix}(1-\alpha)\cdot\Delta\vec{u}\\ \alpha\cdot\Delta\vec{d}\\ \vec{0}\end{pmatrix}=\begin{pmatrix}(1-\alpha)\cdot OR\\ \alpha\cdot DR\\ \beta I\end{pmatrix}\cdot\Delta\vec{\theta} \tag{3}$$

where $\vec{0}$ is a null vector of dimension M, I is a unit matrix of dimension M×M, and β is a kick weight.

The goal of the orbit and dispersion correction tool is to solve the system in equation (3) and get the $\Delta\vec{\theta}$ that minimizes the excursions. To solve the system, the Singular Value Decomposition method (SVD) is used [3].

3. Response matrices and dispersion measurement

In order to be able to implement the correction algorithms, two matrices are needed: the orbit (OR) and dispersion (DR) response matrices. They describe how the changes in the strength of the correctors affect the change of orbit and dispersion. These matrices can be measured directly on the accelerator, by tracking the orbit and dispersion of a particle after acting on a corrector, but this is time consuming. Instead, they can be obtained from the transfer matrices between the correctors and the BPMs. From linear transport theory for a transmission line, the beam motion evolves as:

$$\begin{aligned}x_{final}&=R_{11}\cdot x_0+R_{12}x'_0+R_{13}\cdot y_0+R_{14}\cdot y'_0+R_{16}\cdot\left(\Delta p/p_0\right),\\ y_{final}&=R_{31}\cdot x_0+R_{32}x'_0+R_{33}\cdot y_0+R_{34}\cdot y'_0+R_{36}\cdot\left(\Delta p/p_0\right)\end{aligned} \tag{4}$$

A general matrix notation including first and second-order terms is:

$$u_{i\,final} = \sum_j R_{ij} \cdot u_{0_j} + \sum_{j,k} T_{ijk} \cdot u_{0_j} \cdot u_{0_k} \tag{5}$$

The elements of the OR matrices are then:

$$OR_{ij} = \frac{\Delta \vec{u}_i}{\Delta \vec{\theta}_j} = R_{12_{j\to i}}, R_{34_{j\to i}} \tag{6}$$

And the DR matrix can be obtained as:

$$DR_{ij} = \frac{\Delta \vec{d}_i}{\Delta \vec{\theta}_j} = T_{126_{j\to i}}, T_{346_{j\to i}} \tag{7}$$

where Δu_i and Δd_i are the orbit and dispersion changes in i-th BPM, respectively, caused by a $\Delta\theta_j$ change of j-th corrector strength, R_{12} / R_{34} and T_{126} / T_{346} are first and second order terms for horizontal/vertical plane, correspondingly. In our case, the OR matrix was built using the R_{12} and R_{34} terms provided by the Optics Server [4], but the DR matrix was measured empirically in the machine.

3.1. Dispersion measurement

If the energy of the beam differs from the design energy, its trajectory may deviate from the one of a particle at the right energy. In first order, this deviation is linear in the momentum deviation $\delta = (\Delta p / p_0)$. For a transport line:

$$\Delta x_{final} = R_{16} \cdot \delta \;, \; \Delta y_{final} = R_{36} \cdot \delta \tag{8}$$

R_{16} and R_{36} are often called first-order transverse dispersion.

The dispersion measurement in the tool is based on reading out the orbit for different beam energies [5] that can be changed in modules A1 or AH1 (see figure. 1). The process is as follows:

1. Change the energy in A1.
2. Measure the orbit in the BPMs.
3. Calculate the dispersion as the slope of the second order polynomial fit of the beam position vs energy.

Figure 2 shows the screenshot of the measurement of the dispersion. The top two plots represent the beam orbit for different energies and the lower two plots show the measured dispersion (horizontal and vertical).

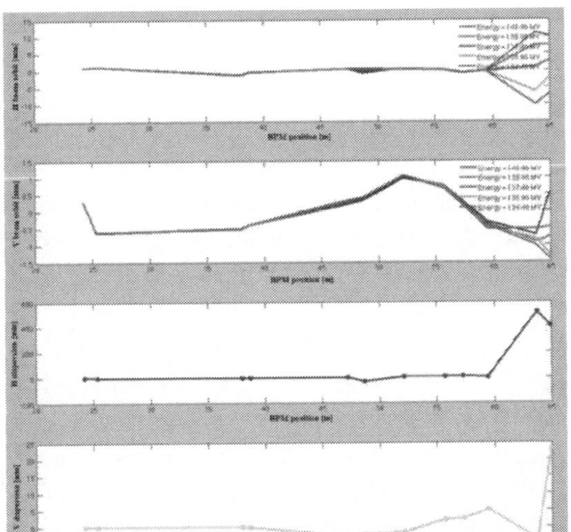

Figure 2. Example of vertical and horizontal dispersion measurements

IOP Conf. Series: Journal of Physics: Conf. Series **874** (2017) 012084 doi:10.1088/1742-6596/874/1/012084

3.2. Dispersion response matrix measurement

As it was said before, the dispersion response matrix was obtained empirically for the XFEL injector with an extra MATLAB tool developed for this purpose. Once the matrix was calculated, it was loaded into the orbit and dispersion correction tool. The process for the measurement is as follows:

1. Measure initial dispersion as described previously.
2. Apply a kick in one corrector upstream A1.
3. Measure dispersion again.
4. Set back the kick to initial value.
5. A new column in the response matrix is obtained by subtracting the measured dispersion minus the initial dispersion.
6. Repeat from point 2.

Figure 3 shows a screenshot of the calculation of the dispersion response matrix. The tool displays the following: The left plots are for the horizontal plane and the right plots for the vertical plane. The lower plots display the final matrices. The upper plots display, for each corrector, the calculated dispersion. The middle plots display how the dispersion in the other plane is affected (crosstalk).

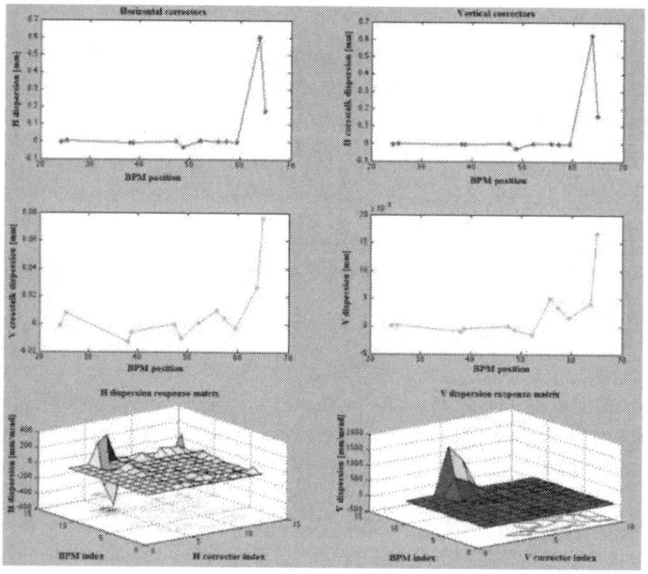

Figure 3. Example of the dispersion response matrix measurements

4. Brief overview of the tool

The orbit and dispersion correction tool was implemented as a GUI using a MATLAB library developed at DESY [6, 7] and interfaces several servers to get the needed information and to communicate with the elements needed for the corrections (Optics Server, Magnets and BPMs servers, etc.). The operator can select the following:

- The elements to include in the correction algorithm.
- The correction of orbit only or dispersion and orbit at the same time.
- The plane to be corrected (horizontal or vertical).
- In case the proposed solution $\Delta\vec{\theta}$ exceeds the limits, the strengths can be reduced by limiting the number of singular values in the SVD solution, by increasing the value of β, or by reducing the size of the calculated angles.
- Dispersion can be measured in the same GUI.

5. First tests and outlook
During the operation of XFEL injector the tool was successfully tested several times. We tried to correct the dispersion and orbit using correctors, including the two last bending magnets in the LH chicane and we succeeded. On the plots the blue lines are the initial, the red ones are the predicted and the green lines are the corrected orbits and dispersions, respectively. In figure 4 the dispersion after and before correction is shown.

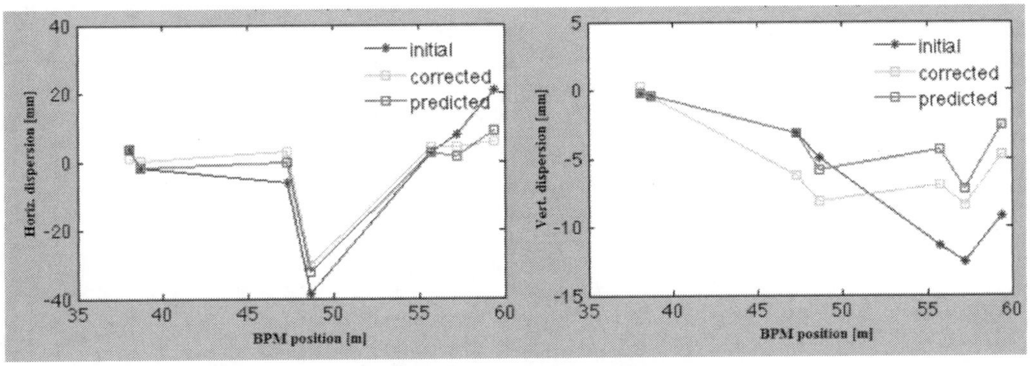

Figure 4. The dispersion screenshot of the correction tool

We managed to improve the dispersion from almost -38.8 mm to around -30 mm at laser heater chicane (the golden dispersion being-30.1mm) and for the rest part we got an improvement of a factor 10.

We managed to improve the orbit from almost -1.5 mm to around -0.5 mm for horizontal and vertical planes. The result for the orbit is shown in figure 5.

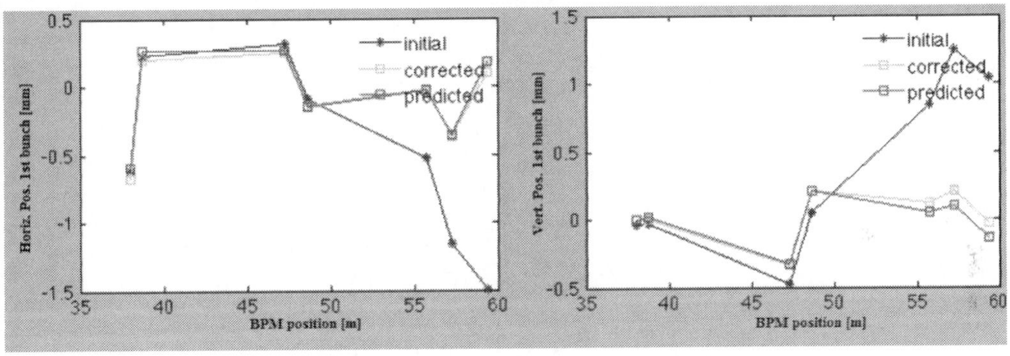

Figure 5. The orbit screenshot of the correction tool

6. Conclusions
The method and the tool for correcting the dispersion and orbit at European XFEL injector, including the dispersion response matrix measurement tools are presented. The tools have been developed and successfully tested by means of simulations and measurements at XFEL injector at DESY. The future all these two tools can be used for the whole beamline as well.

7. References
[1] Rossbach J 1996 A VUV free electron laser at the TESLA test facility at DESY *Nuclear instruments and methods in physics research* section A **vol. 375** pp.269-273
[2] Raubenheimer T *et al.* 1995 Quadrupole alignment and trajectory correction for future linear colliders *4 th International Workshop for Accelerator Alignment* (Tsukuba, Japan)
[3] Gregorici G 2001 *The singular value decomposition and the pseudoinverse* Rep. (Ireland: University College Cork)

IOP Conf. Series: Journal of Physics: Conf. Series **874** (2017) 012084 doi:10.1088/1742-6596/874/1/012084

[4] Prat E *et al.* 2006 Measurement and correction of dispersion in the VUV-FEL *Proc. 10 th European Particle Accelerator Conference* (Edinburgh, Scotland) pp.1951-3
[5] Optics Server, https://xfel-wiki.desy.de/Optics_server
[6] Castro E and Ghazaryan N G 2016 *Orbit and dispersion correction XFEL* Rep. DESY-XFEL (DESY, Hamburg, Germany)
[7] Data GUI library, https://xfel-wiki.desy.de/Data_ GUI_library

Fs level laser-to-RF synchronization at REGAE

M Titberidze[1], M Felber[1], T Lamb[1], R Loch[2], C Sydlo[1] and H Schlarb[1]

[1] Deutsches Elektronen-Synchrotron, Notkestrasse 85, 22607 Hamburg, Germany
[2] Max Planck Institute for the Structure and Dynamics of Matter, Luruper Chaussee 149, 22761 Hamburg, Germany

E-mail: mikheil.titberidze@desy.de

Abstract. The Relativistic Electron Gun for Atomic Exploration (REGAE) is a unique linear accelerator capable of producing ultrashort (~ 10 fs) electron bunches for studying fast processes in matter by means of ultrafast electron diffraction (UED) experiments. Additionally, REGAE is suitable for upcoming external injection experiments for laser wakefield acceleration (LWFA). In order to carry out both mentioned experiments, it is crucial to achieve fs level stability in terms of laser-to-RF synchronization. In this paper we present an advanced laser-to-RF synchronization scheme based on integrated Mach-Zehnder Modulator (MZM). The setup demonstrated the Titanium Sapphire photoinjector laser synchronization with 11 fs (rms) precision in the bandwidth up to 100 kHz. Long term timing drift measurements showed unprecedented peak-to-peak stability of 31 fs (7 fs rms) over 43 hours of measurement time. In addition, AM-PM coefficient of the MZM based laser-to-RF synchronization setup has been evaluated and showed a factor of 10 improved performance compared to conventional direct conversion based laser synchronization setup.

1. Introduction

At REGAE a commercial Titanium Sapphire (Ti:Sa) laser system from Coherent Inc. is used as a photocathode laser to generate electron bunches by means of photoelectric effect. The generated electron bunches are accelerated up to 5 MeV in a S-band ($f = 2.9979$ GHz) 1.5 cell RF gun and further bunched by the 4 cell RF buncher using a technique called ballistic bunching [1, 2, 3, 4] before reaching the UED target chamber. At the target chamber a part of the laser beam together with the electron bunches are used to conduct UED experiments on a femtosecond level in a pump-probe fashion [5]. Therefore, in order to achieve femtosecond level temporal resolution between the laser and electron beams, precise laser-to-RF synchronization is required.

For the past decade, various precise laser-to-RF synchronization methods have been carried out with femtosecond level residual timing instabilities (jitter and drift)[6, 7, 8, 9, 10, 11, 12]. The most recent publication [13] showed very impressive performance of Ti:Sa laser oscillator synchronization to a 2.856 GHz RF source with the residual timing jitter of 3.9 fs (12.2 fs) in the bandwidth up to 100 kHz (1 MHz) and with long term timing drift of 12.5 fs (rms) (> 150 fs peak-to-peak) over 24 hours. The synchronization method in [13] is based on a fibre-loop optical-microwave phase detector (FLOM-PD) [7].

In this paper we present the experimental evaluation of the limitations for the direct conversion based synchronization setup by means of amplitude-to-phase conversion (AM-PM) effects. Experimental results of an alternative laser-to-RF synchronization method based on

IOP Conf. Series: Journal of Physics: Conf. Series **874** (2017) 012085 doi:10.1088/1742-6596/874/1/012085

800 nm integrated MZM [14, 15] is discussed and corresponding out-of-loop residual timing jitter and drift characterization are manifested. In addition, the measurement results of the AM-PM effects for MZM based laser-to-RF synchronization setup are presented in comparison to direct conversion based laser-to-RF synchronization setup.

2. Direct conversion based laser-to-RF synchronization

One of the most common techniques to synchronize the mode-locked lasers to RF reference signal is using a fast photodetector. The idea of employing a photodetection for synchronization purposes is to convert pulsed optical signals to electrical signals. Electrical signals generated by the photodetector are composed of high spectral purity harmonics of the laser repetition rate. The cutoff frequency of the frequency comb is given by the bandwidth (BW) of the photodetector. Typical fast photodetectors are fiber coupled PIN photodiodes (e.g. ET-4000F). Usually, one of the harmonics of the frequency comb can be filtered out using an RF band-pass filter and a low noise RF amplifier in combination with an RF phase detector (e.g. double balanced mixer) can be employed for laser-to-RF synchronization [11, 16]. However, there are several limitations related to the photodetection technique. The two main problems are, AM-PM effects [17, 18], where the optical power fluctuation is converted to phase fluctuation of each frequency component of the generated frequency comb. Second, the power level in the generated frequency comb lines is rather low, leading to the limited signal-to-noise ratio (SNR). Mainly these two effects limit the laser-to-RF synchronization performance.

2.1. Measurement Results of AM-PM Coefficients

Already in our previous work related to REGAE facility [15] we have shown that the AM-PM effects in the direct conversion based laser-to-RF synchronization setup induces significant timing disturbances. Indirect measurement of AM-PM coefficients in [15] revealed that it varies between 160 fs/% to 300 fs/% of input optical power change[1]. Dedicated measurements have been carried out to deduce the AM-PM coefficients of the direct conversion based laser-to-RF synchronization setup. The measurement setup is shown in Figure 1. The optical pulses from Ti:Sa laser oscillator (Micra) with a repetition rate of $f_{\mathrm{rep}} = 83.275$ MHz (multiple of 2.9979 GHz RF reference) are split into two branches. 10 % of its initial power is dedicated for synchronization purposes while the remaining 90 % is used for laser amplifier seeding. The 10 % branch of the laser pulses is further split into two parts using the optical band-pass filter (F1) to operate both in-loop and out-of-loop phase detectors respectively. Fiber coupled collimators are used to couple optical beam from free space to single-mode fibers for both in-loop and out-of-loop phase detectors (yellow, blue lines). A 12 GHz bandwidth fast photodiode (EOT4000F) is used as an in-loop photodetector together with the 3 GHz RF band-pass filter and low noise RF amplifier. The amplified $\approx$3 GHz laser signal is provided to the readout electronics and the phase error information between the RF reference signal from the RF master oscillator (MO) and laser signal is fed to the digital controller (see [11]).

Once the phase locked loop (PLL) was established using the direct conversion based laser-to-RF synchronization setup, the artificial amplitude modulation was introduced to the in-loop photodiode. The amplitude modulation was performed by rotating the free space optical neutral density filter (ND) in front of the in-loop fiber collimator (see Figure 1). Simultaneously, the out-of-loop MZM based laser-to-RF phase detector was used to monitor the phase/timing changes corresponding to the artificial amplitude modulation due to the AM-PM effects. This procedure has been carried out for discrete optical power levels from 1 mW to 8 mW in steps of 0.5 mW and repeated for three different applied bias voltages (3 V, 5 V, 10 V). The measurement results

[1] AM-PM coefficient is defined as $\alpha_{\mathrm{am-pm}} = \frac{\Delta\varphi}{\Delta P/P}$ in units of femtosecond per percent of input optical power change.

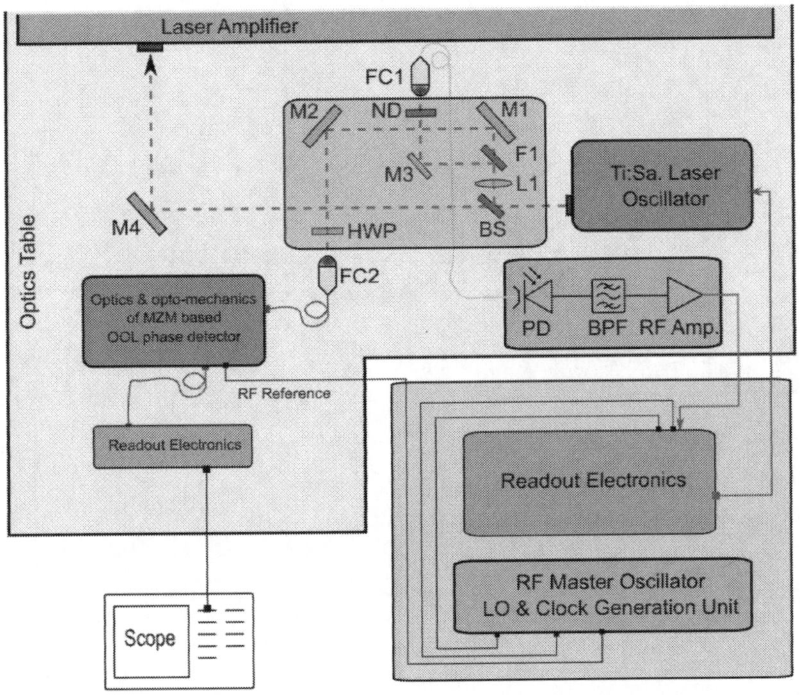

Figure 1. Block diagram of the AM-PM coefficient measurement setup.

are summarized in Figure 2. Figure 2 shows the three curves of the AM-PM coefficients as a function of average optical power for different photodiode bias voltages. At low optical powers the AM-PM coefficient starts out at approximately 300 fs/% of input optical power change and is independent from the photodiode bias voltage. As optical power increases this coefficient starts to change sign and oscillate between ±300 fs/% of input optical power change. At certain optical power levels when the photodiode undergoes the deep over-saturation regime, the AM-PM coefficient becomes very large (see Figure 2, $P > 7\,\text{mW}$, bias voltage 3 V). For the laser synchronization purposes, the most interesting regions in Figure 2 are the zero crossings of the $\alpha_{\text{am-pm}}$, because ideally, $P(\alpha_{\text{am-pm}} = 0)$ is a desired operating point for regular REGAE operation. However, the optical power of the Micra Ti:Sa laser oscillator at REGAE was changing 2 % - 3 % moving the AM-PM coefficient to undesired operating point.

3. MZM based Laser-to-RF synchronization

The main idea of using a MZM based laser-to-RF synchronization setup is to mitigate the problems associated to the direct conversion based laser-to-RF synchronization setup (AM-PM effects). The approach of using a MZM for laser-to-RF synchronization is based on sampling of the RF reference signal zero crossings with optical pulses from a Ti:Sa laser oscillator within the integrated MZM. This allows to convert the relative timing error between these two sources into an amplitude modulation of the optical pulses. For more details about the principle of operation of the setup see [14, 15].

3.1. Measurement Setup

Two identical MZM based laser-to-RF synchronization setups have been commissioned at REGAE. One was used to synchronize the laser oscillator to the RF reference signal ($f = 2.9979\,\text{GHz}$), while the second one was employed as an out-of-loop phase detector to evaluate

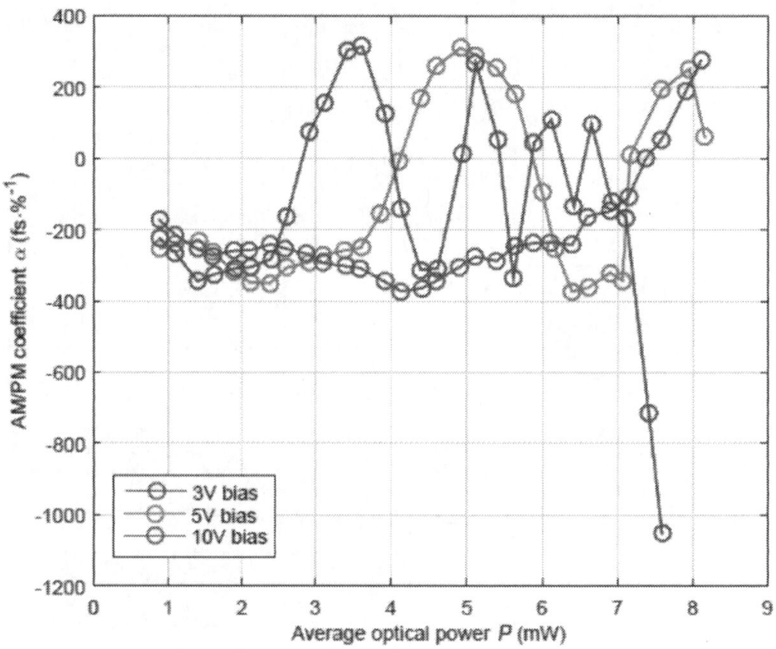

Figure 2. Measured AM-PM coefficients for ET-4000F photodiode.

the relative timing jitter performance between these two sources. The simplified block diagram of the measurement setup is shown in Figure 3. The optical pulses from the laser oscillator are coupled to the fiber collimator and provided to the MZM based synchronization setups via 50/50 fiber splitter. The RF reference signal is transmitted from the RF master oscillator which is further boosted by the high gain RF amplifier. The input optical power of $\approx$4.5 mW and $\approx$+20 dBm of input RF power per setup resulted the in-loop and out-of-loop detector sensitivities of 0.275 mV/fs and 0.43 mV/fs respectively. The relative phase error signal from the in-loop MZM based phase detector is fed to the analog PI controller and the output of the controller is applied to the fast PZT actuator to close the feedback loop.

3.2. Timing Jitter and Drift

While the feedback loop was closed, an out-of-loop MZM based laser-to-RF phase detector was used to measure the residual voltage noise of the phase detector by using a signal source analyzer (SSA) (see Figure 4). Later it was found out that the voltage noise above 40 kHz is dominated by the noise floor of the readout electronics. The noise floor of the readout electronics originates from the thermal noise of the 50Ω load resistor. Since the loop bandwidth is small (few kHz) and significant laser timing jitter contributions beyond 100 kHz are unlikely, the integrated timing jitter from 10 Hz to 100 kHz is evaluated. In this frequency range, the noise floor amounts to 6.85 fs, while the out-of-loop detector measures 13.3 fs about a factor of two higher than the noise floor with strong contributions at 12.5 Hz, 50 Hz and 2.2 kHz. If we quadratically subtract these two numbers, one can calculate that the laser was locked with a precision of $\sim$ 11 fs to the RF source. The 12.5 Hz frequency line is originated from the operation of the DESY II synchrotron which is located in the vicinity of the REGAE facility.

Another important parameter which has been evaluated for the MZM based laser-to-RF synchronization setup is the long term timing drift stability. Figure 6 shows the measurement of the out-of-loop timing drift recorded over 43 hours. The peak-to-peak timing stability resulted

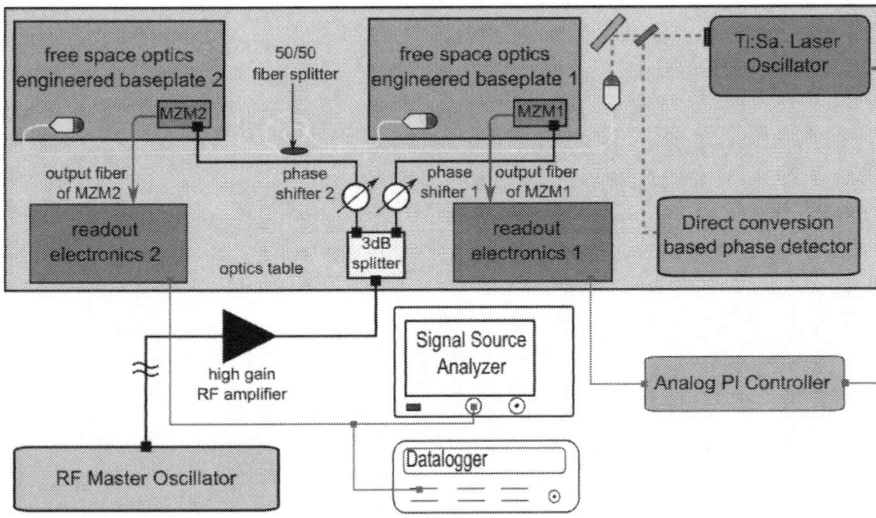

Figure 3. Block diagram of the residual timing jitter and drift measurement setup.

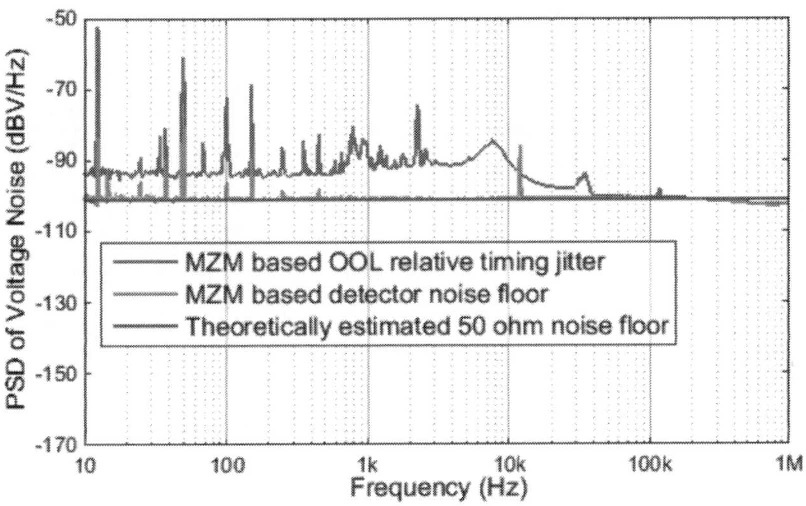

Figure 4. Out-of-loop measured voltage noise.

31 fs while the rms of it amounts only 7 fs which is an excellent long term performance of the phase detector.

3.3. AM-PM Coefficient

The last measurement which has been carried out was the measurement of the AM-PM coefficient for the MZM based laser-to-RF synchronization setup. The input optical power of the in-loop detector was varied using a neutral density filter while monitoring the corresponding timing changes by out-of-loop phase detector. The measurements showed that the mean AM-PM coefficient of the phase detector amounts 34 fs/% of input optical power change which is about factor of 10 smaller value compared to AM-PM coefficients measured for direct conversion based synchronization setup (see Figure 1).

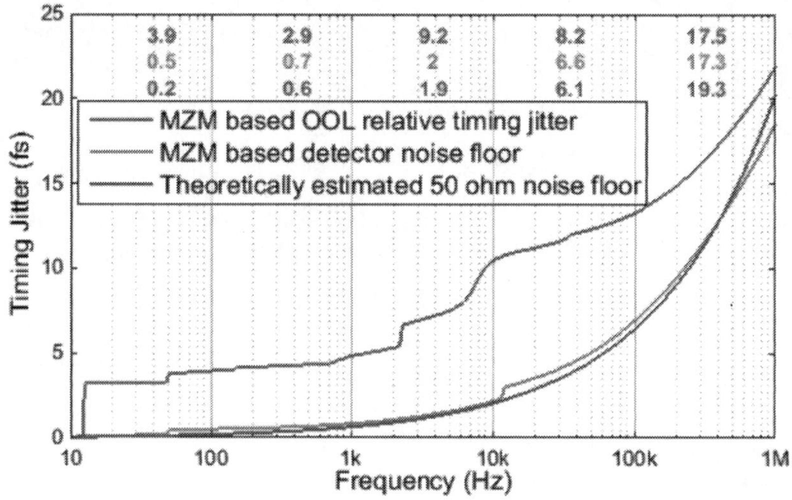

Figure 5. Residual integrated timing jitter.

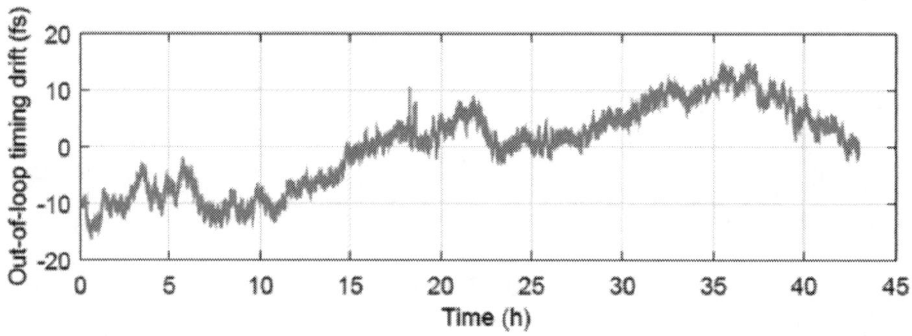

Figure 6. Residual timing drift.

4. Conclusion

In this paper we have shown the limitations of the direct conversion based laser-to-RF synchronization setup. Full experimental evaluation of the MZM based laser-to-RF synchronization setup has been presented with the excellent timing jitter and drift performance of 11 fs rms and 31 fs peak-to-peak respectively. In addition, AM-PM coefficient has been measured for MZM based laser-to-RF synchronization setup with approximately one order of magnitude lower value compared to direct conversion setup. We expect to improve the overall timing performance of the MZM based laser-to-RF synchronization setup by increasing the optical power level which will reduce the noise floor and the corresponding timing jitter. We also expect to reduce the AM-PM coefficient of the MZM based phase detector by removing the DC offsets at the output of the readout electronics.

References

[1] Zeitler B, Dornmair I, Gehrke T, Titberidze M, Maier A R, Hidding B, Floettmann K and Gruener F 2013 in *Proc. SPIE (Prague)* 8779 877904

[2] van Oudheusden T, Pasmans P L E M, van der Geer S B, de Loos M J, van der Wiel M J and Luiten O J 2010 *Phy. Rev. Lett.* **105** 264801

[3] Floettmann K 2014 *Nucl. Instr. Meth.* A **740** 34–38

[4] Zeitler B 2016 *Phase Space Linearization and External Injection of Electron Bunches into Laser-Driven Plasma Wakefields at REGAE* Ph.D. Thesis (Hamburg: University of Hamburg and DESY)
[5] Manz S *et al.* 2015 *Faraday Discussions* **177** 467
[6] Kim J, Cox J, Chen J and Kaertner F X 2008 *Nature Photonics* **2** 733
[7] Jung K and Kim J 2012 *Optics Letters* **37** 2958
[8] Schulz S *et al.* 2015 *Nature Communications* **6** 5938
[9] Lamb T *et al.* 2011 Femtosecond Stable Laser-to-RF Phase Detection Using Optical Modulators in *Proc. FEL11 (Shanghai)* 551
[10] Lamb T *et al.* 2013 Femtosecond Stable Laser-to-RF Phase Detection for Optical Synchronization Systems in *Proc. IBIC13 (Oxford)* 447
[11] Felber M, Hoffmann M, Jalmuzna W, Mavric U, Schulz S and Schlarb H 2012 Laser Synchronization at REGAE using Phase Detection at an Intermediate Frequency in *Proc. IPAC12 (New Orleans)* 2624
[12] Yang H, Jeon C, Jung K, Han B, Jeong Y U, Chung H and Kim J *et al.* 2015 Femtosecond Synchronization of 80-MHz Ti:Sapphire Photocathode Laser Oscillator with S-band RF Oscillator in *Proc. FEL15 (Daejeon)* 105
[13] Yang H, Han B, Shin J, Hou D, Chung H, Baek I H, Jeong Y U and Kim J 2017 *Nature Scientific Reports* **7** 39966
[14] Titberidze M *et al.* 2014 Novel Femtosecond Level Synchronization of Titanium Sapphire Laser and Relativistic Electron Beams in *Proc. IBIC14 (Monterey)* 174
[15] Titberidze M *et al.* 2015 Present and Future Optical-to-Microwave Synchronization Systems at REGAE Facility for Electron Diffraction and Plasma Acceleration Experiments in *Proc. IPAC15 (Richmond)* 833
[16] Kiewiet F B, Kemper A H, Luiten O J, Brussaard G J H and van der Wiel M J 2002 *Nucl. Instr. Meth.* A **484** 619
[17] Ivanov E, Diddams S and Hollberg L 2005 *IEEE Transactions on Ultrasonics, Ferroelectrics, and Frequency Control* **52** 1068
[18] Zhang W, Li T, Lours M, Seidelin S, Santarelli G and Le Coq Y 2012 *Appl. Phys.* B **106** 301

IOP Conf. Series: Journal of Physics: Conf. Series **874** (2017) 012086 doi:10.1088/1742-6596/874/1/012086

First LHC transverse beam size measurements with the beam gas vertex detector*

A Alexopoulos[1], C Barchel[1], A Bay[2], F Blanc[2], E Bravin[1], G Bregliozzi[1], N Chritin[1], B Dehning[1]†, M Ferro-Luzzi[1], S Gianì[2], M Giovannozzi[1], O Girard[2], R Greim[3], G Haefeli[2], P Hopchev[2], R Jacobsson[1], L Jensen[1], O Rhodri Jones[1], V Kain[1], W Karpinski[3], T Kirn[3], A Kuonen[2], R Matev[1], T Nakada[2], M Rihl[1], V Salustino Guimaraes[1], S Schael[3], O Schneider[2], A Schultz von Dratzig[3], G Schwering[3], M Tobin[2], R Veness[1], Q Veyrat[2], S Vlachos[1], M Wlochal[3], B Würkner[1] and Z Xu[2]

[1] CERN, Geneva, Switzerland
[2] EPFL, Lausanne, Switzerland
[3] I. Physikalisches Institut, Aachen, Germany

E-mail: a.alexopoulos@cern.ch

Abstract. The Beam Gas Vertex detector (BGV) is an innovative beam profile monitor based on the reconstruction of beam-gas interaction vertices which is being developed as part of the High Luminosity LHC project. Tracks are identified using several planes of scintillating fibres, located outside the beam vacuum chamber and perpendicular to the beam axis. The gas pressure in the interaction volume is adjusted such as to provide an adequate trigger rate, without disturbing the beam. A BGV demonstrator monitoring one of the two LHC beams was fully installed and commissioned in 2016. First data and beam size measurements show that the complete detector and data acquisition system is operating as expected. The BGV operating parameters are now being optimised and the reconstruction algorithms developed to produce accurate and fast reconstruction on a CPU farm in order to provide real time beam profile measurements to the LHC operators.

1. Introduction

The Beam Gas Vertex detector is a beam profile monitor being developed as part of the high luminosity LHC upgrade [1]. Beam profile measurement based on this method and detector was initially developed at the LHCb experiment [2]. A noble gas is injected in a modified vacuum chamber producing inelastic beam-gas interactions. The charged particles produced in the beam-gas collisions are measured with high precision tracking detectors used to measure the position of the interaction vertices. The BGV allows for non-invasive beam profile and position measurements to be made throughout the full LHC cycle, irrespective of beam energy or luminosity. The detector has been designed to estimate the average transverse beam profile with a precision of about 10% in approximately 5 minutes of integrated beam time [3]. Higher trigger

*Research supported by the HL-LHC project
†Deceased January 14, 2017

IOP Conf. Series: Journal of Physics: Conf. Series **874** (2017) 012086 doi:10.1088/1742-6596/874/1/012086

rates and integration times should allow beam profile measurements on a bunch-by-bunch basis. Based on the beam width σ and given the β-function and dispersion of the magnetic lattice (measured independently), the emittance ϵ can be calculated. The detector can additionally be used for measurements of beam tilt, relative bunch populations and ghost charges (beam intensity in nominally empty bucket slots) [3, 4]. Adding a timing detector would also allow one to measure the longitudinal beam profile.

2. The BGV Demonstrator

2.1. Detector Design

The BGV detector has been installed at Point 4 of the LHC on the beam 2 ring (Figure 1). The detector consists of three main parts: the gas target volume, the tracking detector and the hardware trigger.

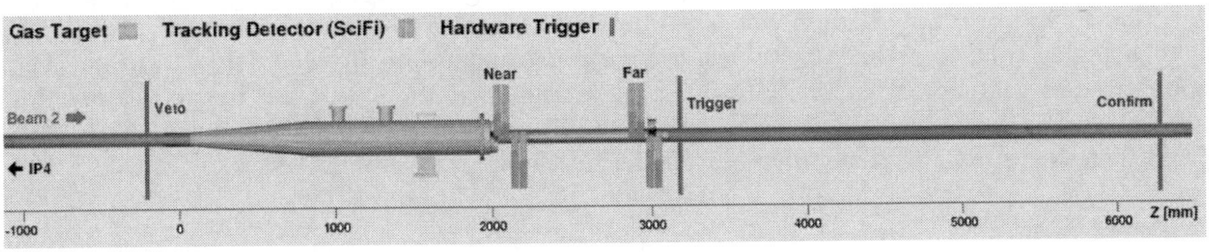

Figure 1. The BGV detector consists of three main parts: the gas target, the tracking detector & the hardware trigger.

The target gas container consists of an aluminium (Al 2219) vacuum chamber of approximately 2 m length and 200 mm maximum diameter. The chamber allows for the injection of (neon) gas giving a local pressure bump of up to about 1×10^{-7} mbar. The BGV vacuum system was optimized to provide a sharp pressure decrease outside the vacuum chamber minimizing the diffusion of gas particles to the beam pipe. In order to reduce multiple scattering of particles produced by the beam-gas interactions, the thickness and density of the exit window towards the tracking detector has been minimized, ranging from 3.25 mm down to 1.15 mm close to the beam pipe. In addition, the beam pipe along the tracking detector has been modified to give a reduced diameter of 52 mm instead of 80 mm and minimum material so that the tracking modules can be as close as possible to the beam to increase the acceptance.

The tracking detector is situated behind the exit window. It is comprised of two stations ('near' and 'far') that are about 1 m apart. Each station has 4 scintillating fibre (SciFi) modules, one pair placed above the beam pipe and the other pair below. Within a pair, the modules have their fibres oriented perpendicular to each other to allow a 2-dimensional measurement.

Each SciFi module contains two fibre mats with 4 (near station) or 5 (far station) layers of scintillating fibres of 250 µm diameter [4]. The two mats are rotated by 2° with respect to each other in order to facilitate pattern recognition. A one-dimensional position measurement resolution of 32 µm has been achieved with a test beam setup [5]. The fibres are read out with silicon photo multipliers (SiPMs) which are cooled with liquid C_6F_{14}. Cooling is needed to improve the signal to noise ratio by reducing the SiPM dark count rate which increases with radiation dose. The radiation dose absorbed by the detector is monitored by a RadMon active detector [6], as well as 6 PIN diodes placed close to the SiPMs.

The BGV trigger consists of scintillator plates with dimensions $30 \times 30\,\text{cm}^2$, arranged in 3 stations. A pair of scintillators is located before the gas target and is used to veto interactions occurring upstream of the target chamber. A second trigger plane is placed downstream of the SciFi stations and provides the trigger signal. A third trigger plane has recently been installed and will be commissioned during the LHC 2017 run. It will be used in coincidence with the second plane to reject background from the other beam and confirm a crossing particle produced in the gas volume by the correct beam.

2.2. Read-out & Timing Control

The read-out chain of the BGV is based on hardware from the LHCb experiment. Each SciFi module is read out by 16 128-channel SiPM arrays operated in Geiger-Müller mode. The Beetle front-end ASIC [7] receives the analogue signals from a SiPM. In order to match the Beetle chip's input current range, the signals are first attenuated by a factor of around 200-500 via an RC circuit. The Beetle has an analogue memory of a programmable maximum length of 160 stages with a serial read-out at the LHC bunch frequency of 40 MHz. Subsequently, the data are time-multiplexed on 4 output ports of 32 channels to form an analogue link (A-link) over which data are forwarded to the VELO repeater boards. At the repeater boards the signal data is amplified and sent over 60 m of cables to the TELL1 boards [8] for digitization.

The timing control of the data acquisition system is supervised by the ODIN board [9]. The board accepts the input from the hardware trigger and then sends a trigger to the Beetle chips to start transmission of the data. Timing studies of the coarse and the fine delays have been performed to ensure that the correct slot of the Beetle memory is retrieved [4].

3. BGV data

Several data taking campaigns were carried out after the commissioning of the detector during the 2016 LHC run, along with studies related to its proper operation and data acquisition.

3.1. Data corrections & clustering

The raw data have to be corrected for effects related to the SiPMs and the read-out electronics before being processed for the final estimation of the beam position and beam size. The corrections are applied in reverse order to how they are introduced in the read-out chain and include: pedestal subtraction, common mode noise suppression and channel correlation corrections. In the future, these steps will be implemented in the FPGAs of the TELL1 boards and performed in real-time. Further studies and analysis of these corrections can be found in [10, 11].

The energy deposited by a traversing particle is dispersed over several detector channels and must be assembled into a cluster. The clustering is performed by groups of 64 channels and uses a three threshold algorithm to suppress noise. Initially, a search is made for a signal over a given threshold. Clusters are formed by including neighbouring channels that themselves contain a significant amount of signal. Finally, the central position of the cluster is calculated as the weighted mean value of the signals inside the cluster [12, 13].

3.2. Analysis Method

Tracks are formed using clusters which are found in all 8 consecutive SciFi layers. For each track two parameters are calculated: the impact parameter d_{xy} which refers to the distance of closest approach of the reconstructed track to the z-axis; the azimuthal angle of the track ϕ is defined as the angle between the x-y projection of the track and the x-axis. These two parameters are related for a given position (x_0, y_0) of the primary vertex:

$$d_{xy} = x_0 \sin(\phi) - y_0 \cos(\phi) \tag{1}$$

The transverse beam width can be calculated using the impact parameter correlation (IPC) method. The event-by-event displacement due to the finite beam width affects all particles of a beam-gas interaction in the same way, thus a correlation is introduced. In the case of an untilted beam ellipse, and assuming the beam width at the BGV location to be $\sigma_x = \sigma_y = \sigma_{beam}$, this correlation can be described as [14, 15]:

$$\langle d_{xy}^{(1)} d_{xy}^{(2)} \rangle = \sigma_{beam}^2 \cos(\phi_1 - \phi_2) \tag{2}$$

where d_{xy} refers to the impact parameter of each track originating from the same vertex and ϕ to its azimuthal angle.

3.3. Results from the 2016 LHC Run

During the 2016 LHC run the detector was operated with a neon gas pressure of approximately 6×10^{-8} mbar which is assumed homogeneous inside the gas targets volume. The SiPMs were cooled to $-25°C$ and the thresholds of the trigger system were set to very low values on the veto side so that the trigger becomes more selective. The results presented are based on non-zero-suppressed data sets taken during LHC fill 5570 over a time period of ~5 minutes and with a rate of about 1 kHz.

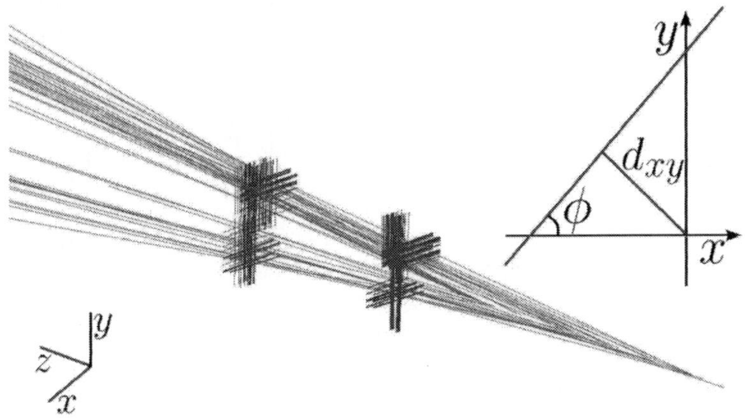

Figure 2. An event display at the BGV detector using the LHCb software *Panoramix*. A representation of the impact parameter d_{xy} and the azimuthal angle ϕ of the track is also displayed.

The distribution of the z-coordinates at which the impact parameter is minimized is displayed in Figure 3. The measurements are well inside the limits of the gas target and are correlated with the expected gas pressure profile.

Plotting the d_0 against ϕ and using equation (1) the beam position can be estimated (Figure 4), in this case giving $x = -0.75$ mm, $y = 0.29$ mm.

Using the IPC estimator (Equation 2) the σ_{beam}^2 is expressed as the slope of the linear fit of the $\langle d_{xy}^{(1)} d_{xy}^{(2)} \rangle$ versus $\cos \Delta\phi$ (Figure 5). The transverse beam width is measured as $\sigma_{beam} = (0.37 \pm 0.13(\text{stat.}))$ mm. Further refinement of the vertexing algorithm along with better event selection through improved triggering will be required to reduce this rather large statistical error and allow a full comparison with other LHC profile measurement devices.

4. Summary & Outlook

The BGV detector has successfully completed its first commissioning steps during the 2016 LHC run. The detector and acquisition system is fully operational and the first data have

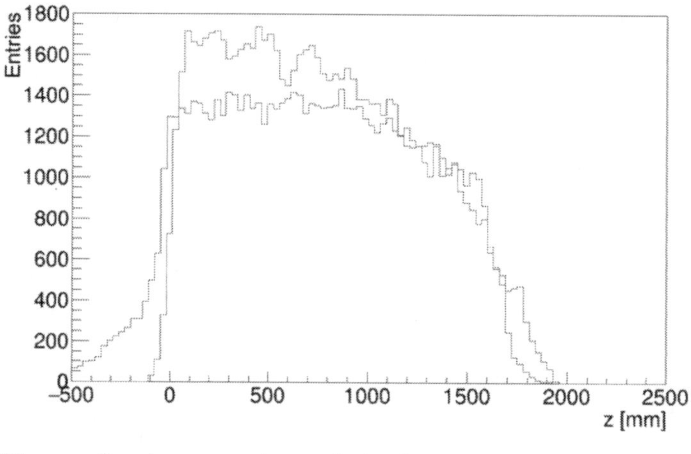

Figure 3. A comparison of the histograms of the points of closest approach to the z-axis between tracks from Monte Carlo (red) and the LHC data set (blue).

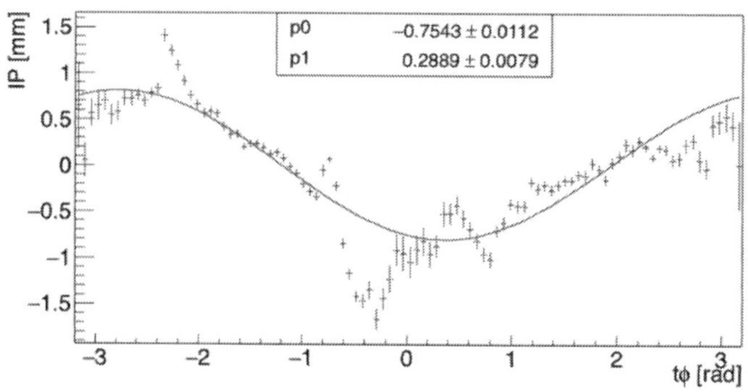

Figure 4. Beam position estimation using the the d_0 vs ϕ method. The fit (red curve) returns a beam position estimation of (−0.75 mm, 0.29 mm) relative to the detector frame.

demonstrated the proper functioning of the complete BGV system. First test measurements result in a transverse beam size estimation with a 0.13 mm statistical error after 5 minutes at a trigger rate of 1 kHz. A larger rate of 100 kHz data will be achieved with the implementation of zero-suppressed acquisition, inserting the data corrections and the clustering in the FPGAs of the read-out electronics, and optimizing the track analysis for real-time execution in a CPU farm [16]. Adding the third plane of the hardware trigger will also drastically improve the selection of events. All of this is expected to reduce the statistical error by a factor of 10. A cross-calibration of the BGV to other instruments during special LHC runs will lead to the determination of any corrections needed. The first real-time beam width results from the BGV demonstrator are expected to be sent to the LHC control room during the 2017 run.

Acknowledgments

We would like to thank the LHCb Collaboration, in particular Clara Gaspar, Niko Neufeld and Markus Frank, for the support in the development of the BGV online control system. We wish

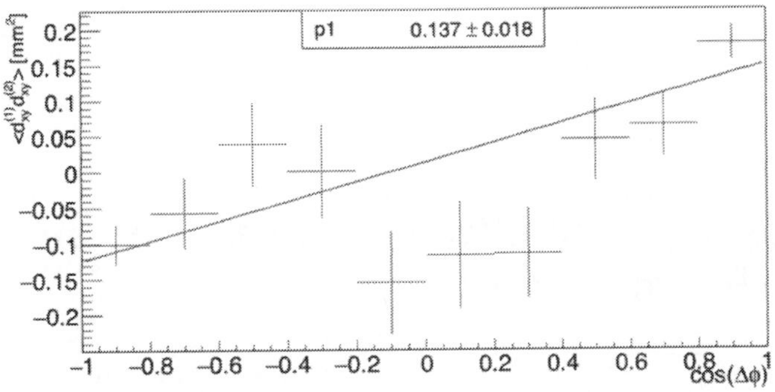

Figure 5. Using the IPC method the beam width is estimated to be $\sigma_{beam} = (0.37 \pm 0.13(\text{stat.}))$ mm.

to express our sorrow over the recent death of Bernd Dehning, a leading collaborator of the BGV project, and dedicate these first results to his memory.

References

[1] 2014 *HL-LHC preliminary design report: Deliverable: D1.5* CERN-ACC-2014-0300 `https://cds.cern.ch/record/1972604`
[2] Ferro-Luzzi M 2005 *Nucl. Instrum. Meth.* A **553** 388–99.
[3] Hopchev P *et al.* 2014 *Proc. 5th Int. Particle Accelerator Conf.* (Dresden) pp. 3680–83.
[4] Rihl M *et al.* 2016 *Proc. 7th Int. Particle Accelerator Conf.* (Busan), pp. 296–8.
[5] Girard O, An L, Kuonen A and Haefeli G 2016 *BGV fibre module: test beam report* (Laussane) `http://lphe.epfl.ch/publications/2016/LPHE-2016-004.pdf`.
[6] Spiezia G *et al.* 2011 *Proc. 10th Int. Conf. on Large Scale Applications and Radiation Hardness of Semiconductor Detectors* (Florence) PoS RD **11** 24
[7] Lchner S and Schmelling M 2005 *The beetle reference manual - chip version 1.3, 1.4 and 1.5* LHCb-2005-105 CERN-LHCb-2005-105.
[8] Haefeli G, Bay A, Gong A, Gong H, Mcke M, Neufeld N and Schneider O 2006 *Nucl. Instrum. Meth.* A **560** 494–502.
[9] Jacobsson R 2002 *The final LHCb readout supervisor "ODIN"* LHCb-TALK-2002-044 `http://lphe.epfl.ch/publications/2016/LPHE-2016-002.pdf`.
[10] Girard O, An L, Kuonen A, Li H and Haefeli G 2016 *BGV fibre module read-out, signal correction and clustering* (Laussane) `http://lphe.epfl.ch/publications/2016/LPHE-2016-002.pdf`.
[11] Aaij R *et al.* 2014 Performance of the LHCb vertex locator *J. Instrumentation* **9** 1405.7808
[12] LHCb Collaboration 2014 *LHCb tracker upgrade technical design report* CERN-LHCC-2014-001
[13] Giani S, Haefeli G, Joram C, Tobin M and Xu Z *Digitization of SiPM signals for the LHCb upgrade SciFi tracker* LHCb-PUB-2014-025
[14] R. Mankel 2010 *Measurement of the transverse beam line width using impact parameter correlations* CMS-AN-2010/196
[15] Donati S and Ristori L 2003 *Measuring beam width and SVX impact parameter resolution* CDF-Note-4189
[16] Frank M, Gaspar C, Jost B and Neufeld N 2015 *J. Phys.: Conf. Series* **664** 082011

Development of wide dynamic range beam loss monitor system for J-PARC main ring

K Satou[1*], T Toyama[1], N Kamikubota[1], S Yamada[1] and S Yoshida[2]

[1]Accelerator Division, Japan Proton Accelerator Research Complex, High Energy Accelerator Research Organization, 2-4 Shirakata, Tokai, Ibaraki 319-1195, Japan
[2] Kanto Imformation Service, 8-21 Bunkyocho, Tsuchiura, Ibaraki 300-0045, Japan

*E-mail: kenichirou.satou@j-parc.jp

Abstract. The new beam loss monitor (BLM) system now in operation at the main ring of J-PARC consists of an isolated front-end current to voltage converter, a VME-based 24 bit ADC system. A dual detector system employs a proportional-type gas chamber (PBLM) and an air-filled ionization chamber (AIC). The system shows a wide dynamic range of 160 dB. It can detect the low level signal that would arise in the case of the detection of residual dose in the ring itself after the beam has been turned off as well as an event such as high level beam loss at the collimators. The signal rise time of the waveform obtained is 17 μs which is fast enough to meet the speed requirement of the Machine Protection System (MPS); which is that the MPS should dump the beam within 100 μs when the beam loss signal exceeds the reference levels set in the ADC system.

1. Introduction

The new BLM system was designed [1] to meet the conditions and the requirements listed below. Some items were modified as we tested how the system worked with various beam tunings.

1. The system is required to detect losses from the weak beams that are passed round the ring during the beam commissioning phase, when the beam intensity is only 1 % of the designed maximum intensity of 4E13 protons per bunch (eight bunches for full).
2. The system must be able to detect an unusual loss of power in the beam and send an alarm to the Machine Protection System (MPS) [2] to dump the beam within 100 μs. Acceptable beam power loss in the main ring (MR) is 0.5 W/m in the arc sections, 2 kW at the collimators, 7.5 kW at the slow extraction (SX) section, and 1.125 kW at the fast extraction (FX) section [3, 4].
3. Independently of loss of power in the beam, the level of activation of the components should be kept within the limits set for workers to perform hands-on maintenance. Each device should display the levels that should not be exceeded for various maintenance operations including unscheduled and urgent repair-works for a malfunction of the device. The worker's daily radiation exposure dose is limited to 0.5 mSv for men and 0.3 mSv for women in J-PARC.
4. The activations have to be measured after the beam has been turned off. We need to be able to detect residual doses as low as 10 μSv/h. On the other hand the expected dose for the detectors in the collimator section was estimated to be 0.053 Gy/Cycle for 2 kW beam power lost uniformly in the collimators [5]. To achieve this, the monitoring system must have a dynamic range wider than 120 dB [1].

5. The bandwidth needed to monitor turn-by-turn beam losses for the system is from DC to 200 kHz. To study the head-tail instability [6], detection of even higher frequency signals of about 100 MHz would be needed.

The new BLM system has been in operation since last summer 2016 with a dynamic range of the system wider than 160 dB. However the system's bandwidth is limited: it goes from DC to 10 kHz. To meet requirement 5, we are currently working on a new detector system. In this paper, the performance of the present BLM system is described.

2. Detectors

We incorporated three types of detector: a cylindrical proportional gas chamber (PBLM), a 1 m long short-type air ionization chamber (sAIC), and a long-type air ionization chamber (longAIC). The maximum gas gain of the PBLM is 2E4 at a bias voltage of -2.0 kV. The 216 PBLMs and the 53 sAICs (only in the straight sections) are mounted at each quadrupole magnet (QM) as shown in figure 1. They stand 360 mm away from the QM surface and 848 mm from the centre of the beam duct. The number labelled on each BLM corresponds to the address number of the MR.

A magnetic shield made of iron suppresses perturbation in the gas gain of the PBLM due to leakage of magnetic field from the QM. Another 19 longAICs were installed on the cable rack running on the outside wall of the MR tunnel to cover whole MR ring. The distance between the detectors and centre of the beam duct is 3 m.

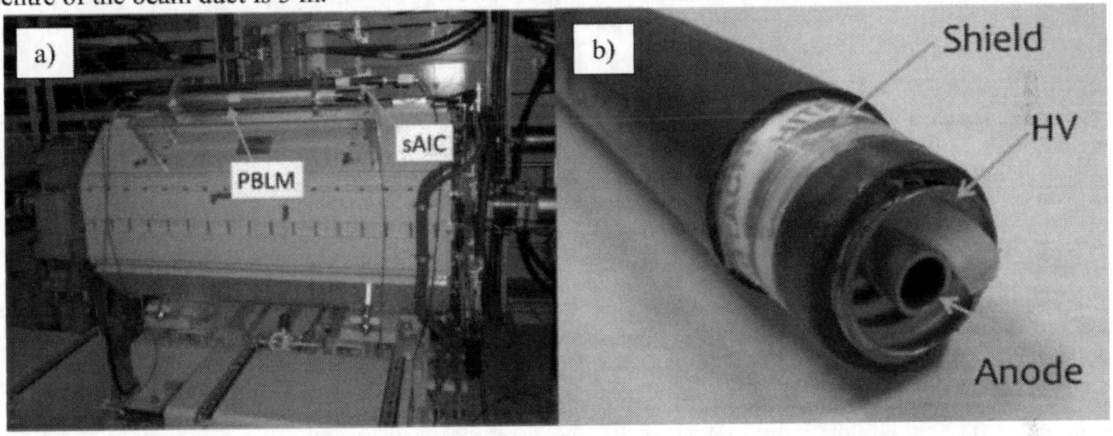

Figure 1. a) Photo of the PBLM and sAIC. b) Cable structure used for AIC.

2.1. *HV system*

To feed a bias voltage to each BLM detector, the MR area was divided into nine sections; these are Injector and Collimator section (#213-20), 1st half of Arc A (#21-44), 2nd half of Arc A (#45-68), SX section (#69-92), 1st half of Arc B (#93-116), 2nd half of Arc B (#117-140), RF & FX section (#141-164), 1st half of Arc C (#165-188), 2nd half of Arc C (#189-212), where (#a-b) means the area from address a to b. In each section, detectors are grouped and the HV cables are daisy-chained so that the PBLMs in each section are thus operated with the same bias voltage. The bias voltages for the AICs are, however, all set at -500 V. During operations, the output currents of the power supplies are less than 0.1 mA which are much less than the maximum output current of 20 mA. The drift of the HV of the power supply is less than 0.03 %.

2.2. *PBLM*

The PBLMs have been used as the main detectors. They offer a wide range in gas gain and a fast impulse response about 0.1 μs. Post-signal processing takes place within the signal rise time of 17 μs, well within the MPS speed requirement. The details of the PBLM are described in refs. [5, 7, 8].

The initial charge induced by the beam loss is amplified by the gas gain. The output current divided by the gas gain can be used as an indication of beam loss signal. By changing the bias setting, a wide dynamic range can be covered. However an increase of the output current decreases the gas gain [9]. For example 1 μA and 10 μA output DC current results in 4 % and 40 % gain decrease in case of the bias voltage of -1.6 kV, respectively. This means that the PBLMs are likely to underestimate the beam power loss especially in the Injector and Collimator section and the SX section where a large beam loss is inevitable.

2.3. sAIC and longAIC

The AIC is an ion chamber that uses a double shielded 20D coaxial cable. A corrugated Cu pipe acts as an inner shield core while spiral polyethylene acts as an insulator as shown in figure 1. The centre Cu tube picks up the charge, while the bias voltage is applied on the corrugated pipe tube and the outermost Cu tape shield is grounded.

The sAIC is a 1 m long AIC. The maximum output current and sensitivity were checked by gamma rays from a strong ^{60}Co gamma source. The data shows linear response up to 10 μA within an uncertainty of about 10 %. The output current corresponds to 3.2 kGy/h [1], so the sensitivity is 11 μC/Gy. This qualifies the sAIC for use in high radiation environment.

The longAICs are, on average, 84 m long. The 19 longAICs cover the whole length of the MR tunnel. They measure average beam power loss to ensure that the beam loss is lower than the criteria of the item 2 listed in the introduction. The calibration was made using intentional beam losses of the injection beam at the collimator units, and in the Arc A and the Arc B. The beam losses were generated by the local bump orbit. The beam energy was fixed as 3 GeV. The number of the lost particles was measured by the DCCT [8]. The longAIC#1-10 outputted 14.1 nC for the beam loss of the 1.1E11±0.2E11 protons at the collimators. The detector covers the area from the injection point of address #1 to the end of the collimator units of address #10. The efficiency, which is the output charge divided by the beam power lost, is thus 1/3.6 μC/kJ. In the Arc sections, it was 1/8.9 μC/kJ on average.

3. The Front-end amp and ADC

A new data acquisition system was developed and installed during the maintenance period last summer 2016. It consists of an isolated current-to-voltage converter that uses photo-couplers (front end amp) and a VME-based 24 bit ADC system. The system is described in ref. [10].

The conversion factor and the -3 dB bandwidth of the front end amp are 1 V/μA and 50 kHz, respectively. The maximum input current is 10 μA. To improve the signal to noise ratio, a low-pass filter of 10 kHz is currently inserted in the front-end amp.

An ADC system with a noise level of 100 μVpp and a total harmonic distortion of 80 dB was developed. The effective number of bits is 16.5. To improve the ground level stability, offset subtraction is performed at each cycle. After the offset subtraction, the ADC digitized at the rate of 1 MS/s. Each digitized point is sequentially compared with a reference level. At the same time, the data is summed and compared with the second reference level. When the data indicates one of the reference levels has been exceeded, the system outputs an alarm to the MPS system within 2 μs. The 1 MS/s waveform data is then down-converted to 1 KS/s data.

4. BLM data

To check the impulse response of the system, the gamma rays from the activated magnet were measured. The 1 MS/s waveform data are shown in figure 2, where the bias voltage was -2.0 kV. The pulses can be seen in the figure. The 10-90 % rise time is 17 μs.

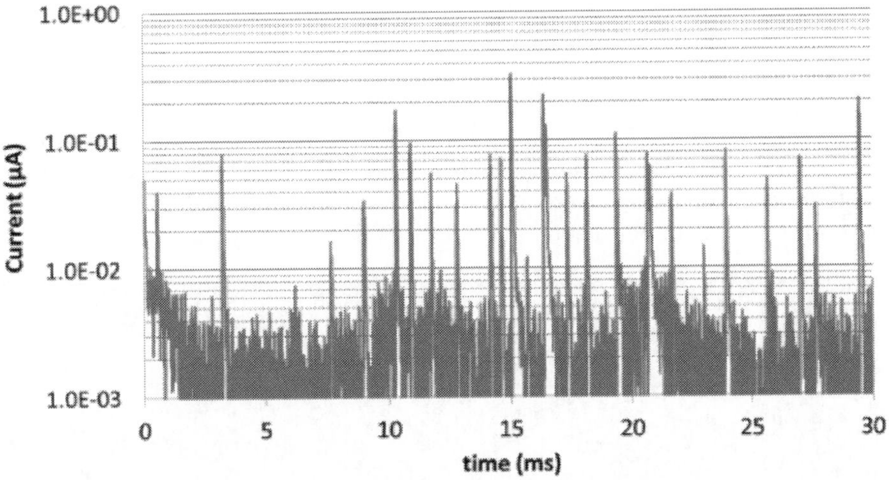

Figure 2. Pulse signal from the PBLM induced by the gamma rays from the activated magnet.

The following data were obtained on the FX mode operation with 460 kW output power. The operation-cycle was 2.48 s long.

The integrated charge data of the PBLMs and sAICs are shown in figure 3. The gas gains of the PBLMs were corrected for the data in this plot. The beam loss signals from PBLMs are shown as blue bars. The bias voltage of the PBLMs was all -1.6 kV except for these in the Injector and Collimator section whose bias voltage was -1.3 kV. The gas gain is 628 and 61, respectively, for bias voltages of -1.6 and -1.3 kV. The sAIC data are shown as red solid circles. The data shows that its detection threshold is about 1E-3 nC/Cycle.

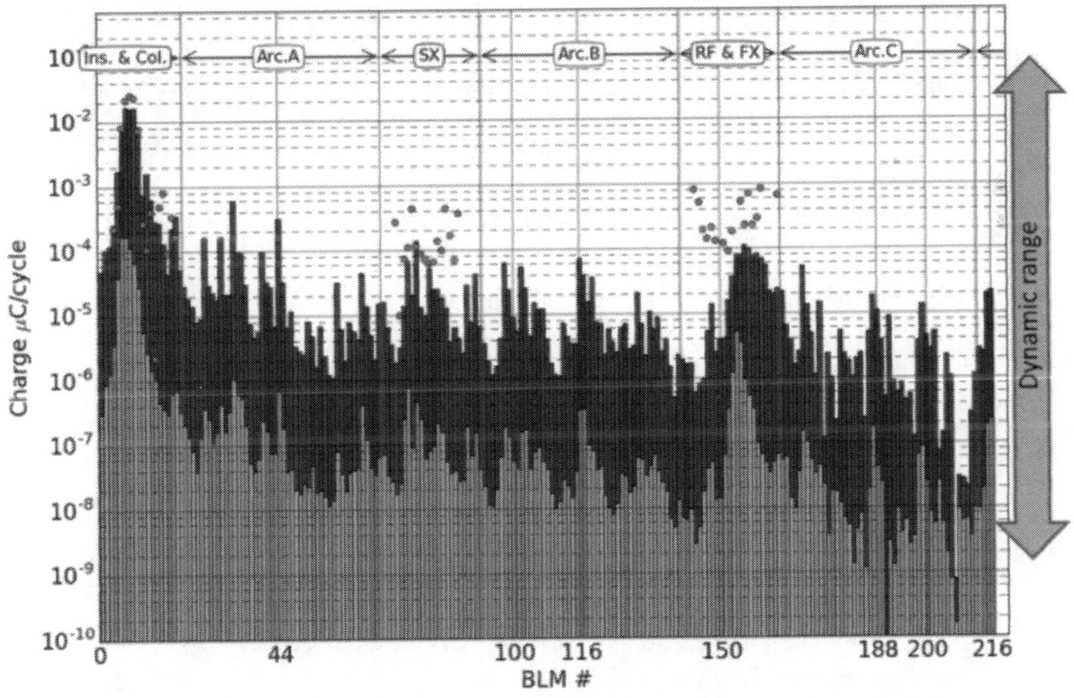

Figure 3. Integrated charge plot. The beam loss signals from PBLMs and sAICs are shown as blue bars and red solid circles, respectively. The yellow bars show the residual dose.

The residual dose data measured five hours after the beam was shut down are shown as yellow bars, where the bias voltage was -2.0 kV. The integration time was 1.6 s. The data were compared with the measurements by using the commercial dose meter, CANBERRA RADIAGEM2000. The output charge of 1.2E-5 μC/Cycle corresponds to the residual dose rate of 100 μSy/h.

The 1 kS/s waveforms of the PBLMs were divided by each gas gain and displayed in a contour plot as shown in figure 4. The post signal processing, a simple averaging process for data points in the time window of 20 ms which is shifted by a 10 ms interval, was adopted to reduce noise due to the fundamental and the higher harmonics coming from the 50 Hz AC supply.

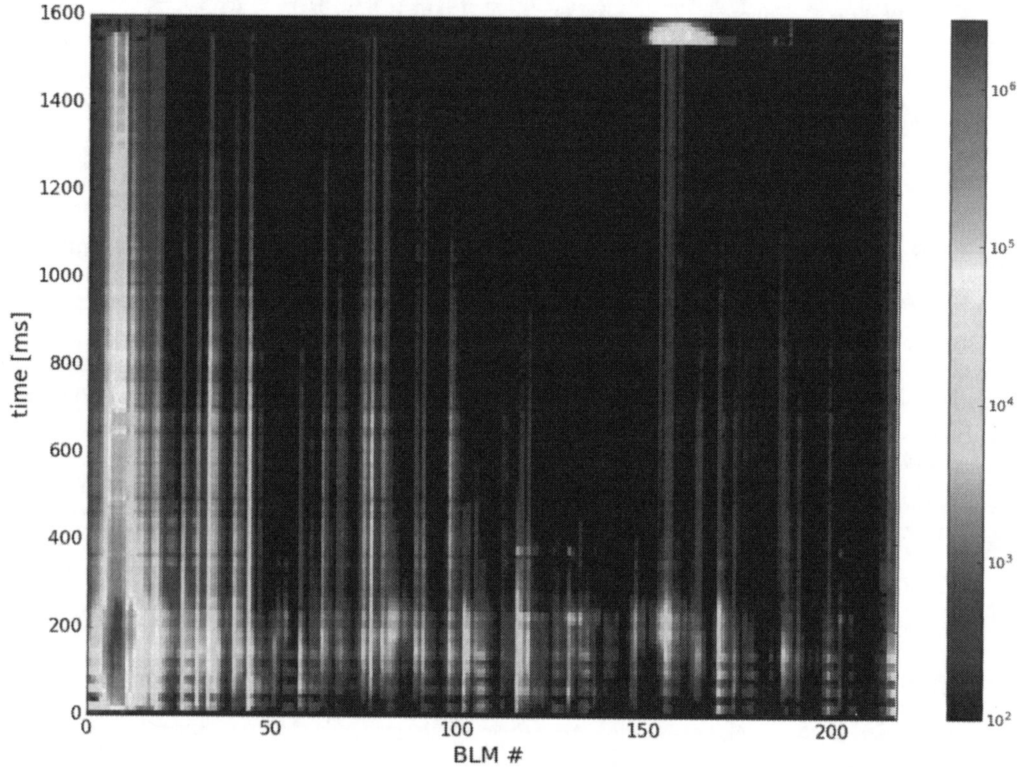

Figure 4. Contour plot of the waveforms from the PBLMs.

We can see in the figure the key details of the beam losses: positions, timings, and intensities. An intense beam loss at the collimators can be seen at the time around 200 ms which is due to the imperfection of the momentum tracking of the bending magnets occurred after the acceleration start. And the extraction beam loss at the FX septum magnet can be seen around BLM#156. Four times losses can also be seen at 0 ms, 40 ms, 80 ms, and 120 ms after the beam injection. These losses were caused by injection miss-matches.

5. Discussion

At the outset of the design process, the target for the rise time was 10 μs for the MPS system to dump the beam within 100 μs. However, it seems to us that the present rise time of 17 μs is fast enough. We took into-consideration the observations that the cable delay is a few μs, the processing time of the ADC system is about 2 μs, and the processing time of the MPS system is about 1 μs.

The 1 KS/s waveforms of PBLM#8 sAIC#8, and longAIC#1-10 are shown in figure 5. The gas gains were not corrected for these waveforms. As shown in the figure, the waveform of the PBLM is saturated at the output current of 10 μA. On the other hand, the waveforms of the sAIC and the longAIC have enough margins to show the whole signal. The integrated charge from the longAIC was 0.31 μC/Cycle which corresponds to the 450 W lost at the collimators.

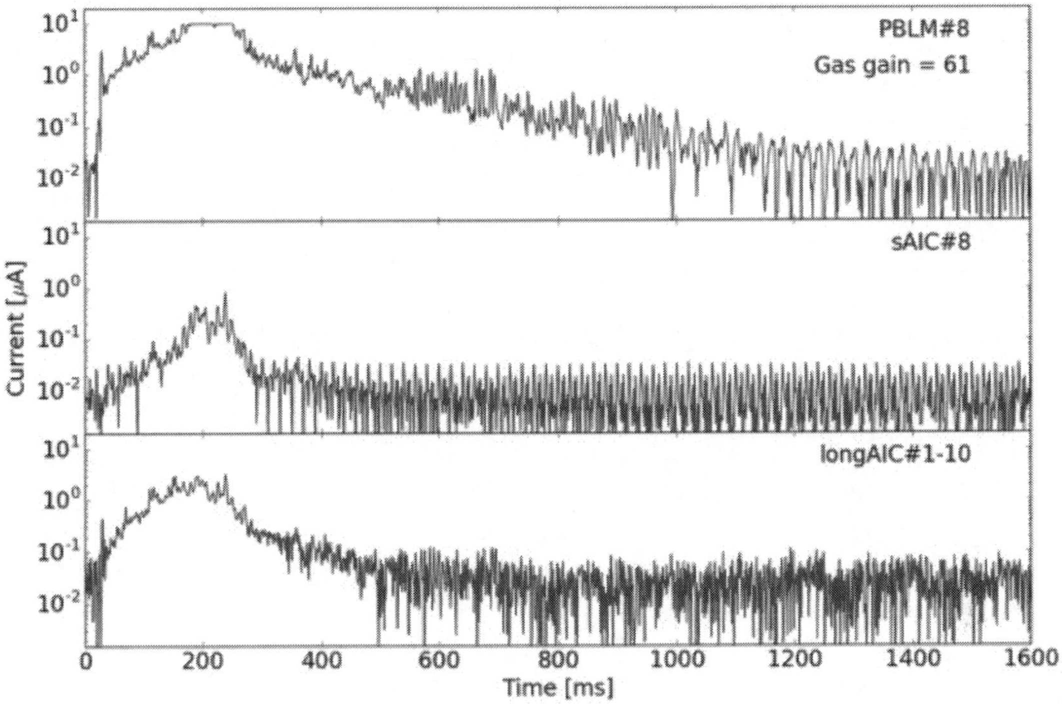

Figure 5. Waveforms of the PBLM, the sAIC, and the longAIC installed at the Injector and Collimator section.

The dual detector system of the PBLM and the sAIC can compensate for the weakness of the PBLM and shows wide dynamic range. At each bias setting the dynamic range is about 80 dB for integrated charge data, and an appropriate bias setting can expands the total dynamic range. The upper limit of the integrated charge of the sAIC can be estimated greater than 0.1 μC/Cycle considering the present margin. Including the data of residual dose measurements, the net dynamic range is now wider than 160 dB.

6. Summary

The new BLM system for the J-PARC MR has been operating successfully since last summer. Thanks to the high gas gain performance of the PBLM and the combined use of the sAIC, the system dynamic range is now wider than 160 dB. The present signal rise time of 17 μs is slower than the initial design goal of 10 μs. However it is likely to be fast enough to ensure that the MPS system will dump the beam within 100 μs. The output signal of the longAIC was calibrated using the controlled beam losses at the Collimators and Arc A and Arc B. The beam power loss in the Injector and Collimator section was 450 W in the FX mode operation which outputs 460 kW beam power.

To improve signal-to-noise ratio of the waveform data obtained, further studies are needed. We are now designing another detector system for a new BLM that will operate with bandwidth of more than 200 kHz.

References

[1] Satou K and Toyama T 2012 *Proc. 3rd Int. Particle Accelerator Conf. 20-25 May 2012 (New Orleans)* p 837

[2] Nakagawa H, Akiyama A and Odagiri J 2009 *Proc. 12th Int. Conf. on Accelerator and Large Experimental Physics Control Systems 12-16 Oct. 2009 (Kobe)* p 406

[3] Yamazaki Y *et al.* 2003 Accelerator Technical Design Report for J-PARC *preprint* KEK Report 2002-13 and JAERI-Tech 2003-044

[4] Shirakata M.J, Igarashi S, Ishii K, Sato Y, and Takano J 2016 *Proc. Advanced Beam Dynamics Workshop on High-Intensity, High Brightness and High Power Hadron Beams* 3-8 July 2016 (Malm) p 543

[5] Toyama T *at al.* 2008 *Proc. Advanced Beam Dynamics Workshop on High-Intensity, High Brightness Hadron Beams 25-29 Aug. 2008 (Nashville)* p 450

[6] Chin Y H 2013 *Proc. North American Particle Accelerator Conf. 29 Sep. – 4 Oct. 2013 (Pasadena)* p 27

[7] Lee S, Toyama T, Kishiro J and Tanaka M 2004 *Proc. 9th European Particle Accelerator Conf. 5-9 Jul. 2004 (Lucerne)* p 2667

[8] Satou K, Toyama T, Arakawa D A, Arinaga A, Hashomoto Y, Igarashi S, Tejima M, Hayashi N and Yamamoto K 2008 *Proc. Advanced Beam Dynamics Workshop on High-Intensity, High Brightness Hadron Beams 25-29 Aug. 2008 (Nashville)* p 472

[9] Sipila H, Vanha-honko V 1978 *Nucl. Instr. And Meth.* **153** 461

[10] Satou K, Kamikubota N, Toyama T and Yoshida S 2014 *Proc. 5th Int. Particle Accelerator Conf. 15-20 Jun. 2014 (Dresden)* p 3547

IOP Conf. Series: Journal of Physics: Conf. Series **874** (2017) 012088 doi:10.1088/1742-6596/874/1/012088

Wire position system to consistently measure and record the location change of girders following ground changes

H J Choi, S B Lee, H G Lee, S Y Back, S H Kim, H S Kang

Pohang Accelerator Laboratory, Pohang, Gyeongbuk 37673, Republic of Korea

Email : choihyo@postech.ac.kr

Abstract. Several parts that comprise the large scientific device should be installed and operated at the accurate three-dimensional location coordinates (X, Y, and Z) where they should be subjected to survey and alignment. The location of the aligned parts should not be changed in order to ensure that the electron beam parameters (Energy 10 GeV, Charge 200 pC, and Bunch Length 60 fs, Emittance X/Y 0.481 μm/0.256 μm) of PAL-XFEL (X-ray Free Electron Laser of the Pohang Accelerator Laboratory) remain stable and can be operated without any problems. As time goes by, however, the ground goes through uplift and subsidence, which consequently deforms building floors. The deformation of the ground and buildings changes the location of several devices including magnets and RF accelerator tubes, which eventually leads to alignment errors (ΔX, ΔY, and ΔZ). Once alignment errors occur with regard to these parts, the electron beam deviates from its course and beam parameters change accordingly.

PAL-XFEL has installed the Hydrostatic Leveling System (HLS) to measure and record the vertical change of buildings and ground consistently and systematically and the Wire Position System (WPS) to measure the two dimensional changes of girders. This paper is designed to introduce the operating principle and design concept of WPS and discuss the current situation regarding installation and operation.

1. Introduction

As shown in Figure 1, the position of the components changes along with the changes in Girder and the location of the ground which a building stands on. When the position of the components moves out of the orbit of the electron beam due to changes in the ground, interaction between the electromagnet field of the component and electron beam also changes. If the frequency of the equipment is higher or the stronger the electromagnetic field or magnetic field, the greater the influence of changes in position.

In Korea, there is a saying that goes "It takes more than pearls to make a necklace." It means that no one will get the performance as he/she wants if he/she does not arrange and position all devices precisely even though each of the devices works well. In PAL-XFEL, the HLS and WPS are installed and operated to observe and record changes in the ground and building through continuous measurement and to quickly arrange devices in the region where significant displacement is shown [1].

IOP Conf. Series: Journal of Physics: Conf. Series **874** (2017) 012088 doi:10.1088/1742-6596/874/1/012088

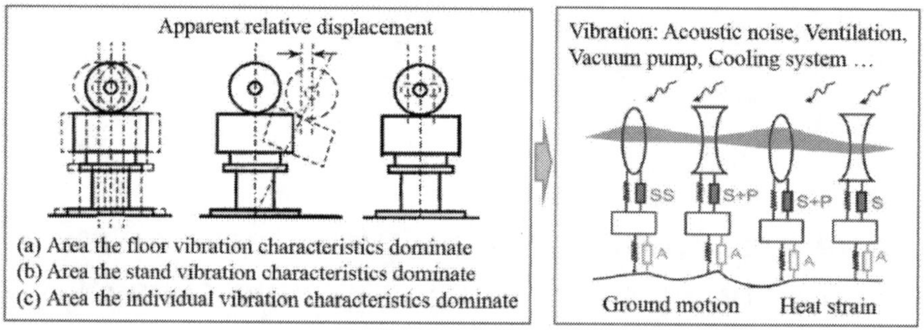

Figure 1. Three vibration patterns which generate relative displacement [2]

2. History of WPS to have used on the accelerator

In 1990, the concept of a wire measurement system was first introduced to observe the changes in the ground of a 10-km-long linear collider [3]. The method for transmitting the RF signal to the wire and comparing the strength of the RF signals at the Pickup which is the same as the beam position monitor (BPM) was also examined in 1993. In this case, impedance around a wire should be kept constant at 50 ohm to transmit the RF signal over a distance over 100 meter [4]. In 1995, the European synchrotron radiation facility (ESRF) succeeded in researching and developing a non-contact mono axe (1D) WPS sensor using carbon wire [5]. After that, FOGALE Nanotech developed and commercialized the two dimension WPS2D sensor.

To establish and operate large scientific instruments including linear colliders, various sensors should be developed for the steady observation of changes in the position of the instruments as a result of changes in the ground [6]. Recently, the Compact Linear Collider (CLIC) studied the method for real-time measurement and recording of changes in the ground of instruments several kilometers in length by using WPS and HLS [7].

3. Electrical characteristics of passive components used as sensors

We should know the characteristics of basic devices, such as resistors (R), condensers (C) and inductors (L) of electronic circuits regarding input power supply in order to design those circuits or to understand the response properties of designed electronic circuits. Input power supply is divided into DC and AC. Whereas the response characteristics of DC circuits can be simply understood by the relationship among the voltage, intensity and resistor, the response characteristics of AC circuits can be understood only after knowing three kinds of resistance components. In AC circuits, the resistor (R) is marked as a real number resistor and the condenser and inductor are marked as imaginary number resistors and called reactance. The condenser (C) is called capacitive reactance (X_C) and the inductor (L) inductive reactance (X_L). An important point is that resistance values of reactance are determined according to the frequency of AC power supply. Equation (1) shows how to calculate the resistance values of R-L-C passive components in relation to AC power supply [8].

$$V = I \cdot Z \tag{1}$$

where

$$Z = R + j \cdot X = \text{resistance } [\Omega]$$
$$X = X_C + X_L = \text{reactance}$$
$$X_C = \frac{1}{j \cdot \omega \cdot C} = \text{capacitor reactance}$$
$$X_L = j \cdot \omega \cdot L = \text{inductor reactance}$$

Various sensors using electric response characteristics of R-L-C passive components are used in diverse areas [9]. The operation principle of WPS sensors is to figure out changes in the X_C resistance values of condensers using AC power supply.

IOP Conf. Series: Journal of Physics: Conf. Series **874** (2017) 012088 doi:10.1088/1742-6596/874/1/012088

4. Working principle of WPS sensor

The working principle of the WPS sensor, which is to measure a distance in a non-contact way, is described in Figure 2. The capacitor AC voltage divider rule, which is widely used as a measurement of distance in various fields, is also applied to WPS. Voltage divider circuits may be constructed from reactive components just as easily as they may be constructed from fixed resistors, and just like resistors, a capacitive voltage divider is not affected by changes in the supply frequency even though they use capacitors which are reactive elements. WPS are used for measuring the displacement of the sensor with respect to a stretched wire, which serves as straightness reference. Horizontal and vertical movements can be measured with these sensors. The capacitive sensors have a working range of 10 mm for each axis with a resolution of minor 0.1 μm and a repeatability of about 1 μm with integrated electronics. The data acquisition rate can be up to 10 Hz. In addition to the sensor, the signal conditioning box and cables compose the calibrated sensor unit. It has two electrodes per axis facing each other, which makes a total of four electrodes. The sensor can be treated as a pair of double capacitors in a differential configuration. FOGALE sets and provides the Voltage AC frequency (4~10 KHz) of electronics. The first WPS have a frequency of 4 KHz and the other WPS have a frequency increased by 0.1 KHz steps.

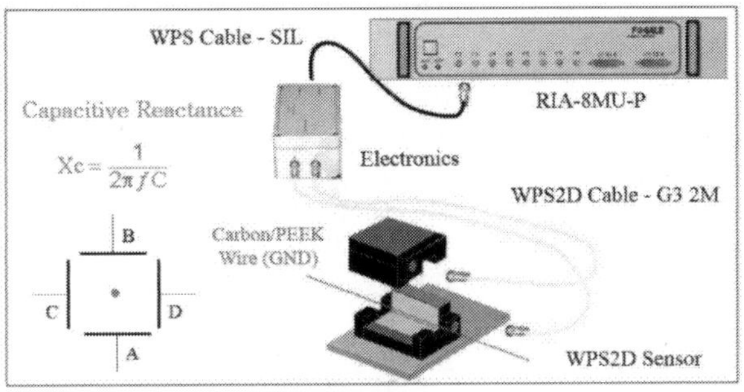

Figure 2. Working principle of WPS sensor

5. WPS reference: wire

The most important thing in the WPS system is a carbon wire which is used as a reference for measurement. As the post which is used to hold up a wire is installed on the ground, changes in the ground where the post is installed and the angle of the post should be measured regularly. Movement of the wire post means a change of reference. Carbon PEEK wire (diameter 0.4 mm, linear mass density 235 g/km) manufactured and marketed in Japan is used with 10~15 kg weights [5][10]. As shown in Figure 3, a precise wire stretcher PDL-WPS/15 provided by FOGALE was installed on the post and wire was connected to it. A wire protection duct is installed to prevent the wire from moving or shaking as a result of air generated from an air handling unit.

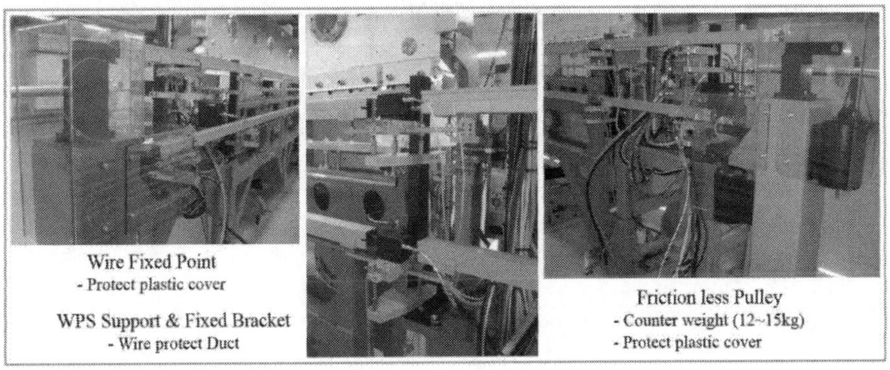

Figure 3. Wire stretched Post and WPS sensor support

To design the support for fixing the WPS sensor and fixed bracket, the wire sag should be calculated. The calculation formula provided by FOGALE Nanotech is shown in Equation (2) and Figure 4. In a measurement performed by PAL-XFEL, the wire sag was about 41 mm in relation to a 135 meter length and 12 kg load wire.

$$C(x) = \frac{\gamma g L}{2T}\left(1+\left(\frac{h}{L}\right)^2\right)\left(x-\frac{x^2}{L}\right) \approx C_{carbon}(x) = 0.03 \times L^2(m)/T(kg) \quad (2)$$

where

γ : linear mass density
g : gravity
L : horizontal distance between A and B
T : tension
h : vertical distance between A and B

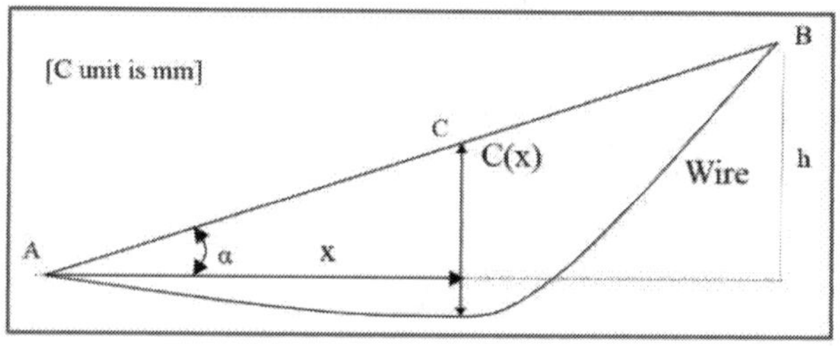

Figure 4. Wire Sag calculation

6. Installation and operation of WPS

A WPS was installed in the inter-section girder of the hard X-ray undulator section as shown in Figure 5. A total of 40 WPS2D-B-10x10 sensors were attached to five RIA-8MU-P racks connected by RS422 communication cables. The length of the WPS2D Cable-G3 connecting electronics and a sensor was reduced to 2 meters, minimizing the cable length. The measurement area of WPS is +/- 5 mm, but the linearly measurable area is +/- 1.5 mm. When installing WPS sensors, WPS fixed brackets should be adjusted to make the WPS measurement value less than +/- 0.5 mm.

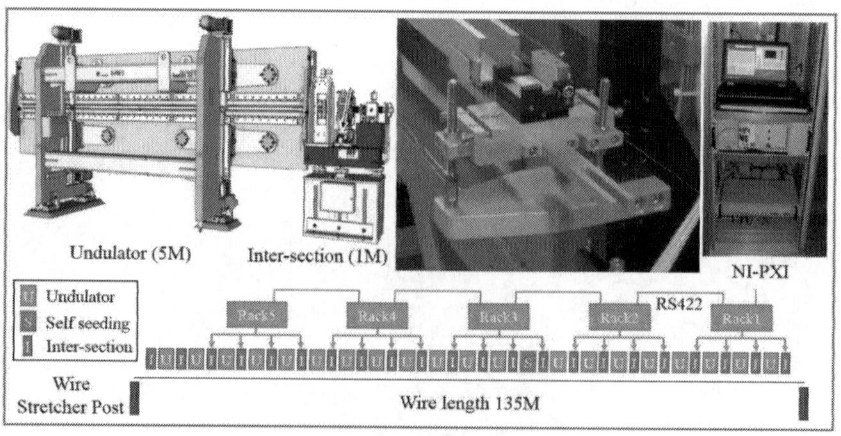

Figure 5. Location of WPS system installation and WPS fixed bracket

Data-acquisition PCs for WPS should be stably operated for a long time. For this, PXIe-8135 produced by National Instrument was selected. The program for measuring WPS was prepared with LabVIEW, which anybody can easily learn and program. The EPICS-type communication for storing WPS measurement values in the database used CA Lab, which was developed by the BEESY research center in Germany, after downloading it. The BEESY distributes CA Lab free of charge. Construction of data-acquisition PXI for WPS is shown in Figure 6 [11].

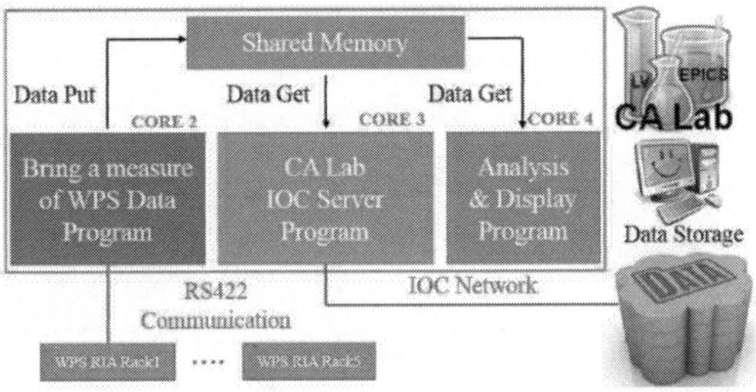

Figure 6. Construction map of a PXI server for WPS

7. Feedbacks for stabilization of PAL-XFEL beams

As various ways have been applied for the stabilization of beams regarding PAL-XFEL, electron beams and photon beams can move following a normal track even if the ground and building floor are changed. Figure 7 shows three methods applied to the PAL-XFEL system to minimize the orbit dispersion of electronic beams taking place due to various factors, including ground change. Beam orbit correction feedbacks are obtained using the strip-line BPM (S-BPM) and quadrupole electric magnetic for stabilization of electron beams in the linac section. In the undulator section, beam based alignment (BBA), a method to adjust the height of a magnet block that is an instrument of the 20 undulators using the screen-monitor and the cavity BPM (C-BPM), is applied. Regarding the beamline section, photon beams were always placed to be projected at the same place of the sample stage using a photon quadrant BPM (Q-BPM) and a mirror [12].

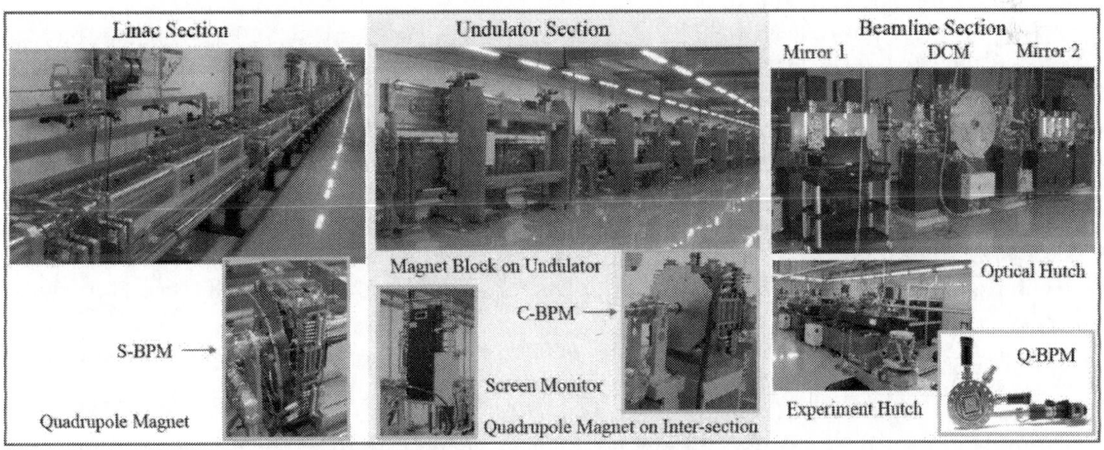

Figure 7. Way to stabilize FEL beams

The extent to which beams can be stabilized using beam stabilization feedbacks is limited. In consequence, the survey and alignment (S/A) team carries out an alignment procedure twice a year to place all equipment at the correct three-dimensional location coordinates (X, Y, and Z) by adjusting their girders and supports based on the results of the measurement of changes in the ground and buildings.

8. Conclusion

By installing and operating WPS, changes in the position of Undulator Hall according to changes in ground can be measured in real time as shown in Figure 8. After detecting undulators that deviated off the alignment by more than 50 μm, operators of PAL-XFEL re-aligns 20 undulators by BBA method, it takes two to three hours, contributes to optimal performance of the undulator.

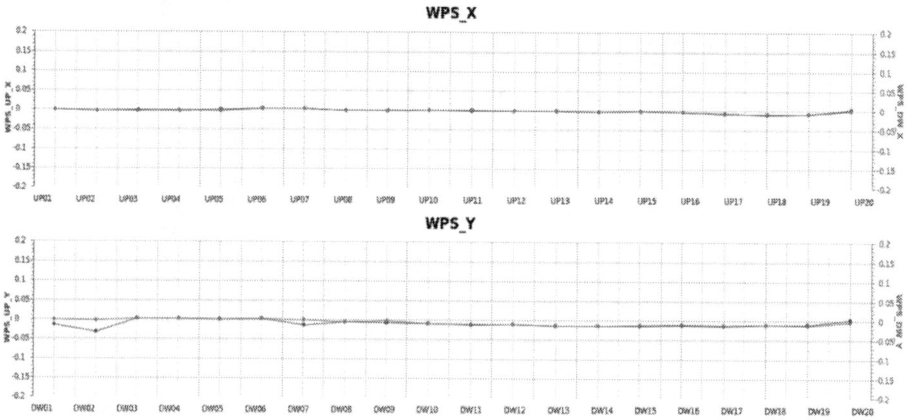

Figure 8. Real-time measurement of WPS

References

[1] Choi H *et al.* 2015 Introduction to BINP HLS to measure vertical changes on PLA_XFEL building and ground *Proc. IPAC2015* Richmond (U.S.A) pp. 994–7

[2] Katayama K *et al.* 1990 The Investigation of Vibration of LINAC at KEK *Proc. IWAA1990* Hamburg (Germany) pp. 333–42

[3] Schwarz W 1990 Wire measurements for the control of the FFTB-magnets *Proc. IWAA1990* Hamburg (Germany) pp. 469–80

[4] Herty A and Mainaud Durand A 2010 Intercomparison Tests with HLS and WPS *Proc. IWAA2010* Hamburg (Germany)

[5] Roux D 1995 Status of ESRF Alignment Facilities: WPS ready for an automatic 2D smoothing of a storage ring *Proc. IWAA1995* Tsukuba (Japan) pp. 159–68

[6] Ruland R E 1995 Some Alignment Considerations for the Next Linear Collider *Proc. IWAA1995* Tsukuba (Japan) pp. 451–62

[7] Coosemans W and Mainaud M 1997 Pre-Alignment of CLIC using The Double-Wie Method *Proc. IWAA1997* Argonne (Illinois, U.S.A)

[8] Gibilisco S 2011 *Teach yourself Electricity and Electronics* (5th edition McGraw-Hill) chapter 13 and chapter 14

[9] Winncy Y D 2014 *Resistive, Capacitive, Inductive, and Magnetic Sensor Technologies* (CRC Press)

[10] Bowden G *Stretched wire mechanics* SLAC-PUB-11465

[11] Choi H *et al.* 2015 LabVIEW EPICS program for measuring BINP HLS of PAL-XFEL *Proc. ICALEPCS2015* Melbourne (Australia) pp. 966–9

[12] Chao A W *et al.* 2013 *Handbook of Accelerator Physics and Engineering* (2nd edition World Scientific)

Acknowledgment
I would like to thank Mr. Jean-Christophe Clenet and Mme. Elodie Bollet of FOGALE Nanotech for their technical advice and for helping us with the process of installing the WPS system in PAL-XFEL.

Possibilities for transverse feedback phase adjustment by means of digital filters

G Kotzian
CERN, Geneva, Switzerland
E-mail: gerd.kotzian@cern.ch

Abstract. In transverse feedback systems a phase adjustment is generally required to convert a beam position signal from a pick-up into a momentum correction signal used by a transverse kicker. In this paper we outline several possibilities for phase adjustments using only single pick-ups or the vector combination of two pick-ups. Analytical expressions are given as a function of the fractional tune and the betatron phase advance between the pick-up location and the kicker. The shortest possible digital filter is formulated, including a notch for closed orbit suppression and a free parameter to adjust for betatron phase. We introduce a novel, fully parametrized digital filter with the feature to be insensitive to variations in fractional tune. Examples are given for the SPS transverse feedback system and compared with measurements.

1. Introduction

In larger synchrotrons pick-ups and kickers can be generally placed at locations such that the betatron phase advance between them allows for the signal to be directly applied for feedback without further phase adjustments. These locations are however not always available for installing a dedicated monitor, or the required phase advance might change in machines with cycle-dependent optics.

The transverse feedback system (TFB) of CERN's Super Proton Synchrotron (SPS) has its pick-ups installed in close proximity to the kickers. The following analysis therefore aims on identifying potential solutions for transverse feedback phase adjustments using short finite impulse response (FIR) digital filters and one or more pick-ups.

2. Simple Beam Model in z-domain

A first-order difference equation for the complex-valued sequence $x[n]$, which describes the linear portion of the particle motion through the magnetic guidance field — we shall consider the case where the amplitude decay is dominated by active damping only, see also [1] — is given by a recurrence formula,

$$x[n+1] = \alpha x[n], \quad n \geq 0; \alpha = re^{j\omega_0} . \tag{1}$$

Introducing the z-Transform, defined by the $\mathcal{Z}\{\cdot\}$-operator (bilateral transform) [2],

$$\mathcal{Z}\{x[n]\} = \sum_{n=-\infty}^{\infty} x[n]z^{-n} = X(z) , \tag{2}$$

and applying it to the recurrence formula yields after some reformulation

$$X(z) = \frac{1}{1 - \alpha z^{-1}} \,. \tag{3}$$

The rational function $X(z)$ has a single complex-valued pole at $z = \alpha$, which corresponds to the fact that the sequence $x[n]$ is complex-valued. Thereby, the convention used for Equation (1) attributes the real part of $x[n]$ as transverse displacement, $y = \Re\{x\}$, and the imaginary part as the trajectory's slope, $y' = \Im\{x\}$.

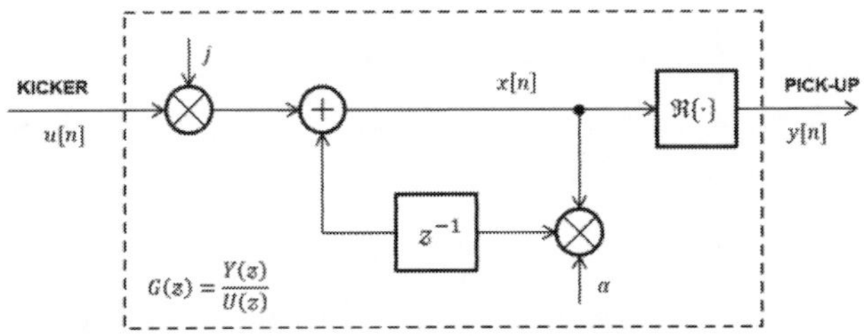

Figure 1. Simple beam model in z-domain.

The action of a kicker only changes a particle's slope. Consequently, in Figure 1 the real-valued input sequence $u[n]$ is first multiplied by the imaginary unit $j = \sqrt{-1}$ and subsequently added to the complex-valued sequence $x[n]$. Beam position monitors detect only transverse displacement, hence $y[n] = \Re\{x[n]\}$, as outlined in Figure 1.

The overall beam transfer function $G(z)$ can therefore be expressed as

$$G(z) = \frac{Y(z)}{U(z)} = j\frac{1}{2}\left(\frac{1}{1 - \alpha z^{-1}} - \frac{1}{1 - \alpha^* z^{-1}}\right) \,. \tag{4}$$

Here, α and α^* are the two conjugate complex poles of $G(z)$. It is worth noting that the second pole seen at the output, $Y(z) = G(z)X(z)$, is the result of the pick-up and its ability to only detect transverse position.

Equation (4) can be expanded and rewritten as

$$G(z) = \frac{r \sin\omega_0 \, z^{-1}}{1 - 2r\cos\omega_0 \, z^{-1} + r^2 \, z^{-2}} \,, \tag{5}$$

which makes it evident that due to the term z^{-1} in the numerator the output sequence is readily delayed by one sample, corresponding to the fact that a response to a transverse deflection at a particular longitudinal position is visible only after one turn.

In Equation (5) the parameter r determines whether the oscillation at frequency ω_0 is stable ($|r| < 1$), steady ($|r| = 1$), or unstable ($|r| > 1$).

Taking into account an arbitrary phase shift ϕ_{PU} between kicker and pick-up, i.e. by extending the pick-up output such that $y[n] = \Re\{x[n] \cdot e^{j\phi_{PU}}\}$ one obtains more generally as beam transfer function

$$G(z, \phi_{PU}) = \frac{r \sin(\omega_0 + \phi_{PU}) \, z^{-1} - r^2 \sin\phi_{PU} \, z^{-2}}{1 - 2r\cos\omega_0 \, z^{-1} + r^2 \, z^{-2}} \,. \tag{6}$$

As can be seen from Equation (6) the introduction of a betatron phase advance between pick-up and kicker has no impact on the position of the poles (the denominator is unaltered). In fact the required phase shift is established solely by an additional zero in the numerator, moving on the real axis of the z-plane as the phase angle changes.

IOP Conf. Series: Journal of Physics: Conf. Series **874** (2017) 012089 doi:10.1088/1742-6596/874/1/012089

3. Pick-up Placement

The ultimate goal could be stated as follows: determine the slope of a particle's trajectory at the position of the kicker, to be able to counteract oscillations by correcting its trajectory (i.e. the transverse momentum p_y) on a turn-by-turn basis.

Given that — at the time of writing — there is no technique known which allows to directly measure transverse momentum, it requires an indirect method to obtain the slope via position measurements.

By recalling that a pick-up measures the real part it becomes obvious that if the sequence is phase rotated, by $-90°$ or $-j$, then the corresponding slope is returned as position, y_p, seen by the pick-up:

$$y_p = \Re\{-jx\} = \Re\{-jy + y'\} = y' . \tag{7}$$

This procedure allows for two possible interpretations to realize the phase rotation: (a) Spatial phase shift, and (b) Temporal phase shift.

3.1. Spatial phase shift

Technically speaking, Equation (7) means nothing else than to physically place a pick-up at betatron phase advance $-90°$ with respect to the kicker.

However, if the installation of a pick-up at this phase advance is not realizable, then a phase rotation φ can be obtained by combining (mixing) the signals of two pick-ups accordingly, that is,

$$y_k = p_1 \cdot \cos\varphi + p_2 \cdot \sin\varphi . \tag{8}$$

Here, $p_1 = \Re\{x \cdot e^{j\psi}\}$ and $p_2 = \Re\{x \cdot e^{j(\psi+\pi/2)}\}$ are two pick-ups at an arbitrary longitudinal position, ψ, but with 90° phase advance between them. The outputs are then mixed together to obtain the sequence y_k for the kicker action.

For the combination of two pick-up signals that have a phase advance different than 90° a more general solution has been already formulated in [3], as *Pick-up Vector Sum.* As detailed therein, the two pick-up mixing coefficients follow from

$$b_{1,2} = -\frac{1}{2}\left(\frac{\cos(\Delta\phi_{Qkm})}{\cos(\Delta\phi/2)} \mp \frac{\sin(\Delta\phi_{Qkm})}{\sin(\Delta\phi/2)}\right) , \tag{9}$$

where $\Delta\phi = \phi_2 - \phi_1$ describes the phase advance between the two pick-ups, and with

$$\Delta\phi_{Qkm} = -3\pi Q_f + \phi_k - \frac{\phi_2 + \phi_1}{2} . \tag{10}$$

The fractional tune is denoted as Q_f, and ϕ_1, ϕ_2, and ϕ_k are the betatron phase advances at pick-up and kicker locations with respect to some fixed reference. Equation (10) is readily taking into account a one-turn delay and the phase shift of the notch filter.

3.2. Temporal phase shift

During one complete revolution a particle exhibits a precise number of oscillations, thus if the fractional tune Q_f is 0.25 then a phase rotation of 90° is achieved between consecutive turns. By reconsidering Equation (7) we can exploit this as

$$y_p[n] = \Re\{jx[n-1]\} = -y'[n-1] , \tag{11}$$

which states that the position y_p at turn n represents the negative slope, $-y'$, of the previous turn, $n-1$.

In reality a fractional tune close to the quarter integer resonance is usually not very practical. Therefore the same Pick-up Vector Sum algorithm as for the two pick-up case can be applied

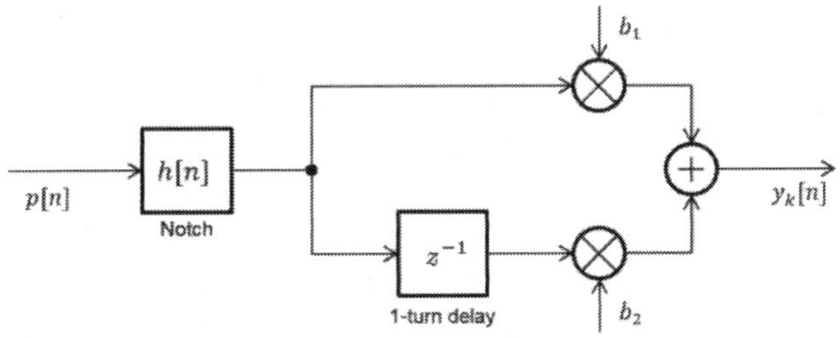

Figure 2. Block diagram of temporal phase shift.

for a single pick-up at subsequent turns, as outlined in Figure 2. In this case we define $\phi_1 = 0$ and $\phi_2 = 2\pi Q_f$, to be used with Equation (9) respectively Equation (10).

Figure 2 suggests that by considering a notch filter using $h[n] = [1, -1]$ the phase shift between $y_k[n]$ and $p[n]$ can be carried out by a short FIR, with the filter coefficients resulting from

$$y_k[n] = b_1 p[n] + (b_2 - b_1)p[n-1] - b_2 p[n-2] \,. \tag{12}$$

With only three taps and including a notch for DC-orbit suppression Equation (12) describes the shortest possible digital filter. The design has two parameters, the fractional tune Q_f and a free parameter ϕ_k which allows for direct phase adjustment.

Optionally, improved noise suppression can be achieved with zeros added at $z = \pm 1$, which can be absorbed into the notch filter, $h[n]$, including an extra phase term, $\Delta\vartheta = -\pi Q_f$ into Equation (10) for every additional tap the notch filter is extended.

By anticipating that the group delay — a measure for linearity of the phase — introduced by a filter lowers the stable phase margin of a closed loop system we made an attempt to compensate this effect by introducing an additional pair of conjugate complex zeros, $c_{1,2} = \zeta e^{\pm j\omega_d}$, at the desired tune frequency, $\omega_d = 2\pi Q_f$, which adds negative group delay as a function of the magnitude, ζ. The group delay as specified in [2] follows for a direct-form FIR transfer function as

$$\tau(\omega) = -\sum_{k=1}^{M} \frac{|c_k|^2 - \Re\{c_k e^{-j\omega}\}}{1 + |c_k|^2 - 2\Re\{c_k e^{-j\omega}\}} \,. \tag{13}$$

By taking into account the secondary phase term resulting from these zeros the overall group delay was compensated to $\tau(\omega) = 0$ at the design tune, $\omega = \omega_d$. Note that a system having zero group delay is able to transmit the signal's envelop without delay.

4. Results

The introduced beam model of transverse oscillations (Equation (6)) allows numerical evaluation of the analytic expressions for the described phase adjustment possibilities. We use root locus plots to study the evolution of the beam's closed-loop poles at selected frequencies, $\omega_0 = 2\pi Q_f$, and as a function of the feedback gain.

The three plots in Figure 3 detail a portion of the upper right quarter of the complex$z$-plane. As an example closed-loop pole trajectories for the SPS vertical plane are shown, assuming a fractional tune of $Q_f = 0.18$ and variations of up to ± 0.04 (blue, solid). Their origin is at the unit circle (black, bold solid) for zero loop gain (corresponding to the undamped open loop response). By increasing the feedback gain the damping time reduces gradually until the trajectories cross the circle at $|z| = 0.95$ (red, dashed) which is the design value of the SPS TFB corresponding to 20 turns. For as long as a trajectory points towards the origin of the z-plane (black, dash-dotted lines) it will follow the desired closed loop negative feedback of $-180°$.

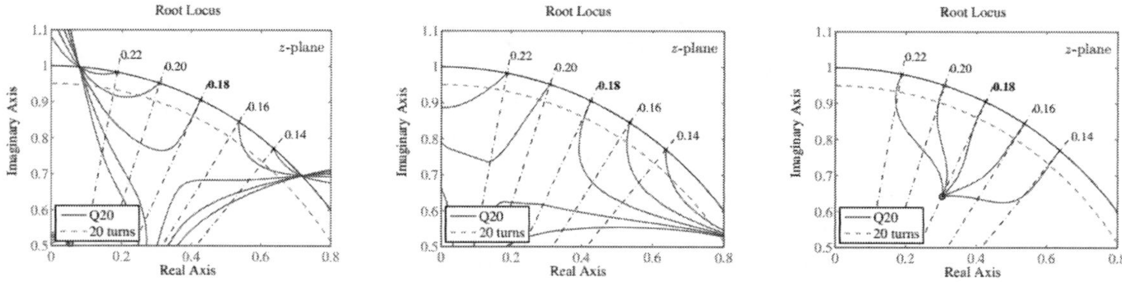

(a) Standard Hilbert phase shifter. (b) Short 3-tap FIR filter. (c) Group delay compensated.

Figure 3. Root locus plots, fractional tune $Q_f = 0.18 \pm 0.04$

Figure 3(a) shows the beam response for a standard FIR Hilbert phase filter, using seven taps for the case of the SPS TFB. As can be seen this filter is working perfectly fine at the target tune of 0.18, however, the feedback phase appears to be sensitive to tune variations, with the active feedback adversely pushing the tune further away the more its value deviates from the desired value. This effect is attributed to the constant group delay of the FIR filter causing the feedback phase to be optimum only for a single frequency and to roll off quickly for long filters.

As already anticipated, shortening the FIR filter length to only 3 taps, as provided by Equation (12), has a positive effect on tune variations, as can be seen by inspecting Figure 3(b), with the closed loop response remaining stable over a larger range of tune values.

Figure 3(c) shows that the loop stability is ultimately improved by carefully compensating the group delay of the digital filter. With its flat phase response around the design tune this filter is robust against changes of the fractional tune in the order of $\pm 22\%$. This filter was tested in the SPS TFB by changing the machine's fractional tune and performing beam transfer function measurements. The results listed in Table 1 confirm the theory of negative feedback over the same range of fractional tunes.

Table 1. Measured loop response vs. tune.

Tune	0.14	0.16	**0.18**	0.20	0.22
Phase	$-167°$	$-179°$	$-182°$	$-182°$	$-187°$

5. Summary and Conclusion

New FIR phase shift filters have been designed and analysed for the SPS TFB based on a simple beam model in z-domain and with root locus plots, to assess their closed loop performance with beam and in presence of active feedback. The obtained results were found to be in good agreement with measurements carried out in the SPS. Limitations on system gain [4], the performance in the presence of noise [5, 6], and the ability to reject disturbances are subject of further studies.

The shortest digital filter described has only 3-taps, including DC suppression by a notch filter (for other realisations of short phase shift filters see for example [7, 8, 9, 10, 11, 12]). Moreover, the tune sensitivity was further compensated with an additional pair of conjugate complex

IOP Conf. Series: Journal of Physics: Conf. Series **874** (2017) 012089 doi:10.1088/1742-6596/874/1/012089

zeros, effectively lowering the filter's group delay to zero at the fractional tune. Compensation of tune-dependent phase variations were reported also in [13].

The described digital filters are reasonable candidates for replacing the Hilbert phase shifter currently in use in the transverse feedback system of the SPS, mostly due to their favourable response over a larger range of tune values.

References

[1] Kotzian G, Höfle W and Vogel E 2008 *Proc. EPAC'08* THPP114

[2] Oppenheim A V, Schafer R W and Buck J R 1999 *Discrete-time Signal Processing - SE* (New Jersey: Prentice Hall)

[3] Baudrenghien P, Höfle W, Kotzian G and Rossi V 2008 *Proc. EPAC'08* THPC122

[4] Teytelmann D 2011 *Proc. PAC'11* WEODN1

[5] Hindi H *et al* 1994 *Proc. EPAC'94* pp. 1622–1624

[6] Yao C Y, Shang H R, Scaminaci A and Di Monte N 2013 *Proc. IPAC'13* WEPME056

[7] Tobiyama M and Kikutani E 1995 *KEK-Proc-96-6* pp. 470–474

[8] Kohaupt R D 1996 *Proc. EPAC'96* MOY02A

[9] Zhabitsky V M 1996 *Proc. EPAC'96* TUP019L

[10] Dehler M, Bulfone D, Kramert R, Lonza M, Pollet P and Schilcher T 2004 *Proc. EPAC'04* THPLT019

[11] Nakamura T, Daté S, Kobayashi K and Ohshima T 2004 *Proc. EPAC'04* THPLT068

[12] Nakamura T, Kobayashi K, Cheng W X, Honda T, Izawa M, Obina T and Tadano M 2006 *Proc. EPAC'06* THPCH02

[13] Bulfone D, Dehler M, Forchi V, Lonza M and Zambon L 2002 *Proc. EPAC'02* THPRI035

IOP Conf. Series: Journal of Physics: Conf. Series **874** (2017) 012090 doi:10.1088/1742-6596/874/1/012090

The design improvement of horizontal stripline kicker in TPS storage ring

P J Chou, C K Chan, C C Chang, K T Hsu, K H Hu, C K Kuan and I C Sheng

Synchrotron Radiation Research Center, Hsinchu 30076, Taiwan

E-mail: pjchou@nsrrc.org.tw

Abstract. We plan to replace the existing horizontal stripline kicker of the transverse feedback system with an improved design. Large reflected power was observed at the downstream port of stripline kicker driven by the feedback amplifier. A rapid surge of vacuum pressure was observed when we tested the high current operation in TPS storage ring in April 2016. A burned feedthrough of the horizontal stripline kicker was discovered during a maintenance shutdown. The improved design is targeted to reduce the reflection of driving power from feedback system and to reduce beam induced RF heating. This major modification of the design is described. The results of RF simulation performed with the electromagnetic code GdfidL are reported as well.

1. Introduction

The designs of stripline kickers of transverse feedbacks used in Taiwan Photon Source (TPS) were adapted from the SLS/ALBA design [1, 2]. Both the horizontal and vertical stripline kickers were installed and used in the commissioning of TPS phase-I insertion devices (IDs) since the 3rd quarter of 2015. The first hint of a problem for the horizontal stripline kicker was noticed when we operated in decay mode and stored 520 mA beam current in the storage ring on 12 Dec. 2015. An abrupt vacuum burst was recorded by an ion gauge nearby the horizontal kicker, with the peak pressure exceeding 6600 nPa. Similar events of vacuum burst recorded by the same ion gauge occurred several times when we tested the high beam current operation of storage ring in the first half of 2016. A hardware inspection was performed during the summer shutdown in 2016. We found the central connector of a feedthrough had brown spots and dielectric filling material of coaxial adapter was burned due to overheating. We also observed large reflected signals traveling towards the drive amplifier, which is connected to the downstream ports of kicker, at beam currents larger than 180 mA. This large reflected signal could cause damage to the drive amplifier. After reviewing all the data, it was decided to replace the horizontal kicker with an improved design which should achieve a better impedance matching and significant reduction of the beam induced RF heating. The detailed design and analysis of improved horizontal stripline kicker is reported and comparisons to the old design will be discussed.

2. Observation of problems

Abrupt vacuum bursts near the horizontal kicker were frequently observed during the high current beam test in TPS storage ring as shown in figure 1. The damaged coaxial feedthrough is shown in figure 2.

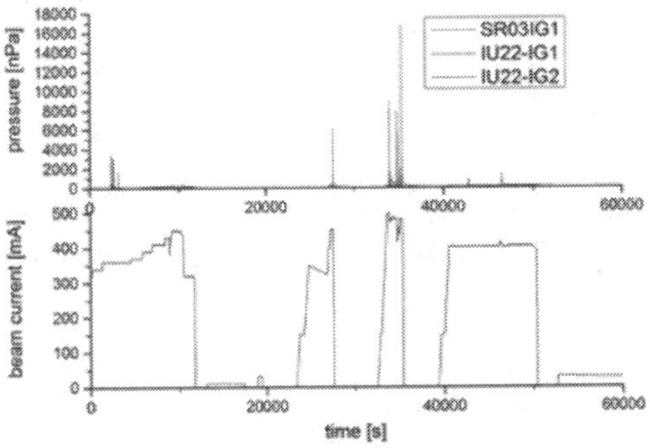

Figure 1. Vacuum pressure recorded by three ion gauges surrounding the horizontal kicker (top), beam current stored in storage ring (bottom).

Figure 2. Burned feedthrough connector (left), damaged coaxial adapter connected to the feedthrough (right).

3. Design improvement

When a particle beam pass through a stripline kicker, both the even and the odd mode are excited by the beam and the drive amplifier, respectively [3]. The revised kicker design should improve the impedance matching to minimize the beam induced signal going downstream toward the drive amplifier and reduce the reflection of driving power applied by drive amplifier. We also made efforts to design the stripline-to-coaxial transition of the kicker which contributed to a large reduction in the loss factor of kicker module.

3.1. Impedance matching

For an ideal stripline pickup/kicker, with the even mode impedance Z_{even} matched to the terminating line impedance (typically 50Ω), the beam induced signal only propagates to the upstream port and there is no beam induced signal at the downstream port [3]. From the viewpoint of an equivalent network, the stripline kicker is composed as a pair of coupled transmission lines with even and odd mode excited simultaneously. The necessary condition for the input impedance of a drive port to be matched to the terminating line impedance is given by the theory of coupled transmission lines [4, 5]:

$$Z_{even} Z_{odd} = Z_0^2 \quad (1)$$

where Z_0 is the terminating line impedance. One should make efforts to match both the impedance of even mode and the input impedance of drive port.

The stripline electrode arrangement has a racetrack cross-section profile to follow that of the beam pipe in the straight sections. The stripline electrodes are enclosed by a circular tube made out of stainless steel. The diameter of circular tube was adjusted to match the input impedance of drive port according to equation (1). To speed up the fabrication process for the scheduled shutdown in Jan. 2015, we settled with a circular tube of diameter 100 mm which could be readily procured off the shelves. The 7/8 EIA vacuum feedthroughs made by Kyocera were used for this design version. The horizontal stripline kicker after assembly is shown in figure 3. The simulated electric field pattern of the odd mode is shown in figure 4. The calculated impedance of both TEM modes of the stripline kicker is listed in Table 1.

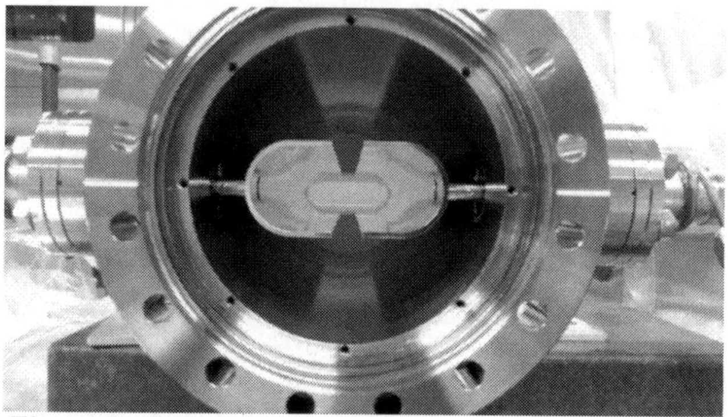

Figure 3. The horizontal stripline kicker after assembly.

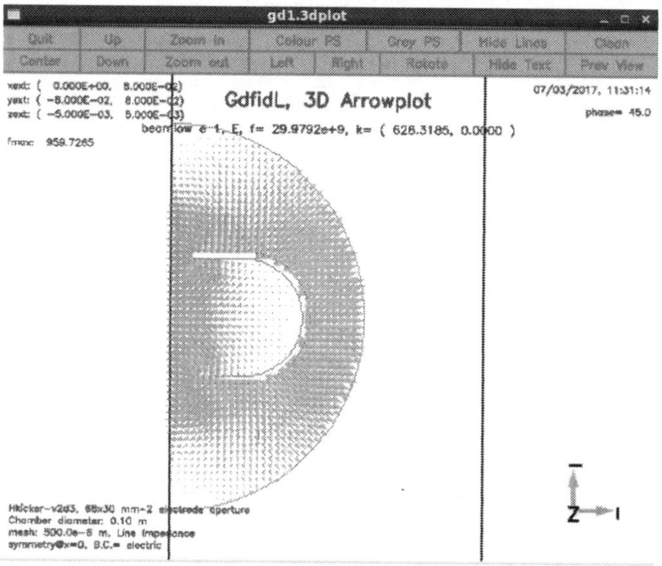

Figure 4. The simulated electric field pattern of the odd mode in the stripline kicker. Only half the structure was depicted.

Table 1. The impedance of TEM modes in the horizontal stripline kicker.

Mode	[Ω]
Z_{even}	66.58
Z_{odd}	34.70

The time domain reflectometry (TDR) was simulated by the 3-D electromagnetic code GdfidL [6] and compared with TDR measurements, see the result in figure 5.

IOP Conf. Series: Journal of Physics: Conf. Series **874** (2017) 012090 doi:10.1088/1742-6596/874/1/012090

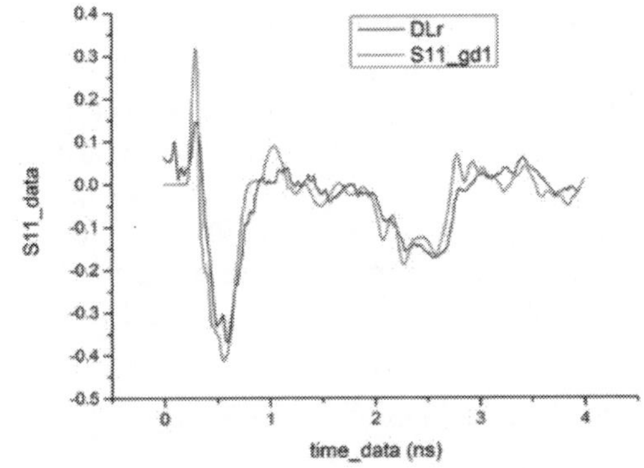

Figure 5. The comparison of simulated TDR (red) and measured data (black). Only one port was excited and the other three ports were terminated with 50 Ω.

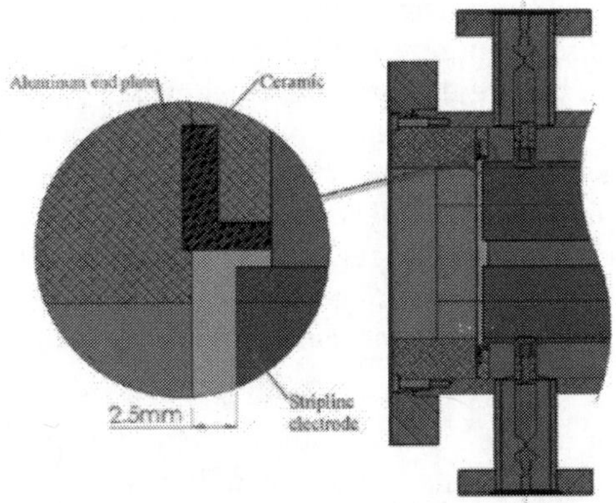

Figure 6. A zoom-in view of end plate with a shallow rectangular indentation used to shield the stripline electrodes.

3.2. Reduction of loss factor of the kicker module

The beam induced RF heating is related to the loss factor of kicker module. We made additional efforts to design the end plates of kicker chamber to effectively reduce the longitudinal and vertical gap as seen by the particle beam. Besides the special end plate design, we also increased the vertical gap of racetrack electrodes from 68x20 mm^2 to 68x30 mm^2. The combined effect of both approaches results in a significant reduction of loss factor for the kicker module. A zoom-in view of the end plate design is shown in figure 6.

A shallow rectangular indentation of depth 4.5 mm was machined on both end plates and the stripline electrodes were extended into the shallow indentation by 2 mm. The vertical gap between end plate and electrode is 2.5 mm. A ceramic washer of 1.5 mm was inserted between the electrodes and the end plates to assure electrical insulation.

The calculated loss factors for the old design and this improved design are listed in Table 2. There are 864 RF buckets in TPS storage ring. In routine operation the storage ring is typically filled with 600 bunches. The average RF heating power at 500 mA total current was estimated approximately by the following expression:

$$\langle P \rangle \approx \frac{\Delta E}{T_b} \tag{2}$$

where ΔE is the parasitic energy loss per bunch, and T_b is the RF period.

Table 2. The Calculated Loss Factor and Average Power of Beam Induced RF Heating at 500 mA Total Current (rms bunch length= 4.5 mm).

Model	Loss factor [V/pC]	RF heating [W]
Old design	1.044	1081.7
Improved design	0.310	321.2

The longitudinal impedance spectrum of the entire kicker module was calculated using a time-domain numerical simulation software (GdfidL), the result is shown in figure 7. The beam pipe and all four coaxial ports terminated to 50 Ω were all included in the GdfidL simulation. There appears a dominant resonant mode at 1.233 GHz. The Q-factor and shunt impedance of this resonant mode were calculated by GdfidL using the eigenmode solver with absorbing boundary conditions [6]. The growth time of longitudinal coupled bunch instabilities driven by this resonant mode can be calculated by the formulas of the theory of bunched beam instabilities [7]. The RF parameters of dominant resonant mode calculated with absorbing boundary conditions are list in Table 3. The longitudinal radiation damping time of TPS storage ring is 6.08 ms, a much lower value than the instability growth time driven by this resonant mode at 500 mA total current. Therefore, the trapped resonant modes in horizontal kicker will not cause longitudinal coupled bunch instabilities.

Table 3. RF parameters of dominant resonant mode calculated by GdfidL with absorbing boundary conditions.

f [GHz]	R/Q [Ω]	Q_{total}	Growth time [ms] at 500 mA total current
1.2327	5.98	412.53	93.1

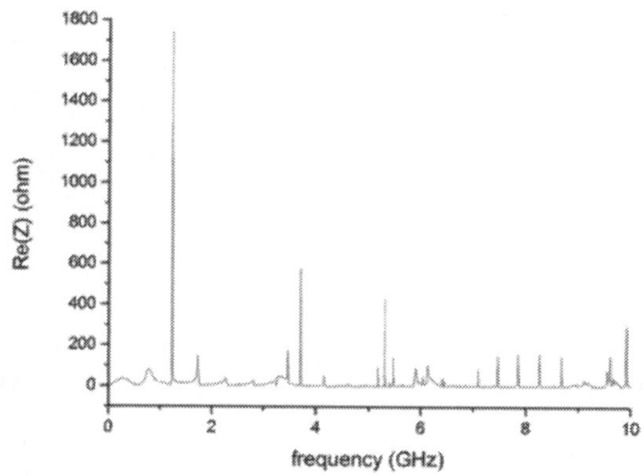

Figure 7. The longitudinal impedance simulated by GdfidL for the entire kicker module. The spectral resolution is 3.12 MHz.

4. Summary

The improved design has decreased the loss factor of horizontal kicker module by a factor of 3.37. Also the beam induced RF heating is reduced substantially. Abrupt vacuum bursts are no longer observed near the location of horizontal kicker. The trapped resonant modes in the horizontal kicker were analysed, but are of no concern for longitudinal coupled bunch instabilities.

5. References

[1] Dehler M *et al.* 2000 *Proc. European Particle Accelerator Conference (Vienna)* p 1894
[2] Iriso U, Günzel T F, and Pérez F 2009 *Proc. European Workshop on Beam Diagnostics and Instrumentation for Particle Accelerators (Basel)* p 86
[3] Goldberg D A and Lambertson G R 1992 Dynamic devices: a primer on pickups and kickers *AIP Conf. Proc. No.249*, ed M Month and M Dienes (New York: AIP) p 537
[4] Brown R G, Sharpe R A, Hughes W L, and Post R E 1973 Matrix representation of transmission-line circuits *Lines, Waves, and Antennas*, 2nd ed. (New York: Wiley) p 136
[5] Pozar D M 2012 Power dividers and directional couplers *Microwave Engineering*, 4th ed. (New York: Wiley) p 317

IOP Conf. Series: Journal of Physics: Conf. Series **874** (2017) 012090 doi:10.1088/1742-6596/874/1/012090

[6] Bruns W *The GdfidL Electromagnetic Field Simulator* (http://www.gdfidl.de)
[7] Suzuki T 2013 Effective impedance *Handbook of Accelerator Physics and Engineering*, 2nd ed. , ed A W Chao, K H Mess, M Tigner, and F Zimmermann (Singapore: World Scientific) p 262

A 166.6 MHz superconducting rf system for the HEPS storage ring

P Zhang, X Hao, T Huang, Z Li, H Lin, F Meng, Z Mi, Y Sun, G Wang, Q Wang and X Zhang

Institute of High Energy Physics, Chinese Academy of Sciences, Beijing, China

E-mail: zhangpei@ihep.ac.cn

Abstract. A superconducting 166.6 MHz quarter-wave beta=1 cavity was recently proposed for the High Energy Photon Source (HEPS), a 6 GeV kilometer-scale light source. Four 166.6 MHz cavities will be used for main acceleration in the newly planned on-axis beam injection scheme realized by a double-frequency RF system. The fundamental frequency, 166.6 MHz, was dictated by the fast injection kicker technology and the preference of using 499.8 MHz SC RF cavity as the third harmonic. Each 166.6 MHz cavity will be operated at 4.2 K providing 1.2 MV accelerating voltage and 150 kW of power to the electron beam. The input coupler will use single-window coaxial type graded up to 200 kW CW power. Each cavity will be equipped with a 200 kW solid-state amplifier and digital low-level RF system. This paper describes the 166.6 MHz RF system with a focus on the design and optimization of the RF cavity and its ancillaries, the LLRF system and the status of the solid-state amplifiers.

1. Introduction

High Energy Photon Source (HEPS) has been initiated by IHEP in Beijing. This is a 6 GeV kilometer-scale light source[1]. Its main beam parameters are listed in Table 1. Prior to its official construction, a test facility namely HEPS-TF has been approved in 2016 to R&D and prototype key technologies and components.

Table 1. Beam parameters of the HEPS storage ring.

Parameter	Value
Energy	6 GeV
Circumference	~1300 m
Current	200 mA
Main RF frequency	166.6 MHz
Harmonic RF frequency	499.8 MHz
Energy loss w/ IDs	2.5 MeV/turn
Total SR power	500 kW

A new injection scheme, namely on-axis accumulation, has been proposed recently to accommodate the relatively small beam dynamic aperture[2]. This however requires a double-frequency RF system, main and third harmonic, to make two stable buckets in order for the injected bunch to be merged into the circulating beam. It needs to be noted that the third harmonic RF system will be active rather than passive.

The recent development of DSRD (Drift Step Recovery Diode) switch makes the total width of the kicker system to be smaller than 10 ns. In addition, in order to use the mature technology of 500 MHz superconducting (SC) cavity as the harmonic cavity, the main RF frequency has thus been chosen to be 166.6 MHz, one third of 499.8 MHz.

This paper describes the newly designed 166.6 MHz RF system with a focus on the SRF cavity and its ancillaries, the LLRF system and the status of the solid-state amplifiers.

2. The RF System

Given the low RF frequency (166.6 MHz), both normal conducting (NC) and superconducting (SC) options have been considered for the main accelerator in the design phase. Their main parameters are listed in Table 2. Compared to the NC option, the SC case will save 23% of AC power due to much lower cavity surface losses, thus requires less RF amplifiers and reduced operation cost, but requires a one-time investment in a larger cryogenic system and its associated electricity bill. These make the SC case marginally cheaper if not considering cavity cost. In addition, the 500 MHz RF system will also use SC cavities due to higher RF voltage required and less longitudinal space needed.

Table 2. NC and SC options for the 166.6 MHz RF system.

Parameter	NC	SC
Frequency	166.6 MHz	
Total SR power	500 kW	
Total RF voltage	3.5 MV	
Number of cavities	6	3~4
Cavity voltage [MV]	0.6	1.2
R/Q [Ω]	400	130
Q_0	4×10^4	1×10^9
Total cavity wall loss [kW]	135	~0.04
Total RF power [kW]	635	501
Total AC power [MW]	1.3	1.0

The double-frequency RF system will make use of two types of SC cavities, both to be in-house designed. Each cavity will equip one power coupler graded above 200 kW CW power. All FPCs will also be in-house designed. A digital low-level RF (LLRF) system will be used for each cavity to obtain high control stability. The parameters of the double-frequency RF system are listed in Table 3. Focus of the R&D is on the 166.6 MHz SRF system.

3. The Cavities

The 166.6 MHz SRF cavity needs to be newly designed. Given that $\beta = 1$, the popular elliptical shape will make the cavity geometry excessively large ($\lambda/2$) thus unpractical for manufacturing. Therefore we have chosen the quarter-wave shape. A proof-of-principle (PoP) cavity has been designed with an optimized peak surface field and manageable transverse size (<400 mm). The RF model, EM field and main RF parameters are shown in Figure 1. Electron bunch with $\beta \sim 1$

Table 3. RF parameters for the HEPS storage ring.

Parameter	Main RF	Harmonic RF
Frequency [MHz]	166.6	499.8
Operating temperature [K]	4.2	
Number of cavities	4	2
Cavity voltage [MV]	1.2	1.7
R/Q [Ω]	~130	~95
E_p, B_p [MV/m, mT]	~30, ~50	~20, ~40
Cavity peak power [kW]	150	200
External Q	~4×10^4	~1×10^5
Number of RF stations	4	2
RF power/station [kW]	200	250
Control technology	digital LLRF	
Stability (amp., phase)	±0.3%, ±0.3°	

is being accelerated through the gap between left beam tube and the inner conductor cone where the electric field peaks. The magnetic field is on the other hand inhabits in the short end of the coaxial-like cavity. A 20% margin on peak surface fields was reserved.

(a)

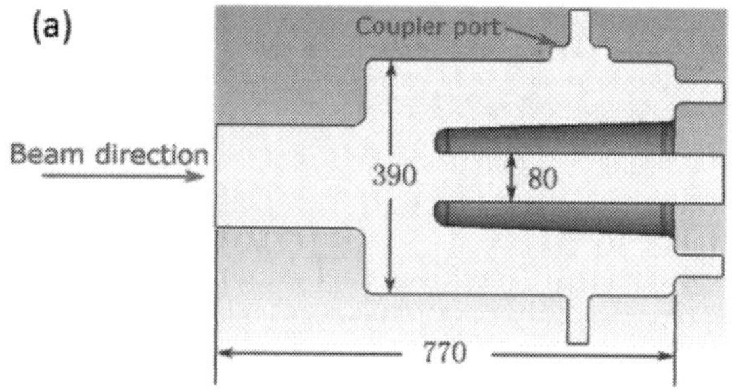

(b)

Parameter	Value
R/Q (=V_c^2/P_c)	136 Ω
G (=$R_s \cdot Q_0$)	54 Ω
V_c	1.5 MV
E_p at V_c=1.5MV	40 MV/m
B_p at V_c=1.5MV	65 mT

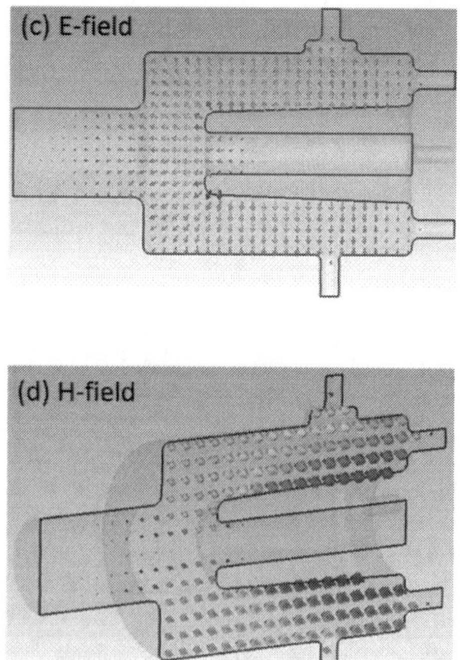

Figure 1. The 166.6 MHz PoP cavity.

The PoP cavity was subsequently fabricated in a domestic company, post processed in Ningxia and IHEP and finally vertical tested in IHEP as shown in Figure 2. The results largely exceed the design goal. The maximum peak field reached 86.5 MV/m and 132 mT respectively with a Q_0 value of $5{\times}10^8$.

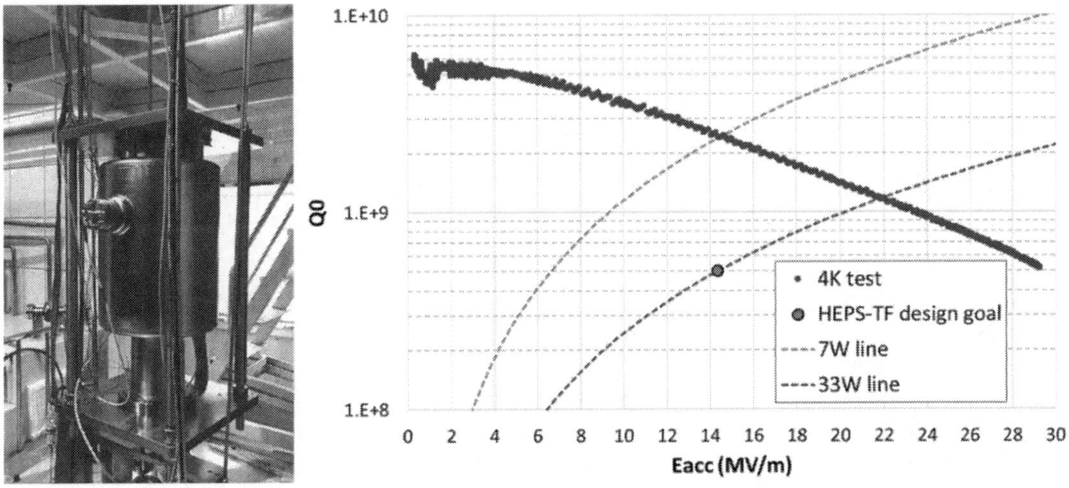

Figure 2. The PoP cavity vertical test.

The impedance threshold of higher order modes (HOM) restricted by multi-bunch instability is calculated[3]. They shall be less than 151 kΩ at 1 GHz for longitudinal direction, 1500 kΩ/m for x-direction and 3100 kΩ/m for y-direction. The R/Q of each HOM below 2 GHz is calculated and shown in Figure 3. Consequently most HOMs need to be damped to a loaded Q value of $10^3{\sim}10^4$. The first few HOMs have rather low frequencies (460 MHz for the 1st monopole HOM and 430 MHz for the 1st dipole mode) and highest R/Q values. Thus a hybrid HOM damping scheme has been conceived. An enlarged beam pipe connected to the cavity will take care of HOMs with higher frequencies subsequently absorbed by a ferrite damper, while a bandpass or high-pass coaxial type HOM coupler will extract the first few HOMs with lower frequencies. The overall HOM power per cavity is less than 500 W during nominal operation. The HOM damping structure is currently being designed. One need to note that there is no HOM damping structure on the PoP cavity.

The 3rd harmonic SC cavity will use 499.8 MHz single-cell elliptical structure, a matured technology popular for light sources. The main RF parameters are listed in Table 3. A RF model is shown in Figure 4.

4. The Power Coupler

The fundamental power coupler (FPC) of the 166.6 MHz SC cavity was newly designed as shown in Figure 5. The bandwidth reached 19 MHz. It uses single-window coaxial structure with electric coupling and is connected to a 9 3/16" coaxial rigid line after transition components. The location of the FPC on the cavity has been optimized to minimize the potential window damage by line-of-sight field emission electrons and to lower the risk of cavity contamination by squeezing the vacuum part into the cryomodule. Two prototype FPCs are currently being fabricated.

5. The LLRF System

A digital LLRF system was chosen for HEPS due to high flexibility. The accelerating field needs to be controlled to be better than ±0.3% in amplitude and ±0.3° in phase with respect to the reference. The system is planned to have digital IQ demodulation, fast data logging

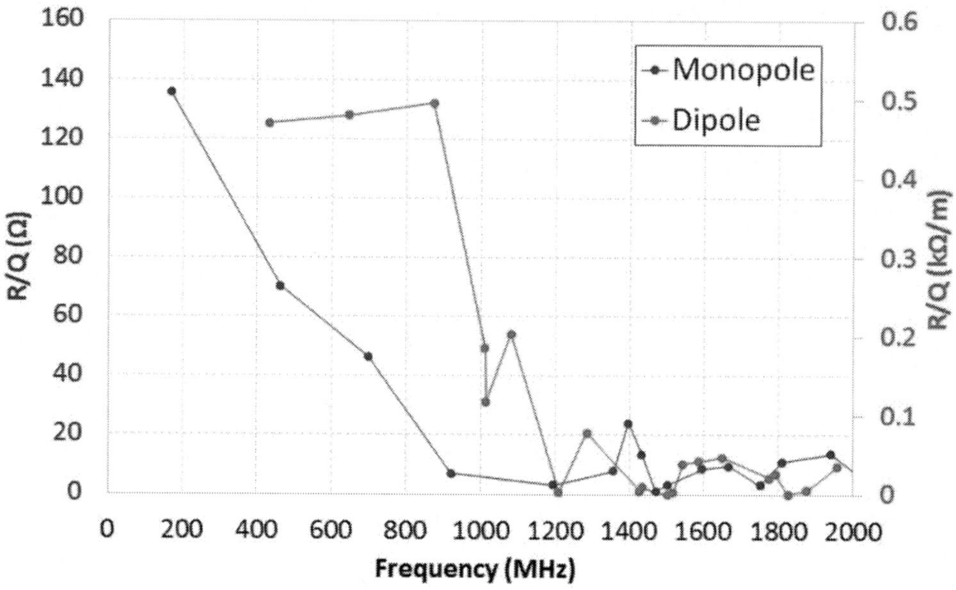

Figure 3. The R/Q value of HOMs calculated on the PoP cavity using CST MWS[4].

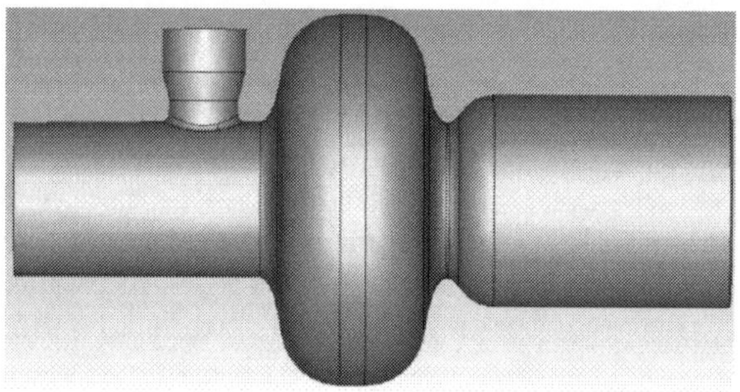

Figure 4. The 499.8 MHz 3rd harmonic cavity.

for post mortem analysis, automatic cavity startup and fast interlock. A first measurement of the prototype system itself was conducted recently. The peak-to-peak residual noise of 0.1% in amplitude and 0.1° in phase w.r.t. the reference were measured as shown in Figure 6.

6. The Amplifier

The required RF power for each 166.6 MHz cavity is less than 150 kW during operation. Reserving 25% of power margin, each RF station will have a 200 kW transmitter using solid-state amplifier (SSA) technology. The power combination scheme is shown in Figure 7. Each transmitter is realized by a cascaded power combination scheme. The smallest unit is a 2 kW power module using LDMOS transistor. Two 25 kW prototype amplifiers are currently being produced in a domestic company.

Recent tests of the 2 kW module showed a power efficiency of 70% at 2 kW output. The suppression of spurious signal, 2nd harmonic and 3rd harmonic was better than -85 dB, -40 dB and -57 dB respectively. The phase noise was measured to be better than -90 dBc/Hz at 1 kHz offset.

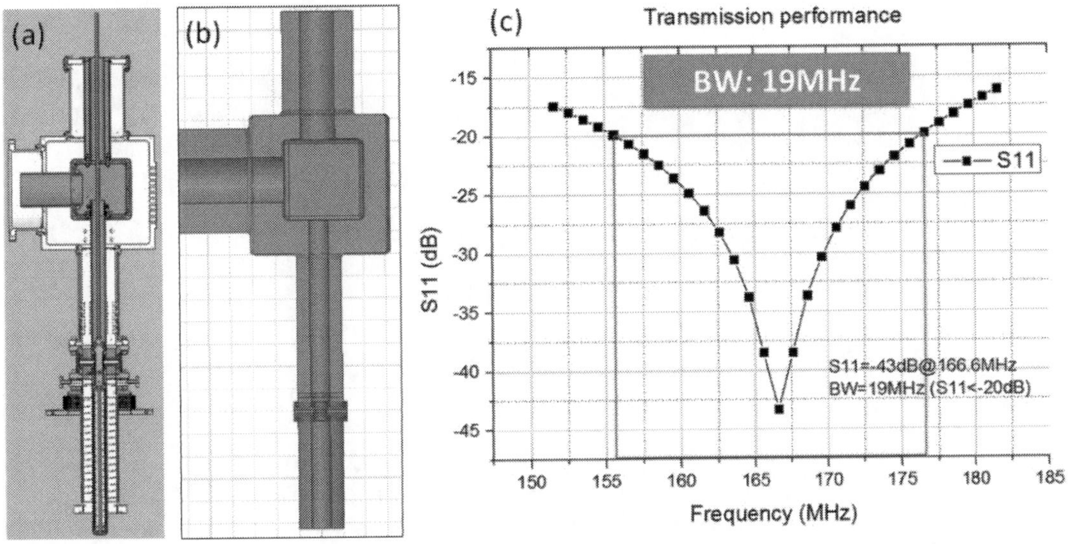

Figure 5. The 166.6 MHz fundamental power coupler.

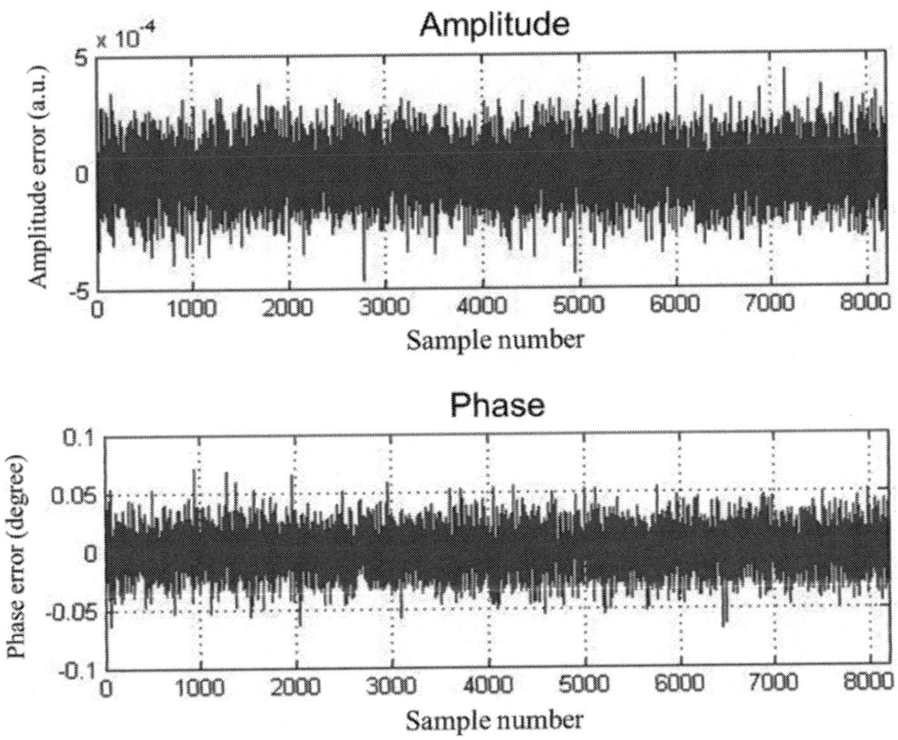

Figure 6. The measurement of the LLRF prototype system.

7. Final Remarks

A 166.6 MHz superconducting RF system has been proposed for HEPS project. A proof-of-principle 166.6 MHz SRF cavity has been designed, fabricated and vertical tested. The cavity

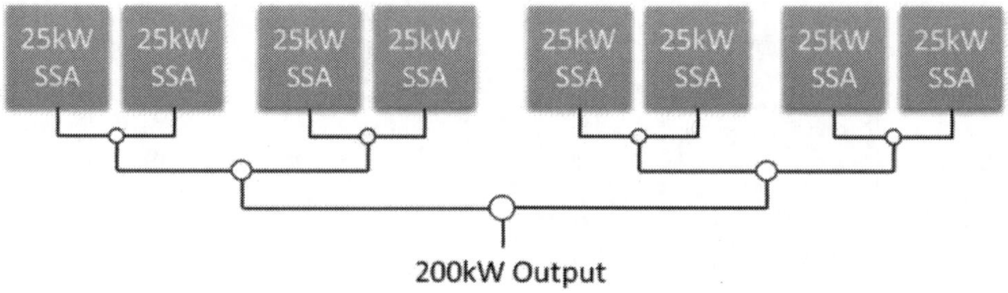

Figure 7. Power combination scheme of the 200 kW 166.6 MHz SSA.

performance largely exceeds the design goal. The cavity HOMs have been evaluated and a hybrid damping structure is being designed. The 200 kW CW power coupler has been designed and the prototypes are being fabricated. A digital LLRF prototype system is under production and initial self-tests presented lower than required residual errors. In summary, the 166.6 MHz SRF system is in the middle stage of its R&D phase and is currently proceeding well. Key components are being prototyped and no fundamental problems have been discovered to date.

8. Acknowledgements

We thank Q Qin and W Pan for their encouragement and constant support. We are grateful to R Rimmer, K Bane, G Burt, T Furuya, J Liu, K Liu, O Napoly, Q Wu and Z Zhao for enlightening discussions. This work has been supported by HEPS-TF project and partly by Pioneer Hundred Talents Program of Chinese Academy of Sciences.

References

[1] Xu G 2017 *Proc. IPAC2017* (Copenhagen Denmark) WEPAB052
[2] Duan Z, Chen J, Guo Y and Xu G 2017 *Proc. IPAC2017* (Copenhagen Denmark) TUPAB069
[3] Wang N, Duan Z, Li X, Shi H, Tian S and Xu G 2017 *Proc. IPAC2017* (Copenhagen Denmark) WEPIK078
[4] *CST Microwave Studio®* ver. 2017 CST AG (Darmstadt Germany)

IOP Conf. Series: Journal of Physics: Conf. Series **874** (2017) 012092 doi:10.1088/1742-6596/874/1/012092

The crab cavities cryomodule for SPS test

C Zanoni[1], A Amorim Carvalho[1], K Artoos[1], S Atieh[1], K Brodzinski[1], R Calaga[1], O Capatina[1], T Capelli[1], F Carra[1], L Dassa[1], T Dijoud[1], K Eiler[1], G Favre[1], P Freijedo Menendez[1], M Garlasché[1], L Giordanino[1], T Jones[2], S A E Langeslag[1], R Leuxe[1], H Mainaud-Durand[1], P Minginette[1], M Narduzzi[1], V Rude[1], M Sosin[1], J S Swieszek[1] and N Templeton[2]

[1] CERN, Geneva, Switzerland

[2] STFC Daresbury Laboratory, Keckwick Lane, Warrington, WA4 4AD, UK

E-mail: carlo.zanoni@cern.ch

Abstract. RF Crab Cavities are an essential part of the HL-LHC upgrade. Two concepts of such systems are being developed: the Double Quarter Wave (DQW) and the RF Dipole (RFD). A cryomodule with two DQW cavities is in advanced fabrication stage for the tests with protons in the SPS. The cavities must be operated at 2 K, without excessive heat loads, in a low magnetic environment and in compliance with CERN safety guidelines on pressure and vacuum systems. A large set of components, such as a thermal shield, a two layers magnetic shield, RF lines, helium tank and tuner are required for the successful and safe operation of the cavities. The sum of all these components with the cavities and their couplers forms the cryomodule. An overview of the design and fabrication strategy of this cryomodule is presented. The main components are described along with the present status of cavity fabrication and processing and cryomodule assembly. The lesson learned from the prototypes and first manufactured systems are also included.

1. Introduction

A cryomodule is by definition an apparatus for maintaining a very low temperature. This condition is a requirement for all superconducting systems in an accelerator. Large sections of the LHC will be modified for its High Luminosity major upgrade [1]. In this frame, novel RF cavities aimed at reducing the crossing angle at the interaction points are foreseen based on 2 different designs, one for vertical (Double Quarter Wave, DQW) and one for horizontal interaction (RF Dipole, RFD). In general, this type of cavity is also called *crab cavity* for the drift motion they impose to the bunches of the beam. 16 crab cavities will be installed, 2 per each beam, on each side of both the ATLAS and CMS experiments.

Successful operation of the SRF crab cavities for HL-LHC requires them not only to stay at a temperature of 2 K, but also the magnetic field to be below 1 μT. Keeping a system at cryogenic temperature is a demanding task in terms of energy, therefore the heat transmitted from the outer environment must be minimized by careful design. This requirement must find a compromise with the need of structural strength, to comply with the mechanical loads and alignment stability.

In this paper, we overview the main systems of the crab cavities cryomodule, previously seen in [2]. The main focus is the cryomodule for the DQW cavity type Figure 1, which is more

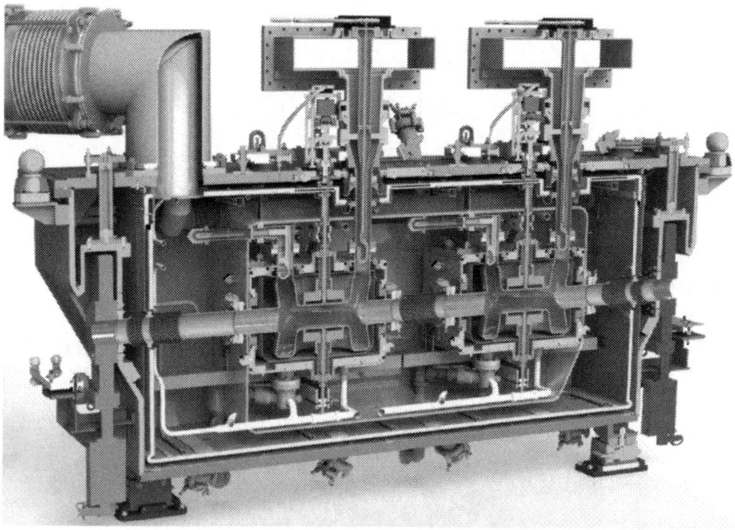

Figure 1. Open view of the cryomodule for the DQW cavity.

advanced in view of tests with high energy protons in SPS.

2. The Dressed Crab Cavity

At the core of the cryomodule there are two cavities. Each of them is enclosed by a set of systems including helium tank, High Order Modes suppressors (HOMS), pickup field antenna, Fundamental Power Coupler (FPC) and cold magnetic shield. The assembly of these systems, with the cavity, is called *dressed cavity* and is shown in Figure 2 [3].

Each crab cavity has a complex geometry (i.e. not axisymmetric) made of 4 mm thick niobium sheets. Bulk niobium was chosen over coated copper, that would be challenging for such geometries. Two cavities have been fabricated at CERN for the SPS tests, Figure 3. The shape is trim tuned before the final electron beam weld. Each cavity is then subjected to a chemical etching of 150-200 μm, heat treated at 650 °C, chemically etched for further 20 μm, rinsed with high pressure water and then conditioned and tested at 2 K.

The *helium tank* is built using thick (25-30 mm) titanium plates. In order to mitigate the deformation induced by the welding, we follow a design approach in which the structural resistance is assured by a large set of bolts at the plates' edges. Thin welds are then performed to provide leak tightness.

This procedure, along with the mechanical strength and leak tightness, was tested on a prototype, i.e. a tank representative of the final design, but without any cavity inside. The tests were successful and showed a good agreement with the simulated deformations. The cavity deformation during tank TIG welding requires careful monitoring, but the measured values were sufficiently low both structurally and in terms of frequency, Figure 4.

The RF performance is not determined by the cavity alone, but also by the HOMS, including the pickup field antenna shown in Figure 5.

The *HOMS* design and fabrication is described in [4]. The mechanical performance is acceptable and with a high safety factor. The fabrication is indeed more critical, in view of the tight tolerances. Five-axes milling machining and careful preparation of electron beam welding by means of tests were carried out.

The *pickup antenna* (that also acts as a HOMS) presents similar challenges, with the difference that there is no liquid helium and therefore the mechanical assessment is straightforward. To assess the thermal performance, the coupling between RF domain and thermal map is performed by means of a lookup table. Even in the worst case, however, RF is not a major source of heat

Figure 2. Open view of the dressed DQW cavity.

and the peak temperature of about 5 K is determined by the static losses through the coaxial cable. The overall picture of the heat load on the cold mass is in [5].

Fabrication tolerances and uncertainty during processing as well as environmental effects during operation impose the use of a system capable of adjusting the cavity RF frequency. This device is called *tuning system* and is detailed in [6] and [7].

3. The Magnetic Shields

The surface resistance of a superconducting cavity depends also on the external magnetic field. In the case of the crab cavities a value below 1 μT is specified, against an environment of $\approx$ 60 μT. To guarantee compliance with such a limit, a two-layers approach is followed. Both shields have been produced and the measurements show an acceptable field reduction factor.

A so called *cold magnetic shield* is installed inside each helium tank. This shield is made of Cryophy® sheets, which provides good permeability also at 2 K [8].

The external layer, or *warm magnetic shield* is located just inside the vacuum vessel and therefore operates at 300 K. The sheets are made of mu-metal. The main challenge is to limit the flux leaks due to the holes and gaps between plates due to several large interfaces. To mitigate this effect, a 100 mm overlap between top plate and bottom assembly is foreseen. Patterns of deformable small plates are mounted at the interfaces that cannot be screwed due

Figure 3. DQW cavity ready for dressing.

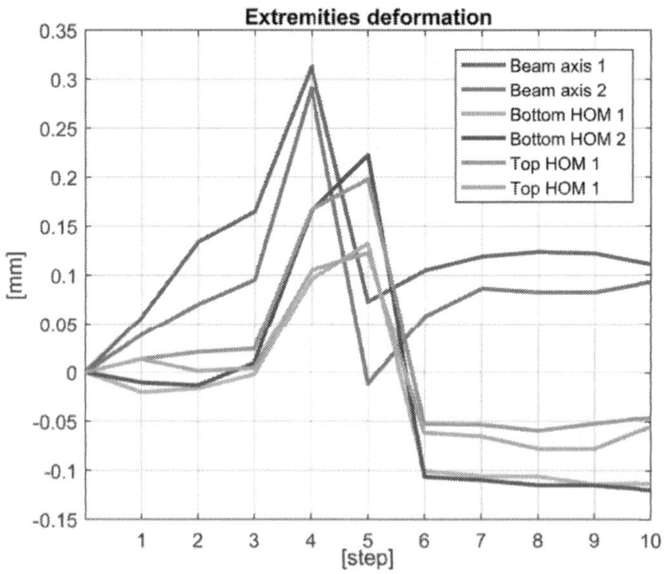

Figure 4. Deformation of tank prototype at the position of the cavity ports. Step 4 is the weld of the covers of the bolts on the plates edges. Step 5-6 are the covers of the bolts around the ports. The subsequent tests do not induce deformation.

to lack of access during assembling. They provide material continuity in those locations. The overall expected magnetic field is shown in Figure 6. We considered the simulations in this domain non-conservative and we therefore aimed at a high margin.

4. The Thermal Shield

The *thermal shield* (see Figure 7) is the layer that prevents the cold mass to receive direct radiation from 300 K and is located inside the magnetic shield. The thermalisations, which mitigate the heat on the cold mass [5], are also attached to the shield that is cooled with helium gas at a temperature between 50 K and 70 K.

Figure 5. Pickup antenna for one of the DQW crab cavities.

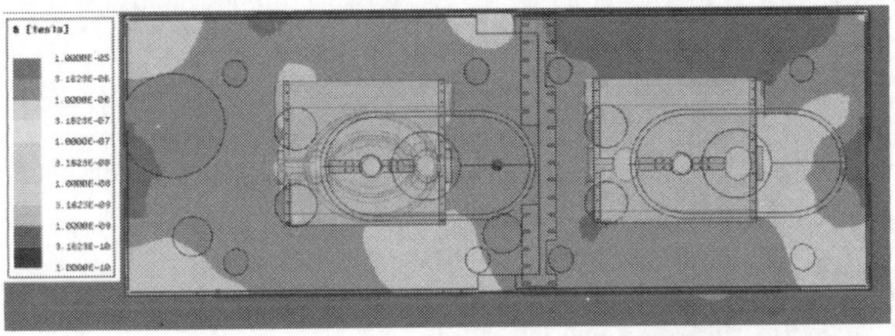

Figure 6. Magnitude of magnetic field on the mid-plane of the beam pipes.

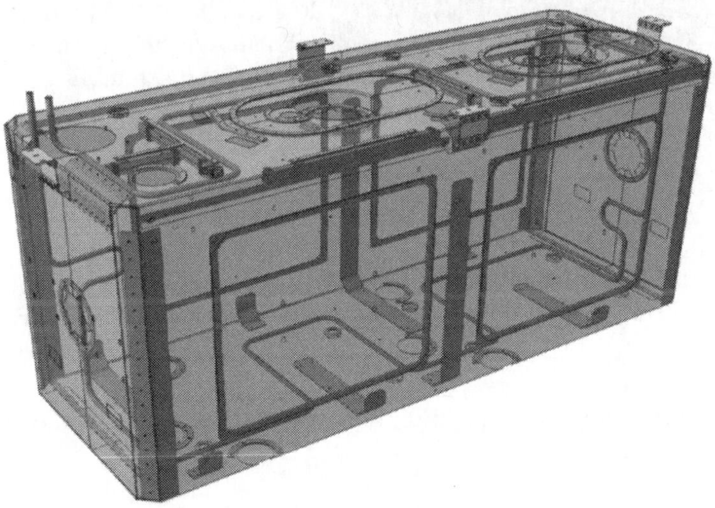

Figure 7. View of the thermal shield.

The design is based on the HIE-ISOLDE experience [9] and employs copper sheets with a copper pipe brazed on them. The system is suspended by the vacuum vessel top plate by means of titanium blades, that allow thermal contraction. A multi-layer insulation is foreseen both between thermal and warm magnetic shield and around each dressed cavity.

5. The Vacuum Vessel and the Alignment Systems

The *vacuum vessel* is the outer layer of the cryomodule and is the system that contains and supports most of the devices and encloses the insulation vacuum.

In order to comply with the tight magnetic and mechanical requirements, the vessel is made from stainless steel 316LN. The thickness of the plates and the configuration of the stiffeners are optimized in order to mitigate the deformation of the top plate, when the vessel is under vacuum. This is aimed at keeping a good alignment of the cavities and accuracy of position monitoring system.

The cavities position is adjusted by means of a plate mounted isostatically at 3 points. Each dressed cavity is rigidly connected to this plate and the vacuum volume is confined by some bellows. A combinations of optical sensor systems [10, 11] provides redundant monitoring of the cavities position and orientation for the SPS installation and tests.

Figure 8 shows the philosophy followed for cryomodule assembling, that foresees for all the systems to be mounted below the top plate and then to lower the sub-assembly in the bottom part of the vacuum vessel.

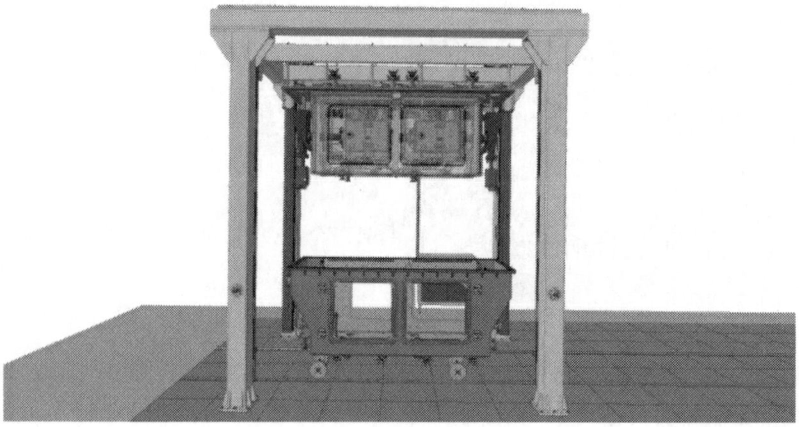

Figure 8. View of the cryomodule during the last steps of the assembling.

6. Conclusion

HL-LHC will include 16 crab cavities to compensate for the crossing angle and contribute to the increase of luminosity. Each pair of cavities will be enclosed in a common cryomodule. A test cryomodule for installation in SPS in 2018 is being prepared and will allow for compact deflecting cavities to be validated with the highest energy protons for the first time in the world.

Acknowledgments

The research and activities for developing and manufacturing the Crab Cavities Cryomodule are supported by the HL-LHC project.

References

[1] Apollinari G, Bjar Alonso I, Brning O, Lamont M and Rossi L 2015 *High-Luminosity Large Hadron Collider (HL-LHC): Preliminary Design Report* CERN Yellow Reports: Monographs (Geneva: CERN) URL `http://cds.cern.ch/record/2116337`

[2] Carra F *et al.* 2015 Crab Cavity and Cryomodule Development for HL-LHC *Proceedings, 17th International Conference on RF Superconductivity (SRF2015): Whistler, Canada, September 13-18, 2015* p FRBA02 URL `http://inspirehep.net/record/1482264/files/frba02.pdf`

[3] Zanoni C *et al.* 2015 Design of Dressed Crab Cavities for the HL-LHC Upgrade *Proceedings, 17th International Conference on RF Superconductivity (SRF2015): Whistler, Canada, September 13-18, 2015* p THPB070 URL `http://inspirehep.net/record/1482224/files/thpb070.pdf`

[4] Zanoni C *et al.* 2015 Engineering design and prototype fabrication of HOM couplers for HL-LHC crab cavities *Proceedings, 17th International Conference on RF Superconductivity (SRF2015): Whistler, Canada, September 13-18, 2015* p THPB069 URL `https://inspirehep.net/record/1482223/files/thpb069.pdf`

[5] Carra F, Jorgen A, Rama C, Ofelia C, Teddy C, Carlo Z and Silvia V A 2017 Assessment of Thermal Loads in the CERN SPS Crab Cavities Cryomodule *Proceedings, 8th International Particle Accelerators Conference (IPAC2017): Copenhagen, Denmark, May 14-19, 2017* p TUPVA008

[6] Artoos K *et al.* 2015 Development of SRF Cavity Tuners for CERN *Proceedings, 17th International Conference on RF Superconductivity (SRF2015): Whistler, Canada, September 13-18, 2015* p THPB060 URL http://inspirehep.net/record/1482215/files/thpb060.pdf

[7] Verd-Andrs S, Artoos K, Ben-Zvi I, Calaga R, Capatina O, Leuxe R, Skaritka J, Wu Q, Xiao B and Zanoni C 2016 Frequency Tuning for a DQW Crab Cavity *Proceedings, 7th International Particle Accelerator Conference (IPAC 2016): Busan, Korea, May 8-13, 2016* p WEPMR038 URL http://inspirehep.net/record/1470198/files/wepmr038.pdf

[8] Masuzawa M, Terashima A, Tsuchiya K and Ueki R 2017 *Superconductor Science and Technology* **30** 034009 URL http://stacks.iop.org/0953-2048/30/i=3/a=034009

[9] Valdarno L, Delruelle N, Leclercq Y, Parma V, Vandoni G and Williams L 2015 *IOP Conf. Ser.* **101** 012045 URL https://cds.cern.ch/record/2145935

[10] Sosin M, Dijoud T, Mainaud Durand H and Rude V 2016 Position Monitoring System for HL-LHC Crab Cavities *Proceedings, 7th International Particle Accelerator Conference (IPAC 2016): Busan, Korea, May 8-13, 2016* p WEPOR018 URL http://inspirehep.net/record/1470302/files/wepor018.pdf

[11] Hashemi K S and Bensinger J 2000 The BCAM Camera Tech. rep. CERN

A compact 10 kW solid-state RF power amplifier at 352 MHz

Dragos Dancila, Long Hoang Duc, Magnus Jobs, Måns Holmberg, Adam Hjort, Anders Rydberg and Roger Ruber

FREIA, Uppsala University, SE-751 21 Uppsala, Sweden

E-mail: dragos.dancila@angstrom.uu.se

Abstract. A compact 10 kW RF power amplifier at 352 MHz was developed at FREIA for the European Spallation Source, ESS. The specifications of ESS for the conception of amplifiers are related to its pulsed operation: 3.5 ms pulse length and a duty cycle of 5%. The realized amplifier is composed of eight kilowatt level modules, combined using a planar Gysel 8-way combiner. The combiner has a low insertion loss of only 0.2 dB, measured at 10 kW peak power. Each module is built around a commercially available LDMOS transistor in a single-ended architecture. During the final tests, a total output peak power of 10.5 kW was measured.

1. Introduction

Solid-state RF high power amplifiers are increasingly used in a large variety of systems including particle accelerators, such as cyclotrons and LINACs and in an increasingly larger variety of applications: radionuclide production, particle therapy for cancer treatment, synchrotron light sources for scientific studies, etc. A pioneering work in using SSA for synchrotron light sources started at SOLEIL [1], with 330 W modules using baluns. The increase in power per transistor allows today more effective designs. The operation in a LINAC using superconducting cavities reduces drastically the bandwidth of operation, which allows narrowband architectures to be implemented and thus simpler designs to be used [2]. In this context, a single-ended architecture has been demonstrated to ensure a high performance for power amplifier at the kilowatt-level [3]. The RF design implements matching networks in a single-ended architecture and avoids the use of baluns. The 50 Ohm input and output impedances are transformed to match the transistors required impedances using wide stepped microstriplines and are fine-tuned using surface mounted capacitors. Note that these impedances are roughly four times smaller than for the push-pull amplifier. DC feed at the drain is realised by connecting the supply voltage via a thick wire air coil and it uses a resistor for biasing at the gate. The kilowatt level module is built around a commercially available LDMOS transistor, the BLF188XR. The measured peak power per module is about 1.25 kW, while operating at 50 V drain voltage with a drain efficiency of 71% [3]. Eight of such modules were combined using a planar 8-way Gysel combiner to achieve above 10 kW output power. A very compact 10 kW amplifier at 352 MHz was realised, see Figure 1. The 10 kW system consists of three stages, first a commercially available amplifier provides around 100 W with a gain of 50 dB. The second stage is composed of one kilowatt level module with a gain of about 20 dB, providing around 15 W input power to each of the eight modules. The third stage is composed of eight kilowatt level modules using a Gysel combiner.

IOP Conf. Series: Journal of Physics: Conf. Series **874** (2017) 012093 doi:10.1088/1742-6596/874/1/012093

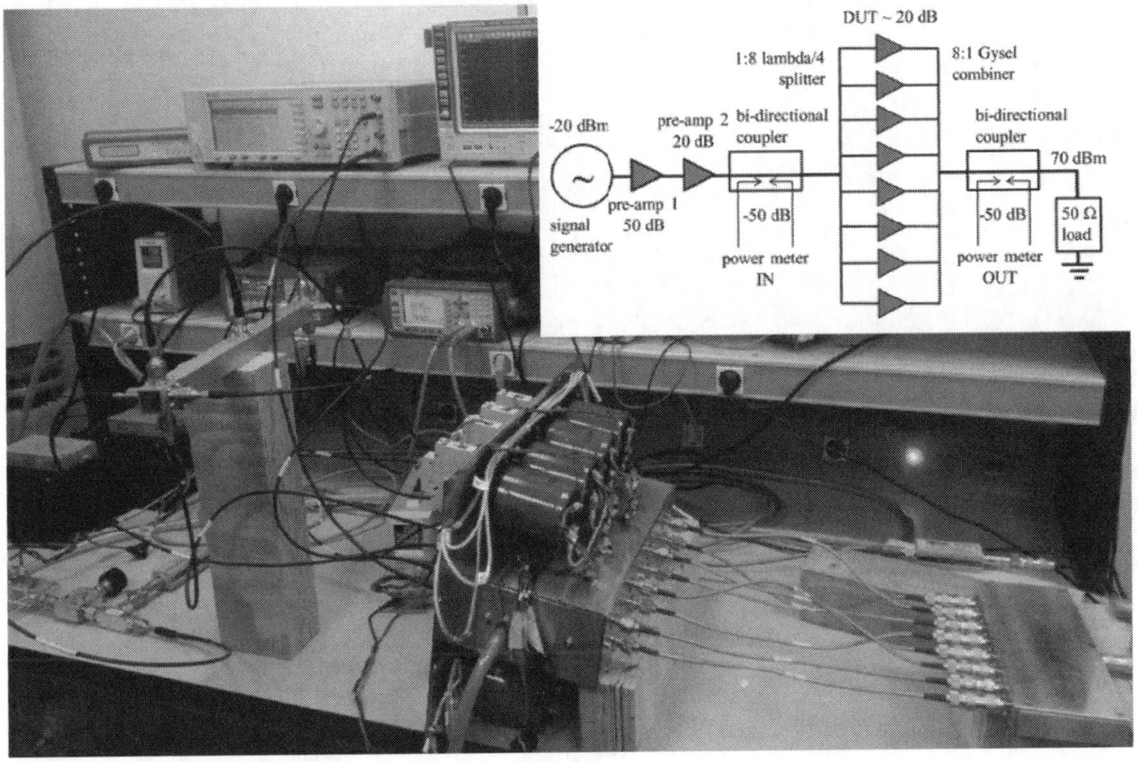

Figure 1. 10 Kilowatt power amplifier at 352 MHz.

2. Kilowatt level module

Each module is built around an LDMOS transistor, the BLF188XR. Input and output matching networks implement a single-ended architecture and are realised on Rogers RO3003 substrate (ε_r = 3; tanδ = 0.001; copper thickness, Cu = 35 m and substrate height, h = 0.76 mm), see Figure 2. The simulated output power with ADS is about 1.3 kW at 1 dB compression for a gain of 20.5 dB with 70% power added efficiency, PAE. This is in good agreement with the measurements in ESS pulsed mode (i.e. duty cycle 5% and pulse duration 3.5 ms) the measured RF output power per module is 1.25 kW with a gain of 19 dB at 1 dB gain compression and a drain efficiency of 71%.

Figure 2. Kilowatt-level power amplifier in a single-ended architecture at 352 MHz.

Assembled in an aluminium casing, the temperature rises for only few degrees, to about 30° C (20° C room temperature) while it is close to 150°C in CW. The heat sink is provided with a water cooling pipe (diameter 10 mm) and a water debit of about 8 l/min is maintained during the whole operation. During CW measurements, the gain compresses by 2.5 dB and the efficiency drops to 61% at 1 kW output power. With improved cooling, the amplifier could perform better in CW operation mode.

The amplifier is operated in Class B, with a quiescent drain current, IDq = 0.1 A and a drain voltage, VDS = 50 V. At nominal output power, i.e. 1.25 kW, the following harmonics levels are monitored: 2nd at -25 dBc, 3rd at -35 dBc, 4th at -44 dBc and the 5th $<$ -60 dBc. Such low harmonic levels are comparable to the harmonic levels of a push-pull amplifier, using e.g. balun transformers. More information could be found in [3].

The eight amplifier modules were individually fine tuned to achieve a nominal power of 1.25 kW in pulsed mode (5% duty cycle and 3.5 ms pulse). The fine tuning resulted in different values and positions of the output matching capacitors. The phase imbalance among the eight amplifiers is small, it is smaller than 5°, see Figure 3. The gain imbalance is also small, it is smaller than 0.5dB at the nominal output power of 1.25 kW, see Figure 4. The low phase and amplitude imbalances allow reducing the combination losses, as we will see next.

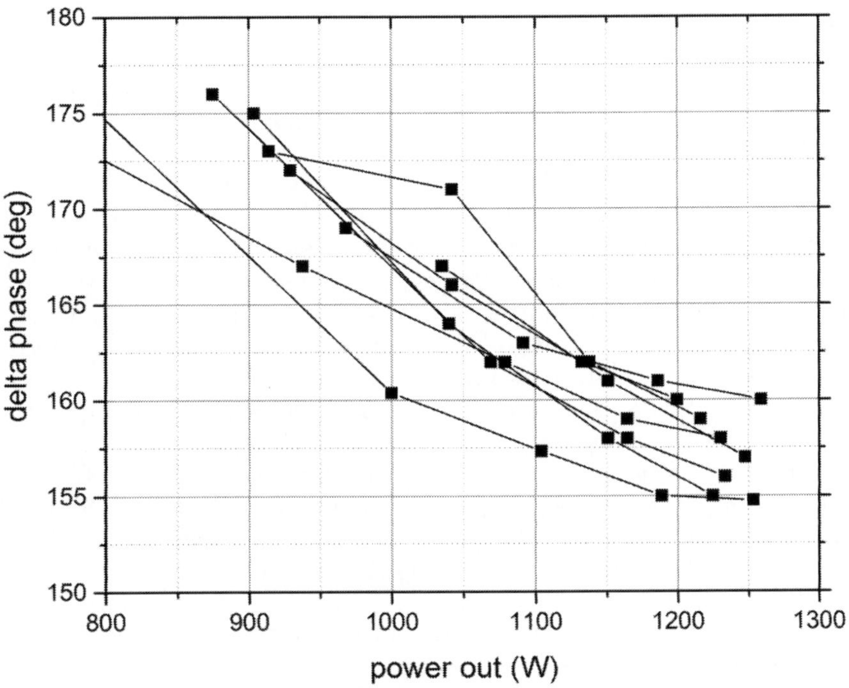

Figure 3. Phase spread of the eight modules.

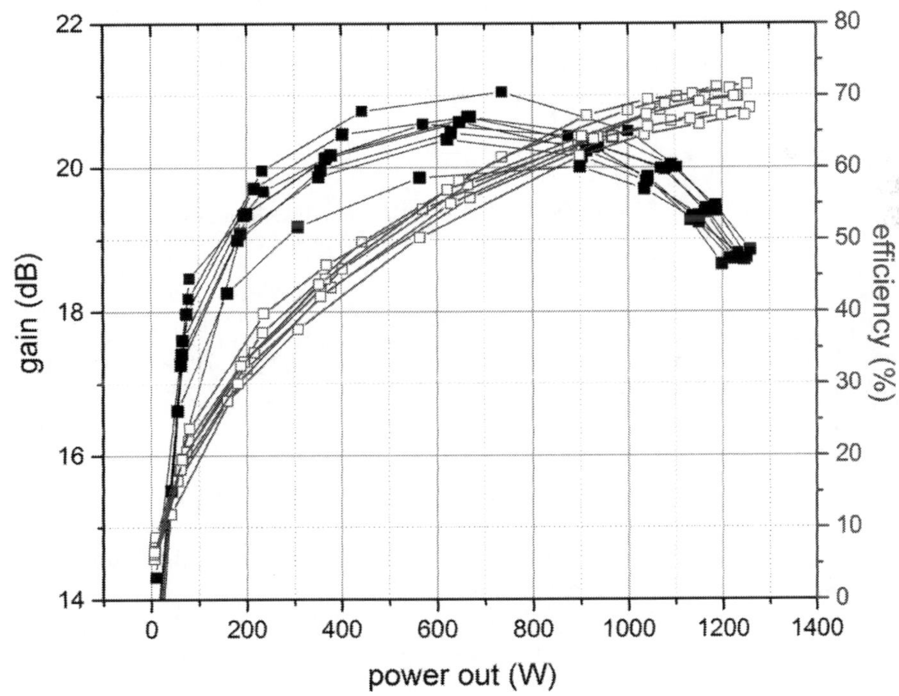

Figure 4. Drain and efficiency spread of the eight modules.

IOP Conf. Series: Journal of Physics: Conf. Series **874** (2017) 012093 doi:10.1088/1742-6596/874/1/012093

3. Planar combiner

A compact 8-way combiner has been developed in a planar technology, suitable for mass fabrication. The combiner allows the combination of eight 1.25 kW modules to a total output power of 10 kW. One of the main challenges in designing a planar combiner for high RF power levels is to ensure that the combiner can dissipate the high power resulting from a mismatch, to avoid dielectric breakdown and to provide a high power combination efficiency.

A topology for power combiners in which the load used to dissipate power in the case of unbalanced inputs was connected directly to ground (i.e unbalanced) instead of between the input lines as is the case of the Wilkinson combiner (i.e balanced) was proposed by Gysel [5]. This architecture removes the restrictions on the electrical length of the load and allows larger loads to be used. By using a single stage 8-1 Gysel type combiner the port to port isolation is maintained while the total insertion loss is kept low [4].

The proposed design is divided into two separate main boards, see Figure 5-a) containing the input port balancing and the dummy loads and Figure 5-b) containing the output matching and combination stages. Two boards were used in order to enable the mounting of dummy loads on the back of the common section as well as to allow the use of two different board thicknesses. The use of multiple separate substrates was needed in order to accommodate the varying line impedances required to achieve a proper matching of the different combiner stages and is related also to power handling capabilities of the substrates. A Rogers TMM3 5.08 mm substrate is used for the output section while a TMM3 1.27 mm substrate is used for the common point configuration. The Rogers TMM3 substrate provide a low loss ($\tan\delta = 0.002$) and high power handling capabilities, suitable to use in high power combiner designs.

The single stage 8-1 Gysel combiner has been tested up to 10 kW peak output power with no observable performance degradation or arcing. The total insertion loss was measured to be 0.2 dB at nominal power. The performance degradation due to parasitic coupling between adjacent lines in the common point section of the combiner was reduced by the use of weakly coupled stubs between adjacent lines which helps to realign the phase balance between transmission lines.

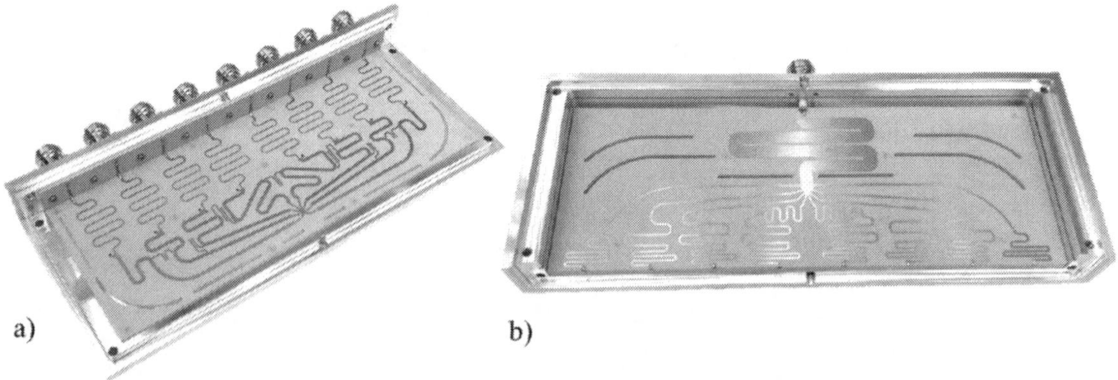

a) b)

Figure 5. The output board (a) mounted on the lower section of the aluminium casing and the common board (b) mounted on the upper section of the aluminium casing, together with output connectors.

4. Time domain measurements

For pulsed operation, a large capacitor (68 mF) is used to supply the energy per pulse. The voltage droop is less than 1 V from the drain operating voltage 50 V or 2% of the drain voltage after 3.5 ms, whereas each module delivers 1.25 kW output power. The power supply recharges the capacitor during the silence between pulses, see Figure 6.

The droop over the pulse is measured around 1 A, see Figure 7. There is a spread in the current used by each module with a maximum current of 42 A and a minimum of 37 A, which reflects into the efficiency per module. The efficiency per module can be evaluated in real time, having available a direct readout of voltage and current per module. Analogue circuits using a microcontroller (ATmega328P) are implemented to monitor simultaneously all eight drain currents and voltages. Each module is provided with a Hall effect current transducer and drain voltage is also monitored. Time-multiplexed measurements are performed resulting in the real time measurements of the voltage and current per module, simultaneously for all modules.

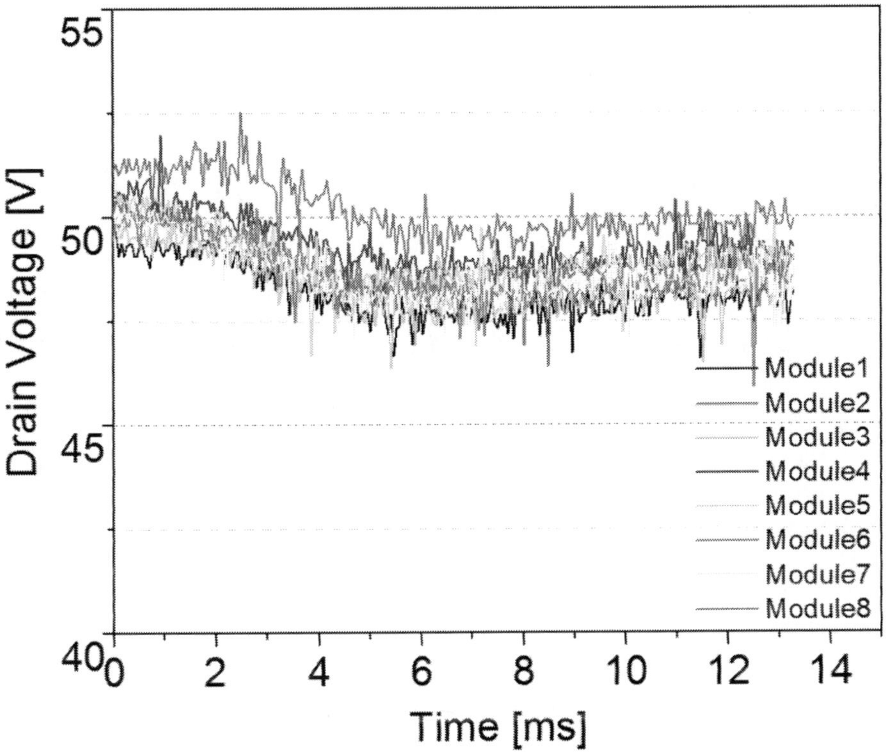

Figure 6. Voltage during pulse for each of the eight modules.

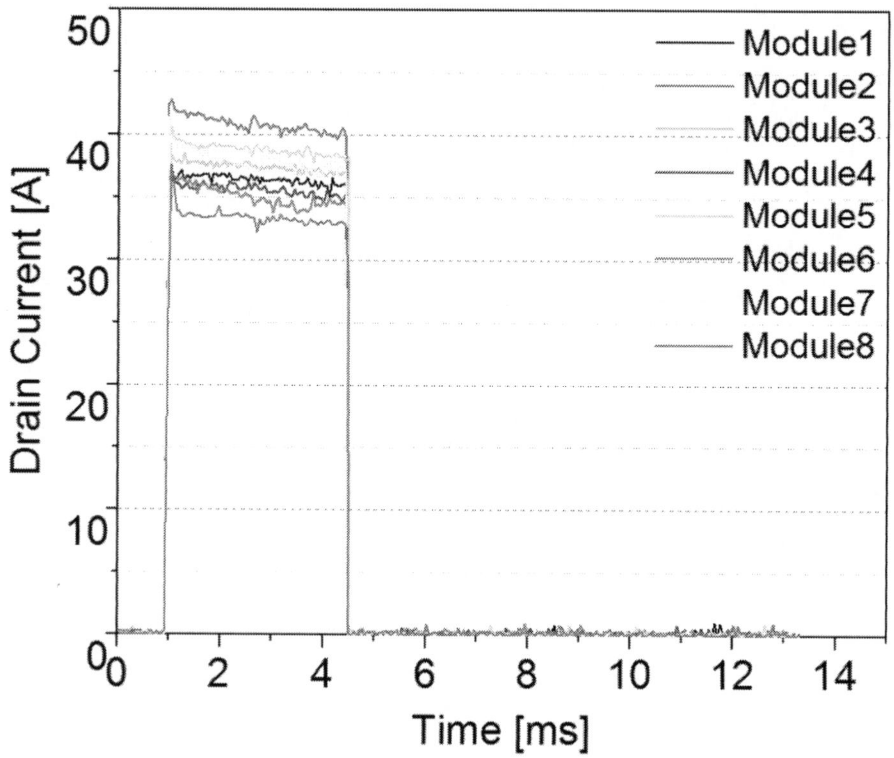

Figure 7. Current during pulse for each of the eight modules.

5. Testing of amplifier at 10 kW

The amplifier is implemented in three gain stages, as shown in Figure 1. The amplifier is made to fit a 19-inch (48 cm) standard rack and has a height of 9 rack units (40 cm) comprising the capacitor bank, with an overall depth of only 20 cm. At the input, a quarter wavelength splitter is used to distribute the input power, around 15 W to each of the eight modules. Power measurements are performed using directional couplers at the input and the output of the combiners. Combiners and cables losses are small (0.4 dB at the input and 0.32 dB at the output). These are not removed from the gain and output power measurements, as presented in Figure 8.

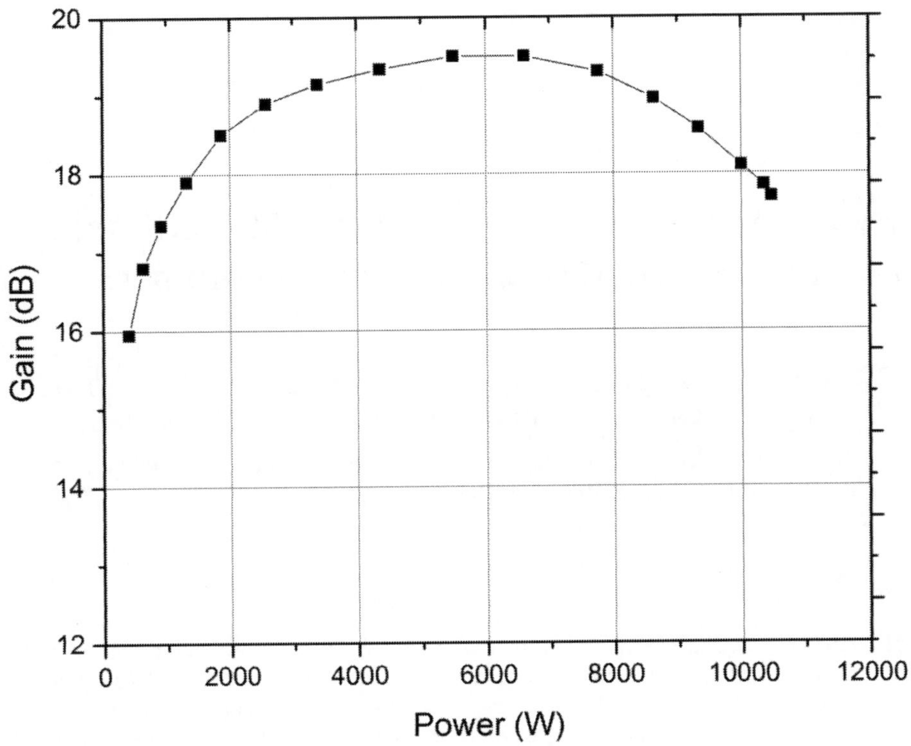

Figure 8. Amplifier gain and output power.

6. Conclusion

A compact 10 kW pulsed peak power amplifier has been developed at 352 MHz using eight modules in a single ended architecture. The modules are robust, easy to manufacture and show an output power of 1.25 kW with a drain efficiency of 71%. Measurements of the variation among the modules show a small variation within phases and gains. A single stage 8-way Gysel combiner is also implemented with an insertion loss of 0.2 dB at nominal power, which allows the combination of the modules to a total measured output power of 10.5 kW. Analogue circuits are used to monitor simultaneously all eight drain currents and voltages using a microcontroller based time-multiplexed measurement scheme. The amplifier, RF modules, combiners and monitoring circuits were all realised in house, at the FREIA laboratory, Uppsala University.

Acknowledgments

This work is in part funded by the Vice-chancellor of Uppsala University and ESS AB. The authors would like to thank Johan Eriksson and Tord Peterson for their help with the practical implementation.

References

[1] Marchand P, Lopes R, Ribeiro F and Ruan T 2011, *Proc. IPAC11* San Sebastian (Spain) MOPC127
[2] Mohania P, Mahawar A, Shrivastava P and Gupta P D 2013 *Rev. Sci. Instrum.* **84(9)** 094703
[3] Haapala L, Eriksson A, Hoang L D and Dancila D 2016 *Electronics Letters* **52** 18
[4] Jobs M, Dancila D, Eriksson J and Ruber R 2017 A 8-1 Single Stage 10 kW Planar Gysel Power Combiner at 352 MHz, *submitted to IEEE Trans. on Components, Packaging and Manufacturing Technology*
[5] Gysel U H 1975 *IEEE-MTT-S International Microwave Symposium*, pp 116-118

IOP Conf. Series: Journal of Physics: Conf. Series **874** (2017) 012094 doi:10.1088/1742-6596/874/1/012094

Quarter wavelength combiner for an 8.5kW solid state amplifier and conceptual study of hybrid combiners

T C Yu, Ch Wang, L H Chang, M S Yeh, M C Lin, C H Lo, M H Tsai, F T Chung, M H Chang, L J Chen, Z K Liu, C L Tsai and F Y Chang

101 Hsin-Ann road, Hsinchu SciencePpark, Hsinchu city, Taiwan, National Synchrotron Radiation Research Center

yu.tc@nsrrc.org.tw

Abstract. Experimental results to combine ten 900W solid state amplifier modules based on typical quarter wavelength 10-way combiners are described for a total of 8.5kW RF power output at 500 MHz. The power gain and phase distribution among the ten modules are measured and calculated to sense the combination efficiency. The combination efficiency of 100 modules differing in power gain and phase distribution is theoretically analysed. Groups of 5, 10, 25, 50 and 100 units are used in 4, 3, 2, and 1-stage power combination for total 100 units and the characteristics are calculated and investigated, including bandwidth, efficiency and even redundancy under various output VSWR levels. To simplify combining complexity and to eliminate the drawbacks of single stage combiners, a multi way 2-stage coaxial to waveguide combiner is thus proposed as an expandable power combiner.

1. Introduction

A project for a homemade 60 kW solid state power amplifier (SSPA) system is in progress at NSRRC. Numerous efforts have been made on the module design including planar balun with better heat dissipation, integrated water cooling plate [1], a temperature measurement circuit and a high directivity PCB directional coupler. After production of some modules, they were used for power combination. These handmade SSPAs were not very identical in gain and phase distribution, but the power combination efficiency still performed satisfactory. To find a relationship between gain and phase distribution on one side and their combination efficiency, circuit models with adjustable power gain and phase distribution on a RF circuit simulator were done based on the measurement data. The range of power and gain distribution among modules seems not to be very narrow at +/- 0.2dB and +/- 5deg, respectively [2]. A Gaussian distribution is applied to groups within +/-0.5 dB &+/-10deg and +/- 1dB & +/-20deg to estimate the efficiency of a 100-unit power combination. In addition to the combination efficiency, combination methodology also affects the system performance as does insertion loss and bandwidth. In general, the SSPAs will be combined first in a basic group as a medium power unit, then basic groups are combined until the required power and quantity is reached. Varying the module numbers in a basic group for 1-stage to 4- stage combinations of 100 units and complete combiner systems are analysed to obtain system bandwidth and redundancy capability under different output VSWR. To reduce the number of combination stages, a multiport coaxial to waveguide combiner is proposed as an expandable 2-stage power combiner.

2. Combination efficiency analysis for an 8.5 kW 10-way SSPA

The maximum power combination efficiency can be obtained when all SSPA units are identical. However, the actual modules cannot be identical due to the dispersion of errors from different chips, passive components, to misaligned soldering etc. Each one contributes some error leading to big differences between modules which can be controlled by trim capacitors, but some errors still remain. How these errors affect the overall power combination efficiency is an interesting and practical point of study. We start from an experimental 8.5kW, 10-way combiner to a theoretical 100-way power combination for different error distribution ranges in this article.

The measured power gain and phase as a function of output power for each handmade module is shown in Figure 1 for a power gain and phase distribution within +/- 0.5dB and +/- 18 degrees of experimental values. According to measured power gain and phase distributions at the saturation point, the estimated and measured combination results are shown in Table 1. The measurement results are close to calculation while the combination efficiency is the ratio of combined output power to the total power of 10 modules. From the result, the combination efficiency seems not to be very sensitive to the range of error distribution.

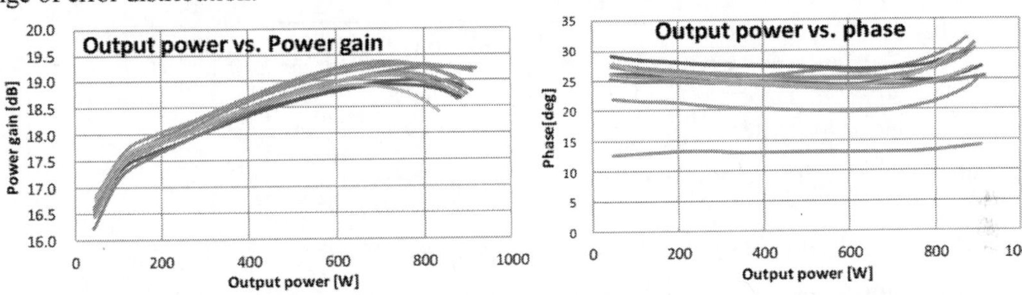

Figure 1. Power gain and phase distribution among the 10 handmade SSPA modules.

Table 1. Calculated and measured combined power measurements of the 10 handmade SSPA modules.

	Calculated	Measured
Combined output power [kW]	8.87kW	8.68kW
Combination efficiency [%]	99.37%	97.24%

2.1. *Combining efficiency under various dispersion*

The available overall SSPA efficiency derives from a combination of a few factors: the AC-DC conversion efficiency ($Eff_{AC\text{-}DC}$), the insertion loss/mismatch loss of passive transmission line components including combiners (IL), the combination efficiency resulting from errors among modules (Eff_{combi}) as well as divider/combiner errors and finally, the DC-RF conversion efficiency of the SSPA module itself ($Eff_{DC\text{-}RF}$). Thus the overall AC to RF conversion efficiency is given by

$$\mathrm{Eff}_{total} = \mathrm{Eff}_{\mathrm{AC-DC}} \times (1 - IL) \times Eff_{combi} \times Eff_{DC-RF} \quad (1)$$

To find the combination efficiency corresponding to specified power gain and phase errors among the modules, 100 SSPA units have been selected randomly in a circuit simulator and assigned to three Gaussian distributions +/-0.2dB&+/-5°, +/-0.5dB&+/-10° and +/-1dB&+/-20° as shown in Figure 2. A few modules seem not to quite fit within their Gaussian distributions, but the resulting combination efficiency is still satisfactory as shown in Table 2. Assuming 90% in $Eff_{AC\text{-}DC}$, 62% in $Eff_{DC\text{-}RF}$ and 2% in IL, the corresponding overall AC to RF conversion efficiency will be affected by less than one percent even under a wide range of errors as in group #3.

IOP Conf. Series: Journal of Physics: Conf. Series **874** (2017) 012094 doi:10.1088/1742-6596/874/1/012094

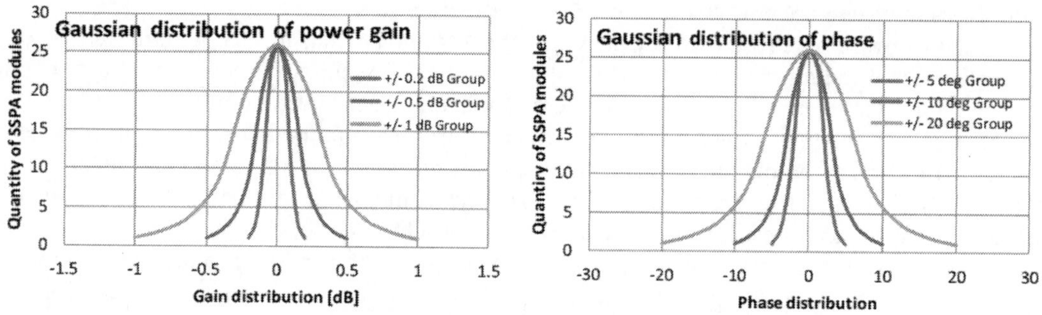

Figure 2. Power gain and phase of 100 SSPA units in three Gaussian distribution ranges

Table 2. Power combination efficiency among three Gaussian distribution ranges.

	Combination efficiency
+/- 0.2dB & +/- 5° (#1)	99.79%
+/-0.5dB & +/-10° (#2)	99.49%
+/-1dB & +/-20° (#3)	98.29%

Table 3. The impact of combination efficiency.

	$Eff_{AC\text{-}DC}$	$Eff_{DC\text{-}RF}$	IL	Eff_{combi}	Eff_{total}
#1	90%	62%	2%	99.79%	54.58%
#2	90%	62%	2%	99.49%	54.41%
#3	90%	62%	2%	98.29%	53.76%

3. The characteristics of combining trees

There are various ways to combine SSPA units to the required power level. Usually, the SSPA will be combined first in a basic group, then, these basic groups will be further combined once or twice or even triple to generate the final output power [3-4]. Examples of a combination tree with a three stage combination of 5x2x2 or a one stage combination like 20x1 for 20 SSPA units are shown in Figure 3. Since combining SSPAs in 1-stage is becoming popular to simplify system complexity [5], some of its properties will be discussed as well.

IOP Conf. Series: Journal of Physics: Conf. Series **874** (2017) 012094 doi:10.1088/1742-6596/874/1/012094

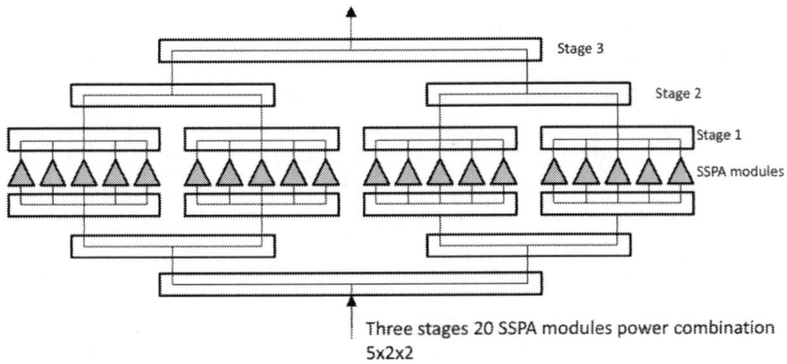

(a)

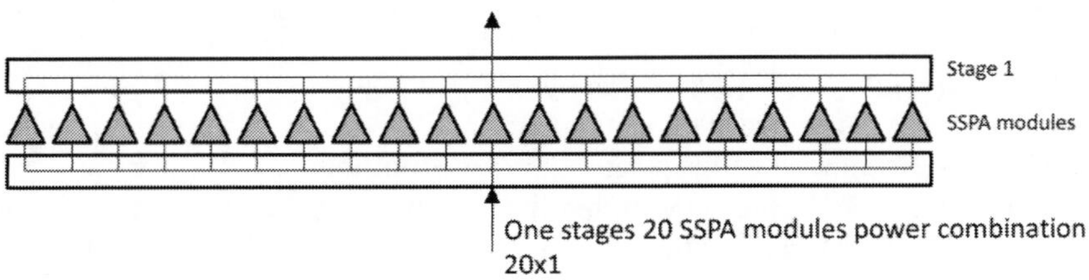

(b)

Figure 3. (a) SSPAs are grouped in sets of five and combined twice in three stages 5x2x2 (b) SSPAs are combined only once as a 20x1 combination

3.1. Bandwidth

Since the RF frequency used in NSRRC is 499.65MHz, the combining trees would be designed accordingly for convenience. The bandwidth of the combination tree can be presented as an S-parameter from the output port to any input port or vice versa. The single stage combination of various quantities is shown in Figure 4 (a) and a multistage combination of a 100 ways input is shown in Figure 4 (b). Obviously, larger combinations in a single stage would let the bandwidth become narrower while multistage benefit a wider bandwidth which is strongly determined by the quantity in the first stage. Since the SSPA module itself usually is a wideband active component, the required bandwidth of the combination tree may dependent on select specifications.

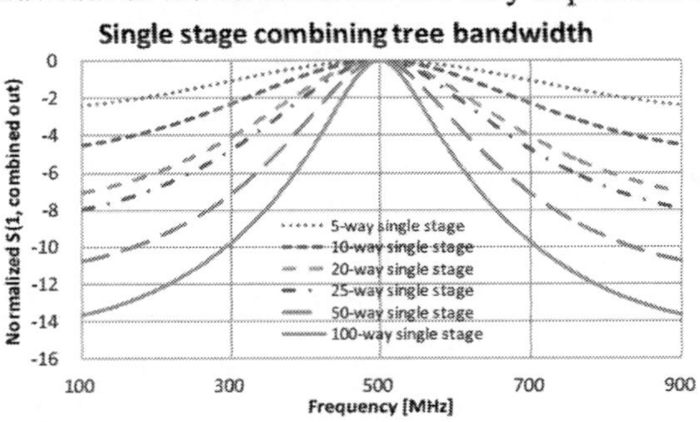

(a)

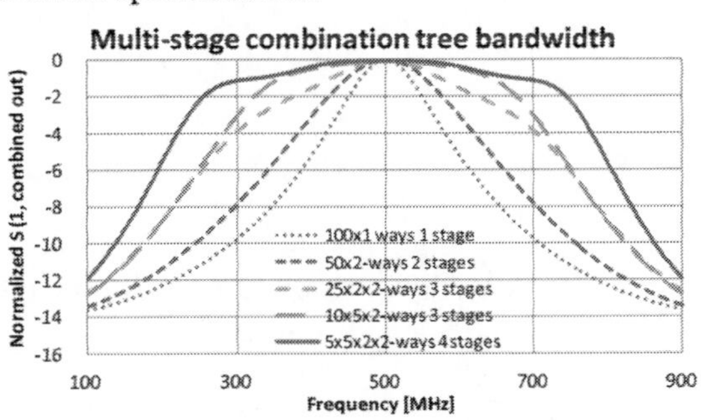

(b)

Figure 4. (a) different 1-stage combinations and (b) different 100 ways combinations

3.2. Redundancy

Besides the property of low DC voltage operation and energy savings, the most attracting feature of the SSPAs is their redundancy. The SSPA can absorb a certain number of module failures without interrupting system operation. Since a high power RF system usually does not operate at its maximum power rating, the power drop caused by a failed module can be compensated immediately by other active unit through the LLRF control system. The available power during some SSPA module failures can be simplified like in Equation (2), where P_{out} is the output power of the final stage, P_n is the nominal power of each module, N_{total} is the total quantity of SSPA modules and N_{fail} is the number of the failed SSPAs. One failing module can reduce the output power by twice its nominal power since other active modules would react by reflecting power approximately equal to P_n. Note, that this equation holds only for $N_{total} > 10 \cdot N_{fail}$, otherwise, the calculated P_{out} will be lower than the actual output power. For example, if we assume five failing modules, we still have at least 90% of the maximum output power as long as the minimum number of total modules is more than 100.

$$P_{out} = P_n \times N_{total} - 2 \times P_n \times N_{fail} \quad (2)$$

3.3. Risk behind the redundancy: reflected power

The SSPA module is usually equipped with a circulator and a load to absorb the reflected power and protect the power transistors as well as keeping matching conditions. The power handling capacity of the circulator and load is usually just a little higher than the maximum saturation power of the module such as 1.2 times P_{sat}. For example, when the combined output is connected to a dummy load (VSWR=1) and some modules in the same group have failed, the reflected power of the basic combination group within a 10x5x2 100 units SSPA is shown in Figure 5. From Figure 5 the reflected power contributed from other active units will not be equally distributed to all ports and the active modules may see higher reflected power than OFF modules when the number of OFF modules is getting larger. For a single failing module, one would see a higher Pr than for two failed units. Therefore, the reflection power of a single failed module should be taken care of since the chance of one module failing is much higher than that of two modules failing at the same time in normal operation.

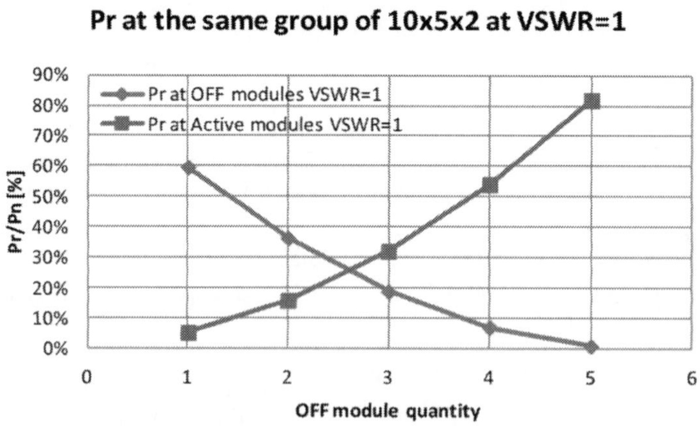

Figure 5. Ratio of reflection power (Pr) to nominal power (Pn) for OFF/active modules in the same group of 10x5x2 SSPA when the output VSWR=1

The highest Pr of SSPAs will appear at the single failing module with output load VSWR=2 and 100 in 5x1, 10x1, 20x1, 25x1, 50x1 and 100x1 1-stage SSPA, which both are analysed in Figure 6. The reflected power on the failing module dramatically increases with output VSWR as well as the number of the SSPA modules in the basic group and is highly dependent on the load phase. The highest Pr is observed at 100x1 1-stage SSPA under VSWR=100 which is almost four times Pn. Even at VSWR=2, the Pr can reach 1.74 times Pn at a 100x1 1-stage SSPA. As the number of SSPA modules get larger within a single stage, the single failed module would have a chance to receive a high reflection power at certain phases. From the analysis, the reflected power could be a risk for the circulator and load of the failed module in a multi way or single stage combiner when the load VSWR is high. The high reflection power could induce RF arcing propagating through the transmission branch. To provide a four to five times power handling capacity for circulator and load at each SSPA module is not feasible, while adding a high power circulator between SSPAs and final output load or fine adjusting the line length/phase between RF output and the actual load could be a solution.

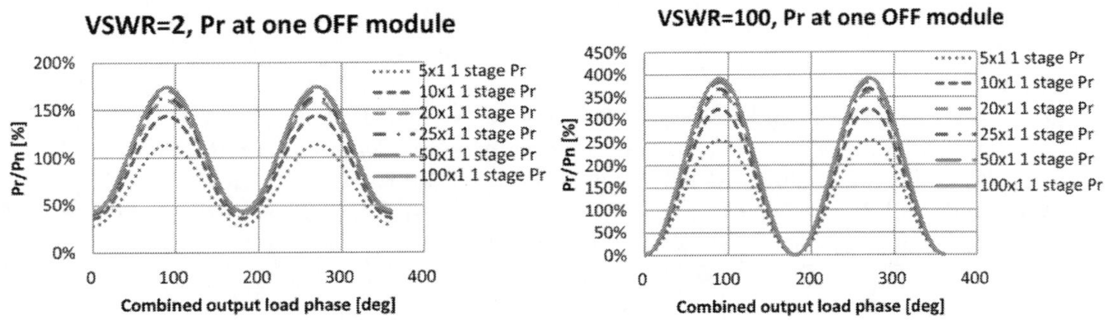

Figure 6. Ratio of reflected power to nominal power for a single OFF module under two VSWRs load conditions and different numbers of 1-stage SSPA units

4. 2-stage expandable combiner

From the analysis discussed above, multistage combination may have higher insertion losses while single stage combination has a narrower bandwidth and a risk of high reflection under some extreme conditions, for which a trade off solution is proposed as a multiport 2-stage waveguide combiner with expansion capability as shown in Figure 7.

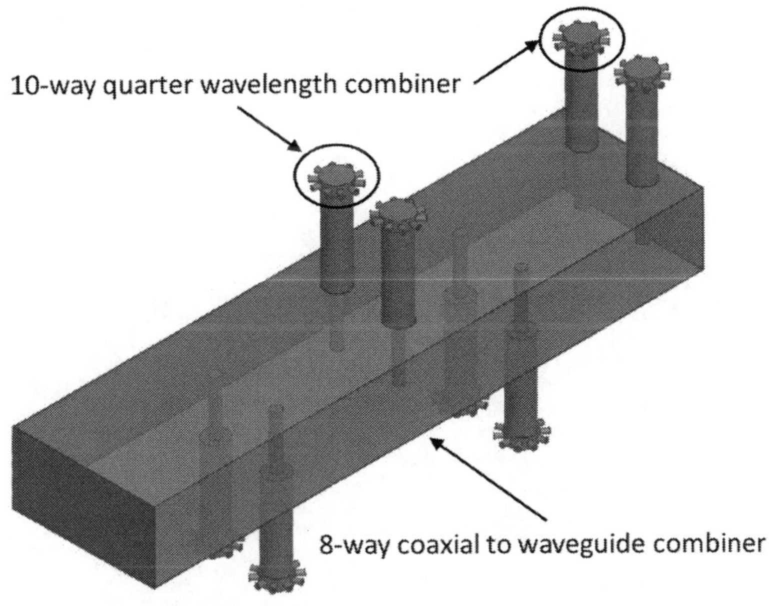

Figure 7. Expandable 2-stage coaxial-to-waveguide power combiner

5. Conclusion

The experimental results of a 10-way SSPA power combiner for 8.5kW have shown acceptable efficiencies with experimental and calculated results matching well. One hundred SSPA random units, in three Gaussian distribution groups, are used for power combination on a RF circuit simulator showing a satisfactory 98% combination efficiency even when gain and phase errors are distributed as wide as +/- 1 dB and +/- 20 ° . The bandwidth of different power combination trees is also discussed. The available output power for a certain number of failing SSPA modules are described as well while the risk behind the redundancy is simply explained by graphs. As a tradeoff between compactness and redundancy, a 2-stage coaxial to waveguide power combiner is proposed as an expandable multiports power combiner.

References

[1] Yu T C, Wang Ch, Chang L H, Yeh M S, Lin M C, Yang T T, Lo C H, Tsai M H, Chung F T, Lin Y H, Chang M H, Chen L J and Liu Z K 2014 *Proc. of the 5th Int. Particle Accelerator Conf.* ed P J G Christine and M Peter et al (Dresden Germany) p 2294-6

[2] Marchand P, EI Ajjouri M E, Lopes R, Ribeiro F and Ruan T 2011 *Proc. of the 2nd Int. Particle Accelerator Conf.* (San Sebastian Spain) p 376-8.

[3] Marchand P, Diop M D, Lopes R L, Polian J, Ribeiro F and Ruan T 2005 *Proc. of 2005 Particle Accelerator Conf.* (Knoxville Tennessee) p 811-3

[4] Gaspar M A and Garvey T 2015, *Proc. of the 6th Int. Particle Accelerator Conf.* (Richmond VA USA) p 3170-2

[5] Jacob J, Farvacque L, Gautier G, Langlois M L, and Mercier J M 2013 *Proc. of the 6th Int. Particle Accelerator Conf.* (Shanghai, China) p 2708-10

The RF distribution system for the ESS

P Aden[1], T R Edgecock[2], D Naeem[1], R Smith[1], A Sunesson[3], N Turner[2] and R Yogi[3]

[1]STFC Daresbury Laboratory, Warrington, WA4 4AD, UK
[2]University of Huddersfield, Queensgate, Huddersfield, HD1 3DH, UK
[3]European Spallation Source ERIC, 223 63 Lund, Sweden

t.r.edgecock@hud.ac.uk

Abstract. The RF distribution system for the super-conducting cavities of the European Spallation Source will be one of the largest systems ever built. It will distribute the power from 146 power sources to the two types of ESS cavity at two different frequencies and will use one line per cavity for resilience. It will consist of a total of around 3.5 km of waveguide and coaxial line and over 1500 hundred bends. It is designed to transport this RF power over a distance of up to 40m per line, while minimising losses, avoiding reflections and allowing the monitoring of performance.

1. Introduction

The European Spallation Source [1] (see figure 1) will be a high brightness source of spallation neutrons for materials and other studies. It is currently under construction in Lund, Sweden and is scheduled to produce its first neutrons in 2019. It is planned to build the facility in two stages, with the first accelerating protons to 1.3 GeV and 3 MW beam power. The second stage will extend this to 2 GeV and 5 MW. It will then deliver a peak neutron flux seven times more powerful than any existing accelerator based spallation source and a peak flux 30 times greater than any reactor based neutron source.

Figure 1. Artist's impression of the ESS after completion.

The ESS will employ a linear accelerator for proton acceleration [2]. The front-end of this, up to 90 MeV, will use normally conducting cavities, while the rest will use two types of super-conducting cavities (see figure 2). There will be 26 spoke cavities running at 352 MHz, 36 medium beta (MB) elliptical cavities at 704 MHz and 84 high beta (HB) elliptical cavities at 704 MHz. Note that for the first stage of the ESS, 40 of the HB cavities will not be used. The RF pulse width will be 3.5 ms and the pulse repetition frequency will be 14 Hz.

2. RF distribution for the super-conducting cavities

The purpose of the RF distribution system (RFDS) is to transport the RF power from the power sources, located above ground in the gallery building, through chicanes called stubs, to the RF cavities below ground in the accelerator tunnel, mainly through waveguides. The ESS requirement for high reliability, 95%, has resulted in a design with one power source per cavity and hence one waveguide run per cavity. As each stub has space for 8 waveguide runs, the RFDS mainly comes in units of 8 (see figure 3). The parameters of the RFDS are shown in table 1.

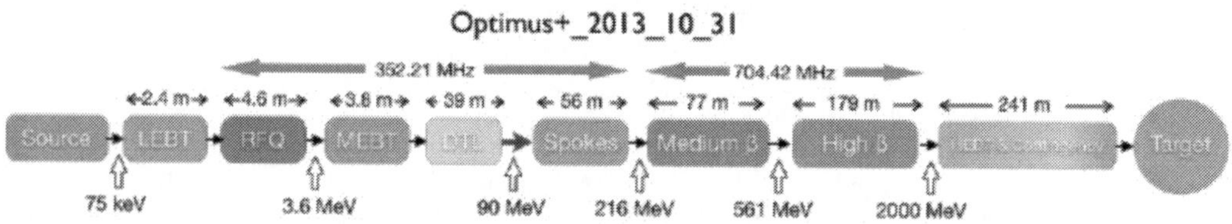

Figure 2. Schematic of the ESS linac.

2.1. Requirements

The requirements for the RFDS are:

- It must deliver power reliably to the cavities (a mean time between failures of 500000 hours), with minimal reflections: return loss (S11) > 30 dB, insertion loss (S21) < 0.05 dB.
- RF average power leakage at 704 MHz must not exceed 0.35 mW/cm^2 and at 352 MHz must not exceed 0.2 mW/cm^2.
- It must be possible to measure the forward and reflected power at various locations.
- The power sources must be protected against reflected power.
- Arcs must be detected and suppressed with a minimal rate of false trips.

The system designed to meet all of these requirements is shown in figure 3.

Table 1. RFDS Parameters for 5 MW operation

Linac module	Frequency (MHz)	Power (kW)		Number of systems
		Peak	Average	
Spoke	352.21	400	20	26
MB	704.42	1200	60	36
HB	704.42	1500	75	84

2.2. RFDS design

Due to the low average power, the spoke systems will employ 6-1/8 coxial cables up to the circulator as this was determined to be cheaper than using waveguides for this section [3]. The rest of the system will use WR2300 half-height (HH) waveguides. The MB and HB systems will use WR1150 waveguides for the entire length. The circulators are three port devices for transporting power from one port to the next. At the ESS, power arriving at port 1 will be transported onto port 2 and from there to the cavities. Reflected power arriving at port 2 will be transported to port 3 and from there to a load, where it will be absorbed. This protects the power sources from reflections due to impedance mis-matches, etc. Directional couplers will be installed between the power source and the circulator, between the circulator and the load and in the gallery waveguide system. These are used to measure the forward and reflected power to monitor that the system is working correctly.

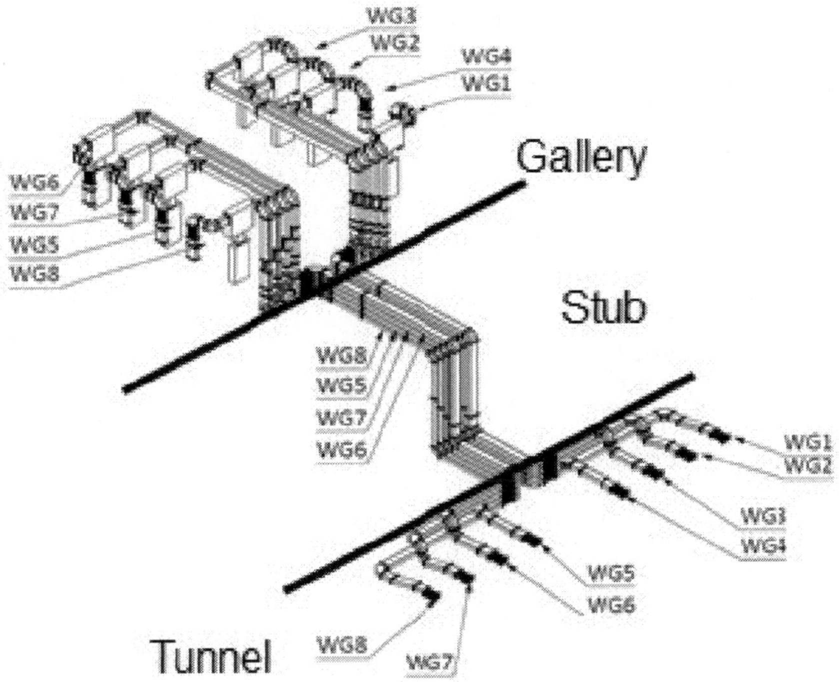

Figure 3. The layout of a group of 8 MB/HB waveguide runs.

Each waveguide run will contain up to 5 flexible waveguides (bellows) to allow for thermal expansion and small misalignments. There will also be so-called shutter switches to allow the testing of the power sources, without delivering power to the cavities. When open, these pass the power as normal. When closed, the power is reflected back to the load via the circulator. Finally, each spoke waveguide run will have 3 arc detectors in the gallery and 2 on the cavity coupler and each MB/HB run will have 4 arc detectors in the gallery and 2 on each cavity coupler. These are used to prevent breakdown in the RFDS. The function of each of these components is described in more detail in the next section.

3. RFDS components

3.1. Waveguides

The waveguide system will consist of standard 6-1/8 inch coaxial straight cable and bends, WR2300 HH straight waveguides and bends and WR1150 straight waveguides and bends. The standard waveguide will be broken into nearly 90 different lengths, ranging from 72 to 4000 mm, with the minimum number of each length required being 1 and the maximum 120. The ESS will also be using a number of non-standard waveguides. As the stubs will be closed at the bottom and top with shielding walls for radio-protection purposes, the air circulation around the waveguides in this region will be limited. As a result, the heating of the waveguides will be sufficient to raise their temperature to approaching 90°C. Although this is fine for the waveguides, it is too hot from the cables that will be

IOP Conf. Series: Journal of Physics: Conf. Series **874** (2017) 012095 doi:10.1088/1742-6596/874/1/012095

above and below them. As a result, the waveguides need to be cooled. This will be done by blowing tunnel air through the waveguides. As this can be activated, it is not permitted to discharge this into the gallery. As a result, the air will be returned down a neighbouring waveguide and expelled back in to the tunnel. Vents will be included in special "air cooling" waveguides to allow sufficient air flow, while maintaining RF leakage below the legal limits. RF windows will also be installed upstream of the vents to prevent activated air entering the gallery through the waveguide.

In addition, a number of waveguide runs will use angled waveguides, to reduce the number of bends required. Each waveguide run will also have one section of "removable" waveguide. This will used as an additional protection, along with the shutter switches, to avoid power being delivered to the cavities while personnel are in the tunnel. They will be equiped with special flanges that will provide a signal to show that the flange is in place.

The total lengths and numbers of waveguides and bends required are shown in Table 2. Mechanical tolerances have been specified for each of these components to ensure they meet the return loss and insertion loss requirements. Further, each must have an internal surface roughness that is better than N6 to minimise reflections. The waveguide system will be supplied by Mega Industries LLC [4].

Table 2. Total lengths and numbers of waveguides and bends required to the ESS

Linac module	Component	Length (mm)	Total length or number required
Spoke	Standard waveguide		497m
	Angled waveguides	733.0	37
	Air cooling waveguides var. A	733.0	7
	Air cooling waveguides var. B	539.1	45
	Removable waveguide	500	26
	H-bends		130
	E-bends		78
	Coax bends		52
	Coax straights		26
MB+HB	Standard waveguide		2583m
	Angled waveguides var. A	459.0	30
	Angled waveguides var. B	1009.0	60
	Air cooling waveguides var. A	305.4	60
	Air cooling waveguides var. B	457.2	180
	Removable waveguide	500	120
	H-bends		600
	E-bends		720

3.2. Circulators

The aim of the circulators is to pass the power onto the next port, with minimal power entering any other port or being reflected. They are included to ensure the reflected power is diverted to the load. The ESS circulators will be three port devices supplied by AFT Microwave GmbH [5] and will use the non-reciprocal nature of ferrite material to function as required. A prototype circulator has been built and observed to meet the requirements of return loss < 30 dB, insertion loss < 0.1 dB and isolation > 26 dB. Each circulator will have an active feedback unit to optimise performance based on the measured temperature and reflected power and an arc detector to avoid sparking.

3.3. Loads
The aim of the loads is to absorb any power reflected through the system via impedance mis-matches, the shutter switches, etc. Due to the energy recovery policy of the ESS, some of the components will be cooled with a water inlet temperature of 50°C or more. Further, the ESS will install the RF loads for MB/HB in series with a klystron collector with an inlet water temperature of 50°C. The water may also be recirculated to improve the heat recovery further. This means that the maximum outlet temperature for the load can be 80°C and they have been designed to work at this temperature. They will again be supplied by AFT Microwave GmbH and a prototype device has been built and tested up to this temperature, though at a low input power.

3.4. Directional couplers
Three types of directional coupler are to be used for spoke and for MB/HB. For spoke, 2 directional couplers will be 6-1/8 inch coaxial type, one with 4 loops and one with 2 loops. The 4 loop types will provide forward and reflected power measurements for the low level RF (LLRF) and diagnostic systems. The remaining type will be WR2300 HH and in the waveguide system. This will again be 4 loop, but with a directivity > 40 dB for more accurate measurements. The MB/HB directional couplers will be the same, except that all will be WR1150.

3.5. Arc detectors
Arc detectors are used to detector sparks within the waveguide system and inhibit the power amplifier before breakdown can occur. If this does occur, the discharge can travel in both directions along the waveguide, damaging this and amplifier and cavity at both ends of the system. The ESS will use at least two types of arc detector in up to 5 different locations. For the spoke waveguides, these will be on the circulator, close to the load and on a bend within the waveguide system. For MB/HB, there will also be a detector close to the power source. For both spoke and MB/HB, there will be arc detectors on the cavity couplers. As these are in a radiation environment, they are likely to be of a different technology from the other detectors.

3.6. Shutter switches
The role of the shutter switch is to allow a waveguide run to be closed so that all the power is reflected back to the load through the circulator. Two types are being considered for the ESS. The first will use a switch which is moved via a level which can be locked in place, to prevent the position of the switch being changed. It will also have micro-switches to indicate what position the switch is in.

The second type is a new one in which the switch is closed by sliding a reflecting plate into the waveguide. A prototype of this has been built and will soon undergo rigorous testing.

4. Conclusions
A complete design of the RF distribution system for the super-conducting cavities of the ESS has been made. A number of prototypes for this system have been built and tested, though not all at high power. The system is now being procured, with deliveries due to start in October 2017 to meet the challenging ESS schedule.

5. References
[1] Lindroos M *et al* 2011 The European Spallation Source *Nucl. Inst. Meth.* B **269** no. 24 pp 3258-60
[2] Danared H *et al* 2010 Status of the ESS accelerator construction project *Proc. of Int. Particle Accelerator Conference (Richmond USA)* pp 3870-2
[3] Yogi R A *et al* 2013 Design of the RF distribution system for the ESS spoke linac *Preprint* FREIA Report 2013/02 *(Uppsala University, Sweden)*
[4] Mega Industries LLC, http://www.megaind.com/
[5] AFT Microwave GmbH, http://www.aft-microwave.com/

IOP Conf. Series: Journal of Physics: Conf. Series **874** (2017) 012096 doi:10.1088/1742-6596/874/1/012096

Design of an Inductive Adder for the FCC injection kicker pulse generator

D Woog, M J Barnes, L Ducimetière, J Holma and T Kramer

CERN, Geneva, Switzerland

david.woog@cern.ch

Abstract. The injection system for a 100 TeV centre-of-mass collider is an important part of the Future Circular Collider (FCC) study. Due to issues with conventional kicker systems, such as self-triggering and long term availability of thyratrons and limitations of HV-cables, innovative design changes are planned for the FCC injection kicker pulse generator. An inductive adder (IA) based on semiconductor (SC) switches is a promising technology for kicker systems. Its modular design, and the possibility of an active ripple suppression are significant advantages. Since the IA is a complex device, with multiple components whose characteristics are important, a detailed design study and construction of a prototype is necessary. This paper summarizes the system requirements and constraints, and describes the main components and design challenges of the prototype IA. It outlines the results from simulations and measurements on different magnetic core materials as well as on SC switches. The paper concludes on the design choices and progress for the prototype to be built at CERN.

1. Pulse Generator

Several injection energies are presently being considered for the FCC-hh injection [1]. Nevertheless a very fast system is required to achieve a high filling factor for the FCC; in addition the system must be reliable to avoid mis-kicking beam and consequent damage to downstream equipment. A transmission line kicker magnet, similar to the one used in the LHC injection system, will be used to achieve the short rise times. Thyratron switches are typically used at CERN for injection kicker systems, however they can occasionally self-trigger, have limited dynamic range and cannot interrupt current: hence, for a thyratron, a line type modulator is required to define the maximum duration of current pulse. Another issue is the high voltage (HV) cables used for Pulse Forming Lines (PFLs). New cable is extremely difficult or impossible to source.

To eliminate the aforementioned problems a new pulse generator design is proposed. Recent developments in power semiconductor switches open the possibility of replacement of a thyratron based pulse generator by a pulse generator based on SC-switches: the latter has the ability to interrupt current. However a drawback of SC-switches is their relatively lower ratings for voltage and current, and hence the need for multiple switches in parallel and series to achieve the required values. The inductive adder (IA), presented in [2], presents one possibility for a SC-based pulse generator design. Thanks to the fundamental design of the IA all MOSFET and trigger electronics are referenced to ground, making the triggering circuitry less complex than a pulse generator with floating gate circuitry. A further advantage of the IA is the modular design, which allows built in redundancy and gives the possibility to use identical modules for applications with different requirements. A SC-based Marx generator is also being considered as a pulse generator for the FCC injection kicker system [3].

Table 1. Injection Kicker Requirements

Parameter	**Injection**
Kinetic Energy [TeV]	3.30
Angle [mrad]	0.18
Pulse length [μs]	2.00
Flat top tolerance [%]	±0.50
Field rise time [μs]	0.43
Voltage [kV]	15.70
Current [kA]	2.50
System impedance [Ω]	6.25

Table 1 shows the FCC injection kicker requirements for a beam energy of 3.3 TeV. To obtain the required field rise and fall times of 430 ns in the kicker magnet, the rise and fall times of the current pulse from the generator need to be <75 ns. A potential drawback of the IA is the output transformer and therefore the risk of saturation of the magnetic core. However with a pulse length of 2.00 μs the requirements for the FCC injection kicker system makes the use of an IA possible. In addition, by employing modulation layers, the flat top tolerance of ±0.5 % is easily achievable with an IA, as these layers allow the generation of pulses with very high stability [4].

Although the available length for the kicker magnets is 150 m, for beam impedance and stability reasons it is desirable to use a significantly shorter length [5], but this increases the current required from the IA. As a compromise a current of 2.5 kA and a magnetic length of ~30 m has initially been chosen: the corresponding magnet deflection angle is 0.18 mrad [5].

2. Inductive Adder principle

The IA consists of a stack of toroidal 1:1 transformer cores: a schematic drawing of the setup can be found in [2]. The rated pulse current of fast switching MOSFETs is typically in the range of 100 to 250 A: hence the primary of each transformer requires multiple parallel branches to provide the required output current of 2.5 kA. The primary windings are located at the outer circumference of the toroidal transformer as shown in [4]. The secondary (stalk) passes through the stacked transformer cores. It encloses the magnetic flux of every "layer" and hence the output voltage is the sum of the output voltages of each layer. The primary voltage per layer is limited by the rated voltage of the SC-switch, which must include a safety margin to ensure good reliability and life time.

To obtain fast rise time the inductance of the primary circuits needs to be low. For this reason the primary winding of the transformer totally encloses the magnetic core. To avoid over voltage across the SC switches when the SC switches are turned off, high voltage freewheeling diodes are included in each primary to provide a path for magnetizing current.

3. Components

Dimensioning the components of the IA requires consideration of many aspects. Realising the required rise and fall times as well as the system impedance of 6.25 Ω are challenging requirements. The characteristic impedance (Z_{IA}) of a layer is defined by the capacitance C_{layer}, the inductance L_{layer} of the coaxial structure and the primary inductance L_{p} [6]. C_{layer} and L_{layer} are formed by the secondary winding (stalk) and the toroidal primary winding. Important factors are the ratio $\frac{d_{\mathrm{o}}}{d_{\mathrm{i}}}$ of the outside diameter of the insulation to its inside diameter, the layer height l and the relative permittivity ε_{r} of this insulation material [7]. The parasitic inductance L_{p} is formed by the primary circuits of a layer. Equation (1) shows the correlation of the values [6]:

$$Z_{IA} = \sqrt{\frac{L_p + L_{layer}}{C_{layer}}} = \sqrt{\frac{L_p}{2\pi\varepsilon l}\ln\frac{d_o}{d_i} + \frac{\mu}{4\pi^2\varepsilon}\ln^2\frac{d_o}{d_i}} \quad (1)$$

In order to achieve the required characteristic impedance with reasonable dimensions, various materials have been considered for the insulation between the primary and secondary windings. The maximum electric field strength, E_{max}, which occurs at the inner diameter (d_i) of the insulation material, must not exceed the breakdown field strength of the insulation including a safety margin. Oil has good insulation and permittivity values. For other insulation materials, e.g. air, the required diameters for the magnetic core end up in the range of 1.5 m which results in very heavy and costly cores [2]. Epoxy has been considered but makes disassembling difficult, for this reason oil has been chosen.

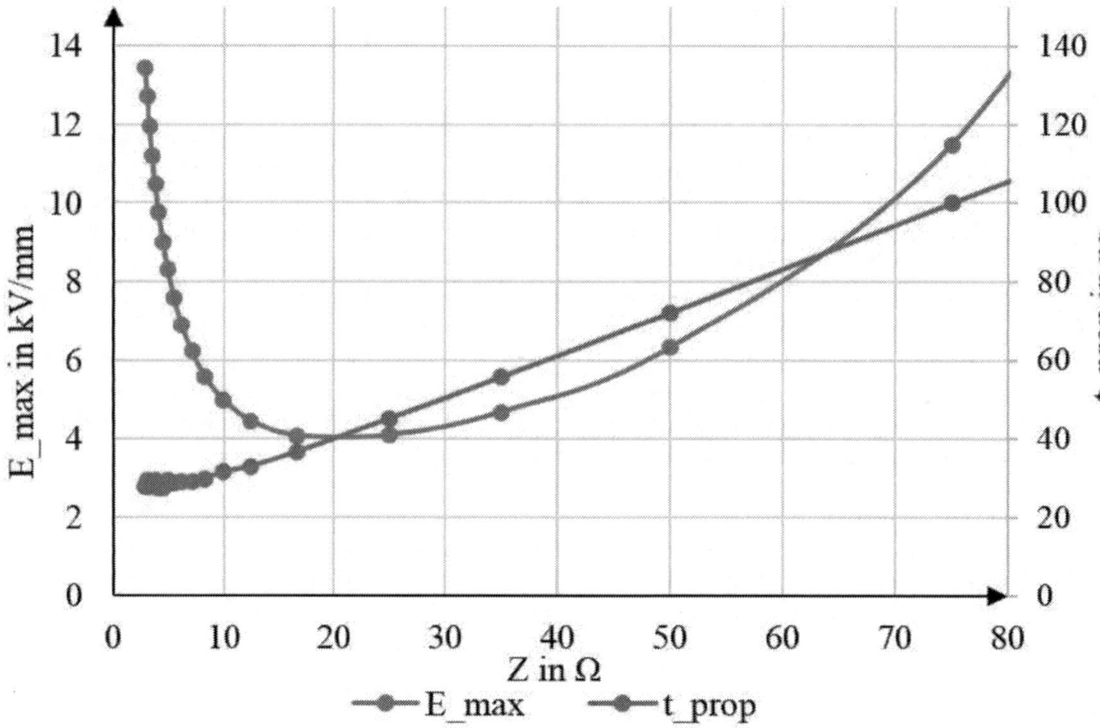

Figure 1. Max. electric field strength versus impedance

Figure 1 shows E_{max} of the insulation material for a relative permittivity of $\varepsilon_r = 2.7$ outer insulation diameter of $d_o = 108$ mm, primary inductivity of $L_p = 5$ nH and output current of $I_{out} = 2.5$ kA. It can be seen that E_{max} has a minimum at around 20 Ω. The two way propagation delay of the IA is also plotted in Figure 1: this delay influences the pulse rise time. In coordination with the kicker magnet design a low (6.25 Ω) impedance was chosen: a major advantage is that a given current can be obtained with a relatively low voltage. For the IA a low impedance is demanding from the insulation point of view, but at the same time reduces the length of the stack and hence makes it easier to obtain short currentrise times for a given current.

3.1. Magnetic Cores

One of the main components of the IA is the magnetic core. The core will be made from nanocrystalline, tape wound steel and has B-H characteristics which can be somewhat chosen during manufacture. The minimum cross sectional area is defined by the voltage-time integral of a layer and the saturation flux density of the material.

$$A_c = \frac{t_{pulse} \cdot V_{layer}}{\Delta B_c} \tag{2}$$

In Equation (2), A_c is the cross section area of the magnetic core per layer, t_pulse is the pulse duration, V_{layer} is the voltage per layer, and ΔB_c the magnetic flux density swing of the core. Equation (2) shows a linear dependence of the core cross section upon the pulse duration. For long pulses the required core cross section becomes very expensive and could exceed the manufacturing capabilities for the cores. Hence the IA is used for pulses of up to a few μs duration. To ensure that the core is not driven into saturation a 40 % safety margin will be considered for the cross section area.

To increase the available flux swing a negative biasing current will be applied to the cores. Biasing means that the working point in the B-H characteristics is set in the third quadrant, before the pulse.

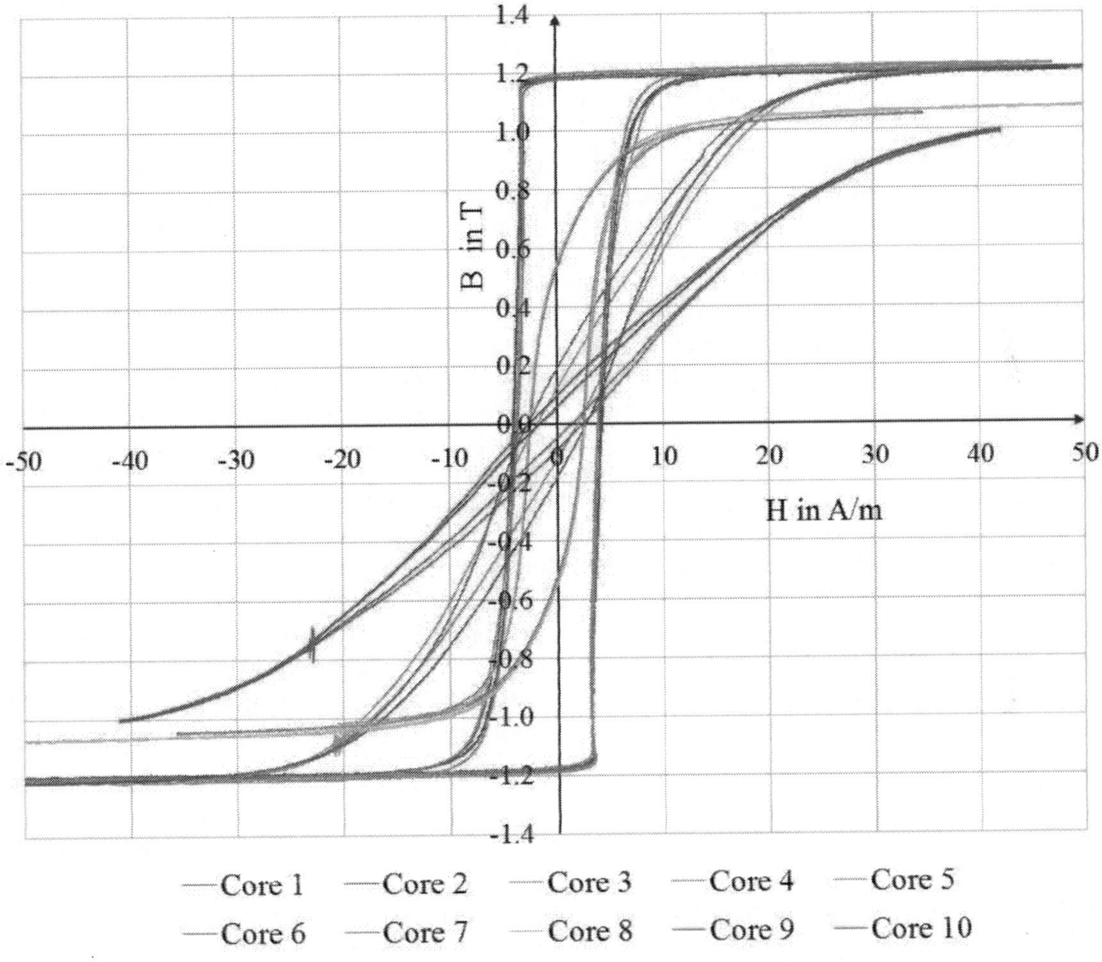

Figure 2. Measured B-H curves of sample cores at 500 Hz

Sample cores from a number of suppliers and with different B-H curves have been ordered and tested, using the technique described in [8]. Furthermore the magnetizing inductance (L_c) as well as the equivalent core loss resistance (R_c) have been measured. Figure 2 shows the measured B-H curves of the sample cores. Magnetizing inductance and core loss resistance, of the different sample cores, are shown in Table 2: cores 1 to 8 have the same dimensions, whereas cores 9 and 10 are slightly smaller.

Table 2. Measured Magnetizing Inductance and Core Loss Resistance of Sample Magnetic Cores

Core	R_c in Ω	L_c in µH	Type of B-H curve
Core 1	55	282	Square
Core 2	50	367	Square
Core 3	65	191	Square
Core 4	75	160	Square
Core 5	150	56	Linear
Core 6	160	42	Linear
Core 7	230	30	Linear
Core 8	200	31	Linear
Core 9	70	28.8	Linear
Core 10	70	28.8	Linear

PSpice simulations with the values shown in Table 2 indicate that all the sample cores are suitable for use in the prototype IA. The simulations show that cores with a square shaped B-H characteristic result in the best performance: even if the core resistance is relatively low the influence is compensated by the higher magnetizing inductance. In addition, the cores with a square shaped B H curve require a relatively low biasing current.

3.2. SC-switches

Many appropriate semiconductor switches, with both turn-on and turn-off capability, exist on the market. IGBTs are available up to several hundred Amps and several kV, but do not have the required fast switching times. Silicon MOSFETs are presently available up to 1200 V and a few hundred Amps: although the switching times are generally fast, the on-state resistance of Si-MOSFETs is still in the range of several hundred mΩ, which causes a relatively high voltage drop for a low impedance system. Silicon Carbide (SiC) MOSFETs are now available with rated voltage of up to 1700 V and pulse currents of more than 100 A. To reach higher current MOSFETs can be connected in parallel. SiC MOSFETs are characterised by a low on-state resistance, usually less than 50 mΩ, and relatively fast switching times. The relatively high voltage rating and low on-state resistance, combined with fast switching, makes SiC MOSFETs the preferred devices for the IA prototype.

SiC MOSFETs with a rated voltage of 1200 V and pulse current values of 190 A or higher have been measured. All tested devices proved capable of switching from 0.5 % to 99.5 % in less than 40 ns, for low currents: measurements for high currents are planned. An important factor to maximize the switching performance is the gate driver, and various drivers are being evaluated.

3.3. Pulse Capacitors

The pulse capacitors require a high capacitance to realise the required pulse current and duration with low droop: thus high energy density is desirable. At the same time the dimensions of the capacitor should be kept small to realize a short layer height, to keep the stack length and hence delay low. In addition, the inductance of the primary circuit is influenced by the parasitic inductance of the capacitor itself. To obtain high energy density with low inductance, custom made sample capacitors were ordered and tested. The parasitic inductance of the capacitors was measured together with its series resistance, using the technique described in [9]: results are shown in Table 3. The preferred capacitor, for its low inductance and relatively high voltage rating and energy density, is labelled Cap1.

Table 3. Measurement Results for Capacitor Samples

Type	Cap1	Cap2	Cap3	Cap4	Cap5
C in uF	25 ±10%	25 ±10%	25 ±10%	30 ±10%	35 ±10%
V in V	1500	1300	1100	1100	1100
L in nH	28	30.6	31.1	39.4	35.5
R in Ω	0.32	0.27	0.22	0.28	0.26

Ed in J/cm^3	0.274	0.212	0.166	0.182	0.212

In Table 3, C is the capacitance, V the rated voltage, Ed the energy density, and L and R are the measured parasitic inductance and resistance, respectively.

4. Conclusion

Design choices for the prototype IA will be based on test results. After the components tests are completed the prototype will be assembled and tested based on the requirements outlined in this paper: the foreseen parameters are shown in Table 4.

Table 4. Design Parameters of the Prototype IA

Parameter	Unit	Value
Nr. of constant voltage layers	-	21
Nr. of modulation layers	-	2
Nr. of branches per layer	-	24
Insulation inner diameter	mm	103.4
Insulation outer diameter	mm	108
Insulation thickness	mm	2.3
Insulation relative permittivity	-	2.7
Characteristic impedance	Ω	6.25
Voltage per layer	V	960
Current per branch	A	105
Core cross section area	cm^2	20.6
Core height	mm	30
Layer height	mm	40
Capacitance per branch	μF	25
Output voltage	kV	15.62
Output current	kA	2.5

Acknowledgement

Many thanks to Silvia Aguilera, Michal Krupa and Patrick Odier from the BE-BI group at CERN for their help with equipment and know how to measure the B-H curves.

References

[1] Stoel L S, Barnes M J, Bartmann W, Burkart F, Goddard B, Herr W, Kramer T, Milanese A, Rumolo G and Shaposhnikova E 2016 High Energy Booster Options for a Future Circular Collider at CERN *proceedings IPAC'16* Busan Korea.

[2] Kramer T, Barna D, Barnes M J, Bartmann W, Burkart F, Ducimetiere L, Goddard B, Senaj V, Stadelbauer T and Woog D 2016 Considerations for the injection and extraction kicker systems of a 100 TeV centre of mass FCC-hh collider *proceedings IPAC'16* Busan Korea.

[3] Barnes M J, Ducimetière L, Fowler T and Goddard B 2016 Pulsed Power at CERN *to be published in the EAPPC2016 proceedings* Lisbon Portugal.

[4] Holma J and Barnes M J 2014 Measurements on prototype inductive adders with ultra-flat-top output pulses for CLIC DR kickers *proceedings IPAC'14* Dresden Germany.

[5] Chmielinska A, Barnes M J, Bartmann W, Burkart F and Goddard B 2017 Preliminary estimate of beam induced power deposition in a FCC-hh injection kicker magnet *proceedings IPAC'17* Copenhagen Denmark.

[6] Zhang W, Eng W, Pai C, Sandberg J, Tan Y and Tian Y 2007 A Simplified Model for Parameter Estimation and Circuit Analysis of Inductive-Adder Modulator *IEEE Transactions on Dielectrics and Electrical Insulation* Vol. **14** Issue: 4.

[7] Holma J and Barnes M J 2012 Pulse power modulator development for the CLIC damping ring kickers *CERN-OPEN-2012-011* CLIC-Note-938 CERN Geneva Switzerland.

IOP Conf. Series: Journal of Physics: Conf. Series **874** (2017) 012096 doi:10.1088/1742-6596/874/1/012096

[8] Holma J, Barnes M J and Ducimetière L 2016 Measurements on magnetic cores for inductive adders with ultra-flat output pulses for CLIC DR kickers *proceedings IPAC'16* Busan Korea.
[9] Holma J and Barnes M J 2012 Evaluation of components for the high precision inductive adder for the CLIC damping rings *proceedings IPAC'12* New Orleans Louisiana.

Study of the suitability of 3D printing for Ultra-High Vacuum applications

Stéphane Jenzer[1], Manuel Alves[1], Nicolas Delerue[1], Alexandre Gonnin[1], Denis Grasset[1], Frederic Letellier-Cohen[1], Bruno Mercier[1], Eric Mistretta[1], Christophe Prevost[1], Alexis Vion[2], Jean-Pierre Wilmes[3]

[1]LAL, Univ. Paris-Sud, CNRS/IN2P3, Université Paris-Saclay, Orsay, France.
[2] BV Proto, Rue de Leupe, Sévenans, France.
[3] AGS Fusion, 35 Route du champ Biolay, Izernore, France.

E-mail: `jenzer@lal.in2p3.fr`, `delerue@lal.in2p3.fr`

Abstract. In the recent year additive manufacturing (3D printing) has revolutionized mechanical engineering by allowing the quick production of mechanical components with complex shapes. So far most of these components are made in plastic and therefore can not be used in accelerator beam pipes. We have investigated samples printed using a metal 3D printer to study their behavior under vacuum. We report on our first tests showing that such samples are vacuum compatible and comparing pumping time.

1. Introduction

The introduction of additive manufacturing has significantly reduced the time required to produce a mechanical part to be used in an accelerator. When a situation requires it a part, even with a complex shape, can be designed and printed within a few hours of the need for it arising. Additive manufacturing also allows the manufacturing of parts with shapes that can not be built using traditional tools.

We have been using 3D printed parts for more than a year in an accelerator environment satisfactorily. Some of these complex parts, made of plastics, have been exposed to large levels of radiation for several months (for example in [1]) without loss of mechanical performances.

However at the moments the benefits of additive manufacturing can not be extended to parts that are inserted in vacuum as the UHV compatibility of 3D printing still has to be established. Some preliminary measurements have already been reported [2] but we are not aware of any systematic campaign of measurement.

2. Method

To tests the suitability of 3D printing to UHV applications we have acquired 130 mm long DN40KF tubes with the same drawing (shown on figure III) from two different manufacturer (BV Proto[1] and AGS Fusion[2]) of metal 3D printed objects and also from a reputable manufacturer of

[1] See `http://bvproto.eu`
[2] See `http://www.ags-fusion.fr`

conventional vacuum equipment. Table 1 lists these tubes and their specific treatment. All these parts have been leak tested with helium and then mounted on a test stand using a synthetic rubber (Viton) gasket. On the test stand the parts have been pumped to measure the limit pressure that could be reached and then left under vacuum without pumping to measure the pressure increase. As described in table 1 some of the 3D printed parts have been tested directly whereas others have been machined on a lathe to improve the surface quality of the flange. The tests have been limited so far to quick flanges (KF) as manufacturing the knife of a conflat flange (CF) is more difficult with conventional tools (and not possible in 3D printing).

The tubes have all been manufactured using Selective Laser Melting (SLM) with a stainless steel 316L powder. They have all been manufactured vertically (horizontal layers). If the tubes had been manufactured horizontally they would have been printed faster but they would probably have sagged under their own weight during the printing or required the addition of many supports (that must then be removed in a time consuming process). The tubes made by BV proto were manufactured using a layer thickness of 0.04 mm and took about 30 hours to print (all 4 tubes together). The tubes made by AGS Fusion were manufactured using a layer thickness of 0.02 mm and took about 60 hours to print (all 4 tubes together).

The raw tubes after printing can be seen on figure I and II.

Figure 1. Raw tubes on their support just after printing at BV proto.

As after printing the tubes are attached to their support it is necessary either to saw them at one end (BV proto) or to wire-cut them out of their support (AGS Fusion).

As expected the surface quality of 3D printed tubes is very different of that obtained from conventional techniques, so for each manufacturer we investigated raw tubes but also tubes with typical surface finishing (bead blasting and lathing). In the case of lathing, we investigated both the case where only the flanges are lathed (see figure III) and the case where both the flanges and the inside are lathed (see figure IV).

Figure 2. Tubes removed from their support at AGS Fusion.

The surface roughness of the raw tubes from BV proto was measured to be Ra = 8.5 μm to 10 μm and for AGS fusion Ra = 6 μm to 7.5 μm.

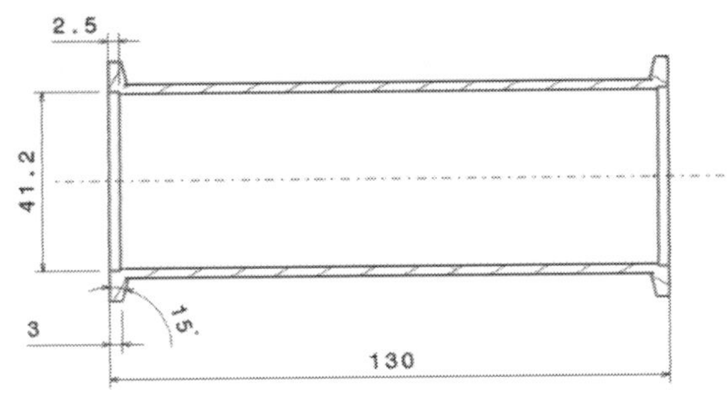

Figure 3. Drawing of the tubes and in red the area that was lathed on the flanges in the case of BV3 and AG3.

3. Results

The results of our measurements are summarized in table 1 and for the tubes for which no leaks were detected on figure V. As can be expected the tubes were a leak had been detected (BV1, BV2, AG1 and AG2) did not stay under static vacuum for very long.

It is important to stress that for all tubes the leak testing did not show any leak on the body of the tubes and the leaks detected for BV1, BV2, AG1 and AG2 were due to the poor quality of the flange surface. These results show that a raw tube produced by selective laser melting (SLM) is not directly vacuum compatible. However once the flanges of the tube have been lathed, such tube had vacuum performances comparable to those of commercial product.

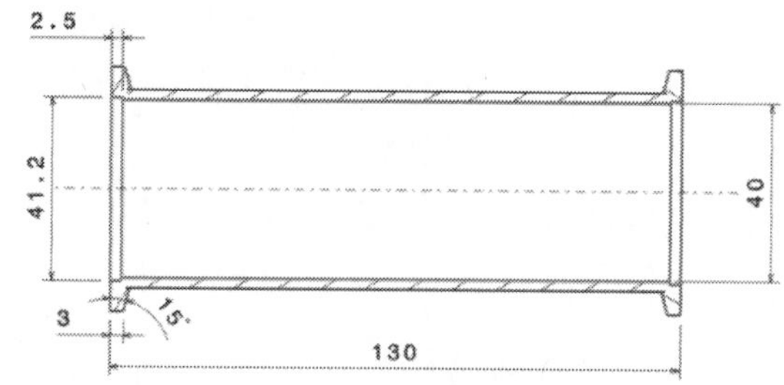

Figure 4. Drawing of the tubes and in red the area that was lathed (flanges and inside) in the case of BV4 and AG4.

The fact that no differences were seen between the two types of lathed tubes shows that the lathing of the inside of the tube is not necessary.

Additive manufacturing is best suited for extruded shapes as such shapes do not require any supports to be realized when built in vertical position (horizontal layers been added one after the other). In the case of the tube presented below (see figure III and IV) the main difficulty is the 15° chamfer required for the tightening clamp. To make this chamfer we used a support (deposited during the printing process) but this required to use tools to remove the support after printing. From our results we conclude that it would have been better to make a 45° chamfer that would then have been rectified during the lathing process.

We plan to continue our study using metal gasket that will allow us to go to lower pressure

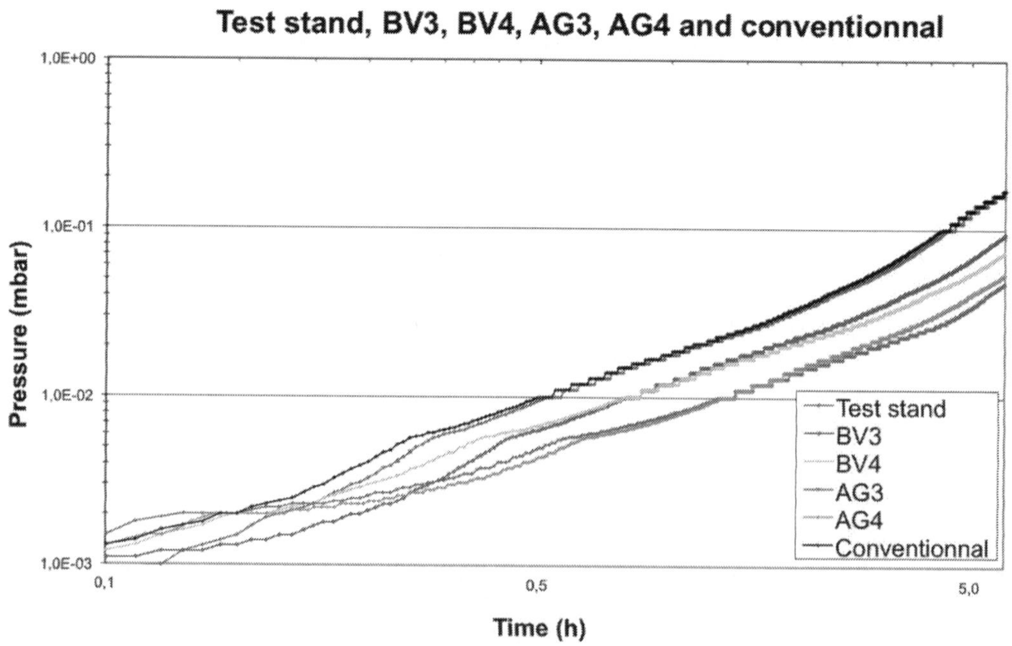

Figure 5. The static vacuum pressure measured as function of time for each of the samples that passed the leak testing. We can see that these sample perform as well as the reference sample in these tests (within measurement uncertainty).

IOP Conf. Series: Journal of Physics: Conf. Series **874** (2017) 012097 doi:10.1088/1742-6596/874/1/012097

Table 1. List of tubes tested in this study and the helium leak test and limit pressure (Penning) test results.

Manufacturer	Part name	Surface finishing	He leak test	Limit pressure
BV Proto	BV1	Sawing at one end	Raw: 1×10^{-7} mbar l/s Sawed: $> 1 \times 10^{-5}$ mbar l/s	1.7×10^{-4} mbar
	BV2	Minor processing with hand tools	$> 1 \times 10^{-5}$ mbar l/s	8.6×10^{-4} mbar
	BV3	Lathing of both flanges	No leak detected	1.2×10^{-5} mbar*
	BV4	Lathing of both flanges and the internal surface	No leak detected	1.2×10^{-5} mbar*
AGS Fusion	AG1	Wire-cutting at one end	Raw: 3×10^{-7} mbar l/s Wire-cut: $> 1 \times 10^{-5}$ mbar l/s	8.5×10^{-4} mbar
	AG2	Wire-cutting at one end	2×10^{-7} mbar l/s Wire-cut: $> 2.8 \times 10^{-7}$ mbar l/s	1.2×10^{-3} mbar
	AG3	Lathing of both flanges	6.2×10^{-8} mbar l/s No leak detected	1.5×10^{-5} mbar*
	AG4	Lathing of both flanges and the internal surface	No leak detected	9.6×10^{-6} mbar*
Vacom	Reference	Conventional	No leak detected	1.8×10^{-5} mbar*

* This is equivalent to the limit pressure of the test stand.

and test these tubes in the Ultra High Vacuum range.

References

[1] Khodnevych V and Delerue N 2017 *IPAC2017* These proceedings p MOPAB026
[2] Rummery P 2015 Tests carried out on st.5t cube private communication

IOP Conf. Series: Journal of Physics: Conf. Series **874** (2017) 012098 doi:10.1088/1742-6596/874/1/012098

Study of the new return yoke for the upgraded Superconducting Cyclotron of INFN-LNS

A Calanna[1], L Calabretta[1], D Rifuggiato[1], G Gallo[1], G Costa[1], L Allegra[1], A D Russo[1], G D'Agostino[1,2]

[1] INFN-LNS, Catania, Italy

[2] Univers. degli Studi di Catania, Dip.to di Fisica e Astronomia, Catania, Italy

E-mail: `calanna@lns.infn.it`

Abstract. The LNS Superconducting Cyclotron (CS) has been working for 20 years making a wide range of ions and energies available. Many experiments are performed each year. In the near future a major upgrade is planned. This will allow to overcome the major limitation of the CS, which is the beam power delivered to the users, that at present does not exceed 100 W. In the new version of the CS, the extracted beam power will be increased up to a factor 100. This improvement will be reached extracting by stripping a specific set of light ions and energies extracted by stripping. Nevertheless, the extraction through the two electrostatic deflectors, providing a beam power limited to 100 W, will be also maintained to fulfil the users requests. The new design could strongly affect the beam dynamics. The iron yoke penetrations do not respect the three folds symmetry of our cyclotron and have a complex shape, due to the double extraction methods and all services entrances. This inhomogeneity produces unwanted field harmonics, which have to be reduced as much as possible to avoid beam precession or second order effects. Here the study accomplished to minimize the perturbation of the non-three fold field symmetry using the current sheet approximation (CSA) is presented, along with the state-of-art configuration of the updated cyclotron.

1. Introduction

The INFN-LNS Superconducting Cyclotron ("Ciclotrone Superconduttore", CS) is an isochronous three-sector compact accelerator equipped with two pairs of superconducting coils and a RF system operating in the range of 15 to 48 MHz These features allow the acceleration of almost all ions up to uranium in a wide range of energies, between 10 and 80 AMeV.

The CS was designed for nuclear experiments with low intensity beams. The beam extraction is performed with through two electrostatic deflectors, which limits the maximum power delivered to the users. Indeed, due to the compactness, the last accelerated orbit is not completely separated from the previous one.

Then, the maximum extraction efficiency is limited to 60%. The non-extracted beam hits the septum on the first electrostatic deflector, causing issues due to thermal deformation of the septum, extra outgassing. The increased dark current on the deflector and the consequent discharges can cause mechanical damages. The present intensity is not sufficient for production of radioactive nuclei or for investigation of rare nuclear reaction, while it is not at all a limit for the exploitation of usual nuclear experiments.

IOP Conf. Series: Journal of Physics: Conf. Series **874** (2017) 012098 doi:10.1088/1742-6596/874/1/012098

The goal of this project is to add a new extraction system dedicated to a selected set of ions and energies, keeping the extraction the other ions and energies allowed by the operating diagram through the present extraction system. The selected ions will be extracted by stripping without the limitation described above and the beam power could reach 10 kW in some cases. See table 1.

At first, the request of high beam power came from the experiment NUMEN [1] and from scientists interested to use the FRIBs facility at LNS [2], which is used to produce radioactive ions with the in-flight method. The selected ions for the extraction by stripping are ^{12}C, ^{18}O, ^{20}Ne at energies between 15 to 70 AMeV, see table 1. This extraction method is based on the sudden change of charge state and the consequently drop of the magnetic rigidity after the stripper foil crossing.

Table 1. List of the Ions to be Extracted by Stripping and Expected Power Extracted.

Ion	Energy	I_{source}	$I_{accelerated}$	$I_{extracted}$	$P_{extracted}$
	AMeV	eμA	eμA	pps	Watt
$^{12}C^{4+}$	45	400	60(4+)	9.4*10^{13}	8100
$^{12}C^{4+}$	60	400	60(4+)	9.4*10^{13}	10800
$^{18}O^{6+}$	29	400	60(6+)	6.2*10^{13}	5220
$^{18}O^{6+}$	45	400	60(6+)	6.2*10^{13}	8100
$^{18}O^{6+}$	60	400	60(6+)	6.2*10^{13}	10800
$^{18}O^{7+}$	70	200	30(7+)	2.7*10^{13}	5400
$^{20}Ne^{7+}$	28	400	60(7+)	5.3*10^{13}	4800
$^{20}Ne^{7+}$	60	400	60(7+)	5.3*10^{13}	10280
$^{40}Ar^{14+}$	60	400	60(14+)	2.7*10^{13}	10280

According to data in [3], for the selected ions at energies bigger than 15 AMeV, the percentage of the particles with q=Z (where q and Z are charge state and atomic number of the ion respectively) after the stripping foils is higher than 99%. That means that low losses are expected inside the machine if the extraction channel is carefully designed. Table 1 summarizes the expected results in term of beam power increase for the cases studied. In table 1, the values of beam current delivered by the ion source and accelerated by the cyclotron are conservative.

Indeed, the limit of 10 kW is posed mainly to maintain the beam losses inside the cyclotron below 100-200 W and to mitigate the problem of machine activation.

For each ion and energy of interest, the position of the stripper foil has been found, as well as the positions of the two magnetic channels needed to provide radial focusing and to slightly steer the beam after the pole radius. See [4], [5] and [6] for more details on the design of the extraction channels. Note that, since the stripping extraction is by definition a multi-turn extraction, an energy spread is introduced. The computations have been done considering an energy spread of +/-0.3% [7].

As showed in [4], [5] and [6] the new extraction system implies major mechanical modifications to the cyclotron.

Firstly, a new cryostat with new superconducting coils [8] is mandatory. The new extraction channel has a larger vertical gap than the existing one. Massachusetts Institute of Technology [8] has done a feasibility study of the cryostat and a call for tenders is at the moment open to select the supplier for the new magnet. Secondly, additional penetrations are necessary to host new magnetic channels and new compensation bars.

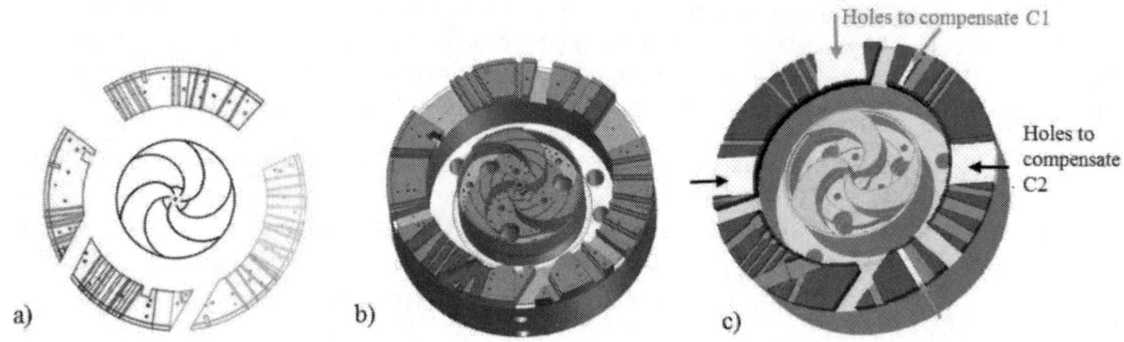

Figure 1. Views of the bottom half of the CS. The first ones, *a* and *b*, are top views of the CS as it is today. In *a*, the existing extraction channel is in the orange sector. In *b* the modifications needed to add the new extraction channel, which is almost 30 degrees away from the old one, are shown in magenta. Finally, *c* shows the results of the optimizations process, see text. The arrows highlight the areas where the iron has been removed.

2. Mechanical modifications to the iron yoke

Since the CS has been designed and constructed more than 30 years ago, the engineering drawings of the machine are only on paper. 3D models were created reconstructing the cyclotron state-of-art.

The introduction of the new extraction channel implies the rotation of the penetrations for the horizontal tie-rods and the addition of new penetrations for the new magnetic channels, the new compensation bars [4], [5] and [6] and the new stripper foil housing. Moreover, other few changes to the penetrations have been optimized. A detailed explanation of all the modifications done by the LNS engineering team working at this project is reported in [9]. The yoke modifications are done to a central ring of the yoke that has a height of 25 cm, 12.5 cm around the median plane, which is the particle acceleration plane.

The drawing of Figure 1*a* shows the state-of-art of this ring. The existing extraction channel is in the orange sector. In the dark blue sector, there are all the penetrations for the first electrostatic deflector and the penetration for the radial injection, which today is not anymore used. The poles and the inner wall of the cryostat are designed as well. The drawing in Figure 1*b* shows in magenta all the modifications needed in the yoke according to the engineering team to add the new extraction line and the necessary subsystems.

However, drilling new non-symmetric penetrations add new field harmonics components, which have to be decreased at least at the level of the existing, working, configuration, Figure 1*a*. This step is mandatory since field harmonics components are responsible for a precession of the beam and for unwanted second order effects on the beam. Therefore, the iron in the central ring of the yoke has to be redistributed, where possible, to guarantee good beam dynamics performance, and, at the same time, have stresses under control.

During the CS construction, the first and second harmonic field components optimization was done keeping off the iron at the right of the violet sector and at the left of the green sector of Figure 1*a*. Today, to achieve a final optimized design, 3D simulations are useful. Note that, in the dark blue sector, there is also the penetration for the radial injection, which today is not anymore used since the beams are axially injected. Then, as a starting point of the optimization process to get the final configuration of the upgraded CS, it is possible to add iron closing the old penetration for the radial injection and closing the holes at the right of the violet sector and at the left of the green sector (keeping only the needed penetrations).

3. Use of CSA to decrease the unwanted harmonics

The method used to define the iron profile in the central ring of the return yoke is the Current Sheet Approximation (CSA), [10], which gives the possibility to study the iron shape using a coil of negligible thickness that occupies the same volume of the iron. This method has been used also to define the profile of the magnetic channels needed in the new extraction channels.

The CSA is valid in the case of full-saturated iron and, as a first approximation; it is possible to consider it reliable for all values of the current flowing in the superconducting coils. Figure 2 shows the central ring after the modifications needed to add the new extraction channel. In blue, there is the outer wall of the cryostat, which is not a part of the return yoke, but it is iron-made and has not a three-fold symmetry. Then, it has an effect of the field harmonics components and it has been taken into account in these magnetostatic computations.

As a starting point, the first and second harmonic of the magnetic field has been evaluated for the configuration with the iron ring as in Figure 2, including only the needed penetrations for the CS and closing the old unused ones as explained above.

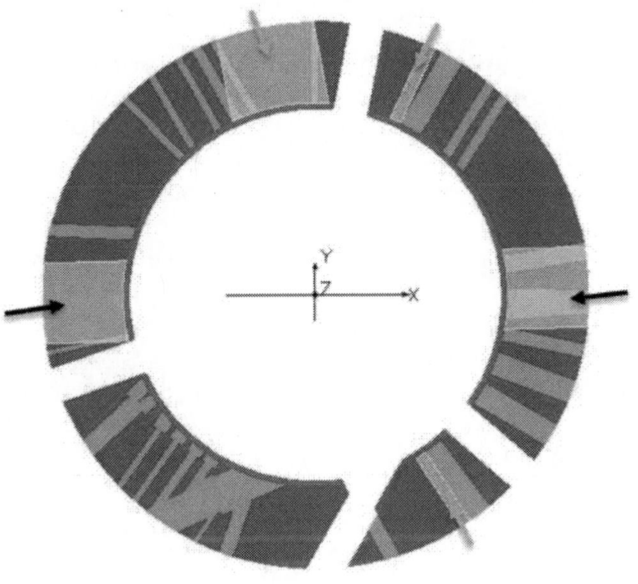

Figure 2. Central iron ring of the yoke of the CS. In navy, there is the iron with all the penetrations needed to guarantee both extraction methods. In translucent yellow, there are the profiles of the new penetrations to add to the central iron ring of the yoke to cancel the first (orange arrows) and second (black arrows) field harmonic components.

Using the FEM code OPERA (Cobham), it is possible to design a parametrized coil combining arc and straight conductors. In this way, it is possible to identify the size and position of iron to be removed to have the field harmonics compensated. The yellow volumes in Figure 2 show the profiles of the holes needed for the corrections using the CSA. The two small iron penetrations and the big one centred at about 100 degrees are needed to cancel the first harmonic component, whose peak is in correspondence of the new extraction channel. The other two pieces to remove are also necessary to compensate the second harmonic component.

A 3D view of the iron yoke after the optimization process is reported in Figure 1*c*.

The first and second harmonic of the magnetic field vs. radius for the initial and of the final configuration achieved by the optimization process, are shown in Figure 3. Along the entire acceleration path the first harmonic stays below 5 Gauss and the second does not reach 1 Gauss.

This result is better than the present performance obtained with configuration in Figure 1*a*.

Although the volume of the central iron ring is similar to the present case and the harmonic compensation is very satisfactory, other solutions are under investigation. The idea is to replace the big hole centred at 100 degrees, with other smaller ones opportunely placed around the ring.

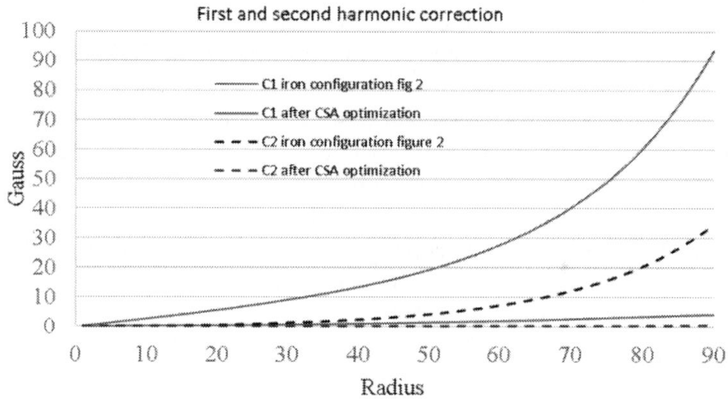

Figure 3. First and second harmonic before and after the optimization process using the CSA method.

This is done in the perspective to distribute as much as possible the compression stresses due to the lack of iron. For the same reason, to compensate the second harmonic component, simulations will help to understand if smaller pieces of iron closer to the acceleration path can avoid the drilling of the big holes in the iron. It worth mentioning, these results are important not only for the high intensity beams, but also for all the accelerated ions included the ones extracted through electrostatic deflectors.

4. Conclusions

The goodness of the optimization process to cancel the first and second field harmonic components through the use of parametrized coils and the CSA approximation has been proven. Further investigations are mandatory to try to reduce the volume of the iron to take off the central iron ring of the yoke of the CS. Moreover, the same calculations will be done for more values of current flowing in the superconducting coils.

References

[1] Capuzzello *et al.* 2015 *Eur. Phys. J.* A **51** 145
[2] Raciti G *et al.* 2010 *Nucl.Phys.* A **834 1-4** 780c-783c
[3] July 1992 *Atomic Data and Nuclear Data Tables* **51** 2 Table 2 187
[4] Calabretta L, Calanna A, Rifuggiato D *et al.* publication process started on March 2017 *MPLA*
[5] Calanna A publication process started on May 2017 *Nuovo Cimento C Colloquia and Communications in Physics* High intensity extraction from the Superconducting Cyclotron at LNS-INFN
[6] D'Agostino G, Calanna A *et al.* 2016 *Proc. Int. Conf. on Cyclotrons* Zurich TUC03
[7] Tomic S S and Samsonov E V 2002 *Phys. Rev* E **65** 036504
[8] Radovinsky A *et al.* 2016 *IEEE Transactions on Applied Superconductivity* **26** Issue 4
[9] Gallo *et al.* 2016 *Proc. Int. Conf. on Cyclotrons* THP09
[10] Gordon M M and Johnson D A 1980 *Particle Accelerators* **10** 217

Improvements of vacuum system in J-PARC 3 GeV synchrotron

J Kamiya[1], Y Hikichi[1], Y Namekawa[1], K Takeishi[1], T Yanagibashi[1], M Kinsho[1], K Yamamoto[1] and A Sato[2]

[1] J-PARC, Japan Atomic Energy Agency, Shirakata 2-4, Tokai, Naka, Ibaraki, Japan
[2] Nippon Advanced Technology Co. Ltd., Muramatsu 3129-45, Tokai, Naka, Ibaraki, Japan

kamiya.junichiro@jaea.go.jp

Abstract. The RCS vacuum system has been upgraded since the completion of its construction towards the objectives of both better vacuum quality and higher reliability of the components. For the better vacuum quality, (1) pressure of the injection beam line was improved to prevent the H- beam from converting to H0; (2) leakage in the beam injection area due to the thermal expansion was eliminated by applying the adequate torque amount for the clamps; (3) new in-situ degassing method of the kicker magnet was developed. For the reliability increase of the components, (1) A considerable number of fluoroelastmer seal was exchanged to metal seal with the low spring constant bellows and the light clamps; (2) TMP controller for the long cable was developed to prevent the controller failure by the severe electrical noise; (3) A number of TMP were installed instead of ion pumps in the RF cavity section as an insurance for the case of pump trouble.

1. Introduction

The 3 GeV Rapid Cycling Synchrotron (RCS) is a keystone of J-PARC facility since it plays a role of both the main accelerator for the Material and Life Science Facility (MLF), and the injector to the Main Ring (MR). The RCS aims to achieve the proton beam power of 1 MW, which corresponds to each cycle 8.3×10^{13} protons accelerated up to 3 GeV at the repetition rate of 25 Hz. The design concept of the RCS vacuum system, the uniquely developed vacuum components, and the basic vacuum characteristics achieved were reported in the previous articles [1, 2]. Since the completion of its construction, the chief purposes of the RCS have been accomplishment of the higher beam power and more stable operation. For both purposes, improvements of the vacuum quality and increase of the vacuum components reliability have been performed in the RCS vacuum system. As an example of such efforts, Figure 1 shows the comparison of the present beam line pressure distribution with past one. The pressure was improved in the several sections. In this article, the upgrades of the vacuum system are shown from the perspectives of the vacuum quality improvement and reliability increase of the vacuum components.

IOP Conf. Series: Journal of Physics: Conf. Series **874** (2017) 012099 doi:10.1088/1742-6596/874/1/012099

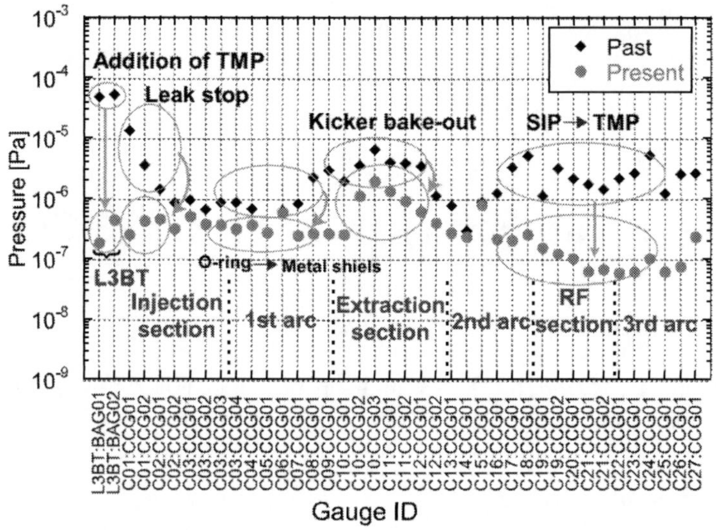

Figure 1. Comparison of the present beam line pressure distribution with the past one. Improvements in 2nd and 3rd arc also owes the replacement of O-ring to the metal seals.

2. Vacuum quality improvements

2.1. Beam loss reduction by lower pressure at the injection beam transport line [3]

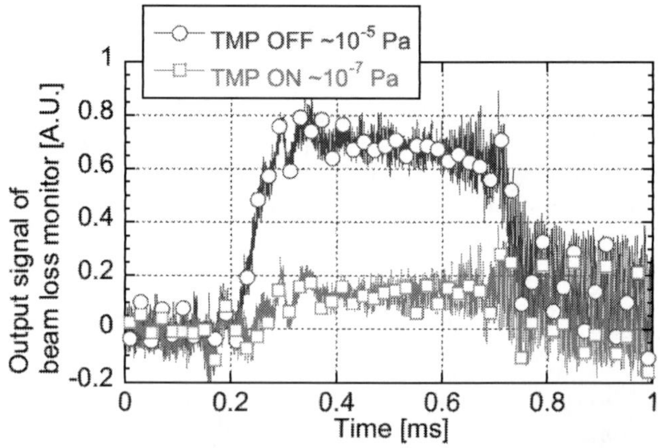

Figure 2. Beam loss suppression in the injection branch area by the vacuum improvement in L3BT line.

The injection line called L3BT line (Linac to 3GeV synchrotron Beam Transport line), was the section, where the high beam line pressure causes the beam loss. In this section, H^- beam from Linac was converted to H^0 by charge stripping due to the interaction between H^- beam and the residual gas molecules. Such H^0 beam was not bended by the injection septum magnets and directly hit the vacuum wall at the injection branch. Additional vacuum pumps were installed in this section to reduce the beam line pressure and beam loss in this section. By installing two turbo molecular pumps (TMP), the beam line pressure was reduced by double-digit from 10^{-5} Pa to 10^{-7} Pa. As a result, the beam loss in the injection branch area due to the charge stripping of the injection beam was successfully suppressed as shown in Figure 2.

2.2. Elimination of vacuum leak in the beam injection area

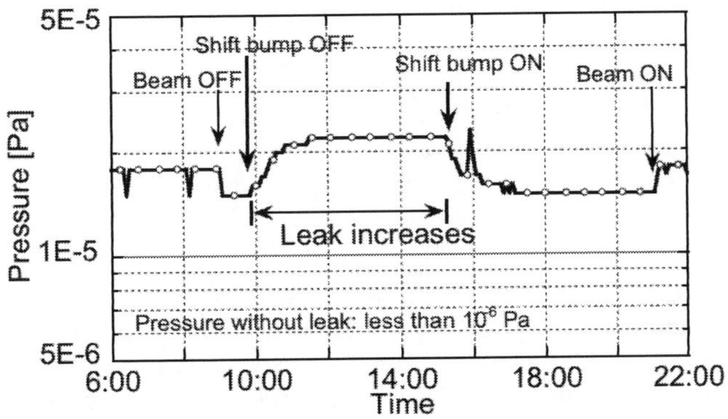

Figure 3. Leak property in the beam injection area. The leak caused by the thermal expansion increases without excitation of the pulse magnets.

Vacuum leak happened in the RCS beam injection area where the normal pressure without leak is less than 10^{-6} Pa. The amount of leak shows unique property as shown in Figure 3. The leak increased when the shift bump magnets, which generates the pulse magnetic field, are not excited, while it decreased with the bump magnets excitation. Such situation would be caused by the thermal expansion of the beam pipe of alumina ceramics and/or the flange of pure titanium, which is installed in the aperture of the shift bump. Measured temperature was about 50 °C for the beam pipes and 100 °C for the flanges, respectively, which was caused by the radiation from the magnet cores and induced current by the pulse magnetic field. Corresponding thermal expansion is 0.4 mm to the beam direction and 0.3 mm to the flange diameter direction, respectively. In the construction period, minimum amount of torque, which guarantees vacuum leak of less than 10^{-11} Pa m^3/s was applied for each flange clamps to prevent the sealing surface of the flange from denting by the metal seal. Actually, the amount of torque was 10-20 N m. However, the repeated thermal expansion would have caused the clamp laxity resulting the gap between the flanges.

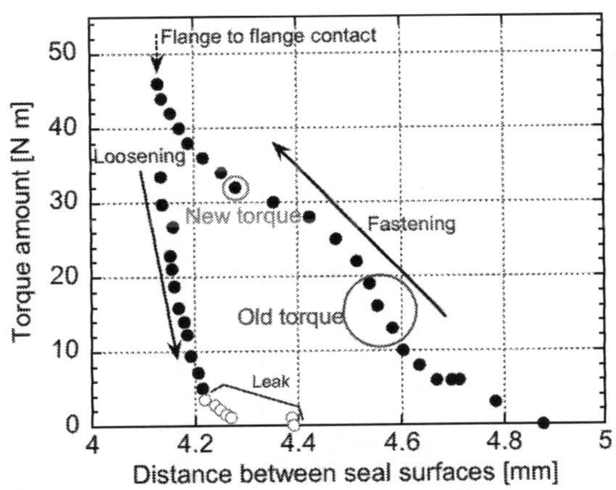

Figure 4. Measured typical compression curve of the metal seal.

IOP Conf. Series: Journal of Physics: Conf. Series **874** (2017) 012099 doi:10.1088/1742-6596/874/1/012099

To search the adequate torque amount, the compression properties of the metal was measured. Figure 4 shows the measured typical compression curve of the metal seal. The torque amount of 10-20 N m is obviously too small because such region is just the start of the compression. The torque amount of 30 N m would be enough because the seal is sufficiently compressed to near the flange and flange contact. Therefore, the torque amount of 30 N m was applied to the clamps in the injection area. As a results the leak has not occurred so far. Furthermore, we are currently measuring the compression curve with the maximum torque amount of 30 N m to demonstrate that such torque amount does not actually cause the vacuum leak if the bolts slack by the repeated thermal expansion.

2.3. In-situ bake-out of the kicker magnets [4]

Degassing of the kicker magnets in vacuum is necessary to prevent the damage of the components, such as ferrite cores and aluminum electric plates, by electric discharge. The usual bake-out method, which involves heating the vacuum chamber by the heaters attached on the outer surface of the chamber, cannot be applied in the beam line due to the thermal expansion of the long chamber with the length of about 5 m. If the chamber was baked at 120 °C, where the main outgassing components of water vapour is effectively desorbed, the thermal expansion would be more than 5 mm and it would break the nearby ceramics pipes.

We have developed a new degassing method based on the principle of the thermal radiation shielding, by which only kicker magnets are baked out without raising the temperature of the vacuum chamber to prevent unwanted thermal expansion of the chamber. The principle of thermal radiation shielding recognizes that the thermal radiation heat flux q is reduced to one-$(n+1)^{th}$ by n shielding plates. For the kicker degassing, by simply installing the heater and thermal radiation shield plates between the kicker magnet and the chamber wall as shown in Figure 5 (a), most of the heat flux from the heater directs towards the kicker magnet. Verification measurement with the R&D kicker demonstrated that each part of the kicker magnet was heated to above 120 °C with a small rise in the vacuum chamber temperature less than 20 °C by using the 5 shield plates. Degassing of the kicker magnets in the RCS beam line was actually performed by the new bake-out method. As a result, with keeping the thermal expansion of the chamber less than 0.2 mm, the beam line pressure was successfully improved as shown in Figure 5 (b).

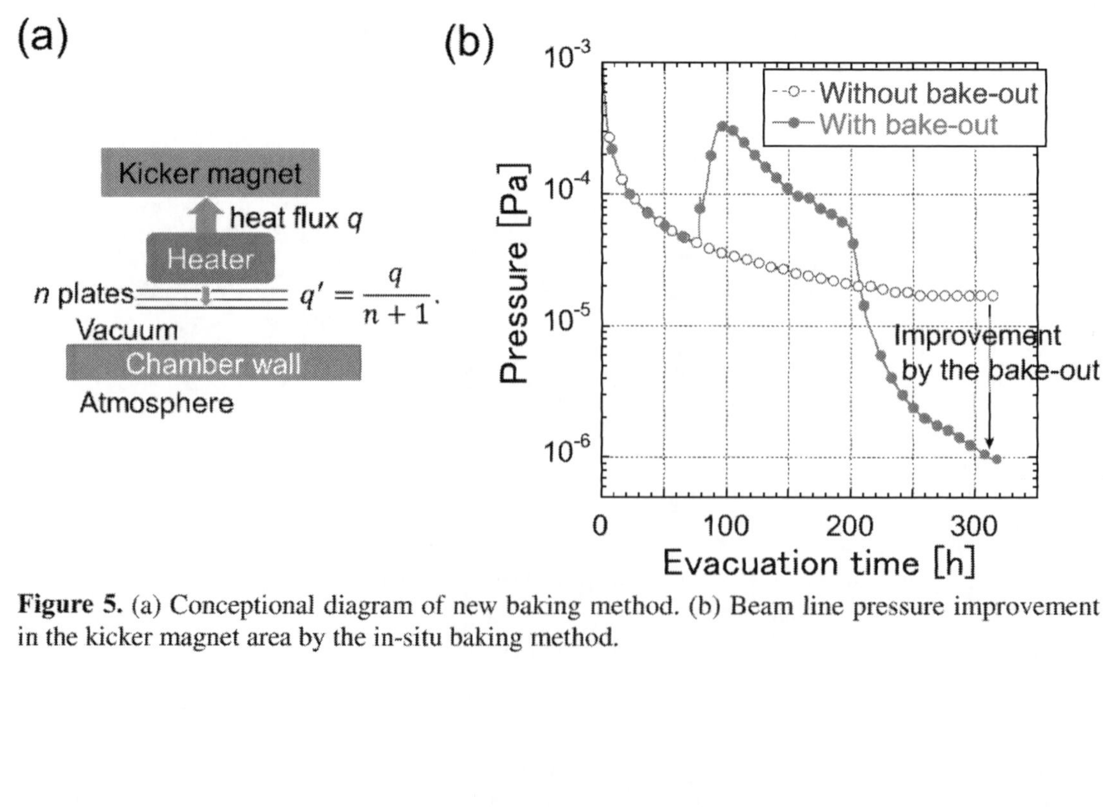

Figure 5. (a) Conceptional diagram of new baking method. (b) Beam line pressure improvement in the kicker magnet area by the in-situ baking method.

3. Reliability increase of the vacuum components

3.1. Replacing O-ring to metal seals along with low spring constant bellows and very light clamps

In the construction period of the RCS vacuum system, O-ring seals of fluoroelastmer were used for about 35 flange joints out of about 450 in total because the leak did not stop with the metal seal. Especially the connection with the beam pipes for the bending magnet tended to leak because it is difficult to parallelize the flanges and align each central axis for the bended beam pipes. However, the O-ring seals should be replaced to the metal ones for higher intensity beam operation because fluoroelastmer would deteriorate with high radiation. Thus, bellows with low spring constants were developed to connect such misaligned flanges. Hydroformed bellows with large apertures generally have high spring constants. The flexible bellows without welding were successfully developed by obtaining a narrow pitch of approximately 3 mm, which was formed by a combination of hydroforming and pressing. A spring constant of less than 25 N/mm was achieved. Universal type, which consists of two bellows and a pipe between them, was applied to connect a large displacement perpendicular to the beam axis of approximately 5 mm. The connection point with developed bellows and clamps are shown in Figure 6. For the flanges joint, the light clamps of aluminum alloy with high mechanical strength was newly manufactured. This clamp consists of only two semi-circular parts for one flange for better handling.

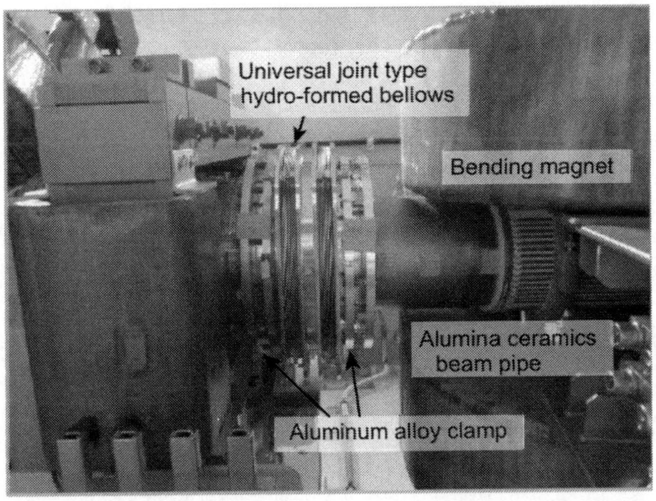

Figure 6. Low spring constant hydro-formed bellows with universal joint type and the light clamp of aluminium alloy.

3.2. TMP controller for long cable

According to the upgrade of the injection beam energy from 181 MeV to 400 MeV, the excitation current for the injection shift bump magnet was increased. After that, the controllers of the TMP often froze due to the electrical noise and rotor touched down with rated rotating speed. This was very big problem because the TMP rotor and stator blades contact with each other, causing the deformation of such blades. Figure 7 shows the solution of this problem. Controllers were placed in the utility tunnel, which was just under the accelerator tunnel, because of the limitation of the cable length between the TMP and its controller. We decided to relocate the controllers to noise-free environment by developing the control method with long cables. By combining the amplifier with the controller, the cable length was successfully lengthened to more than 150 m. The failure of the TMP controller have been eliminated by such countermeasure.

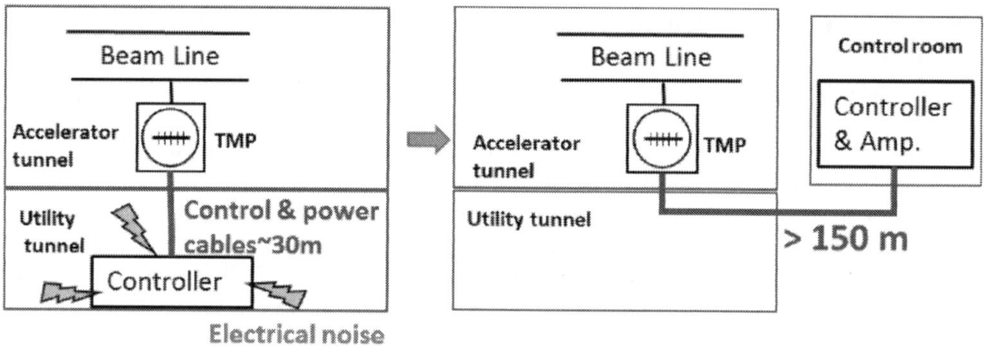

Figure 7. Counter measure against the TMP controller freeze by the electrical noise. The controller with amplifier was newly developed for using the long cables.

3.3. System reliability improvement of RF cavity section

The TMP and spattering ion pumps (SIP) were concurrently used at the first stage of the vacuum system operation. However, the absorbed ions, which might be the residual molecules ionized by the electrical field of the RF cavity, were suddenly re-desorbed as a large amount of outgassing during the beam operation. The SIP had not been used since then. As a result, pumping of the RF cavity section depends on only one TMP. In such situation, failure of the TMP directly leads to the long term stop of the beam operation. Two SIP were replaced by the TMP as shown in Figure 8. The stable operation become insured even if one or two TMP failure in this area with achieving lower beam line pressure as a matter of course.

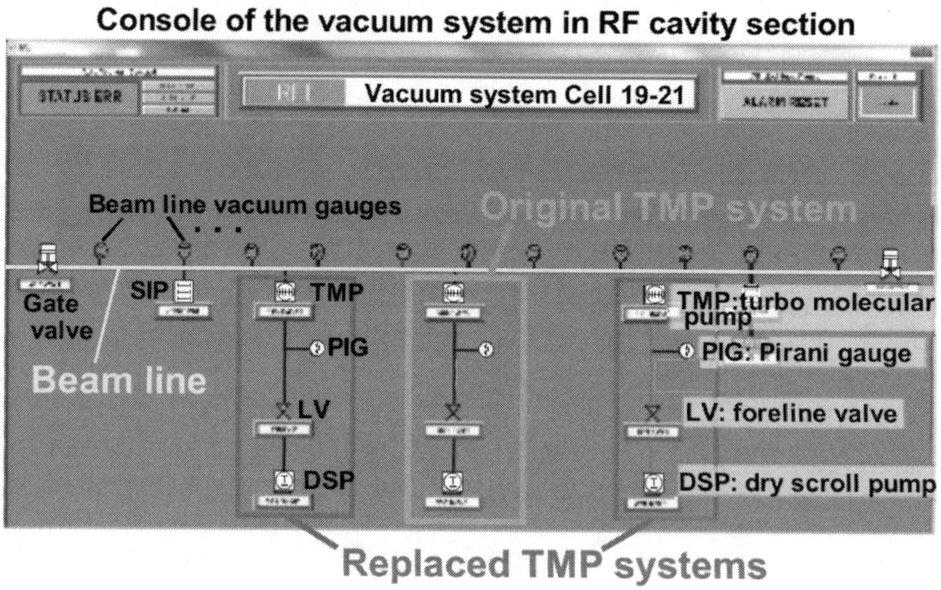

Figure 8. TMP replaced from SIP in the RF cavity section.

4. Conclusion

Continuous upgrades of the RCS vacuum system have been performed for obtaining the better vacuum quality and higher reliability. In consequence, the lower pressure along the entire RCS beam line and the more stable vacuum system operation are achieved.

References

[1] Ogiwara N, Kinsho M, Kamiya J, Yamamoto K, Yoshimoto M, Hikichi Y, Kanazawa K, Mio K, Takiyama Y and Suganuma K 2009 *Vacuum* **84** 723

[2] Kinsho M, Saito Y, Kabeya Z and Ogiwara N 2007 *Vacuum* **81** 808

[3] Kamiya J, Kinsho M, Noshiroya S and Yamamoto K 2014 *Proc. Int. Particle Accelerator Conf (Dresden)* p 2338

[4] Kamiya J, Ogiwara N, Yanagibashi T, Kinsho M and Yasuda Y 2016 *J. Vac. Sci. Technol. A* **34** p.021604

Thermal analysis of the LHC injection kicker magnets

L Vega[1], A Abánades[2], M J Barnes[1], V Vlachodimitropoulos[1] and W Weterings[1]

[1]CERN - European Laboratory for Particle Physics, TE Division, CH-1211, Geneva 23, Switzerland
[2]Departamento de Ingeniería Energética, E.T.S. de Ingenieros Industriales, Universidad Politécnica de Madrid, Madrid, Spain

E-mail: lorena.vega@cern.ch

Abstract. The CERN Large Hadron Collider LHC is equipped with two fast pulsed magnet systems (MKIs) that inject particle beams coming from the injector chain. Operation with high intensity beams for many hours can lead to significant beam induced heating of the ferrite yokes of the MKIs. When the ferrite exceeds the Curie temperature of 125°C it loses its magnetic properties, preventing further injection until the ferrite cools down, potentially causing a delay of several hours. Hence important upgrades of the beam-screen were implemented after Run 1 of LHC. However, the High-Luminosity (HL) LHC will be operated with significantly higher intensity beams and hence additional measures are required to limit the ferrite temperature. These magnets operate under ultra-high vacuum conditions: convection is negligible and, as a result of low emissivity of the inside of the vacuum tanks, thermal radiation is limited. A detailed study of the thermal behaviour of these magnets is reported and compared with measurements. In addition several options to improve cooling of the ferrites are presented and analysed.

1. Introduction

The LHC injection kicker magnets (MKIs) deflect the incoming beam onto the LHC's equilibrium orbits [1]. To deflect the incoming particle beam, these magnets are pulsed at high voltage and current, creating a magnetic field pulse. The field lines are guided using a NiZn ferrite yoke. Although the MKIs only pulse 12 times to fill the LHC, the high intensity beam circulates through the aperture of these magnets constantly for many hours. This leads to significant beam induced heating in the ferrite yokes. If the Curie temperature (~125°C) is reached, the yoke will lose its ferromagnetic properties and the beam will be mis-injected, which can cause quenches of downstream superconducting magnets. In the future the HL-LHC will be operated with significantly higher intensity beams [2], so additional measures are required to avoid heating of the ferrite yokes above the Curie point. Furthermore, because of the ultra-high vacuum (UHV), convection is negligible and, as a result of low emissivity of the inside of the vacuum tanks, thermal radiation is limited [1]. A numerical model has been created to study the thermal behaviour of the MKI magnets. The purpose is to identify solutions to reduce the effect of beam induced heating.

2. Thermal model

A finite element model has been created using ANSYS [3]. The complexity of the MKI magnets is remarkable: there are many components with complex geometries. An MKI magnet has both side-to-side symmetry, in the longitudinal axis, and a periodic structure: each MKI is composed of 33 repeated cells. A cell consists of a U-core ferrite yoke between two HV conducting plates, and two ceramic capacitors sandwiched between a HV plate and a plate connected to ground [4]. Different models of an MKI have been analysed to identify a reasonable trade-off between accuracy and computational cost.

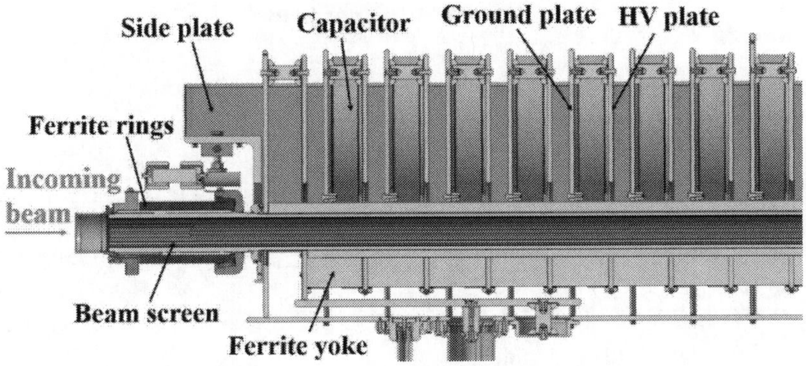

Figure 1. Model of the upstream end of an MKI magnet.

To simplify the model, advantage can be taken of the side-to-side symmetry in the longitudinal axis. The first models were 2D, taking advantage of the periodic structure [4]. However, recent studies indicate that the power is deposited in a non-uniform manner along the longitudinal axis: it is deposited mostly in the upstream end of the magnet, in the ferrite rings located at the beam entrance and in the first two ferrite yokes. From the 3rd to the 33rd ferrite yokes, the power deposition is close-to-uniform and of relatively low magnitude [5, 6]. In order to determine how many cells should be modelled, to obtain accurate predictions, simulations with up to 15 cells have been performed - these simulations allow one to see from which cell the longitudinal heat transfer can be considered adiabatic. For the thermal predictions presented in this paper the power distributions detailed in [6] have been modelled: it is concluded that 10 cells are sufficient, as the temperature difference between the 10th and the 11th ferrite yokes is less than 2%.

3. Validation

The numerical model has been validated by simulating the calculated power for several operational fills of LHC: the predicted temperatures are then compared with measured temperatures in one of the installed MKIs. To compare these temperature it must be taken into account that the PT100 sensors are not installed directly on the ferrite yoke, since it is pulsed to high voltage. The two PT100's for the yoke are located on a side plate: there is one PT100 towards each of the upstream and downstream ends, located at the position of the third yoke from each end of the magnet. In addition there is a PT100 associated with each set of ferrite rings, located on a clamp in contact with the rings (figure 2).

The time dependent values of the power depositions calculated for each of the upstream ferrite rings and each of the first 10 yokes, for a specific fill during LHC operation (in this case from 02-07-2016 00:57hrs to 03-07-2016 15:21hrs), are modelled with ANSYS. The results are shown in figure 3.

In figure 3, Magnet_Up and Tube_Up correspond to the measured temperatures, at the upstream end of the magnet, on the side plate and by the ferrite rings, respectively: both are

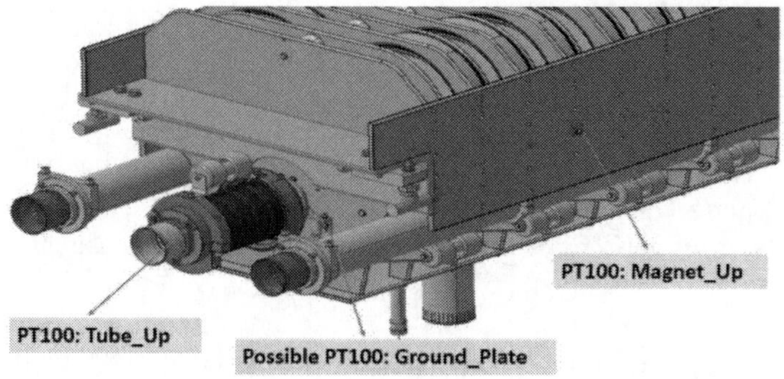

Figure 2. PT100 at the upstream end.

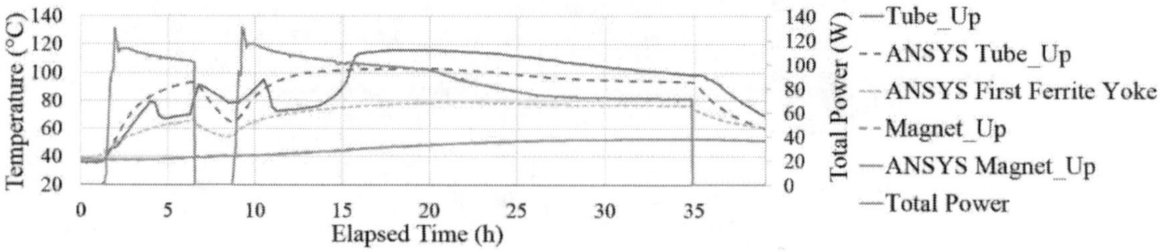

Figure 3. Comparison: measurements and predictions.

for the magnet named MKI8D. The temperatures for the side plate show very good agreement between measurements and predictions: the curves are overlapped.

For the temperature of the clamp next to the ferrite rings (Tube_Up), there is some discrepancy between measurements and predictions: this may be partially due to uncertainty about the value of the thermal contact resistance between the rings and the clamp, which is very difficult to estimate. However the measured temperature Tube_Up (figure 3, continuous green line) shows several rapid changes: these are most likely not realistic for the ferrite rings and the cause is currently under study: one theory is that this might be related to a loss in thermal contact between the ferrite rings and the clamp, due to different thermal expansion. Similar behaviour has been observed for the measured upstream ferrite ring temperatures of all the MKI magnets.

3.1. Optimum location of PT100 sensors

Calculation of the power deposition in the ferrites presents uncertainties [6, 5]. Thus, good agreement between the measured temperatures and thermal predictions also helps to validate the power deposition. Ideally the PT100 sensors should be located in positions with a high sensitivity to the ferrite temperatures, where changes in yoke temperature result in reasonable changes in the measured temperature.

In order to evaluate the quality of the present locations of the PT100s, the same transient simulation used for the validation of the model (figure 3) has been re-run with 25% increased power depositions, and the bake-out jackets that cover the tanks have not been included. The predicted temperature of the first yoke increases by 11%.

Figure 4 shows that the temperature on the side plate has increased by only 4%: hence the present PT100 location on the side-plate is not optimum. Whereas the first Ground Plate (figure 2) temperature, close to the beam aperture increases by 14%. It is planned to install a prototype MKI, with several upgrades [7] at the end of 2017. Moving a side-plate PT100, to

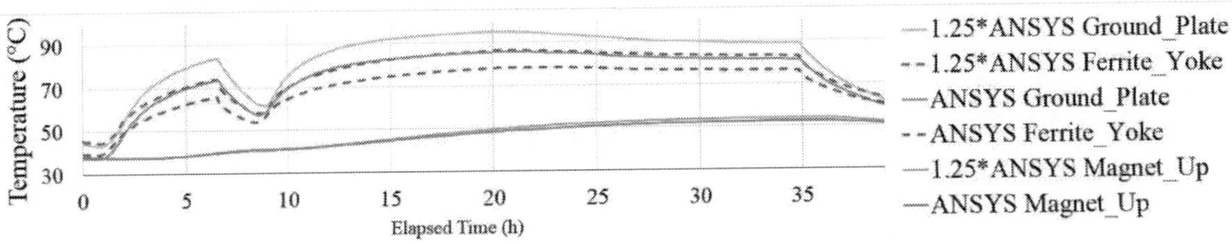

Figure 4. Power input 25% higher and no bake-out jackets.

the first ground plate, is under consideration: however, since design changes, to reduce power deposition in the yoke, are planned [6], if the PT100 is moved, the quantitative comparison to other MKIs is lost.

4. Ferrite yoke temperature

Potentially the most accurate way to predict ferrite yoke temperatures is to run a transient thermal simulation with the power data calculated for a specific fill, as was done for the validation of the model (figure 3). However, typical transient simulations require several days of CPU time and, in addition, maximum temperatures are dependent upon initial conditions which are themselves influenced by previous fills. A simpler and more pessimistic approach is to consider steady-state conditions. Figure 5 shows the relationship between the steady-state power and the temperature of the ferrite yoke and rings.

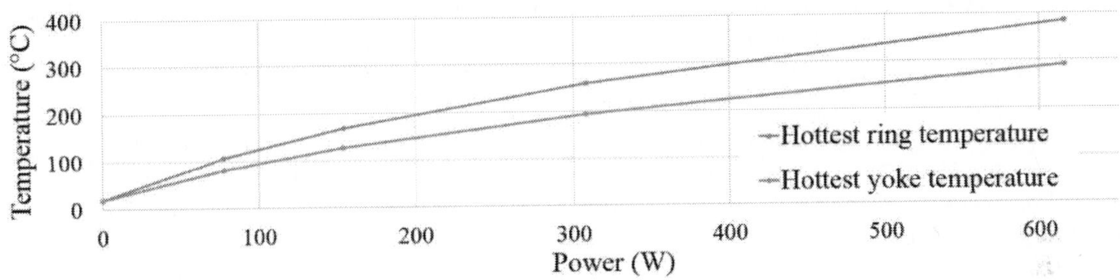

Figure 5. Relationship between steady-state power and maximum temperature of the ferrite yoke and rings.

For estimating ferrite temperature, there are several approaches to choosing a representative steady-state power. Using the above validation fill as an example, the temperature of the 1st ferrite yoke, at the end of the cycle, is 78°C (figure 3, yellow dashed line). The peak power deposition of the fill (120 W) would give a steady-state yoke temperature of 109°C (figure 5). This is a very pessimistic result because this power is only present at the start of the cycle and then reduces due to a decrease in beam intensity. In addition, due to the long thermal-time constant (~20 hours) of the MKIs, many hours are needed to reach steady-state conditions. Alternatively, the average power of the fill (91 W) would give a steady-state yoke temperature of 91°C (figure 5).

4.1. Predicted temperatures for future operation

Based on the long fill of July 2016 (figure 3), and scaling the number of bunches to 2808, the maximum temperature expected during Run 2 is 107°C [7], which is below the Curie point. The

IOP Conf. Series: Journal of Physics: Conf. Series **874** (2017) 012100 doi:10.1088/1742-6596/874/1/012100

predictions presented in this paper for HL-LHC are carried out for 25% more power deposition than expected: this is to give a safety margin. The temperature of the hottest ferrite yoke, for different operation scenarios, can be estimated using figure 5: for future operation of HL-LHC, the expected temperature is 291°C, well above the Curie point. However the power (616 W) corresponds to the peak power expected, which is a very pessimistic approach as it has been commented.

5. Cooling alternatives

5.1. Increased thermal emissivity of the vacuum tank

As a result of the low thermal emissivity of the internal surface of the MKI vacuum tanks, cooling of the ferrite yoke by thermal radiation is presently limited [8]. Extensive research has been carried out to find a high emissivity coating which is compatible with ultra-high vacuum, thermal bake-outs and high voltage and does not peel or flake. Several surface finishes have been considered: laser treatment, carbon coating [9], multilayer optical coatings deposited by magnetron sputtering, developed by Polyteknik AS [10], and coatings deposited by thermal spray (plasma spray and flame spray) [11]. However, although increasing the emissivity reduces both the maximum temperature and the thermal time-constant, so that the yokes cool down more rapidly, the temperatures are not reduced enough (for HL-LHC the highest yoke temperature is reduced from 291°C to 176°C, still above the Curie point). Hence other cooling schemes have been studied.

5.2. Removal of the bake-out jackets

Historically, bake-out jackets are left in place on the MKIs in the LHC tunnel: this is in case there is a need for an in-situ bake-out since there is insufficient space to install them in the tunnel. However the jackets also act to thermally isolate the MKIs vacuum tanks from the local environment. For improving heat extraction from the magnets it is recommended that the jackets are removed: predictions show that the ferrite yoke temperature decreases by 7% for a given power deposition. Alternatively, for a given temperature, the power deposition can be 18% higher without jackets. The following studies, regarding alternative cooling schemes, do not model bake-out jackets. For HL-LHC, the temperature of the hottest yoke would be reduced from 291°C to 267°C.

5.3. Cooling of the ferrite yokes

One alternative is to cool the ferrite yokes with a cold plate. This plate should be a good thermal conductor but also electrically insulating as the ferrite is at pulsed high voltage. A potential material is Aluminium Nitride [12, 13]. Predictions show that if only the first yoke is cooled like this, its temperature will be reduced from 267°C to 110°C but the subsequent yokes approach the Curie temperature. Hence, it would be necessary to cool down the first and the third ferrite yokes to ensure that all of them remain below the 125°C. However, additional cooling is required for the ferrite rings as their temperatures are otherwise predicted to exceed their Curie point (200°C and 250°C for 4M2 and 4B3, respectively [4]).

5.4. Cooling of the ferrite rings

Predictions show that, with the beam screen design implemented during LS1, cooling the rings to 15°C is not enough: the ferrite yokes are still above the Curie temperature (figure 6, blue continuous line). Hence, there are ongoing studies to reduce the power deposition in the ferrite yokes by decreasing the beam screen overlap length [6]: this would potentially reduce the power deposition in the first ferrite yoke by a factor of 9.5. However, the power in the hottest ring is increased by a factor of 31. Without cooling, the design change reduces the predicted first ferrite yoke temperature from 267°C to 216°C (figure 6, orange continuous line). Nevertheless, the high temperature of the first yoke is caused mainly by the heat conducted and radiated from the ferrite rings, where the predicted temperatures are ~500°C. Cooling the rings reduces drastically the temperature of the ferrite yokes too: as shown in figure 6 (orange dotted line), if the rings are cooled at 15°C, the maximum temperature of the ferrite yoke is reduced to 41°C, which is well below the Curie temperature. A design of a water cooling system, for the ferrite rings, is currently under development. A prototype will be constructed for testing the performance before its installation in the MKIs in the LHC. Nevertheless the cooling system will be in a pulsed high voltage and UHV environment: hence one must ensure that the risk of liquid leakage into machine vacuum is negligible.

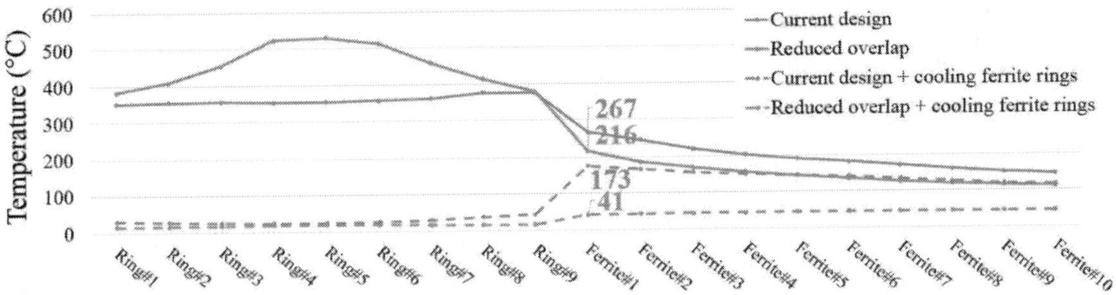

Figure 6. Maximum temperature for two overlaps of beam screen, without and with cooling of ferrite rings.

If the power deposition in the ferrite yokes is not reduced as much as expected, by reducing the overlap length [6], cooling of the rings alone may not be sufficient. Hence direct cooling of the rings combined with a cooling plate for the first ferrite yoke may be required. Considering this case for the present beam screen overlap, the predictions show that all the ferrite yokes are below 107°C.

6. Conclusions

A finite element model has been created using ANSYS in order to analyse the thermal behaviour of the MKI magnets: the model has been validated by comparing predictions with temperature measurements of MKIs in the LHC. Predictions show that for HL-LHC the temperature of the yokes will exceed the Curie point if further measures are not taken. Options have been studied to reduce the effect of beam induced heating: by reducing the beam screen overlap length the power deposited in the ferrite yokes can be decreased- for this scenario, installation of a cooling system on the ferrite rings is predicted to maintain all ferrites below their Curie temperature. In case the power reduction is not as high as expected, the first ferrite yoke must be cooled too. Currently a cooling system for the rings is under development.

References

[1] Barnes M J*et al* 2016 Operational Experience of the Upgraded LHC Injection Kicker Magnets *Proc. of Int. Part. Acc. Conf. (Busan)* p 3623
[2] Apollinari G, Béjar Alonso I, Brning O, Lamont M and Rossi L 2015 High-Luminosity Large Hadron Collider (HL-LHC): Preliminary Design Report *CERN-2015-005 (Geneva)* chapter 1 pp 7–8
[3] ANSYS *http://www.ansys.com/*
[4] Barnes M J *et al* 2013 Beam Induced Ferrite Heating of the LHC Injection Kickers and Proposals for Improved Cooling *Proc. of Int. Part. Acc. Conf. (Shangai)* p 732
[5] Day H, Barnes M J, Ducimitière L, Vega Cid L and Weterings W 2016 Current and Future Beam Thermal Behaviour of the LHC Injection Kicker Magnet *Proc. of Int. Part. Acc. Conf. (Busan)* p 3615

[6] Vlachodimitropoulos V, Barnes M J, Ducimitière , Vega Cid L and Weterings W 2017 Predicted Beam Induced Power Deposition in the LHC injection Kicker Magnets for HL-LHC type Beams *Proc. of Int. Part. Acc. Conf. (Copenhagen)*

[7] Barnes M J *et al* 2017 Operational Experience of the Upgraded LHC Injection Kicker Magnets During Run 2 and Future Plans *Proc. of Int. Part. Acc. Conf. (Copenhagen)*

[8] Sobiech Z *et al* 2014 Cooling of the LHC Injection Kicker Magnet Ferrite Yoke: Measurements and Future Proposals *CERN-ACC-2014-0241 (Geneva)*

[9] Taborelli M, Chiggiato P, Costa Pinto P and Cruikshank P 2016 Nine years of carbon coating development for the SPS upgrade: achievements and heritage *CERN-ACC-2016-0010 (Geneva)*

[10] Polyteknik AS *http://www.polyteknik.com/*

[11] ASM International 2004 *Introduction to Thermal Spray Processing* (Handbook of Thermal Spray Technology (#06994G))

[12] Timmins M, Bertarelli A, Uythoven J and Gaxiola E H 2004 SPS Extraction Kicker Magnet Cooling Design *CERN-AB-2004-038 (Geneva)*

[13] Kumar Sharma B, Kettner M, Biebersmith P and Roldan N 2001 *Aluminum Nitride vs. Beryllium Oxide for High Power Resistor Products* (Microwave Journal)

Operational experience of the upgraded LHC injection kicker magnets during Run 2 and future plans

M J Barnes, A Adraktas, G Bregliozzi, B Goddard, L Ducimetière, B Salvant, J Sestak, L Vega Cid, W Weterings, C Yin Vallgren

CERN, Geneva, Switzerland

E-mail: mike.barnes@cern.ch

Abstract. During Run 1 of the LHC, one of the injection kicker magnets caused occasional operational delays due to beam induced heating with high bunch intensity and short bunch lengths. In addition, there were also sporadic issues with vacuum activity and electrical flashover of the injection kickers. An extensive program of studies was launched and significant upgrades were carried out during Long Shutdown 1 (LS 1). These upgrades included a new design of beam screen to reduce both beam coupling impedance of the kicker magnet and the electric field associated with the screen conductors, hence decreasing the probability of electrical breakdown in this region. This paper presents operational experience of the injection kicker magnets during the first years of Run 2 of the LHC, including a discussion of faults and kicker magnet issues that limited LHC operation. In addition, in light of these issues, plans for further upgrades are briefly discussed.

1. Introduction

The Large Hadron Collider (LHC) is equipped with two injection kicker (MKI) systems, MKI2 and MKI8, for deflecting the incoming particle beams onto the LHC's equilibrium orbits [1]. Counter-rotating beams circulate in two horizontally separated pipes. Each beam pipe is filled by 12 consecutive injections, at 450 GeV. Both MKI2 and MKI8 comprise four systems, named A through D: D is the first to see injected beam. The total deflection by an MKI system is 0.85 mrad, requiring an integrated field strength of 1.3 T·m. To limit beam emittance blow-up due to injection oscillations, reflections and flat top ripple of the field pulse must be less than ±0.5%.

A low impedance (5 Ω) and carefully matched high bandwidth system meets the stringent pulse response requirements. An MKI kicker system consists of a multi-cell Pulse Forming Network (PFN) and a multi-cell travelling wave kicker magnet [2], connected by a matched transmission line and terminated by a matched resistor (TMR). Each travelling wave magnet has 33 cells. A cell consists of a U-core NiZn ferrite sandwiched between high voltage (HV) plates: ceramic capacitors are sandwiched between each HV plate and a plate connected to ground (figure 1). The magnets are operated in vacuum of ~10^{-11} mbar. The complete magnet is baked out at 300°C before HV conditioning and tests.

2. Beam induced heating

With high bunch intensity and short bunch lengths, integrated over many hours of a good physics fill, the impedance of the magnet ferrite yoke can lead to significant beam induced heating. To limit longitudinal beam coupling impedance, while allowing a fast magnetic field rise-time, an alumina tube with screen conductors lodged in its inner wall is placed within the aperture of the magnet [2]. The conductors, which provide a path for the image current of the beam, are connected to the standard LHC vacuum chamber at one end and are capacitively coupled to it at the other end. There is a set of toroidal ferrite rings mounted around each end of the alumina tube, outside of the aperture of the magnet (figure 1), whose original purpose was to damp low-frequency resonances [3]. Each set of nine toroids utilizes two types of NiZn ferrite [4].

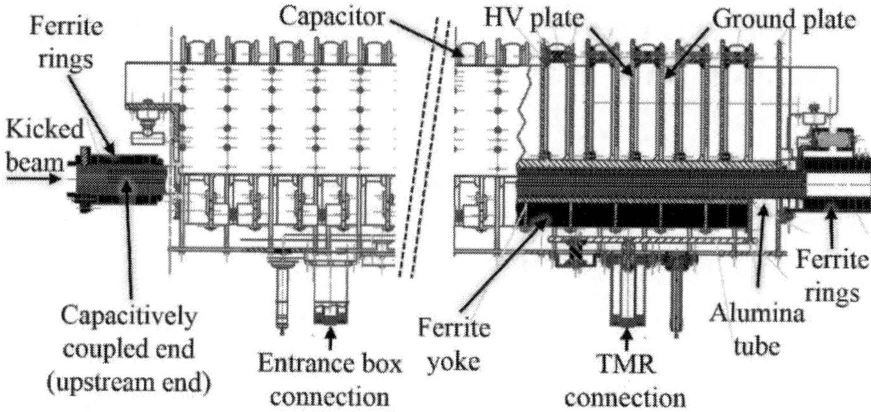

Figure 1. MKI kicker magnet.

During Run 2 the measured temperatures at the upstream (capacitively coupled) end of the MKI magnets have been consistently higher than those measured at the downstream end [5]. CST Particle Studio [6] has been used to study beam coupling impedance and the predictions are in general agreement with observations: the longitudinal distribution of volume losses of the post-LS1 kicker magnets is non-uniform. The ferrite rings at the capacitively coupled end of the structure experience more than 25% of the total power deposition, for a structure that has relatively little volume [7]. The power deposition in the yoke is predominantly at the upstream end of the magnet: the first cell, at the upstream end, experiences ~9% of the total power deposition [8].

Thermal simulations have been carried out to confirm that the calculated power losses for Run 2 result in the temperatures measured during LHC operation. To compare the results with the measurements it must be taken into account that the PT100 temperature sensors for the yoke are located on a side plate of each magnet. There is good agreement between measured and predicted side-plate temperatures [9], validating the power loss calculations and the thermal model:

- Maximum measured side-plate temperature during Run 2 is 57°C, at the upstream end of MKI8D [fill #5069, $1.25x10^{11}$ ppb (protons per bunch), 2076 bunches, 25.5 hrs stable beam];
- Steady-state thermal simulations show that 57°C corresponds to a temperature, in the first cell at the upstream end, of ~80°C and a total power deposition of almost 100 W in the magnet.

Scaling total power deposition linearly with the number of bunches, to the nominal 2808 bunches, and assuming $1.25x10^{11}$ ppb, the total expected power is almost 150 W. The corresponding predicted temperature, in the first cell at the upstream end, is 107°C and an expected side-plate temperature, at the upstream end, is 77°C. The 107°C is below the Curie temperature (125°C), hence no issues with MKI heating are foreseen during Run 2. Jackets are installed on each magnet for the bake-out [4] and are presently left on. If necessary, to improve heat transfer, the jackets could be removed: this would reduce the highest temperature rise in the ferrite yoke, by ~7% [9].

2.1. Beam Induced Heating: Future Plans

For Run 3 operation, with High Luminosity (HL) LHC type beams, the power deposition in the MKIs is expected to be a factor of ~4 greater than for LHC: this would be unacceptably high unless measures

IOP Conf. Series: Journal of Physics: Conf. Series **874** (2017) 012101 doi:10.1088/1742-6596/874/1/012101

are taken. Hence means of reducing heating of the yoke [8] and efficiently removing heat from the magnet [9] are being studied.

3. Electron cloud

Significant pressure rise, due to electron-cloud, occurs in and nearby the MKIs: the predominant gas desorbed from surfaces is H_2. Conditioning of surfaces reduces electron-cloud, and thus dynamic pressure rise, but further conditioning is often required when beam parameters (e.g. bunch spacing, length and intensity) are pushed.

Voltage is induced on the screen conductors during field rise (to 30 kV) and fall (to −17 kV). High pressure, at the capacitively coupled end, can result in breakdown/flashover – hence an interlock (SIS) prevents injection when the pressure is above threshold. The SIS thresholds, for the MKI interconnects, are $5\text{x}10^{-8}$ mbar. During LS1, the vacuum systems on the interconnects between MKI magnets were upgraded: (a) interconnects were NEG coated; (b) a NEG cartridge was integrated to give a nominal pumping speed of 400 l/s for H_2 (prior to LS1, ion pumps provided a nominal 30 l/s for H_2).

The alumina tube of each MKI has a Secondary Electron Yield (SEY) of ~9 when first installed and, together with metallic surfaces facing the beam, requires conditioning with beam. During mid-2016 electron cloud resulted in a factor of ~20 rise in pressure in most MKI8 interconnects: the dynamic pressure rise in the MKI tanks was a factor of ~10. However, electron cloud in the alumina tube of MKI8D resulted in a dynamic pressure rise, measured in the MKI8D-Q5 interconnect, adjacent to MKI8D but outside of the four magnets, being a factor of ~1000: this pressure rise is at the capacitively coupled end of the MKI8D, i.e. where HV is induced. Hence, to avoid flashover, there is an SIS threshold on this interconnect pressure, which is historically set to $6\text{x}10^{-8}$ mbar. Figure 2 shows the pressure in the MKI8D-Q5 interconnect, and beam 2 intensity during June 2016.

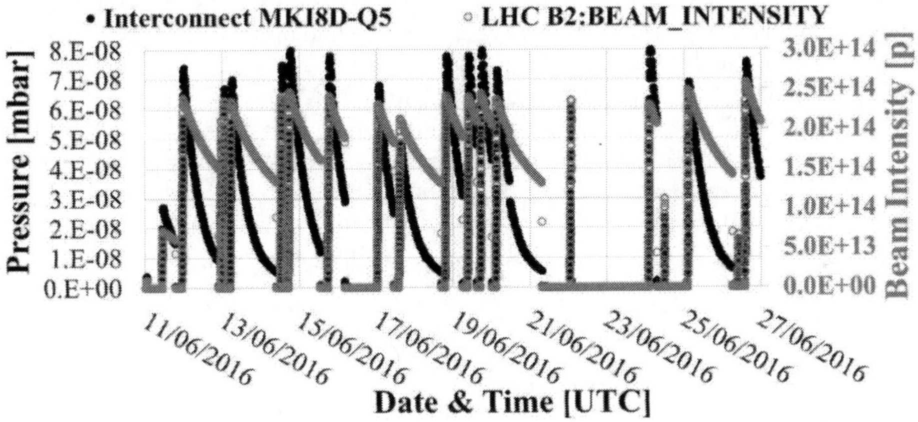

Figure 2. pressure in the MKI8D-Q5 interconnect, and beam 2 intensity during June 2016.

Figure 3 shows pressure normalized to the number of protons, for 25 ns bunch spacing, for interconnect MKI8C-MKI8D, tank MKI8D, and interconnect MKI8D-Q5, from 24 June to 27 June, 2016. Beam 2 intensity is also shown.

The first fill shown in figure 3 had 72 batches per injection (bpi), $1.11\text{x}10^{11}$ ppb and 2040 bunches. Three injections where missing from beam 2 due to the MKI8D-Q5 threshold exceeding $6\text{x}10^{-8}$ mbar. The second long fill shown in figure 3 had 96 bpi (2 x 48 batches, each batch separated by 250 ns), $1.18\text{x}10^{11}$ ppb and 2076 bunches. The third long fill shown in figure 3 had 96 bpi (2 x 48 batches, each batch separated by 250 ns), $1.23\text{x}10^{11}$ ppb and 2076 bunches. The 20% lower normalized pressure for the last two long fills, in comparison with the first fill, is due to the 250 ns gap, in the train, between the two injected 48 bunch batches, reducing electron cloud.

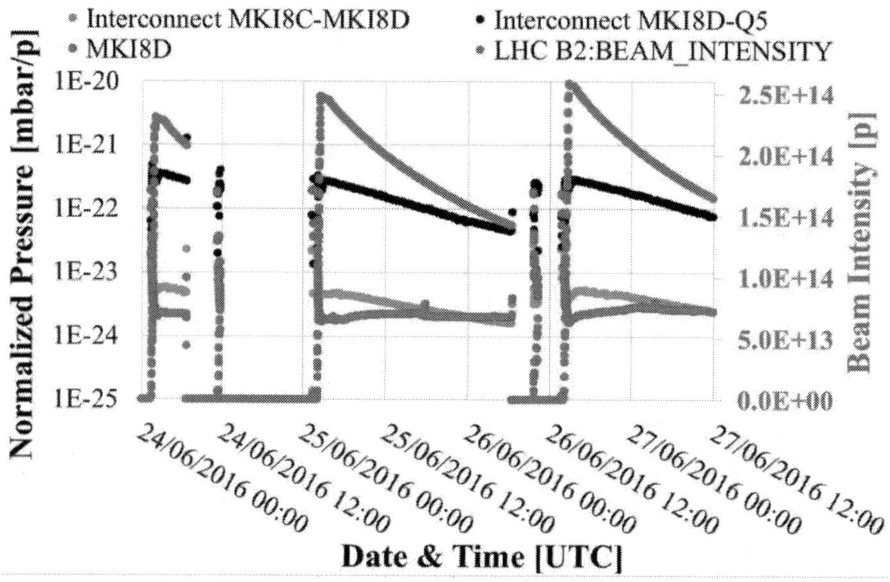

Figure 3. pressure normalized to the number of protons for: interconnect MKI8C-MKI8D, tank MKI8D, and interconnect MKI8D-Q5. Beam 2 intensity is also shown.

During the Extended Year End TS (EYETS), from Dec. 2016, two NEG cartridges, of 400 l/s each for H_2, were integrated in the vacuum sector between MKI8D and Q5. The upgrade locally increases the pumping speed and hence is expected to maintain the dynamic pressure, in this interconnect, below $5x10^{-8}$ mbar, up to the nominal, 2808, 25 ns bunches.

During TS3, Nov. 2016, it was necessary to replace magnet MKI2D (see below). Hence, although electron cloud around MKI2D has not limited injection during Run 2, the alumina tube in the new MKI2D will not have seen high intensity proton beam and will require conditioning after the EYETS. To assist the conditioning, two new NEG cartridges, of 400 l/s each, have also been integrated in the vacuum sector between MKI2D and Q5. However, despite the upgrade to this vacuum sector, the dynamic pressure in the MKI2D-Q5 interconnect is expected to initially limit intensity of LHC beam 1.

3.1. Electron cloud: future plans

For the longer term a coating of Cr_2O_3, applied to the inside of the alumina tube by magnetron sputtering, is a promising means of eliminating electron cloud. Measurements show that naked, high-purity, alumina has a maximum SEY (δ_{max}) of ~9. A Cr_2O_3 coating, applied by magnetron sputtering by Polyteknik [10], reduces δ_{max} to approximately 2.3: bombarding the surface with electrons further reduces δ_{max} to less than 1.4 (figure 4).

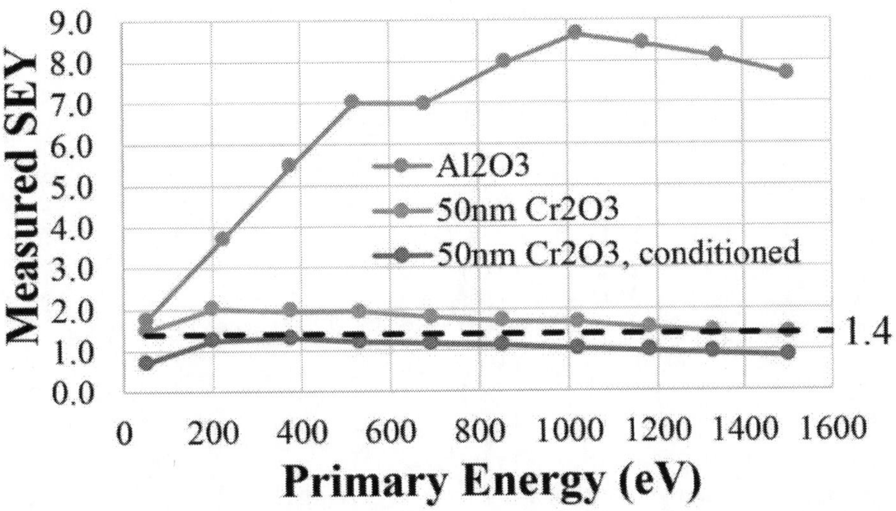

Figure 4. Measured SEY of high purity alumina and alumina with a 50 nm Cr_2O_3 coating: bombarding Cr_2O_3 with electrons reduces δ_{max} to below 1.4 [Measurements courtesy of E. Garcia-Tabares Valdivieso and H. Neupert].

A set of samples has been sputtered with 50 nm of Cr_2O_3 and installed in a setup in the CERN SPS, during the EYETS, for tests with beam.

4. MKI flashovers

Prior to injecting beam into the LHC, in order to re-qualify the MKIs, a short conditioning (SoftStart (SS)) is carried out during which the PFN voltage is ramped up to 8% above the nominal injection voltage. During 2016 (2015) the statistics for MKI flashovers were:

- 4 (5) during "SoftStarts" [~80% of total pulses];
- 3 (0) with injected beam [~20% of total pulses].

There was a disproportionate number of flashovers with high-intensity beam during 2016, in comparison with 2015: this suggests a deconditioning effect of the high-intensity beam upon the alumina. A Cr_2O_3 coating is expected to reduce or eliminate this deconditioning.

5. MKI2D high impedance contact

In Oct. 2016 a high impedance contact, characterized by the magnet not initially carrying current, developed at the input end of the MKI2D magnet. The dashed curves in figure 5 shows currents for magnet MKI2C and solid curves for MKI2D: these two sets of curves should be more or less identical. The MKI2D was exchanged during TS3.

An inspection of the removed magnet showed damage and erosion of a high current contact at the magnet input. However the removed MKI2D magnet had experienced two power-cuts during bake-outs, during LS1. These power-cuts may have resulted in excessive mechanical stress on the contact, due to the tank cooling more rapidly than the magnet – hence giving significant, transient, thermally induced, difference in length between the tank and magnet: this is under study.

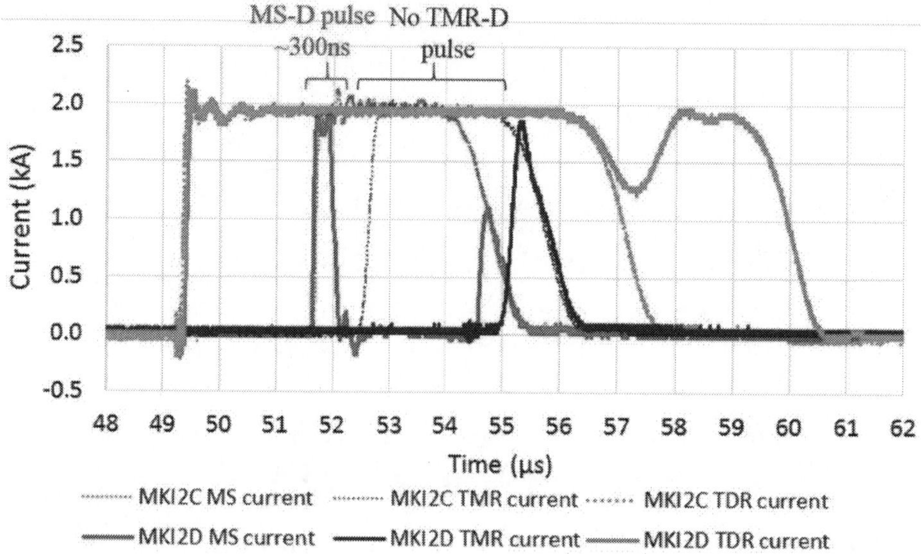

Figure 5. First pulse of a SS (at 20 kV): dashed curves are magnet MKI2C, solid curves are for MKI2D.

6. MKI erratic turn-on

One Main Switch (MS) thyratron erratic (spontaneous turn-on), during resonant charging of the PFN, occurred in September 2016: there were 876 circulating bunches in the LHC, of which 210 were miskicked. Since November 2014 there has been a total, for the two MKI systems, of ~1.4 million MS thyratron pulses. During 2015: 18% of the pulses were for injection; 82% of the pulses were during SS's; almost 60% of the total of the pulses were at or above the nominal injection voltage. There has been a total of three erratics (all on MSs): two during SSs, both at voltages >2 kV above nominal. The erratic in September 2016 was the only one that occurred below nominal voltage.

The observed probability of a MS erratic during injection is ~1.6×10^{-5} per injection. The probability of an erratic occurring is dependent upon the magnitude of the reservoir voltage of the thyratron: hence, during the EYETS, the reservoir voltage of each thyratron was verified.

7. Conclusion

No heating issues are expected for the MKIs during Run 2, with the nominal number of bunches. However, for operation with HL-LHC type beams, methods of reducing heating of the ferrite yoke [8] and efficiently removing heat from the magnet [9] are essential and are being studied. Dynamic pressure rise, due to electron cloud in the MKI8D alumina tube, limited injection of beam 2 during 2016: an upgrade to the vacuum system during the EYETS is expected to solve this issue. Longer term a coating of Cr_2O_3, applied to the inside of the alumina tube by magnetron sputtering, is a promising means of eliminating electron cloud and further reducing electrical flashover in the alumina tube. It is planned to install a prototype MKI, with reduced power deposition and a Cr_2O_3 coated alumina tube, during the Year End TS 2017-18. In addition it is currently planned to install cooling systems during LS2.

8. References

[1] *LHC Design Report*, http://ab-div.web.cern.ch/ab-div/Publications/LHC-DesignReport.html
[2] Barnes M J *et al.* 2013 *Proc. of Int. Part. Acc. Conf. (Shanghai)* pp 735-737
[3] Day H *et al.* 2013 *Proc. of Int. Part. Acc. Conf. (Shanghai)* pp 1649-1651
[4] Barnes M J *et al.* 2013 *Proc. of Int. Part. Acc. Conf. (Shanghai)* pp 732-734
[5] Barnes M J *et al.* 2016 *Proc. of Int. Part. Acc. Conf. (Busan)* pp 3623-3626
[6] CST - Computer Simulation Technology, http://www.cst.com
[7] Day H, Barnes M J, Ducimetière L, Vega Cid L and Weterings W *Proc. of Int. Part. Acc. Conf.*

(Busan), pp 3615-3618

[8] Vlachodimitropoulos V, Barnes M J, Ducimetière L, Vega Cid L and Weterings W 2017 *Proc. of Int. Part. Acc. Conf. (Copenhagen)*

[9] Vega Cid L, Abánades A, Barnes M J, Ducimetière L, Vlachodimitropoulos V and Weterings W 2017 *Proc. of Int. Part. Acc. Conf. (Copenhagen)* "Thermal analysis of the LHC Injection Kickers", paper WEPVA096

[10] Polyteknik AS, DK-9750 Oestervraa, Denmark. http://www.polyteknik.com/

Preliminary Mechanical Design Study of the Hollow Electron Lens for HL-LHC

Carlo Zanoni[1], Giorgia Gobbi[1], Diego Perini[1] and Giulio Stancari[2]
[1] CERN, Engineering Department, Geneva, Switzerland
[2] Fermi National Accelerator Laboratory, PO Box 500, Batavia, Illinois 60510, USA
E-mail: carlo.zanoni@cern.ch

Abstract. A Hollow Electron Lens (HEL) has been proposed in order to improve performance of halo control and collimation in the Large Hadron Collider in view of its High Luminosity upgrade (HL-LHC). The concept is based on a hollow beam of electrons that travels around the protons for a few meters. The electron beam is produced by a cathode and then guided by a strong magnetic field. The first step of the design is the definition of the magnetic field that drives the electron trajectories. The estimation of such trajectories by means of a dedicated MATLAB tool is presented. The influence of the main geometrical and electrical parameters is analyzed and discussed. Then, the main mechanical design choices for the solenoids, cryostats gun and collector are described. The aim of this paper is to provide an overview of the feasibility study of the Electron Lens for LHC. The methods used in this study also serve as examples for future mechanical and integration designs of similar devices.

1. Introduction

Hollow electron collimation is a novel technique [1, 2, 3, 4] based on a magnetically confined beam of electrons traveling around the proton axis. The electrons are emitted by a cathode and then compressed and confined into a long strong solenoid. The electrons dissipate their energy on a metallic collector with an active cooling system. Such a system has recently been proposed for the Large Hadron Collider at CERN [5] and its feasibility is being investigated.

With a HEL, the particles of the protons bunch halo are kicked transversely while the core remains unaltered. The most evident advantage of hollow electron collimation is that the electron beam can have a radius close to the proton beam dimension, avoiding the limitations of mechanical devices.

To avoid stressing the integration at the proposed location in LHC, the final design of the HEL should be as compact as possible, still fulfilling the required 3 m of electrons and protons parallel and centred trajectories [5]. This goal requires frequent iterations between design, solenoids configuration and estimation of electron trajectories. This process has been facilitated by the approach described in the first part of this paper.

The trajectory of the electrons is driven by the magnetic field generated by the solenoids. The solenoid in which the cathode is located is called gun solenoid (GS), the long one that confines the e-cloud is called main solenoid (MS). Intermediate solenoids (BS) are also foreseen in order to keep the electrons on the defined path, Figure 1.

A magnetically confined electron beam closely follows the field lines of the solenoid field [6]. For instance, if the axial field B increases, the requirement of a null magnetic field divergence

IOP Conf. Series: Journal of Physics: Conf. Series **874** (2017) 012102 doi:10.1088/1742-6596/874/1/012102

Table 1. Currents and Magnetic Fields.

Solenoid	Requirement	Derived	r_{int}	Δr	length
MS	$4\ T$	$159.5\ A/mm^2$	99.5 mm	20 mm	3016 mm
BS	$159.5\ A/mm^2$	$2.5\ T$	109 mm	20 mm	150 mm
GSi	$94\ A/mm^2$	$0.4\ T$	110 mm	4.8 mm	200 mm
GSg	$0.2\ T$	$25.9\ A/mm^2$	110 mm	4.8 mm	200 mm

implies that the transverse size r of the beam must decrease to conserve the product Br^2. Therefore, the dimensions of the electron beam in two points along its path follow the equation:

$$\frac{r_0}{r_1} = \sqrt{\frac{B_1}{B_0}} \tag{1}$$

where r_0 and r_1 are the radii of the electron beam in point 0 and 1 and B_0 and B_1 are the magnetic fields in points 0 and 1, respectively. The GS is divided in two parts: a tunable field solenoid around the e-gun cathode (GSc) and a constant field solenoid aligned with the first one (GSi).

Table 1 provides the fields, currents and dimensions of the solenoids. Some solenoids are defined by the current rather than the field. In fact, the BS is constrained to the same current of the MS, that defines a field that depends on the relative position. Table 2 gives the nominal proposed dimensions of the electron beam in the main solenoid and some parameters of the cathode.

Table 2. Dimensions of the Hollow Electron Beam and of the Emitting Cathode.

r_{int} hollow electron beam @ nominal fields	0.9 mm (3 σ)
r_{ext} hollow electron beam @ nominal fields	1.8 mm (6 σ)
Inner diameter of the cathode	8.05 mm
Outer diameter of the cathode	16.1 mm
Nominal current at the cathode	5 A
Nominal energy at the cathode	10 keV

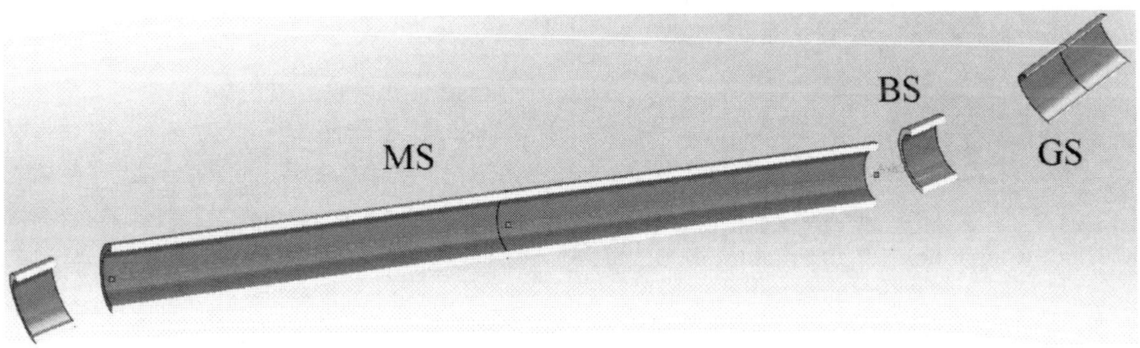

Figure 1. Open view of the solenoids configuration.

2. Estimation of Electron Trajectories

The trajectory of the electrons is estimated in order to obtain a first guess of the mechanical configuration of the solenoids, which has tight constraints to fit in LHC. The sensitivity of the electron trajectory to the main geometrical parameters is also derived. The numerical computations are carried out with MATLAB and occasionally with Comsol 5.2. A full beam dynamics assessment, including self fields, is foreseen for the future.

The electrons are emitted by the cathode and travel toward the MS. In the simulations an emission point is located on the GS axis and we check that the trajectory of the electrons is centred on the axis of the MS. To keep the beam magnetized and to mitigate its space-charge evolution, the magnetic field experienced by the electrons should always be ≥ 0.1 T. The following assumptions are introduced to simplify the computations:

- a single particle instead of the cloud of electrons, is considered. The mutual interaction between particles is then neglected as well as the influence of the proton beam and of the metal vacuum pipe. This particle represents the center of the hollow cloud;
- the calculation is nonrelativistic;
- ambient disturbances are neglected.

This simplified approach allows an accuracy of the solenoids position in the order of the mm that is sufficient for drafting the concept configuration and mechanical design.

The trajectory is calculated by combining the Newton-Lorentz equations with the Biot-Savart law. The motion of the electron is integrated by means of the Boris algorithm [7, 8]. We decided to fix the radii and the nominal currents according to considerations of other nature such as field compression factor, magnetic energy and reasonable space available and focus on the compactness of the design.

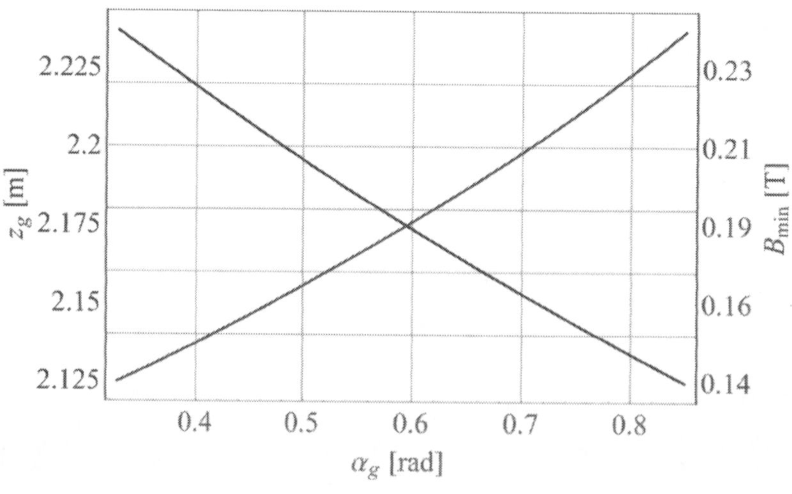

Figure 2. Correlation between distance of the GS and their angle (blue). The minimum field encountered with that configuration is also estimated (red).

The design concept foresees the BS at the inlet and outlet of the MS. These solenoids are tilted and are powered with the same current of the main one in order to bend the electrons with a strong field. The GSg has 0.2 T around the gun, which guarantees the correct compression

Table 3. Comparison of Design Concepts.

Concept	z_g $[m]$	min B $[T]$
with BS	2.19	0.16
without BS	3.48	0.008 T

factor. At this point only the inclination (α_g) and the position along z (z_g) of the gun are free parameters. Parameter z_g is zero in the mid plane of the MS. The simulations suggest that, to guarantee a trajectory that passes in the center of the MS, only the $[\alpha_g, z_g]$ pairs of Figure 2 are allowed.

A value of $\alpha_g = \pi/6$ is chosen as it is considered sufficiently small to avoid excessive bending of the electrons. In fact, it allows the curvature of the trajectory to maintain the same sign between the GS and the BS. Table 3 shows the gun position in the proposed design with and without bending solenoids. It is clear that this second case is not an option. The magnetic field lines as simulated in COMSOL and the electron beam center trajectory are shown in Figure 3 and Figure 4

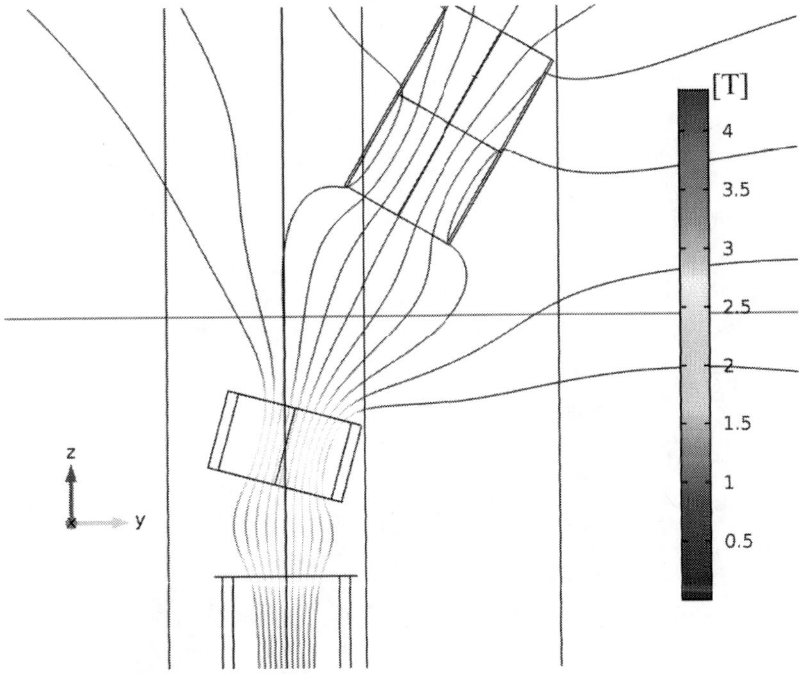

Figure 3. Field lines as estimated in Comsol.

3. Preliminary Mechanical Design

The MS and the BSs are superconducting, with niobium-titanium wires. They are assembled inside cryostats cooled by liquid helium at 4.5 K available in the LHC tunnel. The GS work at a field level that could be obtained using normal resistive coils. A resistive solenoid of 0.4 0.5 T requires copper winding dissipating 10-15 kW power. With liquid helium available in the tunnel

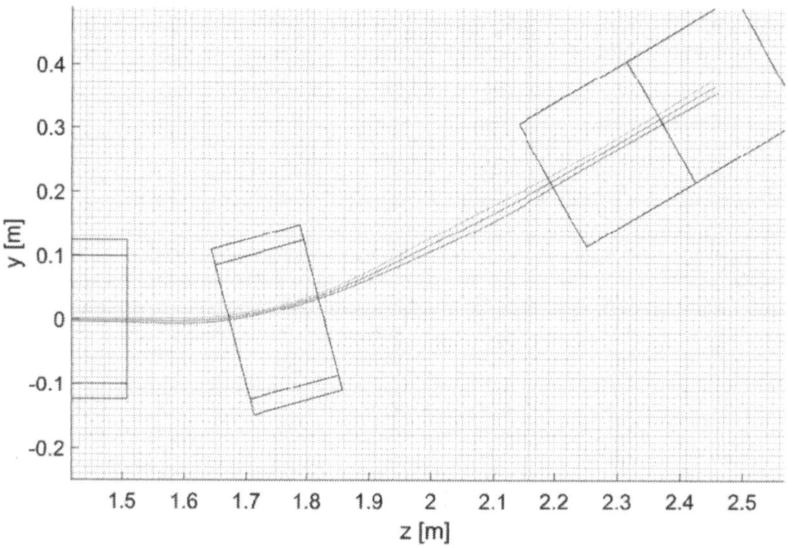

Figure 4. Trajectory estimated with the optimal parameters, zoom on the inlet bending.

and already used in the other coils, we think that superconducting solenoids are the best choice for the GS as well. This also facilitates the compactness of the system.

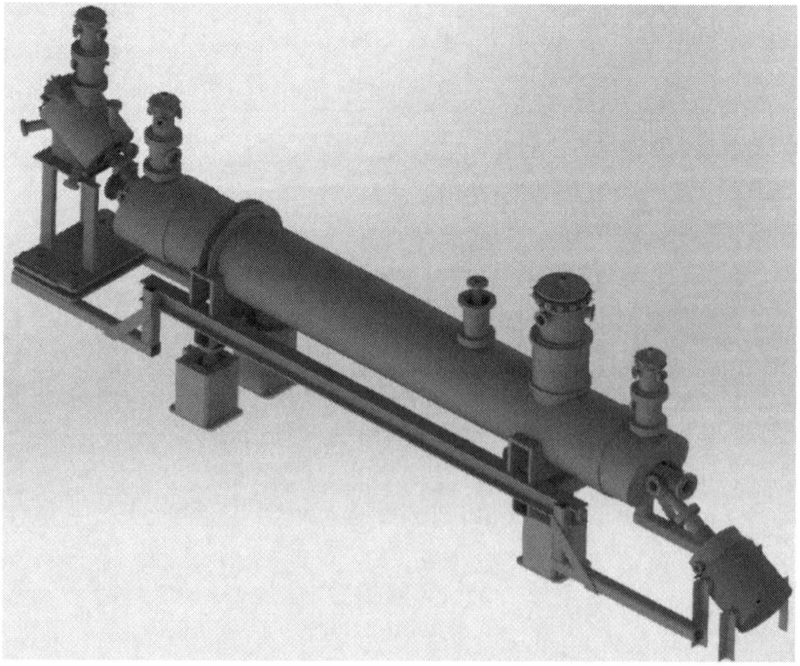

Figure 5. 3D view of the HEL system.

A 3D view of the whole system is shown in Figure 5 [9]. The MS dimensions are a trade off between the minimization of magnetic energy stored and the assemblability (that is facilitated when the solenoid is big). The result is an inner diameter of 199 mm. The solenoid is 3 m long and is divided in 3 equal independent parts for efficient quench protection. The cryostats are in stainless steel and have a central bore to house the room temperature vacuum chamber where the LHC beam and the hollow electron beam travel. Support is provided by G10 short tubes.

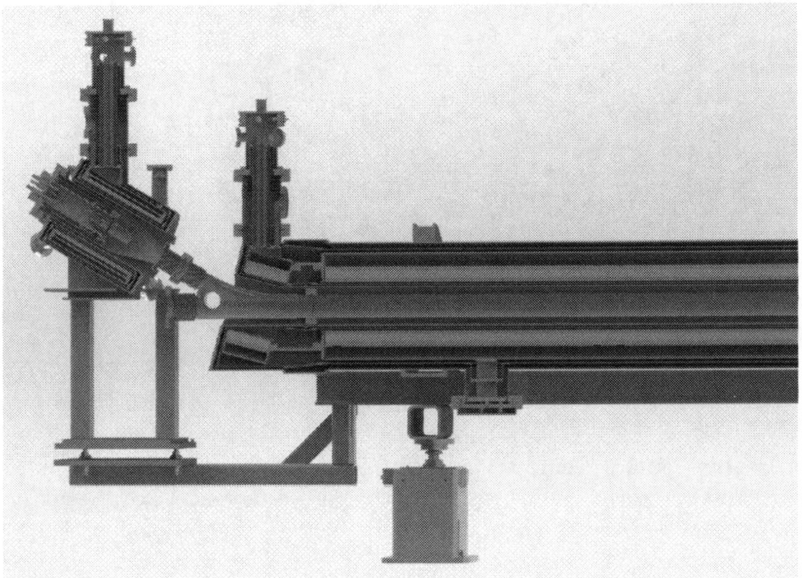

Figure 6. Open view of the HEL system, inlet region.

The BSs are hosted in the same cryostat of the MS, Figure 6. In fact, the force acting between the magnets is above 40 kN. By hosting MS and BSs in the same cold mass, there are no thermal losses through the supports.

In case of continuous mode use of the HEL the maximum power deposited on the collector is 50 kW. In normal operation, this number is reduced by biasing the collector potential. However, the concept feasibility has been performed accounting for all the power. The first draft of the collector is then a 400 mm × ⌀300 mm copper bucket with water cooling on the lateral surfaces. The simulated peak temperature is less than 90 C with a water flow of 8 ls^{-1} (1 ms^{-1} speed). A ferrite shield is foreseen around the collector to open the magnetic field lines and therefore the electron trajectories. In absence of such a shield, all the power would be concentrated on a very small surface, with a power density around 160 Wmm^{-2} instead of 0.55 Wmm^{-2}. Figure 7 shows the temperature obtained with a conceptual geometry of the collector, by means of a coupled fluido-thermal analysis.

4. Conclusion

A Hollow Electron Lens has been proposed to facilitate halo control and collimation for HL-LHC. The conceptual mechanical configuration has been presented. A simplified script that estimates the trajectories position has been prepared to quickly iterate between possible concepts. The system is now considered feasible through a compact concept made of five superconducting solenoids, in which the inlet and outlet field lines of the main 3 m solenoid are bended by two strong solenoids. Future activities include research on high-performance cathodes and the detailed design of all the subsystems.

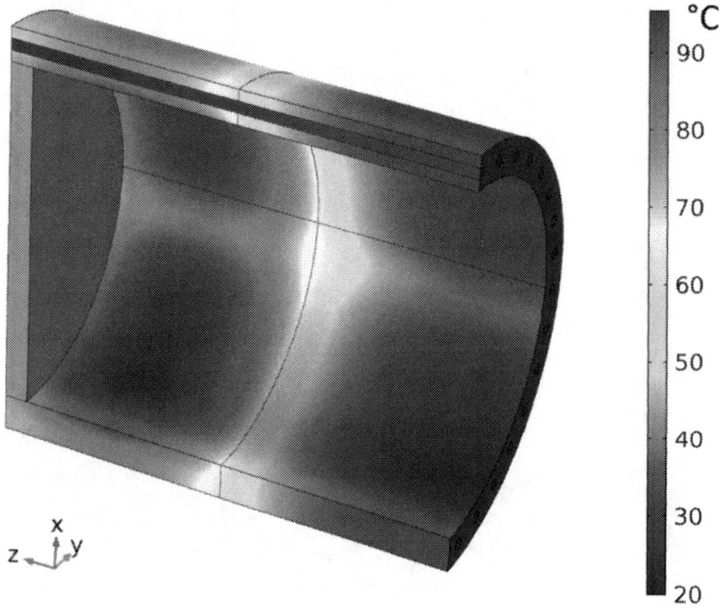

Figure 7. Temperature profile in the collector concept estimated with a coupled fluido-thermal analysis.

References

[1] Shiltsev V 2006 On possible use of electron lenses in LHC *Third CARE HHH APD Workshop towards a roadmap for the upgrade of the LHC and GSI accelerator complex, LHC LUMI 06, Valencia, Spain, 16-20 October 2006* pp 92–96 URL `http://doc.cern.ch/yellowrep/06-505`

[2] Shiltsev V 2007 Electron lenses for compensation of beam-beam effects: Tevatron, RHIC, LHC *CARE-HHH-APD Workshop on Finalizing the Roadmap for the Upgrade of the CERN & GSI Accelerator Complex (BEAM'07) Geneva, Switzerland, October 1-5, 2007* pp 16–18 URL `http://lss.fnal.gov/archive/2007/conf/fermilab-conf-07-699-apc.pdf`

[3] Stancari G, Valishev A, Annala G, Kuznetsov G, Shiltsev V, Still D A and Vorobiev L G 2011 *Phys. Rev. Lett.* **107**(8) 084802 URL `https://link.aps.org/doi/10.1103/PhysRevLett.107.084802`

[4] Shiltsev V D 2016 *Electron Lenses for Super-Colliders* Particle Acceleration and Detection (Springer) ISBN 9781493933150, 9781493933174

[5] Stancari G, Previtali V, Valishev A, Bruce R, Redaelli S, Rossi A and Salvachua Ferrando B 2014 (*Preprint* `https://arxiv.org/abs/1405.2033`)

[6] Reiser M 2007 *Theory and design of charged particle beams* (New York, USA: Wiley-VCH Verlag GmbH) ISBN 9783527407415

[7] Boris J P 1970 Relativistic plasma simulation-optimization of a hybrid code *Proceeding of Fourth Conference on Numerical Simulations of Plasmas* pp 3 – 67

[8] Qin H, Zhang S, Xiao J, Liu J, Sun Y and Tang W M 2013 *Physics of Plasmas* **20** 084503 URL `http://dx.doi.org/10.1063/1.4818428`

[9] Perini D and Zanoni C 2017 Preliminary Design Study of the Hollow Electron Lens for LHC Tech. Rep. CERN-ACC-NOTE-2017-0004 CERN (*Preprint* `https://arxiv.org/abs/1702.00234`) URL `http://cds.cern.ch/record/2242211`

Two year operational experience with the TPS vacuum system

Y C Yang, C K Chan, I C Sheng, I T Huang, J Y Chung and C C Liang

National synchrotron Radiation Research Center, 101 Hsin-Ann Road, Hsinchu Science Park, Hsinchu 30076, Taiwan

E-mail: jolas@nsrrc.org.tw

Abstract. The Taiwan Photon Source (TPS), a 3-GeV third generation synchrotron light source, was commissioned in 2014 December and is now currently operated in top-up mode at 300mA for users. During the past two years, the machine was completed to meet design goals with among others the installation of superconducting cavities (SRF), the installation of insertion devices (ID) and the correction of vacuum chamber structure downstream from the IDs. The design goal of 500mA beam current was achieved with a total accumulated beam dose of more than 1000Ah, resulting in three orders of magnitude reduction of out-gassing. As the beam current was increased, a few vacuum problems were encountered, including vacuum leaks, unexpected pressure bursts, etc. Vacuum related issues including high pressure events, lessons learned and operational experience will be presented and discussed in this paper.

1. Introduction

The TPS is a low-emittance, 3-GeV storage ring synchrotron light source with 518.4m circumference. After 3 years of civil construction, accelerator component installation started in October 2013 followed by the commissioning of the booster synchrotron by December 12th, 2014 to fill the first storage ring beam current on December 31 of the same year. During phase I machine commissioning, 100mA beam current was reached limited by the room temperature PETRA cavities and 35Ah of beam dose was accumulated by March 2015. The average base pressure of the storage ring was 2.8×10^{-8} Pa before commissioning and increased to 1.33×10^{-7} Pa with 100mA beam current. Through beam cleaning an out-gassing rate of 1.33×10^{-7} Pa/mA was reached with 35Ah beam dose [1].

Before phase II commissioning, the machine was shut down for a period of six month for machine upgrades, including the replacement of the PETRA cavities with superconducting cavities, SRF, installation of 10 IDs and associated Front End (FE) systems. The design goal of 500mA beam current was archived on December 12th, 2015 and a beam dose of 194Ah was accumulated by then. While commissioning IDs and Beam lines (BL), the TPS became available to users in 300mA top up mode in September 2016 and 1000Ah accumulated beam dose was reached in November 2016. Table 1 lists operational milestones of the TPS in its first two years. As to the vacuum system, few events like vacuum leakage, pressure bursts and malfunction of vacuum components, occurred while regular shut downs allowed efficiency upgrades as well as correction and optimization of the interlock protection system to be described in more detail below.

Table 1. Milestones for the TPS in its first two years of operation.

Date	Milestone
2014.12	first beam stored
2015.03	100mA beam current archived
2015.12	500mA design goal archived
2016.09	open to user
2016.11	1000 Ah beam dose accumulated

2. Vacuum status

The evolution of beam cleaning in the TPS storage ring is shown in figure 1 where the average pressure raise per beam current (dP-avg/I,) versus accumulated beam dose (Ah) is plotted and the logarithmic slope is -0.84. Totally, four vacuum incidences occurred, as shown in figure 1. The first incidence was the replacement of PETRA cavities with superconductivity cavities and the installation of ten insertion devices in straight sections. The second occurrence was due to the replacement of a bending chamber located in the 02 arc section, SR02B1, to correct an unexpected pressure raise. The third and last incidence was the upgrade of four bending chambers (SR05B1, SR13B1, SR21B1 and SR23B1) located downstream of insertion devices and of a straight section between two EPUs. The goal was to increase the aperture so that synchrotron light from IDs could pass without striking the chamber wall [2].

Based on the design of the TPS vacuum system, more than 70% of synchrotron light was intercepted by crotch absorbers located downstream of bending chambers in the arc section. This explains why the out-gassing dP-avg/I during the first machine shut down is smaller than later ones. The beam cleaning effect continuously decreased, reaching 2.84×10^{-11} Pa/mA after a beam dose of 1000Ah was reached.

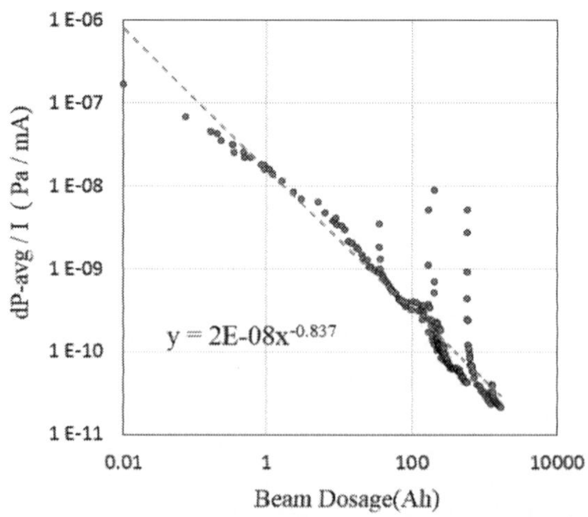

Figure 1. Evolution of beam cleaning.

3. Vacuum events

Several vacuum events occurred during machine commissioning. The two major ones were described below. On 10 February 2015, the vacuum pressure near the septum chamber in the injection section increased by more than 1×10^{-4} Pa. A leak was found near a welding joint between the vacuum tube and a flange in the septum 1 chamber as shown in figure 2. After in-situ repair, the vacuum pressure recovered quickly to less than 1×10^{-7} Pa. Yet, the pressure still increased during beam injection and after the tube was taken out no leak could be detected in the lab with a leakage detector. After reinstallation and activating the septum the problem disappeared without a clear reason. We suspect that out-gassing from the gap between two flanges acted like a virtual leak leading to the rise of pressure.

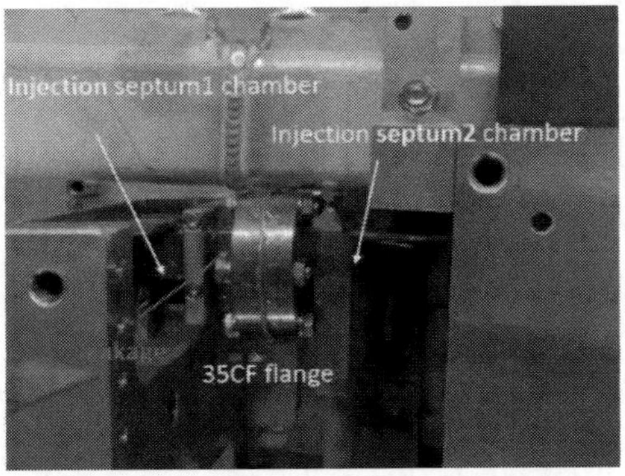

Figure 2. Position of a vacuum leak in the injection section.

After the RF system was upgraded to SRF cavities, the beam current could gradually be increased to meet the 500mA design goal. The first sudden pressure burst, close to 1×10^{-6} Pa, was observed near the SR02B1 region on October 4th, while the beam current was about 190mA. In fact, the normalized dynamic pressure, P/I, increased in the SR02B1 since September 20th, while starting phase II commissioning with more than 100mA beam current. It was not easy to notice, especially in the initial state of commissioning with generally unstable situations. After several leak and temperature checks, no causes were found, but the phenomenon persisted at high current and out-gassing from vacuum components was suspected. First, we replaced the crotch absorber inside the SR02B1 chamber and a photon stopper behind FE02 because they intercepted more than 70% synchrotron radiation from the upstream bending magnet. During the replacement process, the inner aperture of the B1 chamber was checked by endoscopy as well. After replacement, pressure burst still persisted at high beam current operation and the SR02B1 chamber was finally replaced. A quick, in-situ replacement method was developed [2,3]. It took eight day to replace the chamber and recover to ultra-high vacuum. Inside the removed chamber, foreign matter on the surface along the beam orbit was found as shown in figure 3, which explains the pressure behaviour described above. Out-gassing from this matter was caused by image current heating at high beam intensity operation.

After solving this problem, increase of beam current was further pursued with 500mA being the design goal. A maximum of 520 mA was archived on December 12, only 12 days after replacement of the chamber. At 520mA, the average ring pressure increased to 8.6×10^{-8} Pa, whereas in the SR02B1 it was 3.5×10^{-7} Pa due to the new vacuum chamber. Figure 4 illustrates the comparison of normalized dynamic pressure raise in the SR02B1 section, P_{SR02B1} and rest of the storage ring, $P_{average}$. The timeline illustrates: (a) a pressure raise for more than 100mA beam current, (b) replacement of the crotch absorber and photon stopper, (c) replacement of the SR02B1 chamber and (d) operation at 520mA. The ratio P_{SR02B1}/I increased 1.5-15 times for more than 100mA of beam current before replacement of the SR02B1 chamber, whereas the ratio $P_{average}/I$ continued to decrease monotonically and was little affected by P_{SR02B1}.

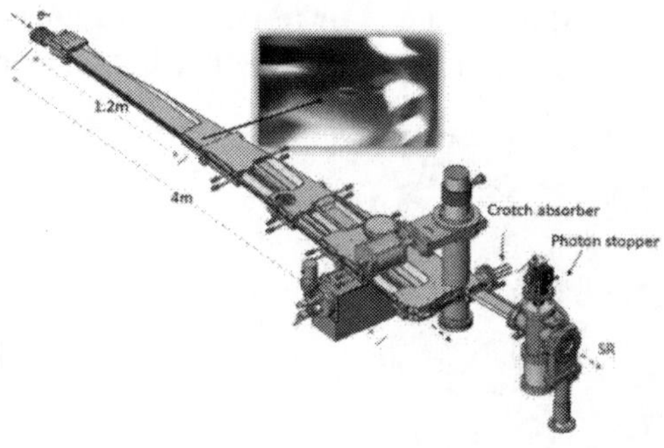

Figure 3. Foreign matter inside the chamber.

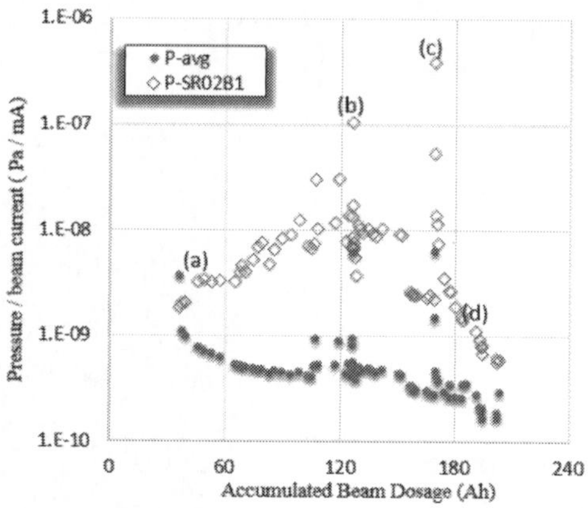

Figure 4. Comparison of the normalized dynamic pressure raise of the P_{SR02B1} and $P_{average}$.

4. Vibration issues

Following design, exhaust pumping systems, including mechanical pumps and turbo molecular pumps (TMP), were used near crotch absorbers to maximize the pumping speed and extend the lifetime of non-evaporable getter pumps (NEG) and sputter ionized pumps (IP) during early commissioning. A 29 Hz vibration was observed in beam position monitors (BPM) and magnets. Turning pumping system on and off, the source of vibration was identified to come from the mechanical pumps [4]. As a consequence all exhaust pumping systems were turned off in March, 2015.

Another source of vibration came from the cooling water in vacuum chambers. The cooling system included two loops: An aluminium (Al) loop for the vacuum chambers and copper (Cu) loop for crotch absorbers and photon stoppers. Originally, the water flow was set to 10 LPM. Because less than 30% synchrotron light from the upstream bending magnet was distributed over the Al chamber wall, a test for cooling water vibrations was considered in arc section 13. From spectral analysis of BPM data, it was found that the power spectral density (PSD) of horizontal orbit oscillations improves 2-3 times under 100Hz when the Al water flow is reduced to 6 LPM, while the vertical orbit oscillations were almost completely eliminated in the range of 40-120Hz. Vacuum pressure and water temperature were not affected during those tests. As a result, the water flow setting of the Al system was reduced to 6 LPM in March 2015, whereas the Cu system retained the original setting to meet the high radiation heat load.

5. Vacuum interlock

The basic vacuum interlock design was described earlier [5]. The vacuum pressure protection function, isolating poor vacuum regions by closing up- and down- stream sector gate valves (SGV), worked well during commissioning. After the in-vacuum ID installation, the interlock logic was updated to include also the ID vacuum components. The principle is to ensure three ionization gauges between two SGVs used in the pressure protection logic. The storage ring interlock protection logic also monitors vacuum signals from FE. Several times during commissioning, false FE vacuum signals caused interrupts from faulty evaluation of signals from FEs and BLs. Another vacuum signal in the storage ring adjacent to the FE system was added and a trip signal is given only when these two signals fault simultaneously.

The temperature monitoring and interlock system were also optimized, since several times during commissioning a trip signal was sent by malfunction of temperature sensors. Originally, plastic covers for the temperature signal cables were used and the sensor readings were affected by radiation. A few sensors were even damaged as shown in figure 5, especially downstream of IDs. These covers were gradually replaced by metal ones.

The temperature of bellows and BPM blocks with no cooling water channels in straight sections were affected by image current heating while the beam current gradually increased. In straight sections, the aperture height of BPM blocks is 16mm, relative to 20mm elsewhere and 38mm in the arc. Cooling fans were installed in all straight sections to reduce the effect of heating. The temperature of air cooling in the SR04S2 section and natural cooling in the SR02S2 section of equal design is compared in figure 6. A 20℃ increase in the SR02S2 bellows with natural cooling was observed at 300mA beam current operation together with a 15% increase in local vacuum pressure.

Figure 5. Temperature sensor cable damage by radiation.

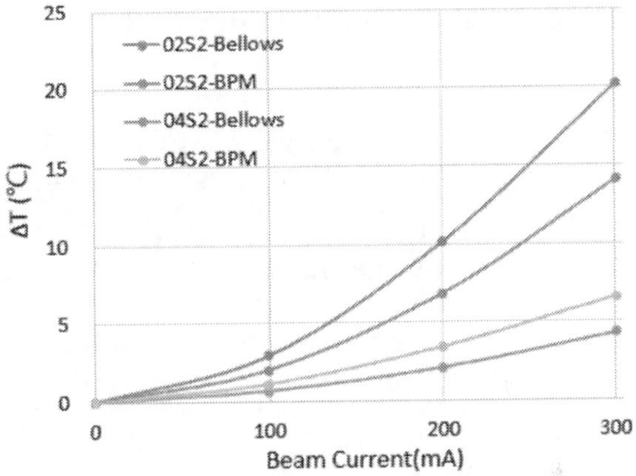

Figure 6. Temperature comparison between air cooling and natural cooling.

6. Conclusions

Two years of vacuum operation experience in the TPS has been presented. Good beam cleaning was obtained with continuing decrease as more beam dose is accumulated. Through machine commissioning, the ability of vacuum diagnostics and recovery was established. Interlock protection and monitoring systems were optimized and expanded to meet machine protection requirements.

References

[1] Yang Y C, Chan C K, Chen B Y, Kuan C K, Sheng I C, Hsiung G Y, Wu L H, Chang C C, Cheng C M, Chung J Y, Liang C C and Chen J R 2015 *Proc. IPAC`15 (Richmond)* pp 3222-4

[2] Sheng I C, Chang C C, Chan C K, Wu L H, Shueh C, Yang Y C, Cheng C M, Huang I T, Chung J Y, Cheng Y T and Hsiao Y M 2016 *Proc. IPAC`16 (Busan)* pp 3676-8

[3] Chan C K, Chang C C, Shueh C, Yang I C, Wu L H, Chen B Y, Cheng C M, Huang Y T, Chung J Y, Cheng Y T, Hsiao Y M and Sheng Albert 2017 *Nuclear Instruments and Methods in Physical Research A* **851** pp 57-61

[4] Huang C H, Liao C Y, Cheng Y S, Chiu P C, Hu K H and Hsu K T 2015 *Proc. ICALEPCS2015 (Melbourne)* pp 277-80

[5] Yang Y C, Hsiung G Y, Chen B Y, Chung J Y, Chen Z W, Chen J R and Lee T Y 2014 *Proc. IPAC`14 (Dresden)* pp 2387-9

Development of 3.95 MeV X-band linac-driven x-ray combined neutron source

J M Bereder[1,5], Y Mitsuya[1,6], Y Takahashi[1,7], K Dobashi[1,8], M Uesaka[1,9], J Kusano[2,10], Y Tanaka[3,11], Y Oshima[4,12], M Ishida[4,13]

[1] The University of Tokyo, 7-3-1, Bunkyou-ku, Tokyo, 113-8654, Japan
[2] Accuthera Inc., 2-22, Shirakata-shirane, Tokai-mura, Naka-gun, Ibaraki, 319-1188, Japan
[3] Institute of Industrial Science, 4-6-1, Komaba, Meguro-ku, Tokyo, 153-8505, Japan
[4] Public Works Research Institute, 1-6, Minamihara, Tsukuba-shi, Ibaraki, 305-8516, Japan

[5] jean-michel-bereder@nuclear.jp
[6] yukimitsuya@nuclear.jp
[7] takahashi@concrete.t.u-tokyo.ac.jp
[8] kdobashi@tokai.t.u-tokyo.ac.jp
[9] uesaka@nuclear.jp
[10] kusano@accuthera.com
[11] yasuxi@iis.u-tokyo.ac.jp
[12] y-ooshima@pwri.go.jp
[13] ishida@pwri.go.jp

Abstract. The existing non-destructive inspection method employed for concrete structures uses high energy X-rays to detect internal flaws in concrete structures and iron reinforcing rods. In addition to this conventional method, the authors are developing an innovative inspection system that uses a mobile compact linac-driven neutron source that utilizes neutron backscattering, to measure the moisture distribution in concrete structures and estimate the corrosion probability distribution of iron reinforcing rods.

1. Introduction

1.1. Research purpose

During the period of rapid economic growth from the 1950s to the 1970s, Japanese governmental investment in infrastructural projects such as bridges and buildings expanded rapidly. However, most of the industrial and social infrastructure projects had an estimated lifetime of only approximately 50 years, so the declining strength of concrete structures has become an issue of national importance.

From a cost-performance point of view, it is better to carry out on-site non-destructive inspections regularly and repairs as and when needed, rather than demolishing or rebuilding. Non-destructive inspection methods for social infrastructures, aimed at detecting internal flaws in concrete structures and iron rods, have therefore been developed[1]. In addition to the conventional high energy X-ray method, the development of a neutron backscatter moisture detection system using a linear accelerator driven neutron source to measure moisture distribution in concrete structures is now under development. By combining the knowledge of the moisture distribution in concrete structures, the corrosion probability distribution of iron reinforcing rods can be estimated. (figure 1)

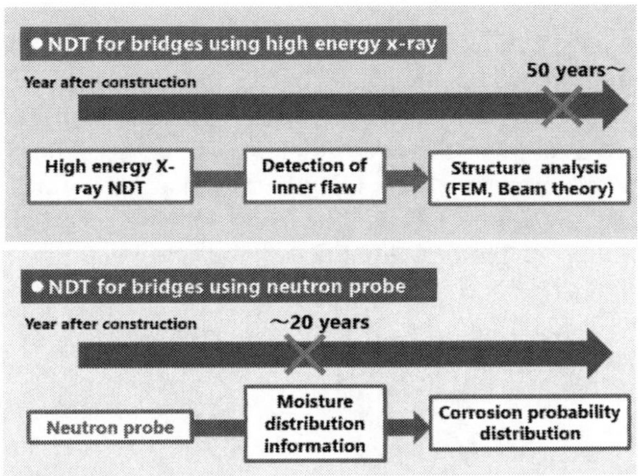

Figure 1. Applications of X-ray source and neutron source in NDT of social infrastructure.

1.2. Moisture detection for PC bridges

About 60% of the bridges, the most important infrastructures, are concrete bridges, and about 40% of them are prestressed concrete bridges (PC bridges).

A PC bridge has cylindrical structure of iron inside the concrete, and steel wires (PC wire) with tensile stress (prestress) are passed through, thereby giving compressive stress to the entire concrete, and strengthening the whole structure. The PC wires are filled with grout filler and placed under basic conditions, so that a passive film of $Ca(OH)_2$ is formed on the surface of the PC wires that prevents internal corrosion. This passive film is destroyed by neutralization due to CO_2 invasion into the concrete. Corrosion of the wire is caused by moisture penetration in the grout unloaded part due to initial construction failure, leading to deterioration of the concrete (figure 2). Therefore, it is important to acquire internal moisture distribution information for the soundness evaluation of concrete structures, but it is not yet realized in non-destructive inspection technology usable on site.

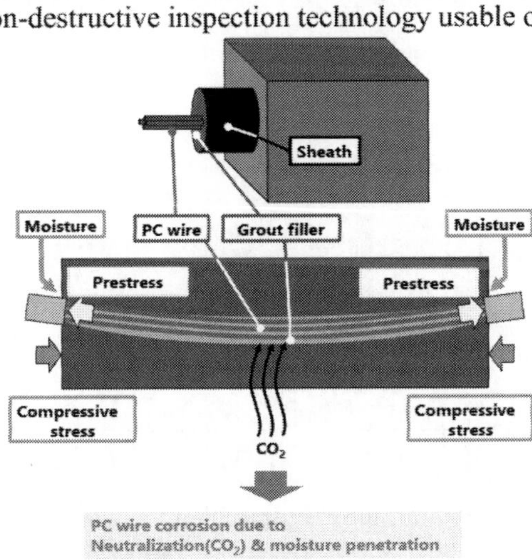

Figure 2. Description of prestressed concrete structure.

2. Description of neutron source

2.1. Target design

Beryllium ^{9}Be, having the lowest threshold energy for photo-nuclear reaction $^{9}Be(\gamma,n)^{8}Be^{*}$, is used in a mobile linac-driven neutron source, as designed by an Italian group in the previous study [2]. A beryllium photo-neutron target has been combined with a lead beam collimator, a boric acid resin layer for neutron shielding, and a lead layer for γ-ray shielding (figure 3). The ^{9}Be target and the 3.95 MV mobile X-ray source together comprise the mobile neutron source system (figure 4). Since mainly fast neutrons are used in the neutron source, a beam line using a high-Z material that does not moderate the neutrons is used. Optimization of the beryllium target size and neutron/γ-ray shielding simulation is performed using the Monte-Carlo code.

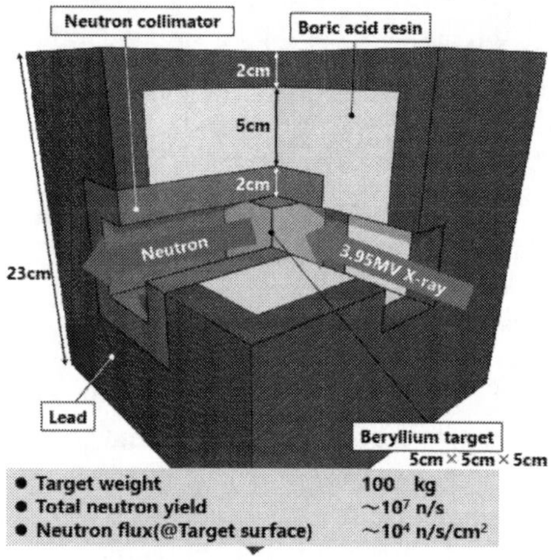

Figure 3. Description of neutron source

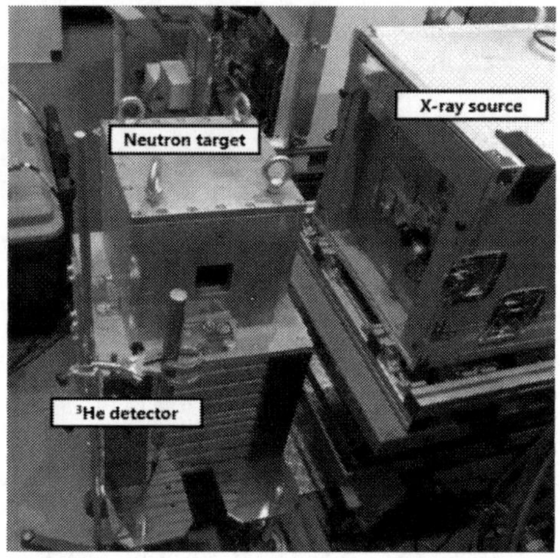

Figure 4. Neutron source setup.

2.2. Neutron yield and energy distribution

The calculated neutron yield in the neutron source is approximately $\sim 10^{7}$n/s. The target weight is about 100 kg, which is manageable by manpower. Radiation safety is satisfactory as well. The distribution of neutron energy produced is shown in figure 5.

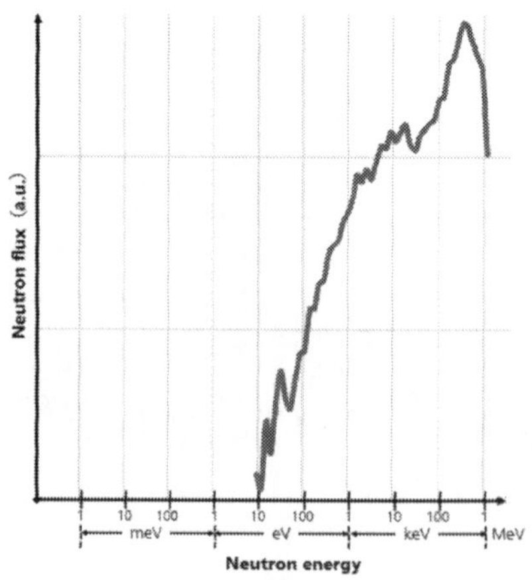

Figure 5. Produced neutron energy distribution.

3. Moisture detection using backscatter neutron

3.1. Detection principle

Moisture detection using neutrons is performed by irradiating concrete with fast neutrons and detecting backscattered moderated neutrons due to multiple elastic scattering with light elements especially hydrogen nuclei. The substance used for neutron detection including ^{3}He has a high reaction cross section with neutrons in the thermal region, and therefore the count of detectors increases with the existence of moisture (figure 6).

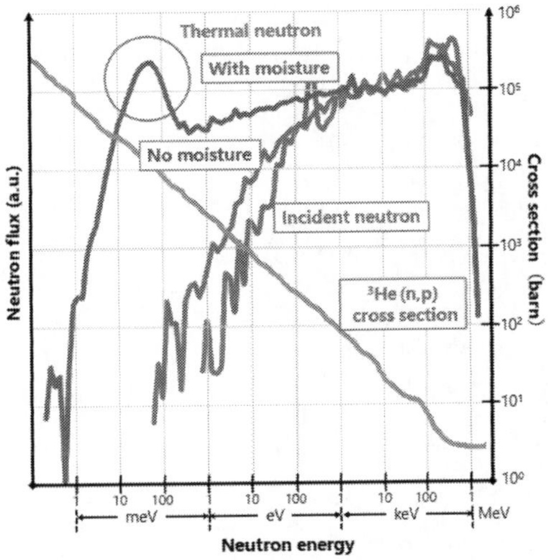

Figure 6. Backscatter neutron energy (with and without water), incident neutron energy and $^{3}He(n,p)$ cross section.

3.2. Moisture detection experiment of PC concrete sample

Moisture detection experiments simulating immersion of PC wire duct in bridges were also conducted (figure 7). 100 g of water was placed in a concrete sample having a cylindrical sheath structure at a depth of 10 cm in the concrete, and the measurement was performed. As a result, the measurement was successful with accuracy of 3σ (figure 8).

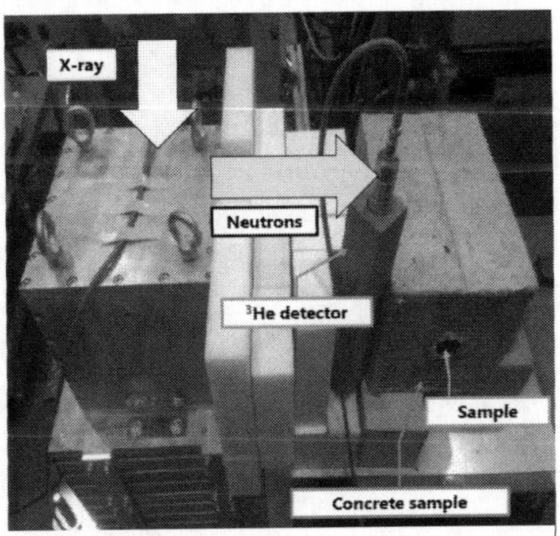

Figure 7. Experiment simulating immersion of PC wire duct.

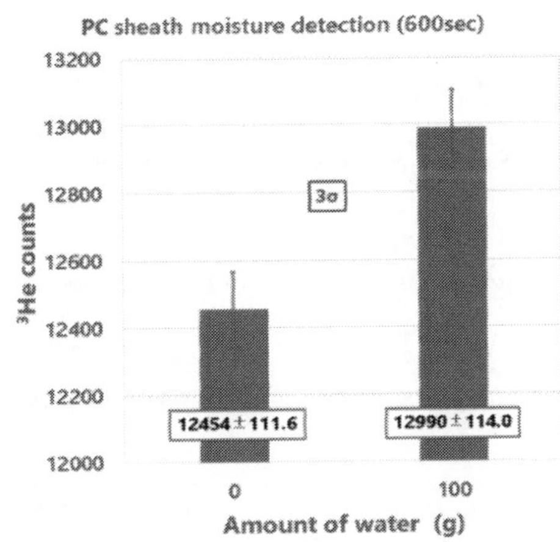

Figure 8. Result of PC sheath moisture detection experiment.

3.3. Concrete moisture content measurement

Concrete moisture content measurement experiment was conducted as a fundamental study for measurement of moisture distribution in concrete structures(figure 9). Each concrete sample with a water cement ratio (W/C ratio) of 36% and 50% was adjusted to 0% moisture content and 100% moisture content and 600 second-measurement were carried out. As a result, even for concrete samples with different W/C ratios, the difference in neutron detector counts between the moisture content 0% case and 100% case is approximately the same, which means the moisture information is successfully acquired regardless of the formulation of concrete sample itself.(figure 10)

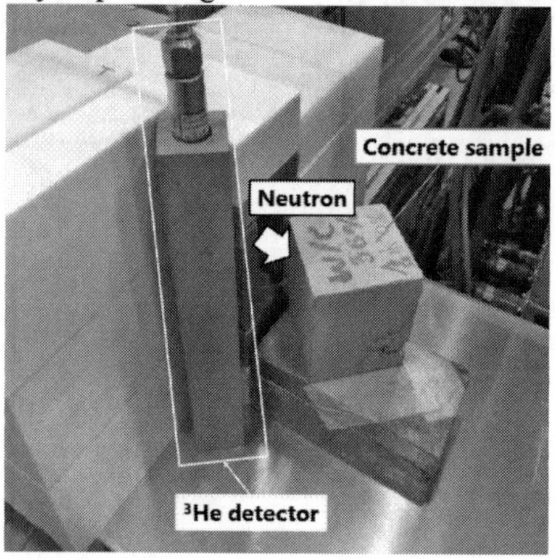

Figure 9. Concrete moisture content measurement setup.

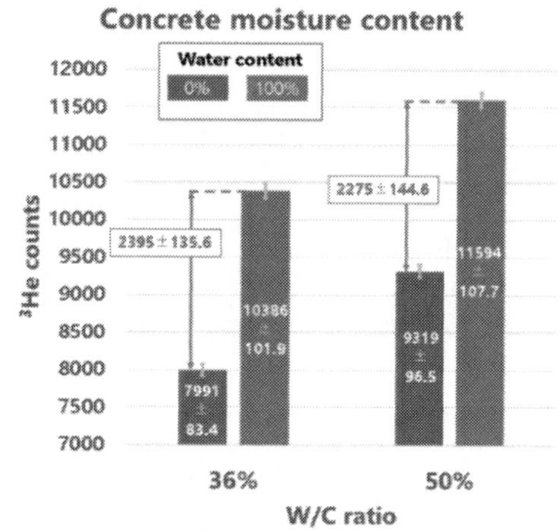

Figure 10. Result of concrete water content measurement experiment.

4. Conclusion

An experiment simulating flood conditions in actual bridges and a measurement experiment of concrete moisture content - using a mobile linac driven compact neutron source - were carried out. The development of a measurement system using neutron energy information by TOF to improve inspection accuracy is now under development. The first outdoor actual bridge test using an accelerator driven neutron source will also be conducted in this fall.

Acknowledgements

This work was supported by Council for Science, Technology and Innovation (CSTI), Cross-Ministerial Strategic Innovation Promotion Program (SIP) (Funding agency: JST).

References

[1] Uesaka M *et al.* 2014 On-site nondestructive inspection by upgraded portable 950 keV/3.95 MeV x-band linac x-ray sources *Journal of Physics* B **47** 234008
[2] Auditore L *et al.* 2005 Study of a 5 MeV electron linac based neutron source *Nuclear Instruments and Methods in Physics Research* B **229** pp 137-43

Structural analysis and evaluation of actual PC bridge using 950 keV/3.95 MeV X-band linacs

H Takeuchi[1], R Yano[1], I Ozawa[1], Y Mitsuya[1], K Dobashi[1], M Uesaka[1], J Kusano[2], Y Oshima[3], M Ishida[3]

[1] Department of Nuclear Engineering and Management, School of Engineering, The University of Tokyo, 7-3-1 Hongo, Bunkyo-ku, Tokyo, Japan
[2] Accuthela Inc., 2-8-22 Kurigi, Asao-ku, Kawasaki, Kanagawa, Japan
[3] Public Works Research Institute, 1-6 Minamihara, Tsukuba, Ibaraki, Japan

E-mail: takeuchi@nuclear.jp

Abstract. In Japan, bridges constructed during the strong economic growth era are facing an aging problem and advanced maintenance methods have become strongly required recently. To meet this demand, we develop the on-site inspection system using 950 keV/3.95 MeV X-band (9.3 GHz) linac X-ray sources. These systems can visualize in seconds the inner states of bridges, including cracks of concrete, location and state of tendons (wires) and other imperfections. At the on-site inspections, 950 keV linac exhibited sufficient performance. But, for thicker concrete, it is difficult to visualize the internal state by 950 keV linac. Therefore, we proceeded the installation of 3.95 MeV linac for on-site bridge inspection. In addition, for accurate evaluation, verification on the parallel motion CT technique and FEM analysis are in progress.

1. Background and objective

In Japan, many concrete bridges are facing aging problems. Potential deteriorations are periodically inspected visually. However, it is necessary to accurately detect imperfection inside concrete for accurate safety assessment. Currently, several inspection methods have been proposed for detecting deterioration of the main cable of the bridge. Transmitted X-ray inspection is one of the effective methods in respect of visualization of internal defects of bridge [1]. So far, low energy X-ray sources, which is up to 300 keV, have been used for on-site concrete structure inspection. However, the applicable thickness of 300 keV X-rays is limited to about 300 mm, and it takes long time to obtain a result even if it can be applied to over 300 mm thickness. Therefore, our research objective is to apply 950 keV/3.95 MeV linac for on-site bridge inspection, which makes it possible to inspect thicker concrete briefly.

950 keV and 3.95 MeV Linac parameters are shown in Table 1 [2]. The X-ray energy generated by the mobile linacs have a spectral distribution with maximum energy of 950 keV and 3.95 MeV, respectively. Both mobile linacs are composed of a high-frequency magnetron source and an accelerating tube part. Both are coupled by X-band flexible waveguide and mutual position is variable. Power supply and water cooling chiller are in stationary boxes. These setting ensures linac's portability and flexibility of X-ray irradiation direction. The weight of the 950 keV X-ray source is about 52 kg and magnetron is 45 kg. As for 3.95 MeV X-ray source, the X-ray source is 140 kg and the magnetron is 62 kg. The 950 keV X-ray source is effective for thin concrete which is 200 to 350

IOP Conf. Series: Journal of Physics: Conf. Series **874** (2017) 012105 doi:10.1088/1742-6596/874/1/012105

mm and X-ray protection chief qualification is required for operation. The 3.95 MeV X-ray source is effective for thicker concrete which is 350 to 740 mm. However, the 3.95 MeV X-ray source require the complex process for operation to follow the Japanese law. Now, we mainly use the 950 keV linac for on-site inspection and installation of 3.95 MeV is now on progress. We will use both linac separately in necessary cases in future.

Table 1. Parameters of Mobile Linac

Parameter	950 keV Linac	3.95 MeV Linac
RF Source	9.3 GHz Magnetron	9.3 GHz Magnetron
RF Power	300 kW	1.5 MW
Repetition	330 pps	200 pps
Beam Energy	950 keV	3.95 MeV
Electron Gun Voltage	20 kV	20 kV
Pulse Width	2.5 μs	4 μs

2. On-site bridge inspection

We applied our inspection system for on-site bridge inspection. The target bridge was a post-tensioned PC box girder bridge. This bridge aged over 40 years, and it will be dismantled in a few years. The purpose of inspection was detecting the corrosion of tendons which are main cables of bridge and play an important role in maintaining the safety of the bridge by introducing compressive stress to the concrete. The experiments using 950 keV linac were conducted to scan the bottom flange of the bridge. Figure 1 shows schematic view of the experiment. In the experiment, X-ray source was placed inside the box, webs and flanges of which can prevent X-ray from spreading outside. In the examination, we irradiated X-ray downward and the IP (Imaging plate, made by Fuji Film) was placed on the opposite side of X-ray source. The linac was operated at a beam current of 120 mA and a repetition was 300 pps.

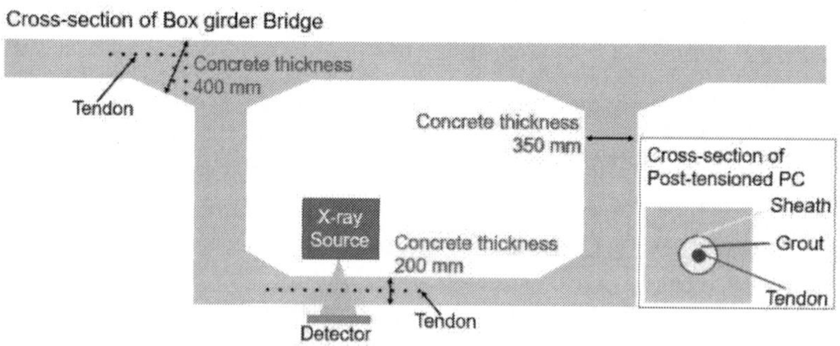

Figure 1. Schematic view of real bridge inspection.

Figure 2 is one of the results in this experiment, and the tendons, sheaths and rebars were successfully visualized by X-ray. As a result, we succeeded in clearly imaging the internal concrete structure, and then assumed the reduced section areas as shown in Table 2 [3].

In general, the thicknesses of concrete to be inspected are often over 400 mm: for instance, web concrete is about 300 mm, connection of web and flange is about 400 mm in this experiment. It is considered difficult to obtain clear results on top flange by 950 keV X-ray source because of the thickness of the concrete. From this consideration, it is necessary to introduce 3.95 MeV linac for a wider range of on-site bridge inspection

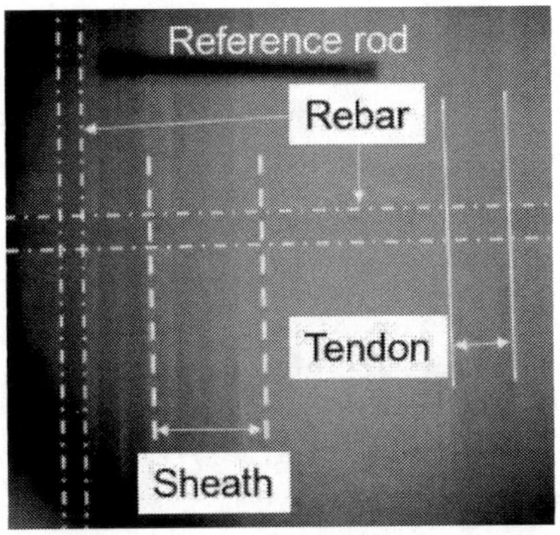

Figure 2. Visualized tendon, sheath and rebars in the bottom flange by X-ray.

Table 2. Assumed reduction rate of cross sectional area of tendon.

Reduction rate of cross sectional area	Number of tendons
0 %	14
0 ~ 30 %	5
30 ~ 60 %	1
60 ~ 90 %	1
100 %	1

3. Benchmark experiment of 3.95 MeV linac

To confirm the feasibility of 3.95 MeV linac for on-site inspection, benchmark test is required by radiation regulatory body. In March 2017, we conducted the benchmark experiment at PWRI (Public Works Research Institute in Japan). In this experiment, the sectional part of a hollow slab cut from an actual bridge was examined by 3.95 MeV in the experimental field. Figure 3 shows the cut-out sample. The picture also shows three examination lines and inspection points in this examination. Point 1 line (P1) has approximately 400 mm, point 2 has 700 mm and point 3 has 800 mm thickness of concrete. Figure 4 shows the 3.95 MeV X-ray source. In this experiment, 3.95 MeV linac was operated at a beam current of 100 mA and a repetition was 200 pps.

For point 1, a FPD (Flat panel detector, XRD1622 made by Perkin Elmer) was placed in a hollow of the bridge. As a result of inspection, the bundle of tendons and some rebars are visualized in 5 seconds (Figure 5).

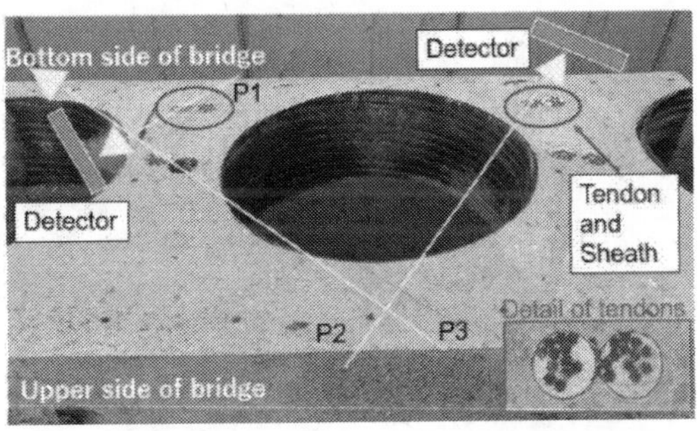

Figure 3. Cut-out sample and inspection points.

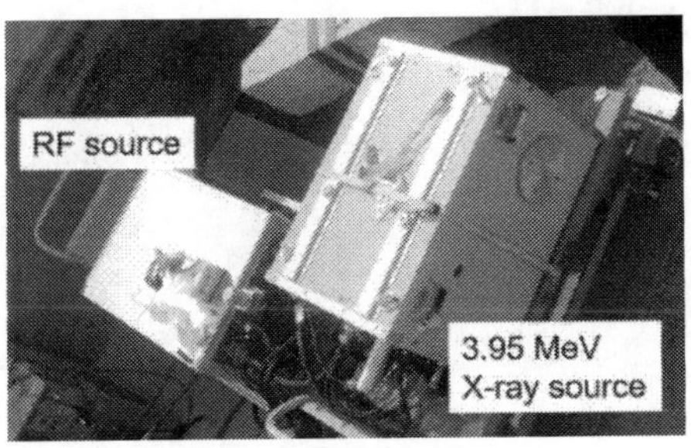

Figure 4. 3.95 MeV linac.

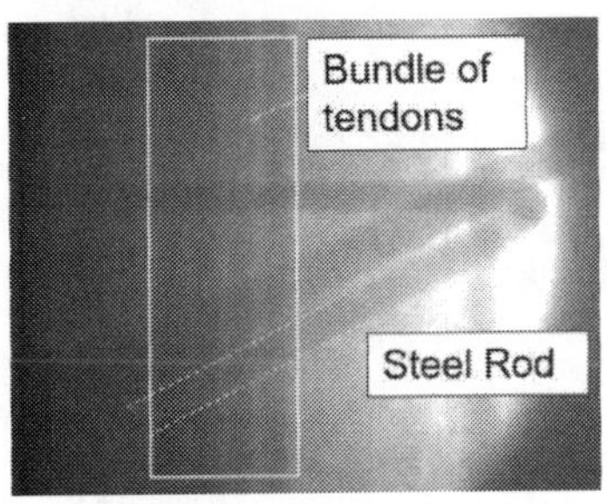

Figure 5. The result of P1.

IOP Conf. Series: Journal of Physics: Conf. Series **874** (2017) 012105 doi:10.1088/1742-6596/874/1/012105

As for point 2 line where X-ray should penetrate the thicker concrete part, we used FPD and IP as detectors. At this inspection point, X-ray crosses two steel plates until the detector. Figure 6 shows the results given by FPD (left side) and IP (right side), respectively. In the result of FPD, several rebars are visualized, but the right part of this figure is blurred due to the steel plate. The integration time of FPD was 20 seconds. In the IP result, in addition to rebar, bundle of tendons is slightly visualized in 1 minute integration time.

On the point 3 line, the internal structure was not clearly visualized. From these result, it is considered that the 3.95 MeV X-ray source is applicable to concrete up to 700 mm thick. However, if obstacle such as steel plate of hollow slabs is included, a long integration time is required for detector to visualize the internal structure.

From this result, however, safety of X-ray inspection executed in the field was confirmed, which enables the permission from the regulatory authority to conduct on-site bridge inspection by 3.95 MeV X-ray source

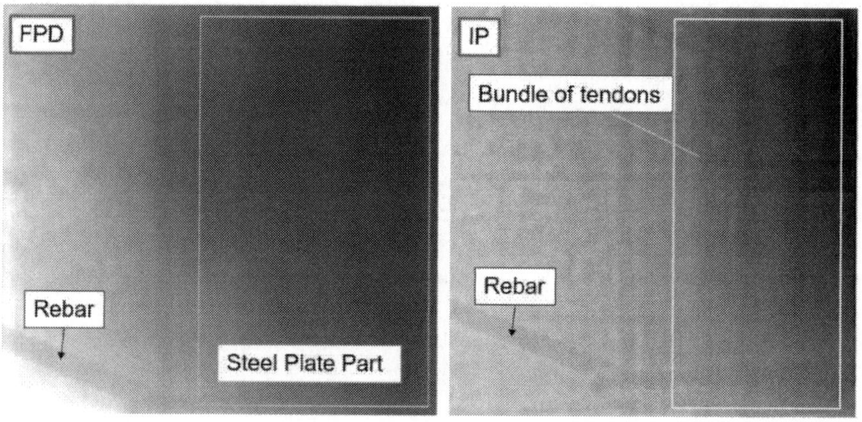

Figure 6. The inspection result of P2.

4. Analysis and evaluation process

4.1. Parallel Motion CT

For evaluation of tendon's corrosion, cross-sectional image is highly required and CT reconstruction technique may open up the possibility to obtain the sectional image. However, it is not practical or impossible to conduct 360 degree rotated CT because rotation is limited. Therefore, we intend to apply the parallel motion CT which can reconstruct the cross-sectional image without rotating the X-ray source and detector [4]. Demonstration of parallel motion CT was conducted by R. Yano, and 950 keV linac was used to radiate X-ray as a cone beam with an angle of 30 degree. In this experiment, target concrete sample has 2 iron rods and a tendon located in the sheath. The thickness of concrete is 200 mm. Figure 7 shows the image reconstructed by parallel motion CT. Due to the limited angle geometry, reconstruction artifacts arise especially in beam direction. However, positional relationship and the diameter are reconstructed. This parallel motion CT technique requires only translational movement and does not need to rotate the X-ray source. Thus, the parallel motion CT is much promising for on-site bridge inspection.

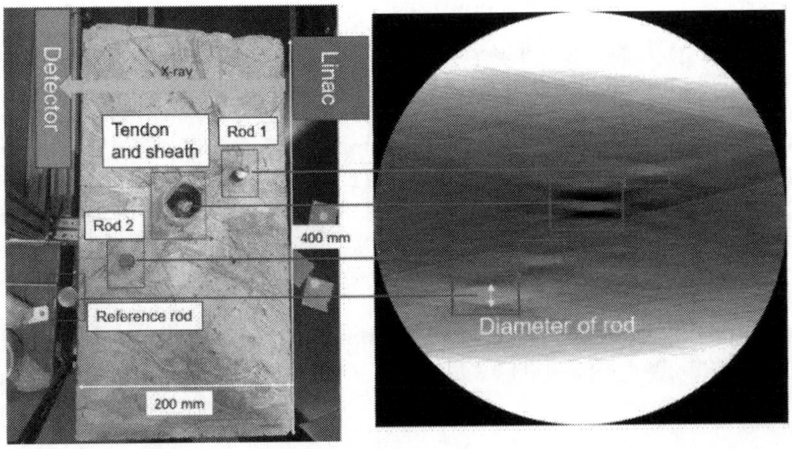

Figure 7. Target sample and reconstructed image.

4.2. Structural Analysis using FEM

Evaluation of cross-sectional reduction of tendon is estimated by the result of X-ray Inspection. This deterioration has a significant influence on the structural safety of the bridge. Therefore, by conducting the structural analysis using finite element method (FEM), the residual bearing capacity of the entire bridge is evaluated from the result of the X-ray inspection. Calculation software named DuCOM-COM3 is used for the FEM program [5]. Figure 8 shows the example of FEM model of bridge. In this software, physical properties of concrete are entered in 3-D meshes and tendon's physical properties and diameter are inputted in 2-D element. Figure 9 shows the strain distribution calculated by the FEM analysis where the physical parameters and boundary condition were properly considered on the basis of on-site inspection result listed in Table 2. The stress and strain distribution on cross section was calculated, confirming 60 μ increase of strain at the bottom part. Through this analysis, the decrease of tendon's sectional area much affect the bearing capacity and X-ray inspection can contribute to structural safety evaluation.

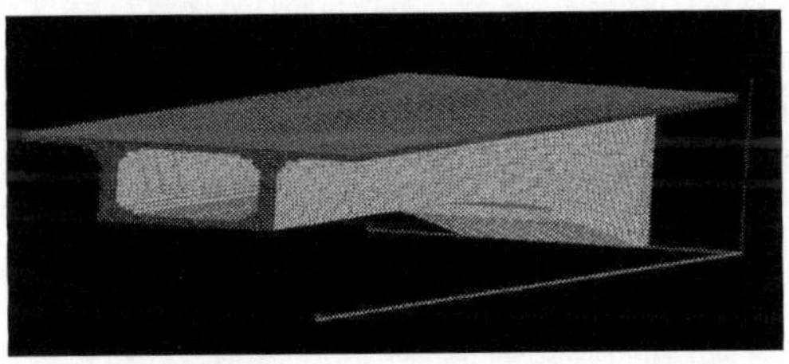

Figure 8. The mesh model of FEM.

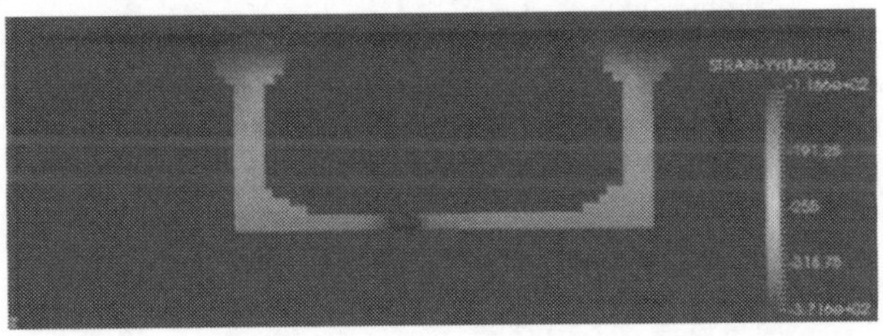

Figure 9. The example of strain distribution analysis.

5. Conclusion

We succeeded in visualizing the internal structure of bridge in on-site inspection using 950 keV linac. For thick concrete inspection, we conducted the benchmark experiment of 3.95 MeV linac. The 3.95 MeV linac succeeded in visualization of internal structure of 700 mm thickness of concrete and it can be applied for on-site bridge inspection in this year. As a post processing of X-ray inspection result, parallel motion CT exhibits high reconstruction performance. It was also suggested that this reconstruction method can be applied to bridge inspection by scanning the bridge using parallel motion stage. It is possible to perform FEM analysis using the result of X-ray inspection and contribute to the safety evaluation of entire bridge.

References

[1] Oshima Y, Ishida M, Uesaka M and Otake Y 2016 *Proc. 8th Int. Conf. on bridge maintenance, safety and management* (Foz Do Iguacu, Brazil) p 185
[2] Ueaska M *et al* 2014 *J. Phys. B: At. Mol. Opt. Phys.* **47** 234008
[3] Uesaka M *et al* 2016 *Inspection Engineering* **21** pp 45-52
[4] Yo M and Yo K 2014 *IECIE Technical Report* **113(499)** pp 103-8
[5] Maekawa K, Ishida T and Kishi T 2009 *Multi-Scale Modelling of Structural Concrete* ed Taylor & Francis (NY, USA)

Acknowledgement

This work was supported by Council for Science, Technology and Innovation(CSTI), Cross-ministerial Strategic Innovation Promotion Program (SIP) (Funding agency: JST).

Focusing and bunching of ion beam in axial injection channel of IPHC cyclotron TR24

T Adam[2], I Ivanenko[1], N Kazarinov[1], F Osswald[2] and E Traykov[2]

[1]Joint Institute for Nuclear Research, Dubna, 141980, Russia
[2] Institut Pluridisciplinaire Hubert Curien/IN2P3/CNRS, University of Strasbourg, 67037, Strasbourg, France

nyk@jinr.ru

Abstract. The CYRCé cyclotron (CYclotron pour la ReCherche et l'Enseignement) is used at IPHC (Institut Pluridisciplinaire Hubert Curien) for the production of radio-isotopes for diagnostics, medical treatments and fundamental research in radiobiology. The TR24 cyclotron produced and commercialized by ACSI (Canada) delivers a 16-25 MeV proton beam with intensity from few nA up to 500 μA. The solenoidal focusing instead of existing quadrupole one is proposed in this report. The changing of the focusing elements will give the better beam matching with the acceptance of the spiral inflector of the cyclotron. The parameters of the focusing solenoid are found. Additionally, the main parameters of the bunching system are evaluated in the presence of the beam space charge. This system consists of the buncher installed in the axial injection beam line of the cyclotron. The using of the grid-less multi harmonic buncher may increase the accelerated beam current and will give the opportunity to new proton beam applications.

1. Introduction

The beam transport and bunching of the H^- ion beam by means of multi-harmonic buncher that may be installed in the axial injection beam line of the TR24 [1] cyclotron have been considered in paper [2]. Using a buncher will give an opportunity to increase the accelerated beam current. In this report, the solenoidal focusing instead of the quadrupole one was calculated. The influence of the space charge on the beam bunching was evaluated also. The simulation of beam transport was carried out by means of 3D version of MCIB04 program code based on momentum method [3].

2. Beam line layout

The scheme of the beam line and the approximate length of the optical elements are shown in Figure 1. This scheme was the basis for simulation of the dynamics of the ion beam.

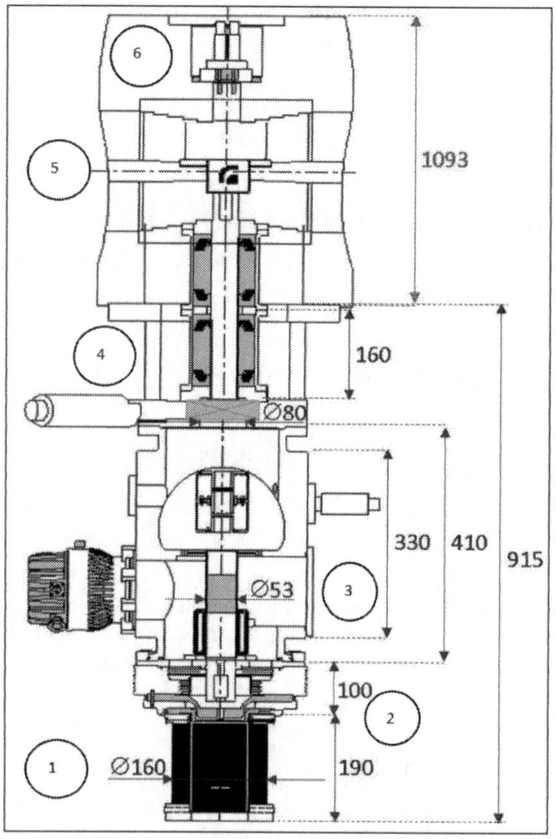

Figure 1. Axial injection beam line of TR24 cyclotron. 1 - CUSP ion source; 2 - extraction electrodes; 3 - EM steering (H/V); 4 - EM quad doublet; 5 - ES deflection; 6 – cyclotron. Dimensions – mm

3. H^- ion beam parameters

H^- ion beam is produced in the CUSP ion source [4] with kinetic energy of 30 keV. The beam emittance is strongly dependent on beam current. For H^- ion beam currents varying from 1 mA to 5 mA the initial beam diameter is equal to 1 cm and the normalized beam emittance is changing within range 0.1÷0.4 π mm×mrad. The main parameters of the H^- ion beam used in the simulation are contained in Table 1.

Table 1. H^- beam parameters

Parameter/notation/unit	Value
Charge/ Z/	1
Mass number/ A/	1
Kinetic energy/ W/ keV	30
Beam diameter/ d/ cm	1
Beam emittance/ ε/ mm×mrad	50 π
Ion beam current/ I/ mA	5
Neutralization factor/ f/	0.3-0.95

4. Simplified beam line scheme

The simplified scheme of the beam line is shown in Figure 2.

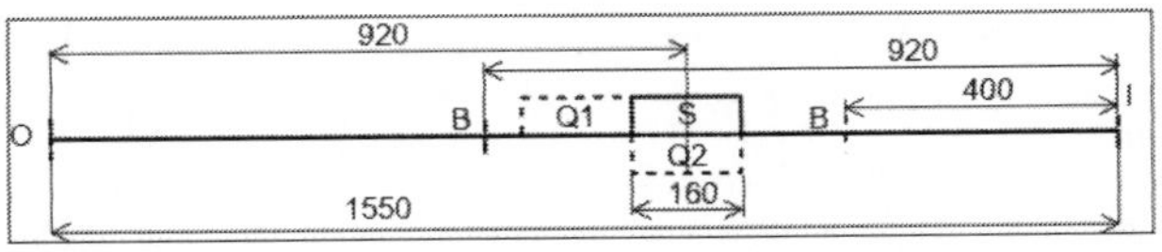

Figure 2. Beam line scheme. O – object point; I – spiral inflector; B – buncher; Q1,2 – existing quadrupole lenses; S – solenoid. Dimensions – mm

The object point O is placed at the edge of the CUSP source (see Figure 1). The solenoidal lens S is installed at the place of the second existing quadrupole lens Q2. The two possible placements of the buncher are shown also.

5. Beam neutralization

The beam current from CUSP ion source may achieve up to 5 mA. The transport of the beam with a big current is impossible without reasonable assumption about beam space charge neutralization. In the simulation the neutralization factor had a different values in the space before f_n (for the non-bunched beam) and after f_b the buncher (for the bunched beam): $f_n = 95\%$; $f_b = (30 \div 95)\%$.

6. Focusing solenoid

The POISSON/SUPERFISH [5] computational model of the solenoid and on-axis magnetic field distribution are shown in Figures 3,4.

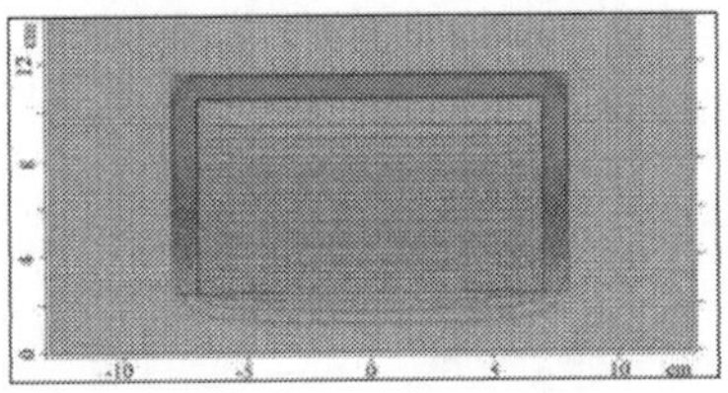

Figure 3. Solenoid computational model

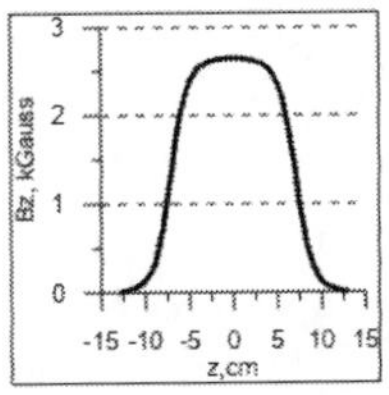

Figure 4. Magnetic field distribution

The parameters of the focusing solenoid are contained in Table 2.

Table 2. Solenoid parameters

Parameter, unit	Value
Overall length, cm	16
Overall diameter, cm	23
Screen width, cm	1
Winding dimension(hor.×vert.), cm	14×8
Number of turn	14×8=112
Diameter of aperture, cm	5
Maximum induction, kGauss	2.96
Maximum current, A	300
Power supply, kW	1.1

7. Beam transport

The maximal magnetic field induction of the solenoidal lens Bm or gradients of the quadrupole lenses were fitted to minimize the amplitude of beam envelopes oscillation at the entrance of the spiral inflector for both degrees of freedom. The fitted value of Bm is approximately equal to 2.3 kGauss.

The results of simulation of beam transport through the axial injection beam line are shown in Figures 5-7.

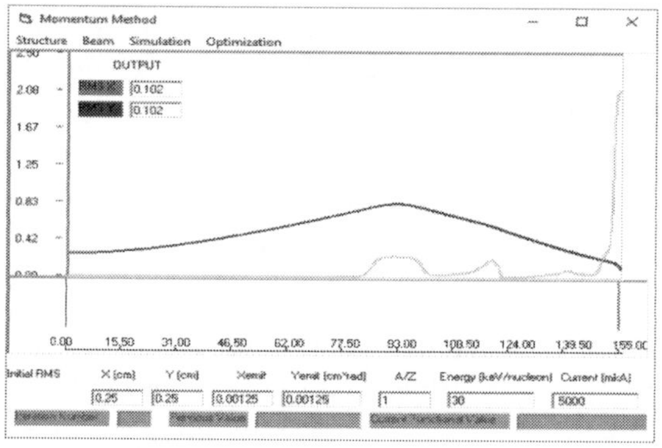

Figure 5. User interface of MCIB04 program (3D momentum method version) with RMS beam envelopes and longitudinal magnetic field distribution

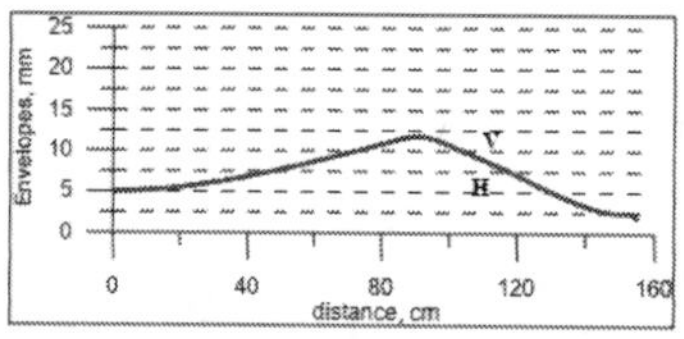

Figure 6. Horizontal (H) and vertical (V) H- ion beam envelopes (2σ) in the case of solenoidal focusing

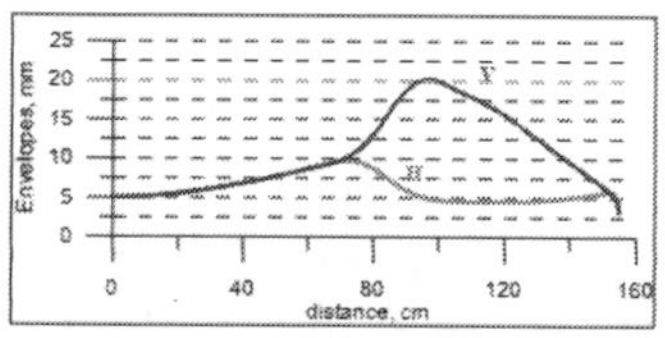

Figure 7. Horizontal (H) and vertical (V) H^- ion beam envelopes (2σ) in the case of quads focusing

As may be seen from Figures 6,7 the beam matching is sufficiently better in the case of the solenoidal focusing. The beam dimensions are rather close to the matched beam radius that is approximately equal to 1 mm. Therefore, the solenoidal focusing is more convenient for the axial injection channel.

8. Beam bunching and space charge

The beam bunching may be realized by multi-harmonic buncher B installed before quadrupole Q1 (see Figure 2) at the distance $L_B = 92$ cm from median plane of cyclotron. The electric field in the buncher produces the modulation of the longitudinal momentum of the ions. After the drift space, this modulation gives the longitudinal modulation of the ion beam density. The grid-less four-harmonic buncher has been successfully used at ATLAS facility [6].

Using of the buncher will give opportunity to double the accelerated beam current in the case $f_n = f_b = 95\%$ and quadrupole focusing of the beam [2]. In fact, the neutralization of the space charge of the beam is not the same for non-bunched and bunched beams. For the non-bunched beam, the neutralization factor f_n is close to 100% and the value $f_n = 95\%$ is a rather good approximation. However, for the bunched beam the neutralization factor f_b may significantly differ from 100%.

Let define the bunching efficiency as the ratio of the number of ions within phase interval 20 degrees of RF field of the cyclotron for bunched beam and non-bunched beam. The dependence of the bunching efficiency on length after the buncher for various values of the bunched beam neutralization factor f_b is shown in Figure 8.

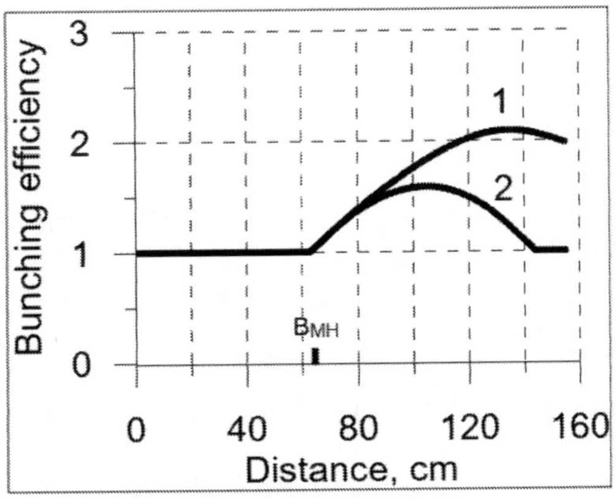

Figure 8. Bunching efficiency. L_B = 92 cm. Curve 1 – f_b = 95%; Curve 2 – f_b = 80%

As may be seen from Figure 8 the bunching efficiency decreases with decreasing of the neutralization factor of the bunched beam f_b and for $f_b \geq 80\%$ the use of the buncher becomes non-effective because of the space charge debunching forces.

The decreasing of the length of bunching L_B leads to decreasing of the lower limit of the neutralization factor f_b.

The dependence of the bunching efficiency on length after the buncher placed at the distance L_B = 40 cm from median plane of cyclotron for various values of the bunched beam neutralization factor f_b is shown in Figure 9.

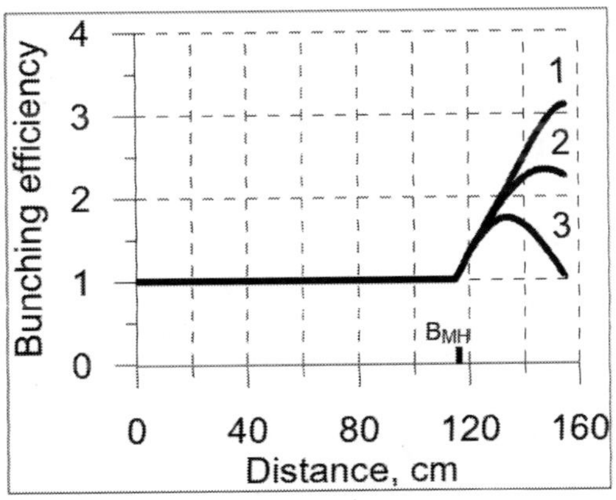

Figure 9. Bunching efficiency. L_B = 40 cm. Curve 1 – f_b = 95%; Curve 2 – f_b = 80%; Curve 3 – f_b = 30%

Comparison of Figures 8,9 shows that the decreasing of the length of bunching L_B leads to increasing of the bunching efficiency (for the same neutralization factor of the bunched beam f_b). The accelerated beam current may be approximately doubled (L_B = 92 cm) or tripled (L_B = 40 cm) in the case of big neutralization of the bunched beam (f_b = 95%). The lower limit of the neutralization of

IOP Conf. Series: Journal of Physics: Conf. Series **874** (2017) 012106 doi:10.1088/1742-6596/874/1/012106

bunched beam when the beam bunching becomes non-effective is smaller for L_B = 40 cm and equal to f_b = 30%.

References

[1] Advanced Cyclotron Systems Inc. http://www.advancedcyclotron.com/cyclotron-solutions/tr24
[2] Kazarinov N, Ivanenko I and Osswald F 2016 Axial Injection Channel of IPHC Cyclotron TR24 and Possibility of Ion Beam Bunching *Proc. of 21st Int. Conf. on Cyclotrons and Their Appl.* (Zurich, Switzerland), pp 224-6 http://www.JACoW.org
[3] Aleksandrov V, Kazarinov N and Shevtsov V 2008 3D Intense Beam Dynamics Simulation Using the Moments Method *Phys. of Part. and Nucl. Lett.* **5** N7 625-8
[4] Jayamanna K, McDonald M, Schmor P W and Yuan D H 1990 The TRIUMF Compact DC H^- /D^- Ion Source *Proc. of Europian Part. Acc. Conf. EPAC'90* (Nice, France) pp 647-9 http://www.JACoW.org
[5] Reference Manual for the POISSON/SUPERFISH Group of Codes 1987 *Los Alamos National Laboratory Report* LA-UR-87-126
[6] Ostroumov P N *et al.* 2007 Beam Test of a Grid-Less Multi-Harmonic Buncher *Proc. of Part. Acc. Conf. PAC'07* (Albuquerque New Mexico USA) pp 2242-4 http://www.JACoW.org

Neutron doses due to beam losses in a novel concept of a proton therapy gantry

V Talanov, D C Kiselev, D Meer, V Rizzoglio, J M Schippers, M Seidel and M Wohlmuther

Paul Scherrer Institut, 5232 Villigen PSI, Switzerland

E-mail: vadim.talanov@psi.ch

Abstract. A novel design of a gantry for proton therapy is investigated in which a degrader and emittance limiting collimators are mounted on the gantry. Due to the interactions of protons in these components there will be an additional neutron dose at the location where a patient is positioned during a proton therapy. The results of numerical study of this additional dose are presented. Neutron prompt dose at the patient position is estimated through the Monte Carlo simulation using the MCNPX 2.7.0 particle transport code. Secondary neutron and photon fluxes from the distinct beam loss points are taken into consideration and the resulting dose is calculated using realistic estimates of beam losses. The dependence of the dose on the beam energy and individual impacts of each loss point on the total dose at the patient position as well as on critical beam line components are estimated and potential design constraints are discussed. It has been found that compared with a conventional gantry the expected additional dose is higher but the optimization of the beam line configuration and additional shielding shall help to reduce the dose to an acceptable value.

1. Novel gantry concept

A recently presented novel concept of a gantry [1] is aimed at the single room solution for the radiation therapy with protons through locating the energy degrader system (EDS) directly on the gantry itself. The existing EDS followed by an energy selection system (ESS), for example the one used at the Center for Proton Therapy at PSI [2] at the COMET cyclotron [3], typically require a space in the facility of about 10 m in length. With the proposed novel gantry concept it is foreseen that the elimination of the ESS and the shift of the EDS shall allow for simplifying and shortening the beam delivery line.

A sketch of the schematic layout for the considered beamline is shown in Figure 1. The proton beam of fixed energy delivered by the accelerator is first deflected by 45° by a normal conducting (NC) dipole and then is directed towards the patient position by a compact superconducting (SC) magnet that provides the final 135° bend. The energy degrader is located on the gantry, right downstream from the NC dipole, to minimize the impact of protons scattered on the degrader on the beam transport. Two additional collimators downstream of the energy degrader shall allow for shaping the beam to match the acceptance of the SC magnet. The primary collimator shall define the beam size and the secondary collimator shall define the beam divergence.

In the proposed solution three major points of beam losses — energy degrader and two collimators — are combined into a single drift space located in the beam waist allowing for compacting the EDS at the ~ 2 m length. However in such design the EDS is shifted closer to

IOP Conf. Series: Journal of Physics: Conf. Series **874** (2017) 012107 doi:10.1088/1742-6596/874/1/012107

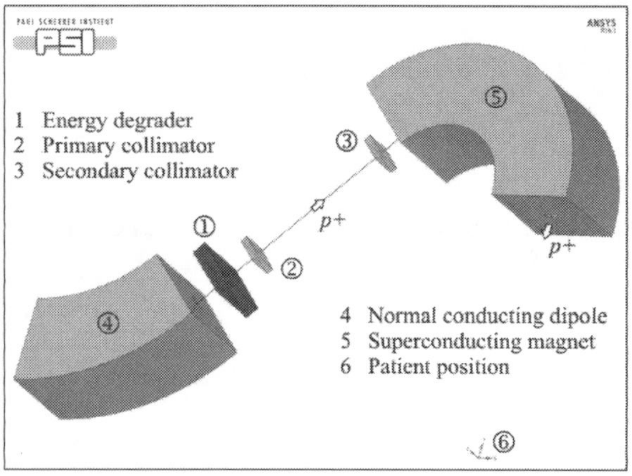

Figure 1. Beamline layout of degrader on gantry concept.

the patient position and so relative and absolute prompt neutron dose at the patient position arising from the losses at the beam transport through the EDS shall be estimated.

2. Estimated beam losses

The losses at the beam transport were calculated in the geometry model of the existing EDS of the COMET cyclotron. The beam envelope of $2\,\sigma_{x,y}$ was simulated with the OPAL charged particle tracking tool [4] as shown in Figure 2 and the reduction of the beam intensity at each EDS component was estimated. For two energy settings of the degrader estimated beam intensity reduction factors and calculated from them absolute beam losses are given in Table 1. To deliver a typical field dose of 1 Gy over a volume of one liter, approximately 10^{11} protons are needed at the patient. Using this number as an initial normalization and the factor of transmission as calculated from the values for the absolute losses, Table 1 gives for each energy setting the estimated number of protons that shall be delivered onto the degrader.

3. Neutron dose simulation

The distribution of secondary neutrons in the proposed beamline layout and resulting prompt neutron dose at the patient position were simulated using the MCNPX 2.7.0 Monte Carlo particle transport code [5]. Geometry models of simulations were designed in SpaceClaim Direct Modeller from the ANSYS Workbench 16.1 [6] as depicted in Figure 3 and then automatically converted

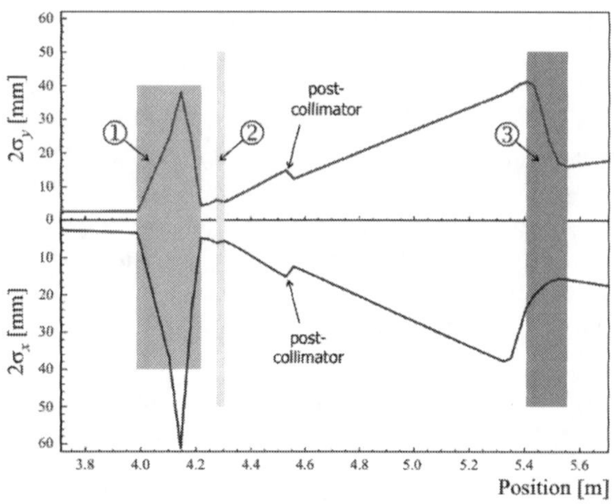

Figure 2. The beam envelope of $2\,\sigma_{x,y}$ simulated for the energy setting of 70 MeV for the EDS layout of the COMET cyclotron. Same labels as in Figure 1. The losses at the post-collimator, downstream from the primary collimator, were included into the losses at the primary collimator.

Table 1. Estimated factors of beam intensity reduction and absolute beam losses for three considered loss points, and calculated factor of transmission and number of protons onto the degrader for a given transmission and a typical field dose, for the energy setting of the degrader of 70 MeV and 230 MeV. Same notation as in Figure 1.

Energy setting	**70 MeV**	**230 MeV**
Beam intensity reduction factor		
1 Energy degrader	0.642	0.939
2 Primary collimator	0.242	0.829
3 Secondary collimator	0.121	0.494
Absolute losses		
1 Energy degrader	35.80 %	6.10 %
2 Primary collimator	48.66 %	16.06 %
3 Secondary collimator	13.66 %	39.39 %
Factor of transmission	1.88 %	38.45 %
Protons onto degrader	5.32×10^{12}	2.6×10^{11}

to the MCNPX geometry description using the SuperMC simulation software system which is developed by INEST-FDS Team [7, 8].

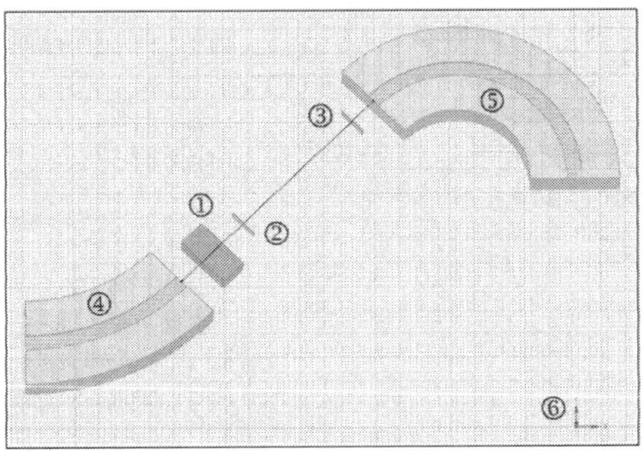

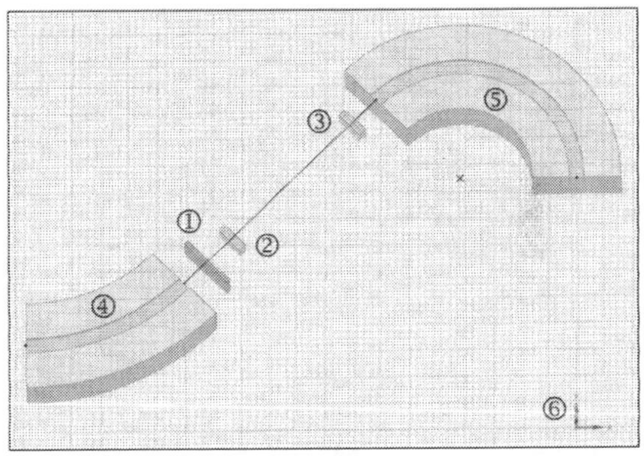

Figure 3. Simulation models for the energy setting of 70 MeV (left) and 230 MeV (right). Same labels as in Figure 1.

The primary proton beam energy was set to 250 MeV. Two cases of the energy setting of the degrader of 70 MeV and 230 MeV were simulated. The dimensions of the degrader were taken according to the given energy setting using the stopping-power and range data for protons from the PSTAR database [9]. As at the time the simulations have been performed the technical design of the collimators was not yet fixed, the thickness of primary and secondary collimators in the beam direction corresponded to 110 % of the range of 70 MeV and 230 MeV protons respectively, as shown in Figure 3, so that all protons will be stopped within the collimator material. The energy degrader was simulated as a solid graphite block. The same block geometry using copper

IOP Conf. Series: Journal of Physics: Conf. Series **874** (2017) 012107 doi:10.1088/1742-6596/874/1/012107

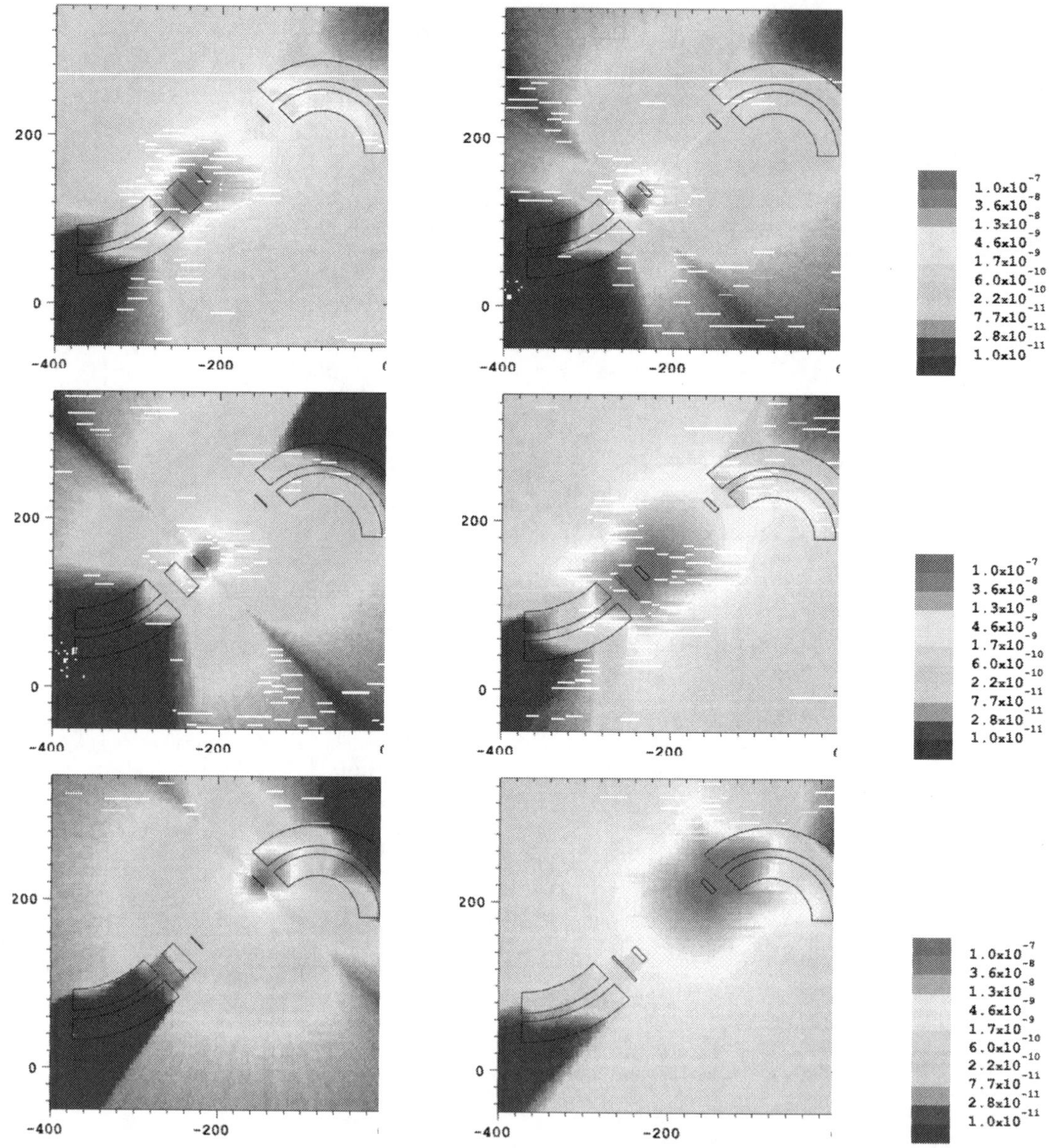

Figure 4. Neutron dose [μSv] per primary proton for the case of the beam impinging onto the degrader (1st row), primary (2nd row) and secondary collimator (3rd row) when degrading to 70 MeV (left column) and to 230 MeV (right column).

as a material was assumed for both collimators. The geometry of both NC dipole and the SC magnet was approximated by a solid curved parallelepiped made out of steel with a rectangular magnet aperture.

4. Results and outlook

The simulated 2D distributions of the neutron dose per primary proton are shown for each individual beam loss point in Figure 4. The neutron dose at the patient position was obtained by normalizing these neutron dose maps by the absolute number of incoming beam particles

IOP Conf. Series: Journal of Physics: Conf. Series **874** (2017) 012107 doi:10.1088/1742-6596/874/1/012107

Table 2. Neutron dose [μSv] at the patient position for a typical field dose and its relative fraction for beam losses at three considered loss points and the energy setting of the degrader of 70 MeV and 230 MeV. Same notation as in Figure 1.

Energy setting		**70 MeV**		**230 MeV**	
1	Energy degrader	764	85 %	9	15 %
2	Primary collimator	100	11 %	12	20 %
3	Secondary collimator	33	4 %	39	65 %
Total dose		897 μSv		60 μSv	

as calculated from the values of absolute losses from Table 1. The resulting partial and total neutron dose for two simulated cases of the energy setting of the degrader is given in Table 2.

The simulations showed a non-negligible additional dose from neutrons at the patient position of an order of 1 mSv for a typical field dose, at the energy setting of 70 MeV. The neutron dose at 230 MeV was found factor 15 lower which is almost proportional to the difference in the primary beam intensity as seen from Table 1. The main contribution to the total neutron dose was found from the thick insertions in the beamline in combination with a relatively high energy of the incoming beam. These thick insertions are the energy degrader in the case of 70 MeV and primary and secondary collimators in the case of 230 MeV, as seen also from the neutron dose maps in Figure 4.

In the final gantry design the thickness of the collimators will be chosen to be fixed according to the highest energy setting of the degrader. At lower energy settings this shall result in additional interactions of secondary particles in the collimators. Supplemental simulations show that the subsequent change of the neutron dose at the patient position will be in the order of only several percents relative to the total dose as listed in Table 2. Also in the present geometry model of simulations only a bare beamline has been introduced which definitely corresponds to an extreme maximal magnitude of the secondary neutron flux at the patient position, produced by the lost protons. One can assume that the appropriate radiation shielding together with further optimization of the beamline layout will certainly allow for reducing the estimated neutron prompt dose.

References

[1] Gerbershagen A, Meer D, Schippers J M and Seidel M 2016 *Z. Med. Phys.* **26** 224–37

[2] *Center for Proton Therapy CPT* Available from: `https://www.psi.ch/protontherapy/` [21 April 2017]

[3] Schippers J M, Dölling R, Duppich J, Goitein G, Jermann M, Mezger A, Pedroni E, Reist H W and Vrankovic V 2007 *Nucl. Instr. and Meth.* B **261** 773–76

[4] Adelmann A, Gsell A, Metzger-Kraus C, Ineichen Y, Russel S, Wang C, Yang J, Sheehy S, Rogers C and Winklehner D 2017 *The OPAL (Object Oriented Parallel Accelerator Library) Framework* Paul Scherrer Institut Report PSI-PR-08-02

[5] *MCNPX Users Manual Version 2.7.0* 2011 ed D B Pelowitz Los Alamos National Laboratory Report LA-CP-11-00438

[6] *ANSYS SpaceClaim: 3-D Modeling Software* Available from: `http://www.ansys.com/products/platform/ansys-spaceclaim/` [21 April 2017]

[7] Wu Y and FDS Team 2009 *Fus. Eng. Des.* **84** 1987–92

[8] Wu Y, Song J, Zheng H, Sun G, Hao L, Long P, Hu L and FDS Team 2015 *Ann. Nucl. En.* **82** 161–8

[9] Berger M J, Coursey J S, Zucker M A and Chang J 2005 *ESTAR, PSTAR, and ASTAR: Computer Programs for Calculating Stopping-Power and Range Tables for Electrons, Protons, and Helium Ions (version 1.2.3)* National Institute of Standards and Technology Available from: `http://physics.nist.gov/Star/` [21 April 2017]

IOP Conf. Series: Journal of Physics: Conf. Series **874** (2017) 012108 doi:10.1088/1742-6596/874/1/012108

Simulations and measurements of proton beam energy spectrum after energy degradation

A Gerbershagen, A Adelmann, R Dölling, D Meer, V Rizzoglio, J M Schippers
Paul Scherrer Institute, Villigen, Switzerland

alexander.gerbershagen@psi.ch

Abstract. At the proton therapy facility PROSCAN of the Paul Scherrer Institute the energy modulation of the cyclotron generated proton beam is performed via material insertion into the beam trajectory. The energy spectrum of the particles propagating forwards after such procedure has been simulated and measured. The current paper summarizes the results of these simulations and measurements and illustrates their significance for the future developments of a gantry for proton therapy at the Paul Scherrer Institute.

1. Introduction

In the cyclotron driven proton therapy facilities the energy of the proton beam is often reduced by an energy degrader [1]. At PSI it consists of six wedges composed of graphite, which are being inserted into the beam trajectory (see Figure 1). In less than 50 ms, the wedges can be moved into the beam path to a smaller or larger degree, providing all the energies in the typical range 230 – 70 MeV.

Figure 1. Depiction of degrader wedges [1].

After the degradation the beam has a momentum width, which can be higher than the momentum acceptance of the beamline and the gantries. Hence, typically an Energy Selection System (ESS) is installed downstream of the degrader. This system comprises two dipoles, a set of quadrupoles and a horizontal collimator slit (see Figure 2).

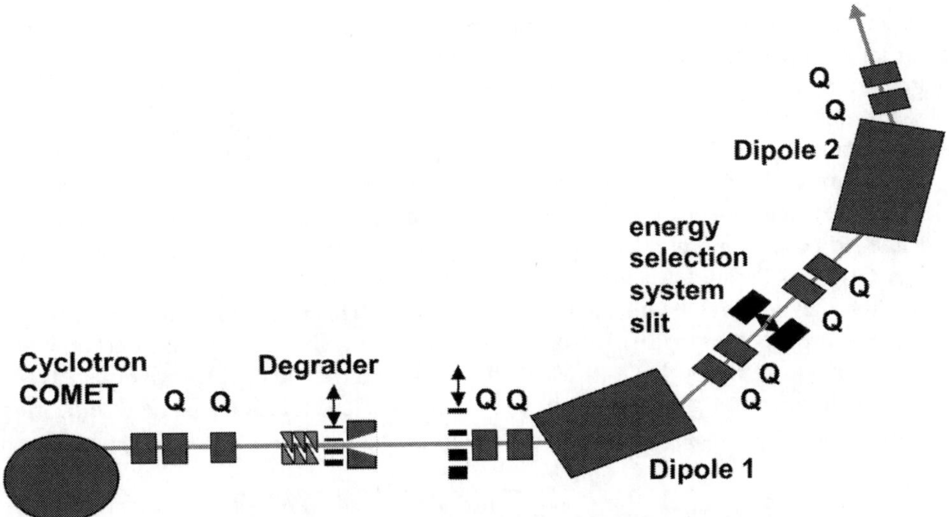

Figure 2. Layout of the first section of PROSCAN beamline with the locations of the cyclotron, the degrader, dipole magnets and energy selection system slit. Quadrupoles are labelled as "Q".

The beam optics in the ESS is arranged so that at the collimator slit location there is a focus in x-plane for a monochromatic beam (see solid red line in Figure 3). The value of horizontal beam dispersion (dotted red line in Figure 3) is high at this location (e.g. 35 mm per percent dp/p in case of PROSCAN). Hence, there is a strong correlation between the particles' horizontal position and their momentum at this location. This makes the horizontal collimator slit an effective energy selection collimator.

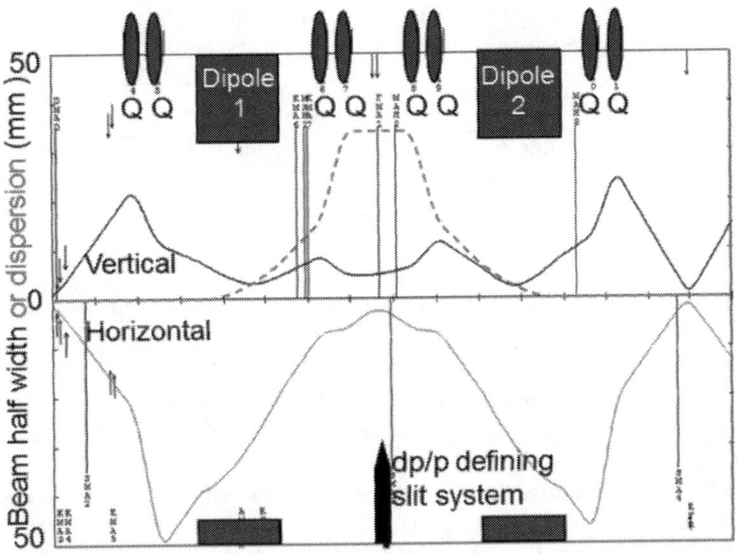

Figure 3. The TRANSPORT [2] envelope of the section presented in Figure 2. The horizontal axis is the z-position along the beam. The vertical axis above zero shows two standard deviations value of the vertical beam size (2σy). The vertical axis below zero shows the value of the horizontal beam size (2σx). The red dotted line represents the dispersion for 1% momentum offset.

Currently, a superconducting gantry including the degrader, but without ESS is being designed at PSI [3]. Such an arrangement makes the exact knowledge of the energy spectrum of a proton beam after the degradation very relevant. In the conventional ESS, as in PROSCAN, the low energy tails are absorbed by the horizontal slit. Without the ESS, as in the superconducting gantry, although the momentum acceptance of the bending section is very large, those low energies particles with strongly deviating momentum will be lost for the most part in the superconducting magnets of the final bend. Figure 4 demonstrates the trajectories of the particles with different momenta passing through the 3D field map of the proposed superconducting bend. The particle tracking has been performed with the OPAL (Object Oriented Parallel Accelerator Library) framework [4]. The five magnets of the bend have been set for the nominal beam energy of 185 MeV. The particles with the nominal energy (purple) pass well through the good field regions of all five magnets. However, particles with lower energies (orange, green and grey for 140 MeV, 90 MeV and 50 MeV, respectively) leave the good field region and will be lost at the aperture of the magnets. These particles can cause heating, which may lead to quenching of the magnet or damage the superconducting wire.

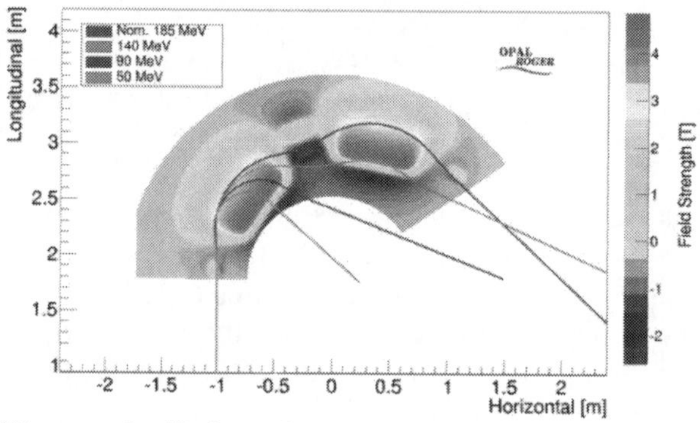

Figure 4. Trajectories of the particles tracked through a 3D field map of the proposed superconducting magnet set for 185 MeV beam.

In order to estimate how problematic this may be, simulation studies and an experiment for the determination of the beam energy spectrum have been performed and are presented in the following chapters.

2. Simulations

The energy spectrum of the proton beam interacting with the graphite degrader has been performed with the Monte Carlo code FLUKA [5]. The simulation uses $5 \cdot 10^5$ protons with an initial energy of 249.5 MeV, which corresponds to the extraction energy of the COMET cyclotron (see Figure 2). At the degrader entrance particles have a Gaussian distribution of 1.2 mm FWHM on both transversal planes and a zero divergence. The beam is assumed to be ideally monochromatic. In the FLUKA simulation, a simplified geometry of the PSI degrader of Figure 1 has been used, where the wedges have been replaced with 6 equivalent slabs of graphite with a density of 1.88 g/cm (see Figure 5). The thickness of the slabs corresponds to the thickness of the degrader wedges for the 80 MeV energy setting. Also the round collimators following the degrader, which limit the transverse emittance, are integrated in the simulation.

IOP Conf. Series: Journal of Physics: Conf. Series **874** (2017) 012108 doi:10.1088/1742-6596/874/1/012108

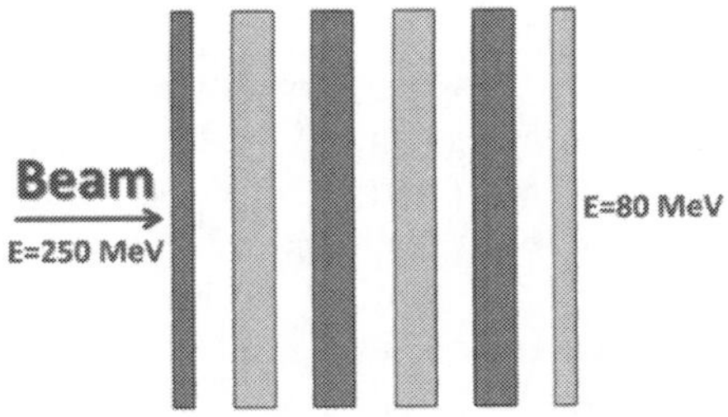

Figure 5. Schematic representation of the multi-slabs degrader geometry implemented in FLUKA.

The result of the simulated energy spectrum is shown in Figure 6, which displays the relative beam intensity as a function of energy. This distribution corresponds to a position approximately 1.5 m downstream of the degrader.

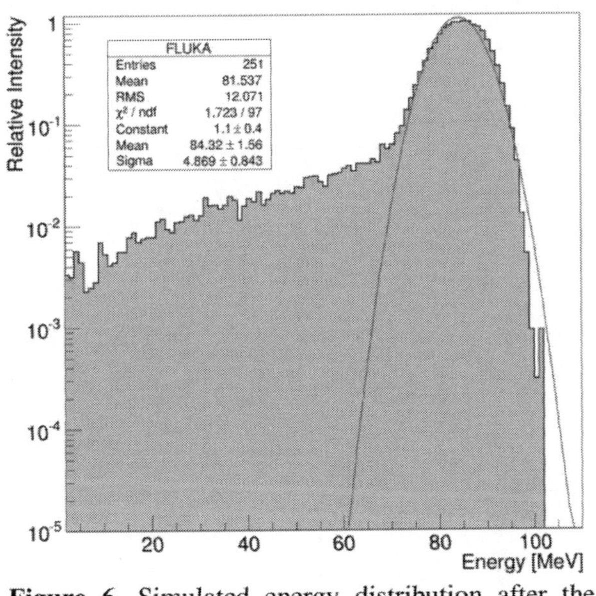

Figure 6. Simulated energy distribution after the beam propagation using the degrader parameters for the energy of 80 MeV (black curve) and Gaussian approximation of the main peak (red curve).

In Figure 6 one can observe an approximately Gaussian peak of beam intensity at approx. 80 MeV. The peak has a sharper fall-off on the high energetic side. The beam intensity above 100 MeV is negligible. However, the simulation predicts an intensity of the beam at the energies below 60 MeV at a level of approximately 0.2-2 % of the intensity maximum. The total share of the beam intensity contained in the "tail" below 70 MeV is 8.86 %. The majority of those protons underwent nuclear interactions with the degrader material.

IOP Conf. Series: Journal of Physics: Conf. Series **874** (2017) 012108 doi:10.1088/1742-6596/874/1/012108

3. Measurement

In order to measure the low energy spectrum experimentally, the degrader has been set to the energy of 80 MeV. The settings of the magnets downstream of the degrader have been adapted to transfer the beam with the lower energies between 80 MeV and 10 MeV (with the steps being 80 / 78 / 76 / 74 / 72 / 70 / 60 / 50 / 40 / 30 / 20 / 10 MeV). For this, the nominal magnet settings for the energy of 80 MeV used for PSI Gantry 3 has been scaled down with the momentum of the beam. The beam profile in front of the ESS slit location has been recorded for each step in the energy using a strip profile monitor.

Due to the high correlation of the horizontal position of the particles and their energy at the location of ESS, the beam intensity measured in the center of the strip monitor corresponds to the beam intensity for a given energy.

The projection of the horizontal profile directly in front of the ESS slit has been measured by a relatively thick ionization profile monitor [6] using an automated procedure [7]. The beam currents to the central 14 strips are derived from the signal currents taking into account the energy dependent amplification and the energy loss of the protons in the chamber foils (Figure 7). At 10 MeV, the beam energy was too low to reach the active volume of the monitor, resulting in a signal below the amplifier threshold of a few pA.

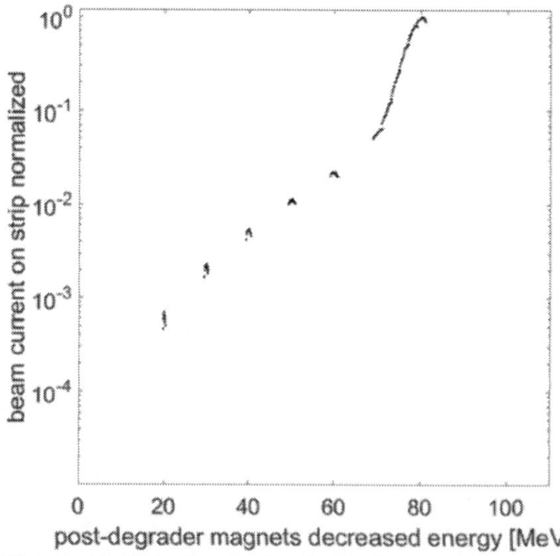

Figure 7. Relative beam intensity as a function of beamline setting energy, measured at the ESS slit location.

In Figure 7 one can recognize the Gaussian peak with a maximum at 80 MeV. At the energies below 70 MeV there is a tail of low energy particles. The transition of Gaussian peak to exponential tail appears at approximately 5 % of the peak intensity. This compares well to the 6 % of the simulation. Also the transition energy of 70 MeV from the Gaussian peak to the tail compares well to the simulation prediction. The general shape of the tail corresponds well to the simulation result displayed in Figure 6. This suggests that the particles populating it correspond to the particles involved into the nuclear interactions in the degrader, as predicted in the simulation.

There is an observed fall-off of the individual profiles to the sides for the energies below 70 MeV. One possible explanation for it is that the particles with a momentum strongly deviating from the beamline setting are lost before the beam arrives at the ESS slit, since in there are two more collimators limiting the beam size horizontally after the first dipole introduces the dispersion. This may also indicate that the measured intensity is too low at low energies.

At lower energies the relative intensity in the simulation results becomes increasingly larger than the measured intensity, e.g. at 20 MeV the difference is about by a factor of 10. We expect that for these low energies the signal of the monitors is too low. Also, the measurement relates to a position, which is approximately 6 m downstream of the position considered in the simulation. Two additional collimators are in between. It is plausible that the transmission over this line is lower for the low energy protons which start with a broader angular distribution.

4. Conclusions and outlook

The results of the FLUKA simulation for the spectrum of the beam after degradation are consistent with the measurement results in the significant, high energy part of the spectrum. This indicates that the effects providing the largest contribution to the beam energy loss, incl. multiple nuclear scattering and the inelastic scattering, are correctly implemented in FLUKA and that FLUKA simulations can be used as a reliable tool for the further studies.

These studies will include the OPAL simulation of the beam with the described spectrum being propagated through the superconducting magnet bend described above (Figure 4). The goal of the simulations is to determine the locations along the bend at which the low energy particles reach the magnet aperture, and hence get lost. The simulations will give a comprehension of what the total power deposited at each bend location will be and how the particles will interact with the vacuum pipe or magnet material. It is expected that the largest share of losses will occur on the inner aperture of the first half of the bend. Also, some protons with a very low energy will take a spiral trajectory transversely to the bending plane and will be lost on the magnet aperture in the non-bending direction at the entrance of the first superconducting dipole.

There are following possible consequences of the beam losses in the superconducting magnet:

a) Long term damage of the superconducting material.
b) Warming up of some locations in the SC magnet and generation of so-called hot spots in the superconductor. The hot spots can increase the probability of quenches. These are very undesirable in the magnet, which must have a very high reliability (and up-time) for the purpose of medical treatments. If the simulations demonstrate that the warming up of the superconducting coil is a problem, the cooling and quench prevention concepts need to be revised.
c) Activation of the gantry material and the area around the patient. Also, the neutron production from the scattered protons must be calculated and the neutron dose the patient will receive due to the lost low energy tail of the proton beam must be estimated.

We are designing a material inlayer at the inner aperture of the bend (in the bending plane), which will absorb the majority of the particles. The material should be selected so that a proton beam scattered on it causes a relatively small activation and a relatively low neutron production.

References

[1] Reist H *et al.* 2003 Concept for Handling and Service of the Proscan Degrader Unit *PSI - Scientific and Technical Report* Volume VI p. 109
[2] Rohrer U PSI Graphic Transport Framework. based on a CERN-SLAC-FERMILAB version by K.L. Brown et al.
[3] Gerbershagen A *et al.* 2016 A novel beam optics concept in a particle therapy gantry utilizing the advantages of superconducting magnets *Zeitschrift für Medizinische Physik* vol. **26** no. 3 p. 224–237

[4] Adelmann A, Gsell A, Metzger-Kraus C, Ineichen Y, Russel S, Wang C, Yang J, Sheehy S, Rogers C and Winklehner D 2008 - 2017 The OPAL (Object Oriented Parallel Accelerator Library) Framework *Technical Report* PSI-PR-08-02
[5] Ferrari A, Sala P, Fasso A and Ranft J 2005 *FLUKA: a multi-particle transport code* CERN-2005-10
[6] Dölling R *et al.* 2007 Beam Diagnostics for the Proton Therapy Facility PROSCAN *Proc. Acc. App.* pp. 152-159
[7] Dölling R 2016 Automated Documentation of Tunes in the Beam Lines of the COMET Cyclotron *Cyclotrons Conference Procedings* Zürich Switzerland THP17

Modelling PET radionuclide production in tissue and external targets using Geant4

T Amin[1], A Infantino[2], C Lindsay[3], R Barlow[1] and C Hoehr[3]

[1] University of Huddersfield, Huddersfield, HD1 3DH, UK
[2] European Organization for Nuclear Research, Geneva 23, Switzerland
[3] TRIUMF, Vancouver, BC V6T 2A3, Canada

Email: tanjilul.amin@hud.ac.uk

Abstract. The Proton Therapy Facility in TRIUMF provides 74 MeV protons extracted from a 500 MeV H^- cyclotron for ocular melanoma treatments. During treatment, positron emitting radionuclides such as ^{11}C, ^{15}O and ^{13}N are produced in patient tissue. Using PET scanners, the isotopic activity distribution can be measured for in-vivo range verification. A second cyclotron, the TR13, provides 13 MeV protons onto liquid targets for the production of PET radionuclides such as ^{18}F, ^{13}N or ^{68}Ga, for medical applications. The aim of this work was to validate Geant4 against FLUKA and experimental measurements for production of the above-mentioned isotopes using the two cyclotrons. The results show variable degrees of agreement. For proton therapy, the proton-range agreement was within 2 mm for ^{11}C activity, whereas ^{13}N disagreed. For liquid targets at the TR13 the average absolute deviation ratio between FLUKA and experiment was 1.9±2.7, whereas the average absolute deviation ratio between Geant4 and experiment was 0.6±0.4. This is due to the uncertainties present in experimentally determined reaction cross sections.

1. Introduction

This paper presents Positron-Emission Tomography (PET) isotope production at TRIUMF from two different medical applications at different energy regimes: proton therapy (PT) at 74 MeV and medical isotope production at 13 MeV. The Monte Carlo (MC) code FLUKA and Geant4 toolkit have long been essential tools in accelerator design and shielding studies; their application at lower energies for medical application is gaining in popularity, and it is therefore important to explore their limitations in this energy regime. The accuracy, relies on the quality of reaction cross section data used by the MC codes.

Compared to traditional X-ray therapy, in proton therapy (PT) the therapeutic protons stop within a defined target, sparing healthy tissue or organs of the cancer patient. Due the rapid dose fall off in proton therapy, patient alignment is extremely critical for the success of the treatment. Unfortunately, the range of protons inside the patient is not always well known, thus treatment plans rely on large error margins. To verify the dose to the patient, secondary particles such as prompt-gammas or positron emitters (e.g. ^{11}C, ^{15}O and ^{13}N) can be measured. Using MC simulations such as MCNP(X), FLUKA or Geant4, the secondaries can be used to trace back the dose deposited. In this work, PET isotope production was simulated in a Poly(methyl methacrylate) (PMMA) phantom with a raw Bragg peak (RBP) and a spread-out Bragg peak (SOBP).

Radioisotopes are an essential component in the diagnosis and treatment of cancer. In the near future, several isotope producing research reactors are scheduled to shut down. This has shifted the attention of the isotope community to accelerator based isotope production. MC codes can be used to simulate the production yields of various isotopes from different targets. This allows the optimization of the target design to maximize the isotope to contaminant ratio [1]. Details of all isotopes investigated in this paper are listed in Table 1.

2. Materials and methods

2.1 Experiments

Experiments were carried out at the TRIUMF PT Facility and the TR13, a 13 MeV medical cyclotron. At the PT facility, isotope production via a 74 MeV proton beam (raw-Bragg Peak - RBP) as well as a 23 mm spread-out proton beam of a maximum energy of 74 MeV (spread-out Bragg peak - SOBP) was measured in a PMMA phantom at the University of British Columbia PET centre. For more details, see [2].

TR13 is a 13 MeV self-shielded cyclotron and accelerates negative hydrogen ions. The H^- ions have their electrons stripped off and are extracted to the target through collimators. The beam enters through an aluminium foil that separates the cooling helium from the vacuum conditions of the cyclotron, and then a HAVAR foil which isolates the target liquid from the cooling helium inside the target assembly. The target body is composed of niobium. Liquids targets of enriched ^{18}O water, of natural water, and of zinc nitrate in water were irradiated, and the resulting isotopes, ^{18}F, ^{13}N and ^{68}Ga respectively, were measured via an ionization chamber and gamma spectroscopy. For more details see [1, 3].

Table 1. Details of PET isotopes investigated [4].

Isotope	**Half Life [mins]**	**$E_{\beta+}$ [keV]**	**Positron yield**
$^{12}C(p,pn)^{11}C$ $^{16}O(p,3(pn))^{11}C$	20	386	100%
$^{16}O(p,2(pn))^{13}N$	10	492	100%
$^{18}O(p,n)^{18}F$	110	250	97%
$^{68}Zn(p,n)^{68}Ga$	68	836	88%

2.2 Monte Carlo Simulations

Geant4 is a MC code implemented in C++ for simulating interactions between particles and matter. In this work Geant4 version 10.1 was used [5]. At TRIUMF's PT facility a PMMA cylindrical phantom of length 55 mm and diameter 22 mm was irradiated with a dose of 50 Gy. The density of the target was 1.2 g/cm^3. The elemental compositions of the TR13 targets are listed in Table 2. The table details enriched ^{18}O and natural water along with the isotopic composition of natural Zn in Zn nitrate hexahydrate salt $^{nat}Zn(NO_3)_2{\cdot}6H_2O$. The Zn solution was prepared by dissolving 0.743, 0.225 and 0.032 weight fractions of zinc salt, water and nitric acid respectively. The Zn solution had a density of 1.56 g/cm^3, whereas the remaining two solutions were of density 1 g/cm^3.

Table 2. Composition of targets at the TR13. The desired target isotopes are highlighted in bold.

^{18}O Water		**Nat. O**		**Nat. Zinc**	
^{16}O	4%	**^{16}O**	**99.757%**	^{64}Zn	48.63%
^{18}O	**96%**	^{17}O	0.038%	^{66}Zn	27.9%
		^{18}O	0.205%	^{67}Zn	4.1%
				^{68}Zn	**18.75%**
				^{70}Zn	0.62%

In Geant4, the saturation yield Y_v of isotope v at the End Of Bombardment (EOB), normalized to beam current in units of Bq/μA was calculated by:

$$Y_v = \frac{N_{iso}}{N_p e} \tag{1}$$

where N_{iso} is the number of isotopes produced, N_p is the number of incident protons per second, and e is the proton electric charge.

In this work, QGSP-BIC-ALLHP was used from the several physics models available in Geant4 to describe hadronic interactions. It is a high precision model that uses TALYS-based Evaluated Nuclear Data Library (TENDL) for isotope production. For electromagnetic interactions, electromagnetic option 1 was used. Isotopic yields have been normalized to incident beam current by simulating 10^9 primaries with 0.01 mm spatial resolution in liquid targets and 0.1 mm in PMMA.

The FLUKA MC package version 2011.2b.6 was used for the isotope production at the medical cyclotron. For more details, see [1, 6]. For the PT simulation, FLUKA version 2011.2c-4 was used, see [7] for more details. Isotope production in FLUKA is always handled internally and the tabulated cross sections used are not accessible to the user for comparison.

3. Results

The MC results for the PT facility are shown in figure 1 and 2. Table 3 lists TR13 experimental saturation yields (Y_{Exp}) and compares them to MC saturation yields of FLUKA and Geant4 (Y_F, Y_G) as ratios. Saturation yield decay corrected to EOB has been normalized for a beam current of 1 μA incident on the liquid target for a 1 hr irradiation. The simulated TENDL cross section (X_T) and an experimental EXFOR cross section (X_{EXFOR}) is also compared. It must be noted that the cross section ratio is limited to only the most significant contributing proton reactions, whereas in simulations incident protons and all secondaries are taken into account.

3.1 PT: ^{11}C and ^{13}N production

For the irradiation with a RBP and a SOBP, the production of ^{11}C inside the PMMA phantom from FLUKA and Geant4 are in good agreement, as illustrated in figure 1. The 50% fall off for the RBP irradiation from FLUKA and Geant4 is 23 mm and 25 mm respectively; for the SOBP irradiation, the fall offs are 17 mm and 18 mm.

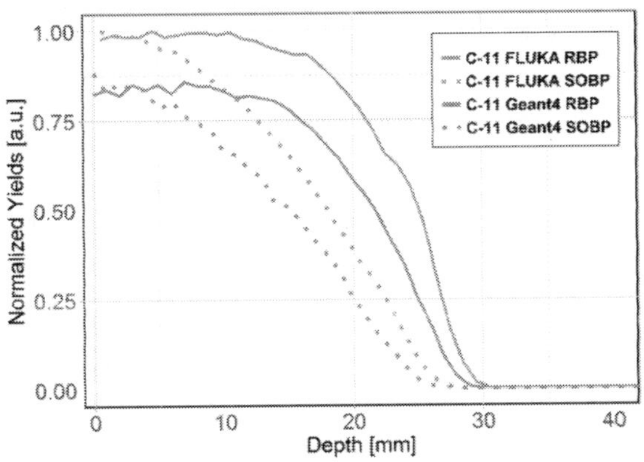

Figure 1. Yields of ^{11}C from a 74 MeV RBP and SOBP from Geant4 and FLUKA.

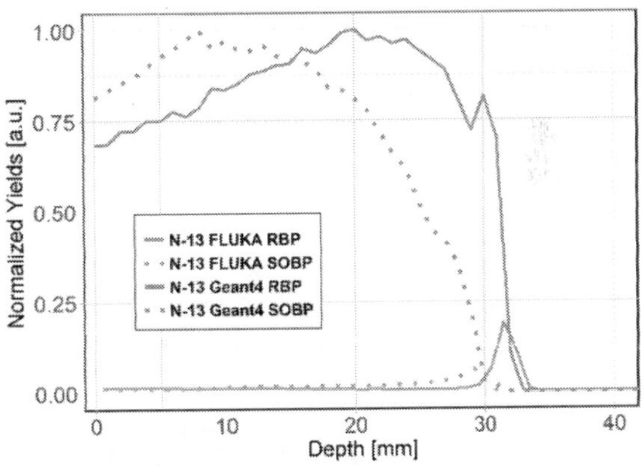

Figure 2. Yields of ^{13}N from a 74 MeV RBP and SOBP from Geant4 and FLUKA.

In figure 2 the ^{13}N yield is significantly larger in Geant4 using TENDL cross sections. For PT, the beam energy inside the PMMA target is 70 to 0 MeV denoted by the blue and red regions in figure 3. Compared to EXFOR cross sections, TENDL grossly overestimates the ^{13}N yield to be even greater than for the ^{11}C production. This overestimation propagates through to the proton range inside the target. At 70 MeV beam energy FLUKA is therefore better able to calculate the ^{13}N yield than Geant4. The 50% fall off from overall PET activity measurements for RBP was 27.9±1.7 mm (Geant4: 30.6 mm, FLUKA:

25.5 mm) and for SOBP was 21.9±1.7 mm (Geant4: 22.9 mm, FLUKA: 18.4 mm) [2]. Consequently, FLUKA better agrees with measurements when comparing the total yield of all isotopes for RBP whereas for a SOBP Geant4 performs better.

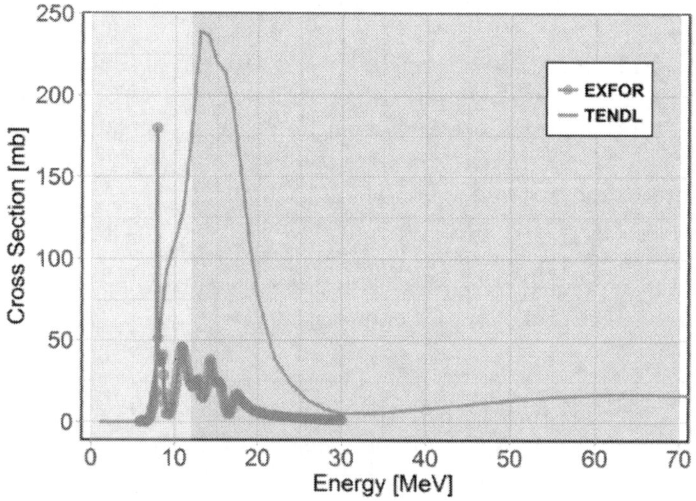

Figure 3. Comparison between IAEA and TENDL cross section for $^{16}O(p,2(pn)))^{13}N$ reaction [8, 9]. The relevant energy range for the TRIUMF PT facility is from 0 MeV to 70 MeV (red and blue), while for the TR13 cyclotron only from 0 MeV to 12 MeV (red).

3.2 TR13: ^{18}F, ^{13}N, and ^{68}Ga production

For the production of ^{18}F, Geant4 underestimates the yield by 53% whereas FLUKA overestimates the yield by 66%, see Table 3. To investigate the cross section data used in Geant4, we compared with experimental cross sections from EXFOR. Even though numerous resonances are present in the EXFOR cross section library, but not taken into account in TENDL, the ratio of cross sections is approximately 1 in this case. At this point, no explanation has been found as to why such a large difference exists between the yields and cross sections ratio. However, at 16.5 MeV, the saturation yield of ^{18}F calculated using FLUKA was in perfect agreement with the recommended IAEA saturation yield. For more details, see [10].

For the ^{13}N production at the TR13, the liquid target is exposed to significantly lower proton energies (13 MeV versus 74 MeV for PT). This greatly reduces the discrepancy between the experimental and TENDL cross section for the $^{16}O(p,2(pn))^{13}N$ reaction. In figure 3 the red region illustrates a fairly acceptable level of agreement, with TENDL and EXFOR having the same threshold energy. The yield ratios from FLUKA and Geant4 are 5.92±0.01 and 2.07±0.01 respectively. Geant4 yield ratio compares well with the cross section ratio of 2.38. Therefore, at energies closer to the production threshold, Geant4 provides a better approximation for ^{13}N yield than FLUKA.

Table 3. Saturation yield and cross section comparison.

Isotope	Y_{Exp} [MBq/μA]	Y_F/Y_{Exp}	Y_G/Y_{Exp}	X_T/X_{EXFOR}
^{18}F	4920±60	1.66±0.01	0.47±0.01	0.98
^{13}N	259±3	5.92±0.01	2.07±0.01	2.38
^{8}Ga	138±2	1.03±0.01	0.69±0.02	0.73

For ^{68}Ga the yield ratio for FLUKA is much more accurate than Geant4. The underestimation in Geant4 can be attributed to the relatively lower TENDL cross sections for the reaction compared to EXFOR. The yield ratio of 0.69±0.02 in Geant4 agrees well with the cross section ratio of 0.73. While both codes calculated ^{68}Ga yield accurately, FLUKA was slightly better with a ratio of 1.03±0.01.

4. Conclusions

The PET isotopes in the medical applications of proton therapy and PET isotope production using a 13 MeV medical cyclotron at TRIUMF have been simulated in Geant4 and FLUKA. For PT, the production of ^{11}C had good agreement between MC codes, whereas ^{13}N calculated using TENDL cross sections showed large discrepancies. This was traced back to deviating experimental cross section from EXFOR and TENDL cross sections used in Geant4. It can be concluded that TENDL cross sections are not suitable for calculating ^{13}N at an energy range of 10 to 70 MeV.

For saturation yields of ^{18}F, ^{13}N, and ^{68}Ga produced at the TR13 medical cyclotron results are mixed. While FLUKA calculated the saturation yield for ^{68}Ga better than Geant4, the situation is reversed for ^{13}N. Overall FLUKA had a mean absolute deviation of 1.9 ± 2.7, whereas Geant4 had a deviation of 0.6 ± 0.4 for the three isotopes here being investigated. This may indicate that saturation yields calculations from Geant4 are slightly closer to measured yields than from FLUKA, despite the MC models not taking into account any thermal effects or density changes in the liquid target or loss in the transfer system. A wider range of isotopes needs to be examined for a better assessment.

By comparing TENDL and EXFOR cross sections, it can be seen that the availability of accurate cross sections greatly affects the isotopic yield calculations. Therefore, in order to improve the accuracy of calculations, the cross sections available should be well known and widely accepted by the community.

Acknowledgement

TRIUMF receives federal funding via a contribution agreement with the National Research Council of Canada.

References

[1] Infantino A *et al.* 2016 Assessment of the production of medical isotopes using the Monte Carlo code FLUKA: Simulations against experimental measurements *Nucl. Instr. Meth.* B vol. **366** p. 117
[2] Amin T *et al.* Proton range verification with PET at small tumour sites: a direct comparison between experiment and simulations with FLUKA and Geant4 submitted for publication
[3] Amin T *et al.* Validating Production of PET Radionuclides in Solid and Liquid Targets: Geant4 versus FLUKA and Measurements unpublished
[4] IAEA Cyclotron Produced Radionuclides: Principles and Practice, *Technical Reports Series* No. 465, IAEA, Vienna (2008)
[5] Geant4-Website A toolkit for particle passage through matter http://geant4.web.cern.ch/geant4//
[6] Ferrari A *et al.* 2005 CERN-2005-10 INFN/TC_05/11 SLAC-R-773
[7] Lindsay C *et al.* Monte Carlo Simulation of Primary and Secondary Dose for the TRIUMF Proton Therapy Facility, submitted for publication
[8] IAEA 2003 Cross-section for $^{16}O(p,x)^{13}N$. https://www.nds.iaea.org/medical/o6p13n0.html
[9] Koning AJ *et al.* 2012 Modern Nuclear Data Evaluation with The TALYS Code System, *Nuclear Data Sheets,* **113** 2841
[10] Infantino A *et al.* 2015 Accurate Monte Carlo modelling of cyclotrons for optimization of shielding and activation calculations in the biomedical field *Radiation Physics and Chemistry vol.* **116** 231–236

Institute of Physics
Dirac House, Temple Back
Bristol BS1 6BE UK

ISSN: 1742-6588
ISBN 978-1-5108-4560-2

CURRAN ASSOCIATES INC.
proceedings.com